PRINCIPLES OF
HEAT TRANSFER

PRINCIPLES OF HEAT TRANSFER

Fourth Edition

Frank Kreith
Mark S. Bohn
Solar Energy Research Institute

HARPER & ROW, PUBLISHERS, New York
Cambridge, Philadelphia, San Francisco,
London, Mexico City, São Paulo, Singapore, Sydney

1817

Sponsoring Editor: Cliff Robichaud
Project Editor: David Nickol
Text Art: J & R Services, Inc.
Production: Delia Tedoff
Compositor: Syntax International Pte. Ltd.
Printer and Binder: The Maple Press

PRINCIPLES OF HEAT TRANSFER, Fourth Edition

Library of Congress Cataloging in Publication Data

Kreith, Frank.
  Principles of heat transfer.

  Includes bibliographical references and index.
  1. Heat—Transmission. I. Bohn, Mark. II. Title.
QC320.K7 1986   621.402′2   86-331
ISBN 0-06-043774-X

HARPER INTERNATIONAL EDITION
ISBN 0-06-350388-3

86 87 88 89 9 8 7 6 5 4 3 2 1

CONTENTS

PREFACE

In preparing a fourth edition, we have attempted to write a book that is useful not only as a text, but also as a convenient reference. To achieve this goal we have infused our personal experience in modern engineering practice into the learning process by using practical and realistic example problems and solving them by methods currently used in thermal design and engineering analysis.

In the presentation we have stressed the physical understanding of heat transfer processes rather than equation plugging. In order to introduce realistic problems early on, we have included in the first chapter sufficient material to make it possible to tackle interesting problems and stimulate the thinking of the reader in terms of overall thermal systems, as well as in terms of specific modes of heat transfer. We have also attempted to show the limits of accuracy of engineering predictions by not merely presenting correlations in the form of an equation or a line, but showing the spread in experimental data to impress on the reader that even under the best of circumstances there are unavoidable limits in accuracy to heat transfer predictions in real-world situations. At the same time, we have updated empirical correlations used in thermal sciences, particularly in convection heat transfer, to reflect the best available data. We have also updated and expanded the list of physical properties of materials used in heat transfer processes, particularly liquid metals, gases, and radiation properties of solids. We have introduced a separate chapter dealing with numerical methods and have eliminated graphical and analogical methods because they are no longer used in practice. With the easy access to computers, numerical analyses can handle more quickly and accurately problems previously solved by graphical and analogical methods. Modern numerical procedures adaptable to any type of computer are emphasized; in particular, matrix inversion, elimination, and iteration procedures are presented in sufficient detail to assist students in their future professional life. Questions of stability and accuracy of numerical solutions of heat transfer problems are also treated.

Our approach to numerical methods assumes that the student has had a course in computer programming and has access to a personal computer. Since the student may have studied any of several programming languages, such as FORTRAN, BASIC, or Pascal, we have tried not to tie our numerical approach to any specific language. In order to obtain numerical answers to simultaneous systems, commonly encountered in numerical analysis of thermal problems, we have provided matrix inversion programs written in BASIC and FORTRAN. It should be emphasized, however, that any other program available to the student or instructor can be used to obtain the numerical answers and that the programs included are merely illustrative and not intended to indicate preferred computer language or approach.

In selecting appropriate topics to update the text, we have introduced material on non-Newtonian fluids and the design of modern types of heat exchange equipment and processes, such as packed beds, heat pipes, and solar collectors. We have also added a section on second law analysis, which is receiving increasing attention in the design of energy-efficient systems.

In previous editions the last chapter was devoted to mass transfer, but feedback from teachers and students indicated that this chapter was hardly ever covered. We have therefore eliminated this chapter and integrated some basic material on mass transfer into the chapters dealing with convection, where the analogy between heat, mass, and moment transfer facilitates the presentation in a logical manner. For a more complete treatment of mass transfer, particularly in relation to the design of equipment, a specialized book on mass transfer should be consulted.

To reflect modern trends, SI units are used in the book. However, we have observed that the change from English to SI units has not been as rapid as was anticipated. Therefore, approximately one-third of the problems and examples are presented in English units, because a practicing engineer must still be fluent in both systems of units in order to communicate properly. Tables of property data in the appendices have been modified to simplify conversion of the data between English and SI units.

We have attempted to make the presentation easy to follow, but at the same time to avoid the tediousness of overly detailed treatment, which adds unnecessary length to a book and at times diminishes the interest of the students. Problems at the end of each chapter are arranged by topics and by increasing complexity to help students and teachers to develop facility in approaching the realistic challenges faced in engineering design. In addition, we have introduced a few problems that are relatively open-ended in order to stimulate the creative capabilities of readers.

The book is sufficiently versatile to be suitable for a single-semester course, a single-quarter course, or a series of two courses. Sections that can be omitted without breaking the continuity of the presentation are marked with an asterisk. If all the sections marked with an asterisk are omitted, the material in the book can be covered in a single quarter. For a semester course, the instructor can select between three and five additional sections. The entire book can be covered in two quarters of undergraduate courses at the junior or senior level.

In an era when energy conservation and efficiency in conversion are of paramount importance to engineers, heat transfer is becoming an increasingly important part of engineering design. It is our hope that by presenting the subject in an interesting and up-to-date manner, we have provided a service to those engineers who will face the challenge of efficient design of heat exchange equipment in their future careers.

When a book goes into its fourth edition, many people have contributed ideas and suggestions, and we thank all of those people for their support. The following have been especially helpful in the preparation of this edition: Dr. A. Bejan of Duke University has stimulated our interest in second law analysis and his writings have influenced our presentation. Dr. A. Bergles of the State University of Iowa, Dr. A. Kirkpatrick of Colorado State University, and Dr. Joseph Mollendorf of the State University of New York at Buffalo have reviewed parts of the manuscript and contributed many constructive suggestions. Mrs. Mary Prescott has served as the technical editor and provided support in obtaining permissions, keeping track of changes in text and illustrations, and spotting errors in spelling and syntax that escaped our attention. Mr. David Nickol provided valuable assistance as the project editor. Mrs. E. Cocciolo and Mrs. P. Rummel have typed the manuscript. Last, but not least, Frank Kreith wants to thank his wife, Marion, for her patience and support in the course of preparing this new edition.

Frank Kreith
Mark S. Bohn

Nomenclature

Symbol	Quantity	English system of units	International system of units
a	velocity of sound; acceleration	ft/s	m/s
a	thermal diffusivity $= k/c\rho$	ft²/h	m²/s
A	area; A_c, cross-sectional area; A_p, projected area of a body normal to the direction of flow; A_q, area through which rate of heat flow is q; A_s, surface area; A_o, outside surface area; A_i, inside surface area	ft²	m²
b	breadth or width	ft	m
c	specific heat; c_p, specific heat at constant pressure; c_v, specific heat at constant volume	Btu/lb$_m$ F	J/kg K
C_A	molar concentration of component A	lb mole/ft³	kg/mole m³
C	constant		
C	thermal capacity	Btu/°F	J/K
$\dot{C}$	hourly heat capacity rate in Chapter 8; $\dot{C}_c$, hourly heat capacity rate of colder fluid in a heat exchanger; $\dot{C}_h$, hourly heat capacity rate of warmer fluid in a heat exchanger	Btu/h °F	W/K
C_D	total drag coefficient		
C_f	skin friction coefficient; C_{fx}, local value of C_f at distance x from leading edge; $\bar{C}_f$, average value of C_f defined by Eq. (4.45)		
D	diameter; D_H, hydraulic diameter; D_o, outside diameter; D_i, inside diameter	ft	m
D_{AB}	mass diffusion coefficient	ft²/h	m²/s
e	base of natural or Napierian logarithm		
e	internal energy per unit mass	Btu/lb$_m$	J/kg
E	internal energy	Btu	J
E	emissive power of a radiating body; E_b, emissive power of blackbody; E_λ, monochromatic emissive power per micron at wavelength λ	Btu/h ft²	W/m²
f	Fanning friction coefficient for flow through a pipe or a duct, defined by Eq. (8.11)		
f'	friction coefficient for flow over banks of tubes		

Symbol	Quantity	English system of units	International system of units
F	force	lb_f	newton
F_T	temperature factor defined by Eq. (9.119)		
F_{1-2}	geometric shape factor for radiation from one blackbody to another		
$\mathcal{F}_{1-2}$	geometric shape and emissivity factor for radiation from one gray body to another		
g	acceleration due to gravity	ft/s^2	m/s^2
g_c	dimensional conversion factor	32.2 ft lb_m/lb_f s^2 or 4.18 $\times 10^8$ ft $lb_m/$ lb_f h^2	1.0 kg m/N s^i
G	mass velocity or flow rate per unit area $(G = \rho V)$	lb_m/h ft^2	kg/m^2 s
G	irradiation incident on unit surface in unit time	Btu/h ft^2	W/m^2
h	enthalpy per unit mass	Btu/lb_m	J/kg
$\bar{h}$	combined unit-surface conductance, $\bar{h} = \bar{h}_c + \bar{h}_r$; h_b, unit-surface conductance of a boiling liquid, defined by Eq. (10.1); h_c, local unit convective conductance; $\bar{h}_c$, average unit convective conductance; $\bar{h}_r$, average unit conductance for radiation	Btu/h ft^2 °F	W/m^2 K
h_{fg}	latent heat of condensation or evaporation	Btu/lb_m	J/kg
h_m	local convective mass transfer coefficient	m/s	ft/s
i	angle between sun direction and surface normal	deg	rad
i'	electric current flow rate	amp	amp
I	intensity of radiation; I_λ, intensity per micron at wavelength λ	Btu/h unit solid angle	W/sr
J	radiosity	Btu/h ft^2	W/m^2
k	thermal conductivity; k_s, thermal conductivity of a solid; k_f, thermal conductivity of a fluid	Btu/h ft °F	W/m^2 K
K	thermal conductance; K_k, thermal conductance for conduction heat transfer; K_c, thermal convective conductance; K_r, thermal conductance for radiation heat transfer	Btu/h °F	W/K

Symbol	Quantity	English system of units	International system of units
log	logarithm to the base 10		
ln	logarithm to the base e		
l	length, general	ft or in.	m
L	length along a heat flow path or characteristic length of a body	ft or in.	m
L_f	latent heat of solidification	Btu/lb$_m$	J/kg
m	mass flow rate	lb$_m$/s or lb$_m$/h	kg/s
M	mass	lb$_m$	kg
N	number in general; number of tubes, etc.		
p	static pressure; p_c, critical pressure; p_A, partial pressure of component A	psi or lb$_f$/ft^2 atm	N/m^2
P	wetted perimeter	ft	m
P	total pressure	atm	N/m^2
q	rate of heat flow; q_k, rate of heat flow by conduction; q_r, rate of heat flow by radiation; q_c, rate of heat flow by convection; q_b, rate of heat flow by nucleate boiling	Btu/h	W
$\dot{q}_G$	rate of heat generation per unit volume	Btu/h ft^3	W/m^3
Q	quantity of heat	Btu	J
$\dot{Q}$	volumetric rate of fluid flow	ft^3/h	m^3/s
r	radius; r_H, hydraulic radius; r_i, inner radius; r_o, outer radius	ft	m
R	thermal resistance; R_c, thermal resistance to convection heat transfer; R_k, thermal resistance to conduction heat transfer; R_r, thermal resistance to radiation heat transfer	h °F/Btu	K/W
R_e	electrical resistance	ohm	ohm
$\mathcal{R}$	perfect gas constant	1545 ft lb$_f$/lb mole °F, or 0.730 ft^3 atm/lb mole °F	8.314 J/K kg mole
S	shape factor for conduction heat flow		
S_L	distance between centerlines of tubes in adjacent longitudinal rows	ft	m
S_T	distance between centerlines of tubes in adjacent transverse rows	ft	m

Symbol	Quantity	English system of units	International system of units
T	temperature; T_b, temperature of bulk of fluid; T_f, mean film temperature; T_s, surface temperature; T_∞, temperature of fluid far removed from heat source or sink; T_m, mean bulk temperature of fluid flowing in a duct; T_s, temperature at surface of a wall; T_{sv}, temperature of saturated vapor; T_{sl}, temperature of a saturated liquid; T_{fr}, freezing temperature; T_l, liquid temperature; T_0, total temperature; T_{as}, adiabatic wall temperature; T_{wb}, wet-bulb temperature	°F or R	K
u	internal energy per unit mass	Btu/lb$_\text{m}$	J/kg
u	time average velocity in x direction; u', instantaneous fluctuating x component of velocity; $\bar{u}$, average velocity	ft/s or ft/h	m/s
U	overall unit conductance, overall heat transfer coefficient	Btu/h ft² °F	W/m² K
U_∞	free-stream velocity	ft/s	m/s
v	specific volume	ft³/lb$_\text{m}$	m³/kg
v	time average velocity in y direction; v', instantaneous fluctuating y component of velocity	ft/s or ft/h	m/s
V	volume	ft³	m³
$\dot{W}$	rate of work output	Btu	W
x	distance from the leading edge; x_c, critical distance from the leading edge where flow becomes turbulent	ft	m
x	coordinate		
y	coordinate		
y	distance from a solid boundary measured in direction normal to surface	ft	m
z	coordinate		
Z	ratio of hourly heat capacity rates in heat exchangers		

Greek Letters

α	absorptance for radiation; α_λ, monochromatic absorptance at wavelength λ		

Symbol	Quantity	English system of units	International system of units
β	temperature coefficient of volume expansion	1/R	1/K
β_k	temperature coefficient of thermal conductivity	1/R	1/K
γ	specific heat ratio, c_p/c_v		
Γ	body force per unit mass	lb_f/lb_m	N/kg
Γ_c	mass rate of flow of condensate per unit breadth $= \dot{m}/\pi D$ for a vertical tube	lb_m/h ft	kg/s m
δ	boundary-layer thickness; δ_h, hydrodynamic boundary-layer thickness; δ_{th}, thermal boundary-layer thickness	ft	m
Δ	difference between values		
ϵ	heat exchanger effectiveness		
ϵ	emittance for radiation; ϵ_λ, monochromatic emittance at wavelength λ; ϵ_ϕ, emittance in direction of ϕ		
ϵ_H	thermal eddy diffusivity	ft^2/s	m^2/s
ϵ_M	momentum eddy diffusivity	ft^2/s	m^2/s
ζ	ratio of thermal to hydrodynamic boundary-layer thickness, δ_{th}/δ_h		
η_f	fin efficiency		
θ	time	h or s	s
λ	wavelength; λ_{max}, wavelength at which monochromatic emissive power $E_{b\lambda}$ is a maximum	micron	μm
λ	latent heat of vaporization	Btu/lb_m	J/kg
μ	absolute viscosity	lb_m/ft s	N s/m^2
ν	kinematic viscosity, μ/ρ	ft^2/s	m^2/s
ν_τ	frequency of radiation	1/s	1/s
ρ	mass density, $1/v$; ρ_l, density of liquid; ρ_v, density of vapor	lb_m/ft^3	kg/m^3
ρ	reflectance for radiation		
τ	shearing stress; τ_s, shearing stress at surface; τ_w, shear at wall of a tube or a duct	lb_f/ft^2	N/m^2
τ	transmittance for radiation		
σ	Stefan-Boltzmann constant	Btu/h ft^2 R^4	W/m^2 K^4
σ	surface tension	lb_f/ft	N/m
ϕ	angle	rad	rad

Symbol	Quantity	English system of units	International system of units
ω	angular velocity	1/s	1/s
ω	solid angle	steradian	sr
χ	quality	percent	percent

Dimensionless Numbers

Bi	Biot number $= \bar{h}L/k_s$ or $\bar{h}r_o/k_s$
Ec	Eckert number $= U_\infty/c_p(T_s - T_\infty)$
Fo	Fourier modulus $= a\theta/L^2$ or $a\theta/r_o^2$
Gz	Graetz number $= \dot{m}c_p/k_f L$
Gr	Grashof number $= \beta g L^3 \Delta T/\nu^2$
j	Colburn j factor for heat transfer $= (Nu/RePr)Pr^{2/3}$; j_M, j factor for mass transfer $= (Sh/ReSc)Sc^{2/3}$
Kn	Knudsen number
Le	Lewis number $= \alpha/D_{AB}$
M	Mach number $= U_\infty/a$
Nu	Nusselt number $= h_c x/k_f$; Nu_x, local value of Nu at point x
$\overline{Nu}$	average value of Nu over surface $= \bar{h}_c L/k_f$; $\overline{Nu}_D$, diameter Nusselt number $= \bar{h}_c D/k_f$
Pe	Peclet number $= RePr$
Pr	Prandtl number $= c_p\mu/k$ or ν/α
Ra	Rayleigh number $= GrPr$
Re	Reynolds number $= U_\infty \rho L/\mu$; $Re_x = U_\infty \rho x/\mu$, local value of Re at a distance x from leading edge; Re_D, diameter Reynolds number; Re_b, bubble Reynolds number
θ	Boundary Fourier modulus $= \bar{h}^2 a\theta/k_s^2$
Sh	Sherwood number $= h_m L/D_{AB}$
Sc	Schmidt number $= \mu/\rho D_{AB}$
St	Stanton number $= h_c/\rho U_\infty c_p$ or $Nu/RePr$

Miscellaneous

$a > b$	a greater than b
$a < b$	a smaller than b
$\propto$	proportional sign
$\simeq$	approximately equal sign
∞	infinity sign
Σ	summation sign

PRINCIPLES OF
HEAT TRANSFER

Chapter 1

Basic Modes of Heat Transfer

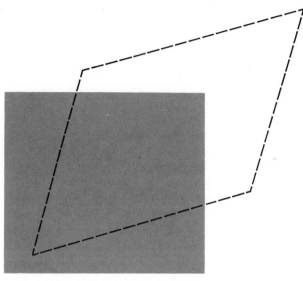

1.1 THE RELATION OF HEAT TRANSFER TO THERMODYNAMICS

Whenever a temperature gradient exists within a system, or when two systems at different temperatures are brought into contact, energy is transferred. The process by which the energy transport takes place is known as *heat transfer*. The thing in transit, called heat, cannot be measured or observed directly, but the effects it produces are amenable to observation and measurement. The flow of heat, as the performance of work, is a process by which the internal energy of a system is changed.

The branch of science which deals with the relation between heat and other forms of energy is called *thermodynamics*. Its principles, like all laws of nature, are based on observations and have been generalized into laws which are believed to hold for all processes occurring in nature, because no exceptions have ever been found. The first of these principles, the first law of thermodynamics, states that energy can be neither created nor destroyed but only changed from one form to another. It governs all energy transformations quantitatively but places no restrictions on the direction of the transformation. It is known, however, from experience that no process is possible whose sole result is the net transfer of heat from a region of lower temperature to a region of higher temperature. This statement of experimental truth is known as the second law of thermodynamics.

All heat transfer processes involve the transfer and conversion of energy. They must therefore obey the first as well as the second law of thermodynamics. At first glance one might therefore be tempted to assume that the principles of heat transfer can be derived from the basic laws of thermodynamics. This, however, would be an erroneous conclusion because classical thermodynamics is restricted primarily to the study of equilibrium states, including mechanical and chemical as well as thermal equilibriums, and is therefore, by itself, of little help in determining quantitatively the transformations that occur from a lack of equilibrium in engineering processes. Since heat flow is the result of temperature nonequilibrium, its quantitative treatment must be based on other branches of science. The same reasoning applies to other types of transport processes such as mass transfer and diffusion.

Limitations of Classical Thermodynamics   Classical thermodynamics deals with the states of systems from a macroscopic view and makes no hypotheses about the structure of matter. To perform a thermodynamic analysis it is necessary to describe the state of a system in terms of gross characteristics, such as pressure, volume, and temperature, which can be measured directly and involve no special assumptions regarding the structure of matter. These variables or thermodynamic properties are of significance for the system as a whole only when they are uniform throughout it, that is, when the system is in equilibrium. Thus, classical thermodynamics is not concerned with the details of a process but rather with equilibrium states and the relations among them. The processes employed in a thermodynamic analysis are idealized processes, devised to give information concerning equilibrium states.

From a thermodynamic viewpoint, the amount of heat transferred during a process simply equals the difference between the energy change of the system and the work done. It is evident that this type of analysis considers neither the mechanism of heat flow nor the time required to transfer the heat. It simply prescribes how much heat to supply to, or reject from, a system during a process between specified end states without taking care of whether, or how, this could be accomplished. The reason for this lack of information obtainable from a thermodynamic analysis is the absence of time as a variable. The question of how long it would take to transfer a specified amount of heat, although it is of great practical importance, does not usually enter into the thermodynamic analysis.

Engineering Heat Transfer   From an engineering viewpoint, the determination of the *rate of heat transfer at a specified temperature difference* is the key problem. To estimate the cost, the feasibility, and the size of equipment necessary to transfer a specified amount of heat in a given time, a detailed heat transfer analysis must be made. The dimensions of boilers, heaters, refrigerators, and heat exchangers depend not only on the amount of heat to be transmitted, but also on the rate at which the heat is to be transferred under given conditions. The successful operation of equipment components such as turbine blades or the walls of combustion chambers depends on the possibility of cooling certain metal parts by removing heat continuously at a rapid rate from a surface. Also, in the design of electric machines, transformers, and bearings, a heat transfer analysis must be made to avoid conditions that will cause overheating and damage the equipment. These examples show that in almost every branch of engineering, heat transfer problems are encountered which are not capable of solution by thermodynamic reasoning alone, but require an analysis based on the science of heat transfer.

In heat transfer, as in other branches of engineering, the successful solution of a problem requires assumptions and idealizations. It is almost impossible to describe physical phenomena exactly, and in order to express a problem in the form of an equation that can be solved it is necessary to make some approximations. In electric circuit calculations, for example, it is usually assumed that the values of the resistances, capacitances, and inductances are independent of the current flowing through them. This assumption simplifies the analysis but may in certain cases severely limit the accuracy of the results.

It is important to keep the assumptions, idealizations, and approximations made in the course of an analysis in mind when the final results are interpreted. Sometimes insufficient information on physical properties makes it necessary to use engineering approximations to solve a problem. For example, in the design of machine parts for operation at elevated temperatures it may be necessary to estimate the proportional limit or the fatigue strength of the material from low-temperature data. To assure satisfactory operation of the part, the designer should apply a factor of safety to the results obtained from the analysis. Similar approximations are also necessary in heat transfer problems. Physical properties, such as the thermal conductivity or the viscosity, change with temperature, but if suitable average values are selected, the calculations can be considerably simplified without introducing an appreciable error in the final result. When heat is transferred from a fluid to a wall, as in a boiler, a scale forms under continued operation and reduces the rate of heat flow. To assure satisfactory operation over a long period of time, a factor of safety must be applied to provide for this contingency.

When it becomes necessary to make an assumption or approximation in the solution of a problem, the engineer must rely on ingenuity and past experience. There are no simple guides to new and unexplored problems, and an assumption valid for one problem may be misleading in another. Experience has shown, however, that the first requirement for making sound engineering assumptions or approximations is a complete and thorough physical understanding of the problem at hand. In the field of heat transfer, this means familiarity not only with the laws and physical mechanisms of heat flow, but also with those of fluid mechanics, physics, and mathematics.

Heat transfer can be defined as the transmission of energy from one region to another as a result of a temperature difference between them. Since differences in temperatures exist all over the universe, the phenomena of heat flow are as universal as those associated with gravitational attractions. Unlike gravity, however, heat flow is governed not by a unique relationship, but rather by a combination of various independent laws of physics.

The literature of heat transfer generally recognizes three distinct modes of heat transmission: *conduction, radiation,* and *convection.* Strictly speaking, only conduction and radiation should be classified as heat transfer processes, because only these two mechanisms depend for their operation on the mere existence of a temperature difference. The last of the three, convection, does not strictly comply with the definition of heat transfer because it depends for its operation on mechanical mass transport also. But since convection also accomplishes transmission of energy from regions of higher temperature to regions of lower temperature, the term "heat transfer by convection" has become generally accepted.

In the next three sections we will survey the basic equations governing each of the three modes of heat transfer. Our initial aim is to obtain a broad perspective of the field without becoming involved in details. We shall, therefore, consider only simple cases. Yet it should be emphasized that in most natural situations heat is transferred not by one, but by several mechanisms operating simultaneously. Hence, we will show in Section 1.5 how to combine the simple relations in situations when several heat transfer modes occur simultaneously. In Section 1.6 we will illustrate how to use the laws of thermodynamics in heat transfer analyses and in the last section we will consider the units used in heat transfer calculations, as well as methods for changing from one system of units to another.

1.2 CONDUCTION

Whenever a temperature gradient exists in a solid medium, heat will flow from the higher-temperature to the lower-temperature region. The rate at which heat is transferred by conduction, q_k, is proportional to the temperature gradient dT/dx times the area A through which heat is transferred, or

$$q_k \propto A \frac{dT}{dx}$$

In this relation $T(x)$ is the local temperature and x is the distance in the direction of the heat flow. The actual rate of heat flow depends on the thermal conductivity k, which is a physical property of the medium. For conduction through a homogeneous medium, the rate of heat transfer is then

$$q_k = -kA \frac{dT}{dx} \tag{1.1}$$

The minus sign is a consequence of the second law of thermodynamics, which requires that heat *must* flow in the direction from higher to lower temperature. The temperature gradient, as shown in Fig. 1.1, will be negative if the temperature decreases with increasing values of x. Therefore, if heat transferred in the positive x direction is to be a positive quantity, a negative sign must be inserted on the right side of Eq. (1.1).

Equation (1.1) defines the thermal conductivity. It is called Fourier's law of conduction in honor of the French scientist J.B.J. Fourier, who proposed it

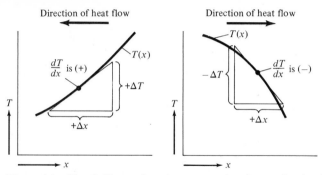

Figure 1.1 Sketch illustrating sign convention for conduction heat flow.

in 1822. The thermal conductivity in Eq. (1.1) is a material property that indicates the amount of heat that will flow per unit time across a unit area when the temperature gradient is unity. In the SI system (see Section 1.7) the area is in square meters (m²), the temperature in kelvin (K), x in meters (m), and the rate of heat flow in watts (W). The thermal conductivity, therefore, has the units of watts per meter per kelvin (W/m K). In the English system, which is still widely used by engineers in the United States, the area is expressed in square feet (ft²), x in feet (ft), the temperature in degrees Fahrenheit (°F), and the rate of heat flow in Btu/h. Thus, k has the units Btu/h ft °F. The conversion constant for k between the SI and English systems is

$$1 \text{ W/m K} = 0.578 \text{ Btu/h ft } °F$$

Orders of magnitude of the thermal conductivity for various types of materials are presented in Table 1.1. Although in general the thermal conductivity varies with temperature, in many engineering problems the variation is sufficiently small to be neglected.

TABLE 1.1 THERMAL CONDUCTIVITIES
OF SOME METALS,
NONMETALLIC SOLIDS,
LIQUIDS, AND GASES

Material	Thermal conductivity at 300 K (W/m K)
Copper	399.0
Aluminum	237.0
Carbon steel, 1% C	43.
Glass	0.81
Plastics	0.2–0.3
Water	0.6
Ethylene glycol	0.26
Engine oil	0.15
Freon (liquid)	0.07
Hydrogen	0.18
Air	0.026

1.2.1 Plane Walls

For the simple case of steady-state heat flow through a plane wall, the temperature gradient and the heat flow do not vary with time and the cross-sectional
area along the heat flow path is uniform. The variables in Eq. (1.1) can then be
separated, and the resulting equation is

$$\frac{q_k}{A} \int_0^L dx = -\int_{T_{hot}}^{T_{cold}} k\, dT = -\int_{T_1}^{T_2} k\, dT$$

The limits of integration can be checked by inspection of Fig. 1.2, where the
temperature at the left face ($x = 0$) is uniform at T_{hot} and the temperature at the
right face ($x = L$) is uniform at T_{cold}.

If k is independent of T, we obtain, after integration, the following expression for the rate of heat conduction through the wall:

$$q_k = \frac{Ak}{L}(T_{hot} - T_{cold}) = \frac{\Delta T}{L/Ak} \tag{1.2}$$

In this equation ΔT, the difference between the higher temperature T_{hot} and
the lower temperature T_{cold}, is the driving potential that causes the flow of heat.
The quantity L/Ak is equivalent to a *thermal resistance* R_k that the wall offers

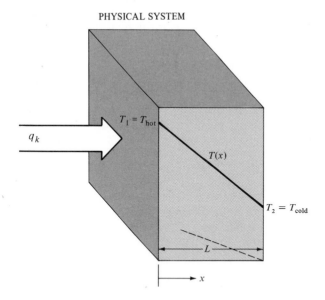

Figure 1.2   Temperature distribution for steady-state conduction through a plane wall.

to the flow of heat by conduction:

$$R_k = \frac{L}{Ak} \tag{1.3}$$

The reciprocal of the thermal resistance is referred to as the *thermal conductance* K_k, defined by

$$K_k = \frac{Ak}{L} \tag{1.4}$$

The ratio k/L in Eq. (1.4), the thermal conductance per unit area, is called the *unit thermal conductance for conduction heat flow*. The subscript k indicates that the transfer mechanism is conduction. The thermal conductance has the units of watts per kelvin temperature difference (Btu/h °F in the engineering system) and the thermal resistance has the units kelvin per watt (h °F/Btu in the engineering system). The concepts of resistance and conductance are helpful in the analysis of thermal systems, where several modes of heat transfer occur simultaneously.

For many materials, the thermal conductivity can be approximated as a linear function of temperature over limited ranges of temperature:

$$k(T) = k_0(1 + \beta_k T) \tag{1.5}$$

where β_k is an empirical constant and k_0 is the value of conductivity at a reference temperature. In such cases, integration of Eq. (1.1) gives

$$q_k = \frac{k_0 A}{L}\left[(T_1 - T_2) + \frac{\beta_k}{2}(T_1{}^2 - T_2{}^2)\right] \tag{1.6}$$

or

$$q_k = \frac{k_{av} A}{L}(T_1 - T_2) \tag{1.7}$$

where k_{av} is the value of k at the average temperature $(T_1 + T_2)/2$.

The temperature distributions for a constant value of thermal conductivity ($\beta_k = 0$) and for thermal conductivity increasing ($\beta_k > 0$) and decreasing ($\beta_k < 0$) with temperature are shown in Fig. 1.3.

EXAMPLE 1.1 ————————————————————————————————

Calculate the thermal resistance and the rate of heat transfer through a pane of window glass ($k = 0.78$ W/m K) 1 m high, 0.5 m wide, and 0.5 cm thick, if the outer-surface temperature is 24°C and the inner-surface temperature is 24.5°C.

Solution. A schematic diagram of the system is shown in Fig. 1.2. Assume that steady state exists and that the temperature is uniform over the inner and outer surfaces. The thermal resistance to conduction R_k is from Eq. (1.3)

$$R_k = \frac{L}{kA} = \frac{0.005 \text{ m}}{0.78 \text{ W/m K} \times 1 \text{ m} \times 0.5 \text{ m}} = 0.0128 \text{ K/W}$$

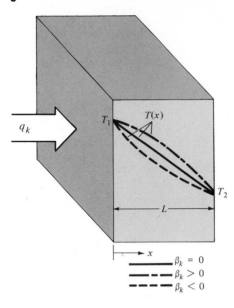

Figure 1.3 Temperature distribution in conduction through a plane wall with constant and variable thermal conductivity.

The rate of heat loss from the interior to the exterior surface is obtained from Eq. (1.2):

$$q_k = \frac{T_1 - T_2}{R_k} = \frac{(24.5 - 24.0)°\text{C}}{0.0128 \text{ K/W}} = 39 \text{ W}$$

Note that a temperature difference of 1°C is equal to a temperature difference of 1 K. Therefore, °C and K can be used interchangeably when temperature differences are indicated. If a temperature level is involved, however, it must be remembered that zero on the Celsius scale (0°C) is equivalent to 273.15 K on the thermodynamic or absolute temperature scale and

$$T(\text{K}) = T(°\text{C}) + 273.15$$

1.2.2 Thermal Conductivity

According to Fourier's law, Eq. (1.1), the thermal conductivity is defined as

$$k \equiv \frac{q_k/A}{dT/dx}$$

For engineering calculations we generally use experimentally measured values of thermal conductivity, although for gases at moderate temperatures the kinetic theory of gases may be used to predict the experimental values accurately. Theories have also been proposed for other materials to calculate thermal conductivities, but in the case of liquids and solids, theories are not adequate to predict the thermal conductivity with satisfactory accuracy (1, 2).

Table 1.1 lists values of thermal conductivity for several materials. Note that the best conductors are pure metals and the poorest ones are gases. In between lie alloys, nonmetallic solids, and liquids.

The mechanism of thermal conduction in a gas can be explained on a molecular level from basic concepts of the kinetic theory of gases. The kinetic energy of a molecule is related to its temperature. Molecules in a high-temperature region have higher velocities than those in a lower-temperature region. But molecules are in continuous random motion, and as they collide with one another they exchange energy as well as momentum. When a molecule moves from a higher-temperature region to a lower-temperature region, it transports kinetic energy from the higher- to the lower-temperature part of the system. Upon collison with slower molecules, it gives up some of this energy and increases the energy of molecules with a lower energy content. In this manner thermal energy is transferred from higher- to lower-temperature regions in a gas by molecular action.

In accordance with the above simplified concept, the faster molecules move, the faster they will transport energy. Consequently, the transport property that we have called thermal conductivity should be dependent on the temperature of the gas. A somewhat simplified analytical treatment [for example, see (3)] indicates that the thermal conductivity of a gas is proportional to the square root of the absolute temperature. At moderate pressures the space between molecules is large compared to the size of a molecule; thermal conductivity of gases is therefore essentially independent of pressure. Figure 1.4 shows how the thermal conductivities of some typical gases vary with temperature.

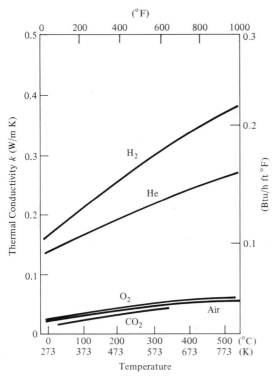

Figure 1.4 Variation of thermal conductivity with temperature for gases. (*Source:* Kreith and J. F. Kreider, *Principles of Solar Engineering*, Hemisphere, Washington, D. C., 1978.)

The basic mechanism of energy conduction in liquids is qualitatively similar to that in gases. However, molecular conditions in liquids are more difficult to describe and the details of the conduction mechanisms in liquids are not as well understood. Figure 1.5 shows the thermal conductivity of some nonmetallic liquids as a function of temperature. For most liquids, the thermal conductivity decreases with increasing temperature, but water is a notable exception. The thermal conductivity of liquids is insensitive to pressure, except near the critical point. As a general rule, the thermal conductivity of liquids decreases with increasing molecular weight. For engineering purposes, values of the thermal conductivity of liquids are taken from tables as a function of temperature in the saturated state. Appendix 2 presents such data for several common liquids. Metallic liquids have much higher conductivities than nonmetallic liquids, and their properties are listed separately in Tables 24 to 26 in Appendix 2.

According to current theories, solid materials consist of free electrons and of atoms in a periodic lattice arrangement. Thermal energy may thus be conducted by two mechanisms: migration of free electrons and lattice vibration. These two effects are additive, but, in general, the transport due to electrons is more effective than the transport due to vibrational energy in the lattice structure. Since electrons transport electric charge in a manner similar to the way in which they carry thermal energy from a higher- to a lower-temperature region, good electrical conductors are usually also good heat conductors, whereas good electrical insulators are poor heat conductors. In nonmetallic solids there is little or no electronic transport and the conductivity is therefore determined

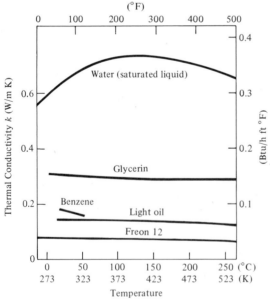

Figure 1.5 Variation of thermal conductivity with temperature for liquids. (*Source:* F. Kreith and J. F. Kreider, *Principles of Solar Engineering*, Hemisphere, Washington, D. C., 1978.)

primarily by lattice vibration. This explains why these materials have a lower thermal conductivity than metals. Thermal conductivities of some typical metals and alloys are shown in Fig. 1.6.

An important group of solid materials for heat transfer design are thermal insulators (4). These materials are solids, but their structure contains air spaces that are sufficiently small to suppress gaseous motion and thus take advantage of the low thermal conductivity of gases in reducing heat transfer. Although we usually speak of a thermal conductivity of thermal insulators, in reality the transport through an insulator is comprised of conduction as well as radiation across the interstices filled with gas. In good insulators the spaces containing

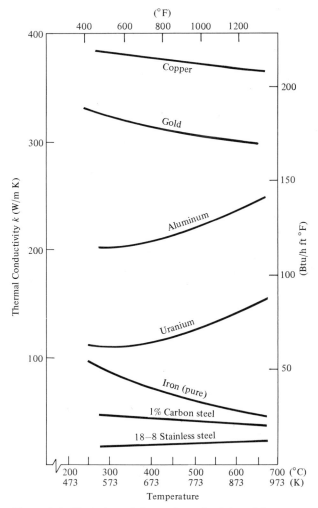

Figure 1.6 Variation of thermal conductivity with temperature for typical metals and alloys. (*Source:* F. Kreith and J. F. Kreider, *Principles of Solar Engineering*, Hemisphere, Washington, D. C., 1978.)

the air are sealed from each other, as in cellular foams made from plastic or glass. The thermal conductivity value of insulation systems is always an effective value that accounts for conduction, radiation, and sometimes also convection within the material. Table 11 in Appendix 2 lists typical values of the effective conductivity for several insulating materials.

1.2.3 Contact Resistance

When different conducting surfaces are placed in contact, as shown in Fig. 1.7, a thermal resistance is present at the interface of the solids. The interface resistance, frequently called the contact resistance, is developed when two materials will not fit tightly together and a thin layer of fluid is trapped between them. Examination of an enlarged view of the contact between the two surfaces shows that the solids touch only at peaks in the surface and that the valleys in the mating surfaces are occupied by a fluid (possibly air), a liquid, or a vacuum.

The interface resistance is primarily a function of surface roughness, the pressure holding the two surfaces in contact, the interface fluid, and the interface temperature. At the interface, the mechanism of heat transfer is complex. Conduction takes place through the contact points of the solid, while heat is transferred by convection and radiation across the trapped interfacial fluid.

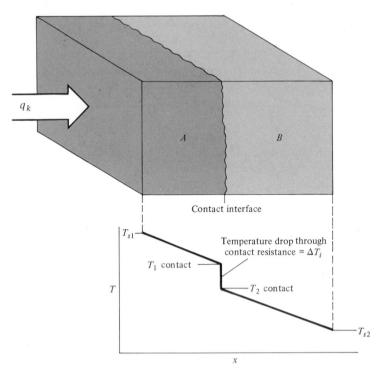

Figure 1.7 Schematic diagram illustrating physical contact and temperature distribution through a contact interface. (*a*) Physical model of contact interface. (*b*) Temperature profile through solids *A* and *B* and contact interface.

TABLE 1.2 APPROXIMATE RANGE OF THERMAL
CONTACT RESISTANCE FOR METALLIC
INTERFACES UNDER VACUUM CONDITIONS (7)

	Thermal resistance, R_i (m^2 K/W $\times$ 10^4)	
Interface material	Contact pressure 100 kN/m^2	Contact pressure 10,000 kN/m^2
Stainless steel	6–25	0.7–4.0
Copper	1–10	0.1–0.5
Magnesium	1.5–3.5	0.2–0.4
Aluminum	1.5–5.0	0.2–0.4

If the heat flux through two solid surfaces in contact is q/A and the temperature difference across the fluid gap separating the two solids is ΔT_i, the interface resistance R_i is defined by

$$R_i = \frac{\Delta T_i}{q/A} \tag{1.8}$$

When two surfaces are in perfect thermal contact, the interface resistance approaches zero and there is no temperature difference across the interface. For imperfect thermal contact, a temperature difference occurs at the interface.

Table 1.2 shows the influence of contact pressure on the thermal contact resistance between metal surfaces under vacuum conditions. It is apparent that an increase in the pressure can reduce the contact resistance appreciably. As shown in Table 1.3, the interfacial fluid also affects the thermal resistance. Putting a viscous liquid such as glycerin on the interface reduces the contact resistance between two aluminum surfaces by a factor of 10 at a given pressure.

Most of the problems at the end of the chapter do not consider interface resistance, even though it exists to some extent whenever solid surfaces are

TABLE 1.3 THERMAL CONTACT
RESISTANCE FOR
ALUMINUM-ALUMINUM
INTERFACE[a]

Fluid	Thermal resistance, R_i (m^2 K/W)
Air	2.75×10^{-4}
Helium	1.05×10^{-4}
Hydrogen	0.720×10^{-4}
Silicone oil	0.525×10^{-4}
Glycerin	0.265×10^{-4}

[a] 10-μm surface roughness under 10^5 N/m^2 contact pressure with different interfacial fluids (7).

mechanically joined. We should therefore always be aware of the existence of the interface resistance and the resulting temperature difference across the interface. Particularly with rough surfaces and low bonding pressures, the temperature drop across the interface can be significant and cannot be ignored. The subject of interface resistance is complex, and no single theory or set of empirical data accurately describes the interface resistance for surfaces of engineering importance. The reader should consult (6) and (7) for more detailed discussions of this subject.

1.3 CONVECTION

The convective mode of heat transfer actually consists of two mechanisms operating simultaneously. The first is the energy transfer due to molecular motion, that is, the conductive mode. But superimposed upon this mode is energy transfer by the macroscopic motion of fluid parcels. The fluid motion is a result of parcels of fluid, each consisting of a large number of molecules, moving by virtue of an extraneous force. This extraneous force may be due to a density gradient, as in natural convection, or due to a pressure difference generated by a pump or a fan, or possibly to a combination of the two.

Figure 1.8 shows a plate at surface temperature T_s and a fluid at temperature T_∞ flowing parallel to the plate. As a result of viscous forces the velocity of the fluid will be zero at the wall and will increase to U_∞ as shown. Since the fluid is not moving at the interface, heat is transferred at that location only by conduction. If we knew the temperature gradient and the thermal conductivity at this interface, we could calculate the rate of heat transfer from Eq. (1.1), or

$$q_c = -k_{\text{fluid}}A\left.\frac{\partial T}{\partial y}\right|_{\text{at } y=0} \tag{1.9}$$

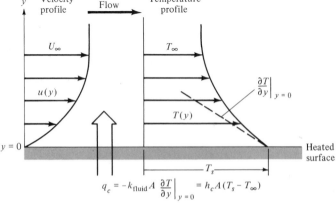

Figure 1.8 Velocity and temperature profile for convection heat transfer from a heated plate with flow over its surface.

But the temperature gradient at the interface depends on the rate at which the macroscopic as well as the microscopic motion of the fluid carries the heat away from the interface. Consequently, the temperature gradient at the fluid-plate interface depends on the nature of the flow field, particularly the free-steam velocity U_∞.

The situation is quite similar in natural convection. The principal difference is that in forced convection the velocity far from the surface approaches the free-stream value imposed by an external force, whereas in natural convection the velocity at first increases with increasing distance from the heat transfer surface and then decreases, as shown in Fig. 1.9. The reason for this behavior is that the action of viscosity diminishes rather rapidly with distance from the surface while the density difference decreases more slowly. Eventually, however, the buoyant force also decreases as the fluid density approaches the value of the unheated surrounding fluid. This interaction of forces will cause the velocity to reach a maximum and then approach zero far from the heated surface. The temperature fields in free and forced convection have similar shapes, and in both cases the heat transfer mechanism at the fluid-solid interface is conduction.

The preceding discussion indicates that the convection heat transfer coefficient will depend on the density, viscosity, and velocity of the fluid as well as on its thermal properties (thermal conductivity and specific heat). Whereas in forced convection the velocity is usually imposed on the system by a pump or a fan and can be directly specified, in free convection the velocity will depend on the temperature difference between the surface and the fluid, the coefficient of thermal expansion of the fluid (which determines the density change per unit temperature difference), and the body force field, which in systems located on the earth is simply the gravitational force.

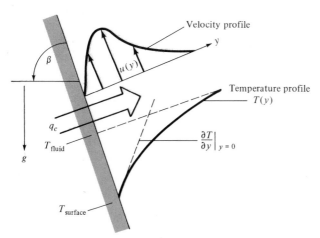

Figure 1.9 Velocity and temperature distribution for free convection over a heated flat plate inclined at angle β from the horizontal. (*Source:* F. Kreith and J. F. Kreider, *Principles of Solar Engineering*, Hemisphere, Washington, D. C., 1978.)

In later chapters we will develop methods for relating the temperature gradient at the interface to the external flow conditions. But for the time being we shall use a simpler approach to calculate the rate of convection heat transfer, as shown below.

Irrespective of the details of the mechanism, the rate of heat transfer by convection between a surface and a fluid may be calculated from the relation

$$q_c = \bar{h}_c A \, \Delta T \tag{1.10}$$

where q_c = rate of heat transfer by convection, W (Btu/h)

      A = heat transfer area, m^2 (ft^2)

    ΔT = difference between the surface temperature T_s and a temperature of the fluid T_∞ at some specified location (usually far away from the surface), K ($°$F)

    $\bar{h}_c$ = average unit thermal convective conductance over the area A (often called the surface coefficient of heat transfer or the convective heat transfer coefficient), W/m^2 K (Btu/h ft^2 $°$F)

The relation expressed by Eq. (1.10) was originally proposed by the British scientist Isaac Newton in 1701. Engineers have used this equation for many years, even though it is a definition of $\bar{h}_c$ rather than a phenomenological law of convection. Evaluation of the convective heat transfer coefficient is difficult because convection is a very complex phenomenon. The methods and techniques available for a quantitative evaluation of $\bar{h}_c$ will be presented in later chapters. At this point it is sufficient to note that the numerical value of $\bar{h}_c$ in a system depends on the geometry of the surface and the velocity, as well as on the physical properties of the fluid and often even on the temperature difference ΔT. In view of the fact that these quantities are not necessarily constant over a surface, the convective heat transfer coefficient may also vary from point to point. For this reason we must distinguish between a local and an average convective heat transfer coefficient. The local coefficient h_c is defined by

$$dq_c = h_c \, dA(T_s - T_\infty) \tag{1.11}$$

while the average coefficient $\bar{h}_c$ can be defined in terms of the local value by

$$\bar{h}_c = \frac{1}{A} \iint_A h_c \, dA \tag{1.12}$$

For most engineering applications, we are interested in average values. For general orientation, typical values of the order of magnitude of average convective heat transfer coefficients encountered in engineering practice are presented in Table 1.4.

Using Eq. (1.10), we can define the *thermal conductance for convective heat transfer* K_c as

$$K_c = \bar{h}_c A \quad \text{(W/K)} \tag{1.13}$$

and the *thermal resistance to convective heat transfer* R_c, which is equal to the

TABLE 1.4 ORDER OF MAGNITUDE OF CONVECTIVE HEAT TRANSFER COEFFICIENTS $\bar{h}_c$

Fluid	W/m² K	Btu/h ft² °F
Air, free convection	6–30	1–5
Superheated steam or air, forced convection	30–300	5–50
Oil, forced convection	60–1,800	10–300
Water, forced convection	300–6,000	50–2,000
Water, boiling	3,000–60,000	500–10,000
Steam, condensing	6,000–120,000	1,000–20,000

reciprocal of the conductance, as

$$R_c = \frac{1}{\bar{h}_c A} \quad (\text{K/W}) \tag{1.14}$$

EXAMPLE 1.2 ————————————————————————————————

Calculate the rate of heat transfer by free convection between a roof of area 10 m × 20 m and ambient air, if the roof surface temperature is 27°C, the air temperature −3°C, and the average unit convective heat transfer coefficient 10 W/m² K.

Solution. Assume that steady state exists and the direction of heat flow is from the air to the roof. The rate of heat transfer by convection from the air to the roof is then given by Eq. (1.10), or

$$q_c = \bar{h}_c A_{roof}(T_{air} - T_{roof})$$
$$= 10 \, (\text{W/m}^2 \, \text{K}) \times 400 \, \text{m}^2 \, (-3 - 27)°\text{C}$$
$$= -120,000 \, \text{W}$$

Note that in using Eq. (1.10) we initially assumed that the heat transfer will be from the air to the roof. But since the heat flow under this assumption turns out to be a negative quantity the *direction of heat flow is actually from the roof to the air*. We could, of course, have deduced this at the outset by applying the second law of thermodynamics, which tells us that heat will always flow from a higher to a lower temperature if there is no external intervention. But as we shall see in a later section, thermodynamic arguments cannot always be used at the outset in heat transfer problems because in many real situations the surface temperature is not known.

1.4 RADIATION

The quantity of energy leaving a surface as radiant heat depends on the absolute temperature and the nature of the surface. A perfect radiator or blackbody*

* A detailed discussion of the meaning of these terms is presented in Chapter 9.

emits radiant energy from its surface at a rate q_r given by

$$q_r = \sigma A_1 T_1^{\,4} \tag{1.15}$$

The heat flow rate q_r will be in watts if the surface area A_1 is in square meters and the surface temperature T_1 is in kelvins; σ is a dimensional constant with a value of 5.67×10^{-8} W/m^2 K^4. (In the engineering system the heat flow rate will be in Btu's per hour if the surface area is in square feet, the surface temperature in degrees Rankine (R), and σ is 0.1714×10^{-8} Btu/h ft^2 R^4.) The constant σ is the Stefan-Boltzmann constant; it was named after two Austrian scientists, J. Stefan, who in 1879 discovered Eq. (1.15) experimentally, and L. Boltzmann, who in 1884 derived it theoretically.

Inspection of Eq. (1.15) shows that any blackbody surface above a temperature of absolute zero radiates heat at a rate proportional to the fourth power of the absolute temperature. While the rate of radiant heat emission is independent of the conditions of the surroundings, a net transfer of radiant heat requires a difference in the surface temperature of any two bodies between which the exchange is taking place. If the blackbody radiates to an enclosure (see Fig. 1.10) that is also black, that is, absorbs all the radiant energy incident upon it, the net rate of radiant heat transfer is given by

$$q_r = A_1 \sigma (T_1^{\,4} - T_2^{\,4}) \tag{1.16}$$

where T_2 is the surface temperature of the enclosure in kelvins.

Real bodies do not meet the specifications of an ideal radiator but emit radiation at a lower rate than blackbodies. If they emit, at a temperature equal

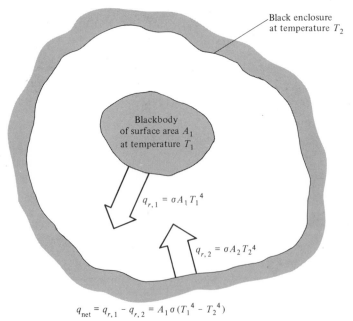

Figure 1.10   Schematic diagram of radiation between body 1 and enclosure 2.

to that of a blackbody, a constant fraction of blackbody emission at each wavelength, they are called gray bodies. A gray body A_1 at T_1 emits radiation at the rate $\epsilon_1 \sigma A_1 T_1{}^4$, and the net rate of heat transfer between a gray body at a temperature T_1 and a surrounding black enclosure at T_2 is

$$q_r = A_1 \epsilon_1 \sigma (T_1{}^4 - T_2{}^4) \tag{1.17}$$

where ϵ_1 is the emittance of the gray surface and is equal to the ratio of the emission from the gray surface to the emission from a perfect radiator at the same temperature.

If neither of two bodies is a perfect radiator and if the two bodies have a given geometric relationship to each other, the net heat transfer by radiation between them is given by

$$q_r = A_1 \mathscr{F}_{1-2} \sigma (T_1{}^4 - T_2{}^4) \tag{1.18}$$

where $\mathscr{F}_{1-2}$ is a dimensionless modulus that modifies the equation for perfect radiators to account for the emittances and relative geometries of the actual bodies. Methods for calculating $\mathscr{F}_{1-2}$ will be taken up in Chapter 9.

In many engineering problems, radiation is combined with other modes of heat transfer. The solution of such problems can often be simplified by using a thermal conductance K_r, or a thermal resistance R_r, for radiation. The definition of K_r is similar to that of K_k, the thermal conductance for conduction. If the heat transfer by radiation is written

$$q_r = K_r (T_1 - T_2') \tag{1.19}$$

the radiation conductance, by comparison with Eq. (1.12), is given by

$$K_r = \frac{A_1 \mathscr{F}_{1-2} \sigma (T_1{}^4 - T_2{}^4)}{T_1 - T_2'} \quad \text{W/K (Btu/h °F)} \tag{1.20}$$

The unit thermal radiation conductance, or *radiation heat transfer coefficient*, $\bar{h}_r$, is then

$$\bar{h}_r = \frac{K_r}{A_1} = \frac{\mathscr{F}_{1-2} \sigma (T_1{}^4 - T_2{}^4)}{T_1 - T_2'} \quad \text{W/m}^2 \text{ K (Btu/h ft}^2 \text{ °F)} \tag{1.21}$$

where T_2' is any convenient reference temperature, whose choice is often dictated by the convection equation, which will be discussed next. Similarly, the *thermal resistance for radiation* is

$$R_r = \frac{T_1 - T_2'}{A_1 \mathscr{F}_{1-2} \sigma (T_1{}^4 - T_2{}^4)} \tag{1.22}$$

EXAMPLE 1.3 ───

A long cylindrical electrically heated rod, 2 cm in diameter, is installed in a vacuum furnace as shown in Fig. 1.11. The surface of the heating rod has an emissivity of 0.9 and is maintained at 1000 K while the interior walls of the furnace are at 800 K. Calculate the net rate at which heat is lost from the rod per unit length and the radiation heat transfer coefficient.

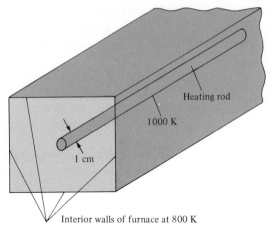

Figure 1.11   Schematic diagram of vacuum furnace with heating rod for Example 1.3.

Solution.   Assume that steady state has been reached. Moreover, note that since the walls of the furnace completely enclose the heating rod, all the radiant energy emitted by the surface of the rod is intercepted by the furnace walls. Thus, for a black enclosure Eq. (1.17) applies and the net heat loss from the rod or surface A_1 is

$$q_r = A\epsilon\sigma(T_1{}^4 - T_2{}^4) = \pi D_1 L\epsilon\sigma(T_1{}^4 - T_2{}^4)$$

$$= \frac{(\pi)(2)(1)}{100}(0.9)(5.67 \times 10^{-8})(1000^4 - 800^4)$$

$$= \frac{2\pi}{100}(0.9)(5.67 \times 10^{-8})(10{,}000 - 4096) \times 10^8$$

$$= 1893 \text{ W}$$

Note that in order for steady state to exist the heating rod must dissipate electrical energy at the rate of 1893 W and the rate of heat loss through the furnace walls must equal the rate of electric input to the system, that is, the rod.

From Eq. (1.17), $\mathscr{F}_{1-2} = \epsilon_1$ and therefore the radiation heat transfer coefficient, according to its definition in Eq. (1.21), is

$$h_r = \frac{\epsilon_1\sigma(T_1{}^4 - T_2{}^4)}{T_1 - T_2} = 150.6 \text{ W/m}^2 \text{ K}$$

1.5 COMBINED HEAT TRANSFER SYSTEMS

In the preceding sections the three basic mechanisms of heat transfer have been treated separately. In practice, however, heat is usually transferred by several of the basic mechanisms occurring simultaneously. For example, in the winter, heat is transferred from the roof of a house to the colder ambient environment not only by convection but also by radiation, while the heat transfer through the roof from the interior to the exterior surface is by conduction. Heat transfer between the panes of a double-glazed window occurs by convection and radiation acting in parallel, while the transfer through the panes of glass is by conduction with some radiation passing directly through the entire window system. In this section we will examine combined heat transfer problems. We will set up and solve these problems by dividing the heat transfer path into sections that can be connected in series, just like an electrical circuit, with heat being transferred in each section by one or more mechanisms acting in parallel. Table 1.5 summarizes the basic relations for the rate equation of each of the three basic heat transfer mechanisms to aid in setting up the thermal circuits for solving combined heat transfer problems.

1.5.1 Plane Walls in Series and Parallel

If heat is conducted through several plane walls in good thermal contact, as through a multilayer wall of a building, the rate of heat conduction is the same

TABLE 1.5 THE THREE MODES OF HEAT TRANSFER

One-dimensional conduction heat transfer through a stationary medium $$q_k = \frac{k_A}{L}(T_1 - T_2)$$ $$R_k = \frac{L}{kA}$$	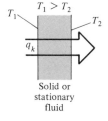 Solid or stationary fluid
Convection heat transfer from a surface to a moving fluid $$q_c = \bar{h}_c A(T_s - T_\infty)$$ $$R_c = \frac{1}{\bar{h}_c A}$$	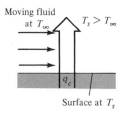 Surface at T_s
Net radiation heat transfer from surface 1 to surface 2 $$q_r = A_1 \mathscr{F}_{1-2}\sigma(T_1{}^4 - T_2{}^4)$$ $$R_r = \frac{T_1 - T_2}{A_1 \mathscr{F}_{1-2}\sigma(T_1{}^4 - T_2{}^4)}$$	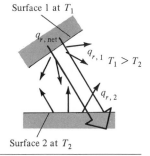 Surface 2 at T_2

PHYSICAL SYSTEM

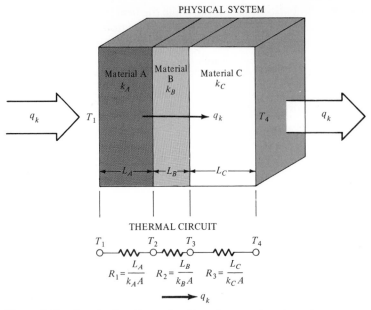

Figure 1.12 Conduction through a three-layer system in series.

through all sections. However, as shown in Fig. 1.12 for a three-layer system, the temperature gradients in the layers are different. The rate of heat conduction through each layer is q_k, and from Eq. (1.1) we get

$$q_k = \left(\frac{kA}{L}\right)_A (T_1 - T_2) = \left(\frac{kA}{L}\right)_B (T_2 - T_3) = \left(\frac{kA}{L}\right)_C (T_3 - T_4) \quad (1.23)$$

Eliminating the intermediate temperatures T_2 and T_3 in Eq. (1.23), q_k can be expressed in the form

$$q_k = \frac{T_1 - T_4}{(L/kA)_A + (L/kA)_B + (L/kA)_C}$$

Similarly, for N layers in series we have

$$q_k = \frac{\Delta T}{(L/kA)_n} = \frac{T_1 - T_{N+1}}{\sum\limits_{n=1}^{n=N} (L/kA)_n} \quad (1.24)$$

where T_1 is the outer-surface temperature of layer 1 and T_{N+1} is the outer-surface temperature of layer N. Using the definition of thermal resistance from Eq. (1.3), Eq. (1.24) becomes

$$q_k = \frac{T_1 - T_{N+1}}{\sum\limits_{n=1}^{n=N} R_{k,n}} = \frac{\Delta T}{\sum\limits_{n=1}^{n=N} R_{k,n}} \quad (1.25)$$

where ΔT is the overall temperature difference, often called the temperature potential. The flow of heat is proportional to the temperature potential.

There is an analogy between the flow of heat and electricity. The flow of electricity is directly proportional to the voltage potential divided by the sum of the electrical resistances in the circuit. This analogy will be found a convenient tool, especially for visualizing more complex situations. The following example illustrates its application to a simple problem.

EXAMPLE 1.4 ───────────────────────────────

Calculate the rate of heat loss from a furnace wall per unit area. The wall is constructed from an inner layer of 0.5-cm-thick steel ($k = 40$ W/m K) and an outer layer of 10-cm zirconium brick ($k = 2.5$ W/m K). The inner-surface temperature is 900 K and the outside surface temperature is 460 K. What is the temperature at the interface?

Solution. Assume that steady state exists, neglect effects at corners and edges of the wall, and assume that the surface temperatures are uniform. The physical system and the corresponding thermal circuit are similar to those in Fig. 1.12, but only two sections or walls are present. The rate of heat loss per unit area can be calculated from Eq. (1.24):

$$\frac{q_k}{A} = \frac{900 - 460}{0.005/40 + 0.1/2.5} = \frac{440}{0.000125 + 0.04} = 10{,}965 \text{ W/m}^2 \simeq 11 \text{ kW/m}^2$$

The interface temperature T_2 is obtained from

$$\frac{q_k}{A} = \frac{T_1 - T_2}{R_1}$$

Solving for T_2 gives

$$T_2 = T_1 - \frac{q_k}{A_1} R_1 = 900 - 10{,}965 \times 0.000125$$

$$= 888.6 \text{ K}$$

Note that the temperature drop across the steel interior wall is only 1.4 K because the thermal resistance of the wall is small compared to the resistance of the brick, across which the temperature drop is many times larger.

A contact or interface resistance can be integrated into the thermal circuit approach. The following example illustrates the procedure.

EXAMPLE 1.5 ───────────────────────────────

Two large aluminum plates, each 1 cm thick, with 10 μm surface roughness are placed on contact under 10^5 N/m^2 pressure in air. The temperatures at the outside surfaces are 395 and 405°C. Calculate (a) the heat flux and (b) the temperature drop due to the contact resistance.

Solution

(a) The rate of heat flow per unit area through the sandwich wall is

$$q'' = \frac{T_{s1} - T_{s3}}{R_1 + R_2 + R_3} = \frac{\Delta T}{(L/k)_1 + R_i + (L/k)_2}$$

From Table 1.3 the contact resistance R_i is 2.75×10^{-4} m² K/W while the other two resistances are equal to

$$(L/k) = (0.01 \text{ m}/240 \text{ W/m K}) = 4.17 \times 10^{-5} \text{ m}^2 \text{ K/W}$$

Hence, the heat flux is

$$q'' = \frac{(405 - 395)°C}{(4.17 \times 10^{-5} + 2.75 \times 10^{-4} + 4.17 \times 10^{-5}) \text{ m}^2 \text{ K/W}}$$
$$= 2.79 \times 10^4 \text{ W/m}^2 \text{ K}$$

(b) The temperature drop in each section of this one-dimensional system is proportional to the resistance. The fraction of the contact resistance is

$$R_i \bigg/ \sum_{n=1}^{3} R_n = 2.75/3.584 = 7.67$$

Hence 7.67°C of the total temperature drop of 10°C is the result of the contact resistance.

Conduction can occur in a section with two different materials in parallel. For example, Fig. 1.13 shows a slab with two different materials of areas A_A and A_B in parallel. If the temperatures over the left and right faces are uniform at T_1 and T_2, we can analyze the problem in terms of the thermal circuit shown to the right of the physical system. Since heat is conducted through the two materials along separate paths between the same potential, the total rate of heat

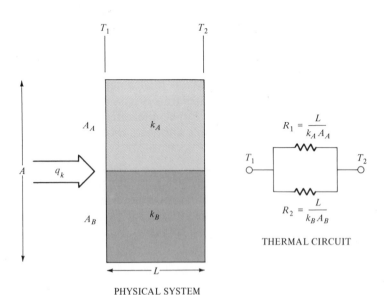

PHYSICAL SYSTEM

Figure 1.13   Heat conduction through a wall section with two paths in parallel.

flow is the sum of the flows through A_1 and A_2:

$$q_k = q_1 + q_2$$
$$= \frac{T_1 - T_2}{(L/kA)_A} + \frac{T_1 - T_2}{(L/kA)_B} = \frac{T_1 - T_2}{R_1 R_2/(R_1 + R_2)} \qquad (1.26)$$

Note that the total heat transfer area is the sum of A_A and A_B and that the total resistance equals the product of the individual resistances divided by their sum, as in any parallel circuit.

A more complex application of the thermal network approach is illustrated in Fig. 1.14, where heat is transferred through a composite structure involving thermal resistances in series and in parallel. For this system the resistance of the middle layer, R_2, in Fig. 1.14, becomes

$$R_2 = \frac{R_B R_C}{R_B + R_C}$$

and the rate of heat flow is

$$q_k = \frac{\Delta T_{\text{overall}}}{\sum\limits_{n=1}^{n=3} R_n} \qquad (1.27)$$

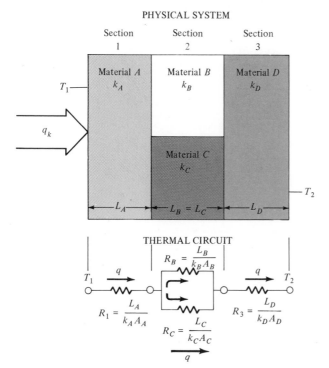

PHYSICAL SYSTEM

Figure 1.14 Conduction through a wall consisting of series and parallel thermal paths.

where N = number of layers in series (three)

R_n = thermal resistance of nth layer

$\Delta T_{\text{overall}}$ = temperature difference across two outer surfaces

By analogy to Eqs. (1.3) and (1.4), Eq. (1.27) can also be used to obtain an overall conductance between the two outer surfaces:

$$K_k = \left(\sum_{n=1}^{n=N} R_n \right)^{-1} \tag{1.28}$$

EXAMPLE 1.6

A layer of 2-in.-thick firebrick (k_b = 1.0 Btu/h ft °F) is placed between two 1/4-in.-thick steel plates (k_s = 30 Btu/h ft °F). The faces of the brick adjacent to the plates are rough, having solid-to-solid contact over only 30 percent of the total area, with the average height of asperities being 1/32 in. If the surface temperatures of the steel plates are 200° and 800°F, respectively, specify the rate of heat flow per unit area.

Solution.   The real system is first idealized by assuming that the asperities of the surface are distributed, as shown in Fig. 1.15. We note that the composite wall is symmetrical with respect to the center plane and therefore only consider half of the system. The overall unit conductance for half the

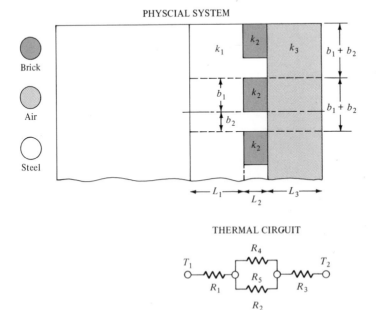

Figure 1.15 Thermal circuit for the parallel-series composite wall in Example 1.6. L_1 = 1 in.; L_2 = 1/32 in.; L_3 = 1/4 in.; T_1 is at the center.

composite wall is then, from Eq. (1.28),

$$K_k = \frac{1}{R_1 + [R_4 R_5/(R_4 + R_5)] + R_3}$$

from an inspection of the thermal circuit.

The thermal resistance of the steel plate R_3 is, on the basis of a unit area, equal to

$$R_3 = \frac{L_3}{k_s} = \frac{1/4}{(12)(30)} = 0.694 \times 10^{-3} \text{ (h ft}^2 \text{ °F/Btu)}$$

The thermal resistance of the brick asperities R_4 is, on the basis of a unit area, equal to

$$R_4 = \frac{L_2}{0.3k_b} = \frac{1/32}{(12)(0.3)(1.0)} = 8.7 \times 10^{-3} \text{ (h ft}^2 \text{ °F/Btu)}$$

Since the air is trapped in very small compartments, the effects of convection are small and it will be assumed that heat flows through the air by conduction. At a temperature of 300°F, the conductivity of air k_a is 0.02 Btu/h ft °F. Then R_5, the thermal resistance of the air trapped between the asperities, is, on the basis of a unit area, equal to

$$R_5 = \frac{L_2}{0.7k_a} = \frac{1/32}{(12)(0.7)(0.02)} = 187 \times 10^{-3} \text{ h ft}^2 \text{ °F/Btu}$$

The factors 0.3 and 0.7 in R_4 and R_5, respectively, represent the percent of the total area for the two separate heat flow paths.

The total thermal resistance for the two paths, R_4 and R_5 in parallel, is

$$R_2 = \frac{R_4 R_5}{R_4 + R_5} = \frac{(8.7)(187) \times 10^{-6}}{(8.7 + 187) \times 10^{-3}} = 8.3 \times 10^{-3} \text{ (h ft}^2 \text{ °F/Btu)}$$

The thermal resistance of half of the solid brick, R_1, is

$$R_1 = \frac{L_1}{k_b} = \frac{1}{(12)(1.0)} = 83.3 \times 10^{-3} \text{ (h ft}^2 \text{ °F/Btu)}$$

and the overall unit conductance is

$$K_k = \frac{1/2 \times 10^3}{83.3 + 8.3 + 0.69} = 5.4 \text{ Btu/h ft}^2 \text{ °F}$$

Inspection of the values for the various thermal resistances show that the steel offers a negligible resistance, while the contact section, although only 1/32 in. thick, contributes 10 percent to the total resistance. From Eq. (1.27), the rate of heat flow per unit area is

$$\frac{q}{A} = K_k \Delta T = 5.4(800 - 200) = 3240 \text{ Btu/h ft}^2$$

1.5.2 Convection and Conduction in Series

In the preceding section we have treated conduction through composite walls when the surface temperatures on both sides are specified. The more common problem encountered in engineering practice, however, is heat being transferred between two fluids separated by a wall with the fluid temperatures specified. In such a situation the surface temperatures are not known, but they can be calculated if the convection heat transfer coefficients on both sides of the wall are known.

Convection heat transfer can easily be integrated into a thermal network. From Eq. (1.14), the thermal resistance for convection heat transfer is

$$R_c = \frac{1}{\bar{h}_c A}$$

Figure 1.16 shows a situation where heat is transferred between two fluids separated by a wall. According to the thermal network shown below the physical system, the rate of heat transfer from the hot fluid h at temperature T_h to the cold fluid c at temperature T_c is

$$q = \frac{T_h - T_c}{\sum\limits_{n=1}^{n=3} R_i} = \frac{\Delta T}{R_1 + R_2 + R_3} \tag{1.29}$$

where　$R_1 = \dfrac{1}{(\bar{h}_c A)_{\text{hot}}}$

$R_2 = \dfrac{L}{kA}$

$R_3 = \dfrac{1}{(\bar{h}_c A)_{\text{cold}}}$

EXAMPLE 1.7 ―――――――――――――――――――――――――――――――

A 0.1-m-thick brick wall ($k = 0.7$ W/m K) is exposed to a cold wind at 270 K through a convection heat transfer coefficient of 40 W/m² K. On the other side is calm air at 330 K, with a free-convection heat transfer

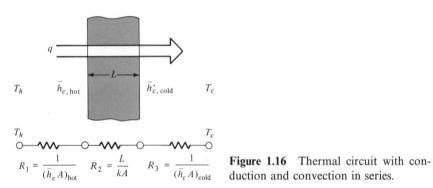

Figure 1.16 Thermal circuit with conduction and convection in series.

coefficient of 10 W/m^2 K. Calculate the rate of heat transfer per unit area (i.e., the heat flux).

Solution.   The three resistances are

$$R_1 = \frac{1}{\bar{h}_{c,\text{hot}}A} = \frac{1}{(10)(1)} = 0.10 \text{ K/W}$$

$$R_2 = \frac{L}{kA} = \frac{0.1}{(0.7)(1)} = 0.143 \text{ K/W}$$

$$R_3 = \frac{1}{\bar{h}_{c,\text{cold}}A} = \frac{1}{(40)(1)} = 0.025 \text{ K/W}$$

and the rate of heat transfer per unit area is from Eq. (1.29)

$$\frac{q}{A} = \frac{\Delta T}{R_1 + R_2 + R_3} = \frac{(330 - 270) \text{ K}}{(0.10 + 0.143 + 0.025) \text{ K/W}} = 223.9 \text{ W}$$

The same approach as used in Example 1.7 can also be used for composite walls, and Fig. 1.17 shows the structure, temperature distribution, and equivalent network for a wall with three layers and convection on both surfaces.

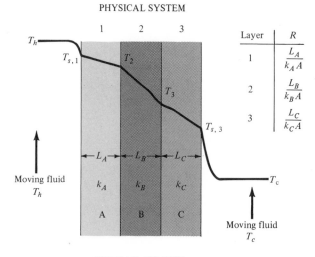

Figure 1.17   Schematic diagram and thermal circuit for composite three-layer wall with convection over both exterior surfaces.

1.5.3 Convection and Radiation in Parallel

In many engineering problems a surface loses or receives thermal energy simultaneously by convection and radiation. For example, the roof of a house heated from the interior is at a higher temperature than the ambient air and thus loses heat by convection as well as radiation. Since both heat flows emanate from the same potential, that is, the roof, they act in parallel. Similarly, the gases in a combustion chamber contain species that emit and absorb radiation. Consequently, the wall of the combustion chamber receives heat by convection as well as radiation. Figure 1.18 illustrates the cocurrent heat transfer from a surface to its surrounding by convection and radiation. The total rate of heat transfer is the sum of the rates of heat flow by convection and radiation, or

$$\begin{aligned} q &= q_c + q_r \\ &= \bar{h}_c A(T_1 - T_2) + h_r A(T_1 - T_2) \\ &= (\bar{h}_c + h_r)A(T_1 - T_2) \end{aligned} \tag{1.30}$$

where $\bar{h}_c$ is the average convection heat transfer coefficient between area A_1 and the ambient air at T_2, and, as shown previously, the radiation heat transfer

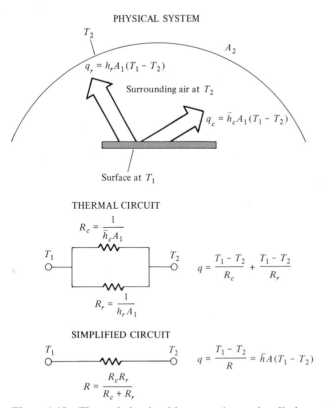

Figure 1.18 Thermal circuit with convection and radiation acting in parallel.

coefficient between A_1 and the surroundings at T_2 is

$$\bar{h}_r = \frac{\epsilon_1 \sigma(T_1{}^4 - T_2{}^4)}{T_1 - T_2} \tag{1.31}$$

The analysis of combined heat transfer, especially at boundaries of a complicated geometry or in unsteady-state conduction, can often be simplified by using an effective unit surface conductance that combines convection and radiation. The combined unit surface conductance, or *unit surface conductance* for short, is defined by

$$\bar{h} = \bar{h}_c + h_r \tag{1.32}$$

The unit surface conductance specifies the average total rate of heat flow between a surface and an adjacent fluid and the surroundings per unit surface area and unit temperature difference between the surface and the fluid. Its units are W/m^2 K.

EXAMPLE 1.8 ————————————————————————————

A 0.5-m-diameter pipe ($\epsilon = 0.9$) carrying steam has a surface temperature of 500 K. The pipe is located in a room at 300 K and the convection heat transfer coefficient between the pipe surface and the air in the room is 20 W/m^2 K. Calculate the combined unit surface conductance and the rate of heat loss per meter of pipe length.

Solution. This problem may be idealized as a small object (the pipe) inside a large black enclosure (the room). Noting that

$$\frac{T_1{}^4 - T_2{}^4}{T_1 - T_2} = (T_1{}^2 + T_2{}^2)(T_1 + T_2)$$

the radiation heat transfer coefficient is, from Eq. (1.31),

$$h_r = \sigma\epsilon(T_1{}^2 + T_2{}^2)(T_1 + T_2) = 13.9 \text{ W/m}^2 \text{ K}$$

The combined unit surface conductance is, from Eq. (1.32),

$$h = \bar{h}_c + h_r = 20 + 13.9 = 33.9 \text{ W/m}^2 \text{ K}$$

and the rate of heat loss per meter is

$$q = \pi D L h(T_{\text{pipe}} - T_{\text{air}}) = \pi(0.5)(1)(33.9)(200) = 10,650 \text{ W}$$

1.5.4 Overall Heat Transfer Coefficient

We noted previously that a common heat transfer problem is to determine the rate of heat flow between two fluids, gaseous or liquid, separated by a wall. If the wall is plane and heat is transferred only by convection on both sides, the rate of heat transfer in terms of the two fluid temperatures is given by Eq. (1.29), or

$$q = \frac{T_h - T_c}{(1/h_c A)_h + (L/kA) + (1/h_c A)_c} = \frac{\Delta T}{R_1 + R_2 + R_3}$$

where the subscripts h and c denote hot and cold and T_h and T_c are the temperatures of the hot and cold fluids, respectively.

In Eq. (1.29) the rate of heat flow is expressed only in terms of an overall temperature potential and the heat transfer characteristics of individual sections in the heat flow path. From these relations it is possible to evaluate quantitatively the importance of each individual thermal resistance in the path. Inspection of the order of magnitudes of the individual terms in the denominator often indicates means of simplifying a problem. When one term dominates quantitatively, it is sometimes permissible to neglect the rest. As we gain facility in the techniques of determining individual thermal resistances and conductances, there will be numerous occasions where such approximations will be illustrated. There are, however, certain types of problems, notably in the design of heat exchangers, where it is convenient to simplify the writing of Eq. (1.29) by combining the individual resistances or conductances of the thermal system into one quantity, called the overall unit conductance, the overall transmittance, or the overall coefficient of heat transfer U. The use of an overall coefficient is a convenience in notation, and it is important not to lose sight of the significance of the individual factors that determine the numerical value of U.

Writing Eq. (1.29) in terms of an overall coefficient gives

$$q = U A \, \Delta T_{\text{total}} \tag{1.33}$$

where

$$U A = \frac{1}{R_1 + R_2 + R_3} = \frac{1}{R_{\text{total}}} \tag{1.34}$$

The overall coefficient U may be based on any chosen area. This becomes particularly important in heat transfer through the walls of tubes in a heat exchanger, and to avoid misunderstandings the area basis of an overall coefficient should always be stated. Additional information about the overall heat transfer coefficient U will be presented in later chapters.

An overall heat transfer coefficient can also be obtained in terms of individual resistances in the thermal circuit when convection and radiation transfer heat to and/or from one or both surfaces of the wall. In general, radiation will not be of any significance when the fluid is a liquid, but can play an important role in convection to or from a gas when the temperatures are high or the convection heat transfer coefficient is small, for instance, in free convection. The integration of radiation into an overall heat transfer coefficient will be illustrated below.

The schematic diagram in Fig. 1.19 shows the heat transfer from hot products of combustion in the chamber of a rocket motor through a wall that is liquid-cooled on the outside by convection. In the first section of this system heat is transferred by convection and radiation in parallel. Hence, the rate of heat flow to the interior surface of the wall is the sum of the two heat flows

$$\begin{aligned}
q &= q_c + q_r \\
&= \bar{h}_c A(T_g - T_{sg}) + h_r A(T_g - T_{sg}) \\
&= (\bar{h}_{c1} + h_{r1}) A(T_g - T_{sg}) = \frac{T_g - T_{sg}}{R_1}
\end{aligned} \tag{1.35}$$

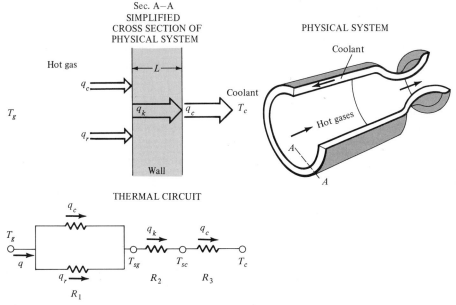

Figure 1.19 Heat transfer from combustion gases to a liquid coolant in a rocket motor.

where $T_g = T_1$ = temperature of the hot gas in the interior

$T_{sg} = T_2$ = temperature of the hot wall surface

$$h_{r1} = \frac{\sigma A(T_g^{\,4} - T_{sg}^{\,4})}{T_g - T_{sg}} = \text{the radiation heat transfer coefficient in the first section (ϵ is assumed unity)}$$

$\bar{h}_{c1}$ = convection heat transfer coefficient from gas to wall

$$R_1 = \frac{1}{(h_r + \bar{h}_{c1})A} = \text{combined thermal resistance of first section}$$

In the steady state, heat is conducted through the shell, the second section of the system, at the same rate as to the surface and

$$q = q_k = \frac{kA}{L}(T_{sg} - T_{sc})$$

$$= \frac{T_{sg} - T_{sc}}{R_2} \tag{1.36}$$

where T_{sc} = surface temperature at wall on coolant side

R_2 = thermal resistance of second section

After passing through the wall, the heat flows through the third section of the system by convection to the coolant. The rate of heat flow in the last step is

$$q = q_c = \bar{h}_{c3}A(T_{sc} - T_c)$$

$$= \frac{T_{sc} - T_c}{R_3} \tag{1.37}$$

where T_c = temperature of coolant

       R_3 = thermal resistance in third section of system

It should be noted that the symbol $\bar{h}_c$ stands for average convection heat transfer coefficient in general, but the numerical values of the convection coefficients in the first, $\bar{h}_{c1}$, and third, $\bar{h}_{c3}$, sections of the system depend on many factors and will, in general, be different. Also, the areas of the three heat flow sections are not equal. But since the wall is very thin, the change in the heat flow area is so small that it can be neglected in this system.

In practice, often only the temperatures of the hot gas and the coolant are known. If intermediate temperatures are eliminated by algebraic addition of Eqs. (1.35), (1.36), and (1.37), the rate of heat flow is

$$q = \frac{T_g - T_c}{R_1 + R_2 + R_3} = \frac{\Delta T_{\text{total}}}{R_1 + R_2 + R_3} \tag{1.38}$$

where the thermal resistances of the three series-connected sections or heat flow steps in the system are defined in Eqs. (1.35), (1.36), and (1.37).

EXAMPLE 1.9 _____

In the design of a heat exchanger for aircraft application (Fig. 1.20), the maximum wall temperature in steady state is not to exceed 800 K. For the conditions tabulated below, determine the maximum permissible unit thermal resistance per square meter of the metal wall between hot gas on the one side and cold gas on the other.

PHYSICAL SYSTEM

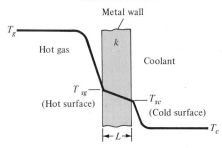

DETAILED THERMAL CIRCUIT

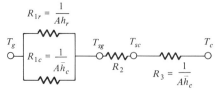

SIMPLIFIED CIRCUIT

Figure 1.20   Physical system and thermal circuit for Example 1.9.

Hot-gas temperature $= T_g = 1300$ K

Combined unit surface conductance on hot side $= \bar{h}_1 = 200$ W/m² K

Combined unit surface conductance on cold side $= \bar{h}_3 = 400$ W/m² K

Coolant temperature $= T_c = 300$ K

Solution.   In the steady state we can write

$\dfrac{q}{A}$ from hot gas to hot side of wall

$$= \frac{q}{A} \text{ from hot side of wall through wall to cold gas}$$

Using the nomenclature in Fig. 1.20, we get

$$\frac{q}{A} = \frac{T_g - T_{sg}}{R_1} = \frac{T_g - T_c}{R_1 + R_2 + R_3}$$

where T_{sg} is the hot-surface temperature. Substituting numerical values for the unit thermal resistances and temperatures yields

$$\frac{1300 - 800}{1/200} = \frac{1300 - 300}{1/200 + R_2 + 1/400}$$

$$\frac{1300 - 800}{0.005} = \frac{1300 - 300}{R_2 + 0.0075}$$

Solving for R_2 gives

$$R_2 = 0.0025 \text{ m}^2 \text{ K/W}$$

Thus, a unit thermal resistance larger than 0.0025 m² K/W for the wall would raise the inner-wall temperature above 800 K. This can place an upper limit on the wall thickness.

1.6* HEAT TRANSFER AND THE LAW OF ENERGY CONSERVATION

In addition to the heat transfer rate equations we shall also often use the first law of thermodynamics, the law of conservation of energy, in analyzing a system. Although, as mentioned previously, a thermodynamic analysis alone cannot predict the rate at which the transfer will occur in terms of the degree of thermal nonequilibrium, the basic laws of thermodynamics must be obeyed and any physical law that must be satisfied by a process or a system provides an equation that can be used for analysis. We have already used the second law of thermodynamics to indicate the direction of heat flow. We will now demonstrate how the first law of thermodynamics can be applied in the analysis of heat transfer problems.

1.6.1 First Law of Thermodynamics

The first law of thermodynamics states that energy cannot be created or destroyed, but may be transformed from one form to another or transferred as heat or work. To apply the law of conservation of energy, we first need to identify a *control volume*. A control volume is a fixed region in space bounded by a *control surface* through which heat, work, and mass can pass. The conservation of energy requirement for an open system in a form useful for heat transfer analysis is:

> The rate at which thermal and mechanical energies enter a control volume plus the rate at which energy is generated within that volume minus the rate at which thermal and mechanical energies leave the control volume must equal the rate at which energy is stored inside this volume.

If the sum of the energy inflow and the generation exceeds the outflow, there will be an increase in the amount of energy stored in the control volume, whereas when the outflow exceeds the inflow and generation there will be a decrease in energy storage. But when there is no generation and the rate of energy inflow is equal to the rate of outflow, steady state exists and there is no change in the energy stored in the control volume.

Referring to Fig. 1.21, the energy conservation requirements may be expressed in the form

$$(e\dot{m})_{\text{in}} + q + \dot{q}_G - (e\dot{m})_{\text{out}} - W_{\text{out}} = \frac{\partial E}{\partial t} \tag{1.39}$$

where $(e\dot{m})_{\text{in}}$ is the rate of energy inflow, $(e\dot{m})_{\text{out}}$ is the rate of energy outflow, q is the *net* rate of heat transfer into the control volume $(q_{\text{in}} - q_{\text{out}})$, W_{out} is the net rate of work output, $\dot{q}_G$ is the rate of energy generation within the control volume, and $\partial E/\partial t$ is the rate of energy storage inside the control volume.

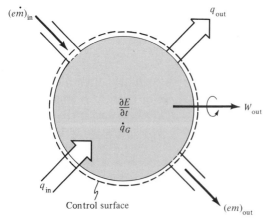

Figure 1.21   Control volume for first law of thermodynamics or conservation of energy.

The specific energy carried by the mass flow, e, across the surface may contain potential and kinetic as well as thermal (internal) forms. But for most heat transfer problems the potential and kinetic energy terms are negligible. The inflow and outflow energy terms may also include work interactions, but these phenomena are of significance only in extremely high-speed flow processes.

Observe that the inflow and outflow rate terms are surface phenomena and are, therefore, proportional to the surface area. The internal energy generation term $\dot{q}_G$ is encountered when a form of energy (such as chemical, electrical, or nuclear energy) is converted to thermal energy within the control volume. The generation term is therefore a volumetric phenomenon and its rate is proportional to the volume within the control surface. Energy storage is also a volumetric phenomenon associated with the internal energy of the mass in the control volume. But the process of energy generation is quite different from that of energy storage, although both will contribute to the rate of energy storage.

Equation (1.39) can be simplified when there is no transport of mass across the boundary. Such a system is called a *closed system*, and Eq. (1.39) for such conditions becomes

$$q + \dot{q}_G - W_{\text{out}} = \frac{\partial E}{\partial t} \tag{1.39a}$$

where the right side represents the rate of energy storage or the rate of increase in internal energy. Note that E is the total internal energy stored in the system and equals the product of the specific internal energy and the mass of the system.

1.6.2 Conservation of Energy Applied to Heat Transfer Analysis

The following two examples demonstrate the use of the energy conservation law in heat transfer analysis. The first example is a steady-state problem in which the storage term is zero, whereas the second example demonstrates the analytic procedure for a problem in which internal energy storage occurs. The latter is called *transient heat transfer*, and a more detailed analysis of such cases will be presented in the next chapter.

EXAMPLE 1.10 ————————————————————————

A house has a black tar, flat horizontal roof. The lower surface of the roof is well insulated, while the upper surface is exposed to ambient air at 300 K through a convective heat transfer coefficient of 10 W/m² K. Calculate the roof equilibrium temperature for the following conditions: (a) a clear sunny day with an incident solar radiation flux of 500 W/m² and the ambient sky at an effective temperature of 50 K, and (b) a clear night with an ambient sky temperature of 50 K.

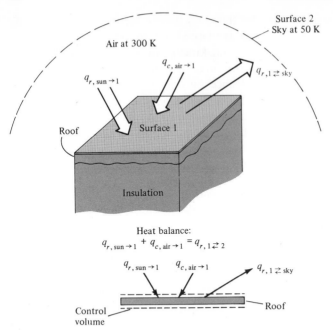

Figure 1.22 Heat transfer by convection and radiation for roof in Example 1.10.

Solution. A schematic sketch of the system is shown in Fig. 1.22. The control volume is the roof. Assume that there are no obstructions between the roof, called surface 1, and the sky, called surface 2, and ʻhat both surfaces are black. The sky behaves as a blackbody because it absorbs all the radiation emitted by the roof and reflects none. Heat is transferred by convection between the ambient air and the roof and by radiation between the sun and the roof and the roof and the sky. This is a closed system in thermal equilibrium and we can express the energy conservation requirement by the conceptual relation

$$\text{Rate of solar radiation heat} \atop \text{transfer } to \text{ roof} \quad + \quad {\text{rate of convection heat} \atop \text{transfer } to \text{ roof}}$$
$$= \quad {\text{net rate of radiation} \atop \text{heat transfer } from \atop \text{roof to ambient sky}}$$

Analytically, this relation can be cast in the form

$$A_1 q_{r,\text{sun}\to\text{roof}} + \bar{h}_c A_1 (T_{\text{air}} - T_{\text{roof}}) = A_1 q_{r,\text{roof}\to\text{ambient sky}}$$

Canceling the roof area A_1 and substituting the Stefan-Boltzmann relation [Eq. (1.17)] for the net radiation from the roof to the ambient sky gives

$$q_{r,\text{sun}\to 1} + \bar{h}_c (300 - T_{\text{roof}}) = \sigma(T_{\text{roof}}^4 - T_{\text{sky}}^4)$$

(a) When the solar radiation to the roof, $q_{r,\,\text{sun}\to1}$, is 500 W/m² and T_{sky} is 50 K, we get

$$500 + 10(300 - T_{\text{roof}}) = 5.67 \times 10^{-8}(T_{\text{roof}}^4 - 50^4)$$

Solving by trial and error for the roof temperature, we get

$$T_{\text{roof}} = 303 = 30°C$$

Note that the convection term is negative because the sun heats the roof to a temperature above the ambient air, so that the roof is not heated but is cooled by convection to the air.

(b) At night the term $q_{r,\,\text{sun}\to1} = 0$ and we get, upon substituting the numerical data in the conservation of energy relation,

$$\bar{h}_c(T_{\text{air}} - T_{\text{roof}}) = \sigma(T_{\text{roof}}^4 - T_{\text{sky}}^4)$$

or

$$10(300 - T_{\text{roof}}) = 5.67 \times 10^{-8}(T_{\text{roof}}^4 - 50^4)$$

Solving this equation for T_{roof} gives

$$T_{\text{roof}} = 270 \text{ K} = -3°C$$

At night the roof is cooler than the ambient air and convection occurs from the air to the roof, which is heated in the process. Observe also that the conditions at night and during the day are assumed to be steady and that the change from one steady condition to the other requires a period of transition in which the energy stored in the roof changes and the roof temperature also changes. The energy stored in the roof increases during the morning hours and decreases during the evening after the sun has set, but this period was not considered in the preceding example.

EXAMPLE 1.11 ────────────────────────────────────

A long, thin copper wire of diameter D and length L has an electrical resistance of ρ_e per unit length. The wire is initially at steady state in a room at temperature T_{air}. At time $t = 0$ an electric current I is passed through the wire. The wire temperature begins to increase due to internal electrical heat generation, but at the same time heat is lost from the wire by convection through a convection coefficient $\bar{h}_c$ to the ambient air.

Set up an equation to determine the change in temperature with time in the wire, assuming that the wire temperature is uniform. This is a good assumption because the thermal conductivity of copper is very large and the wire is thin. We will learn in Chapter 2 how to calculate the transient radial temperature distribution if the conductivity is small.

Solution.   The sketch in Fig. 1.23 shows the wire and the control volume. We shall assume that radiation losses are negligible so that the net rate of heat inflow, q, is negative and equal to the rate of heat loss from the

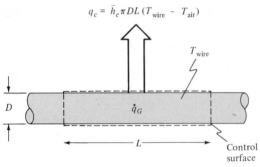

Figure 1.23 Schematic diagram for electric generation system of Example 1.11.

wire, $\dot{q}_{out}$:

$$\dot{q}_{out} = \bar{h}_c A_{surf}(T_{wire} - T_{air}) = \bar{h}_c \pi D L(T_{wire} - T_{air})$$

The rate of energy generation (or electrical dissipation) in the wire-control volume is

$$\dot{q}_G = I^2 R_e = I^2 \rho_e L$$

where $R_e = \rho_e L$, the electrical resistance.

The rate of internal energy storage in the control volume is

$$\frac{\partial E}{\partial t} = \frac{d[(\pi D^2/4)Lc\rho T_{wire}(t)]}{dt}$$

where c is the specific heat and ρ is the density of the wire material.

Applying the conservation of energy relation for a closed system, as given by Eq. (1.39a), to the problem at hand gives

$$\dot{q}_G - q_{out} = \frac{\partial E}{\partial t}$$

since there is no work output and q_{in} is zero.

Substituting the appropriate relations for the three energy terms in the conservation of energy law gives the differential equation

$$I^2 \rho_e L - (\bar{h}_c \pi D L)(T_{wire} - T_{air}) = \left(\frac{\pi D^2}{4} Lc\rho\right)\frac{dT_{wire}(t)}{dt}$$

If the specific heat and density are constant, the solution to this equation for the wire temperature as a function of time, $T(t)$, becomes

$$T_{wire}(t) - T_{air} = C_1(1 - e^{-C_2 t})$$

where $C_1 = \dfrac{I^2 \rho_e}{\bar{h}_c \pi D}$

$C_2 = \dfrac{4\bar{h}_c}{c\rho D}$

Note that as $t \to \infty$, the second term on the right-hand side approaches C_1 and $dT_{\text{wire}}/dt \to 0$. This means physically that *the wire temperature has reached a new equilibrium* value that can be evaluated from the steady-state conservation relation $q_{\text{out}} = \dot{q}_G$ or,

$$(T_{\text{wire}} - T_{\text{air}})\bar{h}_c \pi D L = I^2 \rho_e L$$

Thermodynamics alone, that is, the law of energy conservation, could predict the differences in the internal energy stored in the control volume between the two equilibrium states at $t = 0$ and $t \to \infty$, but it could not predict the rate at which the change occurs. For that calculation it is necessary to use the heat transfer rate analysis shown above.

1.6.3 Boundary Conditions

There are many situations in which the conservation of energy requirement is applied at the surface of a system. In these cases the control surface contains no mass and the volume it encompasses approaches zero, as shown in Fig. 1.24. Consequently, there can be no storage or generation of energy and the conservation requirement reduces to

$$q_{\text{net}} = q_{\text{in}} - q_{\text{out}} = 0 \qquad (1.40)$$

It is important to note that in this form the conservation law holds for steady-state as well as transient conditions and that the heat inflow and outflow may

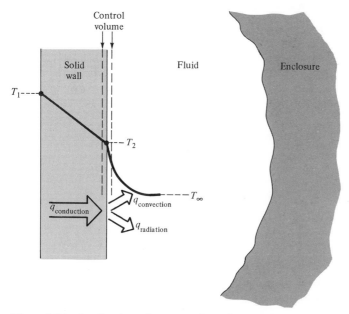

Figure 1.24 Application of conservation of energy law at the surface of a system.

occur by several heat transfer mechanisms in parallel. Applications of Eq. (1.40) to many different physical situations will be illustrated later.

1.7 DIMENSIONS AND UNITS

This section introduces systems of units and defines the system used in this book. It is important not to confuse the meaning of the terms units and dimensions. *Dimensions* are our basic concepts of measurements such as length, time, and temperature. For example, the distance between two points is a dimension called length. Units are the means of expressing dimensions numerically, for instance, meter or foot for length, second or hour for time. Before numerical calculations can be made, dimensions must be quantified by units.

Several different systems of units are in use throughout the world. The SI system (Système International d'Unités) has been adopted by the International Organization for Standardization and is recommended by most U.S. national standard organizations. We will therefore use the SI system of units in this book. But in the United States the English system of units is still widely used. It is, therefore, important to be able to change from one set of units to another. To be able to communicate with engineers who are still in the habit of using the English system, several examples in the book will be worked in the English system.

The basic SI units are those for length, mass, time, and temperature. The unit of force, the newton, is obtained from Newton's second law of motion, which states that force is proportional to the time rate of change of momentum. For a given mass, Newton's law can be written in the form

$$F = \frac{1}{g_c} ma \qquad (1.41)$$

where F is the force, m is the mass, a is the acceleration, and g_c is a constant whose numerical value and units depend on those selected for F, m, and a.

In the SI system the unit of force, the newton, is defined as

$$1 \text{ newton} = \frac{1}{g_c} \times 1 \text{ kg} \times 1 \text{ m/s}^2$$

Thus we see that

$$g_c \equiv 1 \text{ kg m/newtons s}^2$$

In the English system we have the relation

$$1 \text{ lb}_f = \frac{1}{g_c} \times 1 \text{ lb}_m \times g \text{ ft/s}^2$$

The numerical value of the conversion constant g_c is determined by the acceleration imparted to a 1-lb mass by a 1-lb force, or

$$g_c = 32.174 \text{ ft lb}_m/\text{lb}_f \text{ s}^2$$

The weight of a body, W, is defined as the force exerted on the body by gravity. Thus,

$$W = \frac{g}{g_c} m$$

where g is the local acceleration due to gravity. Weight has the dimensions of a force and 1 kg_{mass} will weigh 1 kg_{force} at sea level.

It should be noted that g and g_c are not similar quantities. The gravitational acceleration g depends on the location and the altitude, whereas g_c is a constant whose value depends on the system of units. One of the great conveniences of the SI system is that g_c is identically equal to one and therefore need not be shown specifically. In the English system, on the other hand, the omission of g_c will affect the numerical answer and it is therefore imperative that it be included and clearly displayed in analysis and especially in numerical calculations.

In the SI system with the fundamental units of meter, kilogram, second, and kelvin, the units for both force and energy or heat are derived units. The joule (newton meter) is the only energy unit in the SI system, and the watt (joule per second) is the corresponding unit of power. In the engineering system of units, on the other hand, the Btu is the unit for heat or energy. It is based on thermal phenomena and is defined as the energy required to raise 1 lb_m of water 1°F at 68°F.

The conversions between the Btu and other common energy units are given below:

$$1 \text{ Btu} = 1054.35 \text{ J (or newton meter)}$$
$$1 \text{ Btu} = 778.16 \text{ ft lb}_f$$

The SI unit of temperature is the kelvin, but use of the Celsius temperature scale is widespread and generally considered permissible. The kelvin is based on the thermodynamic scale, while zero on the Celsius scale (0°C) corresponds to the freezing temperature of water and is equivalent to 273.15 K on the thermodynamic scale. Thus, the relation between a temperature in kelvin and degrees Celsius or centigrade is

$$K = 273.15 + °C$$

Note, however, that temperature differences are numerically equivalent in K and °C, since 1 K is equal to 1°C.

In the engineering system of units the temperature is usually expressed in degrees Fahrenheit (°F) or, on the thermodynamic temperature scale, in degrees Rankine (°R). Here 1 K is equal to 1.8°R and conversions for other temperature scales are given below:

$$°F = 1.8°C + 32$$
$$°R = °F + 459.69$$
$$°R = 1.8 K$$
$$°C = \frac{°F - 32}{1.8}$$

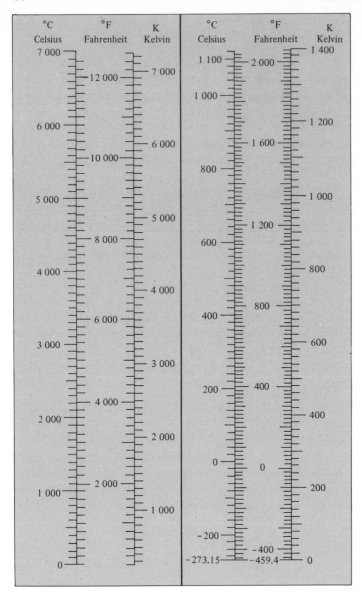

Figure 1.25 Nomograph for temperature conversion.

In order to facilitate conversion, Fig. 1.25 displays the relation between °C, °F, and K in the form of a nomograph.

EXAMPLE 1.12

One surface of a 2-cm-thick copper plate is maintained at 300 K and the other face at 250 K. Calculate the rate of heat transfer per unit area through the plate in SI units and then convert the answer into the English system of units.

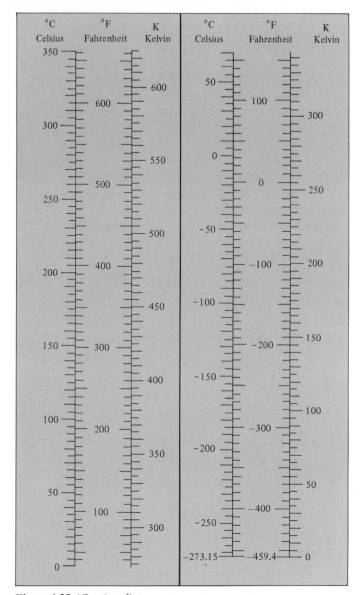

Figure 1.25 (*Continued*)

Solution. From Appendix 2 the thermal conductivity of copper at the average temperature between the two faces is 401 W/m K. From Eq. (1.1)

$$\frac{q}{A} = q'' = k\frac{\Delta T}{\Delta x} = (401 \text{ W/m K})\frac{(300 - 250) \text{ K}}{0.02 \text{ m}}$$
$$= 1,002,500 \text{ W/m}^2 = 1.003 \text{ MW/m}^2$$

To convert to English units, note that 1 Btu/h ft² = 3.413 W/m². Thus,

$$1,002,500 \text{ W/m}^2 \times \frac{1 \text{ Btu/h ft}^2}{3.413 \text{ W/m}^2} = 2.94 \times 10^5 \text{ Btu/h ft}^2$$

To avoid mistakes in changing from one set of units to another, simply treat the units as algebraic symbols and include the units of every conversion factor. The technique is illustrated below.

EXAMPLE 1.13

An empirical relation to determine the average heat transfer coefficient for air flow in a pipe is given by

$$\bar{h}_c = 0.10 \frac{V^{0.3}}{D^{0.7}}$$

where $\bar{h}_c$ = heat transfer coefficient, Btu/h ft² °F

V = velocity, ft/s

D = inside diameter, ft

If $\bar{h}_c$ is to be expressed in watts per square meter per kelvin, what should the constant in place of 0.10 be?

Solution

$$h = \frac{0.1 V^{0.3}}{D^{0.7}} \times \frac{\text{Btu}}{\text{h ft}^2 \, °\text{F}} \times \frac{1054 \, \text{J}}{\text{Btu}} \times \frac{1 \, \text{h}}{3600 \, \text{s}} \left(\frac{1 \, \text{ft}}{12 \, \text{in.}} \right)^2$$

$$\times \left(\frac{1 \, \text{in.}}{0.0254 \, \text{m}} \right)^2 \times \frac{1.8°\text{F}}{1 \, \text{K}}$$

$$= 0.567 \frac{V^{0.3}}{D^{0.7}} \frac{\text{W}}{\text{m}^2 \, \text{K}}$$

It is left as an exercise to determine the constant when V and D are also to be expressed in SI units (see Problem 1.2).

1.8 CLOSURE

This chapter has presented an overview of the field of heat transfer and its relation to thermodynamics. The material in this introductory chapter will be treated in more detail in subsequent chapters, but at this point you should be aware of the basic modes of heat transfer and their physical mechanisms. At the outset you should try to develop a facility for analyzing thermal systems in terms of the relevant transport phenomena and try to visualize the system in terms of its thermal circuit. At the same time, it is important to remember the conservation laws and define the appropriate control volume for a given situation. Table 1.5 presents a summary of the heat transfer rate processes and the appropriate rate equations.

Although there is no simple procedure for analyzing thermal systems that can be applied to all situations, there is a series of steps that engineers have found useful in their professional practice. These steps are outlined below:

1. Carefully read the problem and ask yourself *in your own words* what is known about the system, what information can be obtained from sources such as tables of properties, handbooks, or appendices, and what are the unknowns for which an answer must be found.
2. Draw a schematic diagram of the system, including the boundaries to be used in the application of conservation laws. Identify the relevant heat transfer processes and sketch a thermal circuit for the system.
3. State all the simplifying assumptions that you feel are appropriate for the solution of the problem and flag those that will need to be verified after an answer has been obtained. Pay particular attention to whether the system is in the steady or unsteady state. Also, compile the physical properties necessary for analyzing the system and cite the sources from which they were obtained.
4. Analyze the problem by means of the appropriate conservation laws and rate equations, using, wherever possible, insight into the processes and intuition. As you develop more insights, refer back to the thermal circuit and modify it if appropriate. Perform the numerical calculations in a step-by-step manner so that you can easily check your results by an order-of-magnitude analysis.
5. Comment on the results you have obtained and discuss any questionable points, in particular as they apply to the original assumptions. Then summarize the key conclusions at the end.

As you progress in your studies of heat transfer in subsequent chapters of the book, the procedure outlined above will become more meaningful and you may wish to refer to it as you begin to analyze and design more complex thermal systems.

As a final comment, it should be borne in mind that the subject of heat transfer is in a constant state of evolution and an engineer is well advised to follow the current heat transfer literature if he or she wishes to keep up to date. The most important serial publications that present new findings in heat transfer are listed at the end of the appendices. In addition to serial publications, the engineer will find it useful to refer from time to time to handbooks that periodically summarize the current state of knowledge. One such handbook that has been widely used is the *Handbook of Heat Transfer*, which has recently been updated in its second edition. This handbook contains many details that cannot be covered in an introductory text, and we have made an effort to identify the relations between the material in this text and the more detailed expositions of specific aspects of heat transfer that can be found in the second edition of the *Handbook of Heat Transfer* (5).

PROBLEMS

The problems for this chapter are organized by subject matter as shown below.

Topic	Section	Problem number
Conversion of units	1.7	1.1–1.4
Conduction	1.2	1.5–1.9
Convection	1.3	1.10–1.15
Radiation	1.4	1.16–1.21
Combined heat transfer mechanisms	1.5 and 1.6	1.22–1.31
Thermal circuits	1.5 and 1.6	1.32–1.35

1.1 The heat transfer coefficient between a surface and a liquid is 10 Btu/h ft^2 °F. How many watts per square meter will be transferred in this system if the temperature difference is 10°C?

1.2 Determine the constant in Example 1.13 when V and D are in the SI system of units.

1.3 The thermal conductivity of asbestos at 86°F is 0.025 Btu/h ft °F. What is its value in watts per square centimeter per degree centigrade per centimeter?

1.4 The thermal conductivity of silver at 212°F is 238 Btu/h ft °F. What is the conductivity in watts per meter per kelvin?

1.5 The outer surface of a 0.2-m-thick concrete wall is kept at a temperature of −5°C, while the inner surface is kept at 20°C. The thermal conductivity of the concrete is 1.2 W/m K. Determine the heat loss through a wall 10 m long and 3 m high.

1.6 If the weight, not the space, required for insulation of a plane wall is most significant, show analytically that the lightest insulation for a specified thermal resistance is that insulation which has the smallest product of density times thermal conductivity.

1.7 A furnace wall is to be constructed of bricks having standard dimensions 9 by $4\frac{1}{2}$ by 3 in. Two kinds of material are available. One has a maximum usable temperature of 1900°F and a thermal conductivity of 1 Btu/h ft °F, and the other has a maximum temperature limit of 1600°F and a thermal conductivity of 0.5. The bricks cost the same and can be laid in any manner, but we wish to design the most economical wall for a furnace with a temperature on the hot side of 1900°F and on the cold side of 400°F. If the maximum amount of heat transfer permissible is 300 Btu/h for each square foot of area, determine the most economical arrangements for the available bricks.

1.8 To measure thermal conductivity, two similar 1-cm.-thick specimens are placed in an apparatus shown in the accompanying sketch. Electric current is supplied to the 6-cm by 6-cm. guarded heater, and a wattmeter shows that the power dissipation is 10 watts (W). Thermocouples attached to the warmer and to the cooler surfaces show temperatures of 322 and 300 K, respectively. Calculate the thermal conductivity of the material at the mean temperature in Btu/h ft °F and in W/m K.

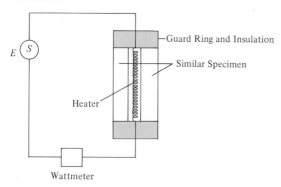

Guard Ring and Insulation

Similar Specimen

Heater

E S

Wattmeter

1.9 To determine the thermal conductivity of a structural material, a large 6-in.-thick slab of the material was subjected to a uniform heat flux of 800 Btu/h ft², while thermocouples embedded in the wall 2 in. apart were read over a period of time. After the system had reached equilibrium, an operator recorded the readings of the thermocouples as shown below for two different environmental conditions:

Distance from surface (in.)	Temperature (°F)
Test 1	
0	100
2	150
4	206
6	270
Test 2	
0	200
2	265
4	335
6	406

From these data, determine an approximate expression for the thermal conductivity as a function of temperature between 100 and 400°F.

1.10 Using Table 1.4 as a guide, prepare a similar table showing the order of magnitudes of the thermal resistances per unit area for convection between a surface and various fluids.

1.11 A thermocouple (0.8-mm-OD wire) is used to measure the temperature of quiescent gas in a furnace. The thermocouple reading is 165°C. It is known, however, that the rate of radiant heat flow per meter length from hotter furnace walls to the thermocouple wire is 1.1 W/m and the unit conductance between the wire and the gas is 6.8 W/m²K. With this information, *estimate* the true gas temperature. State your assumptions and indicate the equations used.

1.12 Water at a temperature of 76.6°C is to be evaporated slowly in a vessel. The water is in a low-pressure container which is surrounded by steam. The steam is condensing at 107°C. The overall heat transfer coefficient between the water and the steam is 1100 W/m²K. Calculate the surface area of the container which would be required to evaporate water at a rate of 0.01 kg/s.

1.13 The heat transfer rate from hot air at 100°C flowing over one side of a flat plate with dimensions 0.1 m by 0.5 m is determined to be 125 W when the surface of the plate is kept at 30°C. What is the average convection heat transfer coefficient between the plate and the air?

1.14 The heat transfer coefficient for a gas flowing over a thin flat plate 3 m long and 0.3 m wide varies with distance from the leading edge according to

$$h_c(x) = 10x^{-1/4} \ \text{W/m}^2 \ \text{K}$$

Calculate the average heat transfer coefficient, the rate of heat transfer between the plate and the gas if the plate is at 170°C and the gas is at 30°C, and the local heat flux 2 m from the leading edge.

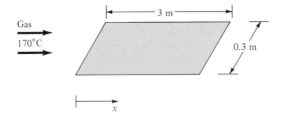

1.15 The heat transfer coefficient for a 1-ft-diameter sphere in still air is 1.2 Btu/h ft² °F. If the air is at 80°F and the surface of the sphere at −297°F, determine the rate of heat transfer by convection.

1.16 Two large parallel plates, having surface conditions which approximate those of a blackbody, are maintained at 1500 and 500°F, respectively. Determine the rate of heat transfer by radiation between the plates in Btu/h ft² and the radiation unit conductance in Btu/h ft² °F and in W/m² K.

1.17 A spherical vessel of 1 ft in diameter is located in a large room whose walls are at 80°F. If the vessel is used to store liquid oxygen at −297°F and the surface of the storage vessel as well as the walls of the room are black, calculate the rate of heat transfer by radiation to the liquid oxygen in Btu/h and in watts.

1.18 Repeat Problem 1.17 but assume that the surface of the storage vessel has an emittance of 0.1. Then determine the rate of evaporation of liquid oxygen in pounds per hour, assuming that convection can be neglected.

1.19 Determine the rate of radiant heat emission in watts per square meter from a blackbody at (*a*) 150°C, (*b*) 600°C, (*c*) 5700°C.

1.20 A spherical satellite orbits about the earth. Assuming there are no internal energy sources, find the equilibrium temperature of the satellite surface (*a*) if the surface is black and (*b*) if the surface is gray and $\epsilon = 0.1$. Assume that the solar incident radiation flux at the orbit is 1340 W/m². *Ans.* (*a*) 4°C.

1.21 Heat is transferred through a plane wall from the inside of a room at 22°C to the outside air at −2°C. The unit-surface conductances at the inside and outside surfaces are 12 and 28 W/m² K, respectively. The thermal resistance of the wall per unit area is 0.5 m² K/W. Determine the temperature at the outer surface of the wall and the rate of heat flow through the wall per unit area.

1.22 Steam is condensing inside a pipe at 134 psia. The unit-surface conductance on the steam side is 500 Btu/h ft^2 °F. The thermal resistance of the pipe per unit area is 0.001 h ft^2 °F/Btu and the unit-surface conductance at the outside of the pipe is 4 Btu/h ft^2 °F. (*a*) Estimate the percent of the overall thermal resistance offered by (1) the steam, (2) the pipe, and (3) the steam and the pipe. (*b*) Determine the temperature at the outer surface of the pipe if the pipe is suspended in a room at 70°F. The values of the unit conductances and the resistance are based on the outside area of the pipe.

1.23 A flat plate placed in the sunlight receives 600 W/m^2 of radiant heat from the sun and the atmosphere. If the air temperature is 27°C and the unit-surface conductance between the plate and the air is 12 W/m^2 K, determine the plate temperature. Neglect heat losses from the bottom of the plate.

1.24 How much fiberglass insulation ($k = 0.035$ W/m K) is needed to enable a guarantee that the outside temperature of a kitchen oven will not exceed 43°C? The maximum oven temperature to be maintained by the conventional type of thermostatic control is 290°C, the kitchen temperature may vary from 15°C to 33°C and the average heat transfer coefficient between the oven surface and the kitchen is 12 W/m^2 K.

1.25 A heat exchanger wall consists of a copper plate $\frac{3}{8}$ in. thick. The surface coefficients on the two sides of the plate are 480 and 1250 Btu/h ft^2 °F, corresponding to fluid temperatures of 200 and 90°F, respectively. Assuming that the thermal conductivity of the wall is 220 Btu/h ft °F, (*a*) draw the thermal circuit, (*b*) compute the surface temperatures in °F, and (*c*) calculate the heat flux in Btu/h ft^2.
Ans. (*b*) 119°F, 124°F; (*c*) 36,300 Btu/h ft^2.

1.26 A horizontal 3-mm-thick flat copper plate, 1 m long and 0.5 m wide, is exposed in air at 27°C to radiation from the sun. If the total rate of incident solar radiation is 300 W and the combined unit-surface conductances on the upper and lower surfaces are 20 and 15 W/m^2 K, respectively, determine the equilibrium temperature of the plate.

1.27 A submarine is to be designed to provide a comfortable temperature for the crew of no less than 70°F. The submarine can be idealized by a cylinder 30 ft in diameter and 200 ft in length. The combined unit-surface conductance on the interior is about 2.5 Btu/h ft^2 °F, while on the outside the unit-surface conductance is estimated to vary from about 10 Btu/h ft^2 °F (not moving) to 150 Btu/h ft^2 °F (top speed). For the following wall constructions, determine the minimum size in kilowatts of the heating unit required if the sea water temperatures varies from 34 to 55°F during operation. The walls of the submarine are (*a*) $\frac{1}{2}$-in. aluminum; (*b*) $\frac{3}{4}$-in. stainless steel with a 1-in. thick layer of fiberglass insulation on the inside; and (*c*) of sandwich construction with a $\frac{3}{4}$-in.-thick layer of stainless steel, a 1-in.-thick layer of fiber-glass insulation, and a $\frac{1}{4}$-in. thickness of aluminum on the inside. What conclusion can you draw? *Ans.* (*a*) 525; (*b*) 57; (*c*) 57.

1.28 A small gray sphere having an emissivity of 0.5 and a surface temperature of 1000°F is located in a blackbody enclosure having a temperature of 100°F. Calculate for this system: (*a*) the net rate of heat transfer by radiation per unit of surface area of the sphere, (*b*) the radiative thermal conductance in Btu/h °F if the surface area of the sphere is 0.1 ft^2, (*c*) the unit thermal resistance for radiation between the sphere and its surroundings, (*d*) the ratio of thermal resistance for radiation to thermal resistance for convection if the unit-surface conductance for convection between the sphere and its surroundings is 2.0 Btu/h ft^2 °F, (e) the total rate of heat transfer from the sphere to the surroundings, and (f) the combined unit-thermal-surface conductance for the sphere.

1.29 A small oven with a surface area of 3 ft^2 is located in a room in which the walls and the air are at a temperature of 80°F. The exterior surface of the oven is at 300°F and the net heat transfer by radiation between the oven's surface and the surroundings is 2000 Btu/h. If the average unit-surface conductance for convection between the oven and the surrounding air is 2.0 Btu/h ft^2 °F, calculate: (a) the net heat transfer between the oven and the surroundings in Btu/h, (b) the thermal resistance at the surface for radiation and convection in h °F/Btu, and (c) the combined unit-thermal-surface conductance in Btu/h ft^2 °F.

1.30 A steam pipe 200 mm in diameter passes through a large basement room. The temperature of the pipe wall is 500°C, while that of the ambient air in the room is 20°C. Determine the heat transfer rate by convection and radiation per unit length of steam pipe, if the emissivity of the pipe surface is 0.8 and the natural convection heat transfer coefficient has been determined to be 10 W/m^2 K.

1.31 A simple solar heater consists of a flat plate of glass below which is located a shallow pan filled with water, so that the water is in contact with the glass plate above it. Solar radiation is passing through the glass at the rate of 156 Btu/h ft^2. The water is at 200°F and the surrounding air is 80°F. If the heat transfer coefficients between the water and the glass and the glass and the air are 5 Btu/h ft^2 °F, respectively, determine the time required to transfer 100 Btu per square foot of surface to the water in the pan. The lower surface of the pan may be assumed insulated. *Ans.* 2.5 h.

1.32 Draw the thermal circuit, determine the rate of heat flow per unit area from a furnace wall, and estimate the exterior surface temperature under the following conditions:
(a) Convective heat transfer coefficient at the interior surface is 15 W/m^2 K.
(b) Rate of heat flow by radiation from hot gases and particles at 2000°C to interior wall surface is 45,000 W/m^2.
(c) Unit thermal conductance of wall (interior surface temperature about 850°C) is 250 W/m^2 K.
(d) Convection from outer surface.

1.33 The inner wall of a combustion chamber receives 50,000 Btu/h ft^2 by radiation from a gas at 5000°F. The convective unit conductance between the gas and the wall is 20 Btu/h ft^2 °F. If the inner wall of the combustion chamber is at a temperature of 1000°F, determine the total unit thermal resistance in h ft^2 °F/Btu. Also, draw the thermal circuit. *Ans.* 0.031.

1.34 A composite refrigerator wall is composed of 2 in. of corkboard sandwiched between a $\frac{1}{2}$-in.-thick layer of oak and a $\frac{1}{32}$-in. thickness of aluminum lining on the inner surface. The average unit-convective-thermal conductances at the interior and exterior wall, respectively, are 2 and 1.5 Btu/h ft^2 °F. (a) Calculate the individual resistance of this composite wall and the resistances at the surface. (b) Calculate the overall conductance per unit area. (c) Draw the thermal circuit. (d) For an air temperature inside the refrigerator at 30°F and outside of 90°F, calculate the rate of heat transfer per unit area. *Ans.* (d) 7.4 Btu/h ft^2.

1.35 Draw the thermal circuit for heat transfer from the sun through a double-glazed window to the air in a room. Identify each circuit element.

REFERENCES

1. P. G. Klemens, "Theory of the Thermal Conductivity of Solids," in R. P. Tye, ed., *Thermal Conductivity*, vol. 1, Academic Press, London, 1969.
2. E. McLaughlin, "Theory of the Thermal Conductivity of Fluids," in R. P. Tye, ed., *Thermal Conductivity*, vol. 2, Academic Press, London, 1969.
3. W. G. Vincenti, and C. H. Kruyer, Jr., *Introduction to Physical Gas Dynamics*, Wiley, New York, 1965.
4. J. F. Mallory, *Thermal Insulation*, Reinhold, New York, 1969.
5. W. M. Rohsenow, J. P. Hartnett, and E. N. Ganic, eds., *Handbook of Heat Transfer*, 2d ed., McGraw-Hill, New York, 1985.
6. T. N. Veizivogen, "Correlation of Thermal Contact Conductance: Experimental Results," in *Progress in Astronautics and Aeronautics*, vol. 20, Academic Press, New York, 1967.
7. E. Fried, "Thermal Conduction Contribution to Heat Transfer at Contacts," in R. P. Tye, ed., *Thermal Conductivity*, vol. 2, Academic Press, London, 1969.

Chapter 2

Conduction

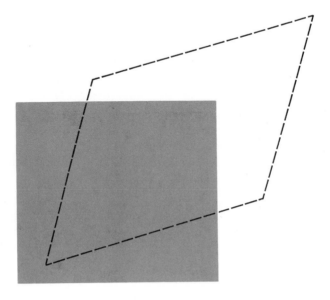

2.1 INTRODUCTION

Conduction heat transfer is the process by which heat flows through a solid. In the conduction mode, heat is transferred by a complex submicroscopic mechanism in which atoms interact by elastic and inelastic collisions to propagate the energy from regions of higher to regions of lower temperature. From an engineering point of view there is no need to delve into the complexities of the mechanism, because the rate of heat propagation can be predicted by Fourier's law, which incorporates the mechanistic features of the process into a physical property known as the *thermal conductivity*.

Although conduction also occurs in liquids and gases, it is rarely the predominante transport mechanism in fluids. The reason for this is that once heat begins to flow in a fluid, even if no external force is applied, density gradients are set up and convective currents are set in motion. In convection, fluid is thus transported on a macroscopic scale as well as on a microscopic scale, and convective currents are generally more effective in transporting heat than conduction, where the motion is limited to submicroscopic transport of energy.

Conduction heat transfer has been a fertile field for applied mathematicians for the past two hundred years. The governing physical relations are partial differential equations, which are susceptible to solution by classical methods (1). Famous mathematicians, including Laplace, Fourier, and others, spent part of their lives seeking and tabulating useful solutions to heat conduction problems. However, the analytic approach to conduction is limited to relatively simple geometric shapes and to boundary conditions that can only approximate the situation in realistic engineering problems. With the advent of the high-speed computer, the situation changed dramatically and a revolution occurred in the field of conduction heat transfer. The computer made it possible to solve, with relative ease, complex problems that closely approximate real conditions. As a result, the analytic approach has nearly disappeared from the engineering scene. The analytic approach is, however, important as background for the next chapter, in which we will show how to solve conduction problems by numerical methods.

2.2 THE CONDUCTION EQUATION

In this section the general conduction equation is derived. A solution of this equation, subject to given initial and boundary conditions, yields the temperature distribution in a solid system. Once the temperature distribution is known, the heat transfer rate in the conduction mode can be evaluated by applying Fourier's law, Eq. (1.1).

The conduction equation is a mathematical expression of the conservation of energy in a solid substance. To derive this equation we perform an energy balance on an elemental volume of material in which heat is being transferred only by conduction. Heat transfer by radiation occurs in a solid only if the material is transparent or translucent.

The energy balance includes the possibility of heat generation in the material. Heat generation in a solid can occur by chemical reactions, electric currents passing through the material, or nuclear reactions. The general form of the conduction equation also accounts for storage of internal energy. Thermodynamic considerations show that when the internal energy of a material increases its temperature also increases. A solid material therefore experiences a net increase in stored energy when its temperature increases with time. If the temperature of the material remains constant, no energy is stored and steady conditions are said to prevail.

Heat transfer problems are classified according to the variables that influence the temperature. If the temperature is a function of time, the problem is classified as *unsteady* or *transient*. If the temperature is independent of time, the problem is called a *steady-state* problem. If the temperature is a function of a single space coordinate, the problem is said to be *one-dimensional*. If it is a function of two or three coordinate dimensions, the problem is *two- or three-dimensional*, respectively. If the temperature is a function of time and only one space coordinate, the problem is classified as *one-dimensional and transient*.

2.2.1 Rectangular Coordinates

To illustrate the analytic method of approach, we will first derive the conduction equation for a one-dimensional, rectangular coordinate system as shown in Fig. 2.1. We will assume that the temperature in the material is only a function of the x coordinate and time, or $T = T(x, t)$, and that the conductivity k, density ρ, and specific heat c of the solid are all constant.

The principle of conservation of energy for the control volume of Fig. 2.1 can be stated as follows:

Rate of heat conduction   rate of heat generation
into control volume $+$ inside control volume $=$

rate of heat conduction   rate of energy storage
out of control volume $+$ inside control volume   (2.1)

We will use Fourier's law to express the two conduction terms and define the symbol $\dot{q}_G$ as the rate of energy generation per unit volume inside the control volume. Then the word Eq. (2.1) can be expressed in mathematical form:

$$-kA \frac{\partial T}{\partial x}\bigg|_x + \dot{q}_G A\,\Delta x = -kA \frac{\partial T}{\partial x}\bigg|_{x+\Delta x} + \rho A\,\Delta x\,c\,\frac{\partial T(x + \Delta x/2, t)}{\partial t} \quad (2.2)$$

Dividing Eq. (2.2) by the control volume $A\,\Delta x$ and rearranging, we obtain

$$k \frac{(\partial T/\partial x)_{x+\Delta x} - (\partial T/\partial x)_x}{\Delta x} + \dot{q}_G = \rho c \frac{\partial T(x + \Delta x/2, t)}{\partial t} \quad (2.3)$$

In the limit as $\Delta x \to 0$, the first term on the left side of Eq. (2.3) can be expressed in the form

$$\frac{\partial T}{\partial x}\bigg|_{x+dx} = \frac{\partial T}{\partial x}\bigg|_x + \frac{\partial}{\partial x}\left(\frac{\partial T}{\partial x}\bigg|_x\right)dx = \frac{\partial T}{\partial x}\bigg|_x + \frac{\partial^2 T}{\partial x^2}\bigg|_x dx \quad (2.4)$$

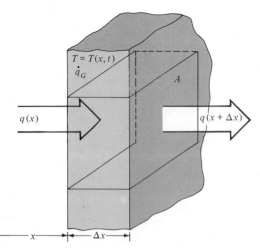

Figure 2.1 Control volume for one-dimensional conduction in rectangular coordinates.

The right side of Eq. (2.3) can be expanded in a Taylor series as

$$\frac{\partial T}{\partial t}\left[\left(x + \frac{\Delta x}{2}\right), t\right] = \frac{\partial T}{\partial t}\bigg|_x + \frac{\partial^2 T}{\partial x \partial t}\bigg|_x \frac{\Delta x}{2} + \cdots$$

Equation (2.2) then becomes, to the order of Δx,

$$k\frac{\partial^2 T}{\partial x^2} + \dot{q}_G = \rho c \frac{\partial T}{\partial t} \qquad (2.5)$$

Physically, the first term on the left side represents the *net rate of heat conduction* into the control volume per unit volume. The second term on the left side is the *rate of energy generation per unit volume* inside the control volume. The right side represents the *rate of increase in internal energy* inside the control volume per unit volume. Each term has dimensions of energy per unit time and volume with the units (W/m^3) in the SI system and $(Btu/h\ ft^3)$ in the engineering system.

Equation (2.5) applies only to unidimensional heat flow because it was derived on the assumption that the temperature distribution is one-dimensional. If this restriction is now removed and the temperature is assumed to be a function of all three coordinates as well as time, or $T = T(x, y, z, t)$, terms like the first one in Eq. (2.5), representing the net rate of conduction per unit volume in the y and z directions, will appear. The three-dimensional form of the conduction equation then becomes (see Fig. 2.2)

$$\frac{\partial^2 T}{\partial x^2} + \frac{\partial^2 T}{\partial y^2} + \frac{\partial^2 T}{\partial z^2} + \frac{\dot{q}_G}{k} = \frac{1}{\alpha}\frac{\partial T}{\partial t} \qquad (2.6)$$

where α is the *thermal diffusivity*, a group of material properties defined as

$$\alpha = \frac{k}{\rho c} \qquad (2.7)$$

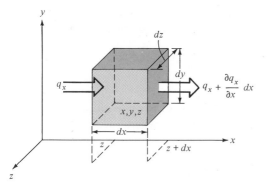

Figure 2.2 Differential control volume for three-dimensional conduction in rectangular coordinates.

The thermal diffusivity has units of (m^2/s) in the SI system and (ft^2/s) in the engineering system. Numerical values of the thermal conductivity, density, specific heat, and thermal diffusivity for several engineering materials are presented in Appendix 2.

Solutions to the general conduction equation in the form of Eq. (2.6) can only be obtained for simple geometric shapes and easily specified boundary conditions. However, as shown in the next chapter, solutions by numerical methods can be obtained quite easily for complex shapes and realistic boundary conditions without a great deal of effort, a procedure used in engineering practice today for the majority of conduction problems. But a basic understanding of analytic solutions is important in writing computer programs, and in the rest of this chapter we will examine problems for which simplifying assumptions can eliminate some terms from Eq. (2.6) and reduce the complexity of the solution.

If the temperature of a material is not a function of time, the system is in the steady state and does not store any energy. The steady form of a three-dimensional conduction equation in rectangular coordinates is

$$\frac{\partial^2 T}{\partial x^2} + \frac{\partial^2 T}{\partial y^2} + \frac{\partial^2 T}{\partial z^2} + \frac{\dot{q}_G}{k} = 0 \tag{2.8}$$

If the system is in the steady state and no heat is generated internally, the conduction equation further simplifies to

$$\frac{\partial^2 T}{\partial x^2} + \frac{\partial^2 T}{\partial y^2} + \frac{\partial^2 T}{\partial z^2} = 0 \tag{2.9}$$

Equation (2.9) is known as the *Laplace equation*, in honor of the French mathematician Pierre Laplace. It occurs in a number of areas in addition to heat transfer, for instance, in diffusion of mass or in electromagnetic fields. The operation of taking the second derivatives of the potential in a field has therefore been given a shorthand symbol, ∇^2, called the Laplacian operator. For the rectangular coordinate system Eq. (2.9) becomes

$$\frac{\partial^2 T}{\partial x^2} + \frac{\partial^2 T}{\partial y^2} + \frac{\partial^2 T}{\partial z^2} = \nabla^2 T = 0 \tag{2.10}$$

Since the operator ∇^2 is independent of coordinate system, the above form will be particularly useful when we want to study conduction in cylindrical and spherical coordinates.

2.2.2 Dimensionless Form

The conduction equation in the form of Eq. (2.6) is dimensional. It is often more convenient to express this equation in a form where each term is dimensionless. In the development of the dimensionless equation we will identify dimensionless groups that govern the heat conduction process. Begin by defining a dimensionless temperature as the ratio

$$\theta = \frac{T}{T_r} \tag{2.11}$$

a dimensionless x coordinate as the ratio

$$\xi = \frac{x}{L_r} \tag{2.12}$$

and a dimensionless time as the ratio

$$\tau = \frac{t}{t_r} \tag{2.13}$$

where the symbols T_r, L_r, and t_r represent a reference temperature, a reference length, and a reference time, respectively. Although the choice of reference quantities is somewhat arbitrary, the values selected should be physically significant. The choice of dimensionless groups varies from problem to problem, but the form of the dimensionless groups should be structured so that they limit the dimensionless variables between convenient extremes, such as zero and one. The value for L_r should therefore be selected as the maximum x dimension of the system for which the temperature distribution is sought. Similarly, a dimensionless ratio of temperature differences that varies between zero and unity is often preferable to a ratio of absolute temperatures.

   If the definitions of the dimensionless temperature, x coordinate, and time are substituted into Eq. (2.5), we obtain the conduction equation in the nondimensional form

$$\frac{\partial^2 \theta}{\partial \xi^2} + \frac{\dot{q}_G L_r{}^2}{k T_r} = \frac{L_r{}^2}{\alpha t_r} \frac{\partial \theta}{\partial \tau} \tag{2.14}$$

The reciprocal of the dimensionless group $(L_r{}^2/\alpha t_r)$ is called the *Fourier number*, designated by the symbol Fo:

$$\text{Fo} = \frac{\alpha t_r}{L_r{}^2} \tag{2.15}$$

In a physical sense, the Fourier number is the ratio of the rate of heat transfer by conduction to the rate of energy storage in the system. It is an important dimensionless group in transient conduction problems and will be encountered frequently. The choice of reference time and length in the Fourier number depends on the specific problem, but the basic form is always a thermal diffusivity multiplied by time divided by the square of a characteristic length.

   The other dimensionless group appearing in Eq. (2.14) is a ratio of internal heat generation per unit time to heat conduction through the volume per unit time. We will use the symbol $\dot{Q}_G$ to represent this dimensionless heat generation number:

$$\dot{Q}_G = \frac{\dot{q}_G L_r{}^2}{k T_r} \tag{2.16}$$

The one-dimensional form of the conduction equation expressed in dimensionless form now becomes

$$\frac{\partial^2 \theta}{\partial \xi^2} + \dot{Q}_G = \frac{1}{\text{Fo}} \frac{\partial \theta}{\partial t} \tag{2.17}$$

If steady state prevails, the right side of Eq. (2.17) becomes zero.

2.2.3 Cylindrical and Spherical Coordinates

Equation (2.6) was derived for a rectangular coordinate system. Although the generation and energy storage terms are independent of the coordinate system, the heat conduction terms depend on geometry and, therefore, on the coordinate system. The dependence on the coordinate system used to formulate the problem can be removed by replacing the heat conduction terms with the Laplacian operator.

$$\nabla^2 T + \frac{\dot{q}_G}{k} = \frac{1}{\alpha}\frac{\partial T}{\partial t} \tag{2.18}$$

The differential form of the Laplacian is different for each coordinate system, and the Laplacian operators in rectangular, cylindrical, and spherical coordinate systems are derived in the appendix.

For a general transient three-dimensional problem in the cylindrical coordinates shown in Fig. 2.3, $T = T(r, \phi, z, t)$ and $\dot{q}_G = \dot{q}_G(r, \phi, z, t)$. If the Laplacian is substituted into Eq. (2.18), the general form of the conduction equation in cylindrical coordinates becomes

$$\frac{1}{r}\frac{\partial}{\partial r}\left(r\frac{\partial T}{\partial r}\right) + \frac{1}{r^2}\frac{\partial^2 T}{\partial \phi^2} + \frac{\partial^2 T}{\partial z^2} + \frac{\dot{q}_G}{k} = \frac{1}{\alpha}\frac{\partial T}{\partial t} \tag{2.19}$$

If the heat flow in a cylindrical shape is only in the radial direction, $T = T(r, t)$, the conduction equation reduces to

$$\frac{1}{r}\frac{\partial}{\partial r}\left(r\frac{\partial T}{\partial r}\right) + \frac{\dot{q}_G}{k} = \frac{1}{\alpha}\frac{\partial T}{\partial t} \tag{2.20}$$

Furthermore, if the temperature distribution does not vary with time, the conduction equation becomes

$$\frac{1}{r}\frac{d}{dr}\left(r\frac{\partial T}{\partial r}\right) + \frac{\dot{q}_G}{k} = 0 \tag{2.21}$$

In this case the equation for the temperature contains only a single variable r and is therefore an ordinary differential equation.

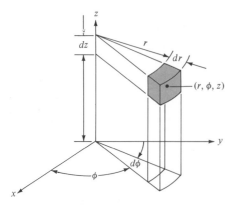

Figure 2.3 Cylindrical coordinate system for the general conduction equation.

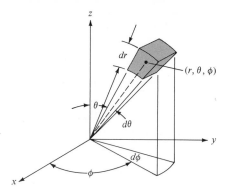

Figure 2.4 Spherical coordinate system for the general conduction equation.

When no internal energy generation is present and the temperature is a function of the radius only, the steady-state conduction equation for cylindrical coordinates is

$$\frac{d}{dr}\left(r\frac{dT}{dr}\right) = 0 \tag{2.22}$$

For spherical coordinates, as shown in Fig. 2.4, the temperature is a function of the three space coordinates r, θ, ϕ and time t, or $T = T(r, \theta, \phi, t)$. The general form of the conduction equation in spherical coordinates is then

$$\frac{1}{r^2}\frac{\partial}{\partial r}\left(r^2\frac{\partial T}{\partial r}\right) + \frac{1}{r^2\sin\theta}\frac{\partial}{\partial\theta}\left(\sin\theta\frac{\partial T}{\partial\theta}\right) + \frac{1}{r^2\sin^2\theta}\frac{\partial^2 T}{\partial\phi^2} + \frac{\dot{q}_G}{k} = \frac{1}{\alpha}\frac{\partial T}{\partial t} \tag{2.23}$$

2.3 STEADY CONDUCTION IN SIMPLE GEOMETRIES

In this section we will demonstrate how to obtain solutions to the conduction equations derived in the preceding section for relatively simple geometric configurations with and without internal heat generation.

2.3.1 Plane Wall with and without Heat Generation

In the first chapter we saw that the temperature distribution for one-dimensional, steady conduction through a wall is linear. We can verify this result by simplifying the more general case expressed by Eq. (2.6). For steady state $\partial T/\partial t = 0$, and since T is only a function of x, $\partial T/\partial y = 0$ and $\partial T/\partial z = 0$.

Furthermore, if there is no internal generation, $\dot{q}_G = 0$, Eq. (2.6) reduces to

$$\frac{d^2 T}{dx^2} = 0 \tag{2.24}$$

Integrating this ordinary differential equation twice yields the temperature distribution

$$T(x) = C_1 x + C_2 \tag{2.25}$$

For a wall with $T(x = 0) = T_1$ and $T(x = L) = T_2$ we get

$$T(x) = \frac{T_2 - T_1}{L} x + T_1 \qquad (2.26)$$

The above relation agrees with the linear temperature distribution deduced by integrating Fourier's law, $q_k = -kA \, dT/dx$.

Next consider a similar problem, but with heat generation throughout the system as shown in Fig. 2.5. If the thermal conductivity is constant and the heat generation is uniform Eq. (2.5) reduces to

$$k \frac{d^2 T(x)}{dx^2} = -\dot{q}_G \qquad (2.27)$$

Integrating this equation once gives

$$\frac{dT(x)}{dx} = -\frac{\dot{q}_G}{k} x + C_1 \qquad (2.28)$$

and a second integration yields

$$T(x) = -\frac{\dot{q}_G}{2k} x^2 + C_1 x + C_2 \qquad (2.29)$$

where C_1 and C_2 are constants of integration whose values are determined by the boundary conditions. The specified conditions require that the temperature at $x = 0$ be T_1 and at $x = L$ be T_2. Substituting these conditions successively

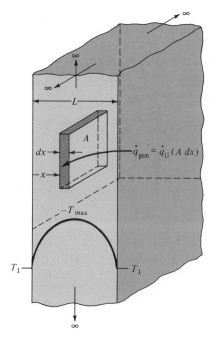

Figure 2.5 Conduction in a plane wall with uniform heat generation. Temperature distribution is for the case $T_1 = T_2$ (see Eq. 2.33).

into the conduction equation gives

$$T_1 = C_2 \qquad (x = 0) \tag{2.30}$$

and

$$T_2 = -\frac{\dot{q}_G}{2k} L^2 + C_1 L + T_1 \qquad (x = L) \tag{2.31}$$

Solving for C_1 and subsituting into Eq. (2.29) gives the temperature distribution

$$T(x) = -\frac{\dot{q}_G}{2k} x^2 + \frac{T_2 - T_1}{L} x + \frac{\dot{q}_G L}{2k} x + T_1 \tag{2.32}$$

Observe that Eq. (2.26) is now modified by two terms containing the heat generation and that the temperature distribution is no longer linear.

If the two surface temperatures are equal, $T_1 = T_2$, the temperature distribution becomes

$$T(x) = \frac{\dot{q}_G L^2}{2k} \left[\frac{x}{L} - \left(\frac{x}{L} \right)^2 \right] + T_1 \tag{2.33}$$

This temperature distribution is parabolic and symmetric about the center plane with a maximum T_{max} at $x = L/2$. At the centerline $dT/dx = 0$, which corresponds to an insulated surface at $x = L/2$. The maximum temperature is

$$T_{max} = T_1 + \frac{\dot{q}_G L^2}{8k} \tag{2.34}$$

For the symmetric boundary conditions the temperature in dimensionless form is

$$\frac{T(x) - T_1}{T_{max} - T_1} = 4(\xi - \xi^2)$$

where $\xi = x/L$.

EXAMPLE 2.1 _____

A long electrical heating element, made of iron, has a cross section of 10 cm × 1.0 cm. It is immersed in a heat transfer oil at 80°C. If heat is generated uniformly at a rate of 1,000,000 W/m³ by an electric current, determine the unit surface conductance necessary to keep the temperature of the heater below 200°C. The thermal conductivity for iron at 200°C, from Table 12 in Appendix 2, is 64 W/m K.

Solution. If we disregard the heat dissipated from the edges, a reasonable assumption since the heater has a width 10 times greater than its thickness, Eq. (2.34) may be used to calculate the temperature difference between the center and the surface, or

$$T_{max} - T_1 = \frac{\dot{q}_G L^2}{8k} = \frac{(1,000,000 \text{ W/m}^3)(0.1 \times 0.01 \text{ m}^2)}{(8)(64) \text{ W/m K}} = 0.2°C$$

The temperature drop from the center to the surface of the heater is so small because the heater material is made of iron, which is a good conductor. We can neglect this temperature drop and calculate the minimum unit convective conductance from a heat balance, or

$$\dot{q}_G \frac{L}{2} = \bar{h}_c(T_1 - T_\infty)$$

Solving for $\bar{h}_c$:

$$\bar{h}_c = \frac{\dot{q}_G(L/2)}{(T_1 - T_\infty)} = \frac{(10^6 \text{ W/m}^3)(0.005 \text{ m})}{120 \text{ K}} = 42 \text{ W/m}^2 \text{ K}$$

Thus, the unit convective conductance to keep the temperature in the heater from exceeding the set limit must be larger than 42 W/m² K.

2.3.2 Cylindrical and Spherical Shapes Without Heat Generation

In this section we will obtain solutions to some problems in cylindrical and spherical systems which are often encountered in practice. Probably the most common case is that of heat transfer through a pipe with a fluid flowing inside. This system can be idealized, as shown in Fig. 2.6, by radial heat flow through

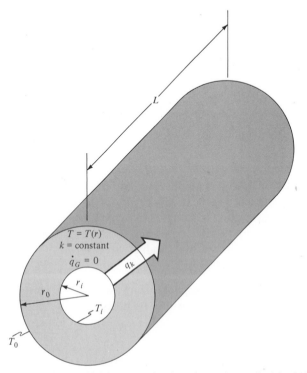

Figure 2.6 Radial heat conduction through a cylindrical shell.

a cylindrical shell. Our problem is then to determine the temperature distribution and the heat transfer rate in a long hollow cylinder of length L if the inner- and outer-surface temperatures are T_i and T_o, respectively, and no internal generation is present. Since the temperatures at the boundaries are constant, the temperature distribution is not a function of time and the appropriate form of the conduction equation is

$$\frac{d}{dr}\left(r\frac{dT}{dr}\right) = 0 \tag{2.35}$$

Integrating once with respect to radius gives

$$r\frac{dT}{dr} = C_1 \quad \text{or} \quad \frac{dT}{dr} = \frac{C_1}{r}$$

A second integration gives

$$T = C_1 \ln r + C_2$$

The constants of integration can be determined from the boundary conditions:

$$T_i = C_1 \ln r_i + C_2 \qquad \text{at } r = r_i$$

Thus, $C_2 = T_i - C_1 \ln r_i$. Similarly, for T_0:

$$T_o = C_1 \ln r_o + T_i - C_1 \ln r_i \qquad \text{at } r = r_o$$

Thus, $C_1 = (T_o - T_i)/\ln(r_o/r_i)$.

The temperature distribution, written in dimensionless form, is therefore

$$\frac{T(r) - T_i}{T_o - T_i} = \frac{\ln(r/r_i)}{\ln(r_o/r_i)} \tag{2.36}$$

The rate of heat transfer by conduction through the cylinder of length L is, from Eq. (1.2),

$$q_k = -kA\frac{dT}{dr} = -k(2\pi rL)\frac{C_1}{r} = 2\pi Lk\frac{T_o - T_i}{\ln(r_o/r_i)} \tag{2.37}$$

In terms of a thermal resistance we can write

$$q_k = \frac{T_o - T_i}{R_{\text{th}}} \tag{2.38}$$

where the resistance to heat flow by conduction through a cylinder of length L, inner radius r_i, and outer radius r_o is

$$R_{\text{th}} = \frac{\ln(r_o/r_i)}{2\pi Lk} \tag{2.39}$$

The principles developed for a plane wall with conduction and convection in series can also be applied to a long hollow cylinder, such as a pipe or a tube. For example, suppose that a hot fluid flows through a tube that is covered by an insulating material, as shown in Fig. 2.7. The system loses heat through a unit convective conductance $\bar{h}_{co}$ to the surrounding air.

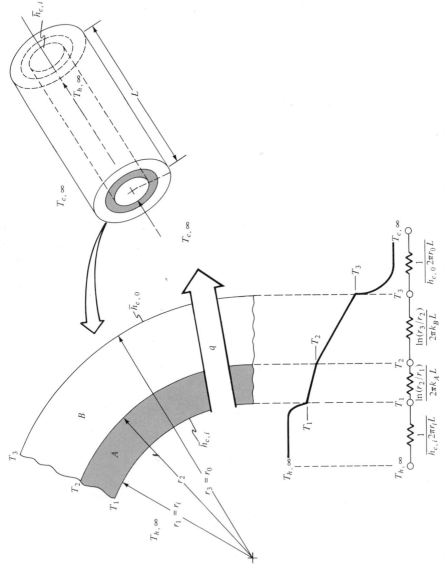

Figure 2.7 Temperature distribution for a composite cylindrical wall with convection at the interior and exterior surfaces.

Using Eq. (2.37) for the thermal resistance of the two cylinders and Eq. (1.14) for the thermal resistance at the inside of the tube and the outside of the insulation, the thermal network is shown below the physical system in Fig. 2.7. Denoting the hot-fluid temperature by $T_{h,\infty}$ and the environmental air temperature by $T_{c,\infty}$, the rate of heat flow is

$$q = \frac{\Delta T}{\sum\limits_{1}^{4} R_{th}} = \frac{T_{h,\infty} - T_{c,\infty}}{\dfrac{1}{\bar{h}_{c,i}2\pi r_1 L} + \dfrac{\ln(r_2/r_1)}{2\pi k_A L} + \dfrac{\ln(r_3/r_2)}{2\pi k_B L} + \dfrac{1}{\bar{h}_{c,o}2\pi r_3 L}} \qquad (2.40)$$

EXAMPLE 2.2

Compare the heat loss from an insulated and an uninsulated copper pipe under the following conditions. The pipe ($k = 400$ W/m K) has an internal diameter of 10 cm and an external diameter of 12 cm. Saturated steam flows inside the pipe at 110°C. The pipe is located in a space at 30°C and the heat transfer coefficient on its outer surface is estimated to be 15 W/m² K. The insulation available to reduce heat losses is 5 cm thick and its conductivity is 0.20 W/m K.

Solution. The uninsulated pipe is depicted by the system in Fig. 2.8. The heat loss per unit length is therefore

$$\frac{q}{L} = \frac{T_s - T_\infty}{R_1 + R_2 + R_3}$$

For the interior surface resistance we can use Table 1.4 to estimate $\bar{h}_{c,i}$. For saturated steam condensing $\bar{h}_{c,i} = 10,000$ W/m² K. Hence we get

$$R_1 = R_i = \frac{1}{2\pi r_i \bar{h}_{c,i}} \simeq \frac{1}{(2\pi)(0.05 \text{ m})(10{,}000 \text{ W/m}^2 \text{ K})} = 0.000318 \text{ m K/W}$$

$$R_2 = \frac{\ln(r_o/r_i)}{2\pi k_{pipe}} = \frac{0.182}{(2\pi)(400 \text{ W/m}^2 \text{ K})} = 0.00007 \text{ m K/W}$$

$$R_3 = R_o = \frac{1}{2\pi r_o \bar{h}_{c,o}} = \frac{1}{(2\pi)(0.06 \text{ m})(15 \text{ W/m}^2 \text{ K})} = 0.177 \text{ m K/W}$$

Since R_1 and R_2 are negligibly small compared to R_3, $q/L = 80/0.177 = 452$ W/m for the uninsulated pipe.

For the insulated pipe the system corresponds to that shown in Fig. 2.7; hence, we must add a fourth resistance between r_1 and r_3:

$$R_4 = \frac{\ln(11/6)}{(2\pi)(0.2 \text{ W/m K})} = 0.482 \text{ m K/W}$$

Also, the outer convective resistance changes to

$$R_o = \frac{1}{(2\pi)(0.11 \text{ m})(15 \text{ W/m}^2 \text{ K})} = 0.096 \text{ m K/W}$$

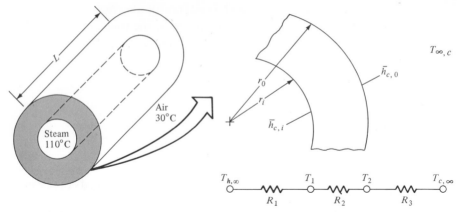

Figure 2.8 Schematic diagram and thermal circuit for a hollow cylinder with convective surface conditions; Example 2.2.

The total thermal resistance per meter length is therefore 0.578 m K/W and the heat loss is $80/0.578 = 138$ W/m. Adding insulation will reduce the heat loss from the steam by 70%.

As shown in Chapter 1 for the case of plane walls with convection resistances at the surfaces, it is often convenient to define an overall heat transfer coefficient by the equation

$$q = UA\,\Delta T_{\text{total}} = UA(T_{\text{hot}} - T_{\text{cold}}) \tag{2.41}$$

Comparing Eqs. (2.40) and (2.41) we see that

$$UA = \frac{1}{\sum\limits_{1}^{4} R_{\text{th}}} = \frac{1}{\dfrac{1}{\bar{h}_{c,i}A_i} + \dfrac{\ln(r_2/r_1)}{2\pi k_A L} + \dfrac{\ln(r_3/r_2)}{2\pi k_B L} + \dfrac{1}{\bar{h}_{c,o}A_o}} \tag{2.42}$$

For plane walls the areas of all sections in the heat flow path are the same, but for cylindrical and spherical systems the area varies with radial distance and the overall heat transfer coefficients can be based on any area in the heat flow path; thus, the numerical value of U will depend on the area selected. Since the outermost diameter is the easiest to measure in practice, $A_o = 2\pi r_3 L$ is usually chosen as the base area. The rate of heat flow is then

$$q = UA_o(T_{\text{hot}} - T_{\text{cold}}) \tag{2.43}$$

and the overall coefficient becomes

$$U = \frac{1}{\dfrac{r_3}{r_1 \bar{h}_{c,i}} + \dfrac{r_3 \ln(r_2/r_1)}{k_A} + \dfrac{r_3 \ln(r_3/r_2)}{k_B} + \dfrac{1}{\bar{h}_{c,o}}} \tag{2.44}$$

EXAMPLE 2.3

A fluid at an average temperature of 200°C flows through a plastic pipe of 4 cm OD and 3 cm ID. The thermal conductivity of the plastic is 0.5 W/m K and the heat transfer coefficient at the inside is 300 W/m² K. The pipe is located in a room at 30°C and the heat transfer coefficient at the outer surface is 10 W/m² K. Calculate the overall heat transfer coefficient and the heat loss per unit length of pipe.

Solution. A sketch of the physical system and the corresponding thermal circuit is shown in Fig. 2.8. The overall heat transfer coefficient from Eq. (2.44) is

$$U_o = \cfrac{1}{\cfrac{r_o}{r_i \bar{h}_{c,i}} + \cfrac{r_o \ln(r_o/r_i)}{k} + \cfrac{1}{\bar{h}_{c,o}}}$$

$$= \cfrac{1}{\cfrac{0.02}{0.015 \times 300} + \cfrac{0.02 \ln(2/1.5)}{0.5} + \cfrac{1}{10}} = 8.62 \text{ W/m}^2 \text{ K}$$

where U_o is based on the outside area of the pipe. The heat loss per unit length is

$$\frac{q}{L} = \frac{T_{\text{hot}} - T_{\text{cold}}}{U A_o} = \frac{200 - 30}{(8.62)(\pi \times 0.04)} = 157 \text{ W/m}$$

For a hollow sphere with uniform temperatures at the inner and outer surfaces (see Fig. 2.9), the temperature distribution without heat generation in the steady state can be obtained by simplifying Eq. (2.23). Under these boundary conditions the temperature is only a function of the radius r and the conduction

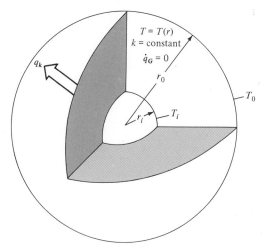

Figure 2.9 Hollow sphere without heat generation and uniform surface temperatures.

equation in spherical coordinates is

$$\frac{1}{r^2}\frac{d}{dr}\left(r^2\frac{dT}{dr}\right) = \frac{1}{r}\frac{d^2(rT)}{dr^2} = 0 \tag{2.45}$$

If the temperature at r_i is uniform and equal to T_i and at r_o equal to T_o, the temperature distribution is

$$T(r) - T_i = (T_o - T_i)\frac{r_o}{r_o - r_i}\left(1 - \frac{r_i}{r}\right) \tag{2.46}$$

The rate of heat transfer through the spherical shell is

$$q_k = -4\pi r^2\frac{\partial T}{\partial r} = \frac{T_o - T_i}{(r_o - r_i)/4\pi k r_o r_i} \tag{2.47}$$

The thermal resistance for a spherical shell is then

$$R_{th} = \frac{r_o - r_i}{4\pi k r_o r_i} \tag{2.48}$$

EXAMPLE 2.4 ——

A spherical, thin-walled metallic container is used to store liquid nitrogen at 77 K. The container has a diameter of 0.5 m and is covered with an evacuated, reflective insulation system composed of silica powder ($k = 0.0017$ W/m K). The insulation is 25 mm thick and its outer surface is exposed to ambient air at 300 K. The latent heat of vaporization h_{fg} of liquid nitrogen is 2×10^5 J/kg. If the convection coefficient is 20 W/m^2 K over the outer surface, determine the rate of liquid boil-off of nitrogen per hour.

Solution. The rate of heat transfer from the ambient air to the nitrogen in the container can be obtained from the thermal circuit shown in Fig. 2.10. We can neglect the thermal resistances of the metal wall and between the boiling nitrogen and the inner wall because that heat transfer coefficient (see Table 1.4) is large. Hence,

$$q = \frac{T_{\infty,\text{air}} - T_{\infty,\text{nitrogen}}}{R_1 + R_2} = \frac{(300 - 77)\text{ K}}{\dfrac{1}{\bar{h}_{c,o}4\pi r_o^2} + \dfrac{r_o - r_i}{4\pi k r_o r_i}}$$

$$= \frac{223\text{ K}}{\dfrac{1}{(20\text{ W/m}^2\text{ K})(4\pi)(0.275\text{ m}^2)} + \dfrac{(0.275 - 0.250)\text{ m}}{4\pi(0.0017\text{ W/m K})(0.275\text{ m})(0.250\text{ m})}}$$

$$= \frac{223\text{ K}}{(0.05 + 17.02)\text{ K/W}} = 13.06\text{ W}$$

Observe that almost the entire thermal resistance is in the insulation. To determine the rate of boil-off we perform an energy balance:

$$\frac{\text{Rate of boil-off}}{\text{of liquid nitrogen}} = \frac{\text{rate of heat transfer}}{\text{to liquid nitrogen}}$$

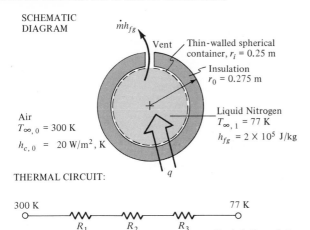

SCHEMATIC
DIAGRAM

THERMAL CIRCUIT:

$R_3 << (R_1 + R_2)$

Figure 2.10   Schematic diagram of spherical container for Example 2.4

or

$$\dot{m}h_{fg} = q$$

Solving for $\dot{m}$ gives

$$\dot{m} = \frac{q}{h_{fg}} = \frac{(13.06 \text{ J/s})(3600 \text{ s/h})}{2 \times 10^5 \text{ J/kg}} = 0.235 \text{ kg/h}$$

2.3.3 Long Solid Cylinder with Heat Generation

A long solid circular cylinder with internal heat generation may be thought of as an idealization of a real system, for example, an electric coil in which heat is generated as a result of the electric current in the wire, or a cylindrical fuel element of uranium 235 in which heat is generated by nuclear fission. The energy equation for an annular element (Fig. 2.11) formed between a fictitious inner cylinder of radius r and a fictitious outer cylinder of radius $r + dr$ is

$$-kA_r \frac{dT}{dr}\bigg|_r + \dot{q}_G L 2\pi r \, dr = -kA_{r+dr} \frac{dT}{dr}\bigg|_{r+dr}$$

where $A_r = 2\pi r L$ and $A_{r+dr} = 2\pi(r + dr)L$. Relating the temperature gradient at $r + dr$ to the temperature gradient at R, we obtain, after simplification,

$$\dot{q}_r = -k\left(\frac{dT}{dr} + r\frac{d^2T}{dr^2}\right) \tag{2.49}$$

Integration of Eq. (2.49) can best be accomplished by noting that

$$\frac{d}{dr}\left(r\frac{dT}{dr}\right) = \frac{dT}{dr} + r\frac{d^2T}{dr^2}$$

and rewriting it in the form

$$\dot{q}_G r = -k\frac{d}{dr}\left(r\frac{dT}{dr}\right)$$

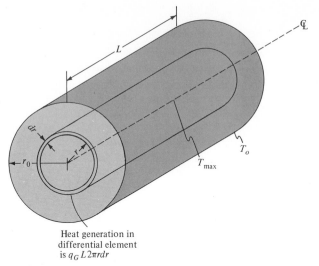

Heat generation in
differential element
is $\dot{q}_G L 2\pi r dr$

Figure 2.11 Nomenclature for heat conduction in a long circular cylinder with internal heat generation.

This is in agreement with the result obtained previously by simplifying the general conduction equation [see Eq. (2.21)]. Integration yields

$$\frac{\dot{q}_G r^2}{2} = -kr\frac{dT}{dr} + C_1$$

from which we deduce that, to satisfy the boundary condition $dT/dr = 0$ at $r = 0$, the constant of integration C_1 must be zero. Another integration yields the temperature distribution

$$T = -\frac{\dot{q} r^2}{4k} + C_2$$

To satisfy the condition that the temperature at the outer surface, $r = r_o$, is T_o, $C_2 = (\dot{q}_G r_o{}^2/4k) + T_o$. The temperature distribution is therefore

$$T = T_o + \frac{\dot{q}_G r_o{}^2}{4k}\left[1 - \left(\frac{r}{r_o}\right)^2\right] \tag{2.50}$$

The maximum temperature at $r = 0$, T_{max}, is

$$T_{\text{max}} = T_o + \frac{\dot{q}_G r_o{}^2}{rk} \tag{2.51}$$

In dimensionless form Eq. (2.50) becomes

$$\frac{T(r) - T_o}{T_{\text{max}} - T_o} = 1 - \left(\frac{r}{r_o}\right)^2 \tag{2.52}$$

For a hollow cylinder with uniformly distributed heat sources and specified surface temperatures the boundary conditions are

$$T = T_i \quad \text{at} \quad r = r_i \text{ (inside surface)}$$
$$T = T_o \quad \text{at} \quad r = r_o \text{ (outside surface)}$$

It is left as an exercise to verify that for this case the temperature distribution is given by

$$T(r) = T_o = \frac{\dot{q}_G}{4k}(r_o^2 - r^2) + \left[(T_i - T_o) + \frac{\dot{q}_G(r_i^2 - r_o^2)}{4k} \ln\left(\frac{r}{r_i}\right) \right] \quad (2.53)$$

If a solid cylinder is immersed in a fluid at a specified temperature T_∞ and the convection heat transfer coefficient at the surface specified and denoted by $\bar{h}_c$, the surface temperature at r_o is not known a priori. The boundary condition for this case requires that the heat conduction from the cylinder equal the rate of convection at the surface, or

$$-k\frac{dT}{dr}\bigg|_{r=r_o} = \bar{h}_c(T_o - T_\infty)$$

Using this condition to evaluate the constants of integration yields for the dimensionless temperature distribution

$$\frac{T(r) - T_\infty}{T_\infty} = \frac{\dot{q}_G r_o}{4\bar{h}_c T_\infty} \left\{ 2 + \frac{\bar{h}_c r_o}{k} \left[1 - \left(\frac{r}{r_o}\right)^2 \right] \right\} \quad (2.54)$$

and for the dimensionless maximum temperature ratio

$$\frac{T_{\max}}{T_\infty} = 1 + \frac{\dot{q}_G r_o}{4\bar{h}_c T_\infty} \left(2 + \frac{\bar{h}_c r_o}{k} \right) \quad (2.55)$$

In the above equations we have two dimensionless parameters of importance in conduction. The first is the heat generation parameter $\dot{q}_G r_o / \bar{h}_c T_\infty$ and the other is the *Biot number*, $\text{Bi} = \bar{h}_c r_o / k$, which appears in problems with simultaneous conduction and convection modes of heat transfer.

Physically, the Biot number is the ratio of a conductive thermal resistance, $R_k = r_o/k$, to a convective resistance, $R_c = 1/\bar{h}_c$. The physical limits on this ratio for the above problem are:

$$\text{Bi} \to 0 \quad \text{when} \quad R_k = \left(\frac{r_o}{k}\right) \to 0 \quad \text{or} \quad R_c = \frac{1}{\bar{h}_c} \to \infty$$

$$\text{Bi} \to \infty \quad \text{when} \quad R_c = \frac{1}{\bar{h}_c} \to 0 \quad \text{or} \quad R_k = \frac{r_o}{k} \to \infty$$

The Biot number approaches zero when the conductivity of the solid or the convective resistance is so large that the solid is practically isothermal and the temperature change is mostly in the fluid at the interface. Conversely, the Biot number approaches infinity when the thermal resistance in the solid predominates and the temperature change occurs mostly in the solid.

EXAMPLE 2.5

Figure 2.12 shows a graphite-moderated nuclear reactor. Heat is generated uniformly in uranium rods of 0.05 m (1.973 in.) OD at the rate of
7.5×10^7 W/m³ (7.24×10^6 Btu/h ft³). These rods are jacketed by an
annulus in which water at an average temperature of 120°C (248°F) is
circulated. The water cools the rods and the average unit convective conductance is estimated to be 55,000 W/m² K (9,700 Btu/h ft² °F). If the
thermal conductivity of uranium is 29.5 W/m K (17.04 Btu/h ft °F), determine the center temperature of the uranium fuel rods.

Solution. Assuming that the fuel rods are sufficiently long that end effects
can be neglected and that the thermal conductivity of uranium does not
change appreciably with temperature, the thermal system can be approximated by that shown in Fig. 2.11. Then the rate of heat flow through the
surface of the rod equals the rate of internal heat generation:

$$2\pi r_o L \left(-k \frac{\partial T}{\partial r} \right)_{r_o} = \dot{q}_G \pi r_o^2 L$$

or

$$-k \frac{\partial T}{\partial r}\bigg|_{r_o} = \frac{\dot{q}_G r_o}{2} = \frac{(7.5 \times 10^7 \text{ W/m}^3)(0.025 \text{ m})}{2}$$

$$= 9.375 \times 10^5 \text{ W/m}^2 \ (2.97 \times 10^5 \text{ Btu/h ft}^2)$$

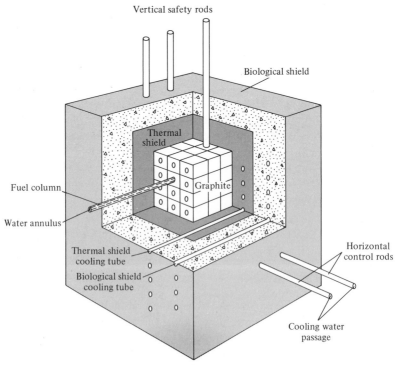

Figure 2.12   Nuclear reactor for Example 2.5. Reprinted from General Electric
Review.

The rate of heat flow by conduction at the outer surface equals the rate of heat flow by convection from the surface to the water:

$$2\pi r_o \left(-k \frac{dT}{dr} \right)\bigg|_{r_o} = 2\pi r_o \bar{h}_o (T_o - T_{\text{water}})$$

from which

$$T_o = \frac{-k(dT/dr)|_{r_o}}{\bar{h}_o} + T_{\text{water}}$$

Substituting the numerical data gives for T_o:

$$T_o = \frac{9.375 \times 10^5 \text{ W/m}^2}{5.5 \times 10^4 \text{ W/m}^2 \text{ K}} + 120°C = 137°C$$

Adding to the surface temperature T_o the temperature difference between the center and the surface of the fuel rods gives the maximum temperature:

$$T_{\text{max}} = T_o + \frac{\dot{q}_G r_o^2}{4k} = 137 + \frac{(7.5 \times 10^7 \text{ W/m}^3)(0.025 \text{ m})^2}{(4)(29.5) \text{ W/m K}}$$

$$= 534°C \ (993.6°F)$$

The same result can be obtained from Eq. (2.55). We observe that most of the temperature drop occurs in the solid because the convective resistance is very small (Bi is about 100).

2.4 EXTENDED SURFACES

The problems considered in this section are encountered in practice when a solid of relatively small cross-sectional area protrudes from a large body into a fluid at a different temperature. Such extended surfaces have wide industrial application as fins attached to the walls of heat transfer equipment in order to increase the rate of heating or cooling.

2.4.1 Fins of Uniform Cross Section

As a simple illustration, consider a pin fin having the shape of a rod whose base is attached to a wall at surface temperature T_s (Fig. 2.13). The fin is cooled along its surface by a fluid at temperature T_∞. The fin has a uniform cross-sectional area A, is made of a material having uniform conductivity k, and the heat transfer coefficient between the surface of the fin and the fluid is $\bar{h}_c$. We will assume that transverse temperature gradients are so small that the temperature at any cross section of the rod is uniform, that is, $T = T(x)$ only. As shown in (2), even in a relatively thick fin the error in a one-dimensional solution is less than 1 percent.

To derive an equation for the temperature distribution, we make a heat balance for a small element of the fin. Heat flows by conduction into the left

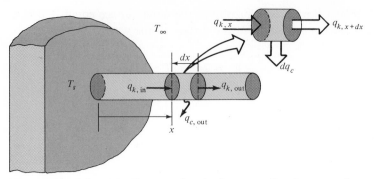

Figure 2.13 Schematic diagram of a pin fin protruding from a wall.

face of the element, while heat flows out of the element by conduction through the right face and by convection from the surface. Under steady-state conditions,

> Rate of heat flow   rate of heat flow by   rate of heat flow by
> by conduction into = conduction out of + convection from surface
>   element at x   element at $x + dx$ between x and $x + dx$

In symbolic form, this equation becomes

$$q_{k,x} = q_{k,x+dx} + dq_c$$

or

$$-kA\left.\frac{dT}{dx}\right|_x = -kA\left.\frac{dT}{dx}\right|_{x+dx} + \bar{h}_c P\,dx[T(x) - T_\infty] \qquad (2.56)$$

where P is the perimeter of the pin and $P\,dx$ is the pin surface area between x and $x + dx$.

If k and $\bar{h}_c$ are uniform, Eq. (2.56) simplifies to the form

$$\frac{d^2T(x)}{dx^2} - \frac{\bar{h}_c P}{kA}[T(x) - T_\infty] = 0 \qquad (2.57)$$

It will be convenient to define an excess temperature of the fin above the environment, $\theta(x) = [T(x) - T_\infty]$, and transform Eq. (2.57) into the form

$$\frac{d^2\theta}{dx^2} - m^2\theta = 0 \qquad (2.58)$$

where $m^2 = \bar{h}_c P/kA$.

Equation (2.58) is a linear, homogeneous, second-order differential equation whose general solution is of the form

$$\theta(x) = C_1 e^{mx} + C_2 e^{-mx} \qquad (2.59)$$

To evaluate the constants C_1 and C_2 it is necessary to specify appropriate boundary conditions. One condition is that at the base ($x = 0$) the fin temperature is equal to the wall temperature, or

$$\theta(0) = T_s - T_\infty = \theta_s$$

The other boundary condition depends on the physical condition at the end of the fin. We will treat the following four cases:

1. The fin is very long and the temperature at the end approaches the fluid temperature, or

$$\theta = 0 \quad \text{at} \quad x \to \infty$$

2. The end of the fin is insulated, or

$$\frac{d\theta}{dx} = 0 \quad \text{at} \quad x = L$$

3. The temperature at the end of the fin is fixed, or

$$\theta = \theta_L \quad \text{at} \quad x = L$$

4. The tip loses heat by convection, or

$$-k\frac{\partial \theta}{\partial x}\bigg|_{x=L} = \bar{h}_{c,L}\theta_L$$

Figure 2.14 illustrates schematically the cases described by these conditions at the tip.

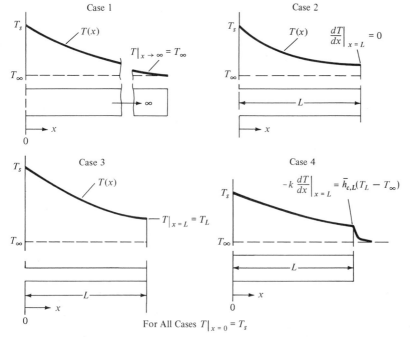

Figure 2.14 Schematic representation of four boundary conditions at the tip of a fin.

For case 1 the second boundary condition can be satisfied only if C_1 in Eq. (2.59) equals zero, or

$$\theta(x) = \theta_s \, e^{-mx} \qquad (2.60)$$

Usually we are interested not only in the temperature distribution, but also in the total rate of heat transfer to or from the fin. The rate of heat flow can be obtained by two different methods. Since the heat conducted across the root of the fin must equal the heat transferred by convection from the surface of the rod to the fluid,

$$q_{\text{fin}} = -kA \left. \frac{dT}{dx} \right|_{x=0} = \int_0^\infty \bar{h}_c P[T(x) - T_\infty] \, dx = \int_0^\infty \bar{h}_c P \theta(x) \, dx \qquad (2.61)$$

Differentiating Eq. (2.60) and substituting the result for $x = 0$ into Eq. (2.61) yields

$$q_{\text{fin}} = -kA[-m\theta(0)e^{(-m)0}] = \sqrt{\bar{h}_c P A k} \; \theta_s \qquad (2.62)$$

The same result is obtained by evaluating the convective heat flow from the surface of the rod

$$q_{\text{fin}} = \int_0^\infty \bar{h}_c P \theta_s \, e^{-mx} \, dx = \frac{\bar{h}_c P}{m} \theta_s \, e^{-mx} \Big|_0^\infty = \sqrt{\bar{h}_c P A k} \; \theta_s$$

Equations (2.60) and (2.62) are reasonable approximations of the temperature distribution and heat flow rate in a finite fin if its length is very large compared to its cross-sectional area. If the rod is of finite length but the heat loss from the end of the rod is neglected, or if the end of the rod is insulated, the second boundary condition requires that the temperature gradient at $x = L$ be zero, or $dT/dx = 0$ at $x = L$. These conditions require that

$$\left(\frac{d\theta}{dx} \right)_{x=L} = 0 = mC_1 \, e^{mL} - mC_2 \, e^{-mL}$$

Solving this equation for condition 2 simultaneously with the relation for condition 1, which required that

$$\theta(0) = \theta_s = C_1 + C_2$$

yields

$$C_1 = \frac{\theta_s}{1 + e^{2mL}} \qquad C_2 = \frac{\theta_s}{1 + e^{-2mL}}$$

Substituting the above relations for C_1 and C_2 into Eq. (2.59) gives the temperature distribution

$$\theta = \theta_s \left(\frac{e^{mx}}{1 + e^{2mL}} + \frac{e^{-mx}}{1 + e^{-2mL}} \right) = \theta_s \frac{\cosh m(L - x)}{\cosh(mL)} \qquad (2.63)^*$$

* The derivation of Eq. (2.63) is left as an exercise for the reader. The hyperbolic cosine, cosh for short, is defined by $\cosh x = (e^x + e^{-x})/2$.

TABLE 2.1 EQUATIONS FOR TEMPERATURE DISTRIBUTION AND RATE OF HEAT TRANSFER FOR FINS OF UNIFORM CROSS SECTION[a]

Case	Tip condition $(x = L)$	Temperature distribution, θ/θ_s	Fin heat transfer rate, q_f
1	Infinite fin $(L \to \infty)$: $\theta(L) = 0$	e^{-mx}	M
2	Adiabatic: $\dfrac{d\theta}{dx}\bigg|_{x=L} = 0$	$\dfrac{\cosh m(L - x)}{\cosh mL}$	$M \tanh mL$
3	Fixed temperature: $\theta(L) = \theta_L$	$\dfrac{(\theta_L/\theta_s) \sinh mx + \sinh m(L - x)}{\sinh mL}$	$M \dfrac{\cosh mL - \theta_L/\theta_s}{\sinh mL}$
4	Convection heat transfer: $\bar{h}\theta(L) = \dfrac{-k\,d\theta}{dx}\bigg|_{x=L}$	$\dfrac{\cosh m(L - x) + (\bar{h}/mk) \sinh m(L - x)}{\cosh mL + (\bar{h}/mk) \sinh mL}$	$M \dfrac{\sinh mL + (\bar{h}/mk) \cosh mL}{\cosh mL + (h/mk) \sinh mL}$

$$^a\theta \equiv T - T_\infty$$
$$\theta_s \equiv \theta(0) = T_s - T_\infty$$
$$m^2 \equiv \frac{\bar{h}_c P}{kA}$$
$$M \equiv \sqrt{\bar{h}_c P k A}\,\theta_s$$

The heat loss from the fin can be found by substituting the temperature gradient at the root into Eq. (2.61). Noting that $\tanh(mL) = (e^{mL} - e^{-mL})/(e^{mL} + e^{-mL})$, we get

$$q_{\text{fin}} = \sqrt{\bar{h}_c P A k}\ \theta_s \tanh(mL) \qquad (2.64)$$

The results for the other two tip conditions can be obtained in a similar manner, but the algebra is more lengthy. For convenience, all four cases are summarized in Table 2.1.

EXAMPLE 2.6 ───────────────────────────────────

A copper pin fin 0.25 cm in diameter protrudes from a wall at 95°C into ambient air at 25°C. The heat transfer is mainly by free convection with a coefficient equal to 10 W/m² K. Calculate the heat loss, assuming that (a) the fin is "infinitely long," and (b) the fin is 2.5 cm long and the coefficient at the end is the same as over the circumference. Finally, (c) how long would the fin have to be for the infinitely long solution to be correct within 5 percent?

Solution. The following assumptions will be made:

1. Thermal conductivity does not change with temperature.
2. Steady state prevails.
3. Radiation is negligible.

4. Convection heat transfer coefficient is uniform over the surface of the fin.
5. Conduction along the fin is one-dimensional.

The thermal conductivity of the copper can be found in Table 12 of Appendix 2. We know that the fin temperature will decrease along its length, but do not know its value at the tip. As an approximation, choose a temperature of 70°C or 343 K. Interpolating the values in Table 12 gives $k = 396$ W/m K.

(a) From Eq. (2.62) the heat loss for the "infinitely long" fin is

$$q_{fin} = \sqrt{\bar{h}_c P k A}(T_s - T_\infty)$$

$$= \left[(10 \text{ W/m}^2 \text{ K})(\pi)(0.0025 \text{ m})(396 \text{ W/m K}) \right.$$

$$\left. \times \left(\frac{\pi}{4} \right) 0.025 \text{ m}^2 \right]^{1/2} (95 - 25)°\text{C}$$

$$= 0.2720 \text{ W}$$

(b) The equation for the heat loss from the finite fin is case 4 in Table 2.1:

$$q_{fin} = \sqrt{\bar{h}_c P k A}(T_s - T_\infty) \frac{\sinh mL + (\bar{h}_c/mk) \cosh mL}{\cosh mL + (\bar{h}_c/mk) \sinh mL}$$

$$= 0.1375 \text{ W}$$

(c) For the two solutions to be within 5 percent, it is necessary that

$$\frac{\sinh mL + (\bar{h}_c/mk) \cosh mL}{\cosh mL + (\bar{h}_c/mk) \sinh mL} \geq 0.05$$

This condition is satisfied when $mL \geq 1.8$ or $L > 9.0$ cm.

2.4.2 Fin Selection and Design

In the preceding section, we developed equations for the temperature distribution and the rate of heat transfer of extended surfaces and fins. Fins are widely used to increase the rate of heat transfer from a wall. As an illustration of such an application, consider a surface exposed to a liquid at temperature T_∞ flowing over the surface. If the wall is bare and the surface temperature T_s is fixed, the rate of heat transfer per unit area from the plane wall is controlled entirely by the heat transfer coefficient $\bar{h}_c$. The coefficient at the plane wall may be increased by increasing the fluid velocity, but this also creates a larger pressure drop and requires increased pumping power.

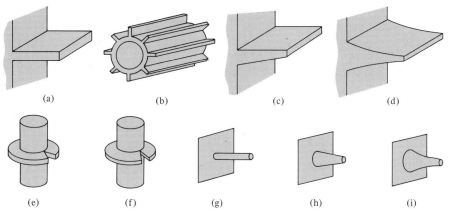

(a)   (b)   (c)   (d)

(e)   (f)   (g)   (h)   (i)

Figure 2.15 Schematic diagrams of different types of fins: (*a*) longitudinal fin of rectangular profile; (*b*) cylindrical tube with fins of rectangular profile; (*c*) longitudinal fin of trapezoidal profile; (*d*) longitudinal fin of parabolic profile; (*e*) cylindrical tube with radial fin of rectangular profile; (*f*) cylindrical tube with radial fin of truncated conical profile; (*g*) cylindrical pin fin; (*h*) truncated conical spine; (*i*) parabolic spine.

In many cases it is thus preferable to increase the rate of heat transfer from the wall by using fins that extend from the wall into the fluid and increase the contact area between the solid surface and the fluid. If the fin is made of a material with high thermal conductivity, the temperature gradient along the fin from base to tip will be small and the heat transfer characteristics of the wall will be greatly enhanced. Fins come in many shapes and forms, some of which are shown in Fig. 2.15. The selection of fins is made on the basis of thermal performance and cost. The selection of a suitable fin geometry requires a compromise among the cost, the weight, the available space, and the pressure drop of the heat transfer fluid, as well as the heat transfer characteristics of the extended surface. From the point of view of thermal performance, the most desirable size, shape, and length of the fin can be evaluated by an analysis such as that outlined below.

The heat transfer effectiveness of a fin is measured by a parameter called the fin efficiency η_f, which is defined as

$$\eta_f = \frac{\text{actual heat transferred by fin}}{\substack{\text{heat that would have been transferred} \\ \text{if entire fin were at the base temperature}}}$$

Using Eq. (2.64), the fin efficiency for a circular pin fin of diameter D and length L with an insulated end is

$$\eta_f = \frac{\tanh \sqrt{4L^2 \bar{h}_c / kD}}{\sqrt{4L^2 \bar{h}_c / kD}} \tag{2.65}$$

whereas for a fin of rectangular cross section (length L and thickness t) the efficiency of a fin with an insulated end is

$$\eta_f = \frac{\tanh \sqrt{\bar{h}PL^2/kA}}{\sqrt{\bar{h}PL^2/kA}} \tag{2.66}$$

If a rectangular fin is long, wide, and thin, $P/A \simeq 2/t$ and the heat loss from the end can be taken into account approximately by increasing L by $t/2$ and assuming that the end is insulated. This keeps the surface area from which heat is lost the same as in the real case, and the fin efficiency then becomes

$$\eta_f = \frac{\tanh \sqrt{2\bar{h}_c L_c^2/kt}}{\sqrt{2\bar{h}_c L_c^2/kt}} \tag{2.67}$$

where $L_c = (L + t/2)$

The error that results from this approximation will be less than 8 percent when

$$\left(\frac{\bar{h}_c t}{2k}\right)^{1/2} \le \frac{1}{2}$$

It is often convenient to use the profile area of a fin, A_m. For a rectangular shape A_m is Lt, whereas for a triangular cross section A_m is $Lt/2$, where t is the base thickness. In Fig. 2.16 the fin efficiencies for rectangular and triangular fins are compared. Figure 2.17 shows the fin efficiency for circumferential fins of rectangular cross section (2, 3).

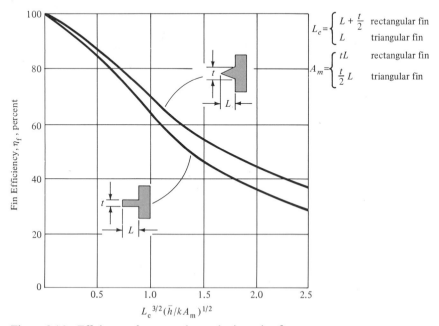

Figure 2.16 Efficiency of rectangular and triangular fins.

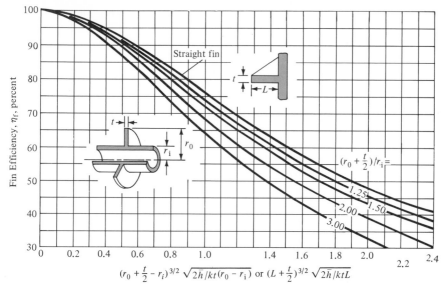

Figure 2.17 Efficiency of circumferential rectangular fins.

EXAMPLE 2.7 _____

To increase the heat dissipation from a 2.5-cm-OD tube, circumferential fins made of aluminum ($k = 200$ W/m K) are soldered to the outer surface. The fins are 0.1 cm thick and have an outer diameter of 5.5 cm. If the tube temperature is 100°C, the environmental temperature is 25°C, and the heat transfer coefficient between the fin and the environment is 65 W/m² K, calculate the rate of heat loss from a fin.

Solution. The geometry of the fin in this problem corresponds to that in Fig. 2.17 and we can therefore use the fin efficiency curve in Fig. 2.17. The parameters required to obtain the fin efficiency are

$$\left(r_0 + \frac{t}{2} - r_i\right)^{3/2} = [(2.75 + 0.05 - 1.25)/100]^{3/2} = 0.00193$$

$$[2\bar{h}_c/kt(r_0 - r_i)]^{1/2} = [2 \times 65/200 \times 0.001 \times 0.015]^{1/2}$$
$$= 208$$

$$\left(r_0 + \frac{t}{2}\right) \Big/ r_i = 2.8/1.25 = 2.24$$

$$\left(r_0 + \frac{t}{2} - r_i\right)^{3/2} [2\bar{h}_c/kt(r_0 - r_i)]^{1/2} = 0.402$$

From Fig. 2.17 the fin efficiency is found to be 91%. The rate of heat loss from a single fin is

$$q_{\text{fin}} = \eta_{\text{fin}}\bar{h}_c A_{\text{fin}}(T_s - T_\infty)$$

$$= \eta_{\text{fin}}\bar{h}_c 2\pi\left[\left(r_0 + \frac{t}{2}\right)^2 - r_i^2\right](T_s - T_\infty)$$

$$= 0.91(65 \text{ W/m}^2 \text{ K})2\pi(7.84 - 1.56) \times 10^{-4} \text{ m}^2 (75 \text{ K}) = 17.5 \text{ W}$$

For a plane surface of area A, the thermal resistance is $1/\bar{h}A$. Addition of fins increases the surface area, but at the same time it introduces a conductive resistance over that portion of the original surface to which the fins are attached. Addition of fins will therefore not always increase the rate of heat transfer. In practice, addition of fins is hardly ever justified unless $\bar{h}A/Pk$ is considerably less than unity.

It is interesting to note that the fin efficiency reaches its maximum value for the trivial case of $L = 0$, or no fin at all. It is therefore not possible to maximize fin performance with respect to fin length. It is normally more important to maximize the efficiency with respect to the quantity of fin material (mass, volume, or cost), because such an optimization has obvious economic significance.

Using the values of the average surface conductances in Table 1.4 as a guide, we can easily see that fins effectively increase the heat transfer to or from a gas, are less effective when the medium is a liquid in forced convection, but offer no advantage in heat transfer to boiling liquids or from condensing vapors. For example, for a 0.3175-cm-diameter aluminum pin fin in a typical gas heater, $\bar{h}A/Pk = 0.00045$, whereas in a water heater, for example, $\bar{h}A/Pk = 0.022$. In a gas heater the addition of fins would therefore be much more effective than in a water heater.

It is apparent from these considerations that when fins are used they should be placed on the side of the heat exchange surface where the heat transfer coefficient between the fluid and the surface is lower. Thin, slender, closely spaced fins are superior to fewer and thicker fins from the heat transfer standpoint. Obviously, fins made of materials having a high thermal conductivity are desirable. Fins are sometimes an integral part of the surface, but there can be a contact resistance at the base of the fin if the fins are mechanically attached.

To obtain the total efficiency of a surface with fins η_t, we combine the unfinned portion of the surface at 100 percent efficiency with the surface area of the fins at η_f, or

$$A_o\eta_t = (A_o - A_b) + A_f\eta_f \tag{2.68}$$

where A_o = total heat transfer area

A_b = base area of the fins

A_f = heat transfer area of the fins

The overall heat transfer coefficient U_o, based on the total outer surface area, for heat transfer between two fluids separated by a wall with fins can then be expressed as

$$U_o = \cfrac{1}{\cfrac{1}{\eta_{to}\bar{h}_o} + R_{k_{\text{wall}}} + \cfrac{A_o}{\eta_{ti}A_i\bar{h}_i}} \tag{2.69}$$

where $R_{k_{\text{wall}}}$ = thermal resistance of the wall to which the fins are attached, W/m² K (outside surface)

A_o = total outer surface area, m²

A_i = total inner surface area, m²

η_{to} = total efficiency for outer surface

η_{ti} = total efficiency for inner surface

$\bar{h}_o$ = average heat transfer coefficient for outer surface, W/m² K

$\bar{h}_i$ = average heat transfer coefficient for inner surface, W/m² K

For tubes with fins on the outside only, the usual case in practice, η_{ti} is unity and $A_i = \pi D_i L$.

In the analysis presented in this chapter, details of the convection heat flow between the fin surface and the surrounding fluid have been omitted. A complete engineering analysis not only requires an evaluation of the fin performance, but must also take the relation between the fin geometry and the convection heat transfer into account. Problems on the convection heat transfer part of the design will be considered in later chapters.

2.5* MULTIDIMENSIONAL STEADY CONDUCTION

In the preceding part of this chapter we dealt with problems in which the temperature and the heat flow can be treated as functions of a single variable. Many practical problems fall into this category, but when the boundaries of a system are irregular or when the temperature along a boundary is nonuniform, a one-dimensional treatment may no longer be satisfactory. In such cases, the temperature is a function of two, and possibly even three, coordinates. The heat flow through a corner section where two or three walls meet, the heat conduction through the walls of a short, hollow cylinder, and the heat loss from a buried pipe are typical examples of this class of problems.

We shall now consider some methods for analyzing conduction in two- and three-dimensional systems. The emphasis will be placed on two-dimensional problems because they are less cumbersome to solve, yet they illustrate the basic methods of analysis for three-dimensional systems. Heat conduction in two- and three-dimensional systems can be treated by analytic, graphic, analogic, and numerical methods. For some cases, "shape factors" are also available. We will consider in this chapter the analytic, graphic, and shape-factor methods of solution. The numerical approach will be taken up in Chapter 3. The analytic treatment in this chapter is limited to an illustrative example, and for more extensive coverage of analytic methods the reader is referred to (1, 4–6). The analogic method is presented in (7), but is omitted here because it is no longer used in practice.

2.5.1 Analytic Solution

The objective of any heat transfer analysis is to predict the rate of heat flow, the temperature distribution, or both. According to Eq. (2.10), in a two-dimensional system without heat sources the general conduction equation governing the temperature distribution in the steady state is

$$\frac{\partial^2 T}{\partial x^2} + \frac{\partial^2 T}{\partial y^2} = 0 \qquad (2.70)$$

if the thermal conductivity is uniform. The solution of Eq. (2.70) will give $T(x, y)$, the temperature as a function of the two space coordinates x and y. The rate of heat flow per unit area in the x and y directions, respectively, can then be obtained from Fourier's law:

$$q_x'' = \left(\frac{q}{A}\right)_x = -k\frac{\partial T}{\partial x}$$

$$q_y'' = \left(\frac{q}{A}\right)_y = -k\frac{\partial T}{\partial y}$$

It should be noted that whereas the temperature is a scalar, the heat flux depends on the temperature gradient and is therefore a vector. The total rate of heat flow at a given point x, y is the resultant of the components q_x and q_y at that point and is directed perpendicular to the isotherm as shown in Fig. 2.18. Thus, if the temperature distribution in a system is known, the rate of heat flow can easily be calculated. Therefore, heat transfer analyses usually concentrate on determining the temperature field.

An analytic solution of a heat conduction problem must satisfy the heat conduction equation as well as the boundary conditions specified by the physical

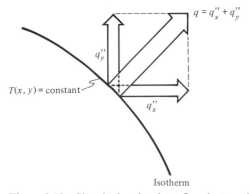

Figure 2.18 Sketch showing heat flow in two dimensions.

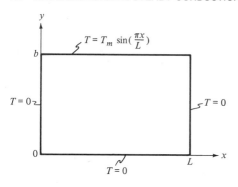

Figure 2.19 Rectangular adiabatic plate with sinusoidal temperature distribution on one edge.

conditions of the particular problem. The classical approach to an exact solution of the Fourier equation is the separation-of-variables technique. We shall illustrate this approach by applying it to a relatively simple problem. Consider a thin rectangular plate, free of heat sources and insulated at the top and bottom surfaces (Fig. 2.19). Since $\partial T/\partial z$ is assumed to be negligible, the temperature is a function of x and y only. If the thermal conductivity is uniform, the temperature distribution must satisfy Eq. (2.70). This is a linear and homogeneous partial differential equation which can be integrated by assuming a product solution for $T(x, y)$ of the form

$$T = XY \tag{2.71}$$

where $X = X(x)$, a function of x only, and $Y = Y(y)$, a function of y alone. Substituting Eq. (2.71) into Eq. (2.70) yields

$$-\frac{1}{X}\frac{d^2X}{dx^2} = \frac{1}{Y}\frac{d^2Y}{dy^2} \tag{2.72}$$

The variables are now separated. The left-hand side is a function of x only, while the right-hand side is a function of y alone. Since neither side can change as x and y vary, both must be equal to a constant, say λ^2. We have, therefore, the two ordinary differential equations

$$\frac{d^2X}{dx^2} + \lambda^2 X = 0 \tag{2.73}$$

and

$$\frac{d^2Y}{dy^2} - \lambda^2 Y = 0 \tag{2.74}$$

The general solution to Eq. (2.73) is

$$X = A\cos\lambda x + B\sin\lambda x$$

and the general solution to Eq. (2.74) is

$$Y = Ce^{-\lambda y} + De^{\lambda y}$$

and therefore, from Eq. (2.71),

$$T = XY = (A\cos\lambda x + B\sin\lambda x)(Ce^{-\lambda y} + De^{\lambda y}) \tag{2.75}$$

where A, B, C, and D are constants to be evaluated from the boundary conditions. As shown in Fig. 2.19, the boundary conditions to be satisfied are

$$T = 0 \quad \text{at} \quad y = 0$$
$$T = 0 \quad \text{at} \quad x = 0$$
$$T = 0 \quad \text{at} \quad x = L$$
$$T = T_m \sin(\pi x/L) \quad \text{at} \quad y = b$$

Substituting these conditions into Eq. (2.75) for T, we get from the first condition

$$(A \cos \lambda x + B \sin \lambda x)(C + D) = 0$$

from the second condition

$$A(Ce^{-\lambda y} + De^{\lambda y}) = 0$$

and from the third condition

$$(A \cos \lambda L + B \sin \lambda L)(Ce^{-\lambda y} + De^{\lambda y}) = 0$$

The first condition can be satisfied only if $C = -D$, and the second if $A = 0$. Using these results in the third condition, we obtain

$$(B \sin \lambda L)(Ce^{-\lambda y}) = 2BC \sin \lambda L \sinh \lambda y = 0$$

To satisfy this condition, $\sin \lambda L$ must be zero or $\lambda = n\pi/L$, where $n = 1, 2, 3$, etc.* There exists therefore a different solution for each integer n and each solution has a separate integration constant C_n. Summing these solutions, we get

$$T = \sum_{n=1}^{\infty} C_n \sin \frac{n\pi x}{L} \sinh \frac{n\pi y}{L} \tag{2.76}$$

The last boundary condition demands that, at $y = b$,

$$\sum_{n=1}^{\infty} C_n \sin \frac{n\pi x}{L} \sinh \frac{n\pi b}{L} = T_m \sin \frac{\pi x}{L}$$

so that only the first term in the series solution with $C_1 = T_m/\sinh(\pi b/L)$ is needed. The solution therefore becomes

$$T(x, y) = T_m \frac{\sinh(\pi y/L)}{\sinh(\pi b/L)} \sin \frac{\pi x}{L} \tag{2.77}$$

The corresponding temperature field is shown in Fig. 2.20. The solid lines are isotherms and the dashed lines are heat flow lines. It should be noted that lines indicating the direction of heat flow are perpendicular to the isotherms.

When the boundary conditions are not as simple as in the illustrative problem, the solution is obtained in the form of an infinite series. For example, if the temperature at the edge $y = b$ is a function of x, say $T(x, b) = F(x)$, then

---

* The value $n = 0$ is excluded because it would give the trivial solution $T = 0$.

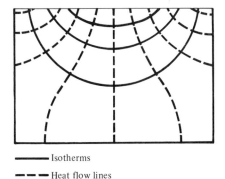

——— Isotherms

– – – Heat flow lines

Figure 2.20 Isotherms and heat flow lines for the plate shown in Fig. 2.19.

the solution, as shown in (1), is the infinite series

$$T = \frac{2}{L} \sum_{n=1}^{\infty} \frac{\sinh(n\pi/L)y}{\sinh n\pi(b/L)} \sin \frac{\pi n}{L} x \int_0^L F(x) \sin \frac{n\pi}{L} x \, dx \qquad (2.78)$$

which is quite laborious to evaluate quantitatively.

The separation-of-variables method can be extended to three-dimensional cases by assuming $T = XYZ$, substituting this expression for T in Eq. (2.9), separating the variables, and integrating the resulting total differential equations subject to the given boundary conditions. Examples of three-dimensional problems are presented in (1, 4, 5).

2.5.2 Graphic Method and Shape Factors

The graphic method presented in this section can yield rapidly a reasonably good estimate of the temperature distribution and heat flow in geometrically complex two-dimensional systems, but its application is limited to problems with isothermal and insulated boundaries. The object of a graphic solution is to construct a network consisting of isotherms (lines of constant temperature) and constant-flux lines (lines of constant heat flow). The flux lines are analogous to streamlines in a potential fluid flow; that is, they are tangent to the direction of heat flow at any point. Consequently, no heat can flow across constant-flux lines and a constant amount of heat flows between any two of them. The isotherms are analogous to constant-potential lines and heat flows perpendicular to them. Thus, lines of constant temperature and lines of constant heat flux intersect at right angles. To obtain the temperature distribution one first prepares a scale model and then draws isotherms and flux lines freehand, by trial and error, until they form a network of curvilinear squares. The procedure is illustrated in Fig. 2.21 for a corner section of unit depth ($\Delta z = 1$) with faces ABC at temperature T_1, faces FED at temperature T_2, and faces CD and AF insulated. Figure 2.21a shows the scale model and Fig. 2.21b presents the curvilinear network of isotherms and flux lines. It should be noted that the flux lines emanating from isothermal boundaries are perpendicular to the boundary, except when they come from a corner. Flux lines leading to or from a corner

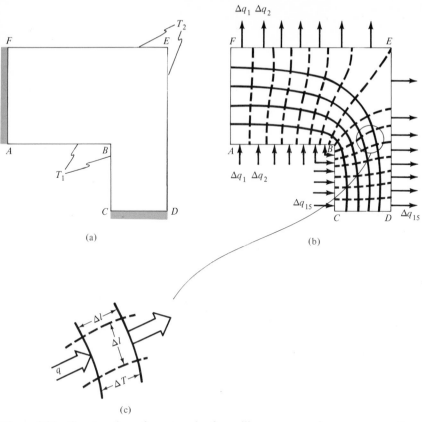

Figure 2.21 Construction of a network of curvilinear squares for a corner section: (*a*) scale model; (*b*) flux plot; (*c*) typical curvilinear square.

of an isothermal boundary bisect the angle between the surfaces forming the corner.

A graphic solution, like an analytic solution of a heat conduction problem described by the Laplace equation and the associated boundary conditions, is unique. Therefore, any curvilinear network, irrespective of the size of the squares, which satisfies the boundary conditions represents the correct solution. Taking any curvilinear square, such as Fig. 2.21*c*, the rate of heat flow is given by Fourier's law:

$$\Delta q = -k(\Delta l \times 1)\frac{\Delta T}{\Delta l} = -k\,\Delta T$$

This heat flow will remain the same across any square within any one heat flow lane from the boundary at T_1 to the boundary at T_2. The ΔT across any one element in the heat flow lane is therefore

$$\Delta T = \frac{T_2 - T_1}{N}$$

where N is the number of temperature increments between the two boundaries at T_1 and T_2. The total rate of heat flow from the boundary at T_2 to the boundary at T_1 equals the sum of the heat flow through all the lanes. According to the above relations, the heat flow rate is the same through all lanes since it is independent of the size of the squares in a network of curvilinear squares. The total rate of heat transfer can therefore be written

$$q = \sum_{n=1}^{n=M} \Delta q_n = \frac{M}{N} k(T_2 - T_1) = \frac{M}{N} k \Delta T_{\text{overall}} \tag{2.79}$$

where Δq_n is the rate of heat flow through the mth lane, and M is the number of heat flow lanes.

Thus, to calculate the rate of heat transfer we need only construct a network of curvilinear squares in the scale model and count the number of temperature increments and heat flow lanes. Although the accuracy of the method depends a good deal on the skill and patience of the person sketching the curvilinear square network, even a crude sketch can give a reasonably good estimate of the temperature distribution, which, if desired, can be refined by the numerical method described in the next chapter.

In any two-dimensional system in which heat is transferred from one surface at T_1 to another at T_2 the rate of heat transfer per unit depth depends only on the temperature difference $T_1 - T_2 = \Delta T_{\text{overall}}$, the thermal conductivity k, and the ratio M/N. This ratio depends on the shape of the system and is called the *shape factor* S. The rate of heat transfer can thus be written

$$q = kS \, \Delta T_{\text{overall}} \tag{2.80}$$

when the grid consists of curvilinear squares. Values of S for several shapes of practical significance (6–9) are summarized in Table 2.2.

EXAMPLE 2.8 ——————————————————————————————

A long, 10-cm-OD pipe is buried with its centerline 60 cm below the surface in soil having a thermal conductivity of 0.4 W/m K. (*a*) Prepare a curvilinear square network for this system and calculate the heat loss per meter length if the pipe surface temperature is 100°C and the soil surface is at 20°C. (*b*) Compare the result from part (*a*) with that obtained with the appropriate shape factor S.

Solution. (*a*) The curvilinear square network for the system is shown in Fig. 2.22. Because of symmetry, only half of this heat flow field needs to be plotted. There are 18 heat flow lanes leading from the pipe to the surface, and each lane consists of 8 curvilinear squares. The shape factor is therefore

$$S = \frac{18}{8} = 2.25$$

and the rate of heat flow per meter is, from Eq. (2.80),

$$q = (0.4)(2.25)(100 - 20) = 72 \text{ W/m}$$

TABLE 2.2 CONDUCTION SHAPE FACTOR S FOR VARIOUS SYSTEMS
$[q_k = Sk(T_1 - T_2)]$

Description of system	Symbolic sketch	Shape factor S
Conduction through a homogeneous medium of thermal conductivity k between an isothermal surface and a sphere buried a distance z below		$\dfrac{2\pi D}{1 - D/4z}$
Conduction through a homogeneous medium of thermal conductivity k between an isothermal surface and a horizontal cylinder of length L buried with its axis a distance z below the surface		$\dfrac{2\pi L}{\cosh^{-1}(2z/D)}$ if $z/L \ll 1$
Conduction through a homogeneous medium of thermal conductivity k between an isothermal surface and an infinitely long cylinder buried a distance z below (per unit length of cylinder)		$\dfrac{2\pi}{\cosh^{-1}(2z/D)}$
Conduction through a homogeneous medium of thermal conductivity k between an isothermal surface and a vertical circular cylinder of length L		$\dfrac{2\pi L}{\ln(4L/D)}$ if $D/L \ll 1$
Horizontal thin circular disk buried far below an isothermal surface in a homogeneous material of thermal conductivity k		$\dfrac{4.45D}{1 - D/5.67z}$
Conduction through a homogeneous material of thermal conductivity k between two long parallel cylinders a distance L apart (per unit length of cylinders)		$2\pi/\cosh^{-1}\left(\dfrac{L^2 - 1 + r^2}{2Lr}\right)$ $+\cosh^{-1}\left(\dfrac{L + 1 - r}{2L}\right)$ $(r = r_1/r_2 \text{ and } L = l/r_2)$
Conduction through two plane sections and the[a] edge section of two walls of thermal conductivity k—inner- and outer-surface temperatures uniform		$\dfrac{al}{\Delta x} + \dfrac{bl}{\Delta x} + 0.54l$
Conduction through the corner section C of three[a] homogeneous walls of thermal conductivity k— inner- and outer-surface temperatures uniform		$0.15\Delta x$ if Δx is small compared to the lengths of walls

[a] Sketch illustrating dimensions for use in calculating three-dimensional shape factors.

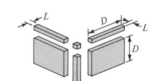

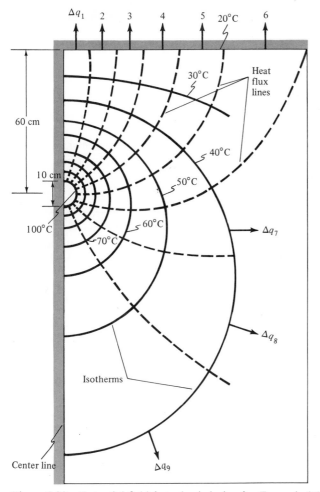

Figure 2.22 Potential field for a buried pipe for Example 2.8.

(b) From Table 2.2

$$S = \frac{2\pi(1)}{\cosh^{-1}(120/10)} = \frac{2\pi}{3.18} = 1.98$$

and the rate of heat loss per meter length is

$$q = (0.4)(2.1)(100 - 20) = 63.3 \text{ W/m}$$

The reason for the difference is that the potential field in Fig. 2.22 has a finite number of flux lines and isotherms and is therefore only approximate.

For a three-dimensional wall, as in a furnace, separate shape factors are used to calculate the heat flow through the edge and corner sections. When all the interior dimensions are greater than one-fifth of the wall thickness,

$$S_{wall} = \frac{A}{L}; \qquad S_{edge} = 0.54D; \qquad S_{corner} = 0.15L$$

where A = area of wall
        L = wall thickness
        D = length of edge

These dimensions are illustrated in Table 2.2. Note that the shape factor per unit depth is given by the ratio M/N when the curvilinear-squares method is used for calculations.

EXAMPLE 2.9

A small cubic furnace 50 × 50 cm on the inside is constructed of fireclay brick ($k = 1.04$ W/m °C) with a wall thickness of 10 cm. The inside of the furnace is maintained at 500°C and the outside at 50°C. Calculate the heat lost through the walls.

Solution.   We compute the total shape factor by adding the shape factors for the walls, edges, and corners.

Walls:

$$S = \frac{A}{L} = \frac{(0.5)(0.5)}{0.1} = 2.5 \text{ m}$$

Edges:

$$S = 0.54D = (0.54)(0.5) = 0.27 \text{ m}$$

Corners:

$$S = 0.15L = (0.15)(0.1) = 0.015 \text{ m}$$

There are 6 wall sections, 12 edges, and 8 corners, so that the total shape factor is

$$S = (6)(2.5) + (12)(0.27) + (8)(0.015) = 18.36 \text{ m}$$

and the heat flow is calculated as

$$q = kS\,\Delta T = (1.04)(18.36)(500 - 50) = 8.59 \text{ kW}$$

2.6* TRANSIENT HEAT CONDUCTION

So far we have only dealt with steady-state conduction in this chapter. But before steady-state conditions are reached, some time must elapse after the heat transfer process is initiated. During this transient period the temperature changes and the analysis must take into account changes in the internal energy. Example 1.10 in Chapter 1 illustrates this phenomenon for a simple case. In the remainder of this chapter we will deal with methods for analyzing more complex unsteady heat flow problems, because transient heat flow is of great practical importance in industrial heating and cooling.

In addition to unsteady heat flow when the system undergoes a transition from one steady state to another, there are also engineering problems involving periodic variations in heat flow and temperature. Examples of such cases are the periodic heat flow in a building between day and night and the heat flow in an internal combustion engine.

We shall first analyze problems that can be simplified by assuming that the temperature is only a function of time and is uniform throughout the system at any instant. This type of analysis is called the lumped-heat-capacity method. In subsequent sections of this chapter we shall consider methods for solving problems of unsteady heat flow when the temperature not only depends on time, but also varies in the interior of the system. Throughout this chapter we shall not be concerned with the mechanisms of heat transfer by convection or radiation. Where these modes of heat transfer affect the boundary conditions of the system, an appropriate value for the unit-surface conductance will simply be specified.

2.6.1 Systems with Negligible Internal Resistance

Even though no materials in nature have an infinite thermal conductivity, many transient heat flow problems can be readily solved with acceptable accuracy by assuming that the internal conductive resistance of the system is so small that the temperature within the system is substantially uniform at any instant. This simplification is justified when the external thermal resistance between the surface of the system and the surrounding medium is so large compared to the internal thermal resistance of the system that it controls the heat transfer process.

A measure of the relative importance of the thermal resistance within a solid body is the ratio of the internal to the external resistance, called the Biot number Bi, which has been defined by the equation

$$\text{Bi} = \frac{R_{\text{internal}}}{R_{\text{external}}} = \frac{\bar{h}L}{k_s} \tag{2.81}$$

where $\bar{h}$ is the average unit-surface conductance, L is a significant length dimension obtained by dividing the volume of the body by its surface area, and k_s is the thermal conductivity of the solid body. In bodies whose shape resembles a plate, a cylinder, or a sphere, the error introduced by the assumption that the temperature at any instant is uniform will be less than 5 percent when the internal resistance is less than 10 percent of the external surface resistance, that is, when $\bar{h}L/k_s < 0.1$.

As a typical example of this type of transient heat flow, consider the cooling of a small metal casting or a billet in a quenching bath after its removal from a hot furnace. Suppose that the billet is removed from the furnace at a uniform temperature T_0 and is quenched so suddenly that we can approximate the environmental temperature change by a step. Designate the time at which the cooling begins as $t = 0$, and assume that the heat transfer coefficient $\bar{h}$ remains constant during the process and that the bath temperature T_∞ at a distance far removed from the billet does not vary with time. Then, in accordance with the assumption that the temperature within the body is substantially uniform at any instant, an energy balance for the billet over a small time interval $d\theta$ is

$$\begin{array}{c} \text{Change in internal energy} \\ \text{of the billet during } dt \end{array} = \begin{array}{c} \text{net heat flow from the} \\ \text{billet to the bath during } dt \end{array}$$

or

$$-c\rho V\, dT = \bar{h}A_s(T - T_\infty)\, dt \tag{2.82}$$

where c = specific heat of billet, J/kg K

ρ = density of billet, kg/m^3

V = volume of billet, m^3

T = average temperature of billet, K

$\bar{h}$ = average heat transfer coefficient, W/m^2 K

A_s = surface area of billet, m^2

dT = temperature change (K) during time interval dt (s)

The minus sign in Eq. (2.82) indicates that the internal energy decreases when $T > T_\infty$. The variables T and t can be readily separated and, for a differential time interval dt, Eq. (2.82) becomes

$$\frac{dT}{T - T_\infty} = \frac{d(T - T_\infty)}{(T - T_\infty)} = -\frac{\bar{h}A_s}{c\rho V}\, dt \tag{2.83}$$

where it is noted that $d(T - T_\infty) = dT$, since T_∞ is constant. With an initial temperature of T_0 and a temperature at time t of T as limits, integration of Eq. (2.83) yields

$$\ln \frac{T - T_\infty}{T_0 - T_\infty} = -\frac{\bar{h}A_s}{c\rho V}\, t$$

or

$$\frac{T - T_\infty}{T_0 - T_\infty} = e^{-(\bar{h}A_s/c\rho V)t} \tag{2.84}$$

where the exponent $\bar{h}A_s t/c\rho V$ must be dimensionless. The combination of variables in this exponent is the product of two dimensionless groups we encountered previously:

$$\frac{\bar{h}A_s t}{c\rho V} = \left(\frac{\bar{h}L}{k_s}\right)\left(\frac{\alpha t}{L^2}\right) = \text{Bi Fo} \tag{2.85}$$

where the characteristic length L is the volume of the body V divided by its surface area A_s.

An electrical network analogous to the thermal network for a lumped-single-capacity system is shown in Fig. 2.23. In this network the capacitor is initially "charged" to the potential T_0 by closing the switch S. When the switch is opened, the energy stored in the capacitance is discharged through the resistance $1/\bar{h}A_s$. The analogy between this thermal system and an electrical system is apparent. The thermal resistance is $R = 1/\bar{h}A_s$ and the thermal capacitance is $C = c\rho V$, while R_e and C_e are the electrical resistance and capacitance, respectively. To construct an electrical system that would behave exactly like the thermal system we would only have to make the ratio $\bar{h}A_s/c\rho V$ equal $1/RC$. In the thermal system internal energy is stored, while in the electrical system electric charge is stored. The flow of energy in the thermal system is heat, and the flow of charge is electric current. The quantity $c\rho V/\bar{h}A$ is called the *time constant* of the system since it has the dimensions of time. Its value is indicative

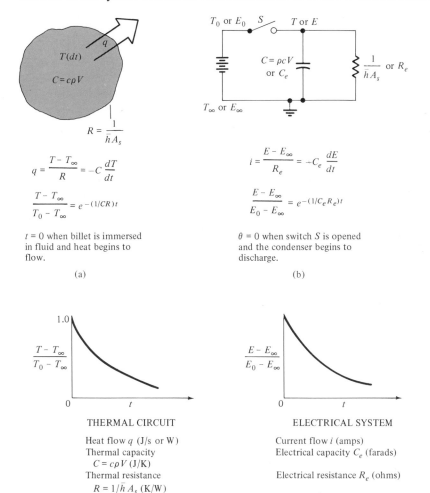

$$q = \frac{T - T_\infty}{R} = -C\frac{dT}{dt}$$

$$\frac{T - T_\infty}{T_0 - T_\infty} = e^{-(1/CR)t}$$

$t = 0$ when billet is immersed in fluid and heat begins to flow.

(a)

$$i = \frac{E - E_\infty}{R_e} = -C_e\frac{dE}{dt}$$

$$\frac{E - E_\infty}{E_0 - E_\infty} = e^{-(1/C_e R_e)t}$$

$\theta = 0$ when switch S is opened and the condenser begins to discharge.

(b)

THERMAL CIRCUIT

Heat flow q (J/s or W)
Thermal capacity
$C = c\rho V$ (J/K)
Thermal resistance
$R = 1/\bar{h} A_s$ (K/W)
Thermal potential $(T - T_\infty)$ (K)

ELECTRICAL SYSTEM

Current flow i (amps)
Electrical capacity C_e (farads)

Electrical resistance R_e (ohms)

Electrical potential $(E - E_\infty)$ (volts)

Figure 2.23   Network and schematic of transient lumped-capacity system.

of the rate of response of a single-capacity system to a sudden change in the environmental temperature. Observe that when the time $t = c\rho V/\bar{h}A_s$ the temperature difference $T - T_\infty$ is equal to 36.8 percent of the initial difference $T_0 - T_\infty$.

EXAMPLE 2.10

Determine the temperature response of a 0.10-cm-diameter copper wire originally at 150°C when suddenly immersed in (a) water at 40°C ($\bar{h}_c =$ 80 W/m² K) and (b) air at 40°C ($\bar{h}_c = 10$ W/m² K).

Solution. From Table 12, Appendix 2 we get

$$k_s = 374 \text{ W/m K}$$
$$c = 383 \text{ J/kg K}$$
$$\rho = 8930 \text{ kg/m}^3$$

The surface area A_s and the volume of the wire per unit length are

$$A_s = \pi D = (\pi)(0.001 \text{ m}^2) = 3.14 \times 10^{-3} \text{ m}^2$$
$$V = \frac{\pi D^2}{4} = (\pi)(0.001^2/4 \text{ m}^2) = 7.85 \times 10^{-7} \text{ m}^3$$

The Biot number in air is

$$\text{Bi} = \frac{\bar{h}_c D}{4k_s} = \frac{(10)(0.001)}{(4)(374)} \ll 1$$

Hence, the internal resistance may be neglected for both cases and Eq. (2.84) applies. From Eq. (2.85):

$$\text{Bi Fo} = \frac{\bar{h}A}{c\rho V}t = \frac{4\bar{h}}{c\rho D}t$$

From the property values we obtain:

$$\text{Bi Fo} = \frac{4(80 \text{ J/s m}^2 \text{ K})}{(383 \text{ J/kg K})(8930 \text{ kg/m}^3)(0.001 \text{ m})}$$
$$= 0.0936\ t \quad \text{for water}$$

$$\text{Bi Fo} = \frac{4(10 \text{ J/s m}^2 \text{ K})}{(383 \text{ J/kg K})(8930 \text{ kg/m}^3)(0.001 \text{ m})}$$
$$= 0.0117\ t \quad \text{for air}$$

The temperature response is given by Eq. (2.84):

$$\frac{T - T_\infty}{T_0 - T_\infty} = e^{-\text{Bi Fo}}$$

The results are plotted in Fig. 2.24. Note that the time required for the temperature of the wire to reach 67°C is more than 20 min in air but only 15 seconds in water. A thermocouple 0.1 cm in diameter would therefore

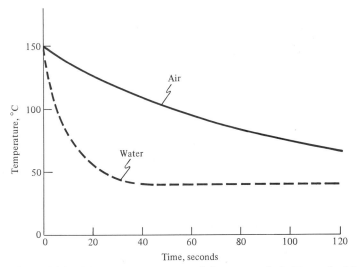

Figure 2.24 Temperature response of thermocouple in Example 2.10 after immersion in air and water.

lag considerably if it were used to measure rapid changes in air temperature, and it would be advisable to use wire of smaller diameter to reduce this lag.

The same general method can also be used to estimate the temperature-time history and internal energy change of a well-stirred fluid in a container suddenly immersed in a medium at a different temperature. If the walls of the container are so thin that their heat capacity is negligible, the temperature-time history of the fluid is given by a relation similar to Eq. (2.84):

$$\frac{T - T_\infty}{T_0 - T_\infty} = e^{-(UA_s/c\rho V)t}$$

where U is the overall heat transfer coefficient between the fluid and the surrounding medium, V is the volume of the fluid in the container, A_s is its surface area, and c and ρ are the specific heat and density of the fluid, respectively.

The lumped-capacity method of analysis can also be applied to composite systems or bodies. For example, if the walls of the container shown in Fig. 2.25 have a substantial thermal capacitance $(c\rho V)_2$, the unit thermal conductance at A_1, the inner surface of the container, is $\bar{h}_1$, the unit thermal conductance at A_2, the outer surface of the container, is $\bar{h}_2$, and the thermal capacitance of the fluid in the container is $(c\rho V)_1$, the temperature-time history of the fluid $T_1(t)$ is obtained by solving simultaneously the energy balance equations

Fluid:

$$-(c\rho V)_1 \frac{dT_1}{dt} = \bar{h}_1 A_1 (T_1 - T_2) \qquad (2.86a)$$

PHYSICAL SYSTEM

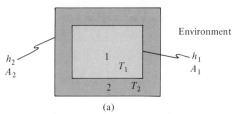

(a)

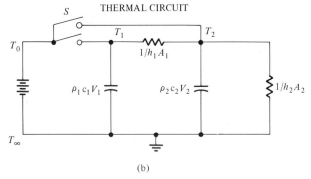

(b)

Figure 2.25 Schematic diagram and thermal network for a two-lump heat capacity system.

Container:

$$-(c\rho V)_2 \frac{dT_2}{dt} = \bar{h}_2 A_2(T_2 - T_\infty) - \bar{h}_1 A_1(T_1 - T_2) \qquad (2.86b)$$

where T_2 is the temperature of the walls of the container.

The above two simultaneous linear differential equations can be solved for the temperature history in each of the bodies. If the fluid and the container are initially at T_0, the boundary conditions for the system are

$$T_1 = T_2 = T_0 \quad \text{at} \quad t = 0$$

which implies that at $t = 0$, $dT_1/dt = 0$ from Eq. (2.86a).

Equations (2.86a) and (2.86b) may be rewritten in operator form as

$$\left(D + \frac{\bar{h}_1 A_1}{\rho_1 c_1 V_1}\right) T_1 - \left(\frac{\bar{h}_1 A_1}{\rho_1 c_1 V_1}\right) T_2 = 0$$

$$-\left(\frac{\bar{h}_1 A_1}{\rho_2 c_2 V_2}\right) T_1 + \left(D + \frac{\bar{h}_1 A_1 + \bar{h}_2 A_2}{\rho_2 c_2 V_2}\right) T_2 = \frac{\bar{h}_2 A_2}{\rho_2 c_2 V_2} T_\infty$$

where the symbol D denotes differentiation with respect to time. For convenience let

$$K_1 = \frac{\bar{h}_1 A_1}{\rho_1 c_1 V_1} \qquad K_2 = \frac{\bar{h}_1 A_1}{\rho_2 c_2 V_2} \qquad K_3 = \frac{\bar{h}_2 A_2}{\rho_2 c_2 V_2}$$

then

$$(D + K_1)T_1 - K_1 T_2 = 0$$
$$-K_2 T_1 + (D + K_2 + K_3)T_2 = K_3 T_\infty$$

Solving the equations simultaneously, we get a differential equation involving only T_1:

$$[D^2 + (K_1 + K_2 + K_3)D + K_1 K_3]T_1 = K_1 K_3 T_\infty$$

The general solution of this equation is

$$T = T_\infty + M e^{m_1 t} + N e^{m_2 t}$$

where m_1 and m_2 are given by

$$m_1 = \frac{-(K_1 + K_2 + K_3) + [(K_1 + K_2 + K_3)^2 - 4K_1 K_3]^{1/2}}{2}$$

$$m_2 = \frac{-(K_1 + K_2 + K_3) - [(K_1 + K_2 + K_3)^2 - 4K_1 K_3]^{1/2}}{2}$$

The arbitrary constants M and N may be obtained by applying the initial conditions

$$T_1 = T_0 \quad \text{at} \quad t = 0$$

and

$$\frac{dT_1}{dt} = 0 \quad \text{at} \quad t = 0$$

This leads to the two equations

$$T_0 = T_\infty + M + N$$
$$0 = m_1 M + m_2 N$$

The final solution for T_1, in dimensionless form, is

$$\frac{T_1 - T_\infty}{T_0 - T_\infty} = \frac{m_2}{m_2 - m_1} e^{m_1 t} - \frac{m_1}{m_2 - m_1} e^{m_2 t} \tag{2.87}$$

The solution for $T_2(t)$ is obtained by substituting the relation for T_1 from Eq. (2.87) into Eq. (2.86a).

The network analogy for the two-lump system is shown in Fig. 2.25. When the switch S is closed, the two thermal capacitances are charged to the potential T_0. At time zero the switch is opened and the capacitances discharge through the two thermal resistances shown.

2.6.2* Infinite Slab

In the remainder of this chapter we will consider some transient conduction problems in which the temperature of the system interior is not uniform. An example of such a problem is transient heat flow in an infinite slab, as shown

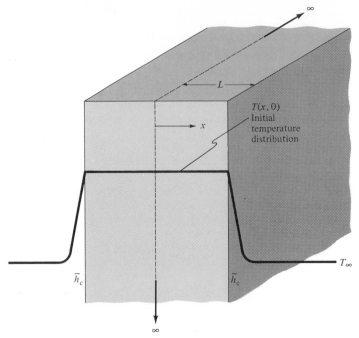

Figure 2.26 Nomenclature for analytical solution of a slab, initially at uniform temperature, subjected at time zero to a change in environmental temperature through a unit
surface conductance $\bar{h}_c$.

in Fig. 2.26. If the temperatures over the two surfaces are uniform, the problem
is one-dimensional and transient. If, furthermore, there are no internal heat
sources and the physical properties of the slab are constant, the general heat
conduction equation reduces to the form

$$\frac{1}{\alpha}\frac{\partial T}{\partial t} = \frac{\partial^2 T}{\partial x^2} \qquad 0 \leq x \leq L \tag{2.88}$$

The thermal diffusivity α, which appears in all unsteady heat conduction problems, is a property of the material, and the time rate of temperature change
depends on its numerical value. Qualitatively we observe that, in a material
that combines a low thermal conductivity with a large specific heat (small α),
the rate of temperature change will be slower than in a material with a large
thermal diffusivity.

Since the temperature T must be a function of time t and distance x, we
begin by assuming a product solution

$$T(x, t) = X(x)\Theta(t)$$

Note that

$$\frac{\partial T}{\partial t} = X\frac{\partial \Theta}{\partial t} \quad \text{and} \quad \frac{\partial^2 T}{\partial x^2} = \Theta\frac{\partial^2 X}{\partial x^2}$$

Substituting these partial derivatives into Eq. (2.88) yields

$$\frac{1}{\alpha} X \frac{\partial \Theta}{\partial t} = \Theta \frac{\partial^2 X}{\partial x^2}$$

We can now separate the variables, that is, bring all functions that depend on x to one side of the equation and all functions that depend on θ to the other. By dividing both sides by $X\Theta$, we obtain

$$\frac{1}{\alpha\Theta} \frac{\partial \Theta}{\partial t} = \frac{1}{X} \frac{\partial^2 X}{\partial x^2}$$

Now observe that the left-hand side is a function of t only and therefore is independent of x, whereas the right-hand side is a function of x only and will not change as t varies. Since neither side can change as t and x vary, both sides are equal to a constant, which we will call μ. Hence, we have two ordinary and linear differential equations with constant coefficients:

$$\frac{d\Theta(t)}{dt} = \alpha\mu\Theta(t) \tag{2.89}$$

and

$$\frac{d^2 X}{dx^2} = \mu X(x) \tag{2.90}$$

The general solution for Eq. (2.89) is

$$\Theta(t) = C_1 e^{\alpha\mu t}$$

If μ were a positive number, the temperature of the slab would become infinitely high as t increased, which is physically impossible. Therefore, we must reject the possibility that $\mu > 0$. If μ were zero, the slab temperature would be a constant. Again, this possibility must be rejected because it would not be consistent with the physical conditions of the problem. We therefore conclude that μ must be a negative number, and for convenience we let $\mu = -\lambda^2$. The time-dependent function, then, becomes

$$\Theta(t) = C_1 e^{-\alpha\lambda^2 t} \tag{2.91}$$

Next we direct attention to the equation involving x, Eq. (2.90). Its general solution can be written in terms of a sinusoidal function. Since this is a second-order equation, there must be two constants of integration in the solution. In convenient form, the solution to the equation

$$\frac{d^2 X(x)}{dx^2} = -\lambda^2 X(x)$$

can be written as

$$X(x) = C_2 \cos \lambda x + C_3 \sin \lambda x \tag{2.92}$$

The temperature, as a function of distance and time in the slab, is given by

$$T(x, t) = C_1 e^{-\alpha\lambda^2 t}(C_2 \cos \lambda x + C_3 \sin \lambda x)$$
$$= e^{-\alpha\lambda^2 t}(A \cos \lambda x + B \sin \lambda x) \tag{2.93}$$

where $A = C_1 C_2$ and $B = C_1 C_3$ are constants that must be evaluated from the boundary and initial conditions. In addition, we must determine the value of the constant λ in order to complete the solution.

The boundary and initial conditions are:

1. At $x = 0$, $\partial T/\partial x = 0$.
2. At $x = \pm L$, $-(\partial T/\partial x)|_{x = \pm L} = (\bar{h}/k_s)(T_{x = \pm L} - T_\infty)$.
3. At $t = 0$, $T = T_i$.

Boundary condition 1 requires that

$$\left.\frac{\partial T}{\partial x}\right|_{x=0} = e^{-\alpha\lambda^2 t}(-A\lambda \sin \lambda x + B\lambda \cos \lambda x)\bigg|_{x=0} = 0$$

Now $\sin 0 = 0$, but the second term in the parentheses, involving $\cos 0$, can be zero only if $B = 0$ or $\lambda = 0$. Since $\lambda = 0$ gives a trivial solution we reject it and the solution for $T(x, t)$ becomes, therefore,

$$T(x, t) = e^{-\alpha\lambda^2 t} A \cos \lambda x$$

In order to satisfy the second boundary condition, namely that the heat flow by conduction at the interface must be equal to the heat flow by convection, the equality

$$-\frac{\partial T}{\partial x}\bigg|_{x=L} = e^{-\alpha\lambda^2 t} A\lambda \sin \lambda L = \frac{\bar{h}}{k_s}(T_{x=L} - 0) = \frac{\bar{h}}{k_s} e^{-\alpha\lambda^2 t} A \cos \lambda L$$

must hold for all values of t, which gives

$$\frac{\bar{h}}{k_s} \cos \lambda L = \lambda \sin \lambda L \quad \text{or}$$

$$\cot \lambda L = \frac{k_s}{\bar{h}L} \lambda L = \frac{\lambda L}{\text{Bi}} \tag{2.94}$$

Equation (2.94) is *transcendental*, and there is an infinite number of values of λ, called characteristic values, that will satisfy it. The simplest way to determine the numerical values of λ is to plot $\cot \lambda L$ and $\lambda L/\text{Bi}$ against λL. The values of λ at the points of intersection of these curves are the characteristic values and satisfy the second boundary condition. Figure 2.27 is a plot of these curves, and if $L = 1$ we can read off the first few characteristic values as $\lambda_1 = 0.86\text{Bi}$, $\lambda_2 = 3.43\text{Bi}$, $\lambda_3 = 6.44\text{Bi}$, etc. The value $\lambda = 0$ is disregarded because it leads to the trivial solution $T = 0$. A particular solution of Eq. (2.94) corresponds to each value of λ. Therefore, we shall adopt a subscript notation to identify the correspondence

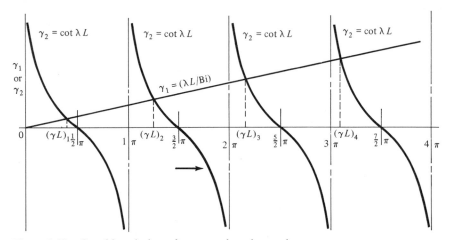

Figure 2.27 Graphic solution of transcendental equation.

between A and λ. For instance, A_1 corresponds to λ_1 or, in general, A_n to λ_n. The complete solution is formed as the sum of the solutions corresponding to each characteristic value, or

$$T(x, t) = \sum_{n=1}^{\infty} e^{-\alpha \lambda_n^2 t} A_n \cos \lambda_n x \qquad (2.95)$$

Each term of this infinite series contains a constant. These constants are evaluated by substituting the initial condition into Eq. (2.95):

$$T(x, 0) = T_i = \sum_{n=1}^{\infty} A_n \cos \lambda_n x \qquad (2.96)$$

It can be shown that the characteristic functions $\cos \lambda_n x$ are orthogonal between $x = 0$ and $x = L$ and therefore*

$$\int_0^L \cos \lambda_n x \cos \lambda_m x \, dx \begin{cases} = 0 & \text{if} \quad m \neq n \\ \neq 0 & \text{if} \quad m = n \end{cases} \qquad (2.97)$$

where λ_m may be any characteristic value of λ. To obtain a particular value of A_n, we multiply both sides of Eq. (2.96) by $\cos \lambda_m x$ and integrate between 0 and

---

* This can be verified by performing the integration, which yields

$$\int_0^L \cos \lambda_n x \cos \lambda_m x \, dx = \frac{\lambda_n \sin L\lambda_n \cos L\lambda_m - \lambda_m \sin L\lambda_m \cos L\lambda_n}{2L(\lambda_m^2 - \lambda_n^2)}$$

when $m \neq n$. However, from Eq. (2.94) we have

$$\frac{\cot \lambda_m L}{\lambda_m} = \frac{k}{h} = \frac{\cot \lambda_n L}{\lambda_n}$$

or

$$\lambda_n \cos \lambda_m L \sin \lambda_n L = \lambda_m \cos \lambda_n L \sin \lambda_m L$$

Therefore, the integral is zero when $m \neq n$.

L. In accordance with Eq. (2.97), all terms on the right-hand side disappear except the one involving the square of the characteristic function, $\cos \lambda_n x$, and we obtain

$$\int_0^L (T_i - T_\infty) \cos \lambda_n x \, dx = A_n \int_0^L \cos^2 \lambda_n x \, dx$$

From standard integral tables (10) we get

$$\int_0^L \cos^2 \lambda_n x \, dx = \frac{1}{2} x + \frac{1}{2\lambda_n} \sin \lambda_n x \cos \lambda_n x \Big|_0^L = \frac{L}{2} + \frac{1}{2\lambda_n} \sin \lambda_n L \cos \lambda_n L$$

and

$$\int_0^L \cos \lambda_n x \, dx = \frac{1}{\lambda_n} \sin \lambda_n L$$

whence the constant A_n is

$$A_n = \frac{2\lambda_n}{L\lambda_n + \sin \lambda_n L \cos \lambda_n L} \frac{(T_i - T_\infty) \sin \lambda_n L}{\lambda_n}$$

$$= \frac{2(T_i - T_\infty) \sin \lambda_n L}{L\lambda_n + \sin \lambda_n L \cos \lambda_n L} \tag{2.98}$$

As an illustration of the general procedure outlined above, let us determine A_1 when $\bar{h} = 1$, $k_s = 1$, and $L = 1$. From the graph of Fig. 2.27, the value of λ_1 is 0.86 radians or 49.2°. Then we have

$$A_1 = (T_i - T_\infty) \frac{2 \sin 49.2}{(1)(0.86) + \sin 49.2 \cos 49.2} = (T_i - T_\infty) \frac{(2)(0.757)}{0.86 + (0.757)(0.653)}$$

$$= 1.12(T_i - T_\infty)$$

Similarly, we obtain

$$A_2 = -0.152(T_i - T_\infty) \quad \text{and} \quad A_3 = 0.046(T_i - T_\infty)$$

The series converges rapidly and, for Bi $= 1$, three terms represent a fairly good approximation for practical purposes.

To express the temperature in the slab in terms of conventional dimensionless moduli, we let $\lambda_n = \delta_n/L$. The final form of the solution, obtained by substituting Eq. (2.98) into Eq. (2.96), is then

$$\frac{T(x, t) - T_\infty}{T_i - T_\infty} = \sum_{n=1}^{\infty} e^{-\delta_n^2(t\alpha/L^2)} 2 \frac{\sin \delta_n \cos(\delta_n x/L)}{\delta_n + \sin \delta_n \cos \delta_n} \tag{2.99}$$

The time dependence is now contained in the dimensionless Fourier modulus, Fo $= t\alpha/L^2$. Furthermore, if we write the second boundary condition in terms of δ_n, we obtain from Eq. (2.94)

$$\cot \delta_n = \frac{k_s}{hL} \delta_n \tag{2.100}$$

or

$$\delta_n \tan \delta_n = \frac{\bar{h}L}{k_s} = \text{Bi}$$

Since δ_n is a function only of the dimensionless Biot modulus, $\text{Bi} = \bar{h}L/k_s$, the temperature $T(x, t)$ can be expressed in terms of the three dimensionless quantities $\text{Fo} = t\alpha/L^2$, $\text{Bi} = \bar{h}L/k_s$, and x/L.

The rate of internal energy change of the slab per unit area of the surface of the slab, dQ/dt, is given by

$$\frac{dQ}{dt} = \frac{q}{A} = -k_s \frac{\partial T}{\partial x}\bigg|_{x=L} \tag{2.101}$$

The temperature gradient can be obtained by differentiating Eq. (2.99) with respect to x for a given value of t, or

$$\frac{\partial T}{\partial x}\bigg|_{x=L} = -\frac{2(T_0 - T_\infty)}{L} \sum_{n=1}^{\infty} e^{-\delta_n^2 \text{Fo}} \frac{\delta_n \sin^2 \delta_n}{\delta_n + \sin \delta_n \cos \delta_n} \tag{2.102}$$

Substituting Eq. (2.102) into Eq. (2.101) and integrating between the limits of $t = 0$ and t gives the change in internal energy of the slab during the time t, which is equal to the amount of heat Q absorbed by (or removed from) the slab. After some algebraic simplification, we obtain

$$Q = 2(T_0 - T_\infty)Lc\rho \sum_{n=1}^{\infty} (1 - e^{-\delta_n^2 \text{Fo}}) \frac{\sin^2 \delta_n}{\delta_n^2 + \delta_n \sin \delta_n \cos \delta_n} \tag{2.103}$$

In order to make Eq. (2.103) dimensionless, we note that $c\rho L T_0$ represents the initial internal energy per square foot of the slab. If we denote $c\rho L(T_0 - T_\infty)$ by Q_0, we get

$$\frac{Q}{Q_0} = \sum_{n=1}^{\infty} \frac{2 \sin^2 \delta_n}{\delta_n^2 + \delta_n \sin \delta_n \cos \delta_n} (1 - e^{-\delta_n^2 \text{Fo}}) \tag{2.104}$$

The temperature distribution and the amount of heat transferred at any time may be determined from Eqs. (2.99) and (2.104), respectively. The final expressions are in the form of infinite series. These series have been evaluated, and the results are available in the form of charts. Use of the charts for the problem treated in this section as well as for other cases of practical interest will be taken up in the following section. A complete understanding of the methods by which the mathematical solutions have been obtained, although helpful, is not necessary for using the charts.

2.6.3* Semi-Infinite Solid

Another simple geometric configuration for which analytic solutions are available is the semi-infinite solid. Such a solid extends to infinity in all but one direction, and can therefore be characterized by a single surface (Fig. 2.28). A semi-infinite solid approximates many practical problems. It may be used to

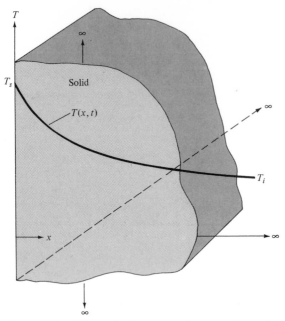

Figure 2.28 Schematic diagram and nomenclature for transient conduction in a semi-infinite solid.

estimate transient heat transfer effects near the surface of the earth or to approximate the transient response of a finite solid, such as a thick slab, during the early portion of a transient when the temperature in the slab interior is not yet influenced by the change in surface conditions.

   If a thermal change is suddenly imposed at this surface, a one-dimensional temperature wave will be propagated by conduction within the solid. The appropriate equation for transient conduction in a semi-infinite solid is Eq. (2.88) in the domain $0 \leq \alpha \leq \infty$. To solve this equation we must specify two boundary conditions and the initial temperature distribution. For the initial condition we shall specify that the temperature inside the solid is uniform at T_i, that is, $T(x, 0) = T_i$. For one of the two required boundary conditions we postulate that far from the surface the interior temperature will not be affected by the temperature wave, that is, $T(\infty, t) = T_i$, with the above specifications.

   Closed-form solutions have been obtained for three types of changes in surface conditions, instantaneously applied at $t = 0$. These three cases are illustrated in Fig. 2.29 and are (5):

1. a sudden change in surface temperature, $T_s \neq T_i$;
2. a sudden application of a specified heat flux q_0'', as, for example, exposing the surface to radiation; and
3. a sudden exposure of the surface to a fluid at a different temperature through a uniform and constant heat transfer coefficient $\bar{h}$.

The solutions are summarized below.

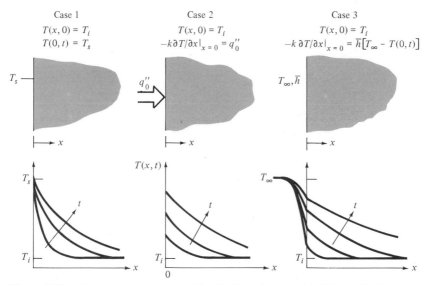

Figure 2.29 Transient temperature distributions in a semi-infinite solid for three surface conditions: (1) constant surface temperature, (2) constant surface heat flux, and (3) surface convection.

Case 1. Change in surface temperature:

$$T(0, t) = T_s$$

$$\frac{T(x, t) - T_s}{T_i - T_s} = \text{erf}\left(\frac{x}{2\sqrt{\alpha t}}\right) \tag{2.105}$$

$$q_s''(t) = -k \left.\frac{\partial T}{\partial x}\right|_{x=0} = \frac{k(T_s - T_i)}{\sqrt{\pi \alpha t}}$$

Case 2. Constant surface heat flux:

$$q_s'' = q_0''$$

$$T(x, t) - T_i = \frac{2q_0''(\alpha t/\pi)^{1/2}}{k_s} \exp\left(\frac{-x^2}{4\alpha t}\right) - \frac{q_0'' x}{k_s} \text{erfc}\left(\frac{x}{2\sqrt{\alpha t}}\right) \tag{2.106}$$

Case 3. Surface convection:

$$-k \left.\frac{\partial T}{\partial x}\right|_{x=0} = \bar{h}[T_\infty - T(0, t)]$$

$$\frac{T(x, t) - T_i}{T_\infty - T_i} = \text{erfc}\left(\frac{x}{2\sqrt{\alpha t}}\right) - \exp\left(\frac{\bar{h}x}{k} + \frac{\bar{h}^2 \alpha t}{k^2}\right) \text{erfc}\left(\frac{x}{2\sqrt{\alpha t}} + \frac{\bar{h}\sqrt{\alpha t}}{k}\right) \tag{2.107}$$

Note that the quantity $\bar{h}^2 \alpha t/k^2$ equals the product of the Biot number squared (Bi = $\bar{h}x/k$) times the Fourier number (Fo = $\alpha t/x^2$).

The function erf appearing in Eq. (2.105) is the *Gaussian error function*, which is encountered frequently in engineering and is defined as

$$\text{erf}\left(\frac{x}{2\sqrt{\alpha t}}\right) = \frac{2}{t} \int_0^{x/2\sqrt{\alpha t}} e^{-2} \, d\eta \tag{2.108}$$

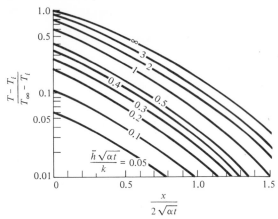

Figure 2.30 Dimensionless transient temperatures for a semi-infinite solid with surface convection.

Values of this function are tabulated in the appendix. The complementary error function, erfc(w), is defined as

$$\text{erfc}(w) = 1 - \text{erf}(w) \tag{2.109}$$

Temperature histories for the three cases are illustrated qualitatively in Fig. 2.29. For case 3, the specific temperature histories computed from Eq. (2.107) are plotted in Fig. 2.30. The curve corresponding to $\bar{h} = \infty$ is equivalent to the result that would be obtained for a sudden change in the surface temperature to $T_s = T(x, 0)$ because when $\bar{h} = \infty$ the second term on the right-hand side of Eq. (2.107) is zero, and the result is equivalent to Eq. (2.105) for case 1.

EXAMPLE 2.11 ─────────────────────────────────────

Estimate the minimum depth x_m at which one must place a water main below the surface to avoid freezing. The soil is initially at a uniform temperature of 20°C. Assume that under the worst conditions anticipated it is subjected to a surface temperature of $-15°C$ for a period of 60 days. Use the following properties for soil (300 K):

$$\rho = 2050 \text{ kg/m}^3 \qquad k = 0.52 \text{ W/m K} \qquad c = 1840 \text{ J/kg K}$$

$$\alpha = \frac{k}{\rho c} = 0.138 \times 10^{-6} \text{ m}^2/\text{s}$$

A sketch of the system is shown in Fig. 2.31.

Solution. To simplify the problem assume that

1. conduction is one-dimensional;
2. the soil is a semi-infinite medium; and
3. the soil has uniform and constant properties.

The prescribed conditions correspond to those of case 1 of Fig. 2.29, and the transient temperature response of the soil is governed by Eq. (2.105).

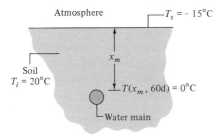

Figure 2.31 Schematic diagram for Example 2.11.

At the time $t = 60$ days after the change in surface temperature, the temperature distribution in the soil is

$$\frac{T(x_m, t) - T_s}{T_i - T_s} = \text{erf}\left(\frac{x_m}{2\sqrt{\alpha t}}\right)$$

or

$$\frac{0 - (-15°C)}{20°C - (-15°C)} = 0.43 = \text{erf}\left(\frac{x_m}{2\sqrt{\alpha t}}\right)$$

From Table 43 we find by interpolation that when $x_m/2\sqrt{\alpha t} = 0.4$, erf(0.4) = 0.43 to satisfy the above. Thus

$$x_m = (0.4)(2\sqrt{\alpha t})$$
$$= 0.8[(0.138 \times 10^{-6} \text{ m}^2/\text{s})(60)(24)(3600)]^{1/2} = 0.68 \text{ m}$$

To use Fig. 2.30 first calculate $[T(x, t) - T_s]/(T_\infty - T_s) = (0 - 20)/(-15 - 20) = 0.57$, then enter the curve for Bi = ∞ and obtain $x/2\sqrt{\alpha t} = 0.4$, the same result as above.

2.7 CHARTS FOR TRANSIENT HEAT CONDUCTION

For transient heat conduction in several simple shapes, subject to boundary conditions of practical importance, the temperature distribution and the heat flow have been calculated and the results are available in the form of charts or tables (4, 6, 11–13). In this section we shall illustrate the application of some of these charts to typical problems of transient heat conduction in solids having Biot moduli larger than 0.1.

2.7.1 One-Dimensional Solutions

Three simple geometries for which results have been prepared in graphic form are:

1. an infinite plate of width $2L$ (see Fig. 2.32);
2. an infinitely long cylinder of radius r_0 (see Fig. 2.33); and
3. a sphere of radius r_0 (see Fig. 2.34).

The boundary conditions and the initial conditions for all three geometries are similar. One boundary condition requires that the temperature gradient at the midplane of the plate, the axis of the cylinder, and the center of the sphere be equal to zero. Physically, this corresponds to no heat flow at these locations.

The other boundary condition requires that the heat conducted to or from the surface be transferred by convection to or from a fluid at temperature T_∞ through a uniform and constant heat transfer coefficient $\bar{h}_c$, or

$$\bar{h}_c(T_s - T_\infty) = -k \left.\frac{\partial T}{\partial n}\right|_s \tag{2.110}$$

where the subscript s refers to conditions at the surface and n to the coordinate direction normal to the surface. It should be noted that the limiting case of $\text{Bi} \to \infty$ corresponds to a negligible thermal resistance at the surface ($\bar{h}_c \to \infty$) so that the surface temperature is specified as equal to T_∞ for $t > 0$.

The initial conditions for all three chart solutions require that the solid be initially at a uniform temperature T_i and that when the transient begins at time zero ($t = 0$) the entire surface of the body is contacted by fluid at T_∞.

The solution for all three cases are plotted in terms of dimensionless parameters. The forms of the dimensionless parameters are summarized in Table 2.3. Use of the graphic solutions is discussed below.

For each geometry there are three graphs, the first two for the temperatures and the third for the heat flow. The dimensionless temperatures are presented in the form of two interrelated graphs for each shape. The first set of graphs, Figs. 2.32a for the plate, 2.33a for the cylinder, and 2.34a for the sphere, gives the dimensionless temperature at the center or midpoint as a function of the Fourier number, that is, dimensionless time, with the inverse of the Biot number as the constant parameter. The dimensionless center or midpoint temperature for these graphs is defined as

$$\frac{T(0, t) - T_\infty}{T_i - T_\infty} \equiv \frac{\theta(0, t)}{\theta_0} \tag{2.111}$$

To evaluate the local temperature as a function of time the second temperature graph must be used. The second set of graphs, Figs. 2.32b for a plate, 2.33b for a cylinder, and 2.34b for a sphere, give the ratio of the local temperature to the center or midpoint temperature as a function of the inverse of the Biot number for various values of the dimensionless distance parameter, x/L for the slab and r/r_0 for the cylinder and the sphere. For the infinite plate this temperature ratio is

$$\frac{T(x, t) - T_\infty}{T(0, t) - T_\infty} = \frac{\theta(x, t)}{\theta(0, t)} \tag{2.112}$$

For the cylinder and the sphere the expressions are similar, but x is replaced by r.

TABLE 2.3 SUMMARY OF DIMENSIONLESS PARAMETERS FOR USE WITH TRANSIENT HEAT CONDUCTION CHARTS IN FIGS. 2.32, 2.33, AND 2.34

Situation	Infinite plate, width 2L	Infinitely long cylinder, radius r_0	Sphere, radius r_0
Geometry			
Dimensionless position	$\dfrac{x}{L}$	$\dfrac{r}{r_0}$	$\dfrac{r}{r_0}$
Biot number	$\dfrac{\bar{h}_c L}{k}$	$\dfrac{\bar{h}_c r_0}{k}$	$\dfrac{\bar{h}_c r_0}{k}$
Fourier number	$\dfrac{\alpha t}{L^2}$	$\dfrac{\alpha t}{r_0^2}$	$\dfrac{\alpha t}{r_0^2}$
Dimensionless centerline temperature $\dfrac{\theta(0, t)}{\theta_0}$	Fig. 2.32a	Fig. 2.33a	Fig. 2.34a
Dimensionless local temperature $\dfrac{\theta(x, t)}{\theta(0, t)}$ or $\dfrac{\theta(r, t)}{\theta(0, t)}$	Fig. 2.32b	Fig. 2.33b	Fig. 2.34b
Dimensionless heat transfer $\dfrac{Q(t)}{Q_0}, \dfrac{Q'(t)}{Q_0'}, \dfrac{Q''(t)}{Q_0''}$	$Q_0'' = \rho c L(T_0 - T_\infty)$	$Q_0' = \rho c \pi r_0^2(T_0 - T_\infty)$	$Q_0 = \rho c \,\dfrac{4}{3}\pi r_0^{\,3}(T_0 - T_\infty)$

To determine the local temperature at any time 0, form the product

$$\frac{T(x, t) - T_\infty}{T_i - T_\infty} = \left[\frac{T(0, t) - T_\infty}{T_i - T_\infty}\right]\left[\frac{T(x, t) - T_\infty}{T(0, t) - T_\infty}\right]$$

$$= \frac{\theta(0, t)}{\theta_0}\,\frac{\theta(x, t)}{\theta(0, t)} \tag{2.113}$$

for the plate and

$$\frac{T(r, t) - T_\infty}{T_i - T_\infty} = \left[\frac{T(0, t) - T_\infty}{T_i - T_\infty}\right]\left[\frac{T(r, t) - T_\infty}{T(0, t) - T_\infty}\right] \tag{2.114}$$

for the cylinder and the sphere.

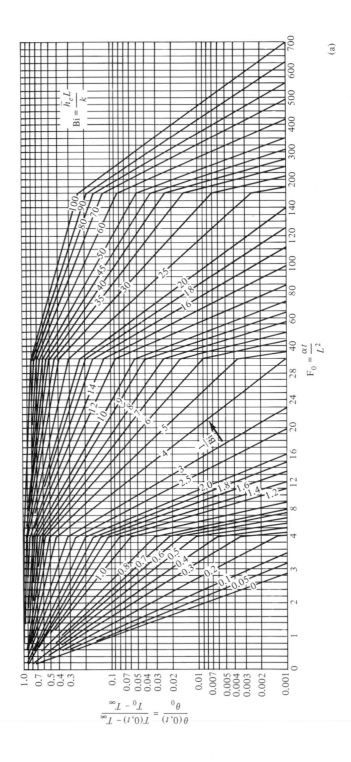

(a)

$$\text{Bi} = \frac{\bar{h}_c L}{k}$$

$$F_0 = \frac{\alpha t}{L^2}$$

$$\frac{\theta(0,t)}{\theta_0} = \frac{T(0,t) - T_\infty}{T_0 - T_\infty}$$

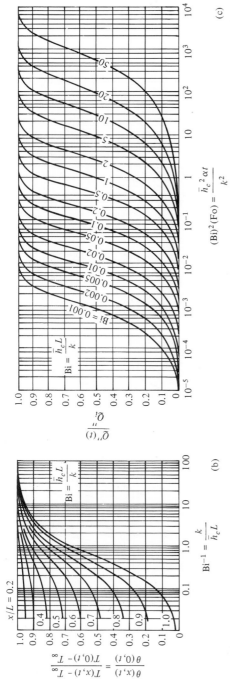

Figure 2.32 Dimensionless transient temperatures and heat flow in an infinite plate of width $2L$.

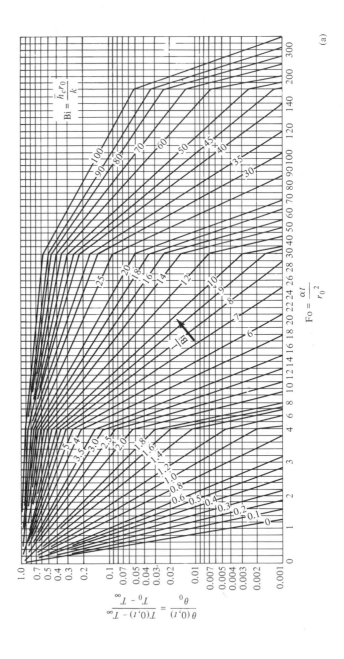

(a)

$$\frac{\theta(0,t)}{\theta_0} = \frac{T(0,t) - T_\infty}{T_0 - T_\infty}$$

$$Fo = \frac{\alpha t}{r_0^2}$$

$$Bi = \frac{\bar{h}_c r_0}{k}$$

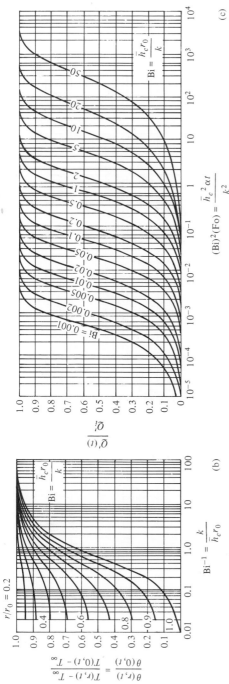

Figure 2.33 Dimensionless transient temperatures and heat flow for a long cylinder.

117

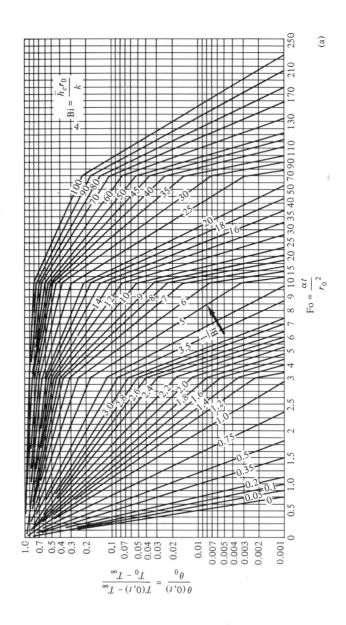

(a)

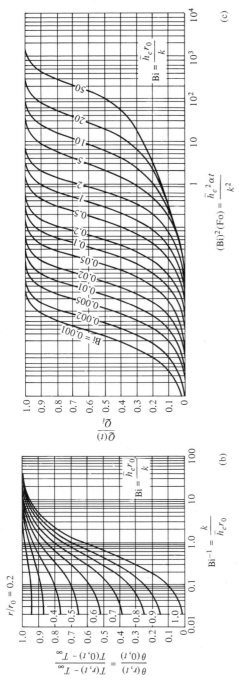

Figure 2.34 Dimensionless transient temperatures and heat flow for a sphere.

The instantaneous rate of heat transfer to or from the surface of the solid can be evaluated from Fourier's law once the temperature distribution is known. The change in internal energy between time $t = 0$ and $t = t$ can be obtained by integrating the instantaneous heat transfer rates, as shown for the slab by Eqs. (2.101) and (2.103). Denoting by $Q(t)$ the internal energy relative to the fluid at time t, and by Q_i the initial internal energy relative to the fluid, the ratios $Q(t)/Q_i$ are plotted against $Bi^2Fo = \bar{h}^2t/k^2$ for various values of Bi in Fig. 2.32c for the plate, Fig. 2.33b for the cylinder, and Fig. 2.34c for the sphere.

Each heat transfer value $Q(t)$ is the total amount of heat that is transferred from the surface to the fluid during the time from $t = 0$ to $t = t$. The normalizing factor Q_i is the initial amount of energy in the solid at $t = 0$ when the reference temperature for zero energy is T_∞. The values for Q_i for each of the three geometries are listed in Table 3.2 for convenience. Since the volume of the plate is infinite, the dimensionless heat transfer for this geometry, per unit surface area, is designated by the ratio $Q''(t)/Q''_i$. The volume of an infinitely long cylinder is also infinite, so the dimensionless heat transfer ratio is written, per unit length, as $Q'(t)/Q'_i$. The sphere has a finite volume, so the heat transfer ratio is simply $Q(t)/Q_i$ for that geometry. If the value of $Q(t)$ is positive, heat flows from the solid into the fluid; that is, the body is cooled. If it is negative, the solid is heated by the fluid.

Two general classes of transient problems can be solved by using the charts. One class of problem involves knowing the time while the local temperature at that time is unknown. In the other type of problem, the local temperature is the known quantity and the time required to reach that temperature is the unknown. The first class of problems can be solved in a straightforward fashion by use of the charts. The second class of problem occasionally involves a trial-and-error procedure. Both types of solutions will be illustrated by the following examples.

EXAMPLE 2.12 ──

A 2.0-m-long, 0.2-m-diameter steel cylinder ($k = 40$ W/m K, $\alpha = 1.0 \times 10^{-5}$ m^2/s), initially at 400°C, is suddenly immersed in water at 50°C. If the unit surface conductance is 100 W/m^2 K, calculate 20 min after immersion

1. the center temperature;
2. the surface temperature; and
3. the heat transferred to the water during the initial 20 min.

Solution. Since the cylinder has a length 10 times the diameter we can neglect end effects. To determine whether the internal resistance is negligible, we calculate first the Biot number

$$Bi = \frac{\bar{h}_c r_0}{k} = \frac{(100)(0.2)}{40} = 0.5$$

Since the Biot number is larger than 0.1 the internal resistance is significant and we cannot use the lumped-capacitance method. To use the chart solution we calculate the appropriate dimensionless parameters according to Table 2.3:

$$\text{Fo} = \frac{\alpha t}{r_0^2} = \frac{(1 \times 10^{-5} \text{ m/s})(20 \text{ min})(60 \text{ s/min})}{0.1^2 \text{ m}^2} = 1.2$$

and

$$\text{Bi}^2 \text{ Fo} = (0.5^2)(1.2) = 0.3$$

The initial amount of internal energy stored in the cylinder per unit length is

$$Q_i = c\rho\pi r_0^2 (T_i - T_\infty) = \left(\frac{k}{\alpha}\right)\pi r_0^2 (T_i - T_\infty)$$

$$= \frac{400 \text{ W/m K}}{1 \times 10^{-5} \text{ m}^2/\text{s}}(\pi)(0.1^2 \text{ m}^2)(350 \text{ K}) = 4.4 \times 10^8 \text{ W s/m}$$

The dimensionless centerline temperature for $1/\text{Bi} = 2.0$ and $\text{Fo} = 1.2$ from Fig. 2.33a is

$$\frac{T(0, t) - T_\infty}{T_i - T_\infty} = 0.48$$

Since $T_i - T_\infty$ is specified as 350°C and $T_\infty = 50$°C, $T(0, t) = (0.48)(350) + 50 = 218$°C.

The surface temperature at $r/r_0 = 1.0$ and $t = 1200$ s is obtained from Fig. 2.32b in terms of the centerline temperature, or

$$\frac{T(r_0, t) - T_\infty}{T(0, t) - T_\infty} = 0.8$$

The surface temperature ratio is thus

$$\frac{T(r_0, t) - T_\infty}{(T_i - T_\infty)} = 0.8 \frac{T(0, t) - T_\infty}{T_i - T_\infty} = (0.8)(0.48) = 0.384$$

and the surface temperature after 20 min is

$$T(r_0, t) = (0.384)(350) + 50 = 184.8°\text{C}$$

Then the amount of heat transferred from the steel rod to the water can be obtained from Fig. 2.33c. Since $Q(t)/Q_i = 0.61$,

$$Q(t) = (0.61)\frac{(2 \text{ m})(4.4 \times 10^7 \text{ W s/m})}{3600 \text{ s/h}} = 14.8 \text{ kW h}$$

EXAMPLE 2.13 _____

A large concrete wall 50 cm thick is initially at 60°C. One side of the wall is insulated. The other side is suddenly exposed to hot combustion gases at 900°C through a unit surface conductance of 25 W/m² K. Determine (a) the time required for the insulated surface to reach 600°C, (b) the temperature distribution in the wall at that instant, and (c) the heat transferred during the process. The following average physical properties are given:

$$k_s = 1.25 \text{ W/m K}$$
$$c = 837 \text{ J/kg K}$$
$$\rho = 500 \text{ kg/m}^3$$
$$\alpha = 0.30 \times 10^{-5} \text{ m}^2/\text{s}$$

Solution. Note that the wall thickness is equal to L since the insulated surface corresponds to the center plane of a slab of thickness $2L$ when both surfaces experience a thermal change. The temperature ratio $(T_s - T_\infty)/(T_0 - T_\infty)$ for the insulated face at the time sought is

$$\frac{T_s - T_\infty}{T_0 - T_\infty}\bigg|_{x=0} = \frac{600 - 900}{60 - 900} = 0.357$$

and the reciprocal of the Biot number is

$$\frac{k_s}{\overline{h}L} = \frac{1.25 \text{ W/m K}}{(25 \text{ W/m}^2 \text{ K})(0.5 \text{ m})} = 0.10$$

From Fig. 2.32a we find that for the above conditions the Fourier number $\alpha t/L^2 = 0.70$ at the midplane. Therefore,

$$t = \frac{(0.7)(0.5^2 \text{ m}^2)}{0.3 \times 10^{-5} \text{ m}^2/\text{s}}$$
$$= 58{,}333 \text{ s} = 16.2 \text{ h}$$

The temperature distribution in the wall 16 h after the transient was initiated can be obtained from Fig. 2.32b for various values of x/L, as shown below

$\dfrac{x}{L}$	1.0	0.8	0.6	0.4	0.2
$\dfrac{T\left(\dfrac{x}{L}\right) - T_\infty}{T(0) - T_\infty}$	0.13	0.41	0.64	0.83	0.96

From the above dimensionless data we can obtain the temperature distribution as a function of distance from the insulated surface:

x, m	0.5	0.4	0.3	0.2	0.1	0
$T_\infty - T(x)$, °C	39	123	192	249	288	300
$T(x)$, °C	861	777	708	651	612	600

The heat transferred to the wall per square meter of surface area during the transient can be obtained from Fig. 2.32c. For Bi = 10, $Q(t)/Q_0$ at $Bi^2Fo = 70$ is 0.70. Thus we get

$$Q(t) = c\rho L(T_0 - T_\infty) = (837 \text{ J/kg K})(500 \text{ kg/m}^3)(0.5 \text{ m})(-840 \text{ K})$$
$$= -1.756 \times 10^8 \text{ J/m}^2$$

The minus sign indicates that the heat was transferred into the wall and the internal energy increased during the process.

2.7.2* Multidimensional Systems[†]

The use of the one-dimensional transient charts can be extended to two- and three-dimensional problems (14). The method involves using the product of multiple values from the one-dimensional charts: Figs. 2.32, 2.33, and 2.34. The basis for obtaining two- and three-dimensional solutions from one-dimensional charts is the manner in which partial differential equations can be separated into the product of two or three ordinary differential equations. A proof of the method can be found in Arpaci (Ref. 15, section 5-2).

The product solution method can best be illustrated by an example. Suppose we wish to determine the transient temperature at point P in a cylinder of finite length, as shown in Fig. 2.35. The point P is located by the two coordinates (x, r), where x is the axial location measured from the center of the cylinder and r is the radial position. The initial condition and boundary conditions are the same as those that apply to the transient one-dimensional charts. The cylinder is initially at a uniform temperature T_0. At time $t = 0$ the entire surface is subjected to a fluid with constant ambient temperature T_∞, and the convective heat transfer coefficient between the cylinder surface area and fluid is a constant value h.

The radial temperature distribution for an infinitely long cylinder is given in Fig. 2.33. For a cylinder with finite length the radial and axial temperature distribution is given by the product solution of an infinitely long cylinder and

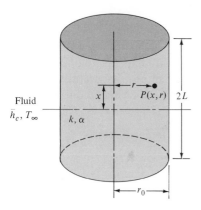

Figure 2.35 Geometry for a short-cylinder product solution.

[†] This subsection is adapted from reference 14 with permission.

infinite plate:

$$\frac{\theta_p(r, x)}{\theta_0} = C(r)P(x)$$

where the symbols $C(r)$ and $P(x)$ are the dimensionless temperatures of the infinite cylinder and infinite plate, respectively:

$$C(r) = \frac{\theta(r, t)}{\theta_0}$$

$$P(x) = \frac{\theta(x, t)}{\theta_0}$$

The solution for $C(r)$ is obtained from Fig. 2.33, a and b, while the value for $P(x)$ is obtained from Fig. 2.32, a and b.

Solutions for other two- and three-dimensional geometries may be obtained in a procedure similar to the one illustrated for the finite cylinder. Three-dimensional problems involve the product of three solutions, while two-dimensional problems can be solved by taking the product of two solutions.

Two-dimensional geometries that have chart solutions are summarized in Table 2.4. Three-dimensional solutions are outlined in Table 2.5. The symbols used in the two tables represent the following solutions:

$$S(x) = \frac{\theta(x, t)}{\theta_0} \qquad \text{for a semi-infinite solid, Fig. 2.28}$$

$$P(x) = \frac{\theta(x, t)}{\theta_0} \qquad \text{for an infinite plate, Fig. 2.32, } a \text{ and } b$$

$$C(r) = \frac{\theta(r, t)}{\theta_0} \qquad \text{for an infinite cylinder, Fig. 2.32, } a \text{ and } b$$

The extension of the one-dimensional charts to two- and three-dimensional geometries allows us to solve a large variety of transient conduction problems.

EXAMPLE 2.14

A 10-cm-diameter, 16-cm-long cylinder with properties $k = 0.5$ W/m K and $\alpha = 5 \times 10^{-7}$ m²/s is initially at a uniform temperature of 20°C. The cylinder is placed in an oven where the ambient air temperature is 500°C and $\bar{h}_c = 30$ W/m² K. Determine the minimum and maximum temperatures in the cylinder 30 min after it has been placed in the oven.

Solution.   The Biot number based on the cylinder radius is

$$\text{Bi} = \frac{\bar{h}_c r_0}{k} = \frac{(30)(0.05)}{0.5} = 3.0$$

The problem cannot be solved by using the simplified approach assuming negligible internal resistance; a chart solution is necessary.

TABLE 2.4   SCHEMATIC DIAGRAMS AND NOMENCLATURE FOR PRODUCT SOLUTIONS
TO TRANSIENT CONDUCTION PROBLEMS WITH FIGS. 2.32, 2.33, AND 2.34
FOR TWO-DIMENSIONAL SYSTEMS

	Geometry	Dimensionless temperature at point P
Semi-infinite plate		$\dfrac{\theta_p(x_1, x_2)}{\theta_0} = P(x_1)S(x_2)$
Infinite rectangular bar		$\dfrac{\theta_p(x_1, x_2)}{\theta_0} = P(x_1)P(x_2)$
One-quarter infinite solid		$\dfrac{\theta_p(x_1, x_2)}{\theta_0} = S(x_1)S(x_2)$
Semi-infinite cylinder		$\dfrac{\theta_p(x, r)}{\theta_0} = S(x)C(r)$
Finite cylinder		$\dfrac{\theta_p(x, r)}{\theta_0} = P(x)C(r)$

TABLE 2.5 SCHEMATIC DIAGRAMS AND NOMENCLATURE FOR PRODUCT SOLUTIONS
TO TRANSIENT CONDUCTION PROBLEMS WITH FIGS. 2.32, 2.33, AND 2.34
FOR THREE-DIMENSIONAL SYSTEMS

	Geometry	Dimensionless temperature at point P
Semi-infinite rectangular bar		$\dfrac{\theta_p(x_1, x_2, x_3)}{\theta_0} = S(x_1)P(x_2)P(x_3)$
Rectangular parallelepiped		$\dfrac{\theta_p(x_1, x_2, x_3)}{\theta_0} = P(x_1)P(x_2)P(x_3)$
One-quarter infinite plate		$\dfrac{\theta_p(x_1, x_2, x_3)}{\theta_0} = S(x_1)S(x_2)P(x_3)$
One-eighth infinite plate		$\dfrac{\theta_p(x_1, x_2, x_3)}{\theta_0} = S(x_1)S(x_2)S(x_3)$

Table 2.4 indicates that the temperature distribution in a cylinder of finite length can be determined by the product of the solution for an infinite plate and an infinite cylinder. At any time the minimum temperature is at the geometric center of the cylinder and the maximum temperature is at the outer circumference at each end of the cylinder. Using the co-

ordinates for the finite cylinder shown in Fig. 2.35, we have

$$\text{Minimum temperature at:} \quad x = 0 \quad r = 0$$
$$\text{Maximum temperature at:} \quad x = L \quad r = r_0$$

The calculations are summarized in the tables below.

Infinite plate			
		$P(0) = \dfrac{\theta(0, t)}{\theta_0}$	$P(L) = \dfrac{\theta(L, t)}{\theta_0}$
$\text{Fo} = \dfrac{\alpha t}{L^2}$	$\text{Bi}^{-1} = \dfrac{k}{h_c L}$	[Fig. 2.32a]	[Fig. 2.32, a and b]
$\dfrac{(5 \times 10^{-7})(1800)}{(0.08)^2} = 0.14$	$\dfrac{0.5}{(30)(0.08)} = 0.21$	0.90	$(0.90)(0.27) = 0.249$

Infinite cylinder			
$\text{Fo} = \dfrac{\alpha t}{r_0^2}$	$\text{Bi}^{-1} = \dfrac{k}{h_c r_0}$	$C(0) = \dfrac{\theta(0, t)}{\theta_0}$	$C(r_0) = \dfrac{\theta(r_0, t)}{\theta_0}$
		[Fig. 2.33a]	[Fig. 2.33, a and b]
$\dfrac{(5 \times 10^{-7})(1800)}{(0.05)^2} = 0.36$	$\dfrac{0.5}{(30)(0.05)} = 0.33$	0.47	$(0.47)(0.33) = 0.155$

The minimum cylinder temperature is

$$\frac{\theta_{min}}{\theta_0} = P(0)C(0) = (0.90)(0.47) = 0.423$$
$$T_{min} = 0.423(20 - 500) + 500 = 297°C$$

The maximum cylinder temperature is

$$\frac{\theta_{max}}{\theta_0} = P(L)C(r_0) = (0.249)(0.155) = 0.039$$
$$T_{max} = 0.039(20 - 500) + 500 = 481°C$$

2.8 CLOSURE

In this chapter, we have considered methods of analyzing heat conduction problems in the steady and unsteady states. Problems in the steady state are divided into one-dimensional and multidimensional geometries. For one-dimensional problems, solutions are available in the form of simple equations that can incorporate various boundary conditions by using thermal circuits. For problems of heat conduction in more than one dimension, solutions can be obtained by analytic, graphic, and numerical means. The analytic approach is recommended only for situations involving systems with a simple geometry and

simple boundary conditions. It is accurate and lends itself readily to parameter-
ization, but when the boundary conditions are complex, the analytic approach
usually becomes too involved to be practical, and for complex geometries it is
impossible to obtain a closed-form solution.

Systems of complex geometry, but with isothermal and insulated boundaries, are readily amenable to graphic solutions. The graphic method, however,
becomes unwieldy when the boundary conditions involve heat transfer through
a surface conductance. For such cases the numerical approach to be considered
in the next chapter is recommended because it can easily be adapted to all
kinds of boundary conditions and geometric shapes.

Conduction problems in the unsteady state can be subdivided into those
that can be handled by the lumped-capacity method and those in which the
temperature is a function not only of time, but also of one or more spatial
coordinates. In the lumped-capacity method, which is a good approximation
for conditions in which the Biot number is larger than one-tenth, it is assumed
that internal conduction is sufficiently large that the temperature throughout
the system can be considered uniform at any instance of time. When this approximation is not permissible it is necessary to set up and solve partial differential equations, which generally require series solutions and are attainable
only for simple geometric shapes. However, for spheres, cylinders, slabs, plates,
and other simple geometric shapes, the results of analytic solutions have been
presented in the form of charts, which are relatively easy and straightforward
to use. As in the case of steady-state conduction problems, when the geometries
are complex and when the boundary conditions vary with time or have other
complex features, it is necessary to obtain the solution by numerical means, as
discussed in the next chapter.

PROBLEMS

The problems for this chapter are organized by subject matter as shown below.

Topic	Section	Problem number
Conduction equation	2.2	2.1–2.4
Steady conduction in simple geometries	2.3	2.5–2.29
Extended surfaces	2.4	2.30–2.43
Multidimensional steady conduction	2.5	2.44–2.63
Transient conduction (analytical solutions)	2.6	2.64–2.80
Transient conduction (chart solutions)	2.7	2.81–2.97

2.1   The heat conduction equation in cylindrical coordinates is

$$\rho c \frac{\partial T}{\partial t} = k \left(\frac{\partial^2 T}{\partial r^2} + \frac{1}{r} \frac{\partial T}{\partial r} + \frac{1}{r^2} \frac{\partial^2 T}{\partial \phi^2} + \frac{\partial^2 T}{\partial z^2} \right) + \dot{q}_G$$

(a) Simplify this equation by eliminating terms equal to zero for the case of steady-state heat flow without sources or sinks around a right-angle corner such as the one shown in the accompanying sketch. It may be assumed that the corner extends to infinity in the direction perpendicular to the paper. (b) Solve the resulting equation for the temperature distribution by substituting the boundary conditions. (c) Determine the rate of heat flow from T_1 to T_2. Assume $k = 1$ W/m K and unit depth perpendicular to the paper.

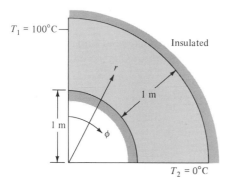

2.2 Derive Eq. (2.19) starting with an energy balance.

2.3 Write Eq. (2.20) in a dimensionless form similar to Eq. (2.17).

2.4 Derive the heat conduction equation in spherical coordinates and simplify it for (a) steady state and (b) uniform surface temperature without internal generation.

2.5 A plane wall, 7.5 cm thick, generates heat internally at the rate of 10^5 W/m³. One side of the wall is insulated, and the other side is exposed to an environment at 93°C. The convection heat transfer coefficient between the wall and the environment is 567 W/m² K. If the thermal conductivity of the wall is 0.12 W/m K, calculate the maximum temperature in the wall.

2.6 A small dam, which may be idealized by a large slab 1.2 m thick, is to be completely poured in a short period of time. The hydration of the concrete results in the equivalent of a distributed source of constant strength of 100 W/m³. If both dam surfaces are at 16°C, determine the maximum temperature to which the concrete will be subjected, assuming steady-state conditions. The thermal conductivity of the wet concrete may be taken as 8.4×10^{-3} W/m K.

2.7 Two large steel plates at temperatures of 93° and 71°C are separated by a steel rod 0.3 m long and 2.5 cm in diameter. The rod is welded to each plate. The space between the plates is filled with insulation, which also insulates the circumference of the rod. Because of the voltage difference between the two plates, current flows through the rod, dissipating electrical energy at a rate of 11.7 W. Determine the maximum temperature in the rod and the heat flux at each end. Check your results by comparing the net heat flow rate at the two ends with the total rate of heat generation.   *Ans.* 103°C, −4.2 W, 7.5 W.

2.8 The shield of a nuclear reactor can be idealized by a large 10-in.-thick flat plate having a thermal conductivity of 2 Btu/h ft °F. Radiation from the interior of the reactor penetrates the shield and produces heat generation in the shield which decreases exponentially from a value of 10 Btu/h in.³ at the inner surface to a value of 1.0 Btu/h in.³ at a distance 5 in. from the interior surface. If the exterior surface is kept cooling at 100°F by forced convection, determine the temperature at the inner surface of the shield. *Hint:* First set up the differential equation for a system in which the heat generation rate varies according to $q(x) = q(0)e^{-Cx}$.

2.9 Derive an expression for the temperature distribution in an infinitely long rod of uniform cross section within which there is uniform heat generation at the rate of 1 W/m. Assume that the rod is attached to a surface at T_s and is exposed through a unit-surface conductance $\bar{h}$ to a fluid at T_∞.

2.10 Derive an expression for the temperature distribution in a plane wall in which there are uniformly distributed heat sources which vary according to the linear relation

$$\dot{q}_G = \dot{q}_w[1 - \beta(T - T_w)]$$

where $\dot{q}_w$ is a constant equal to the heat generated per unit volume at the wall temperature T_w. Both sides of the plate are maintained at T_w and the plate thickness is $2L$.

2.11 A plane wall of thickness $2L$ has internal heat sources whose strength varies according to

$$\dot{q}_G = \dot{q}_0 \cos(ax)$$

where $\dot{q}_0$ is the heat generated per unit volume at the center of the wall ($x = 0$) and a is a constant. If both sides of the wall are maintained at a constant temperature of T_w, derive an expression for the total heat loss from the wall per unit surface area.

2.12 Suppose that a pipe carrying a hot fluid with an external temperature of T_i and outer radius r_i is to be insulated with an insulation material of thermal conductivity k and outer radius r_o. Show that if the outer surface conductance is $\bar{h}$ and the environmental temperature is T_∞, the addition of insulation can actually increase the rate of heat loss if $r_o < k/\bar{h}$ and that maximum heat loss occurs when $r_o = k/\bar{h}$. This radius, r_c is often called the *critical radius*. However, this concept is based on idealizations and assumptions and the student is referred to (16) for a more realistic analysis of this problem.

2.13 A solution whose boiling point is 180°F boils on the outside of a 1-in. tube with a No. 14 BWG gauge wall. On the inside of the tube flows saturated steam at 60 psia. The surface heat transfer coefficients are on the steam side 1500 and on the exterior surface 1100 Btu/h ft² °F. Calculate the increase in the rate of heat transfer for a copper over a steel tube. *Ans.* 15%.

2.14 Steam having a quality of 98% at a pressure of 1.37×10^5 N/m² is flowing at a velocity of 1 m/s through a steel pipe of 2.7 cm OD and 2.1 cm ID. The heat transfer coefficient at the inner surface, where condensation occurs, is 567 W/m² K. A dirt film at the inner surface adds a unit thermal resistance of 0.18 m² K/W. Estimate the rate of heat loss per foot length of pipe if (a) the pipe is bare, (b) the pipe is covered with a 5-cm layer of 85% magnesia insulation. For both cases assume that the unit-surface conductance at the outer surface is 11 W/m² K and that the environmental temperature is 21°C. Also estimate the change in quality per 3-m length of pipe in both cases.

2.15 Estimate the rate of heat loss per unit length from a 2-in.-ID, $2\frac{3}{8}$-in.-OD steel pipe covered with asbestos insulation ($3\frac{3}{8}$ in. OD). Steam flows in the pipe. It has a quality of 99 percent and is at 300°F. The unit thermal resistance at the inner wall is 0.015 h ft^2 °F/Btu, the heat transfer coefficient at the outer surface is 3.0 Btu/h ft^2 °F, and the ambient temperature is 60°F.

2.16 The rate of heat flow per unit length q/L through a hollow cylinder of inside radius r_i and outside radius r_o is

$$\frac{q}{L} = \frac{\bar{A}k\,\Delta T}{r_o - r_i}$$

where $\bar{A} = 2\pi(r_o - r_i)/\ln(r_o/r_i)$. Determine the percent error in the rate of heat flow if the arithmetic mean area $\pi(r_o + r_i)$ is used instead of the logarithmic mean area $\bar{A}$ for ratios of inside to outside diameters (D_o/D_i) of 1.5, 2.0, and 3.0. Plot the results.

2.17 Show that the rate of heat conduction per unit length through a long hollow cylinder of inner radius r_i and outer radius r_o, made of a material whose thermal conductivity varies linearly with temperature, is given by

$$\frac{q_k}{L} = \frac{T_i - T_o}{(r_o - r_i)/k_m\bar{A}}$$

where T_i = temperature at the inner surface

T_o = temperature at the outer surface

$\bar{A} = 2\pi(r_o - r_i)/\ln(r_o/r_i)$

$k_m = k_o[1 + \beta_k(T_i + T_o)/2]$

L = length of cylinder

2.18 A long hollow cylinder is constructed from a material whose thermal conductivity is a function of temperature according to $k = 0.060 + 0.00060\,T$, where T is in °F and k is in Btu/h ft °F. The inner and outer radii of the cylinder are 5 and 10 in., respectively. Under steady-state conditions, the temperature at the interior surface of the cylinder is 800°F and the temperature at the exterior surface is 200°F. (a) Calculate the rate of heat transfer per foot length, taking into account the variation in thermal conductivity with temperature. (b) If the surface heat transfer coefficient on the exterior surface of the cylinder is 3 Btu/h ft^2 °F, calculate the temperature of the air on the outside of the cylinder.

Ans. (a) 1959 Btu/h; (b) 75.3°F.

2.19 A 2.5-cm-OD, 2-cm-ID copper pipe carries brine at -7°C and 0.04 m^3/min. The ambient air is at 21°C and has a dew point of 10°C. How much insulation with a conductivity of $k = 0.002$ W/m K is needed to prevent condensation on the exterior of the insulation if $\bar{h}_c + \bar{h}_r = 17$ W/m^2 K on the outside?

2.20 A hollow sphere with inner and outer radii of R_1 and R_2, respectively, is covered with a layer of insulation having an outer radius of R_3. Derive an expression for the rate of heat transfer through the insulated sphere in terms of the radii, the thermal conductivities, the heat transfer coefficients, and the temperatures of the interior and the surrounding medium of the sphere.

$$\text{Ans. } q = 4\pi\,\Delta T \bigg/ \left(\frac{1}{h_1 R_1{}^2} + \frac{1}{h_3 R_3{}^2} + \frac{R_2 - R_1}{k_{12}R_1 R_2} + \frac{R_3 - R_2}{k_{23}R_2 R_3}\right).$$

2.21 A rocket motor has the shape of a sphere with one-sixth of it cut out to allow for the insertion of a nozzle. The chamber has a 14.5-in. ID. The wall consists of an inner layer of refractory $\frac{1}{4}$ in. thick ($k = 1$ Btu/h ft^2 °F), followed by a 1-in.-thick layer of steel having a thermal conductivity given by

$$k = 18(1 + 0.0011T) \text{ Btu/h ft °F}$$

where T is in °F. In steady-state operation the gases inside the chamber are at 5800°F, and the unit-surface conductance at the inner wall is 12 Btu/h ft^2 °F. Determine the amount of heat transferred through the wall under steady-state operation in 10 s if the outer surface temperature is 200°F. *Ans.* 806 Btu.

2.22 The thermal conductivity of a material may be determined in the following manner. Saturated steam at 2.41×10^5 N/m^2 is condensed at the rate of 0.68 kg/h inside a hollow iron sphere that is 1.3 cm thick and has an internal diameter of 51 cm. The sphere is coated with the material whose thermal conductivity is to be evaluated. The thickness of the material to be tested is 10 cm and there are two thermocouples embedded in it, one 1.3 cm from the surface of the iron sphere and one 1.3 cm from the exterior surface of the system. If the inner thermocouple indicates a temperature of 110°C and the outer thermocouple a temperature of 57°C, calculate (a) the thermal conductivity of the material surrounding the metal sphere, (b) the temperatures at the interior and exterior surfaces of the test material, and (c) the overall heat transfer coefficient based on the interior surface of the iron sphere, assuming the thermal resistances at the surfaces, as well as at the interface between the two spherical shells, are negligible.
Ans. (a) 3.3×10^{-3} W/m K; (b) 51° and 122°C; (c) 7.1 W/m^2 K.

2.23 A cylindrical liquid oxygen (LOX) tank has a diameter of 4 ft, a length of 20 ft, and hemispherical ends. The boiling point of LOX is -297°F. An insulation is sought which will reduce the boil-off rate in the steady state to no more than 25 lb/h. The heat of vaporization of LOX is 92 Btu/lb. If the thickness of this insulation is to be no more than 3 in., what would the value of its thermal conductivity have to be? *Ans.* ~ 0.005 Btu/h ft °F.

2.24 The addition of insulation to a cylindrical surface, such as a wire, may sometimes increase the rate of heat dissipation to the surroundings (see Problem 2.12). (a) For a No. 10 wire (0.26 cm in diameter), what is the thickness of rubber insulation ($k = 9.6 \times 10^{-4}$ W/m^2 K) that will maximize the rate of heat loss if the unit-surface conductance is 0.04 W/m^2 K? (b) If the current-carrying capacity of this wire is considered to be limited by the insulation temperature, what percent increase in capacity is realized by addition of the insulation? State your assumptions.

2.25 A standard 4-in. steel pipe (ID = 4.026 in., OD = 4.500 in.) carries superheated steam at 1200°F in an enclosed space where a fire hazard exists, limiting the outer-surface temperature to 100°F. In order to minimize the insulation cost, two materials are to be used; first a high temperature insulation (relatively expensive) applied to the pipe and then magnesia (a less expensive material) on the outside. The maximum temperature of the magnesia is to be 600°F. The following constants are known:

Steam side coefficient	$\bar{h} = 100$ Btu/h ft^2 °F
High-temperature insulation conductivity	$k = 0.06$ Btu/h ft °F

Magnesia conductivity	$k = 0.045$ Btu/h ft °F
Outside heat transfer coefficient	$\bar{h} = 2.0$ Btu/h ft °F
Steel conductivity	$k = 25$ Btu/h ft °F
Ambient temperature	$T_\infty = 70$ °F

(a) Specify the thickness for each insulating material. (b) Calculate the overall conductance based on the pipe OD. (c) What fraction of the total resistance is due to (1) steam-side resistance, (2) steel pipe resistance, (3) insulation (combination of the two), and (4) outside resistance? (d) How much heat is transferred per hour, per foot length of pipe?

2.26 For the system outlined in Problem 2.20, determine an expression for the critical radius of the insulation in terms of the thermal conductivity of the insulation and the surface coefficient between the exterior surface of the insulation and the surrounding fluid. Assume that the temperature difference, R_1, R_2, the heat transfer coefficient on the interior, and the thermal conductivity of the material of the sphere between R_1 and R_2 are constant.   *Ans.* $R_{3,\text{crit}} = 2k_{23}/\bar{h}_3$.

2.27 Show that the temperature distribution in a sphere of radius r_o, made of a homogeneous material in which energy is released at a uniform rate per unit volume $\dot{q}_G$, is

$$T(r) = T_o + \frac{\dot{q}_G r_o^{\,2}}{6k}\left[1 - \left(\frac{r}{r_o}\right)^2\right]$$

2.28 In a cylindrical fuel rod of a nuclear reactor, heat is generated internally according to the equation

$$\dot{q}_G = \dot{q}_1\left[1 - \left(\frac{r}{r_o}\right)^2\right]$$

where $\dot{q}_G$ = local rate of heat generation per unit volume at r

r_o = outside radius

$\dot{q}_1$ = rate of heat generation per unit volume at the centerline

Calculate the temperature drop from the center line to the surface for a 1-in.-OD rod having a thermal conductivity of 15 Btu/h ft °F if the rate of heat removal from its surface is 500,000 Btu/h ft^2.

2.29 An electrical heater capable of generating 10,000 W is to be designed. The heating element is to be a stainless steel wire, having an electrical resistivity of 80 × 10^{-6} ohms per centimeter length per square centimeter area. The operating temperature of the stainless steel is to be no more than 1260°C. The heat transfer coefficient at the outer surface is expected to be no less than 1702 W/m^2 K in a medium whose maximum temperature is 93°C. A transformer capable of delivering current at 9 and 12 V is available. Determine a suitable size for the wire, the current required, and discuss what effect a reduction in the heat transfer coefficient could have. *Hint:* Demonstrate *first* that the temperature drop between the center and the surface of the wire is independent of the wire diameter, and determine its value.

2.30 The tip of a soldering iron consists of a 0.6-cm-OD copper rod, 7.6 cm long. If the tip must be 204°C, what is the required minimum temperature of the base and the heat flow, in Btu's per hour and in watts, into the base? Assume that $\bar{h}_c =$ 22.7 W/m² K and $T_{air} = 21°C$.

2.31 One end of a 0.3-m-long steel rod is connected to a wall at 204°C. The other end is connected to a wall which is maintained at 93°C. Air is blown across the rod so that a heat transfer coefficient of 17 W/m² K is maintained over the entire surface. If the diameter of the rod is 5 cm and the temperature of the air 38°C, what is the net rate of heat loss to the air?

2.32 Both ends of a 0.6-cm copper U-shaped rod, as shown in the accompanying sketch, are rigidly affixed to a vertical wall, the temperature of which is maintained at 93°C. The developed length of the rod is 0.6 m and it is exposed to air at 38°C. The combined radiative and convective unit-thermal conductance for this system is 34 W/m² K. (*a*) Calculate the temperature of the midpoint of the rod. (*b*) What will the heat transfer from the rod be?

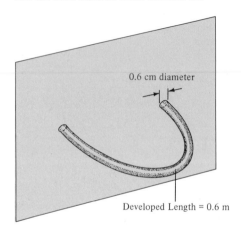

0.6 cm diameter

Developed Length = 0.6 m

2.33 (*a*) Derive an expression for the temperature rise at the center of a current-carrying wire relative to the surface as a function of the current, the diameter, and the electrical and thermal conductivities. (*b*) Compare the temperature differences between center and surface for No. 14 wires (0.064 in. diameter) of copper and nichrome when both are carrying 15 A. (*c*) Compare the surface temperature rise of these wires if the unit-surface conductance is 2 Btu/h ft °F for both. Assume thermal conductivities for copper and nichrome of 220 and 8 Btu/h ft °F, respectively, and electrical conductivities of 1.47×10^6 and 3.76×10^3 ohm/in., respectively.

2.34 A circumferential fin of rectangular cross section 0.6 cm long and 0.3 cm thick surrounds a 2.5-cm-diameter tube. The fin is constructed of mild steel and air blowing over the fin produces a heat transfer coefficient of 28.4 W/m² K. If the temperatures of the base of the fin and the air are 260° and 38°C, respectively, calculate the heat transfer rate from the fin.

2.35 Derive a differential equation for the temperature distribution in a straight triangular fin of length L and base width t. For convenience take the coordinate axis as shown in the sketch below and assume one-dimensional heat flow. If you are familiar with Bessel functions, solve the equations assuming you know the base temperature, the environmental temperature, and the heat transfer coefficient, and present your results in dimensionless coordinates.

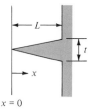

$x = 0$

2.36 A 0.3-cm-thick aluminum plate has rectangular fins on one side, 0.16×0.6 cm, spaced 0.6 cm apart. The finned side is in contact with low-pressure air at 38°C and the average unit-surface conductance is 28.4 W/m². On the unfinned side water flows at 93°C and the unit-surface conductance is 283.7 W/m² K. (a) Calculate the efficiency of the fins; (b) calculate the rate of heat transfer per unit area of wall; and (c) comment on the design if the water and air were interchanged. *Hint:* Show first that the fin efficiency is tanh mL/mL.

2.37 A turbine blade 6.3 cm long, with cross-sectional area $A = 4.6 \times 10^{-4}$ m² and perimeter $P = 0.12$ m, is made of stainless steel ($k = 0.18$ W/m K). The temperature of the root, T_s, is 482°C. The blade is exposed to a hot gas at 871°C, and the unit-surface conductance $\bar{h}$ is 454 W/m² K. Determine the temperature distribution and the rate of heat flow at the root of the blade. Assume that the tip is insulated.

2.38 To determine the thermal conductivity of a long, solid 2.5-cm-diameter rod, one-half was inserted into a furnace while the other half was projecting into air at 27°C. After steady state had been reached, the temperatures at two points 7.6 cm apart were measured and found to be 126° and 91°C, respectively. The heat transfer coefficient over the surface of the rod exposed to the air was estimated to be 22.7 W/m² K. What is the thermal conductivity of the rod?

2.39 The tip of a soldering iron consists of a 0.6-cm-OD copper rod 7.6 cm long. If the tip must be 204°C, what is the temperature of the base and the heat flow, in Btu's per hour and in watts, into the base? Assume that $\bar{h}_c = 22.7$ W/m² K and $T_{air} = 21$°C.

2.40 Heat is transferred from water to air through a brass wall ($k = 5.4$ W/m K). The addition of rectangular brass fins, 0.08 cm thick and 2.5 cm long, spaced 1.25 cm apart, is contemplated. Assuming a water-side heat transfer coefficient of 170 W/m² K and an air-side heat transfer coefficient of 17 W/m² K, compare the gain in heat transfer rate achieved by adding fins to (a) the water side, (b) the air side, and (c) both sides. (Neglect temperature drop through the wall.)

2.41 The wall of a heat exchanger has a surface area on the liquid side of 1.8 m² (0.6 m × 30 m) with a unit-surface conductance of 255 W/m² K. On the other side of the heat exchanger wall flows a gas, and the wall has 96 thin steel fins 0.05 cm wide and 1.25 cm high ($k = 3$ W/m K). The fins are 3 m long and the unit-surface conductance on the gas side is 57 W/m² K. Assuming that the thermal resistance of the wall is negligible, determine the rate of heat transfer if the overall temperature difference is 38°C. *Ans.* 12,556 W.

2.42 The top of a 12-in. I-beam is maintained at a temperature of 500°F, while the bottom is at 200°F. The thickness of the web is $\frac{1}{2}$ in. Air at 500°F is blowing along the side of the beam so that $\bar{h} = 7$ Btu/h ft² °F. The thermal conductivity of the steel may be assumed constant and equal to 25 Btu/h ft °F. Find the temperature distribution along the web from top to bottom and plot the result.

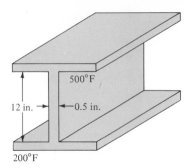

2.43 The handle of a ladle used for pouring molten lead is 30 cm long. Originally the handle was made of 1.9- by 1.25-cm mild-steel bar stock. To reduce the grip temperature, it is proposed to form the handle of tubing 0.15 cm thick to the same rectangular shape. If the average unit-surface conductance over the handle surface is 14 W/m² K, estimate the reduction of the temperature at the grip in air at 21°C.

2.44 Show that for a semi-infinite plate of width L, having the boundary condition for $T(x, y)$

$$T(0, y) - T_1 = 0$$
$$T(L, y) - T_1 = 0$$
$$T(x, \infty) - T_1 = 0$$
$$T(x, 0) - T_1 = (T_2 - T_1)$$

the temperature distribution is

$$\frac{T - T_1}{T_2 - T_1} = \frac{4}{\pi}\left[e^{-(\pi/L)y} \sin\frac{\pi x}{L} + \frac{1}{3}e^{-(3\pi/L)y}\sin\frac{3\pi x}{L} + \cdots \right]$$

For $T_1 = 0$ and $T_2 = 100°C$, plot isotherms of 25, 50, and 75°C.

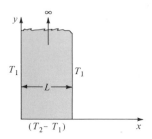

2.45 Show that the temperature distribution in a long rectangular bar, shown in the accompanying sketch, with the bottom side at a uniform temperature of 100°C and the other sides at 0°C is

$$T(x, y) = \frac{400}{\pi} \sum_{n=0}^{\infty} \frac{1}{2n + 1} \sin\frac{(2n + 1)\,\pi x}{a}$$

$$\times \sinh\frac{(b - y)(2n + 1)\pi}{a} \operatorname{cosech}\frac{(2n + 1)\,\pi b}{a}$$

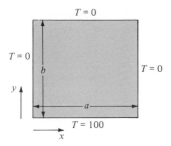

2.46 A rectangular plate 1 ft wide in the x direction, infinite in the y direction, has a temperature distribution given by $T(x, 0) = 100 \sin \pi x$ imposed on the $y = 0$ edge. Determine the temperature distribution $T(x, y)$.

2.47 In a long rectangular bar, $0 \leq x \leq a, 0 \leq y \leq b$, heat is generated uniformly at the rate $\dot{q}$ W/m³. The boundaries at $x = 0$ and $y = 0$ are insulated while at the boundaries at $x = a$ and $y = b$ heat is convected into a surrounding at zero temperature through a unit surface conductance $\bar{h}$. Set up the differential equation with the appropriate boundary condition and obtain an expression for the temperature distribution.

2.48 Compare the rate of heat flow from the top to the bottom in the aluminum structure shown in the sketch with the rate of heat flow through a solid slab. The top is at $-10°C$, the bottom at $0°C$. The holes are filled with insulation which does not conduct heat appreciably.

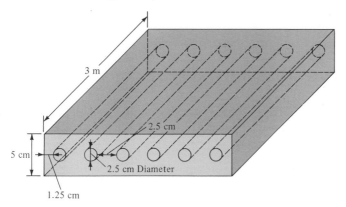

2.49 Determine by means of flux plot the temperatures and heat flow per unit depth in the ribbed insulation shown in the accompanying sketch.

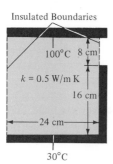

2.50 By means of a flux plot, estimate the rate of heat flow through the object ($k =$ 15 W/m K) shown in the sketch. Assume that no heat is lost from the sides.

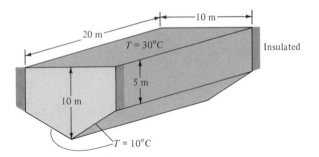

2.51 Determine the rate of heat transfer per foot length from a 5-cm-OD pipe at 150°C placed eccentrically within a larger cylinder of rock wool as shown in the sketch. The outside diameter of the larger cylinder is 15 cm. and the surface temperature 50°C.

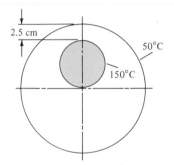

2.52 Determine the rate of heat flow per foot length from the inner to the outer surface of the molded asbestos insulation shown in the accompanying sketch ($k =$ 0.1 Btu/h ft °F).

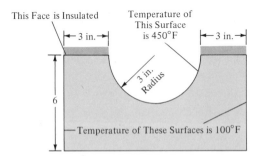

2.53 A long 1-cm-diameter electric copper cable is embedded in the center of a 25-cm square concrete block. If the outside temperature of the concrete is 25°C and the rate of electrical energy dissipation in the cable is 150 W per meter length, determine the temperatures at the outer surface and at the center of the cable.

2.54 A large number of 1.5-in.-OD pipes carrying hot and cold liquids are embedded in concrete stone in an equilateral staggered arrangement with center lines 4.5 in. apart as shown in the sketch. If the pipes in rows A and C are at 60°F while the pipes in rows B and D at 150°F, determine the rate of heat transfer per foot length from pipe X in row B.

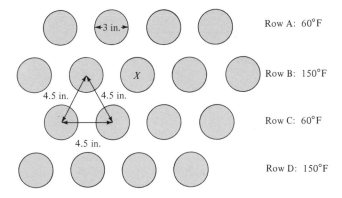

2.55 A plane insulating wall has alternatingly spaced ribs extending from both surfaces as shown in the accompanying sketch. The ribs are made of aluminum so that their temperature is essentially the same as that of the surface to which they are attached. If the upper surface is at 120°C, the lower surface at 50°C, and the space between the surfaces is filled with powdered diatomaceous earth, estimate the rate of heat transfer per square foot of wall. What would be the percentage reduction in the rate of heat transfer if the ribs were removed?

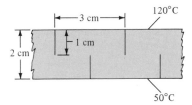

2.56 A long 1-cm.-diameter electric cable is imbedded in a concrete wall ($k = 0.60$ Btu/h ft °F) which is 1 m by 1 m, as shown in the sketch below. If the lower surface is insulated, the surface of the cable is 100°C and the exposed surface of the concrete is 25°C. Estimate the rate of energy dissipation per meter of cable.

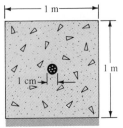

Insulated Surface

2.57 Determine the temperature distribution and the heat flow rate per length meter in a long stone concrete block having the shape shown below. The cross-sectional area of the block is square and the hole is centered.

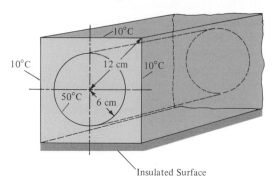

2.58 A 1-ft-OD pipe with a surface temperature of 200°F carries steam over a distance of 300 ft. The pipe is buried with its center line at a depth of 3 ft, the ground surface is 20°F, and the mean thermal conductivity of the soil is 0.4 Btu/h ft °F. Calculate the heat loss per day, and the cost, if steam is worth $0.14 per 10^6 Btu Also, estimate the thickness of 85 percent magnesia insulation necessary to achieve the same insulation with a total unit-surface conductance of 4 Btu/h ft² °F on the outside of the insulation.

2.59 Two long pipes, one having a 10-cm. OD and a surface temperature of 300°C, the other having a 5-cm. OD and a surface temperature of 100°C, are buried deeply in dry sand with their centerlines 15 cm apart. Determine the rate of heat flow from the larger to the smaller pipe per foot length.

2.60 A radioactive sample is to be stored in a protective box with 4-in.-thick walls having interior dimensions of 4 by 4 by 12 cm. The radiation emitted by the sample is completely absorbed at the inner surface of the box, which is made of concrete. If the outside temperature of the box is 25°C, but the inside temperature is not to exceed 50°C, determine the maximum permissible radiation rate from the sample, in watts.

2.61 A 6-in.-OD pipe is buried with its centerline 50 in. below the surface of the ground (k of soil is 0.20 Btu/h ft °F). An oil having a specific gravity of 0.8 and a specific heat of 0.5 Btu/lb °F flows in the pipe at 100 gpm. Assuming a ground-surface temperature of 40°F and a pipe-wall temperature of 200°F, estimate the length of pipe in which the oil temperature decreases by 10°F.

2.62 A 1-in.-OD hot-steam line at 212°F runs parallel to a 2-in.-OD cold-water line at 60°F. The pipes are 2 in. center to center and deeply buried in concrete with k = 0.50 Btu/h ft °F. What is the heat transfer per foot of pipe between the two pipes? If the steam were used to heat cold water at 60°F to 100°F, how many gallons per hour could be heated in a 1000-ft length of pipe?

2.63 Calculate the rate of heat transfer between a 6-in.-OD pipe at 250°F and a 4-in.-OD pipe at 100°F. The two pipes are 1000 ft long; they are buried in sand (k = 0.19 Btu/h ft °F) 4 ft below the surface (T_s = 80°F); they are parallel and separated by 9 in., center to center distance.   *Ans.* 74,700 Btu/h.

2.64 A 0.6 cm.-diameter mild-steel rod at 38°C is suddenly immersed in a liquid at 93°C with $\bar{h}_c$ = 113.5 W/m². Determine the time required for the rod to warm to 88°C.

2.65 A spherical shell satellite (0.3 m OD, 1.25 cm. wall thickness, made of stainless steel) reenters the atmosphere from outer space. If its orginal temperature is 38°C, the effective average temperature of the atmosphere is 1093°C, and the effective heat transfer coefficient is 113.5 W/m², estimate the temperature of the shell after reentry, assuming the time of reentry is 10 min and the interior of the shell is evacuated.

2.66 A thin-wall cylindrical vessel (3 ft in diameter) is filled to a depth of 4 ft with water at an initial temperature of 60°F. The water is well stirred by a mechanical agitator. Estimate the time required to heat the water to 120°F if the tank is suddenly immersed into oil at 220°F. The overall heat transfer coefficient between the oil and the water is 50 Btu/h ft² °F, and the effective heat transfer surface area is 45 ft².

2.67 A thin-wall jacketed tank, heated by condensing steam at 10×104 N/m² psia, contains 91 kg of agitated water (assume uniform water temperature). The heat transfer area of the jacket is 0.9 m² and the overall conductance $U = 227$ W/m² K based on that area. Determine the heating time required for an increase in temperature from 16° to 60°C. *Ans.* 18 min.

2.68 The heat transfer coefficients for the flow of 26.6°C air over a 1.25 cm. diameter sphere are measured by observing the temperature-time history of a copper ball of the same dimension. The temperature of the copper ball ($c = 376$ J/kg K $\rho = 8928$ kg/m³) was measured by two thermocouples, one located in the center, and other near the surface. Both of the thermocouples registered, within the accuracy of the recording instruments, the same temperature at a given instant. In one test run the initial temperature of the ball was 66°C and in 1.15 min the temperature decreased by 7°C. Calculate the heat transfer coefficient for this case.

2.69 A spherical stainless steel vessel at 93°C contains 45 kg of water at the same temperature. If the entire system is suddenly immersed in ice water, determine (*a*) the time required for the water in the vessel to cool to 16°C, and (*b*) the temperature of the walls of the vessel at that time. Assume that the unit-surface conductance at the inner surface is 17 W/m² K, the unit-surface conductance at the outer surface is 22.7 W/m² K, and the wall of the vessel is 2.5 cm. thick.

2.70 A copper wire, $\frac{1}{32}$ in. OD, 2 in. long, is placed in an air stream whose temperature rises as $T_{air} = (50 + 25t)$°F, where t is the time in seconds. If the initial temperature of the wire is 50°F, determine its temperature after 2 s, 10 s, and 1 min. The unit-surface conductance between the air and the wire is 7 Btu/h ft² °F.

2.71 Ball bearings are to be hardened by quenching them in a water bath at a temperature of 100°F. Suppose you are asked to devise a continuous process in which the balls could roll from a soaking oven at a uniform temperature of 1600°F into the water, where they are carried away by a rubber conveyer belt. The rubber conveyer belt would, however, not be satisfactory if the surface temperature of the balls leaving the water is above 200°F. If the surface coefficient of heat transfer between the balls and the water may be assumed to be equal to 104 Btu/h ft² °F, (*a*) find an approximate relation giving the minimum allowable cooling time in the water as a function of the ball radius for balls up to $\frac{1}{2}$ in. in diameter, (*b*) calculate the cooling time, in seconds, required for a ball having a 1-in. diameter, and (*c*) calculate the total amount of heat in Btu/h which would have to be removed from the water bath in order to maintain its temperature uniform if 100,000 balls of 1-in. diameter are to be quenched per hour.

2.72 A mercury thermometer in a cylindrical thermometer well, filled with oil, is subjected to a sudden temperature rise of 38°C. Derive the equation describing the response of the *thermometer*, lumping the heat capacity of the well and the thermometer separately. For a unit-surface conductance at the outer surface of the well of 68 W/m² K and a thermal resistance between the inner surface of the well and the mercury thermometer of 1.7×10^{-3} m² K/W plot the response of the thermometer as a function of time.

2.73 Estimate the time required to heat the center of a 1.5 kg roast in a 204°C oven to 149°C. State your assumptions carefully and compare your results with cooking instructions in a standard cookbook.

2.74 A large 2.54 cm.-thick copper plate is placed between two air streams. The unit-surface conductance on the one side is 28 W/m² K and on the other side is 57 W/m² K. If the temperature of both streams is suddenly changed from 38° to 93°C, determine how long it will take for the copper plate to reach a temperature of 82°C.

2.75 A 1.4 kg aluminum household iron has a 500-W heating element. The surface area is 0.046 m². The ambient temperature is 21°C and the surface heat transfer coefficient is 11 W/m² K (assumed constant). How long after the iron is plugged in will its temperature reach 104°C?

2.76 A slab of material having a thermal diffusivity of 4.6×10^{-2} m²/h is 5 cm. thick and of relatively large breadth and width. The slab is being held at a mean temperature of 538°C in a gas stream having a mean temperature 538°C. The gas temperature is controlled by an on-off controller, which produces an essentially triangular gas-temperature variation of 14°C amplitude and 10-min period. Presuming the film conductance to be 113.5 W/m² K and the heat transfer to be convective only, comment on the adequacy of the control system if the slab temperature should not depart from the mean value of 538°C by more than 3°C at any point in the slab.

2.77 Estimate the depth in moist soil at which the annual temperature variation will be 10 percent of that at the surface.

2.78 A small aluminum sphere of diameter D, initially at a uniform temperature T_0, is immersed in a liquid whose temperature, T_∞, varies sinusoidally according to

$$T_\infty - T_m = A \sin \omega \theta$$

where T_m = time-averaged temperature of the liquid

A = amplitude of the temperature fluctuation

ω = frequency of the fluctuations

If the heat transfer coefficient between the fluid in the sphere, $\bar{h}_a$, is constant and the system may be treated as a "lumped capacity," derive an expression for the sphere temperature as a function of time.

2.79 A wire of perimeter P and cross-sectional area A emerges from a die at a temperature T above ambient and with a velocity U. Specify the temperature distribution along the wire in the steady state if the exposed length downstream from the die is quite long. State clearly and try to justify all assumptions.

2.80 A long, slender metal rod of length L is attached at its base to a wall at 0°F. The curved surface the rod is insulated. The end of the rod is in contact with a fluid at 0°F, where the unit-surface conductance at the interface $\bar{h}$ is constant and uniform. If the initial temperature in the rod is given by $T(x, 0) = f(x)$, show that the temperature distribution after time θ is

$$T(x, \theta) = \sum_{n=1}^{\infty} C_n e^{-a\lambda_n^2\theta} \sin \lambda_n x$$

where

$$C_n = \frac{2\lambda_n L}{(\lambda_n L - \sin \lambda_n L \cos \lambda_n L)L} \int_0^L f(x) \sin \lambda_n x \, dx$$

Calculate the temperature at the end ($x = L$) of a 0.1-in. diam, 2-in.-long stainless steel rod as a function of time, if the initial temperature distribution is linear, with 100°F at the end, and $\bar{h} = 10$ Btu/h ft^2 °F at the end.

2.81 A stainless steel cylindrical billet ($k = 0.12$ W/m K, $a = 1.4 \times 10^{-3}$ m^2) is heated to 593°C preparatory to a forming process. If the minimum temperature permissible for forming is 482°C, how long may the billet be exposed to air at 38°C if the average unit-surface conductance is 85 W/m^2 K? The shape of the billet is shown in the accompanying sketch.

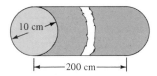

2.82 In the vulcanization of tires, the carcass is placed into a jig, and steam at 149°C is admitted suddenly to both sides. If the tire thickness is 2.5 cm and the initial temperature 21°C, estimate the time required for the central layer to reach 132°C.

2.83 A long copper cylinder 0.6 m in diam and initially at a uniform temperature of 38°C is placed in a water bath at 93°C. Assuming that the heat transfer coefficient between the copper and the water is 1248 W/m^2 K, calculate the time required to heat the center of the cylinder to 66°C. As a first approximation neglect the temperature gradient within the cylinder; then repeat your calculation without this simplifying assumption and compare your results.   *Ans.* 4.8 min; 8.1 min.

2.84 A steel sphere with a diameter of 3 in. is to be hardened by first heating it to a uniform temperature of 1600°F and then quenching it in a large bath of water at a temperature of 38°C. The following data apply:

Surface heat transfer coefficient $\bar{h} = 590$ W/m^2 K

Thermal conductivity of steel = 0.3 W/m^2 K

Specific heat of steel = 628 J/kg K

Density of steel = 7840 kg/m^3

Calculate: (a) time elapsed in cooling the surface of the sphere to 204°C and (b) time elapsed in cooling the center of the sphere to 204°C.

2.85 A fireproof safe is to be constructed. Its walls consist of two 0.16 cm. steel sheets with a layer of asbestos board between them. Using the chart for a slab, estimate the thickness of asbestos required to give 1 h of fire protection on the basis that, for an outside temperature of 816°C, the inside temperature is not to rise above 121°C during this period. The heat transfer coefficient at the exterior surface is 28.4 W/m² K.

2.86 A 2.5 cm.-thick sheet of plastic initially at 21°C is placed between two heated steel plates that are maintained at 138°C. The plastic is to be heated just long enough for its midplane temperature to reach 132°C. If the thermal conductivity of the plastic is 1.1×10^{-3} W/m K, the thermal diffusivity is 2.7×10^{-4} m², and the thermal resistance at the interface between the plates and the plastic is negligible, calculate: (a) the required heating time, (b) the temperature at a plane 0.6 cm. from the steel plate at the moment the heating is discontinued, and (c) the time required for the plastic to reach a temperature of 132°C 0.6 cm from the steel plate.

Ans. (a) 49 min; (b) 134°C; (c) 43 min.

2.87 A turnip (assume spherical) weighing 0.45 kg is dropped into water boiling at atmospheric pressure. If the initial temperature of the turnip is 17°C, how long does it take to reach 92°C at the center? Assume that:

$$\bar{h}_c = 1702 \text{ W/m}^2 \text{ K} \qquad c_p = 3.9 \times 10^3 \text{ J/kg K}$$
$$k = 0.519 \text{ W/m K} \qquad \rho = 1040 \text{ kg/m}^3$$

2.88 An egg, which for the purposes of this problem can be assumed to be a 5-cm.-diameter sphere having the thermal properties of water, is initially at a temperature of 4°C. It is immersed in boiling water at 100°C for 15 min. The heat transfer coefficient from the water to the egg may be assumed to be 0.5674 W/m² K. What is the temperature of the egg center at the end of the cooking period?

2.89 A long wooden rod 2.5 cm. OD is placed at 38°C into an airstream at 816°C. The unit-surface conductance between the rod and air is 28.4 W/m² K. If the ignition temperature of the wood is 427°C, $\rho = 800$ kg/m³, k = 0.173 W/m K, and c = 2500 J/kg K, determine the time between initial exposure and ignition of the wood.

2.90 In the inspection of a sample of meat intended for human consumption, it was found that certain undesirable organisms were present. In order to make the meat safe for consumption, it is ordered that the meat be kept at a temperature of at least 121°C for a period of at least 20 min during the preparation. Assume that a slab of this meat, 2.5 cm. thick, is originally at a uniform temperature of 27°C; that it is to be heated from both sides in a constant temperature oven; and that the maximum temperature meat can withstand is 154°C. Assume furthermore that the surface coefficient of heat transfer remains constant and is .048 W/m K. The following data may be taken for the sample of meat: specific heat = 4184 J/kg K; density = 1280 kg/m³; thermal conductivity = 4.8×10^{-3} W/m K. Calculate the *minimum total time* of heating required to fulfill the safety regulation.

2.91 A frozen-food company freezes its spinach by first compressing it into large slabs and then exposing the slab of spinach to a low-temperature cooling medium. The large slab of compressed spinach is initially at a uniform temperature of 21°C; it must be reduced to an average temperature over the entire slab of −34°C. The temperature at any part of the slab, however, must never drop below −51°C. The cooling medium which passes across both sides of the slab is at a constant temperature of −90°C. The following data may be used for the spinach: density = 80 kg/m³; thermal conductivity = 6×10^{-3} W/m K; specific heat = 2100 J/kg K. Present a detailed analysis outlining a method to estimate the maximum thickness of the slab of spinach that can be safely cooled in 60 min.   *Ans.* 12 in.

2.92 To determine the heat transfer coefficient between a heated steel ball and cooler ground or crushed mineral solids experimentally, a series of SAE 1040 steel balls were heated to a temperature of 700°C and the center temperature-time history of each was measured with a thermocouple while it was cooling in a bed of crushed iron ore, which was placed in a steel drum rotating horizontally at about 30 rpm. For a 5-cm.-diameter ball, the time required for the temperature difference between the ball center and the surrounding ore to decrease from 500° to 250°C was found to be 64, 67, and 72 s, respectively, in three different test runs. Determine the average unit-surface conductance between the ball and the ore. Compare the results obtained by assuming the thermal conductivity to be infinite with those obtained by taking the internal thermal resistance of the ball into account.

Ans. 306 W/m² K.

2.93 A mild-steel cylindrical billet, 25 cm. in diameter, is to be raised to a minimum temperature of 760°C by passing it through a 6 m-long strip type furnace. If the furnace gases are at 1538°C and the over-all unit-surface conductance on the outside of the billet is 68 W/m² K, determine the maximum speed at which a continuous billet entering at 204°C can travel through the furnace.

2.94 A solid lead cylinder 0.6 m in diameter and 0.6 m long, initially at a uniform temperature of 121°C, is dropped into a 21°C liquid bath in which the unit-surface conductance $\bar{h}_c$ is 1135 W/m² K. Plot the temperature-time history of the center of this cylinder and compare it with the time histories of a 0.6-m-diameter, infinitely long lead cylinder and a lead slab 0.6 m thick.

2.95 A long, 0.6-m-OD solid steel ($k = 0.14$ W/m K) cylindrical billet at 16°C room temperature is placed in an oven where the temperature is 260°C. If the average unit-surface conductance is 170 W/m² K, estimate the time required for the center temperature to increase to 232°C by (*a*) using the appropriate chart and (*b*) dividing the solid into *two* equal lumped thermal capacities with appropriate thermal resistances between them. Also, (*c*) determine the instantaneous surface heat fluxes when the center temperature is 232°C.

2.96 Repeat Problem 2.95(*a*), but assume that the billet is only 1.2 m long with the average unit-surface conductance at both ends equal to 136 W/m² K.

2.97 A large billet of steel originally at 260°C is placed in a radiant furnace where the surface temperature is held at 1200°C. Assuming the billet infinite in extent, compute the temperature at point *P* shown in the accompanying sketch after 25 min have elapsed. The average properties of steel are: $k = 0.28$ W/m K, $\rho = 7360$ kg/m³, and $c = 500$ J/kg K. *Ans.* 1040°C.

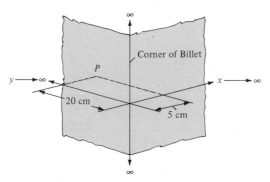

REFERENCES

1. H. S. Carslaw and J. C. Jaeger, *Conduction of Heat in Solids*, 2d ed., Oxford University Press, London, 1959.
2. K. A. Gardner, "Efficiency of Extended Surfaces," *Trans. ASME*, vol. 67, pp. 621–631, 1945.
3. W. P. Harper and D. R. Brown, "Mathematical Equation for Heat Conduction in the Fins of Air-Cooled Engines," NACA Rep. 158, 1922.
4. P. J. Schneider, *Conduction Heat Transfer*, Addison-Wesley, Cambridge, Mass., 1955.
5. M. N. Ozisik, *Boundary Value Problems of Heat Conduction*, International Textbook Co., Scranton, Pa., 1968.
6. L. M. K. Boelter, V. H. Cherry, and H. A. Johnson, *Heat Transfer Notes*, 3d ed., University of California Press, Berkeley, 1942.
7. C. F. Kayan, "An Electrical Geometrical Analogue for Complex Heat Flow," *Trans. ASME*, vol. 67, pp. 713–716, 1945.
8. I. Langmuir, E. Q. Adams, and F. S. Meikle, "Flow of Heat through Furnace Walls," *Trans. Am. Electrochem. Soc.*, vol. 24, pp. 53–58, 1913.
9. O. Rüdenberg, "Die Ausbreitung der Luft und Erdfelder um Hochspannungsleitungen besonders bei Erd-und Kurzschlüssen," *Electrotech. Z.*, vol. 46, pp. 1342–1346, 1925.
10. B. O. Pierce, *A Short Table of Integrals*, Ginn, Boston, 1929.
11. M. P. Heisler, "Temperature Charts for Induction and Constant Temperature Heating," *Trans. ASME*, vol. 69, pp. 227–236, 1947.
12. H. Gröber, S. Erk, and U. Grigull, *Fundamentals of Heat Transfer*, McGraw-Hill, New York, 1961.
13. P. J. Schneider, *Temperature Response Charts*, Wiley, New York, 1963.
14. F. Kreith and W. Z. Black, *Basic Heat Transfer*, Harper & Row, New York, 1980.
15. V. Arpaci, *Conduction Heat Transfer*, Addison-Wesley, Reading, Mass., 1966.

Chapter 3

Numerical Analysis of Heat Conduction

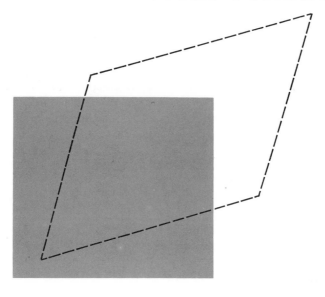

3.1 INTRODUCTION

Equations describing the transport of heat by conduction were developed in Chapter 2 and analytical solutions for several examples were presented. Analytical solutions of heat transfer problems are generally possible only for simple geometries, simple boundary conditions, or extreme values of certain parameters—for instance, asymptotic solutions. Nevertheless, such solutions play an important role in heat transfer analysis by providing insight into complex problems which can be simplified under certain conditions, by providing exact solutions to new heat transfer problems which involve simple geometries, and as a partial check on other solution schemes for complex problems.

Many practical heat transfer problems are not amenable to complete solution by analytical techniques and must be approached by the method of numerical analysis. In some cases problems can be solved more quickly by numerical methods and changes in problem parameters can be made conveniently. This last feature allows an engineer to model an experiment numerically and take "data" much more rapidly and more inexpensively than if the actual experimental apparatus had been assembled and operated.

In the past ten years the cost of computation with large digital computers has dropped, while at the same time the storage capacity and computational speed have increased significantly. These developments have led to the use of increasingly complex computer codes for numerical heat transfer analysis by industry. More recently, the increased sophistication and dramatically reduced cost of desk-top computers have made relatively powerful numerical techniques available to many more practicing engineers. In addition, the solution of many engineering problems begins with a review of the literature, more and more of which includes numerical methods. Whether the engineer is involved with developing sophisticated codes, use of codes, interaction with those who develop such codes, or the solution of less complex heat transfer problems on a desk-top computer, the need for understanding the fundamentals of numerical heat transfer analysis is clear.

One of the most widely used numerical techniques for solving differential equations is the finite-difference method. Analytical methods, such as those described in Chapter 2, provide a solution at every point in time and space within the problem boundaries. The finite-difference method provides a solution only at discrete points within the boundaries and therefore is only an approximation of the exact solution. However, representing the problem at only a finite number of points simplifies the solution procedure to one of making several arithmetic calculations rather than one complex calculation. The finite-difference method transforms the problem into one that can be solved by computers. Generally speaking, if more discrete points are used in the finite-difference calculation, the solution will be closer to the exact solution but more arithmetic calculations will be required. To complete the finite-difference solution we must also represent the boundary conditions and initial conditions at discrete points. Finally, we must be aware that the method used to carry out the simple calculations introduces errors or, in fact, may not give a meaningful solution at all.

Some heat transfer problems do not require the solution of a differential equation. For example, numerical integration may be required in a problem that generates an integral equation. Most schemes for numerical integration (trapezoidal rule, Simpson's rule, etc.) are taught in introductory numerical analysis courses and subroutines are generally available on most computers for performing numerical integration.

It is the purpose of this chapter to introduce the student to methods of solving the differential equations of heat conduction with a computer. For this reason, we consider only linear differential equations. We have placed minimal emphasis on the actual writing of computer programs. Any computer language or version of a language used to explain coding of finite-difference equations is almost certain to differ from that which the student may be familiar with, or may use in the future. Rather, we are stressing the fundamentals of finite-difference methods that will be applicable for any computer or computer language the student may use.

3.2 FINITE-DIFFERENCE REPRESENTATION

In Chapter 2 we derived the differential heat conduction equation, Eq. (2.6). To cast this equation in finite-difference form we must be able to express the first time derivative and first and second space derivatives in terms of temperatures at discrete points. The most straightforward method of doing this is by writing a Taylor series expansion of the temperature at a point in time and space. For simplicity, we first assume that temperature depends only on one space variable. In Cartesian coordinates

$$T(x + \Delta x, t) = T(x, t) + \frac{\partial T}{\partial x}(x, t)\Delta x + \frac{\partial^2 T}{\partial x^2}(x, t)\frac{\Delta x^2}{2!}$$

$$+ \frac{\partial^3 T}{\partial x^3}(x, t)\frac{\Delta x^3}{3!} + O(\Delta x^4) \tag{3.1}$$

$$T(x - \Delta x, t) = T(x, t) - \frac{\partial T}{\partial x}(x, t)\Delta x + \frac{\partial^2 T}{\partial x^2}(x, t)\frac{\Delta x^2}{2!}$$

$$- \frac{\partial^3 T}{\partial x^3}(x, t)\frac{\Delta x^3}{3!} + O(\Delta x^4) \tag{3.2}$$

We use Δx to represent the distance between the discrete grid points. By neglecting terms Δx^2 and smaller we can write from Eq. (3.1)

$$\frac{\partial T}{\partial x}(x, t) = \frac{T(x + \Delta x, t) - T(x, t)}{\Delta x} + O(\Delta x) \tag{3.3}$$

or from Eq. (3.2)

$$\frac{\partial T}{\partial x}(x, t) = \frac{T(x, t) - T(x - \Delta x, t)}{\Delta x} + O(\Delta x) \tag{3.4}$$

By subtracting Eq. (3.2) from Eq. (3.1) and neglecting terms Δx^3 and smaller we get

$$\frac{\partial T}{\partial x}(x, t) = \frac{T(x + \Delta x, t) - T(x - \Delta x, t)}{2\Delta x} + O(\Delta x^2) \tag{3.5}$$

These three representations of $\partial T/\partial x$ are known as forward-difference, backward-difference, and centered-difference approximations, Eqs. (3.3), (3.4), (3.5), respectively. We see that for small Δx the centered-difference approximation has an error proportional to Δx^2, while the forward- and backward-difference approximations have an error proportional to Δx. These truncation errors are the deviation from an exact representation of the derivatives by the finite-difference representation.

Adding Eqs. (3.1) and (3.2) and neglecting terms in Δx^4 and smaller yields an expression for the second derivative with truncation error Δx^2:

$$\frac{\partial^2 T}{\partial x^2}(x, t) = \frac{T(x + \Delta x, t) - 2T(x, t) + T(x - \Delta x, t)}{\Delta x^2} + O(\Delta x^2) \tag{3.6}$$

In general, a more accurate numerical solution will result from a smaller increment size Δx. However, more computer storage and more calculation steps will be required for smaller increment size. Use of more calculation steps increases the cost of the solution by using more computing time. It is often desirable to repeat a numerical calculation with a smaller and smaller increment size until the change in the solution is negligible. A secondary consideration is accumulation of round-off error as the number of calculation steps increases. Round-off error is due to the fact that the computer solution is carried out with a finite number of digits.

EXAMPLE 3.1 _____

Compute the error in the finite-difference representation of the second derivative at $x = L/2$ for the temperature distribution

$$T = T_0 + T_1 \sin \frac{\pi x}{L} \qquad 0 \le x \le L$$

for values of $\Delta x = 0.01L$, $0.1L$, and $0.25L$.

Solution.   Differentiating the temperature distribution twice, we find

$$\frac{d^2 T}{dx^2} = -T_1 \left(\frac{\pi}{L}\right)^2 \sin \frac{\pi x}{L}$$

or

$$\left.\frac{d^2 T}{dx^2}\right|_{x=L/2} = -T_1 \left(\frac{\pi}{L}\right)^2$$

for the exact value of the second derivative. From Eq. (3.6) we need to compute the temperature at $x = L/2 \pm \Delta x$ for the specified values of Δx. Let $\Delta x = \alpha L$ ($\alpha = 0.01, 0.10, 0.25$); we have

$$T - T_0 = T_1 \sin \frac{\pi}{L} \left(\frac{L}{2} \pm \alpha L\right)$$

$$= T_1 \left(\sin \frac{\pi}{2} \cos \alpha \pi \pm \sin \alpha \pi \cos \frac{\pi}{2}\right)$$

$$= T_1 \cos \alpha \pi$$

Applying Eq. (3.6), we find

$$\left.\frac{d^2 T}{dx^2}\right|_{x=L/2} = \frac{2T_1 \cos \alpha \pi - 2T_1}{(\alpha L)^2}$$

$$= \frac{2T_1(\cos \alpha \pi - 1)}{(\alpha L)^2}$$

The ratio of the exact value to the finite-difference value is

$$\frac{(\pi/L)^2 (\alpha L)^2}{2(1 - \cos \alpha \pi)} = \frac{(\alpha \pi)^2}{2(1 - \cos \alpha \pi)}$$

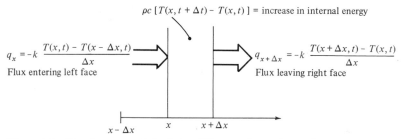

$$\rho c\,[T(x, t + \Delta t) - T(x, t)\,] = \text{increase in internal energy}$$

$$q_x = -k\,\frac{T(x, t) - T(x - \Delta x, t)}{\Delta x}$$

Flux entering left face

$$q_{x + \Delta x} = -k\,\frac{T(x + \Delta x, t) - T(x, t)}{\Delta x}$$

Flux leaving right face

$$x - \Delta x \qquad x \qquad x + \Delta x$$

Figure 3.1 Development of the finite-difference form of one-dimensional, unsteady heat conduction equation.

which is 1.0001 for $\alpha = 0.01$, 1.0083 for $\alpha = 0.10$, and 1.053 for $\alpha = 0.25$. Hence a step size of $L/100$ will yield a 0.01 percent accurate approximation in the second derivative while a step size of $L/4$ produces an error of about 5.3 percent.

Applying these finite-difference equations to Eq. (2.6) demonstrates the physical significance of the finite-difference approximations. For a transient, one-dimensional heat conduction problem with no internal heat generation, Eq. (2.6) reduces to

$$\rho c\,\frac{\partial T}{\partial t} = k\,\frac{\partial^2 T}{\partial x^2} \tag{3.7}$$

Using a forward difference for the time derivative and Eq. (3.6) for the space derivative we find

$$\rho c\,[T(x, t + \Delta t) - T(x, t)] = k\,\frac{\Delta t}{\Delta x^2}\,[T(x + \Delta x, t) - 2T(x, t) + T(x - \Delta x, t)] \tag{3.8}$$

Consider the element from x to $x + \Delta x$; the right side of Eq. (3.8) represents the difference in heat flux out of the face $x + \Delta x$ and into the face x, Fig. 3.1. The left side of Eq. (3.8) represents the increase in internal energy of the element during the time period Δt due to a net heat flux into the element. Figure 3.1 closely resembles Fig. 2.1, which was used to develop Eq. (2.5) by allowing Δx and Δt to become infinitesimally small. In Eq. (3.8) we have kept Δx and Δt of finite size.

3.3 DIFFERENCE EQUATIONS

3.3.1 One-Dimensional Steady Conduction

For a steady one-dimensional problem, Eq. (3.8) simplifies to

$$T(x + \Delta x) - 2T(x) + T(x - \Delta x) = 0 \tag{3.9}$$

that is, the temperature at any point is just the average of the temperature at the two surrounding points.

Consider a steady, one-dimensional conduction problem in the domain $0 \le x \le 1$. We divide this domain into four equal-length segments, $\Delta x = 0.25$. Suppose the boundary conditions are $T(0) = T_a$, $T(1) = T_b$. Then the five simultaneous equations which express the temperature of each node ($x = 0$, Δx, $2\,\Delta x$, $3\,\Delta x$, 1) from Eq. (3.9) are

$$
\begin{aligned}
T(0) &= T_a \\
T(0) - 2T(\Delta x) + T(2\,\Delta x) &= 0 \\
T(\Delta x) - 2T(2\,\Delta x) + T(3\,\Delta x) &= 0 \\
T(2\,\Delta x) - 2T(3\,\Delta x) + T(1) &= 0 \\
T(1) &= T_b
\end{aligned}
$$

By writing the five equations in this way, we see that they can be expressed in matrix form:

$$
\begin{bmatrix}
1 & & & & \\
1 & -2 & 1 & & \\
& 1 & -2 & 1 & \\
& & 1 & -2 & 1 \\
& & & & 1
\end{bmatrix}
\begin{bmatrix}
T(0) \\
T(\Delta x) \\
T(2\,\Delta x) \\
T(3\,\Delta x) \\
T(1)
\end{bmatrix}
=
\begin{bmatrix}
T_a \\
0 \\
0 \\
0 \\
T_b
\end{bmatrix}
\qquad (3.10)
$$

Blank spaces in the matrix represent zeros, a convention we shall follow throughout this chapter.

Equation (3.10) may be written

$$\mathbf{AT = C} \qquad (3.11)$$

where $\mathbf{A}$ is the matrix of coefficients, $\mathbf{T}$ is the vector of nodal temperatures, and $\mathbf{C}$ is the vector of boundary conditions.

If we could find the inverse of $\mathbf{A}$ we could write the solution of Eq. (3.11) as

$$\mathbf{T = A^{-1}C} \qquad (3.12)$$

Because many numerical solution schemes can be reduced to finding the solution of a set of simultaneous equations such as Eq. (3.12), we will postpone discussion of this step until Section 3.5.

3.3.2 One-Dimensional, Unsteady Conduction

A rather arbitrary choice of difference schemes led to Eq. (3.8). We might have chosen a centered representation for the time derivative, or we might have chosen to calculate the space derivative at $t + \Delta t$ rather than at time t. The choice of difference schemes used to assemble the difference equation not only affects the method one uses to solve the equation but also can determine whether one finds a solution at all.

If we let $r = \alpha \, \Delta t / \Delta x^2$ (where α is the thermal diffusivity $k/\rho c$), Eq. (3.8) can be written

$$T(x, t + \Delta t) = rT(x + \Delta x, t) + (1 - 2r)T(x, t) + rT(x - \Delta x, t) \qquad (3.13)$$

so the temperature distribution at any time can be easily calculated from the temperature distribution at an earlier time. This is known as an explicit method and it is first-order accurate in time and second-order accurate in space, that is, the truncation error is $O(\Delta t + \Delta x^2)$.

It can be shown (1) that Eq. (3.13) will converge to the exact solution as Δx and Δt tend to zero only if $0 < r \le 1/2$. Physically, this means that the temperature at a point cannot cause a decrease in the temperature at that point at a later time. We note also that the characteristic diffusion time for the problem is $t \sim L^2/\alpha$ (where L is the length scale for the problem). The restriction on r, therefore, requires that the numerical time step be less than or equal to one-half of the diffusion time. This is demonstrated in Example 3.2.

EXAMPLE 3.2 ⎯⎯⎯⎯⎯⎯⎯⎯⎯⎯⎯⎯⎯⎯⎯⎯⎯⎯⎯⎯⎯⎯⎯⎯⎯⎯

Consider the problem of determining the minimum depth a water main must be buried to avoid freezing, Example 2.11. We determined in that example that a depth of 0.68 m was required. Compute the temperature at a depth of 0.7 m as a function of time, showing the effect of $r > 1/2$ on stability of the numerical solution.

Solution. The finite-difference equation is Eq. (3.13), where $r = \alpha \, \Delta t / \Delta x^2$, and from Appendix 2, Table 11, $\alpha = 0.138 \times 10^{-6} \text{ m}^2/\text{s}$ for dry soil at 300 K. Divide the depth into grid points separated by 0.1 m. The time step is given by

$$\Delta t = \frac{r \Delta x^2}{\alpha} = 0.8387r \text{---days}$$

We must decide how deep the calculation should go, that is, how many elements N must be in the vector $T(N \, \Delta x)$. Obviously, at a very large depth, one would see no change in temperature for the entire period of the calculation, 60 days. This suggests that we should choose N large enough so that $T(N \, \Delta x, t = 0) = T(N \, \Delta x, t = 60 \text{ days})$. To some degree this may be a trial-and-error process. For this problem $N = 50$ will be assumed.

To carry out the solution of Eq. (3.13) it is convenient to use two arrays, one T_j to represent the spatial temperature distribution at a time t, and another T'_j to represent the spatial temperature distribution at time $t + \Delta t$, that is, the left side of Eq. (3.13). The subscript j takes values of $1, 2, \ldots N = 50$ and denotes the depth by $x = \Delta x(j - 1)$.

Having assigned the desired value to r, a solution procedure might be:

1. $T_j = 20$ $j = 2, \ldots, 50$ Set the initial soil temperature

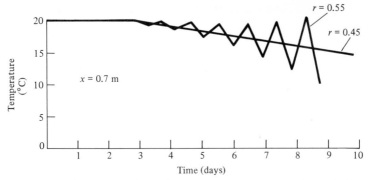

Figure 3.2 Illustration of a stable and an unstable numerical solution of a differential equation. Illustration for Example 3.2.

2. $T_1 = -15$	Set the surface temperature
3. $t = 0$	Initialize time
4. $t = t + \Delta t$	Increment time
5. $T'_j = rT_{j+1} + (1 - 2r)T_j + rT_{j-1}$ $\qquad j = 2, 3, \ldots, 49$	Calculate temperature distribution at the new value of time
6. $T_j = T'_j \qquad j = 2, \ldots, 50$	Replace the old distribution with the new one
7. print $t, T_j \qquad j = 1, 2, \ldots, 50$	Print the distribution for time t
8. go to step 4	Repeat loop until $t = 60$ days

The results of this procedure for $r = 0.45$ and $r = 0.55$ are shown in Fig. 3.2. For $r = 0.45$ the temperature at 0.7 m depth decreases smoothly with time and is within $\pm 0.1°C$ of the exact solution. For $t = 60$ days the temperature at this depth is $0.42°C$, just above freezing. The curve for $r = 0.55$ oscillates around the $r = 0.45$ curve with an increasing amplitude of oscillation. This latter solution is unstable.

By restricting $r \le 1/2$, we are limited in the size of time step we can use in the numerical solution, often to a time step much smaller than we would choose only on the basis of accuracy. This restriction is characteristic of explicit numerical solutions of parabolic partial differential equations like Eq. (3.7).

If Eq. (3.13) is modified so that it involves more than one value at $t + \Delta t$, the difference scheme is called an implicit scheme. That is, the temperature at a point is not completely determined by the temperature distribution at an earlier time. This leads to a more difficult solution; however, implicit finite-difference schemes yield a stable solution for any size of time step. This does not necessarily mean that the stable solution is the correct solution for arbitrarily large time steps.

If we evaluate the right-hand side of Eq. (3.8) at $t + \Delta t$ rather than at time t we get a fully implicit scheme:

$$T(x, t + \Delta t) - T(x, t) = r[T(x + \Delta x, t + \Delta t) - 2T(x, t + \Delta t)$$
$$+ T(x - \Delta x, t + \Delta t)] \tag{3.14}$$

which, like Eq. (3.13), is also first-order accurate in time and second-order accurate in space.

A method known as the *Crank-Nicholson scheme* replaces the right-hand side of Eq. (3.8) with the average of values at time t and time $t + \Delta t$:

$$T(x, t + \Delta t) - T(x, t) = \frac{r}{2}[T(x + \Delta x, t + \Delta t) - 2T(x, t + \Delta t) + T(x - \Delta x, t + \Delta t)$$
$$+ T(x + \Delta x, t) - 2T(x, t) + T(x - \Delta x, t)] \tag{3.15}$$

This scheme increases the accuracy in time to second order; space accuracy remains second order.

EXAMPLE 3.3

Consider a 1/2 in. thick aluminum plate undergoing an annealing process, Fig. 3.3. The slab is initially at $T_i = 100°F$ when the surfaces are suddenly reduced to $T_s = 0°F$. Determine the implicit finite-difference equations for the problem.

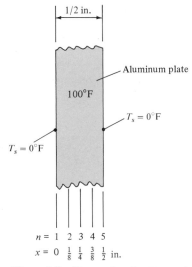

Figure 3.3 Illustration for Example 3.3.

Solution. Divide the plate into four elements, 1/8 in. thick, $x = (0.5 \text{ in.})$ $\times (n - 1)/4$, $n = 1, 2, 3, 4, 5$. From Eq. (3.14) the five simultaneous equations are

$n = 1$; $x = 0$:

$$T(0, t + \Delta t) - T(0, t) = 0$$

$n = 2$; $x = 0.125$ in.:

$$T(0.125, t + \Delta t) - T(0.125, t) = r[T(0.250, t + \Delta t) - 2T(0.125, t + \Delta t) + T(0, t + \Delta t)]$$

$n = 3$; $x = 0.250$ in.:

$$T(0.250, t + \Delta t) - T(0.250, t) = r[T(0.375, t + \Delta t) - 2T(0.250, t + \Delta t) + T(0.125, t + \Delta t)]$$

$n = 4$; $x = 0.375$ in.:

$$T(0.375, t + \Delta t) - T(0.375, t) = r[T(0.5, t + \Delta t) - 2T(0.375, t + \Delta t) + T(0.250, t + \Delta t)]$$

$n = 5$; $x = 0.5$ in.:

$$T(0.5, t + \Delta t) - T(0.5, t) = 0$$

The initial condition is $T(x, 0) = 100$. Note that the first and last equations simply express the boundary conditions.

It is clear that Eq. (3.15) leads to a system of N simultaneous equations, where N is the number of spatial grid points. It is not at all clear, however, how one should go about solving these equations, as Example 3.3 demonstrates.

In order to proceed with the solution of the set of equations generated in Example 3.3, it is best to write the system in matrix form. Replacing the numerical values of x with $n = 1, 2, 3, 4,$ or 5, a general form of each equation is

$$(1 + 2r)T(n, t + \Delta t) - rT(n + 1, t + \Delta t) - rT(n - 1, t + \Delta t) = T(n, t)$$

Forming one vector from the $T(n, t + \Delta t)$ and another from the $T(n, t)$, the matrix form of the above equation is

$$\begin{bmatrix} 1 & & & & \\ -r & 1+2r & -r & & \\ & -r & 1+2r & -r & \\ & & -r & 1+2r & -r \\ & & & & 1 \end{bmatrix} \begin{bmatrix} T(1, t + \Delta t) \\ T(2, t + \Delta t) \\ T(3, t + \Delta t) \\ T(4, t + \Delta t) \\ T(5, t + \Delta t) \end{bmatrix}$$

$$= \begin{bmatrix} 0 & & & & \\ & 1 & & & \\ & & 1 & & \\ & & & 1 & \\ & & & & 0 \end{bmatrix} \begin{bmatrix} T(1, t) \\ T(2, t) \\ T(3, t) \\ T(4, t) \\ T(5, t) \end{bmatrix} \qquad (3.16a)$$

For the Crank-Nicholson scheme, this would become

$$
\begin{bmatrix}
1 & & & & \\
-\dfrac{r}{2} & 1+r & -\dfrac{r}{2} & & \\
 & -\dfrac{r}{2} & 1+r & -\dfrac{r}{2} & \\
 & & -\dfrac{r}{2} & 1+r & -\dfrac{r}{2} \\
 & & & 1 &
\end{bmatrix}
\begin{bmatrix}
T(1, t + \Delta t) \\
T(2, t + \Delta t) \\
T(3, t + \Delta t) \\
T(4, t + \Delta t) \\
T(5, t + \Delta t)
\end{bmatrix}
$$

$$
=
\begin{bmatrix}
0 & & & & \\
\dfrac{r}{2} & 1-r & \dfrac{r}{2} & & \\
 & \dfrac{r}{2} & 1-r & \dfrac{r}{2} & \\
 & & \dfrac{r}{2} & 1-r & \dfrac{r}{2} \\
 & & & 0 &
\end{bmatrix}
\begin{bmatrix}
T(1, t) \\
T(2, t) \\
T(3, t) \\
T(4, t) \\
T(5, t)
\end{bmatrix}
\qquad (3.16b)
$$

Equation (3.16a or b) may be written in matrix notation

$$
\mathbf{A}\mathbf{T}(t + \Delta t) = \mathbf{B}\mathbf{T}(t) + \mathbf{C} \tag{3.17}
$$

where $\mathbf{T}(t)$ is the vector of temperatures at time t and is known. In general, the vector $\mathbf{C}$ will be needed to represent the boundary conditions. Given the initial conditions the right side of Eq. (3.17) can then be simplified to a known vector $\mathbf{D}$, giving

$$
\mathbf{A}\mathbf{T}(\Delta t) = \mathbf{D} \tag{3.18}
$$

The matrix $\mathbf{A}$ is called a tridiagonal matrix because only elements on the main diagonal and in adjacent diagonals are nonzero. This is significant because, as will be shown in Section 3.5, a system such as Eq. (3.18) with a tridiagonal matrix can be solved quite easily.

In deriving the matrix representation for a finite-difference equation it is helpful to visualize the computation scheme in x, t space by means of the computational grid. The computational grid shows the numerical weighting applied to each point in space-time associated with each application of the difference equation. Figure 3.4 shows the computational grid for Eq. (3.16b). Each row of matrix $\mathbf{A}$ consists of two zeros and the weighting factors shown in Fig. 3.4 at time $t + \Delta t$. Similarly, matrix $\mathbf{B}$ consists of two zeros and the weighting factors at time t. The computational grid shows that the zeros in the matrices $\mathbf{A}$ and $\mathbf{B}$ result because the temperature at any point in space is affected only by neighboring points.

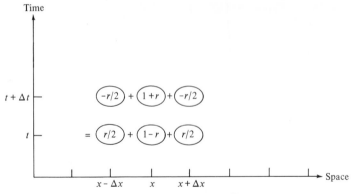

Figure 3.4   Computational grid for Eq. (3.16b).

Using the given initial value $T(x, 0) = 100°F$ from Example 3.3, we have

$$\mathbf{T}(0) = \begin{bmatrix} 100 \\ 100 \\ 100 \\ 100 \\ 100 \end{bmatrix} \qquad \mathbf{C} = \begin{bmatrix} 0 \\ 0 \\ 0 \\ 0 \\ 0 \end{bmatrix} \qquad (3.19)$$

We then use Eq. (3.18) to find $\mathbf{T}(\Delta t)$. We then replace $\mathbf{T}(0)$ by $\mathbf{T}(\Delta t)$ and find $\mathbf{T}(2\,\Delta t)$, $\mathbf{T}(3\,\Delta t)$, etc., in a similar fashion.

3.3.3 Multidimensional Unsteady Conduction

To find the difference equation for a transient, multidimensional problem, consider:

$$\rho c \frac{\partial T}{\partial t} = k \left(\frac{\partial^2 T}{\partial x^2} + \frac{\partial^2 T}{\partial y^2} \right) \qquad (3.20)$$

An expression like Eq. (3.6) may be derived for the y derivative, yielding

$$\frac{\partial^2 T}{\partial y^2}(x, y, t) = \frac{T(x, y + \Delta y, t) - 2T(x, y, t) + T(x, y - \Delta y, t)}{\Delta y^2} + O(\Delta y^2) \qquad (3.21)$$

The explicit finite-difference representation of Eq. (3.20) is then

$$\rho c [T(x, y, t + \Delta t) - T(x, y, t)]$$
$$= k\,\Delta t \left[\frac{T(x + \Delta x, y, t) - 2T(x, y, t) + T(x - \Delta x, y, t)}{\Delta x^2} \right.$$
$$\left. + \frac{T(x, y + \Delta y, t) - 2T(x, y, t) + T(x, y - \Delta y, t)}{\Delta y^2} \right] \qquad (3.22)$$

We again have the convenience of solving for the temperature at time $t + \Delta t$ with only known information (at time t). However, we also have the problem of stability. Equation (3.22) is stable only if

$$\alpha \, \Delta t \left(\frac{1}{\Delta x^2} + \frac{1}{\Delta y^2} \right) \leq \frac{1}{2} \tag{3.23}$$

which is even more restrictive than the one-dimensional equivalent. To demonstrate this let $\Delta x = \Delta y$ in Eq. (3.23); this gives $\alpha \, \Delta t \leq \Delta x^2/4$, compared to $\alpha \, \Delta t \leq \Delta x^2/2$ in one dimension.

The next logical step is to cast Eq. (3.22) in its implicit form. For example, the Crank-Nicholson scheme gives

$$
\begin{aligned}
T(x, y, t + \Delta t) - T(x, y, t) = \frac{r_x}{2} \big[& T(x + \Delta x, y, t + \Delta t) - 2T(x, y, t + \Delta t) \\
& + T(x - \Delta x, y, t + \Delta t) + T(x + \Delta x, y, t) \\
& - 2T(x, y, t) + T(x - \Delta x, y, t) \big] \\
+ \frac{r_y}{2} \big[& T(x, y + \Delta y, t + \Delta t) - 2T(x, y, t + \Delta t) \\
& + T(x, y - \Delta y, t + \Delta t) + T(x, y + \Delta y, t) \\
& - 2T(x, y, t) + T(x, y - \Delta y, t) \big]
\end{aligned} \tag{3.24}
$$

where $r_x = \alpha \, \Delta t / \Delta x^2$ and $r_y = \alpha \, \Delta t / \Delta y^2$. This approximation is second-order accurate in time and space. However, Eq. (3.24) is not usually solved as written. The matrix representation of Eq. (3.24) is no longer tridiagonal and this leads to excessive computation time or storage requirements. The approximation Eq. (3.24) can be modified to give a tridiagonal matrix by a method known as the Alternating Direction Implicit (ADI) method. The method produces tridiagonal matrices by breaking each time iteration into two half-steps (in a two-coordinate domain), each step comprised of only one directional derivative being evaluated at the new time step. For example, the pair of equations could be written

$$
\begin{aligned}
\frac{\rho c}{k} & \frac{T(x, y, t + \Delta t/2) - T(x, y, t)}{\Delta t/2} \\
& = \frac{T(x + \Delta x, y, t + \Delta t/2) - 2T(x, y, t + \Delta t/2) + T(x - \Delta x, y, t + \Delta t/2)}{\Delta x^2} \\
& \quad + \frac{T(x, y + \Delta y, t) - 2T(x, y, t) + T(x, y - \Delta y, t)}{\Delta y^2}
\end{aligned}
$$

$$
\begin{aligned}
\frac{\rho c}{k} & \frac{T(x, y, t + \Delta t) - T(x, y, t + \Delta t/2)}{\Delta t/2} \\
& = \frac{T(x + \Delta x, y, t + \Delta t/2) - 2T(x, y, t + \Delta t/2) + T(x - \Delta x, y, t + \Delta t/2)}{\Delta x^2} \\
& \quad + \frac{T(x, y + \Delta y, t + \Delta t) - 2T(x, y, t + \Delta t) + T(x, y - \Delta y, t + \Delta t)}{\Delta y^2}
\end{aligned} \tag{3.25}
$$

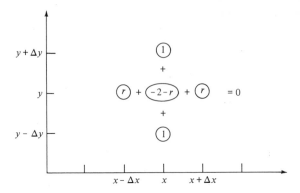

Figure 3.5 Computational grid for Eq. (3.26).

Equation (3.25) is stable for any time step and is second-order accurate in time and space. Organization of these two equations into their matrix form is left as an exercise (see Problem 3.27).

For the two-dimensional steady problem, Eq. (3.22) simplifies to

$$\frac{T(x + \Delta x, y) - 2T(x, y) + T(x - \Delta x, y)}{\Delta x^2}$$

$$+ \frac{T(x, y + \Delta y) - 2T(x, y) + T(x, y - \Delta y)}{\Delta y^2} = 0 \qquad (3.26)$$

If we let $r = (\Delta y / \Delta x)^2$, the computational grid is shown in Fig. 3.5.

EXAMPLE 3.4 _____

Consider steady conduction in a square domain $0 \le x \le 1, 0 \le y \le 1$, Fig. 3.6. Let three of the boundaries be held at temperature T_a and the

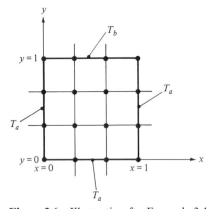

Figure 3.6 Illustration for Example 3.4.

remaining boundary be held at T_b. Determine the matrix representation of the finite-difference equation.

Solution. In practice, the number of grid points would probably be at least 121 ($\Delta x = \Delta y = 0.1$), but such a large matrix is difficult to show here. Letting $\Delta x = \Delta y = 1/3$ produces 16 grid points and a matrix of 16×16 elements. This is sufficient to demonstrate the pattern to allow one to write a computer program for a larger matrix. Assuming that the top surface is at T_b, the grid would be as shown in Fig. 3.6.

The computational grid for Eq. (3.26) is given in Fig. 3.5 and simplifies the construction of the matrix. Since only four grid points are not at boundaries, the computational grid will apply only at these four points. Temperatures at the remaining 12 grid points are fixed at the specified boundary temperatures.

Let any grid point be identified by $x = (i - 1)\Delta x$ and $y = (j - 1)\Delta y$, $i, j = 1, 2, 3, 4$. We then construct a vector of the temperatures $T(i, j)$ and apply the computational grid at the four interior points (2, 2), (3, 2), (2, 3), and (3, 3). The other 12 equations establish the boundary conditions:

$$
\begin{bmatrix}
1 \\
 & 1 \\
 & & 1 \\
 & & & 1 \\
 & & & & 1 \\
1 & & & & 1 & -4 & 1 & & & 1 \\
 & 1 & & & & 1 & -4 & 1 & & & 1 \\
 & & & & & & & 1 \\
 & & & & & & & & 1 \\
 & & & & & 1 & & & 1 & -4 & 1 & & & 1 \\
 & & & & & & 1 & & & 1 & -4 & 1 & & & 1 \\
 & & & & & & & & & & & 1 \\
 & & & & & & & & & & & & 1 \\
 & & & & & & & & & & & & & 1 \\
 & & & & & & & & & & & & & & 1 \\
 & & & & & & & & & & & & & & & 1
\end{bmatrix}
\begin{bmatrix}
T(1,1) \\
T(2,1) \\
T(3,1) \\
T(4,1) \\
T(1,2) \\
T(2,2) \\
T(3,2) \\
T(4,2) \\
T(1,3) \\
T(2,3) \\
T(3,3) \\
T(4,3) \\
T(1,4) \\
T(2,4) \\
T(3,4) \\
T(4,4)
\end{bmatrix}
=
\begin{bmatrix}
T_a \\
T_a \\
T_a \\
T_a \\
T_a \\
0 \\
0 \\
T_a \\
T_a \\
0 \\
0 \\
T_a \\
T_a \\
T_b \\
T_b \\
T_a
\end{bmatrix}
$$

$$\tag{3.27}$$

Note that we have set the top corner temperatures ($j = 4$, $i = 1$ and 4) to T_a. This seems to violate the boundary condition for the top surface. The temperature at these corners is actually indeterminate. However, the assumption $T = T_a$ for these corners will not seem so implausible when we increase the number of grid points.

3.4 BOUNDARY CONDITIONS

To completely formulate the numerical solution of a differential equation we must consider the boundary conditions. For simplicity, the examples presented in previous sections utilized isothermal boundary conditions in rectangular domains. We must now consider boundary conditions other than for the simple isothermal case and for more complex geometries. We will do so in a formal way for a few simple problems to provide a methodology for handling more complex problems.

Consider first a one-dimensional, steady problem with a prescribed heat flux q'' at the boundary $x = 0$, Fig. 3.7. We wish to express the nodal temperatures in terms of the prescribed surface heat flux. Setting the left-hand side of Eq. (3.7) to zero and integrating the resulting differential equation from $-\Delta x/2$ to 0 yields

$$\frac{dT}{dx}(0) - \frac{dT}{dx}\left(\frac{-\Delta x}{2}\right) = 0 \tag{3.28}$$

We know from the prescribed heat flux $(dT/dx)(0) = -q/k$. Replacing the second term in Eq. (3.28) with the finite-difference representation, Eq. (3.5), we find

$$-\frac{q''}{k} - \frac{T(0) - T(-\Delta x)}{2(\Delta x/2)} = 0$$

or

$$T(0) = T(-\Delta x) - \frac{q'' \Delta x}{k} \tag{3.29}$$

By integrating from $-\Delta x/2$ to 0 we can evaluate the temperature gradient at $-\Delta x/2$ in terms of the temperature at node points via Eq. (3.5), a centered-difference approximation. Although this is a straightforward mathematical exercise, it has a reasonable physical interpretation. The surface temperature is a good representation of the half-node from $-\Delta x/2$ to 0 just as the other node temperatures represent a region of thickness Δx centered on each node. Since the energy content of this half-node is constant, the flux entering the left face must equal that leaving the right face, according to Eq. (3.28). The flux entering the left face can be evaluated from the node temperatures $x = 0$, $x = -\Delta x$.

Boundary condition examples include an adiabatic wall, $q'' = 0$, for which

$$T(0) = T(-\Delta x) \tag{3.30}$$

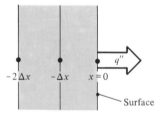

Figure 3.7 Illustration of prescribed heat flux boundary condition.

or a wall exposed by convective heat transfer to a fluid at temperature T_f,

$$q'' = h_c[T(0) - T_f]$$ (3.31)

From Eq. (3.29) this may be written

$$T(0) = \frac{T(-\Delta x) + (h_c \Delta x/k) T_f}{1 + h_c \Delta x/k}$$ (3.32)

Note that as $h_c \Delta x/k \to \infty$ the surface temperature is forced to equal the fluid temperature T_f. Also, as $h_c \Delta x/k \to 0$ we have the adiabatic condition $T(0) = T(-\Delta x)$.

Rather than specifying $T(x = 1) = T_b$ in the problem represented by Eq. (3.10), suppose we incorporate a convective heat transfer condition there. This would modify Eq. (3.10) to

$$\begin{bmatrix} 1 & & & & \\ 1 & -2 & 1 & & \\ & 1 & -2 & 1 & \\ & & 1 & -2 & 1 \\ & & -1 & 1 + \dfrac{h_c \Delta x}{k} \end{bmatrix} \begin{bmatrix} T(0) \\ T(\Delta x) \\ T(2\,\Delta x) \\ T(3\,\Delta x) \\ T(1) \end{bmatrix} = \begin{bmatrix} T_a \\ 0 \\ 0 \\ 0 \\ \dfrac{h_c \Delta x}{k} T_f \end{bmatrix}$$ (3.33)

which would be solved in exactly the same way as Eq. (3.10).

For a transient, one-dimensional problem we must consider the thermal capacitance near the boundary. Integrating Eq. (3.7) from $-\Delta x/2$ to 0 we find

$$\frac{\rho c}{k} \frac{\partial}{\partial t} \int_{-\Delta x/2}^{0} T(x, t)\, dx = \frac{\partial T}{\partial x}(0, t) - \frac{\partial T}{\partial x}\left(-\frac{\Delta x}{2}, t\right)$$ (3.34)

We have interchanged the order of integration and differentiation in the left side of Eq. (3.34). In the domain $-\Delta x/2 \le x \le 0$ we write the integrand as

$$T(x, t) = T(0, t) + \frac{\partial T}{\partial x}(0, t)x + \cdots$$

Performing the integration yields

$$\int_{-\Delta x/2}^{0} T(x, t)\, dx = T(0, t)\frac{\Delta x}{2} - \frac{\partial T}{\partial x}(0, t)\left(\frac{\Delta x}{2}\right)^2 \frac{1}{2} + \cdots$$

We use Eq. (3.5) for the last term in Eq. (3.34), that is,

$$\frac{\partial T}{\partial x}\left(\frac{-\Delta x}{2}, t\right) = \frac{T(0, t) - T(-\Delta x, t)}{\Delta x} + O(\Delta x^2)$$

After deleting terms $O(\Delta x^2)$ and smaller, Eq. (3.34) becomes

$$\frac{\rho c}{k} \frac{\partial}{\partial t}\left[\frac{\Delta x}{2} T(0, t)\right] = \frac{-q''(t)}{k} - \frac{T(0, t) - T(-\Delta x, t)}{\Delta x}$$ (3.35)

Using a forward difference for the time derivative on the left side of Eq. (3.35) gives

$$\frac{\rho c \, \Delta x}{k \quad 2} \left[\frac{T(0, t + \Delta t) - T(0, t)}{\Delta t} \right] = \frac{-q''(t)}{k} - \frac{T(0, t) - T(-\Delta x, t)}{\Delta x}$$

which may be rearranged to give the desired boundary temperature:

$$T(0, t + \Delta t) = T(0, t) \left[1 - 2 \frac{k}{\rho c} \frac{\Delta t}{\Delta x^2} \right] + 2 \frac{k}{\rho c} \frac{\Delta t}{\Delta x^2} T(-\Delta x, t) - 2 \frac{\Delta t}{\rho c \, \Delta x} q''(t)$$

$$(3.36)$$

Calculation of the desired boundary temperature for the adiabatic boundary or the boundary exposed to convective or radiative heat transfer can be easily derived from Eq. (3.36). An implicit version of Eq. (3.36) would be

$$\frac{\rho c \, \Delta x}{k \quad 2} \left[\frac{T(0, t + \Delta t) - T(0, t)}{\Delta t} \right] = \frac{-q''(t + \Delta t)}{k} - \frac{T(0, t + \Delta t) - T(-\Delta x, t + \Delta t)}{\Delta x}$$

$$(3.37)$$

The expression for the boundary temperature can then be incorporated into Eq. (3.16a) or (3.16b) in a straightforward manner.

EXAMPLE 3.5 ——

Find the matrix representing the difference equations for Example 3.3 if the boundary condition at one surface of the aluminum sheet is replaced with $q'' = 0$, that is, an insulated face, Fig. 3.8.

Solution. For the boundary condition Eq. (3.37) gives

$$T(0.5, t + \Delta t) = \frac{T(0.5, t)}{1 + 2r} + \frac{2r}{1 + 2r} T(0.375, t + \Delta t) \qquad (3.38)$$

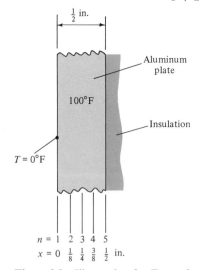

Figure 3.8 Illustration for Example 3.5.

Eq. (3.16b) becomes

$$
\begin{bmatrix}
\vdots & \vdots \\
\dfrac{-2r}{1+2r} & 1
\end{bmatrix}
\begin{bmatrix}
\vdots \\
T(4, t+\Delta t) \\
T(5, t+\Delta t)
\end{bmatrix}
=
\begin{bmatrix}
\vdots & \vdots \\
\dfrac{1}{1+2r}
\end{bmatrix}
\begin{bmatrix}
\vdots \\
T(4, t) \\
T(5, t)
\end{bmatrix}
\qquad (3.39)
$$

We have only shown the portions of Eq. (3.16a) that have changed.

The above procedure may be used to develop boundary condition expressions in two dimensions. For example, consider the plane wall in Fig. 3.7. A two-dimensional grid is shown in Fig. 3.9.

We could integrate Eq. (3.20) from $-\Delta x/2 \leq x \leq 0$ and $-\Delta y/2 \leq y \leq \Delta y/2$, that is, the shaded area in Fig. 3.9. However, an alternate procedure may be less cumbersome for multidimensional systems. If we consider that the temperature at $x = y = 0$ represents that of the shaded area in Fig. 3.9, then the time rate of change of energy in that area is $(\rho c\,\Delta y\,\Delta x/2)\,(\partial/\partial t)T(0, 0, t)$, which must equal the net flux across the four boundaries of the shaded area. This yields

$$
\underbrace{\frac{\rho c\,\Delta x\,\Delta y}{2}\frac{\partial}{\partial t}T(0, 0, t)}_{\text{heat stored}} = \underbrace{-k\,\Delta y\,\frac{\partial T}{\partial x}\left(\frac{-\Delta x}{2}, 0, t\right)}_{\substack{\text{conduction from} \\ \text{left}}} - \underbrace{q''(0, t)\,\Delta y}_{\substack{\text{specified heat} \\ \text{flux at surface}}}
$$

$$
+ \underbrace{k\,\frac{\Delta x}{2}\frac{\partial T}{\partial y}\left(0, \frac{\Delta y}{2}, t\right)}_{\substack{\text{conduction from} \\ \text{above}}} - \underbrace{k\,\frac{\Delta x}{2}\frac{\partial T}{\partial y}\left(0, \frac{-\Delta y}{2}, t\right)}_{\substack{\text{conduction from} \\ \text{below}}} \qquad (3.40)
$$

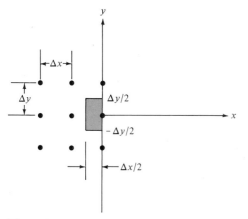

Figure 3.9 Boundary condition at a plane wall, two dimensions.

For each of the three interior faces, the spatial gradient has been evaluated at the point on the face crossing an axis. The specified heat flux $q''(y, t)$ was evaluated at $y = 0$ as if it acted over an area Δy of the face. Using a forward difference in time and centered differences for spatial gradients we find

$$
\begin{aligned}
T(0, 0, t + \Delta t) - T(0, 0, t) = 2\, \frac{k}{\rho c}\, \frac{\Delta t}{\Delta x^2} \Big\{ &T(-\Delta x, 0, t) - T(0, 0, t) \\
&- \frac{q''(0, t)\Delta x}{k} + \frac{1}{2}\left(\frac{\Delta x}{\Delta y}\right)^2 [T(0, \Delta y, t) \\
&- 2T(0, 0, t) + T(0, -\Delta y, t)] \Big\}
\end{aligned} \tag{3.41}
$$

Again using $r = \Delta t k/(\rho c\, \Delta x^2)$ and letting $\beta \equiv (\Delta x/\Delta y)^2$, this reduces to

$$
\begin{aligned}
T(0, 0, t + \Delta t) - T(0, 0, t) = &-2r\Delta x\, \frac{q''(0, t)}{k} - 2r(1 + \beta)T(0, 0, t) + 2rT(-\Delta x, 0, t) \\
&+ \beta r T(0, \Delta y, t) + \beta r T(0, -\Delta y, t)
\end{aligned} \tag{3.42}
$$

For the exterior corner, Fig. 3.10, we find

$$
\begin{aligned}
T(0, 0, t + \Delta t) - T(0, 0, t) = &\frac{-2r}{k} [\beta q_y(0, t)\Delta y + q''_x(0, t)\Delta x] - 2r(1 + \beta)T(0, 0, t) \\
&+ 2rT(-\Delta x, 0, t) + 2\beta r T(0, -\Delta y, t)
\end{aligned} \tag{3.43}
$$

The specified flux at the x axis is $q''_y(x, t)$ and the specified flux at the y axis is $q''_x(y, t)$.

For an interior corner, Fig. 3.11

$$
\begin{aligned}
T(0, t + \Delta t) - T(0, 0, t) = &-\frac{2}{3}\frac{r}{k}[-\beta q_y(0, t)\Delta y + q_x(0, t)\Delta x] \\
&- 2r(1 + \beta)T(0, 0, t) + \frac{4}{3}rT(-\Delta x, 0, t) \\
&+ \frac{2}{3}rT(\Delta x, 0, t) \\
&+ \frac{4}{3}\beta r T(0, \Delta y, t) + \frac{2}{3}\beta r T(0, -\Delta y, t)
\end{aligned} \tag{3.44}
$$

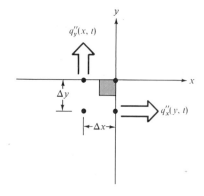

Figure 3.10 Boundary condition at an exterior corner.

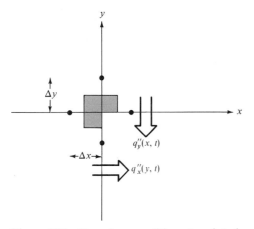

Figure 3.11 Boundary condition at an interior corner.

Implicit forms of Eqs. (3.42), (3.43), and (3.44) may be derived by evaluating the right side of each equation at $t + \Delta t$ instead of t. Derivation of the boundary condition at a curved boundary is left as an exercise (see Problem 3.10).

3.5 SOLUTION OF THE DIFFERENCE EQUATIONS

3.5.1 Introduction

In Section 3.3 we saw that difference equations for several problems could be cast in the form of matrix equations representing a system of simultaneous equations. These equations arise because the temperature at any grid point depends on the temperature at a few neighboring grid points. The matrix equations may be solved by inverting the coefficient matrix and then carrying out any required matrix multiplication. For example, the solution of Eq. (3.18) would be

$$\mathbf{T}(\Delta t) = \mathbf{A}^{-1}\mathbf{D} \tag{3.45}$$

Most computers have subroutines for matrix inversion and multiplication, and in some cases this will be the most direct route to the solution. For large systems, matrix inversion is not economical from the standpoint of storage requirements or computing time. If the matrix $\mathbf{A}$ has $N \times N$ (where N is large) elements, all nonzero, $N^3 + N^2$ multiplications are required to invert $\mathbf{A}$ and multiply $\mathbf{A}^{-1}$ by $\mathbf{D}$. A direct elimination method to be discussed in the next section requires only $N^3/3 + N^2$ multiplications. For matrices with some elements equal to zero (sparse matrices) the argument against matrix inversion is stronger.

Besides matrix inversion, two methods for solving the set of simultaneous equations are elimination methods and iterative methods. In elimination, the set of equations is modified by a series of operations so that ultimately one

TABLE 3.1 COMPARISON OF THREE SOLUTION METHODS

Solution method	Storage required	Number of operations
Matrix inversion	N^2	$N^3 + N^2$
Elimination		
Full matrix	N^2	$N^3/3 + N^2$
Tridiagonal	$3N$	$8N$
Iteration	N^2	$2N^2 - N$ per iteration

equation has only one unknown, one equation has two unknowns (one of which is the unknown from the first equation), and so on. The equations are solved by starting with the simplest equation and working through the set of equations in order of increasing number of unknowns. Iterative schemes involve guessing at the solution and iteratively generating more and more accurate solutions. Table 3.1 compares each solution method on the basis of the required number of operations and storage for a system with N unknowns.

3.5.2 Elimination

The most straightforward elimination technique is called Gaussian elimination and is the method generally used for solution of a few simultaneous equations. The procedure is best shown in an example.

EXAMPLE 3.6 ───

An intermediate-stage power amplifier that dissipates 3 watts is tested by operating it in a closed chamber to eliminate radio-frequency interference, Fig. 3.12. To eliminate any electrical noise generated by a fan, passive cooling is provided by conduction along two copper rods passing through the chamber. One rod ends in another testing chamber, and the end of the other rod is exposed to ambient air. A third copper rod connects the second testing chamber to ambient. The three copper rods are 1 cm in diameter, 15 cm long, and are soldered together to form a triangle. The surfaces of the rods are insulated, except near the ends.

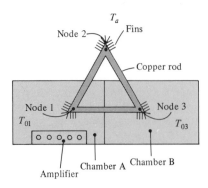

Figure 3.12 Illustration for Example 3.6.

Cooling fins are provided at each apex of the triangle to provide an effective ⌄ooling area of 100 cm². Assume a thermal conductivity of 382 W/m °C for the copper and a convective conductance of 10 W/m² °C at the cooling fins. Can the testing chamber be kept below 100°C if the second chamber is kept open to ambient at 20°C? Can the testing chamber be kept below 100°C if a second amplifier is tested simultaneously in the second chamber? Assume steady-state operation.

Solution. Let each apex of the triangle be considered a node numbered according to the diagram. A heat balance on each node gives

node 1:

$$\frac{kA_\rho}{l}(T_1 - T_3) + \frac{kA_\rho}{l}(T_1 - T_2) = hA_3(T_{01} - T_1)$$

node 2:

$$\frac{kA_\rho}{l}(T_2 - T_1) + \frac{kA_\rho}{l}(T_2 - T_3) = hA_3(T_a - T_2)$$

node 3:

$$\frac{kA_\rho}{l}(T_3 - T_1) + \frac{kA_\rho}{l}(T_3 - T_2) = hA_3(T_{03} - T_3)$$

where $\quad k$ = thermal conductivity of copper rod, 382 W/m °C

$\quad\quad A_\rho$ = cross-sectional area of copper rod, 0.785 cm²

$\quad\quad l$ = length of copper rod, 15 cm

$\quad\quad h$ = convective conductance at fins, 10 W/m² °C

$\quad\quad A_3$ = effective fin surface area, 100 cm²

$\quad\quad T_{01}$ = temperature in chamber A, °C

$\quad\quad T_a$ = ambient temperature, °C

$\quad\quad T_{03}$ = temperature in chamber B, °C

T_1, T_2, T_3 = node temperatures, °C

Substituting these values into the three equations and rearranging produces three equations in the three unknown temperatures:

$$5T_1 - 2T_2 - 2T_3 = T_{01}$$
$$-2T_1 + 5T_2 - 2T_3 = T_a \qquad (3.46)$$
$$-2T_1 - 2T_2 + 5T_3 = T_{03}$$

We may now use Gaussian elimination on this set of equations.
Subtract the third equation from the second:

$$5T_1 - 2T_2 - 2T_3 = T_{01}$$
$$0 + 7T_2 - 7T_3 = T_a - T_{03} \qquad (3.47)$$
$$-2T_1 - 2T_2 + 5T_3 = T_{03}$$

Add twice the first equation to five times the third equation:

$$5T_1 - 2T_2 - 2T_3 = T_{01}$$
$$7T_2 - 7T_3 = T_a - T_{03} \tag{3.48}$$
$$-14T_2 + 21T_3 = 2T_{01} + 5T_{03}$$

Add twice the second equation to the third equation:

$$5T_1 - 2T_2 - 2T_3 = T_{01}$$
$$7T_2 - 7T_3 = T_a - T_{03} \tag{3.49}$$
$$7T_3 = 2T_{01} + 2T_a + 3T_{03}$$

We may now solve the equations in reverse order:

$$T_3 = \frac{1}{7}(2T_{01} + 2T_a + 3T_{03})$$

$$T_2 = \frac{1}{7}(2T_{01} + 3T_a + 2T_{03})$$

$$T_1 = \frac{1}{7}(3T_{01} + 2T_a + 2T_{03})$$

In the first case, where only one amplifier is under test, $q = hA_3(T_{01} - T_1) = 3.0$ W. This gives $T_1 = T_{01} - 30$. Inserting this into the solution for T_1 we find $T_{01} = 72.5°$C. Therefore, the test chamber can be kept below $100°$C.

In the second case, where both amplifiers are under test, set $T_{01} = T_{03} = 100°$C and $T_a = 20°$C, giving $T_1 = 77.14°$C. The rate of heat rejection is only $q = 2.29$ W. Therefore the test chambers cannot be kept below $100°$C.

We now generalize the procedure presented in Example 3.6 (2). For a system of N equations in N unknowns we have

$$a_{11}x_1 + a_{12}x_2 + \cdots + a_{1N}x_N = d_1$$
$$a_{21}x_1 + a_{22}x_2 + \cdots + a_{2N}x_N = d_2 \tag{3.50}$$
$$\vdots \qquad \vdots \qquad \qquad \vdots$$
$$a_{N1}x_1 + a_{N2}x_2 + \cdots + a_{NN}x_N = d_N$$

First we eliminate the x_1 term in each equation except the first. We do this by subtracting a_{n1}/a_{11} times the first equation from the nth equation. This produces a new set of equations, which we denote by using a superscript (1):

$$a_{11}x_1 + a_{12}x_2 + \cdots + a_{1N}x_N = d_1$$
$$a_{22}^{(1)}x_2 + \cdots + a_{2N}^{(1)}x_N = d_2^{(1)}$$
$$\vdots \qquad \vdots \qquad \qquad \vdots$$
$$a_{N2}^{(1)}x_2 + \cdots + a_{NN}^{(1)}x_N = d_N^{(1)}$$

where

$$a_{ij}^{(1)} = a_{ij} - a_{1j}\frac{a_{i1}}{a_{11}} \qquad d_i^{(1)} = d_i - d_1\frac{a_{i1}}{a_{11}} \qquad i, j = 2, 3, \ldots, N$$

The superscript indicates that one elimination operation has been performed on the row.

We now repeat this step to eliminate the x_2 term in each equation except the first and second. This is done by subtracting $a_{n2}^{(1)}/a_{22}^{(1)}$ times the second equation from the nth equation. The procedure is repeated $N - 1$ times to produce a triangular system:

$$
\begin{aligned}
a_{11}x_1 + a_{12}x_2 + \cdots + a_{1N}x_N &= d_1 \\
a_{22}^{(1)}x_2 + \cdots + a_{2N}^{(1)}x_N &= d_2^{(1)} \\
&\;\;\vdots \\
a_{NN}^{(N-1)}x_N^{(N-1)} &= d_N^{(N-1)}
\end{aligned}
\tag{3.51}
$$

where

$$a_{ij}^{(k+1)} = a_{ij}^{(k)} - a_{k+1,j}^{(k)} \frac{a_{i,k+1}^{(k)}}{a_{k+1,k+1}^{(k)}}$$

$$d_i^{(k+1)} = d_i^{(k)} - d_{k+1}^{(k)} \frac{a_{i,k+1}^{(k)}}{a_{k+1,k+1}^{(k)}}$$

$$i, j = k + 2, k + 3, \ldots, N$$

We now solve the equations starting with the Nth equation:

$$
\begin{aligned}
x_N &= \frac{d_N^{(N-1)}}{a_{NN}^{(N-1)}} \\
x_{N-1} &= \frac{d_{N-1}^{(N-2)} - a_{N-1}x_N}{a_{N-1,N-1}^{(N-2)}} \\
&\;\;\vdots \\
x_1 &= \frac{d_1 - a_{12}x_2 - \cdots - a_{1N}x_N}{a_{11}}
\end{aligned}
\tag{3.52}
$$

Note that the procedure assumes that all divisors are nonzero. Equations can always be interchanged (for a nonsingular system) to assure that this condition is met.

Tridiagonal matrices such as **A** in Eq. (3.16) arise because the temperature at any interior node point depends only on the temperature at adjacent node points. These systems can be solved by the preceding elimination method; however, the reduction procedure can be simplified significantly relative to that for full matrices. In some mainframe computers, algorithms are available that have been optimized to take advantage of tridiagonal systems. Such a procedure is given in Appendix 2. Efficient elimination algorithms for sparse matrices other than tridiagonal matrices are presented in (2, 4).

3.5.3 Iteration

Systems of equations more densely populated than tridiagonal systems can often be solved economically by iteration. Storage is minimized since only the coefficients of each equation must be stored. In an iteration procedure, we first guess the solution and then use the algebraic equations to compute a better guess. Repeating the procedure generally produces a more accurate solution with each iteration. If the coefficients of the N simultaneous equations are all nonzero, then an iterative solution will be more efficient than an elimination solution if satisfactory convergence can be achieved in fewer than $N/3$ iterations.

The Jacobi iterative method uses the first equation in the system of equations to solve for the first unknown by using values of unknowns determined from the previous step. Gauss-Seidel iteration is similar except that values of unknowns are used as soon as they become available. The advantages of Gauss-Seidel iteration relative to Jacobi iteration are somewhat faster convergence and smaller storage requirements.

An initial guess is necessary to begin the first iteration. Obviously, the better the first guess, the more quickly one will converge to the solution. Any intuition that can be applied to generate a good first guess will help. For example, in a conduction problem one can easily construct a linear temperature variation between specified boundary temperatures.

The procedure for either iteration method may be generalized as follows. Starting with the system Eq. (3.50), the kth iteration ($k = 0$ is the initial guess) of x_i, that is, $x_i^{(k)}$, is

$$x_i^{(k)} = \frac{d_i - a_{i1}x_1^{(k)} - \cdots - a_{i,i-1}x_{i-1}^{(k)} - a_{i,i+1}x_{i+1}^{(k-1)} - \cdots - a_{i,N}x_N^{(k-1)}}{a_{ii}} \tag{3.53}$$

for Gauss-Seidel iteration. For Jacobi iteration, all x_i on the right side of Eq. (3.53) would be evaluated from values determined at step $k - 1$. Equation (3.53) simply states that the ith unknown is determined from the ith equation by substituting known values for all remaining unknowns.

Convergence of either the Gauss-Seidel or the Jacobi iteration is guaranteed if

$$|a_{ii}| \geq \sum_{j=1}^{N} |a_{ij}| \qquad i = 1, 2, \ldots, N \quad \text{with} \quad j \neq i \tag{3.54}$$

If this condition is met, the matrix of elements a_{ij} is said to be diagonally dominant. The condition can generally be met by rearranging the order of equations. The iteration may converge, however, even if the a_{ij} are not diagonally dominant.

EXAMPLE 3.7

Consider the system of equations developed in Example 3.6 for the case $T_{01} = 100$, $T_a = 20$, $T_{03} = 20$.

$$5T_1 - 2T_2 - 2T_3 = 100$$
$$-2T_1 + 5T_2 - 2T_3 = 20$$
$$-2T_1 - 2T_2 + 5T_3 = 20$$

According to Eq. (3.53) we rewrite these as

$$T_1^{(k)} = \frac{1}{5}(100 + 2T_2^{(k-1)} + 2T_3^{(k-1)})$$

$$T_2^{(k)} = \frac{1}{5}(20 + 2T_1^{(k)} + 2T_3^{(k-1)})$$

$$T_3^{(k)} = \frac{1}{5}(20 + 2T_1^{(k)} + 2T_2^{(k)})$$

A reasonable initial guess would be $T_2^{(0)} = T_3^{(0)} = 20$. We use these values in the first equation to find $T_1^{(1)}$, which we substitute in the second equation to find $T_2^{(1)}$ and so forth:

Iteration	$T_1^{(k)}$	$T_2^{(k)}$	$T_3^{(k)}$
$k = 0$	—	20.00	20.00
1	36.00	26.40	28.96
2	42.14	32.44	33.83
3	46.51	36.14	37.06
4	49.28	38.54	39.13
5	51.06	40.08	40.46
Exact solution (Elimination method)	54.29	42.86	42.86

The error in calculated heat flux would be about 7 percent if we stop after five iterations. Had we chosen a poorer first guess such as $T_2^{(0)} = T_3^{(0)} = 0$ the first five iterations would be

Iteration	$T_1^{(k)}$	$T_2^{(k)}$	$T_3^{(k)}$
$k = 0$	—	0	0
1	20.00	12.00	16.80
2	24.80	20.64	22.18
3	37.13	27.72	29.94
4	43.06	33.20	34.51
5	47.08	36.64	37.49

which is about two iterations "behind" our previous attempt with the better initial guess.

For systems of practical size, that is, $N \gg 3$, Gauss-Seidel and Jacobi iteration are rarely used because convergence is too slow. It is possible, however, to accelerate the convergence of either method. Note from Eq. (3.53) that we produce $x_i^{(k)}$ by adding a correction to $x_i^{(k-1)}$

$$x_i^{(k)} = x_i^{(k-1)}$$
$$+ \left(\frac{d_i - a_{i1}x_1^{(k)} - \cdots - a_{i,i-1}x_{i-1}^{(k)} - a_{i,i+1}x_{i+1}^{(k-1)} - \cdots - a_{i,N}x_N^{(k-1)}}{a_{ii}} - x_i^{(k-1)} \right)$$

$$\tag{3.55}$$

If we "amplify" the correction by multiplying the term in parentheses by ω ($1 < \omega < 2$), we will accelerate convergence. If $\omega < 1$ we decelerate convergence. When $\omega = 1$, we have Eq. (3.53) again. The method of accelerating Gauss-Seidel iteration is known as successive over-relaxation (SOR). Under-relaxation ($\omega < 1$) is often used in iterative solutions of nonlinear equations. For convergence, it is necessary that $\omega < 2$, but there is no general procedure for choosing the amplification factor ω.

EXAMPLE 3.8 ───────────────────────────────────────

Condensing steam is to be used to maintain a room at 20°C. The steam flows through pipes maintaining the pipe surface at 100°C. To increase heat transfer from the pipes, stainless steel pin fins, 10 cm long and 0.5 cm in diameter, are welded to the pipe surface, Fig. 3.13. A fan forces room air over the pipe and fins, giving a heat transfer coefficient of 50 W/m² °C at the base of the fins. However, because of the air flow distribution, the heat transfer coefficient increases to 75 W/m² °C at the fin tip. The variation of heat transfer coefficient may be approximated as parabolic from the base to the tip of the fin. Calculate the rate of heat dissipation of each fin, using the SOR method to determine the fin temperature distribution. Determine the effect of $\omega = 1.0, 1.25$, and 1.50 on accelerating convergence.

Solution. An energy balance at any location x in the fin gives

$$\frac{\pi d_f^2}{4} k_f \frac{d^2 T}{dx^2} = (T - T_\infty) h(x) \pi d_f$$

From the diagram, divide the fin length into seven equal-length segments, $\Delta x = 10/7$ cm, with node 1 at the base of the fin and node 8 at the tip of the fin.

In finite-difference form the energy balance is

$$k_f \frac{T_{j+1} - 2T_j + T_{j-1}}{\Delta x^2} = (T_j - T_\infty) h_j \frac{4}{d_f} \qquad j = 2, 3, \ldots, 7$$

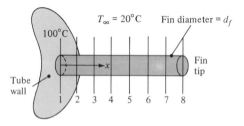

Figure 3.13 Illustration for Example 3.8.

Define $\theta_j = T_j - T_\infty$; then

$$\theta_{j+1} - 2\theta_j + \theta_{j-1} = \theta_j\left(\frac{4h_j\Delta x^2}{d_f k_f}\right)$$

for nodes 2 through 7. The heat transfer coefficient at node j that fits the prescribed parabolic form is

$$h_j = h_0\left\{1 + \frac{[(j-1)10/7]^2}{200}\right\}$$

where h_0 is the value at the base, that is, $h_0 = 50$ W/m^2 °C. At the tip of the fin, node $j = 8$, an energy balance is

$$-\frac{\pi d_f^2}{4}k_f\frac{dT}{dx}\bigg|_{\text{node 8}} = \frac{3}{2}h_0(T_8 - T_\infty)\left(\frac{\pi d_f\Delta x}{2} + \frac{\pi d_f^2}{4}\right)$$

The last term in parentheses is the surface area associated with node 8. Using a backward-difference approximation for the left side of this equation we find

$$T_7 - T_8 = \frac{6h_0\Delta x}{\pi d_f^2 k_f}(T_8 - T_\infty)\left(\frac{\pi d_f\Delta x}{2} + \frac{\pi d_f^2}{4}\right)$$

or

$$\theta_8 = \frac{\theta_7}{1 + (6h_0\Delta x/k_f)(\Delta x/2d_f + 1/4)}$$

Using a thermal conductivity for stainless steel, $k_f = 21$ W/m °C we have

$$\theta_1 = 80$$

$$\theta_j = \frac{\theta_{j+1} + \theta_{j-1}}{2 + C_1[1 + (j-1)^2 C_2]} \qquad \begin{array}{l} j = 2, 3, \ldots, 7 \\ C_1 = 0.3887, C_2 = 0.0102 \end{array}$$

$$\theta_8 = 0.7448\theta_7$$

To calculate the kth iteration of θ_j, that is, $\theta_j^{(k)}$, we have available $\theta_j^{(k-1)}$, $\theta_{j-1}^{(k)}$, and $\theta_{j+1}^{(k-1)}$. To accelerate convergence we write

$$\theta_j^{(k)} = \theta_j^{(k-1)} + \omega\left\{\frac{\theta_{j+1}^{(k-1)} + \theta_{j-1}^{(k)}}{2 + C_1[1 + (j-1)^2 C_2]} - \theta_j^{(k-1)}\right\} \qquad j = 2, 3, \ldots, 7$$

To initiate the iteration procedure, we make a first guess that the fin temperature profile decreases linearly to ambient temperature at the tip, that is, for $k = 1$

$$\theta_j^{(1)} = 91.4286 - 11.4286j \ °C \qquad j = 1, 2, \ldots, 8$$

For the three values of ω we show in tabular form the temperature at node 4 for the first 10 iterations:

	$\omega = 1$	1.25	1.50
$k = 2$	31.816	25.744	18.273
3	23.461	16.789	10.088
4	18.466	13.085	9.517
5	15.519	11.792	10.760
6	13.987	11.889	12.497
7	13.147	11.952	12.274
8	12.671	11.983	12.000
9	12.395	11.986	11.895
10	12.232	11.988	11.985

For the accelerated ($\omega > 1$) convergence, we notice that some oscillation in the solution for θ_4 occurs. This oscillation is more pronounced in the $\omega = 1.5$ case and presumably would become unstable if $\omega = 2$. In both cases, it is clear that the convergence is more rapid than for $\omega = 1$.

The heat dissipated from each node is

$$q_j = \theta_j h_j A_j$$

where A_j is the heat transfer area associated with each node and we take θ_j from the $\omega = 1.25$ calculation, 10 iterations.

j	θ_j (°C)	h_j (W/m² °C)	A_j (cm²)	q_j (W)
1	80	50.00	1.122	0.449
2	42.991	50.51	2.244	0.487
3	22.865	52.04	2.244	0.207
4	11.988	54.59	2.244	0.147
5	6.200	58.16	2.244	0.081
6	3.215	62.76	2.244	0.045
7	1.799	68.37	2.244	0.028
8	1.340	75.00	1.318	0.013
			Total =	1.457 W

3.6 CLOSURE

In this chapter we have presented numerical methods for solving conduction heat transfer problems. When applying such techniques one must be aware not only of the types of differencing schemes used to represent the differential equation, but also of questions of stability, accuracy, and boundary condition representation. In addition, one may need to be concerned with the cost of carrying out the solution on the computer, especially if the program will be used many times. This latter constraint may determine the type of solution method one uses for the difference equations. Table 3.1 in Section 3.5 provides a guideline for choosing the solution method, but that information must be considered in light of the type of computer to be used. For example, on a mainframe computer computational time may be the primary cost driver, while on a microcomputer, storage requirements may need to be minimized.

Finally, one should be aware that many computer programs have been written which provide solutions for a variety of heat transfer problems. Many such programs are in the public domain, and one would do well to survey their capabilities before undertaking the development of a computer program. In either case, the material presented in this chapter will provide a foundation needed for numerical analysis work.

The problems for this chapter are related to the subject matter as shown below.

Topic	Section	Problem Number
Finite difference representation	3.2	3.1–3.3
One-dimensional, steady conduction	3.3	3.4–3.5
One-dimensional, unsteady conduction	3.3	3.6–3.8
Two-dimensional, unsteady conduction	3.3	3.9
Stability	3.3	3.10
Boundary conditions	3.4	3.11–3.16
Solution of difference equations	3.5	3.17–3.30

PROBLEMS

3.1 Determine the Seebeck coefficient (dV/dT) for a type K (chromel-alumel) thermocouple at $1000°F$. Compare the results of a backward-difference, forward-difference, and centered-difference approximation. Use the following table, which gives junction millivolts as a function of temperature:

T °C	V (mV)
998	22.203
999	22.227
1000	22.251
1001	22.274
1002	22.298

3.2 Calculate the first derivative of the function

$$T(x) = T_1 \cos\left(\frac{\pi x}{L}\right)$$

using the first-order accurate difference equation, Eq. (3.3), and the second-order accurate difference equation, Eq. (3.5), at $x = L/2$. Use a step size of $\Delta x = 0.01\,L$. Compare your results with the exact first derivative and explain your result.

3.3 The temperature of an object placed in a furnace has been recorded as a function of time with a period of 1 second. It is desired to calculate the rate of change of temperature of the object from the data. Write an algorithm which could accomplish this. Comment on the ability of your method to handle data with random noise.

3.4 Develop the finite-difference equations for one-dimensional steady conduction in a fin with variable cross-sectional area $A(x)$ and perimeter $P(x)$. The unit surface conductance from the fin to ambient is a constant, h_0.

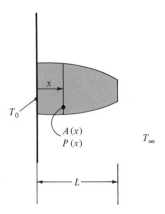

3.5 Using your results from Problem 3.4, find the heat flow at the base of the fin for the following conditions:

$$k = 20 \text{ Btu/h ft } ^\circ\text{F}$$
$$L = 2 \text{ in.}$$
$$A(x) = 0.5\left(1 - \frac{1}{3}\sinh\left(\frac{x}{L}\right)\right) \text{ in.}^2$$
$$P(x) = [A(x)]^{1/2}$$
$$h_0 = 20 \text{ Btu/h ft}^2 \, ^\circ\text{F}$$
$$T_0 = 200^\circ\text{F}$$
$$T = 80^\circ\text{F}$$

Use a grid spacing of 0.2 in.

3.6 Derive an implicit difference equation for one-dimensional, unsteady heat flow in a solid with a volumetric heat source. Assume that the heat source varies in both time and location.

3.7 Write an explicit finite-difference representation of the one-dimensional, unsteady conduction equation with variable thermal conductivity. Apply your results to Example 3.8, using appropriate data for the variation of thermal conductivity of stainless steel with temperature. How much does the variation of thermal conductivity affect the heat flow?

3.8 Sketch the computational grid for the explicit [Eq. (3.13)], implicit [Eq. (3.14)], and Crank-Nicholson [Eq. (3.15)] representations of the one-dimensional unsteady conduction equation. Note that the Crank-Nicholson scheme is an "average" of the explicit and implicit schemes. Derive a general difference equation for the one-dimensional unsteady conduction equation which includes a parameter s such that for $s = 0$ the scheme is explicit, for $s = 1$ the scheme is implicit, and for $s = 1/2$ the scheme is equivalent to the Crank-Nicholson scheme.

3.9 Show that Eq. (3.25) reduces to a tridiagonal matrix.

3.10 The stability criterion for a one-dimensional explicit scheme, Eq. (3.13), $0 \le r \le 1/2$, was justified on the basis of the coefficient of $T(x, t)$ being equal to or greater than zero. Use the same reasoning to derive the stability criterion for the two-dimensional explicit scheme, Eq. (3.23). Extend to three dimensions.

3.11 Given the temperature distribution near the surface of a solid, how would you calculate the heat flux at the surface? Give an equation for the heat flux in terms of material properties and nodal temperatures.

3.12 In deriving Eq. (3.28) how do you justify integration over half of the end grid element? Try integrating over the entire element and explain your results.

3.13 Consider two-dimensional steady conduction near a curved boundary. Determine the finite-difference equation for node (m, n) based on temperatures at nodes $(m, n + 1)$, $(m + 1, n)$, 1, and 2.

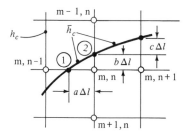

Interior Node Near Curved Boundary

3.14 How would you treat radiation heat transfer at the surface of a one-dimensional slab of material? Assume that the heat flux at the surface is $q = \epsilon\sigma(T_s^4 - T^4)$, where T_s is the temperature of the surface node and T is the temperature of an enclosure surrounding the surface.

3.15 How should the numerical method be altered at an interface between two materials of different thermal conductivities? Illustrate by a simple example.

3.16 How would you include contact resistance between the two materials in Problem 3.15? Derive the finite-difference equation for one-dimensional steady flow between two solids in contact.

3.17 From the elimination procedure developed in Section 3.5.2, Eqs. (3.51) and (3.52), develop a computer program to solve an $N \times N$ system of equations.

3.18 Compare the solution of Example 3.8 by Gaussian elimination, matrix inversion, Gauss-Seidel iteration, Jacobi iteration, and the SOR method. Prepare a table showing the computation time (using the same computer for each method) to achieve a solution accurate to 1/2 percent. Use 10 grid points instead of 8. Repeat for 50 grid points.

3.19 A 3-m-long steel rod is initially at 20°C and insulated completely except for its end faces. One end is suddenly exposed to the flow of combustion gases at 1000°C through a unit surface conductance of 250 W/m² °C and the other end is held at 20°C. How long will it take for the exposed end to reach 700°C? How much energy will the rod have absorbed if its diameter is 3 cm?

3.20 A Trombe wall is a masonry wall frequently used in passive solar homes to store solar energy. Suppose such a wall, composed of 8 in. of solid concrete, is initially at 60°F in equilibrium with the room in which it is located. It is suddenly exposed to sunlight and absorbs 150 Btu/h ft² on the exposed face. The wall is coupled to the room air through a unit surface conductance of 1 Btu/h ft² °F. Compute the temperature distribution in the Trombe wall as a function of time, neglecting radiation. Compute the rate of heat transfer to the room as a function of time.

3.21 Suppose the Trombe wall in Problem 3.20 is exposed to a heat flux that varies sinusoidally in time. Let the heat flux vary from a maximum of 150 Btu/h ft² to zero over a 12-h half-period. Assume that the wall, initially at 60°F throughout, is suddenly exposed to the solar energy at its peak value. Predict the development of the temperature distribution in the wall. How long will it take to achieve a steady-state temperature distribution, that is, one that has only a periodic variation in time? What is the phase relationship between the peak in the input energy and the energy released to the room? Is the Trombe wall an effective energy storage device?

3.22 A turbine blade $2\frac{1}{2}$ in. long, with cross-sectional area $A = 0.005$ ft² and perimeter $P = 0.40$ ft, is made of stainless steel ($k = 15$ Btu/h ft °F). The temperature of the root T_w is 900°F. The blade is exposed to a hot gas at 1600°F, and the unit surface conductance h is 80 Btu/h ft² °F. Using the grid network shown in the accompanying sketch, estimate the temperature distribution and the rate of heat transfer by an elimination method. Assume that the tip is insulated.

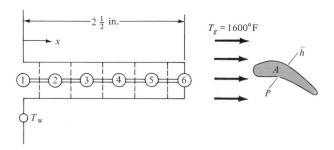

3.23 An interior wall of a cold furnace, initially at 20°C, is suddenly exposed to a radiant flux of 5 kW/m² at start-up. The outer surface of the wall is exposed to ambient air at 20°C through a unit surface conductance of 5 W/m² °C. The wall is composed

of calcium silicate insulation and is 30 cm thick. Determine the temperature profile
in the wall after start-up. At what time will it be necessary to consider reradiation
from the inner surface of the wall?

3.24 In a long, 30-cm-square bar shown in the accompanying sketch the left face is maintained at 40°C, the top face at 250°C, and the other two faces in contact with a
fluid at 40°C through a unit surface conductance of 60 W/m² °C. If the thermal
conductivity of the bar is 20 W/m °C, calculate the temperatures at points 1 through
9 by Gaussian elimination.

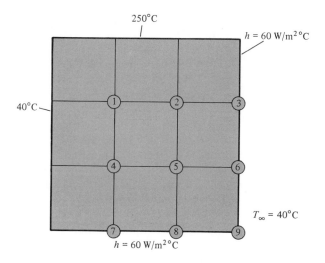

3.25 Repeat Problem 3.24 if the temperature distribution on the top surface of the bar
varies sinusoidally from 40°C at the left face to a maximum of 250°C in the center
and back to 40°C at the right face.

3.26 Determine (*a*) the temperatures at the 16 equally spaced points shown in the accompanying sketch to an accuracy of three significant figures and (*b*) the rate of
heat flow per meter thickness. Assume two-dimensional heat flow, $k = 1$ W/m °C,
and make use of the symmetry of the system.

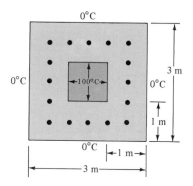

3.27 A long concrete bar of triangular cross section having the dimensions shown in the accompanying sketch is initially at a temperature of 60°F. It is suddenly placed in an environment at 160°F and is being heated through a unit surface conductance of 5 Btu/h ft² °F over its two upper surfaces, while the lower surface is insulated. Using a 1 × 1 ft grid, estimate the temperature distribution after 1 h of heating and also the time required for the minimum temperature in the slab to reach 80°F.

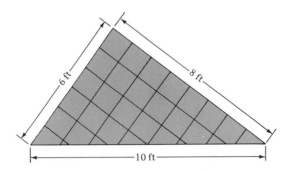

3.28 A tall chimney ($k = 0.2$ W/m °C) has the cross section shown in the accompanying sketch. If the interior surface temperature is 250°C and the exterior surface temperature is 40°C, calculate the rate of heat loss and the temperature distribution, using the Gauss-Seidel method with a 0.2-m grid spacing.

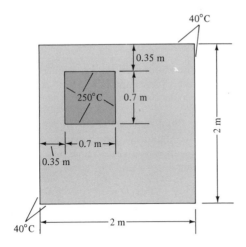

3.29 Repeat Problem 3.28, but assume that the values of unit surface conductance over the interior and exterior surfaces are 30 and 12 W/m² °C, respectively, and that the hot gases are at 250°C while the air outside is at 20°C.

3.30 A solid stainless steel 1-m cube ($k = 15$ W/m °C), is fully insulated except for two touching faces. One of the faces is held at 0°C and the other at 100°C. For N grid points in each spatial direction, how many simultaneous equations must be solved to determine the temperature distribution in the cube? How many of these equations only express boundary conditions, that is, are explicit? What would be a reasonable way to treat the nodes that lie on the boundary between the two isothermal faces? Compute the heat flow through the cube.

REFERENCES

1. W. F. Ames, *Numerical Methods for Partial Differential Equations*, Barnes and Noble. New York, 1969.
2. J. M. Ortega and W. G. Poole, Jr., *An Introduction to Numerical Methods for Differential Equations*, Pitman Publishing, Marshfield, Mass., 1981.
3. S. V. Patankar, *Numerical Heat Transfer and Fluid Flow*, Hemisphere Publishing Corp. Washington, D.C., 1980.
4. A. Jennings, *Matrix Computation for Engineers and Scientists*, Wiley, New York, 1977.

Chapter 4

Analysis of Convection Heat and Mass Transfer

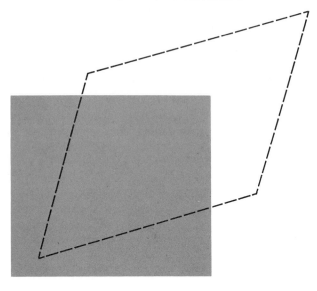

4.1 INTRODUCTION

In the preceding chapters convection has been considered only to the extent that it provides boundary conditions when the surface of a body is in contact with a fluid at a different temperature. However, from the illustrative problems it has probably already become apparent that there are hardly any practical problems that can be treated without a knowledge of the mechanism by which heat is transferred between the surface of a body and the surrounding medium. In this chapter we shall therefore extend our treatment of convection to gain a better understanding of the mechanism and some of the key parameters that influence it.

The convection mechanism can transfer heat as well as mass. In many situations the heat and mass transfer occur simultaneously and the processes are interlinked. In these cases it is necessary to consider both processes in order to arrive at the correct answer. Since the transport mechanisms of these two phenomena are similar, there are also similarities in the method of analysis. Mass transfer by convection transports a species, such as water vapor in air when a concentration gradient exists, while heat transfer by convection transfers heat when a temperature gradient exists. The similarities and interactions between the two processes will be treated in this introductory chapter on convection. Processes with phase change (such as boiling and condensation) will be treated later.

4.2 CONVECTION HEAT TRANSFER

Before attempting to calculate a heat transfer or mass transfer coefficient, we shall examine the convection process in some detail and relate the convection of heat and mass to the flow of the fluid. Figure 4.1 shows a heated flat plate cooled by a stream of air flowing over it. Also shown are the velocity and the temperature distributions. The first point to note is that the velocity decreases in the direction toward the surface as a result of viscous forces acting in the fluid. Since the velocity of the fluid layer adjacent to the wall is zero, the heat transfer between the surface and this fluid layer by conduction alone must be:

$$q_c'' = -k_f \left.\frac{\partial T}{\partial y}\right|_{y=0} = h_c(T_s - T_\infty) \tag{4.1}$$

Although this viewpoint suggests that the process can be viewed as conduction, the temperature gradient at the surface $(\partial T/\partial y)_{y=0}$ is determined by the rate at which the fluid farther from the wall can transport the energy into the mainstream. Thus the temperature gradient at the wall depends on the flow field, with higher velocities being able to produce larger temperature gradients and higher rates of heat transfer. At the same time, however, the thermal conductivity of the fluid plays a role. For example, the value of k_f for water is an order of magnitude larger than that for air; thus, as shown in Table 1.2, the convection heat transfer coefficient for water is larger than that for air.

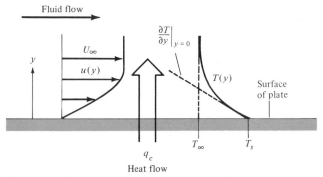

Figure 4.1 Temperature and velocity distributions in laminar forced convection flow over a heated flat plate at temperature T_s.

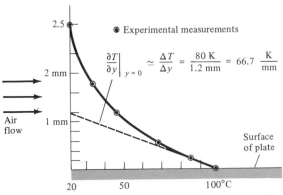

Figure 4.2 Experimental data on temperature distribution for Example 4.1.

EXAMPLE 4.1 _____

Air at 20°C is flowing over a flat plate whose surface temperature is 100°C. At a certain location the temperature is measured as a function of distance from the surface of the plate; the results are plotted in Fig. 4.2. From these data determine the convection heat transfer coefficient at this location.

Solution. From Eq. (4.1) the heat transfer coefficient can be expressed in the form

$$h_c = \frac{-k_f(\partial T/\partial y)_{y=0}}{T_s - T_\infty}$$

From Table 27 in Appendix 2 the thermal conductivity of air, evaluated at the average temperature between plate and fluid stream of 60°C, is 0.028 W/m K. The temperature gradient $\partial T/\partial y$ at the surface is obtained graphically by drawing the tangent to the measured temperature data as shown in Fig. 4.2. We thus obtain $(\partial T/\partial y)_{y=0} \approx -66.7$ K/mm. Substituting for the gradient at the heated surface of the plate in Eq. (4.1) yields

$$h_c = \frac{-(0.028 \text{ W/m K})(-66.7 \text{ K/mm})}{(100 - 20) \text{ K}} \times 10^3 \text{ mm/m}$$

$$= 23.3 \text{ W/m}^2 \text{ K}$$

The situation is quite similar in natural convection, as shown in Fig. 4.3. The principal difference is that in forced convection the velocity approaches the free-stream value imposed by an external force, whereas in natural convection the velocity at first increases with increasing distance from the plate, because the action of viscosity diminishes rather rapidly while the density difference decreases more slowly. Eventually, however, the buoyant force decreases as the fluid density approaches the value of the surrounding fluid; this will cause the velocity to reach a maximum and approach zero far away from the heated surface. The temperature fields in natural and forced convection have similar shapes, and in both cases the heat transfer mechanism at the fluid/solid interface is conduction.

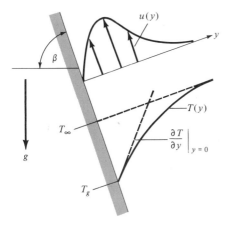

Figure 4.3 Temperature and velocity distributions in natural convection over a heated plate inclined at an angle β from the horizontal.

The preceding discussion indicates that the convection heat transfer coefficient will depend on the density, viscosity, and velocity of the fluid as well as on its thermal properties. Whereas in forced convection the velocity is usually imposed on the system by a pump or a fan and can be directly specified, in natural convection the velocity will depend on the temperature difference between the surface and the fluid, the coefficient of thermal expansion of the fluid, which determines the density change per unit temperature difference, and the force field, which in systems located on the earth is simply the gravitational force.

4.3* CONVECTION MASS TRANSFER

The transfer of mass by convection can be treated in a similar manner. Consider a fluid with a molar concentration of species A, $C_{A,\infty}$, flowing over a surface at which the concentration is maintained at a value $C_{A,s}$, as shown in Fig. 4.4.

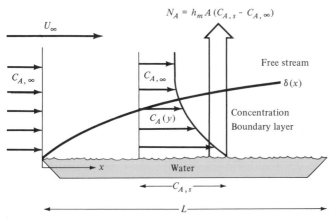

Figure 4.4 Concentration and velocity distributions in laminar forced convection flow over a flat surface with species concentration C_{As}.

An example of such a system is dry air flowing over the surface of a lake. In this situation species A is water vapor and it is transferred by convection into the air.

The relationship between species transfer rate and concentration gradient, known as Fick's law, is

$$N_A'' = -D_{AB} \frac{\partial C_A}{\partial y} \tag{4.2}$$

where N_A'' = molar flux of species A in the direction y, kmol/s m^2

$\partial C_A/\partial y$ = concentration gradient, kmol/m^3/m

D_{AB} = property known as the *binary diffusion coefficient*, m^2/s

Observe that Eq. (4.2) is analogous to Fourier's law in conduction heat transfer. We can therefore apply the physical nature of convection heat transfer to convection mass transfer. As shown in Fig. 4.4, when a fluid with species concentration $C_{A,\infty}$ in the bulk flows over a surface with species concentration $C_{A,s}$, a concentration boundary layer develops and a concentration distribution, similar to a temperature distribution, is set up in the concentration boundary layer. Since there is no fluid motion at the interface, the mass transfer at $y = 0$ can occur only by diffusion and we can define a local mass transfer coefficient h_m as

$$h_m = \frac{-D_{AB}(\partial C_A/\partial y)_{y=0}}{C_{A,s} - C_{A,\infty}} \tag{4.3}$$

Obviously, the flow conditions that influence the temperature gradient will exert a similar influence on the concentration gradient and the convection mass transfer coefficient. This physical similarity is the basis of the heat and mass transfer analogy to be developed later in this chapter.

We can also express the molar flux of species A as the product of a transfer coefficient and a potential difference, similar to the transfer of heat. However, in heat transfer the potential difference is a temperature difference, while in mass transfer the concentration difference is the driving potential, or

$$N_A = \bar{h}_m A(C_{A,s} - C_{A,\infty}) \tag{4.4}$$

where N_A = molar transfer rate of A, kmol/s

$\bar{h}_m$ = average convection mass transfer coefficient, m/s

A = area of transfer, m^2

C_A = molar concentration, kmol/m^3

As in heat transfer, we must take care to distinguish between the local mass transfer coefficient h_m and the average coefficient $\bar{h}_m$ over a surface area A_s. The two are related by

$$\bar{h}_m = \frac{1}{A_s} \iint_{A_s} h_m \, dA_s \tag{4.5}$$

For the flat surface shown in Fig. 4.4

$$\bar{h}_m = \frac{1}{L} \int_0^L h_m(x)\,dx \tag{4.6}$$

In many cases it is convenient to know the mass transfer rate n_A (kg/s) rather than the species rate N_A. The two are related by the molecular weight of species A, $\mathcal{M}_A$ (kg/kmol), or

$$n_A = N_A \mathcal{M}_A = \bar{h}_m A_s (\rho_{A,s} - \rho_{A,\infty}) \tag{4.7}$$

where ρ_A is the mass density of A (kg/m^3).

To calculate the rate of mass transfer one must first determine the concentration potential. The concentration at the surface can be determined under normal conditions from thermodynamic relations. Since equilibrium must exist at the interface between the gas and the liquid or solid surface over which it is flowing, the temperature of the vapor there is equal to the surface temperature T_s. Moreover, the vapor at the interface must be saturated, and thermodynamic tables such as Table 13 in Appendix 2 for water, can be used to obtain the density ρ_s corresponding to the surface temperature T_s. To a first approximation, the molar concentration at the surface can also be determined from the vapor pressure, assuming that the vapor approximates a perfect gas. Then,

$$C_{A,s} = \frac{P_{\text{sat}}(T_s)}{R T_s} \tag{4.8}$$

where $P_{\text{sat}}(T_s)$ is the vapor pressure corresponding to saturation at T_s and R is the universal gas constant.

EXAMPLE 4.2 ───

Air at 32°C containing water vapor with a partial pressure of 0.2 atm is flowing over a swimming pool whose temperature is 27°C. If the pool is 5×5 m in area and the average mass transfer coefficient is 0.002 m/s, calculate the rate of mass transfer, that is, the evaporation rate, from the pool.

Solution. Assume that the pool is at a uniform temperature $T_s = 27 + 273 = 300$ K. From the thermophysical properties of saturated water the saturation pressure $P_{A,s}$ is 0.0353 bar, and the corresponding specific volume 39.13 m^3/kg. Assuming that the vapor can be approximated as a perfect gas

$$C_{A,s} = \frac{P_{\text{sat}}(T_s)}{R T_s} = \frac{0.0353 \text{ bar}}{(8.314 \times 10^{-2} \text{ m}^3 \text{ bar/kmol K})(300 \text{ K})}$$

$$= 1.45 \times 10^{-6} \text{ kmol/m}^3$$

$$C_{A,\infty} = \frac{P_{A,\infty}}{R T_\infty} = \frac{0.2 \text{ atm}}{(8.2 \times 10^{-2} \text{ m}^3 \text{ atm/kmol K})(305 \text{ K})}$$

$$= 8.0 \times 10^{-7} \text{ kmol/m}^3$$

From Eq. (4.7) the rate of mass transfer is

$$n_A = \bar{h}_m A \mathcal{M}_A (C_{A,s} - C_{A,\infty})$$
$$= (0.002 \text{ m/s})(25 \text{ m}^2)(18 \text{ kg/kmol})(1.45 - 0.8) \times 10^{-6} \text{ kmol/m}^3$$
$$= 5.85 \times 10^{-6} \text{ kg/s}$$

4.4 BOUNDARY-LAYER FUNDAMENTALS

To gain an understanding of the parameters significant in forced convection we shall examine the flow field in more detail. Figure 4.5 shows the velocity distribution at various distances from the leading edge of a plate. Starting at the leading edge, a region develops in the flow where viscous forces cause the fluid to slow down. These viscous forces depend on the shear stress τ. In flow over a flat plate the fluid velocity parallel to the plate can be used to define this stress as

$$\tau = \mu \frac{du}{dy} \qquad (4.9)*$$

where du/dy is the velocity gradient and the constant of proportionality μ is called the dynamic viscosity. If the shear stress is expressed in newtons per square meter and the velocity gradient in (seconds)$^{-1}$, then μ has the units newton-seconds per square meter (N s/m^2).

The region in the flow near the plate where the velocity of the fluid is decreased by viscous forces is called the boundary layer. The distance from the plate at which the velocity reaches 99 percent of the free-stream velocity is

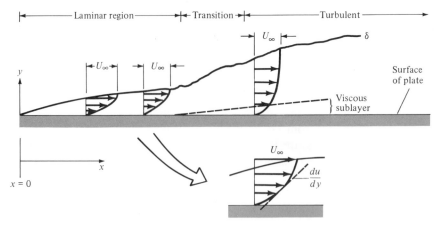

Figure 4.5 Velocity profiles in laminar, transition, and turbulent boundary layers in flow over a flat plate.

* Note that μ can also be expressed in the units kg/ms. With these units it is necessary to use the conversion constant g_c for dimensional consistency and Eq. 4.9 would read $\tau = (\mu/g_c) \, du/dy$.

arbitrarily designated as the boundary-layer thickness, and the region beyond this point is called the undisturbed free stream or potential flow regime.

Initially, the flow in the boundary layer is completely laminar. The boundary thickness grows with increasing distance from the leading edge, and at some critical distance x_c the inertial effects become sufficiently large compared to the viscous damping action that small disturbances in the flow begin to grow. As these disturbances become amplified, the regularity of the viscous flow is disturbed and a transition from laminar to turbulent flow takes place. In the turbulent-flow region macroscopic chunks of fluid move across streamlines and vigorously transport thermal energy as well as momentum. As shown in books on fluid mechanics [e.g., (1)], the dimensionless parameter that quantitatively relates the viscous and inertial forces and whose value determines the transition from laminar to turbulent flow is the Reynolds number Re_x, defined as

$$Re_x = \frac{U_\infty x}{v_f} \tag{4.10}$$

where U_∞ = free-stream velocity

x = distance from the leading edge

$v_f = \mu_f/\rho_f$ = kinematic viscosity of the fluid

The critical value of Re_x at which transition occurs, $Re_{c,x}$, depends on the surface roughness and the level of turbulent activity—the turbulence level—in the mainstream. When large disturbances are present in the main flow, transition begins when $Re_x = 10^5$, but in less disturbed flow fields it will not start until $Re_x = 2 \times 10^5$ (1, 2). The transition regime extends to a Reynolds number about twice the value at which transition began, and beyond this point the boundary layer is turbulent.

Approximate shapes of the velocity profiles in laminar and turbulent flow are sketched in Fig. 4.5. In the laminar range the boundary-layer velocity profile is approximately parabolic. In the turbulent range there exists a thin layer near the surface, called the viscous sublayer, across which the velocity profile is nearly linear. Outside this layer the velocity profile is flat compared to the laminar profile.

4.5 CONSERVATION OF MASS, MOMENTUM, ENERGY, AND SPECIES EQUATIONS FOR LAMINAR FLOW OVER A FLAT PLATE

In the classical approach to convection one derives differential equations for the momentum, energy, and species balance in the boundary layer and then solves these equations for the temperature or concentration gradient in the fluid at the fluid/wall interface to evaluate the convection heat transfer or the convection mass transfer coefficient. A somewhat simpler, but practically more useful, approach is to derive integral instead of differential equations and use an approximate analysis to obtain a solution. In this section the differential equations governing the flow of a fluid over a flat plate will be derived to illustrate the similarity between heat, mass, and momentum transfer and to introduce appropriate dimensionless parameters relating the processes. Then the integral equations for flow over a flat surface will be derived and solved to illustrate an analytical approach that will also be used to obtain the heat transfer and mass transfer boundary-layer coefficients in turbulent flow.

To derive the *conservation of mass* or continuity equation, consider a control volume within the boundary layer as shown in Fig. 4.6 and assume that steady-state conditions prevail. There are no gradients in the z direction (perpendicular to the plane of the sketch), and the fluid is incompressible. Then the rates of mass flow into and out of the control volume, respectively, in the x direction are

$$\rho u \, dy \quad \text{and} \quad \rho \left(u + \frac{\partial u}{\partial x} \, dx \right) dy$$

Thus the net mass flow into the element in the x direction is

$$-\rho \frac{\partial u}{\partial x} \, dx \, dy$$

Similarly, the net mass flow into the control volume in the y direction is

$$-\rho \frac{\partial v}{\partial y} \, dy \, dx$$

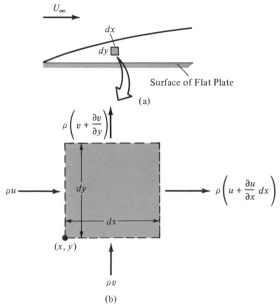

(a)

(b)

Figure 4.6 Control volume $(dx \, dy \cdot 1)$ for conservation of mass in an incompressible boundary layer in flow over a flat plate.

Since the net mass flow rate out of the control volume must be zero, we obtain

$$-\rho\left(\frac{\partial u}{\partial x} + \frac{\partial v}{\partial y}\right) dx\, dy = 0$$

from which it follows that in two-dimensional steady flow, conservation of mass requires that

$$\frac{\partial u}{\partial x} + \frac{\partial v}{\partial y} = 0 \qquad (4.11)$$

The *conservation of momentum* equation is obtained from application of Newton's second law of motion to the element. Assuming that the flow is Newtonian, that there are no pressure gradients in the y direction, and that viscous shear in the y direction is negligible, the rates of momentum flow in the x direction for the fluid flowing across the left- and right-hand vertical faces (see Fig. 4.7) are $\rho u^2\, dy$ and $\rho[u + (\partial u/\partial x)\, dx]^2\, dy$. It should be noted, however, that flow across the horizontal faces will also contribute to the momentum balance in the x direction. The x-momentum flow entering through the bottom face is $\rho uv\, dx$, and the momentum flow per unit width leaving through the upper face is

$$\rho\left(v + \frac{\partial v}{\partial y}\, dy\right)\left(u + \frac{\partial u}{\partial x}\, dx\right) dx$$

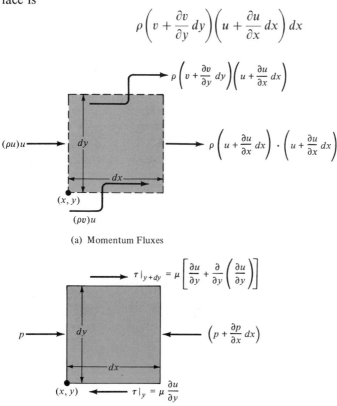

(a) Momentum Fluxes

(b) Forces

Figure 4.7 Differential control volume for conservation of momentum in a two-dimensional incompressible boundary layer.

The viscous shear force at the bottom face is $\tau|_y = -\mu(\partial u/\partial y)\,dx$ and over the top face is

$$\tau\Big|_{y+dy} = \mu\,dx\left[\frac{\partial u}{\partial y} + \frac{\partial}{\partial y}\left(\frac{\partial u}{\partial y}\right)dy\right]$$

Thus the net viscous shear in the x direction is $\mu\,dx(\partial^2 u/\partial y^2)\,dy$.

The pressure force over the left face is $p\,dy$ and over the right face is $-[p + (dp/dx)\,dx]\,dy$. Thus the net pressure force in the direction of motion is $-(\partial p/\partial x)\,dx\,dy$. Equating the sum of the forces to the momentum flow rate out of the control volume in the x direction gives

$$\rho uv\,dx + \rho\frac{\partial v}{\partial y}u\,dy\,dx + \rho v\frac{\partial u}{\partial x}\,dx + \rho\frac{\partial v}{\partial y}\frac{\partial u}{\partial x}\,dy\,dx = \left(\mu\frac{\partial^2 u}{\partial y^2} - \frac{\partial p}{\partial x}\right)dy\,dx$$

Neglecting second-order differentials and using the conservation of mass equation, the conservation of momentum equation reduces to

$$\rho\left(u\frac{\partial u}{\partial x} + v\frac{\partial u}{\partial y}\right) = \mu\frac{\partial^2 u}{\partial y^2} - \frac{\partial p}{\partial x} \qquad (4.12)$$

The *conservation of energy* equation will be derived on the assumption that all physical properties are temperature-independent and that the flow velocity is sufficiently small that the frictional shear work may be neglected. Figure 4.8a shows the rate at which energy will be conducted and convected into and out of the control volume. There are four convective terms in addition to the conductive terms derived in Chapter 2. An energy balance requires that the net rate of conduction and convection be zero. This yields

$$k\,dx\,dy\left(\frac{\partial^2 T}{\partial x^2} + \frac{\partial^2 T}{\partial y^2}\right) - \left[\rho c_p\left(u\frac{\partial T}{\partial x} + \frac{\partial u}{\partial x}T + \frac{\partial u}{\partial x}\frac{\partial T}{\partial x}\,dx\right)\right]dx\,dy$$
$$-\left[\rho c_p\left(v\frac{\partial T}{\partial y} + \frac{\partial v}{\partial y}T + \frac{\partial v}{\partial y}\frac{\partial T}{\partial y}\,dy\right)\right]dx\,dy = 0$$

Using the conservation of mass equation and neglecting second-order terms, as in the derivation of the conservation of momentum equation, gives the following expression for the energy equation:

$$u\frac{\partial T}{\partial x} + v\frac{\partial T}{\partial y} = \alpha\left(\frac{\partial^2 T}{\partial x^2} + \frac{\partial^2 T}{\partial y^2}\right) \qquad (4.13)$$

When the fluid in the boundary layer is a binary mixture, such as water vapor in moist air, relative transport of the species will occur if there is a species concentration. Then *species conservation* must be satisfied along with mass, momentum, and energy conservation. To derive the differential equation governing species conservation in steady incompressible boundary-layer flow, consider the control volume shown in Fig. 4.8b. Species A may be transported by convection with the mean velocity of the mixture and by diffusion relative to the mean motion in each direction. We will neglect the possibility of species generation, which may occur in a chemical reaction.

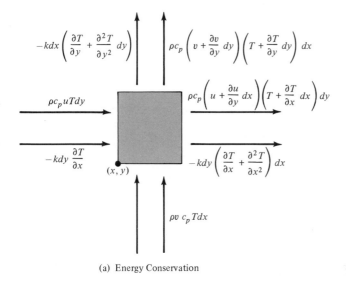

(a) Energy Conservation

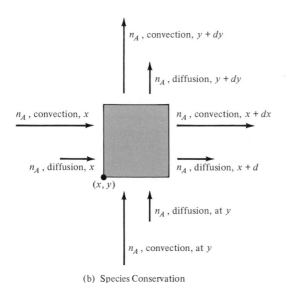

(b) Species Conservation

Figure 4.8 Differential control volume for conservation of energy and conservation of species.

The net rate at which species A enters the two-dimensional control volume in the x direction by convection is

$$n_{A,\,\text{convection},\,x} - n_{A,\,\text{convection},\,x+dx} = (\rho_A u)\,dy - \left[\rho_A u + \frac{\partial(\rho_A u)}{\partial x}\right]dy$$

$$= -\frac{\partial(\rho_A u)}{\partial x}\,dx\,dy$$

The net rate at which species A enters the control volume by diffusion is obtained by applying Fick's law:

$$n_{A, \text{diffusion}, x} - n_{A, \text{diffusion}, x+dx} = \left(-D_{AB} \frac{\partial \rho_A}{\partial x} \right) dy$$

$$- \left[\left(-D_{AB} \frac{\partial \rho_A}{\partial x} + \frac{\partial}{\partial x} \left(D_{AB} \frac{\partial \rho_A}{\partial x} \right) dx \right] dy$$

$$= \frac{\partial}{\partial x} \left(D_{AB} \frac{\partial \rho_A}{\partial x} \right) dx \, dy$$

Similar relations can be written for the net rates at which species A enters in the y direction. Combining all four of these net terms yields

$$\frac{\partial(\rho_A u)}{\partial x} + \frac{\partial(\rho_A u)}{\partial y} = \frac{\partial}{\partial x} \left(D_{AB} \frac{\partial \rho_A}{\partial x} \right) + \frac{\partial}{\partial y} \left(D_{AB} \frac{\partial \rho_A}{\partial y} \right) \quad (4.14)$$

By expanding the first and second terms on the left side and using the continuity equation, Eq. (4.10), we can arrive at a more convenient form for Eq. (4.14):

$$u \frac{\partial \rho_A}{\partial x} + v \frac{\partial \rho_A}{\partial y} = \frac{\partial}{\partial x} \left(D_{AB} \frac{\partial \rho_A}{\partial x} \right) + \frac{\partial}{\partial y} \left(D_{AB} \frac{\partial \rho_A}{\partial y} \right) \quad (4.15a)$$

In molar form, Eq. (4.15a) reads

$$u \frac{\partial C_A}{\partial x} + v \frac{\partial C_A}{\partial y} = \frac{\partial}{\partial x} \left(D_{AB} \frac{\partial C_A}{\partial x} \right) + \frac{\partial}{\partial y} \left(D_{AB} \frac{\partial C_A}{\partial y} \right) \quad (4.15b)$$

Since a boundary layer is quite thin, under normal conditions $\partial T/\partial y \gg \partial T/\partial x$ and $\partial C_A/\partial y \gg \partial C_A/\partial x$. Also, the pressure term in the momentum equation is zero for flow over a flat plate since $(\partial U_\infty/\partial x) = 0$. Then the similarity between the momentum, energy, and mass conservation equations becomes apparent:

$$u \frac{\partial u}{\partial x} + v \frac{\partial u}{\partial y} = v \left(\frac{\partial^2 u}{\partial y^2} \right) \quad (4.16a)$$

$$u \frac{\partial T}{\partial x} + v \frac{\partial T}{\partial y} = \alpha \left(\frac{\partial^2 T}{\partial y^2} \right) \quad (4.16b)$$

$$u \frac{\partial C_A}{\partial x} + v \frac{\partial C_A}{\partial y} = D_{AB} \left(\frac{\partial^2 C_A}{\partial y^2} \right) \quad (4.16c)$$

In the preceding relations, v is the kinematic viscosity, equal to μ/ρ, often called momentum diffusivity. The ratio v/α is equal to $(\mu/\rho)/(k/\rho c_p)$, which is the Prandtl number Pr:

$$\text{Pr} = \frac{c_p \mu}{k} = \frac{v}{\alpha} \quad (4.17)$$

If v equals α, then Pr is 1 and the momentum and energy equations are identical. For this condition, nondimensional solutions of $u(y)$ and $T(y)$ are identical.

Thus it is apparent that the Prandtl number, which is the ratio of fluid properties, controls the relation between velocity and temperature distributions.

If in the preceding relations the momentum diffusivity v equals the binary diffusion coefficient D_{AB}, the mass and momentum conservation equations are identical. The ratio v/D_{AB} is the Schmidt number Sc:

$$\text{Sc} = \frac{v}{D_{AB}} \tag{4.18}$$

The Schmidt number controls the relation between the velocity and concentration distributions. It provides a measure of the relative effectiveness of momentum and mass transport by diffusion in the boundary layer.

The Prandtl and Schmidt numbers are related by the Lewis number Le, defined as

$$\text{Le} = \frac{\alpha}{D_{AB}} = \frac{\text{Sc}}{\text{Pr}} \tag{4.19}$$

The Lewis number is important in situations where heat and mass transfer occur simultaneously.

4.6 DIMENSIONLESS BOUNDARY-LAYER EQUATIONS AND SIMILARITY PARAMETERS

Solutions to Eqs. (4.16a–c), the so-called laminar boundary-layer equations for low-speed forced convection, will yield the velocity, temperature, and concentration profiles. In general these solutions are quite complicated, and the reader is referred to (1–3) for a treatment of the mathematical procedures, which are beyond the scope of this text. However, considerable additional insight into the physical aspects of boundary-layer flow as well as the form of similarity parameters governing the transport processes can be gained by non-dimensionalizing the governing equations, even without solving them.

Figure 4.9 shows the development of the velocity, thermal, and concentration boundary layers in flow over a flat surface of arbitrary shape. To express

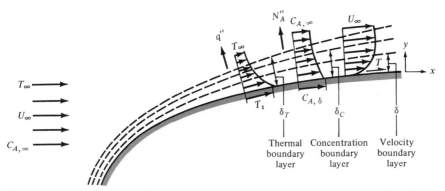

Figure 4.9 Development of the velocity, thermal, and concentration boundary layers in flow over a flat surface of arbitrary shape.

the boundary-layer equations in dimensionless form define the following dimensionless variables similar to Sec. 2.2:

$$x^* = \frac{x}{L} \qquad p^* = \frac{p}{\rho_\infty U_\infty^2}$$

$$y^* = \frac{y}{L} \qquad T^* = \frac{T - T_s}{T_\infty - T_s}$$

$$u^* = \frac{u}{U_\infty} \qquad C_A^* = \frac{C_A - C_{A,s}}{C_{A,\infty} - C_{A,s}}$$

$$v^* = \frac{v}{U_\infty}$$

where L is a characteristic length dimension such as the length of a plate and U_∞ is the free-stream velocity.

Substituting the above dimensionless variables into the dimensional Eqs. (4.10), (4.11), (4.12), and (4.14) yields the corresponding boundary-layer equations.

$$\frac{\partial u^*}{\partial x^*} + \frac{\partial v^*}{\partial y^*} = 0 \tag{4.20a}$$

$$u^* \frac{\partial u^*}{\partial x^*} + v^* \frac{\partial u^*}{\partial y^*} = -\frac{dp^*}{dx^*} + \frac{1}{\mathrm{Re}_L} \frac{\partial^2 u^*}{\partial y^{*2}} \tag{4.20b}$$

$$u^* \frac{\partial T^*}{\partial x^*} + v^* \frac{\partial T^*}{\partial y^*} = \frac{1}{\mathrm{Re}_L \mathrm{Pr}} \frac{\partial^2 T^*}{\partial y^{*2}} \tag{4.20c}$$

$$u^* \frac{\partial C_A^*}{\partial x^*} + v^* \frac{\partial C_A^*}{\partial y^*} = \frac{1}{\mathrm{Re}_L \mathrm{Sc}} \frac{\partial^2 C_A^*}{\partial y^{*2}} \tag{4.20d}$$

Observe that by nondimensionalizing the boundary-layer equations we have cast them into a form in which the dimensionless similarity parameters Re_L, Pr, and Sc appear. These similarity parameters permit us to apply solutions from one system to another geometrically similar system provided the similarity parameters have the same value in both. For example, if the Reynolds number is the same, the dimensionless velocity distribution for air, water, or glycerin flowing over a flat plate will be the same at given values of x^*.

Inspection of Eq. (4.20a) shows that v^* is related to u^*, y^*, and x^* or

$$v^* = f_1(u^*, y^*, x^*) \tag{4.21}$$

and that from Eq. (4.20b) the solution for u^*, accordingly, can be expressed in the form

$$u^* = f_2\left(x^*, y^*, \mathrm{Re}_L, \frac{dp^*}{dx^*}\right) \tag{4.22}$$

The pressure distribution over the surface of a body is determined by its shape. Hence, dp^*/dx^* can be obtained independently and represents the influence of the shape on the velocity distribution in the free stream just outside the boundary layer.

4.6.1 Friction Coefficient

From Eq. (4.9) the surface shear stress τ_s is given by

$$\tau_s = \mu \frac{\partial u}{\partial y}\bigg|_{y=0} = \frac{\mu U_\infty}{L} \frac{\partial u^*}{\partial y^*}\bigg|_{y^*=0} \tag{4.23}$$

Defining the local frictional drag coefficient C_f as

$$C_f = \frac{\tau_s}{\rho U_\infty^2/2} \tag{4.24}$$

and substituting Eq. (4.23) for τ_s gives

$$C_f = \frac{2}{\mathrm{Re}_L} \frac{\partial u^*}{\partial y^*}\bigg|_{y^*=0} \tag{4.25}$$

From Eq. (4.22) it is apparent that the dimensionless velocity gradient $\partial u^*/\partial y^*$ at the surface ($y^* = 0$) depends only on x^*, Re_L, and dp^*/dx^*. But, since dp^*/dx^* is entirely determined by the geometric shape of a body, for bodies of similar shape Eq. (4.25) reduces to the form

$$C_f = \frac{2}{\mathrm{Re}_L} f_3(x^*, \mathrm{Re}_L) \tag{4.26}$$

The above relation implies that for flow over bodies of similar shape the local frictional drag coefficient is related to x^* and Re_L by a universal function that is independent of the fluid or the free-stream velocity.

The average frictional drag over a body $\bar{\tau}$ can be determined by integrating the local shear stress τ over the surface of the body. Hence, $\bar{\tau}$ must be independent of x^* and the average friction coefficient $\bar{C}_f$ depends only on the value of the Reynolds number for flow over geometrically similar bodies, or

$$\bar{C}_f = \frac{\bar{\tau}}{\rho U_\infty^2/2} = \frac{2}{\mathrm{Re}_L} f_4(\mathrm{Re}_L) \tag{4.27}$$

EXAMPLE 4.3 ───

For flow over a slighty curved surface, the local shear stress is given by the relation

$$\tau_s(x) = 0.3 \left(\frac{\rho\mu}{x}\right)^{0.5} U_\infty^{1.5}$$

Obtain from this dimensional equation non-dimensional relations for the local and average friction coefficients.

Solution. From Eq. (4.24), the local friction coefficient is

$$C_{fx} = \frac{\tau_s(x)}{\frac{1}{2}\rho U_\infty^2} = 0.6 \left(\frac{\rho\mu}{x}\right)^{0.5} \frac{U_\infty^{1.5}}{\rho U_\infty^2}$$

$$= 0.6 \left(\frac{\mu}{\rho U_\infty x}\right)^{1/2} = \frac{0.6}{\mathrm{Re}_x^{0.5}} = \frac{0.6 x^{*1/2}}{\mathrm{Re}_L^{1/2}}$$

Integrating the local value and dividing by the area per unit width ($L \times 1$) gives the average shear $\bar{\tau}$:

$$\bar{\tau} = \frac{1}{L} \int_0^L 0.6 \left(\frac{\mu}{\rho U_\infty}\right)^{1/2} x^{1/2}\, dx$$

and the average friction coefficient is therefore

$$\bar{C}_f = \frac{\bar{\tau}}{\rho U_\infty^2/2} = 0.3 \mathrm{Re}_L^{-1/2}$$

4.6.2 Nusselt Number

In convection heat transfer the key unknown is the heat transfer coefficient. In terms of the dimensionless parameters we obtain

$$h_c = -\frac{k_f}{L}\left[\frac{(T_\infty - T_s)}{(T_s - T_\infty)}\right]\frac{\partial T^*}{\partial y^*}\bigg|_{y^*=0} = +\frac{k_f}{L}\frac{\partial T}{\partial y^*}\bigg|_{y^*=0} \tag{4.28}$$

Inspection of this equation suggests that the appropriate dimensionless form of the heat transfer coefficient is the so-called Nusselt number Nu, defined by

$$\mathrm{Nu} = \frac{h_c L}{k_f} \equiv \frac{\partial T^*}{\partial y^*}\bigg|_{y^*=0} \tag{4.29}$$

From Eqs. (4.20a) and (4.20c) it is apparent that for a prescribed geometry the Nusselt number depends only on x^*, Re_L, and Pr or

$$\mathrm{Nu} = f_5(x^*, \mathrm{Re}_L, \mathrm{Pr}) \tag{4.30}$$

Once this functional relation is known, either from an analysis or from experiments with a particular fluid, it can be used to obtain the value of Nu for other fluids and any value of U_∞ and L. Moreover, from the local value of Nu, the local value of h_c can be obtained and then an average value of the heat transfer coefficient $\bar{h}_c$ and an average Nusselt number $\overline{\mathrm{Nu}}_L$. Since the average heat transfer coefficient is obtained by integrating over the heat transfer surface of a body, it is independent of x^*, and the average Nusselt number is a function of only Re_L and Pr, or

$$\overline{\mathrm{Nu}}_L = \frac{\bar{h}_c L}{k_f} = f_6(\mathrm{Re}_L, \mathrm{Pr}) \tag{4.31}$$

4.6.3 Sherwood Number

Similar reasoning can be applied to mass transfer. From its definition

$$h_m = -\frac{D_{AB}}{L}\left[\frac{C_{A,\infty} - C_{A,s}}{C_{A,s} - C_{A,\infty}}\right]\frac{\partial C^*}{\partial y^*}\bigg|_{y^*=0} \equiv \frac{D_{AB}}{L}\frac{\partial C^*}{\partial y^*}\bigg|_{y^*=0} \tag{4.32}$$

The appropriate dimensionless form of the mass transfer coefficient, called the Sherwood number Sh, is defined by

$$Sh = \frac{h_m L}{D_{AB}} = \frac{\partial C_A^*}{\partial y^*}\bigg|_{y^*=0} \qquad (4.33)$$

The local value of Sh for a specified geometry depends on x^*, Re_L, and Sc, or

$$Sh = f_7(x^*, Re_L, Sc) \qquad (4.34)$$

and the average value of Sh is a function of only Re_L and Sc, or

$$\overline{Sh}_L = \frac{\bar{h}_m L}{D_{AB}} = f_8(Re_L, Sc) \qquad (4.35)$$

4.7 EVALUATION OF CONVECTION HEAT AND MASS TRANSFER COEFFICIENTS

Five general methods are available for the evaluation of convection heat and mass transfer coefficients:

1. Dimensionless analysis combined with experiments
2. Exact mathematical solutions of the boundary-layer equations
3. Approximate analyses of the boundary-layer equations by integral methods
4. The analogy between heat, mass, and momentum transfer
5. Numerical analysis

All five of these techniques have contributed to our understanding of convective heat transfer. Yet no single method can solve all the problems because each one has limitations that restrict its scope of application.

Dimensional analysis is mathematically simple and has found a wide range of application (3, 4). The chief limitation of this method is that the results obtained are incomplete and quite useless without experimental data. It contributes little to our understanding of the transfer process, but facilitates the interpretation and extends the range of experimental data by correlating them in terms of dimensionless groups.

There are two different methods for determining dimensionless groups suitable for correlating experimental data. The first of these methods, discussed in the following section, requires only listing of the variables pertinent to a phenomenon. This technique is simple to use, but if a pertinent variable is omitted, erroneous results ensue. In the second method the dimensionless groups and similarity conditions are deduced from the differential equations describing the phenomenon. This method is preferable when the phenomenon can be described mathematically, but the solution of the resulting equations is often too involved to be practical. The technique was presented in Section 4.6.

Exact mathematical analyses require simultaneous solution of the equations describing the fluid motion and the transfer of energy or species in the moving fluid (5). The method presupposes that the physical mechanisms are sufficiently well understood to be described in mathematical language. This preliminary requirement limits the scope of exact solutions because complete mathematical equations describing the fluid flow and the heat and mass transfer mechanisms can be written only for laminar flow. Even for laminar flow the equations are quite complicated, but solutions have been obtained for a number of simple systems such as flow over a flat plate or a circular cylinder (5).

Exact solutions are important because the assumptions made in the course of the analysis can be specified accurately and their validity can be checked by experiment. They also serve as a basis of comparison and as a check on simpler, but approximate, methods. Furthermore, the development of electronic computers has increased the range of problems amenable to mathematical solution, and results of computations for different systems are continually being published in the literature.

Approximate analysis of the boundary layer avoids the detailed mathematical description of the flow in the boundary layer. Instead, a plausible but simple equation is used to describe the velocity and temperature distributions in the boundary layer. The problem is then analyzed on a macroscopic basis by applying the equation of motion and the energy equation to the aggregate of the fluid particles contained within the boundary layer. This method is relatively simple; moreover, it yields solutions to problems that cannot be treated by an exact mathematical analysis. In instances where other solutions are available, they agree within engineering accuracy with the solutions obtained by this approximate method. The technique is not limited to laminar flow, but can also be applied to turbulent flow.

The analogy between heat, mass, and momentum transfer is a useful tool for analyzing turbulent transfer processes. Our knowledge of turbulent-exchange mechanisms is insufficient for us to write mathematical equations describing the temperature distribution directly, but the transfer mechanism can be described in terms of a simplified model. According to one such model, which has been widely accepted, a mixing motion in a direction perpendicular to the mean flow accounts for the transfer of momentum as well as energy. The mixing motion can be described on a statistical basis by a method similar to that used to picture the motion of gas molecules in the kinetic theory. There is by no means general agreement that this model corresponds to conditions actually existing in nature, but for practical purposes its use can be justified by the fact that experimental results are substantially in agreement with analytical predictions based on the hypothetical model.

Numerical methods can solve in an approximate form the exact equations of motion (6). The approximation results from the need to express the field variables (temperature, velocity, pressure) at discrete points in time and space rather than continuously. However, the solution can be made sufficiently accurate if care is taken in discretizing the exact equations. One of the most important advantages of numerical methods is that once the procedure has been programmed, solutions for different boundary conditions, property variables, and so on can be easily computed. Generally, numerical methods can handle complex boundary conditions easily.

4.8 DIMENSIONAL ANALYSIS

Dimensional analysis differs from other methods of approach in that it does not yield equations that can be solved. Instead, it combines several variables into dimensionless groups, such as the Nusselt number, which facilitate the interpretation and extend the range of application of experimental data. In practice, convective heat transfer coefficients are generally calculated from empirical equations obtained by correlating experimental data with the aid of dimensional analysis.

The most serious limitation of dimensional analysis is that it gives no information about the nature of a phenomenon. In fact, to apply dimensional analysis it is necessary to know beforehand what variables influence the phenomenon, and the success or failure of the method depends on the proper selection of these variables. It is therefore important to have at least a preliminary theory or a thorough physical understanding of a phenomenon before a dimensional analysis can be performed. However, once the pertinent variables are known, dimensional analysis can be applied to most problems by a routine procedure, which is outlined below.*

4.8.1 Primary Dimensions and Dimensional Formulas

The first step is to select a system of primary dimensions. The choice of the primary dimensions is arbitrary, but the dimensional formulas of all pertinent variables must be expressible in terms of them. In the SI system the primary dimensions of length L, time t, temperature T, and mass M are used.

The dimensional formula of a physical quantity follows from definitions or physical laws. For instance, the dimensional formula for the length of a bar is $[L]$ by definition.† The average velocity of a fluid particle is equal to a distance divided by the time interval taken to traverse it. The dimensional formula of velocity is therefore $[L/t]$ or $[Lt^{-1}]$ (i.e., a distance or length divided by a time). The units of velocity could be expressed in meters per second, feet per second, or miles per hour, since they all are a length divided by a time. The dimensional formulas and the symbols of physical quantities occurring frequently in heat transfer problems are given in Table 4.1.

---

* The algebraic theory of dimensional analysis will not be developed here. For a rigorous and comprehensive treatment of the mathematical background, chapters 3 and 4 of (3) are recommended.

† Square brackets indicate that the quantity has the dimensional formula stated within the brackets.

TABLE 4.1 IMPORTANT HEAT AND MASS
TRANSFER PHYSICAL QUANTITIES AND
THEIR DIMENSIONS

Quantity	Symbol	Dimensions in $MLtT$ System
Length	L, x	L
Time	t	t
Mass	M	M
Force	F	ML/t^2
Temperature	T	T
Heat	Q	ML^2/t^2
Velocity	u, U_∞	L/t
Acceleration	a, g	L/t^2
Work	W	ML^2/t^2
Pressure	p	$M/t^2 L$
Density	ρ	M/L^3
Internal energy	e	L^2/t^2
Enthalpy	h	L^2/t^2
Specific heat	c	$L^2/t^2 T$
Absolute viscosity	μ	M/Lt
Kinematic viscosity	$v = \mu/\rho$	L^2/t
Thermal conductivity	k	$ML/t^3 T$
Thermal diffusivity	α	L^2/t
Thermal resistance	R	Tt^3/ML^2
Coefficient of expansion	β	$1/T$
Surface tension	σ	M/t^2
Shear per unit area	τ	M/Lt^2
Heat transfer coefficient	$\bar{h}$	$M/t^3 T$
Mass flow rate	$\dot{m}$	M/t

4.8.2 Buckingham π Theorem

To determine the number of independent dimensionless groups required to obtain a relation describing a physical phenomenon, the Buckingham π theorem may be used.* According to this rule, the required number of independent dimensionless groups that can be formed by combining the physical variables pertinent to a problem is equal to the total number of these physical quantities n (e.g., density, viscosity, heat transfer coefficient) minus the number of primary dimensions m required to express the dimensional formulas of the n physical quantities. If we call these groups π_1, π_2, etc., the equation expressing the relationship among the variables has a solution of the form

$$F(\pi_1, \pi_2, \pi_3, \ldots) = 0 \qquad (4.36)$$

---

* A more rigorous rule, proposed by Van Driest (4), shows that the π theorem holds as long as the set of simultaneous equations formed by equating the exponents of each primary dimension to zero is linearly independent. If one equation in the set is a linear combination of one or more of the other equations (i.e., if the equations are linearly dependent), the number of dimensionless groups is equal to the total number of variables n minus the number of independent equations.

In a problem involving five physical quantities and three primary dimensions, $n - m$ is equal to two and the solution either has the form

$$F(\pi_1, \pi_2) = 0 \qquad (4.37)$$

or the form

$$\pi_1 = f(\pi_2) \qquad (4.38)$$

Experimental data for such a case can be presented conveniently by plotting π_1 against π_2. The resulting empirical curve reveals the functional relationship between π_1 and π_2, which cannot be deduced from dimensional analysis.

For a phenomenon that can be described in terms of three dimensionless groups (i.e., if $n - m = 3$), Eq. (4.36) has the form

$$F(\pi_1, \pi_2, \pi_3) = 0 \qquad (4.39)$$

but can also be written as

$$\pi_1 = f(\pi_2, \pi_3)$$

For such a case, experimental data can be correlated by plotting π_1 against π_2 for various values of π_3. Sometimes it is possible to combine two of the π's in some manner and to plot this parameter against the remaining π on a single curve.

4.8.3 Determination of Dimensionless Groups

A simple method for determining dimensionless groups will now be illustrated by applying it to the problem of correlating experimental convection heat transfer data for a fluid flowing across a heated tube. Exactly the same approach could be used for heat transfer in flow through a tube or over a plate.

From the description of the convective heat transfer process, it is reasonable to expect that the physical quantities listed in Table 4.2 are pertinent to the problem.

There are seven physical quantities and four primary dimensions. We therefore expect that three dimensionless groups will be required to correlate the data. To find these dimensionless groups, we write π as a product of the

TABLE 4.2 PERTINENT PHYSICAL QUANTITIES IN CONVECTIVE HEAT TRANSFER

Variable	Symbol	Dimensions
Tube diameter	D	$[L]$
Thermal conductivity of fluid	k	$[ML/t^3T]$
Free-stream velocity of fluid	U_∞	$[L/t]$
Density of fluid	ρ	$[M/L^3]$
Viscosity of fluid	μ	$[M/Lt]$
Specific heat at constant pressure	c_p	$[L^2/t^2T]$
Heat transfer coefficient	$\bar{h}_c$	$[M/t^2T]$

variables, each raised to an unknown power

$$\pi = D^a k^b U_\infty^d \rho^d \mu^e c_p^{\,f} \bar{h}_c^{\,g} \tag{4.40}$$

and substitute the dimensional formulas

$$\pi = [L]^a \left[\frac{ML}{t^3 T}\right]^b \left[\frac{L}{t}\right]^c \left[\frac{M}{L^3}\right]^d \left[\frac{M}{Lt}\right]^e \left[\frac{L^2}{t^2 T}\right]^f \left[\frac{M}{t^3 T}\right]^g \tag{4.41}$$

For π to be dimensionless, the exponents of each primary dimension must separately add up to zero. Equating the sum of the exponents of each primary dimension to zero, we obtain the set of equations

$$b + d + e + g = 0 \quad \text{for} \quad M$$
$$a + b + c - 3d - e + 2f = 0 \quad \text{for} \quad L$$
$$-3b - c - e - 2f - 3g = 0 \quad \text{for} \quad t$$
$$-b - f - g = 0 \quad \text{for} \quad T$$

Evidently any set of values of a, b, c, d, and e that simultaneously satisfies this set of equations will make π dimensionless. There are seven unknowns, but only four equations. We can therefore choose values for three of the exponents in each of the dimensionless groups. The only restriction on the choice of the exponents is that each of the selected exponents be independent of the others. An exponent is independent if the determinant formed with the coefficients of the remaining terms does not vanish (i.e., is not equal to zero).

Since $\bar{h}_c$, the convective heat transfer coefficient, is the variable we eventually want to evaluate, it is convenient to set its exponent g equal to unity. At the same time we let $c = d = 0$ to simplify the algebraic manipulations. Solving the equations simultaneously, we obtain $a = 1$, $b = -1$, $e = f = 0$, and the first dimensionless group is

$$\pi_1 = \frac{\bar{h}_c D}{k}$$

which we recognize as the *Nusselt number* $\overline{\text{Nu}}_D$.

For π_2 we select g equal to zero, so that $\bar{h}_c$ will not appear again, and let $a = 1$ and $f = 0$. Simultaneous solution of the equations with these choices yields $b = 0$, $c = d = 1$, $e = -1$, and

$$\pi_2 = \frac{U_\infty D \rho}{\mu}$$

This dimensionless group is a *Reynolds number* Re_D, with the tube diameter as the length parameter.

If we let $e = 1$ and $c = g = 0$, we obtain the third dimensionless group

$$\pi_3 = \frac{c_p \mu}{k}$$

which is the *Prandtl number* Pr.

We observe that, although the heat transfer coefficient is a function of six variables, with the aid of dimensional analysis the seven original variables have been combined into three dimensionless groups. According to Eq. (4.39), the functional relationship can be written

$$\overline{Nu}_D = f(Re_D, Pr)$$

and experimental data can now be correlated in terms of three variables instead of the original seven. The importance of this reduction in the variables becomes apparent when we attempt to correlate experimental data.

4.8.4 Correlation of Experimental Data

Suppose that, in a series of tests with air flowing over a 1-in.-OD pipe, the heat transfer coefficient has been measured experimentally at velocities ranging from 0.5 to 100 ft/s. This range of velocities corresponds to Reynolds numbers based on the diameter $D\rho U_\infty/\mu$ ranging from 250 to 50,000. Since the velocity was the only variable in these tests, the results are correlated in Fig. 4.10a by plotting the heat transfer coefficient $\overline{h}_c$ against the velocity U_∞. The resulting curve permits a direct determination of $\overline{h}_c$ at any velocity for the system used in the tests, but it cannot be used to determine the heat transfer coefficients for

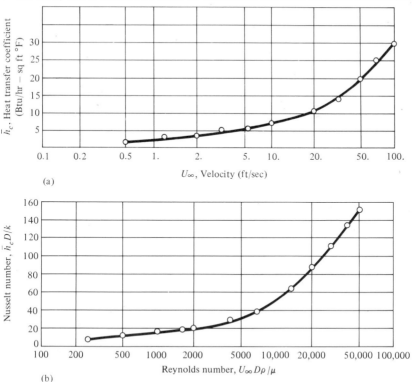

Figure 4.10 Variation of Nusselt number with Reynolds number for cross-flow of air over a pipe or a long cylinder.

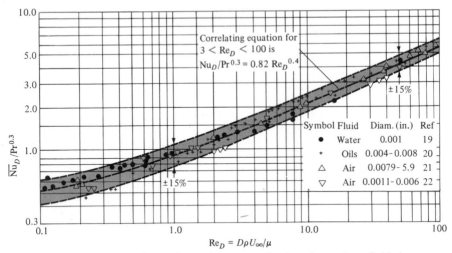

Figure 4.11   Correlation of experimental heat transfer data for various fluids in cross-flow over pipes and cylinders.

cylinders that are larger or smaller than the one used in the tests. Neither could the heat transfer coefficient be evaluated if the air were under pressure and its density were different from that used in the tests. Unless experimental data could be correlated more effectively, it would be necessary to perform separate experiments for every cylinder diameter, every density, etc. The amount of labor would obviously be enormous.

With the aid of dimensional analysis, however, the results of one series of tests can be applied to a variety of other problems. This is illustrated by Fig. 4.10b, where the data of Fig. 4.10a are replotted in terms of pertinent dimensionless groups. The abscissa in Fig. 4.10b is the Reynolds number $U_\infty D\rho/\mu$, and the ordinate is the Nusselt number $\bar{h}_c D/k$. This correlation of the data permits the evaluation of the heat transfer coefficient for air flowing over any size of pipe or wire as long as the Reynolds number of the system falls within the range covered in the experiment.

Experimental data obtained with air alone do not reveal the dependence of the Nusselt number on the Prandtl number, since the Prandtl number is a combination of physical properties whose value does not vary appreciably for gases. To determine the influence of the Prandtl number it is necessary to use different fluids. According to the preceding analysis, experimental data with several fluids whose physical properties yield a wide range of Prandtl numbers are necessary to complete the correlation.

In Fig. 4.11 the experimental results of several independent investigations for heat transfer between air, water, and oils in cross-flow over a tube or a wire are plotted for a wide range of temperatures, cylinder sizes, and velocities. The ordinate in Fig. 4.11 is the dimensionless quantity* $\overline{Nu}_D/Pr^{0.3}$ and the abscissa is Re_D. An inspection of the results shows that all of the data follow a single line reasonably well, so that they can be correlated empirically.

4.8.5 Principle of Similarity

The remarkable result of Fig. 4.11 can be explained by the principle of similarity. According to this principle, often called the model law, the behavior of two systems will be similar if the ratios of their linear dimensions, forces, velocities, etc., are the same. Under conditions of forced convection in geometrically similar systems, the velocity fields will be similar provided the ratio of inertia forces to viscous forces is the same in both fluids. The Reynolds number is the ratio of these forces, and consequently we expect similar flow conditions in forced convection for a given value of the Reynolds number. The Prandtl number is the ratio of two molecular transport properties, the kinematic viscosity $v = \mu/\rho$, which affects the velocity distribution, and the thermal diffusivity $k/\rho c_p$, which affects the temperature profile. In other words, it is a dimensionless group that relates the temperature distribution to the velocity distribution. Hence, in geometrically similar systems having the same Prandtl and Reynolds numbers, the temperature distribution will be similar. The Nusselt number is equal to the ratio of the temperature gradient at a fluid-to-surface interface to a reference-temperature gradient. We expect therefore that, in systems having similar geometries and similar temperature fields, the numerical values of the Nusselt numbers will be equal. This fact is borne out by the experimental results in Fig. 4.11.

Dimensional analyses have been performed for numerous heat and mass transfer systems, and Table 4.3 summarizes the most important dimensionless groups used in design.

4.9* ANALYTIC SOLUTION FOR LAMINAR BOUNDARY-LAYER FLOW OVER A FLAT PLATE[†]

In the preceding section we determined dimensionless groups for correlating experimental data of heat transfer by forced convection. We found that the Nusselt number depends on the Reynolds number and the Prandtl number, i.e.,

$$Nu = \phi(Re)\psi(Pr) \tag{4.42}$$

---

* Combining the Nusselt number with the Prandtl number for plotting the data is simply a matter of convenience. As mentioned previously, any combination of dimensionless parameters is satisfactory. The selection of the most convenient parameter is usually made on the basis of experience by trial and error with the aid of experimental results, although sometimes the characteristic groups are suggested by the results of analytic analyses.

[†] In the remainder of this chapter the mathematical details may be omitted in an introductory course without breaking the continuity of the presentation.

TABLE 4.3 DIMENSIONLESS GROUPS OF IMPORTANCE FOR HEAT AND MASS
TRANSFER

Group	Definition	Interpretation
Biot number (Bi)	$\dfrac{hL}{k_s}$	Ratio of internal thermal resistance of a solid body to its surface thermal resistance
Drag coefficient (C_f)	$\dfrac{\tau_s}{\rho U_\infty^2/2}$	Ratio of surface shear stress to free-stream kinetic energy
Eckert number (Ec)	$\dfrac{U_\infty^2}{c_p(T_s - T_\infty)}$	Kinetic energy of flow relative to boundary-layer enthalpy difference
Fourier number (Fo)	$\dfrac{\alpha t}{L^2}$	Dimensionless time; ratio of rate of heat conduction to rate of internal energy storage in a solid
Friction factor (f)	$\dfrac{\Delta p}{(L/D)(\rho U_m^2/2)}$	Dimensionless pressure drop for internal flow
Grashof number (Gr_L)	$\dfrac{g\beta(T_s - T_\infty)L^3}{\nu^2}$	Ratio of buoyancy to viscous forces
Colburn j factor (j_H)	$St Pr^{2/3}$	Dimensionless heat transfer coefficient
Lewis number (Le)	$\dfrac{\alpha}{D_{AB}}$	Ratio of molecular thermal diffusivity to mass diffusivity
Nusselt number (Nu_L)	$\dfrac{h_c L}{k_f}$	Dimensionless heat transfer coefficient; ratio of convection heat transfer to conduction in a fluid layer of thickness L
Peclet number (Pe_L)	$Re_L Pr$	Product of Grashof and Prandtl numbers
Prandtl number (Pr)	$\dfrac{c_p\mu}{k} = \dfrac{\nu}{\alpha}$	Ratio of molecular momentum diffusivity to thermal diffusivity
Rayleigh number (Ra)	$Gr_L Pr$	Product of Grashof and Prandtl numbers
Reynolds number (Re_L)	$\dfrac{U_\infty L}{\nu}$	Ratio of inertia and viscous forces
Schmidt number (Sc)	$\dfrac{\nu}{D_{AB}}$	Ratio of molecular momentum diffusivity to mass diffusivity
Sherwood number (Sh_L)	$\dfrac{h_m L}{D_{AB}}$	Ratio of convection mass transfer to diffusion in a slab of thickness L
Stanton number	$\dfrac{h_c}{\rho U_\infty c_p} = \dfrac{Nu_L}{Re_L Pr}$	Dimensionless heat transfer coefficient

In the next few sections we shall consider analytical methods for determining
the functional relations in Eq. (4.42) for low-speed flow over a flat plate. This
system has been selected primarily because it is the simplest to analyze. But
the results have many practical applications; for instance, they are good approximations for flow over the surfaces of streamlined bodies such as airplane wings
or turbine blades.

In view of the differences in the flow characteristics, the frictional forces
as well as the heat and mass transfer are governed by different relations for

laminar and turbulent flow. We will first treat the laminar boundary layer, which is amenable to an exact and an approximate method of solution. The turbulent boundary layer is taken up in Section 4.11.

To determine the forced-convection heat transfer coefficient and the friction coefficient for incompressible flow over a flat surface we must satisfy the continuity, momentum, and energy equations simultaneously. These relations were derived in Section 4.5 and are for convenience repeated below.

Continuity:

$$\frac{\partial u}{\partial x} + \frac{\partial v}{\partial y} = 0 \tag{4.11}$$

Momentum:

$$\rho\left(u\frac{\partial u}{\partial x} + v\frac{\partial u}{\partial y}\right) = \mu\frac{\partial^2 u}{\partial y^2} - \frac{\partial p}{\partial x} \tag{4.12}$$

Energy:

$$u\frac{\partial T}{\partial x} + v\frac{\partial T}{\partial y} = \alpha\frac{\partial^2 T}{\partial y^2} \tag{4.13}$$

4.9.1 Boundary-Layer Thickness and Skin Friction

Equation (4.12) must be solved simultaneously with the continuity equation, Eq. (4.11), in order to determine the velocity distribution, boundary-layer thickness, and friction force at the wall. These equations are solved by first defining a stream function $\psi(x, y)$ which automatically satisfies the continuity equation, or

$$u = \frac{\partial \psi}{\partial y} \quad \text{and} \quad v = -\frac{\partial \psi}{\partial x}$$

Introducing the new variable

$$\eta = y\sqrt{\frac{U_\infty}{vx}}$$

we can let

$$\psi = \sqrt{vxU_\infty}\, f(\eta)$$

where $f(\eta)$ denotes a dimensionless stream function. In terms of $f(\eta)$, the velocity components are

$$u = \frac{\partial \psi}{\partial y} = \frac{\partial \psi}{\partial \eta}\frac{\partial \eta}{\partial y} = U_\infty\frac{d[f(\eta)]}{d\eta}$$

and

$$v = -\frac{\partial \psi}{\partial x} = \frac{1}{2}\sqrt{\frac{vU_\infty}{x}}\left\{\frac{d[f(\eta)]}{d\eta}\eta - f(\eta)\right\}$$

Expressing $\partial u/\partial x$, $\partial u/\partial y$, and $\partial^2 u/\partial y^2$ in terms of η and inserting the resulting expressions in the momentum equation yields the ordinary, nonlinear, third-order differential equation

$$f(\eta)\frac{d^2[f(\eta)]}{d\eta^2} + 2\frac{d^3[f(\eta)]}{d\eta^3} = 0$$

which can be solved subject to the three boundary conditions that

$$f(\eta) = 0 \qquad \frac{d[f(\eta)]}{d\eta} = 0 \qquad \text{at } \eta = 0$$

and

$$\frac{d[f(\eta)]}{d\eta} = 1 \qquad \text{at } \eta = \infty$$

The solution to this differential equation was obtained numerically by Blasius, in 1908 (7). The significant results are shown in Figs. 4.12 and 4.13.

In Fig. 4.12 the Blasius velocity profiles in the laminar boundary on a flat plate are plotted in dimensionless form together with experimental data obtained by Hansen (8). The ordinate is the local velocity in the x direction u divided by the free-stream velocity u_∞, and the abscissa is a dimensionless distance parameter $(y/x)\sqrt{(\rho U_\infty x)/\mu}$. We note that a single curve is sufficient to correlate the velocity distributions at all stations along the plate. The velocity u reaches 99

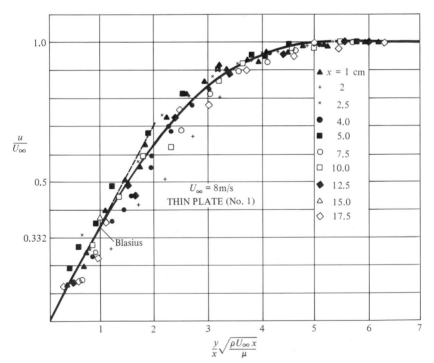

Figure 4.12 Velocity profile in a laminar boundary layer according to Blasius with experimental data of Hansen (8). Courtesy of the National Advisory Committee for Aeronautics, NACA TM 585.

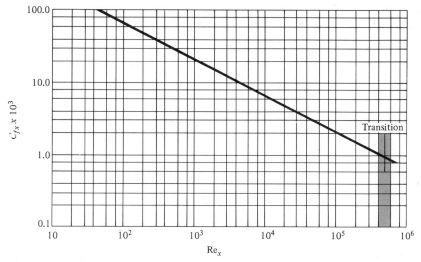

Figure 4.13   Local friction coefficient versus Reynolds number based on distance from leading edge for laminar flow over a flat plate.

percent of the free-stream value U_∞ at $(y/x)\sqrt{(\rho U_\infty x)/\mu} = 5.0$. If we define the hydrodynamic boundary-layer thickness as that distance from the surface at which the local velocity u reaches 99 percent of the free-stream value u_∞, the boundary-layer thickness δ becomes

$$\delta = \frac{5x}{\sqrt{Re_x}} \tag{4.43}$$

where $Re_x = \rho U_\infty x/\mu$, the local Reynolds number. Equation (4.43) satisfies the qualitative description of the boundary-layer growth, δ being zero at the leading edge $(x = 0)$ and increasing with x along the plate. At any station, that is, given value of x, the thickness of the boundary layer is inversely proportional to the square root of the local Reynolds number. Hence, an increase in velocity will result in a decrease of the boundary-layer thickness.

The shear force at the wall can be obtained from the velocity gradient at $y = 0$ in Fig. 4.12. We see that

$$\frac{\partial(u/U_\infty)}{\partial(y/x)\sqrt{Re_x}}\bigg|_{y=0} = 0.332$$

and thus at any specified value of x the velocity gradient at the surface is

$$\frac{\partial u}{\partial y}\bigg|_{y=0} = 0.332\frac{U_\infty}{x}\sqrt{Re_x}$$

Substituting this velocity gradient in the equation for the shear, the wall shear per unit area τ_s becomes

$$\tau_s = \mu\frac{\partial u}{\partial y}\bigg|_{y=0} = 0.332\,\mu\frac{U_\infty}{x}\sqrt{Re_x} \tag{4.44}$$

We note that the wall shear near the leading edge is very large and decreases with increasing distance from the leading edge.

For a graphic presentation it is more convenient to use dimensionless coordinates. Dividing both sides of Eq. (4.44) by the velocity pressure of the free stream $\rho U_\infty^2/2$, we obtain

$$C_{fx} = \frac{\tau_s}{\rho U_\infty^2/2} = \frac{0.664}{\sqrt{Re_x}} \tag{4.45}$$

where C_{fx} is the dimensionless local drag or friction coefficient. Figure 4.13 is a plot of C_{fx} against Re_x and shows the variation of the local friction coefficient graphically. The average friction coefficient is obtained by integrating Eq. (4.45) between the leading edge $x = 0$ and $x = L$:

$$\bar{C}_f = \frac{1}{L} \int_0^L C_{fx}\, dx = 1.33 \sqrt{\frac{U_\infty \rho L}{u}} \tag{4.46}$$

Thus, for laminar flow over a flat plate the average friction coefficient $\bar{C}_f$ is equal to twice the value of the local friction coefficient at $x = L$.

4.9.2 Convection Heat and Mass Transfer

The energy and species conservation equations for a laminar boundary layer are

$$u \frac{\partial T}{\partial x} + v \frac{\partial T}{\partial y} = \alpha \frac{\partial^2 T}{\partial y^2} \tag{4.16b}$$

$$u \frac{\partial C_A}{\partial x} + v \frac{\partial C_A}{\partial y} = D_{AB} \frac{\partial^2 C_A}{\partial y^2} \tag{4.16c}$$

The velocities in the energy and species conservation equations, u and v, have the same values at any point x, y as in the fluid dynamic equation, Eq. (4.12). For the case of the flat plate, Pohlhausen (9) used the velocities calculated previously by Blasius (7) to obtain the solution of the heat transfer problem. Without considering the details of this mathematical solution, we can obtain significant results by comparing Eq. (4.16) with Eq. (4.12), the momentum equation. The three equations are similar; in fact $u(x, y)$ is also a solution for the temperature distribution $T(x, y)$ if $v = \alpha$ and if the temperature of the plate T_s is constant. We can easily verify this by replacing the symbol T in Eq. (4.16b) by the symbol u and noting that the boundary conditions for both T and u are identical. If we use the surface temperature as our datum and let the variable in Eq. (4.16b) be $(T - T_s)/(T_\infty - T_s)$, then the boundary conditions are

$$\frac{T - T_s}{T_\infty - T_s} = 0 \quad \text{and} \quad \frac{u}{U_\infty} = 0 \qquad \text{at } y = 0$$

$$\frac{T - T_s}{T_\infty - T_s} = 1 \quad \text{and} \quad \frac{u}{U_\infty} = 1 \qquad \text{at } y \to \infty$$

where T_∞ is the free-stream temperature.

The condition that $v = a$ corresponds to a Prandtl number of unity since

$$\text{Pr} = \frac{c_p \mu}{k} = \frac{v}{\alpha}$$

For $\text{Pr} = 1$ the velocity distribution is therefore identical to the temperature distribution. An interpretation in terms of physical processes is that the transfer of momentum is analogous to the transfer of heat when $\text{Pr} = 1$. The physical properties of most gases are such that they have Prandtl numbers ranging from 0.65 to 1.0, and the analogy is therefore satisfactory. Liquids, on the other hand, have Prandtl numbers considerably different from unity, and the preceding analysis cannot be applied directly (10).

Similarly, $u(x, y)$ is also a solution for the species concentration $C_A(x, y)$ when the concentration at the surface is constant and $D_{AB} = \lambda$, a condition that corresponds to the Schmidt number being equal to unity, that is, $\text{Sc} = \lambda/D_{AB} = 1$. In other words, the velocity and concentration distributions are identical in boundary layer for these conditions.

Using the analytical results of Pohlhausen's work, the temperature distribution in the laminar boundary layer for $\text{Pr} = 1$ can be modified empirically to include fluids having Prandtl numbers different from unity. In Fig. 4.14, theoretically calculated temperature profiles in the boundary layer are shown for values of Pr of 0.6, 0.8, 1.0, 3.0, 7.0, 15, and 50. We now define a thermal boundary-layer thickness δ_{th} as the distance from the surface at which the temperature difference between the wall and the fluid reaches 99 percent of the free-stream value. Inspection of the temperature profiles shows that the thermal boundary layer is larger than the hydrodynamic boundary layer for fluids having Pr less than unity, but smaller when Pr is greater than unity. According to Pohlhausen's calculations, the relationship between the thermal and hydrodynamic boundary layer is approximately

$$\delta_{th} = \frac{\delta}{\text{Pr}^{1/3}} \tag{4.47}$$

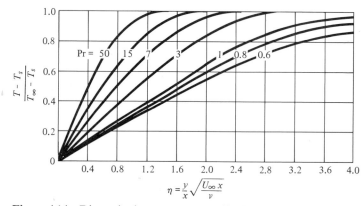

Figure 4.14 Dimensionless temperature distributions in a fluid flowing over a heated plate for various Prandtl numbers.

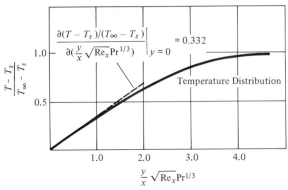

Figure 4.15 Correlation of dimensionless temperature distribution for laminar flow over a heated plate at uniform temperature.

Using the same correction factor, $Pr^{1/3}$, at any distance from the surface, the curves of Fig. 4.14 are replotted in Fig. 4.15. The new abscissa is $Pr^{1/3}(y/x)\sqrt{Re_x}$ and the ordinate is the dimensionless temperature $(T - T_s)/(T_\infty - T_s)$, where T is the local temperature of the fluid, T_s the surface temperature of the plate, and T_∞ the free-stream temperature. This modification of the ordinate brings the temperature profiles for a wide range of Prandtl numbers together on a single line, the curve for $Pr = 1$.

4.9.3 Evaluation of the Convective Heat Transfer Coefficient

The rate of heat transfer by convection and the convective heat transfer coefficient can now be determined. The dimensionless temperature gradient at the surface (at $y = 0$) is

$$\frac{\partial[(T - T_s)/(T_\infty - T_s)]}{\partial[(y/x)\sqrt{Re_x}\ Pr^{1/3}]}\bigg|_{y=0} = 0.332$$

Therefore, at any specified value of x

$$\frac{\partial T}{\partial y}\bigg|_{y=0} = 0.332\ \frac{Re_x^{1/2}Pr^{1/3}}{x}(T_\infty - T_s) \tag{4.48}$$

and the local rate of heat transfer by convection per unit area becomes, on substituting $\partial T/\partial y$ from Eq. (4.48),

$$q_c'' = -k\frac{\partial T}{\partial y}\bigg|_{y=0} = -0.332k\ \frac{Re_x^{1/2}Pr^{1/3}}{x}(T_\infty - T_s) \tag{4.49}$$

The total rate of heat transfer from a plate of width b and length L, obtained by integrating q_c'' from Eq. (4.49) between $x = 0$ and x L, is

$$q = 0.664kRe_L^{1/2}Pr^{1/3}b(T_s - T_\infty) \tag{4.50}$$

The local convective heat transfer coefficient is

$$h_{cx} = \frac{q_c''}{(T_s - T_\infty)} = 0.332\frac{k}{x}\ Re_x^{1/2}Pr^{1/3} \tag{4.51}$$

and the corresponding local Nusselt number is

$$\text{Nu}_x = \frac{h_{cx}x}{k} = 0.332\text{Re}_x^{1/2}\text{Pr}^{1/3} \qquad (4.52)$$

The average Nusselt number $\bar{h}_c L/k$ is obtained by integrating the right-hand side of Eq. (4.51) between $x = 0$ and $x = L$ and dividing the result by L to obtain $\bar{h}_c$, the average value of h_{cx}; multiplying $\bar{h}_c$ by L/k gives

$$\overline{\text{Nu}}_L = 0.664\text{Re}_L^{1/2}\text{Pr}^{1/3} \qquad (4.53)$$

The average value of the Nusselt number over a length L of the plate is therefore twice the local value of Nu_x at $x = L$. It can easily be verified that the same relation between the average and local value holds also for the heat transfer coefficient, that is,

$$\bar{h}_c = 2h_{c(x=L)} \qquad (4.54)$$

In practice, the physical properties in Eqs. (4.47) to (4.53) vary with temperature, while for the purpose of analysis it was assumed that the physical properties are constant. Experimental data have been found to agree satisfactorily with the results predicted analytically if the properties are evaluated at a mean temperature halfway between that of the wall and the free-stream temperature.

EXAMPLE 4.4 _____

Air at 60°F and at a pressure of 1 atm is flowing over a plate at a velocity of 10 ft/s. If the plate is 1 ft wide and at 140°F, calculate the following quantities at $x = 1$ ft and $x = x_c$ in English and SI units:

(a) Boundary-layer thickness
(b) Local friction coefficient
(c) Average friction coefficient
(d) Local drag or shearing stress due to friction
(e) Thickness of thermal boundary layer
(f) Local convective heat transfer coefficient
(g) Average convective heat transfer coefficient
(h) Rate of heat transfer by convection

Solution. Properties of air at 100°F in English units are:

$$\rho = 0.071 \text{ lb}_m/\text{ft}^3$$
$$c_p = 0.240 \text{ Btu/lb}_m \text{ °F}$$
$$\mu = 1.285 \times 10^{-5} \text{ lb}_m/\text{ft s}$$
$$k = 0.0154 \text{ Btu/h ft °F}$$
$$\text{Pr} = 0.72$$

The local Reynolds number at $x = 1$ ft (0.305 m) is

$$\text{Re}_{x=1} = \frac{U_\infty \rho x}{\mu} = \frac{(10 \text{ ft/s})(0.071 \text{ lb}_m/\text{ft}^3)(1 \text{ ft})}{1.285 \times 10^{-5} \text{ lb}_m/\text{ft s}} = 55{,}200$$

TABLE 4.4 RESULTS FOR EXAMPLE 4.4

Part	Symbol	English units	Equation used	Result (x = 1 ft)	Result (x = 9 ft)	SI units	Result (x = 0.305 m)	Result (x = 2.74 m)
a	δ	ft	(4.43)	0.0212	0.064	m	0.00657	0.0195
b	C_{fx}		(4.45)	0.00282	0.00094	—	0.00282	0.00094
c	$\overline{C}_f$		(4.46)	0.00564	0.00188	—	0.00564	0.00188
d	τ_s	lb_f/ft^2	(4.44)	3.12×10^{-4}	1.04×10^{-4}	N/m^2	0.0149	0.00498
e	δ_{th}	ft	(4.47)	0.0236	0.0715	m	0.0072	0.0218
f	h_{cx}	Btu/h ft °F	(4.51)	1.03	0.36	W/m^2 K	5.85	2.046
g	$\overline{h}_c$	Btu/h ft² °F	(4.54)	2.06	0.72	W/m^2 K	11.7	4.092
h	q_c	Btu/h	(4.50)	206	648	W	60.36	189.9

Assuming that the critical Reynolds number is 5×10^5, the critical distance is

$$x_c = \frac{5 \times 10^5 \, \mu}{U_\infty \rho} = \frac{(5 \times 10^5)(1.285 \times 10^{-5} \, \text{lb}_\text{m}/\text{ft s})}{(10 \, \text{ft/s})(0.071 \, \text{lb}_\text{m}/\text{ft}^3)} = 9 \, \text{ft} \, (2.74 \, \text{m})$$

The desired quantities are determined by substituting appropriate values of the variable into the pertinent equations. The results of the calculations are shown in Table 4.4, and it is suggested that the reader verify them.

A useful relation between the local Nusselt number Nu_x and the corresponding friction coefficient C_{fx} is obtained by dividing Eq. (4.52) by $\text{Re}_x \text{Pr}^{1/3}$, or

$$\left(\frac{\text{Nu}_x}{\text{Re}_x \text{Pr}} \right) \text{Pr}^{2/3} = \frac{0.322}{\text{Re}_x^{1/2}} = \frac{C_{fx}}{2} \tag{4.55}$$

The dimensionless ratio $\text{Nu}_x/\text{Re}_x \text{Pr}$ is known as the *Stanton number* St_x. According to Eq. (4.55) the Stanton number times the Prandtl number raised to the two-thirds power is equal to one-half the value of the friction coefficient. This relation between heat transfer and fluid friction was proposed by Colburn (11) and illustrates the interrelationship of the two processes.

4.10* APPROXIMATE INTEGRAL BOUNDARY-LAYER ANALYSIS

To circumvent the problems involved in solving the partial differential equations of the boundary layer, an integral approach can be used. For that purpose let us consider an elemental control volume that extends from the wall to beyond the limit of the boundary layer in the y direction, is dx thick in the x direction, and has a unit width in the z direction, as shown in Fig. 4.16. To obtain a relationship for the net momentum in-flow and the net energy transport,

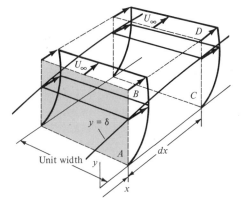

Figure 4.16 Control volume for integral conservation of momentum analysis.

we proceed in a manner similar to that used to derive the boundary-layer equations in the preceding section.

The momentum flow across the face, AB in Fig. 4.16, will be

$$\int_0^\delta \rho u^2 \, dy$$

Similarly, the momentum flow across the face CD will be

$$\int_0^\delta \rho u^2 \, dy + \frac{d}{dx} \int_0^\delta \rho u^2 \, dy \, dx$$

However, fluid also enters the control volume across face BD at the rate

$$\frac{d}{dx} \int_0^\delta \rho u \, dy \, dx$$

This quantity is the difference between the rate of flow leaving across face CD and that entering across face AB. Since the fluid entering across BD has a velocity component in the x direction equal to the free-stream velocity U_∞, the flow of x momentum into the control volume across the upper face is

$$U_\infty \frac{d}{dx} \int_0^\delta \rho u \, dy \, dx$$

Adding up the x-momentum components gives

$$\frac{d}{dx} \int_0^\delta \rho u^2 \, dy \, dx - U_\infty \frac{d}{dx} \int_0^\delta \rho u \, dy \, dx = -\frac{d}{dx} \int_0^\delta \rho u(U_\infty - u) \, dy$$

There will be no shear across the face BD, since this face is outside the boundary layer, where du/dy is equal to zero. There is, however, a shear force τ_w acting at the fluid-solid interface, and there will be pressure forces acting on faces AB and CD. Writing out the net forces acting on the control volume and adding them yields the relation

$$p_x \delta - \left(p_x + \frac{dp_x}{dx} dx \right) \delta - \tau_w \, dx = -\delta \frac{dp_x}{dx} dx - \tau_w \, dx \qquad (4.56)$$

For flow over a flat plate the pressure gradient in the x direction may be neglected and the momentum equation can then be written in the form

$$\frac{d}{dx} \int_0^\delta \rho u(U_\infty - u) \, dy = \tau_w \qquad (4.57)$$

The integral energy equation may be derived in a similar fashion. In this case, however, a control volume extending beyond the limits of both the temperature and velocity boundary layer must be used in the derivation (see Fig. 4.17). The first law of thermodynamics demands that energy in the form of enthalpy, kinetic energy, and heat, as well as shear work, should be considered. For low velocities, however, the kinetic energy terms and the shear work are small compared to the other quantities and may be neglected. Then, the rate

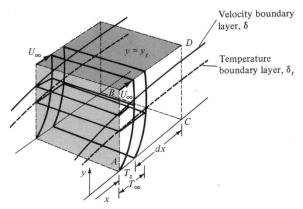

Figure 4.17 Control volume for integral conservation of energy analysis.

at which enthalpy enters across face *AB* is given by

$$\int_0^{y_s} c_p \rho u T \, dy$$

whereas the rate of enthalpy flow across face *CD* is

$$\int_0^{y_s} c_p \rho u T \, dy + \frac{d}{dx} \int_0^{y_s} c_p \rho u T \, dy \, dx$$

The enthalpy carried into the control volume across the upper face is given by

$$c_p T_s \frac{d}{dx} \int_0^{y_s} \rho u \, dy \, dx$$

Finally, heat will be conducted across the interface between the fluid and solid surface at the rate

$$-k \, dx \left(\frac{\partial T}{\partial y} \right)_{y=0}$$

Adding up all the energy quantities yields the integral equation for the conservation of energy in the form

$$c_p T_\infty \frac{d}{dx} \int_0^{y_s} \rho u \, dy \, dx - \frac{d}{dx} \int_0^{y_s} \rho c_p T u \, dy \, dx - k \, dx \left(\frac{\partial T}{\partial y} \right)_{y=0} = 0 \quad (4.58)$$

It should be noted, however, that outside the limit of the temperature boundary layer, the temperature equals the free-stream temperature, T_∞, so that integration need only be taken up to $y = \delta_t$. Equation (4.58) therefore can be simplified to the form

$$\frac{d}{dx} \int_0^{\delta_t} (T_\infty - T) u \, dy - \alpha \left(\frac{\partial T}{\partial y} \right)_{y=0} = 0 \quad (4.59)$$

which is usually known as the integral energy equation of the laminar boundary layer for low-speed flow.

4.10.1 Evaluation of Heat Transfer and Friction Coefficients in Laminar Flow

In the approximate integral method the first step is to assume velocity and temperature contours in the form of polynomials. Then, the coefficient in the polynomial will be evaluated to satisfy the boundary conditions. Assuming a four-term polynomial for the velocity distribution (12)

$$u(y) = a + by + cy^2 + dy^3 \tag{4.60}$$

the constants are evaluated by applying the boundary conditions

$$\text{at } y = 0: \quad u = 0 \quad \text{and therefore} \quad a = 0$$

$$u = v = 0 \quad \text{and therefore} \quad \frac{\partial^2 u}{\partial y^2} = 0$$

$$y = \delta: \quad u = U_\infty \quad \text{and} \quad \frac{\partial u}{\partial y} = 0$$

The conditions above provide four equations for the evaluation of the four unknown coefficients in terms of the free-stream velocity and the boundary-layer thickness. It can easily be verified that the coefficients that satisfy these boundary conditions are

$$a = 0 \qquad b = \frac{3}{2}\frac{U_\infty}{\delta} \qquad c = 0 \qquad d = -\frac{U_\infty}{2\delta^3}$$

Substituting these coefficients in Eq. (4.60) and dividing through by the free-stream velocity U_∞ to nondimensionalize the result yields

$$\frac{u}{U_\infty} = \frac{3}{2}\frac{y}{\delta} - \frac{1}{2}\left(\frac{y}{\delta}\right)^3 \tag{4.61}$$

Substituting Eq. (4.61) for the velocity distribution in the integral momentum equation [Eq. (4.57)] yields

$$\frac{d}{dx}\int_0^\delta \rho U_\infty^2 \left[\frac{3}{2}\frac{y}{\delta} - \frac{1}{2}\left(\frac{y}{\delta}\right)^3\right]\cdot\left[1 - \frac{3}{2}\frac{y}{\delta} + \frac{1}{2}\left(\frac{y}{\delta}\right)^3\right]dy = \tau_w = \mu\left(\frac{du}{dy}\right)_{y=0} \tag{4.62}$$

The wall shear stress τ_w can be obtained by evaluating the velocity gradient from Eq. (4.61) at $y = 0$. Substituting for τ_w and performing the integration in Eq. (4.62) yields

$$\frac{d}{dx}\left(\rho U_\infty^2 \frac{39\delta}{280}\right) = \frac{3}{2}\mu\frac{U_\infty}{\delta} \tag{4.63}$$

Equation (4.63) may be rearranged and integrated to obtain the boundary-layer thickness in terms of the viscosity, distance from the leading edge, and free-stream velocity distribution:

$$\frac{\delta^2}{2} = \frac{140\nu x}{13U_\infty} + C \tag{4.64}$$

Since $\delta = 0$ at the leading edge (i.e., at $x = 0$), the coefficient C in the preceding relation must equal 0 and

$$\delta^2 = \frac{280 v x}{13 U_\infty}$$

or

$$\frac{\delta}{x} = \frac{4.64}{Re_x^{1/2}} \tag{4.65}$$

To evaluate the friction coefficient substitute Eq. (4.61) into Eq. (4.62)

$$\tau_w = \mu \frac{du}{dy}\bigg|_{y=0} = \mu \frac{3}{2} \frac{U_\infty}{\delta}$$

Substituting for δ from Eq. (4.65) gives

$$\tau_w = \frac{3}{9.28} \frac{\mu U_\infty}{x} Re_x^{1/2}$$

and the friction coefficient C_{fx} is

$$C_{fx} = \frac{\tau_{wx}}{\frac{1}{2}\rho U_\infty^2} = \frac{0.647}{Re_x^{1/2}} \tag{4.66}$$

We next turn to the energy equation and propose a temperature distribution in the boundary layer of the same form as the velocity distribution:

$$T(y) = e + fy + gy^2 + hy^3 \tag{4.67}$$

The boundary conditions for the temperature distribution are that at $y = 0$, $T = T_s$; at $y = \delta_t$ (the thickness of the temperature boundary layer), $T = T_\infty$, and $dT/dy = 0$. Also, from Eq. (4.16a), d^2T/dy^2 at $y = 0$ must be zero because both u and v are zero at the interface. From these conditions it follows that the constants are

$$e = T_s \qquad f = \frac{3}{2} \frac{T_\infty}{\delta_t} \qquad g = 0 \qquad h = -\frac{T_\infty}{2\delta_t^2}$$

If the variable in the energy equation is taken as the temperature in the fluid minus the wall temperature, the temperature distribution can be written in the dimensionless form

$$\frac{T - T_s}{T_\infty - T_s} = \frac{3}{2}\frac{y}{\delta_t} - \frac{1}{2}\left(\frac{y}{\delta_t}\right)^3 \tag{4.68}$$

Using Eqs. (4.68) and (4.61) for $T - T_s$ and u, respectively, the integral in Eq. (4.59) can be written as

$$\int_0^{\delta_t} (T_\infty - T)u\,dy = \int_0^{\delta_t} [(T_\infty - T_s) - (T - T_s)]u\,dy$$

$$= (T_\infty - T_s)U_\infty \int_0^{\delta_t} \left[1 - \frac{3}{2}\frac{y}{\delta_t} \div \frac{1}{2}\left(\frac{y}{\delta_t}\right)^3\right]\left[\frac{3}{2}\frac{y}{\delta} - \frac{1}{2}\left(\frac{y}{\delta}\right)^3\right]dy$$

Performing the multiplication under the integral sign, we obtain

$$(T_\infty - T_s)U_\infty \int_0^{\delta_t} \left(\frac{3}{2\delta}y - \frac{9}{4\delta\delta_t}y^2 + \frac{3}{4\delta\delta_t{}^3}y^4 - \frac{1}{2\delta^3}y^3 + \frac{3}{4\delta_t\delta^3}y^4 - \frac{1}{4\delta_t{}^3\delta^3}y^6 \right) dy$$

which yields, after integration,

$$(T_\infty - T_s)U_\infty \left(\frac{3}{4}\frac{\delta_t{}^2}{\delta} - \frac{3}{4}\frac{\delta_t{}^2}{\delta} + \frac{3}{20}\frac{\delta_t{}^2}{\delta} - \frac{1}{8}\frac{\delta_t{}^4}{\delta^3} + \frac{3}{20}\frac{\delta_t{}^4}{\delta^3} - \frac{1}{28}\frac{\delta_t{}^4}{\delta^3} \right)$$

If we let $\zeta = \delta_t/\delta$, the expression above can be written

$$(T_\infty - T_s)U_\infty \delta \left(\frac{3}{20}\zeta^2 - \frac{3}{280}\zeta^4 \right)$$

For fluids having a Prandtl number equal to or larger than unity, ζ is equal to or less than unity and the second term in the parentheses can be neglected compared to the first.* Substituting this approximate form for the integral in Eq. (4.59), we obtain

$$\frac{3}{20}U_\infty(T_s - T_\infty)\zeta^2 \frac{\partial\delta}{\partial x} = \alpha \left. \frac{\partial T}{\partial y} \right|_{y=0} = \frac{3}{2}\alpha \frac{T_\infty - T_s}{\delta\zeta}$$

or

$$\frac{1}{10}U_\infty\zeta^3\delta \frac{\partial\delta}{\partial x} = \alpha$$

From Eq. (4.65) we obtain

$$\delta \frac{\partial\delta}{\partial x} = 10.75 \frac{\nu}{U_\infty}$$

and with this expression we get

$$\zeta^3 = \frac{10}{10.75}\frac{\alpha}{\nu}$$

or

$$\delta_t = 0.976\delta \mathrm{Pr}^{-1/3} \tag{4.69}$$

Except for the numerical constant (0.976 compared with 1.0), the foregoing result is in agreement with the exact calculations of Pohlhausen (9).

The rate of heat flow by convection from the plate per unit area is, from Eqs. (4.1) and (4.68),

$$q_c'' = -k \left. \frac{\partial T}{\partial y} \right|_{y=0} = -\frac{3}{2}\frac{k}{\delta_t}(T_\infty - T_s)$$

Substituting Eqs. (4.65) and (4.69) for δ and δ_t yields

$$q'' = -\frac{3}{2}\frac{k}{x}\frac{\mathrm{Pr}^{1/3}\mathrm{Re}_x{}^{1/2}}{(0.976)(4.64)}(T_\infty - T_s) = 0.33 \frac{k}{x}\mathrm{Re}_x{}^{1/2}\mathrm{Pr}^{1/3}(T_s - T_\infty) \tag{4.70}$$

* For liquid metals, which have $\mathrm{Pr} \ll 1$, $\zeta > 1$ and the second term cannot be neglected.

and the local Nusselt number Nu_x is

$$Nu_x = \frac{h_{cx}x}{k} = \frac{q_c''}{(T_s - T_\infty)}\frac{x}{k} = 0.33 Re_x^{1/2} Pr^{1/3} \tag{4.71}$$

This result is in excellent agreement with the result of the exact analysis of (9).

The foregoing example illustrates the usefulness of the approximate boundary-layer analysis. Guided by a little physical insight and intuition, this technique yields satisfactory results without the mathematical complications inherent in the exact boundary-layer equations. The approximate method has been applied to many other problems, and the results are available in the literature.

4.11* ANALOGY BETWEEN MOMENTUM, HEAT, AND SPECIES TRANSFER IN TURBULENT FLOW OVER A FLAT SURFACE

In a majority of practical applications the flow in the boundary layer is turbulent rather than laminar. Qualitatively, the exchange mechanism in turbulent flow can be pictured as a magnification of the molecular exchange in laminar flow. In steady laminar flow, fluid particles follow well-defined streamlines. Heat and momentum are transferred across streamlines only by molecular diffusion and the cross flow is so small that when a colored dye is injected into the fluid at some point, it follows a streamline without appreciable mixing. In turbulent flow, on the other hand, the color will be distributed over a wide area a short distance downstream from the point of injection. The mixing mechanism consists of rapidly fluctuating eddies that transport fluid particles in an irregular manner. Groups of particles collide with each other at random, establish cross flow on a macroscopic scale, and effectively mix the fluid. Since the mixing in turbulent flow is on a macroscopic scale with groups of particles transported in a zigzag path through the fluid, the exchange mechanism is many times more effective than in laminar flow. As a result, the rates of heat and momentum transfer in turbulent flow and the associated friction and heat-transfer coefficients are many times larger than in laminar flow.

If turbulent flow at a point is averaged over a long period of time (as compared with the period of a single fluctuation) the time-mean properties and the velocity of the fluid are constant if the average flow remains steady. It is therefore possible to describe each fluid property and the velocity in turbulent flow in terms of a *mean value* that does not vary with time and a *fluctuating component* which is a function of time. To simplify the problem, consider a two-dimensional flow (Fig. 4.18) in which the mean value of velocity is parallel to the x direction. The instantaneous velocity components u and v can then be expressed in the form

$$u = \bar{u} + u'$$
$$v = v' \tag{4.72}$$

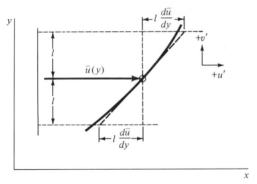

Figure 4.18   Mixing length for momentum transfer in turbulent flow.

where the bar over a symbol denotes the temporal mean value, and the prime denotes the instantaneous deviation from the mean value. According to the model used to describe the flow,

$$\bar{u} = \frac{1}{\theta^*} \int_0^{\theta^*} u\, dt \tag{4.73}$$

where θ^* is a time interval large compared with the period of the fluctuations. Figure 4.19 shows qualitatively the time variation of u and u'. From Eq. (4.73) or from an inspection of the graph it is apparent that the time average of u' is zero (i.e., $\bar{u}' = 0$). A similar argument shows that $\bar{v}'$ and $\overline{(\rho v)}'$ are also zero.

The fluctuating velocity components continuously transport mass, and consequently momentum, across a plane normal to the y direction. The instantaneous rate of transfer in the y direction of x momentum per unit area at any point is

$$-(\rho v)'(\bar{u} + u')$$

where the minus sign, as will be shown later, takes account of the statistical correlation between u' and v'.

The time average of the x-momentum transfer gives rise to an *apparent* turbulent shear or Reynolds stress τ_t, defined by

$$\tau_t = -\frac{1}{\theta^*} \int_0^{\theta^*} (\rho v)'(\bar{u} + u')\, dt \tag{4.74}$$

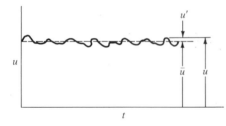

Figure 4.19   Time variation of instantaneous velocity in turbulent flow.

Breaking this term up into two parts, the time average of the first is

$$\frac{1}{\theta*} \int_0^{\theta*} (\rho v)' \bar{u} \, dt = 0$$

since $\bar{u}$ is a constant and the time average of $(\rho v)'$ is zero. Integrating the second term, Eq. (4.74) becomes

$$\tau_t = -\frac{1}{\theta*} \int_0^{\theta*} (\rho v)' u' \, dt = -\overline{(\rho v)' u'} \tag{4.75}$$

or, if ρ is constant,

$$\tau_t = -\rho(\overline{v'u'}) \tag{4.76}$$

where $(\overline{v'u'})$ is the time average of the product of u' and v'.

It is not difficult to visualize that the time averages of the mixed products of velocity fluctuations, such as $\overline{v'u'}$, differ from zero. From Fig. 4.18 we can see that the particles that travel upward $(v' > 0)$ arrive at a layer in the fluid in which the mean velocity $\bar{u}$ is larger than in the layer from which they come. Assuming that the fluid particles preserve on the average their original velocity $\bar{u}$ during their migration, they will tend to slow down other fluid particles after they have reached their destination and thereby give rise to a negative component u'. Conversely, if v' is negative, the observed value of u' at the new destination will be positive. On the average, therefore, a positive v' is associated with a negative u', and vice versa. The time average of $\overline{u'v'}$ is therefore on the average not zero but a negative quantity. The turbulent shearing stress defined by Eq. (4.76) is thus positive and has the same sign as the corresponding laminar shearing stress,

$$\tau_l = \mu \frac{d\bar{u}}{dy} = \rho v \frac{d\bar{u}}{dy}$$

It should be noted, however, that the laminar shearing stress is a true stress, whereas the apparent turbulent shearing stress is simply a concept introduced to account for the effects of the momentum transfer by turbulent fluctuations. This concept allows us to express the total shear stress in turbulent flow as

$$\tau = \frac{\text{viscous force}}{\text{area}} + \text{turbulent momentum flux} \tag{4.77}$$

To relate the turbulent momentum flux to the time-average velocity gradient $d\bar{u}/dy$ we postulate that fluctuations of macroscopic fluid particles in turbulent flow are, on the average, similar to the motion of molecules in a gas [i.e., they travel on the average a distance l perpendicular to $\bar{u}$ (Fig. 4.18) before coming to rest in another y plane]. This distance l is known as *Prandtl's mixing length* (13,14) and corresponds qualitatively to the mean free path of a gas molecule. Assuming that the fluid particles retain their identity and physical properties during the cross motion and that the turbulent fluctuation arises chiefly from the difference in the time-mean properties between y planes spaced a distance l apart, if a fluid particle travels from a layer y to a layer $y + l$,

$$u' \simeq l \frac{d\bar{u}}{dy} \tag{4.78}$$

With this model the turbulent shearing stress τ_t in a form analogous to the laminar shearing stress is

$$\tau_t = -\rho \overline{v'u'} = \rho \epsilon_M \frac{d\bar{u}}{dy} \tag{4.79}$$

where the symbol ϵ_M is called the *eddy viscosity* or the turbulent exchange coefficient for momentum. The eddy viscosity ϵ_M is formally analogous to the kinematic viscosity v, but whereas v is a physical property, ϵ_M depends on the dynamics of the flow. Combining Eqs. (4.78) and (4.79) shows that $\epsilon_M = -\overline{v'l}$, and Eq. (4.77) gives the total shearing stress in the form

$$\tau = \rho(v + \epsilon_M)\frac{d\bar{u}}{dy} \tag{4.80}$$

In turbulent flow ϵ_M is much larger than v and the viscous term may therefore be neglected.

The transfer of energy as heat in a turbulent flow can be pictured in an analogous fashion. Let us consider a two-dimensional time-mean temperature distribution as shown in Fig. 4.20. The fluctuating velocity components continuously transport fluid particles and the energy stored in them across a plane normal to the y direction. The instantaneous rate of energy transfer per unit area at any point in the y direction is

$$(\rho v')(c_p T) \tag{4.81}$$

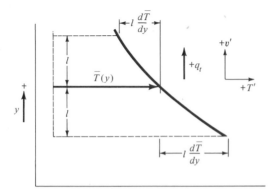

Figure 4.20   Mixing length for energy transfer in turbulent flow.

where $T = \bar{T} + T'$. Following the same line of reasoning that led to Eq. (4.76), the time average of energy transfer due to the fluctuations, called the *turbulent rate of heat transfer* q_t, is

$$q_t = A\rho c_p \overline{v'T'} \qquad (4.82)$$

Using Prandtl's concept of mixing length, we can relate the temperature fluctuation to the time-mean temperature gradient by the equation

$$T' \simeq l\frac{d\bar{T}}{dy} \qquad (4.83)$$

This means physically that, when a fluid particle migrates from the layer y to another layer a distance l above or below, the resulting temperature fluctuation is caused chiefly by the difference between the time-mean temperatures in the layers. Assuming that the transport mechanisms of temperature (or energy) and velocity are similar, the mixing lengths in Eqs. (4.78) and (4.83) are equal. The product $\overline{v'T'}$, however, is positive on the average because a positive v' is accompanied by a positive T', and vice versa.

Combining Eqs. (4.82) and (4.83), the turbulent rate of heat transfer per unit area becomes

$$q_t'' = \frac{q_t}{A} = c_p\rho\overline{v'T'} = -c_p\rho\overline{v'l}\frac{d\bar{T}}{dy} \qquad (4.84)$$

where the minus sign is a consequence of the second law of thermodynamics (see Chapter 1). To express the turbulent heat flux in a form analogous to the Fourier conduction equation, we define ϵ_H, a quantity called the *turbulent exchange coefficient for temperature, eddy diffusivity of heat,* or *eddy heat conductivity,* by the equation $\epsilon_H = \overline{v'l}$. Substituting ϵ_H for $\overline{v'l}$ in Eq. (4.84) gives

$$q_t'' = -c_p\rho\epsilon_H\frac{d\bar{T}}{dy} \qquad (4.85)$$

The total rate of heat transfer per unit area normal to the mean stream velocity can then be written as

$$q'' = \frac{q}{A} = \frac{\text{molecular conduction}}{\text{unit area}} + \frac{\text{turbulent transfer}}{\text{unit area}}$$

or in symbolic form as

$$q'' = -c_p\rho(\alpha + \epsilon_H)\frac{d\bar{T}}{dy} \qquad (4.86)$$

where $\alpha = k/c_p\rho$, the molecular diffusivity of heat. The contribution to the heat transfer by molecular conduction is proportional to α, and the turbulent contribution is proportional to ϵ_H. For all fluids except liquid metals, ϵ_H is much larger than α in turbulent flow. The ratio of the molecular kinematic viscosity to the molecular diffusivity of heat v/α has previously been named the Prandtl number. Similarly, the ratio of the turbulent eddy viscosity to the eddy diffusivity ϵ_M/ϵ_H could be considered a turbulent Prandtl number Pr_t. According to the Prandtl mixing-length theory, the turbulent Prandtl number is unity, since $\epsilon_M = \epsilon_H = \overline{v'l}$.

Although this treatment of turbulent flow is oversimplified, experimental results indicate it is at least qualitatively correct. Isakoff and Drew (15) found that Pr_t for the heating of mercury in turbulent flow inside a tube may vary from 1.0 to 1.6, and Forstall and Shapiro (16) found that Pr_t is about 0.7 for gases. The latter investigators also showed that Pr_t is substantially independent of the value of the laminar Prandtl number as well as of the type of experiment. Assuming that Pr_t is unity, the turbulent heat flux can be related to the turbulent shear stress by combining Eqs. (4.79) and (4.85):

$$q_t'' = -\tau_t c_p \frac{d\bar{T}}{d\bar{u}} \tag{4.87}$$

This relation was originally derived in 1874 by the British scientist Osborn Reynolds and is called the *Reynolds analogy*. It is a good approximation whenever the flow is turbulent and can be applied to turbulent boundary layers as well as to turbulent flow in pipes or ducts. However, the Reynolds analogy does not hold in the viscous sublayer (17). Since this layer offers a large thermal resistance to the flow of heat, Eq. (4.87) does not, in general, suffice for a quantitative solution. Only for fluids having a Prandtl number of unity can it be used directly to calculate the rate of heat transfer. This special case will now be considered.

4.12 REYNOLDS ANALOGY FOR TURBULENT FLOW OVER PLANE SURFACES

To derive a relation between the heat transfer and the skin friction in flow over a plane surface for a Prandtl number of unity, the reader will recall that the laminar shearing stress τ is

$$\tau = \mu \frac{du}{dy}$$

and the rate of heat flow per unit area across any plane perpendicular to the y direction is

$$q'' = -k \frac{dT}{dy}$$

Combining the equations above yields

$$q'' = -\tau \frac{k}{\mu} \frac{dT}{du} \tag{4.88}$$

An inspection of Eqs. (4.87) and (4.88) shows that if $c_p = k/\mu$ (i.e., for $\text{Pr} = 1$), the same equation of heat flow applies in the laminar and turbulent layers.

To determine the rate of heat transfer from a flat plate to a fluid with $\text{Pr} = 1$ flowing over it in turbulent flow, we replace k/μ by c_p and separate the variables in Eq. (4.88). Assuming that q'' and τ are constant, we get

$$\frac{q_s''}{\tau_s c_p} du = -dT \tag{4.89}$$

where the subscript s is used to indicate that both q'' and τ are taken at the surface of the plate. Integrating Eq. (4.89) between the limits $u = 0$ when $T = T_s$, and $u = U_\infty$ when $T = T_\infty$, yields

$$\frac{q_s''}{\tau_s c_p} U_\infty = (T_s - T_\infty) \tag{4.90}$$

But since by definition the local heat-transfer and friction coefficients are

$$h_{cx} = \frac{q_s''}{(T_s - T_\infty)} \quad \text{and} \quad \tau_{sx} = C_{fx} \frac{\rho U_\infty^2}{2}$$

Equation (4.90) can be written

$$\frac{h_{cx}}{c_p \rho U_\infty} = \frac{\text{Nu}_x}{\text{Re}_x \text{Pr}} = \frac{C_{fx}}{2} \tag{4.91}$$

Equation (4.91) is satisfactory for gases in which Pr is approximately unity. Colburn (11) has shown that Eq. (4.91) can also be used for fluids having Prandtl numbers ranging from 0.6 to about 50 if it is modified in accordance with experimental results to read

$$\frac{\text{Nu}_x}{\text{Re}_x \text{Pr}} \text{Pr}^{2/3} = \text{St}_x \text{Pr}^{2/3} = \frac{C_{fx}}{2} \tag{4.92}$$

where the subscript x denotes the distance from the leading edge of the plate.

To apply the analogy between heat transfer and momentum transfer, in practice it is necessary to know the friction coefficient C_{fx}. For turbulent flow over a plane surface the empirical equation for the local friction coefficient

$$C_{fx} = 0.0576 \left(\frac{U_\infty x}{\nu}\right)^{-1/5} \tag{4.93}$$

is in good agreement with experimental results in the Reynolds number range between 5×10^5 and 10^7 as long as no separation occurs. Assuming that the turbulent boundary layer starts at the leading edge, the average friction coefficient over a plane surface of length L can be obtained by integrating Eq. (4.93):

$$\bar{C}_f = \frac{1}{L} \int_0^L C_{fx} dx = 0.072 \left(\frac{U_\infty L}{\nu}\right)^{-1/5} \tag{4.94}$$

4.13 MIXED BOUNDARY LAYER

In reality, a laminar boundary layer precedes the turbulent boundary layer between $x = 0$ and $x = x_c$. Since the local frictional drag of a laminar boundary layer is less than the local frictional drag of a turbulent boundary layer at the same Reynolds number, the average drag calculated from Eq. (4.94) without correcting for the laminar portion of the boundary layer is too large. The actual drag can be closely estimated, however, by assuming that, behind the point of transition, the turbulent boundary layer behaves as though it had started at the leading edge.

Adding the laminar friction drag between $x = 0$ and $x = x_c$ to the turbulent drag between $x = x_c$ and $x = L$ gives, per unit width,

$$\bar{C}_f = \frac{[0.072\mathrm{Re}_L^{-1/5}L - 0.072\mathrm{Re}_{x_c}^{-1/5}x_c + 1.33\mathrm{Re}_{x_c}^{-1/2}x_c]}{L}$$

For a critical Reynolds number of 5×10^5 this yields

$$\bar{C}_f = 0.072\left(\mathrm{Re}_L^{-1/5} - \frac{0.0464x_c}{L}\right) \tag{4.95}$$

Substituting Eq. (4.93) for C_{fx} in Eq. (4.92) yields the local Nusselt number at any value of x larger than x_c:

$$\mathrm{Nu}_x = \frac{h_{cx}x}{k} = 0.0288\mathrm{Pr}^{1/3}\left(\frac{U_\infty x}{\nu}\right)^{0.8} \tag{4.96}$$

We observe that the local heat transfer coefficient h_{cx} for heat transfer by convection through a turbulent boundary layer decreases with the distance x as $h_{cx} \propto 1/x^{0.2}$. Equation (4.96) shows that, in comparison with laminar flow, where $h_{cx} \propto 1/x^{1/2}$, the heat transfer coefficient in turbulent flow decreases less rapidly with x and that the turbulent heat transfer coefficient is much larger than the laminar heat transfer coefficient at a given value of the Reynolds number.

The average conductance in turbulent flow over a plane surface of length L can be calculated to a first approximation by integrating Eq. (4.96) between $x = 0$ and $x = L$:

$$\bar{h}_c = \frac{1}{L}\int_0^L h_{cx}\, dx$$

In dimensionless form we get

$$\overline{\mathrm{Nu}}_L = \frac{\bar{h}_c L}{k} = 0.036\mathrm{Pr}^{1/3}\mathrm{Re}_L^{0.8} \tag{4.97}$$

Equation (4.97) neglects the existence of the laminar boundary layer and is therefore valid only when $L \gg x_c$. The laminar boundary layer can be included in the analysis if Eq. (4.71) is used between $x = 0$ and $x = x_c$, and Eq. (4.96) between $x = x_c$ and $x = L$ for the integration of h_{cx}. This yields, with $\mathrm{Re}_c = 5 \times 10^5$,

$$\overline{\mathrm{Nu}}_L = 0.036\mathrm{Pr}^{1/3}(\mathrm{Re}_L^{0.8} - 23{,}200) \tag{4.98}$$

EXAMPLE 4.5

The crankcase of an automobile is approximately 0.6 m long, 0.2 m wide, and 0.1 m deep. Assuming that the surface temperature of the crankcase is 350 K, estimate the rate of heat flow from the crankcase to atmospheric air at 276 K at a road speed of 30 m/s. Assume that the vibration of the engine and the chassis induce the transition from laminar to turbulent flow so near to the leading edge that, for practical purposes, the boundary layer is turbulent over the entire surface. Neglect radiation and use for the front and rear surfaces the same average convective heat transfer coefficient as for the bottom and sides.

Solution. Using the properties of air at 313 K, the Reynolds number is

$$\text{Re}_L = \frac{\rho U_\infty L}{\mu} = \frac{1.092 \times 30 \times 0.6}{19.123 \times 10^{-6}} = 1.03 \times 10^6$$

From Eq. (4.97) the average Nusselt number is

$$\overline{\text{Nu}}_L = 0.036\, \text{Pr}^{1/3}\text{Re}_L^{0.8}$$
$$= 0.036(0.71)^{1/3}(1.03 \times 10^6)^{0.8}$$
$$= 2075$$

and the average convective heat transfer coefficient becomes

$$\bar{h}_c = \frac{\overline{\text{Nu}}_L k}{L} = \frac{2075 \times 0.0265}{0.6} = 91.6 \text{ W/m}^2 \text{ K}$$

The surface area that dissipates heat is 0.28 m^2 and the rate of heat loss from the crankcase is therefore

$$q = \bar{h}_c A(T_s - T_\infty) = (9.16)(0.28)(350 - 276)$$
$$q = 1898 \text{ W}$$

4.14* HEAT AND MASS TRANSFER ANALOGY

The same reasoning used to relate turbulent heat and momentum transfer for a Prandtl number of unity can be applied to mass transfer of species A when the Schmidt number is unity. This leads to the relation

$$\frac{h_{mx}}{c_p \rho U_\infty} = \frac{\text{Sh}_x}{\text{Re}_x \text{Sc}} = \frac{C_{fx}}{2} \qquad (4.99)$$

This relation can also be extended empirically to Schmidt numbers between 0.6 and 3000 (18) by the equation

$$\frac{\text{Sh}_x}{\text{Re}_x \text{Sc}} \text{Sc}^{2/3} = \frac{C_{fx}}{2} \qquad (4.100)$$

This is often referred to as the analogy between mass and momentum transfer.

EXAMPLE 4.6 ——

Estimate the rate of evaporation from an open water storage reservoir 12×24 m in size in kilograms per day. Typical weather conditions for the location are:

ambient air temperature: 25°C
ambient relative humidity: 50 percent
wind speed: 2 m/s

The relative humidity is defined as the ratio of the water vapor density under ambient conditions to the water vapor density at saturation at the ambient temperature.

Solution.   Assume that the water temperature is the same as the air temperature. Neglect any effects due to atmospheric turbulence and ripples on the water surface. Assume that the wind blows perpendicular to the long side of the reservoir and that the analogy between heat and mass transfer is applicable.

The following property data are taken from the appendices as noted:

air viscosity at 25°C (Table 27), $v = 16.2 \times 10^{-6}$ m²/s

air-water vapor diffusivity (Table 38), $D_{AB} = 2.6 \times 10^{-5}$ m²/s

water vapor density, $\rho_{A,\,\text{sat}} = v_g^{-1} = 0.023$ kg/m³

First we determine the Reynolds number to ascertain the nature of the flow:

$$\text{Re}_L = \frac{U_\infty L}{v} = \frac{(2 \text{ m/s})(12 \text{ m})}{16.2 \times 10^{-6} \text{ m}^2/\text{s}} = 1.48 \times 10^6$$

Hence the flow will be turbulent at the trailing edge, but a transition will occur at a distance x_c corresponding to the critical value $\text{Re}_c = 5 \times 10^5$, or

$$x_c = \frac{\text{Re}_c}{\text{Re}_L} L = \frac{5 \times 10^5}{1.48 \times 10^6} (12) \simeq 4 \text{ m}$$

The flow over the first 4 m will be laminar, over the rest of the water surface turbulent. For this mixed boundary layer Eq. (4.98) gives the average Nusselt number and, using the analogy, the average Sherwood number is

$$\overline{\text{Sh}}_L = 0.036 S_c^{1/3} (\text{Re}_L^{0.8} - 23{,}200)$$

$$= (0.036) \left(\frac{1.6}{2.6} \right)^{1/3} (86{,}340 - 23{,}200)$$

$$= 1920$$

The mass transfer coefficient is thus

$$\bar{h}_{m,L} = \overline{Sh}_L \frac{D_{AB}}{L} = 1920 \left(\frac{2.6 \times 10^5 \ m^2/s}{12 \ m} \right)$$
$$= 4.33 \times 10^{-3} \ m/s$$

The evaporation rate is from Eq. (4.7)

$$n_A = \bar{h}_m A(\rho_{A,s} - \rho_{A,\infty})$$

The relative humidity ϕ_∞ is defined by

$$\phi_\infty = \frac{\rho_{A,\infty}}{\rho_{A,sat}(T_\infty)}$$

Since $T_s = T_\infty = 25°C$, we get

$$n_A = \bar{h}_m A[\rho_{A,sat}(T_s) - \phi_\infty \rho_{A,sat}(T_\infty)]$$
$$= \bar{h}_m A \rho_{A,sat}(25°C)(1 - \phi_\infty)$$
$$= (4.33 \times 10^{-3} \ m/s)(144 \ m^2)(0.023 \ kg/m^2)(0.5)(86,400) \ s/day$$
$$= 620 \ kg/day$$

4.15 CLOSURE

In this chapter we have studied the principles of heat and mass transfer by forced convection. We have seen that the transfer of heat and mass by convection is intimately related to the mechanics of the fluid flow, particularly to the flow in the vicinity of the transfer surface. We have also observed that the nature of heat and mass transfer, as well as flow phenomena, depends greatly on whether the fluid far away from the surface is in laminar or in turbulent flow.

To become familiar with the basic principles of boundary-layer theory and forced-convection heat transfer, we have considered the problem of convection in flow over a flat plate in some detail. This system is geometrically simple, but it illustrates the most important features of forced convection. In subsequent chapters we shall treat heat and mass transfer by convection in geometrically more complicated systems. In the next chapter we shall examine natural-convection phenomena. In Chapter 6, heat transfer by convection to and from fluids flowing inside pipes and ducts will be taken up. In Chapter 7, forced convection in flow over the exterior surfaces of bodies such as cylinders, spheres, tubes, and tube bundles will be considered. The application of the principles of forced-convection heat transfer to the selection and design of heat transfer equipment will be taken up in Chapter 8.

For the convenience of the reader, a summary of the equations used to calculate the heat transfer, mass transfer, and friction coefficients in low-speed flow of gases and liquids over flat, or only slightly curved, plane surfaces is presented in Table 4.5. For additional information the reader is referred to Refs. 5, 6 and 23.

TABLE 4.5 SUMMARY OF USEFUL EMPIRICAL EQUATIONS FOR CALCULATING FRICTION AND HEAT AND MASS TRANSFER COEFFICIENTS IN FLOW OVER FLAT SURFACES AT ZERO ANGLE OF ATTACK[a]

Coefficient	Equation	Conditions
	LAMINAR FLOW	$Re_x < 5 \times 10^5$
Local friction coefficient	$C_{fx} = 0.664 Re_x^{-0.5}$	
Local Nusselt number at distance x from leading edge	$Nu_x = 0.332 Re_x^{0.5} Pr^{0.33}$ $Nu_x = 0.565 (Re_x Pr)^{0.5}$	$Pr > 0.1, Re_x < 5 \times 10^5$ $Pr < 0.1, Re_x < 5 \times 10^5$
Local Sherwood number	$Sh_x = 0.332 Re_x^{0.5} Sc^{0.3}$	$Sc > 0.1, Re_x < 5 \times 10^5$
Average friction coefficient	$C_f = 1.33 Re_L^{-0.5}$	$Re_L < 5 \times 10^5$
Average Nusselt number between $x = 0$ and $x = L$	$\overline{Nu}_L = 0.664 Re_L^{0.5} Pr^{0.33}$	$Pr < 0.1, Re_L < 5 \times 10^5$
Average Sherwood number	$\overline{Sh}_L = 0.664 Re_L^{0.5} Sc^{0.33}$	$Sc > 0.1, Re_L < 5 \times 10^5$
	TURBULENT FLOW	
Local friction coefficient	$C_{fx} = 0.0576 Re_x^{-0.2}$	$Re_x > 5 \times 10^5, Pr > 0.5$
Local Nusselt number at distance x from leading edge	$Nu_x = 0.0288 Re_x^{0.8} Pr^{0.33}$	
Local Sherwood number	$Sh_x = 0.0288 Re_x^{0.8} Sc^{0.33}$	$Re_x > 5 \times 10^5, Sc > 0.5$
Average friction coefficient	$\overline{C}_f = 0.072[Re_L^{-0.2} - 0.0464(x_{cr}/L)]$	$Re_L > 5 \times 10^5, Pr > 0.5$
Average Nusselt number between $x = 0$ and $x = L$ with transition at $Re_{x,cr} = 5 \times 10^5$	$\overline{Nu}_L = 0.036 Pr^{0.33}[Re_L^{0.8} - 23,200]$	
Average Sherwood number	$\overline{Sh}_L = 0.036 Sc^{0.33}[Re_L^{0.8} - 23,200]$	$Re_L > 5 \times 10^5, Sc > 0.5$

[a] Applicable to low-speed flow (Mach number < 0.5) of gases and liquids with all physical properties at the mean film temperature, $T_f = (T_s + T_\infty)/2$.

$$C_{fx} = \tau_s/(\rho u_\infty^2/2g_c) \qquad \overline{C}_f = (1/L) \int C_{fx} dx \qquad Pr = c_p \mu/k$$
$$Nu_x = h_c x/k \qquad \overline{Nu} = \overline{h}_c L/k \qquad \overline{h}_c = (1/L) \int_0^L h_c(x) dx$$
$$Re_x = \rho u_\infty x/\mu \qquad Re_L = \rho u_\infty L/\mu \qquad Sc = \nu/D_{AB}$$
$$Sh_x = h_m x/D_{AB} \qquad \overline{Sh} = \overline{h}_m L/D_{AB} \qquad \overline{h}_m = (1/L) \int_0^L h_m(x) dx$$

PROBLEMS

The problems for this chapter are organized by subject matter as shown below.

Topic	Section	Problem number
Dimensionless numbers	all	4.1 to 4.5
Dimensionless analysis and similarity	4.6	4.6 to 4.15
Heat transfer in flow over a flat plate	4.7 and 4.11	4.16 to 4.27
Boundary layers	4.9 and 4.10	4.28 to 4.31
Thermal design	all	4.32 to 4.36
Mass transfer	4.3 and 4.14	4.37 to 4.40

4.1 Evaluate the Reynolds number from the following data:

$$D = 6 \text{ in.}$$
$$U_\infty = 1.0 \text{ ft/sec.}$$
$$\rho = 300 \text{ kg/m}^3$$
$$\mu = 90 \text{ lb}_m/\text{ft hr}$$

4.2 Evaluate the Prandtl number from the data below:

$$c_p = 0.5 \text{ Btu/lb}_m\,^\circ\text{F}$$
$$k = 2 \text{ Btu/ft hr}^\circ\text{F}$$
$$\mu = 0.5 \text{ Ns/m}^2$$

4.3 Evaluate the Nusselt number for the following condition:

$$D = 6 \text{ in.}$$
$$k = 0.2 \text{ W/mK}$$
$$\bar{h}_c = 18 \text{ Btu/ft}^2 \text{ hr}^\circ\text{F}$$

4.4 Evaluate the Stanton number for the data below:

$$D = 10 \text{ cm}$$
$$U_\infty = 12 \text{ ft/sec.}$$
$$\rho = 1280 \text{ lb}_m/\text{ft}^3$$
$$\mu = 6 \times 10^4 \text{ Ns/m}^2$$
$$c_p = 0.95 \text{ Btu/lb}_m$$
$$\bar{h}_c = 3.0 \text{ Btu/hr ft}^2{}^\circ\text{F}$$

4.5 Evaluate the dimensionless groups $\bar{h}_c D/k$, $U_\infty D\rho/\mu$, $c_p\mu/k$, and $\bar{h}_c/c_p\, G$ for water, ethyl alcohol, mercury, hydrogen, air, and saturated steam over as wide a temperature range as possible and plot the results vs. temperature. For the purpose of These calculations let $D = 1$ m, $U_\infty = 1$ m/sec, and $\bar{h}_c = 1$ W/m² K.

4.6 Replot the data points of Fig. 4.9 on log-log paper and find an equation approximating the best correlation line. Compare your results with Fig. 4.10. Then, suppose steam at 1 atm and 100°C is flowing across a 5 cm-OD pipe at a velocity of 10 m/s. Using the data in Fig. 4.11 estimate the Nusselt number, the heat transfer coefficient, and the rate of heat transfer per meter length of pipe if the pipe is at 200°C and compare with predictions from your correlation equation.

4.7 The average Reynolds number for air passing in turbulent flow over a 2 m-long flat plate is 2.4×10^6. Under these conditions the average Nusselt number was found to be equal to 4150. Determine the average heat-transfer coefficient for an oil having thermal properties similar to those of Table A-17 at 30°C at the same Reynolds number in flow over the same plate.

4.8 The dimensionless ratio $U_\infty/\sqrt{Lg}$, called Froude number, is a measure of similarity between the shapes of the waves produced by a ship model and by its prototype. A 500 ft long cargo ship is designed to run at 20 knots, and a 5 ft. geometrically similar model is towed in a water channel to study the wave resistance. What should be the towing speed in m s^{-1}?

4.9 The torque due to the frictional resistance of the oil film between a rotating shaft and its bearing is found to be dependent on the force F normal to the shaft, the speed of rotation N of the shaft, the dynamic viscosity μ of the oil, and the shaft diameter D. Establish a correlation between the variables by using dimensional analysis. *Ans.* $[T/(F^3/N\mu)^{1/2} = \phi(N\mu D^2/F)$

4.10 When a sphere falls freely through a homogeneous fluid, it reaches a terminal velocity at which the weight of the sphere is balanced by the buoyant force and the frictional resistance of the fluid. Make a dimensional analysis of this problem and indicate how experimental data for this problem could be correlated. Neglect compressibility effects and the influence of surface roughness.

4.11 Experiments have been performed on the temperature distribution in a homogeneous long cylinder (0.1 m diameter, thermal conductivity of 0.2 W m^{-1} K^{-1}) with uniform internal heat generation. By dimensional analysis determine the relation between the steady-state temperature at the center of the cylinder T_c the diameter, the thermal conductivity, and the rate of heat generation. Take the temperature at the surface as your datum. What is the equation for the center temperature if the difference between center and surface temperature is 30°C when the heat generation rate is 3000 W m^{-3}?

4.12 The convection equations relating the Nusselt, Reynolds, and Prandt numbers can be rearranged to show that for gases the heat-transfer coefficient $\bar{h}_c$ depends on the absolute temperature T and the group $\sqrt{U_\infty}/x$. This formulation is of the form $\bar{h}_{c,x} = CT^n\sqrt{U_\infty}/x$, where n and C are constants. Indicate clearly how such a relationship could be obtained for the laminar flow case from $\mathrm{Nu}_x = 0.332\,\mathrm{Re}_x^{0.5}\mathrm{Pr}^{0.333}$ for the condition $0.5 < \mathrm{Pr} < 5.0$. State restrictions on method if necessary.

4.13 Experimental pressure-drop data obtained in a series of tests in which water was heated while flowing through an electrically heated tube of 0.527 in. ID 38.6 in. long, are tabulated below.

Mass flow rate $\dot{m}$ (lb/sec)	Fluid bulk temperature T_b (°F)	Surface temperature T_s (°F)	Pressure drop with heat transfer Δp_{ht} (psi)
3.04	90	126	9.56
2.16	114	202	4.74
1.82	97	219	3.22
3.06	99	248	8.34
2.15	107	283	4.45

Isothermal pressure-drop date for the same tube are given in terms of the dimensionless friction coefficient $f = (\Delta p/\rho\bar{u}^2)(D/2L)g_c$ and the Reynolds number based on the pipe diameter, $\mathrm{Re}_D = \bar{u}D/\nu = 4\dot{m}/\pi D\mu$ below.

Re_D	1.71×10^5	1.05×10^5	1.9×10^5	2.41×10^5
f	0.00472	0.00513	0.00463	0.00445

By comparing the isothermal with the nonisothermal friction coefficients at similar bulk Reynolds numbers, derive a dimensionless equation for the nonisothermal friction coefficients in the form

$$f = \text{constant} \times \mathrm{Re}_D{}^n(\mu_s/\mu_b)^m$$

where μ_s = viscosity at surface temperature;

μ_b = viscosity at bulk temperature;

n and m = empirical constants.

4.14 Tabulated below are some experimental data obtained by passing *n*-butyl alcohol at a bulk temperature of 15°C over a heated flat plate (0.3 m long, 0.9 m wide, surface temperature of 60°C). Correlate the experimental data by appropriate dimensionless numbers and compare the line which best fits the data with Eq. 4.53.

Velocity (m/s)	0.089	0.305	0.488	1.14
Unit-Surface Conductance (W/m²°C)	64.6	130.4	196.2	391.2

4.15 Tabulated below are reduced test data from measurements made to determine the heat-transfer coefficient inside tubes at Reynolds numbers only slightly above transition and at relatively high Prandtl numbers (as associated with oils). Tests were made in a double-tube exchanger with a counterflow of water to provide the cooling. The pipe used to carry the oils was $\frac{5}{8}$-in. OD, 18 BWG, 121 in. long. Correlate the data in terms of appropriate dimensionless parameters.

Test No.	Fluid	$\bar{h}_c$	$\rho\bar{u}$	c_p	k	μ_b	μ_f
11	10C oil	87.0	1,072,000	0.471	0.0779	13.7	19.5
19	10C oil	128.2	1,504,000	0.472	0.0779	13.3	19.1
21	10C oil	264.8	2,460,000	0.486	0.0776	9.60	14.0
23	10C oil	143.8	1,071,000	0.495	0.0773	7.42	9.95
24	10C oil	166.5	2,950,000	0.453	0.0784	23.9	27.3
25	10C oil	136.3	1,037,000	0.496	0.0773	7.27	11.7
36	1488 pyranol	140.7	1,795,000	0.260	0.0736	12.1	16.9
39	1488 pyranol	133.8	2,840,000	0.260	0.0740	23.0	29.2
45	1488 pyranol	181.4	1,985,000	0.260	0.0735	10.3	12.9
48	1488 pyranol	126.4	3,835,000	0.260	0.0743	40.2	53.5
49	1488 pyranol	105.8	3,235,000	0.260	0.0743	39.7	45.7

where $\bar{h}_c$ = mean surface heat-transfer coefficient, based on the mean temperature difference, Btu/hr sq ft °F;

$\rho\bar{u}$ = mass velocity, lb_m/hr sq ft;

c_p = specific heat, Btu/lb_m°F;

k = thermal conductivity, Btu/hr ft °F (based on average bulk temperature);

μ_b = viscosity, based on average bulk (mixed mean) temperature, lb_m/hr ft;

μ_f = viscosity, based on average film temperature, lb_m/hr ft.

Hint: Start by correlating $\overline{Nu}$ and Re_D irrespective of the Prandtl numbers, since the influence of the Prandtl number on the Nusselt number is expected to be relatively small. By plotting $\overline{Nu}$ vs. Re on log-log paper, one can guess the nature of the correlation equation, $\overline{Nu} = f_1(Re)$. A plot of $\overline{Nu}/f_1(Re)$ vs. Pr will then reveal the dependence upon Pr. For the final equation, the influence of the viscosity variation should also be considered.

One possible answer: $\overline{Nu}_D = 0.0067 \dfrac{\rho V D}{\mu_b} \left(\dfrac{c_p \mu_b}{k_b}\right)^{0.2} \left(\dfrac{\mu_b}{\mu_f}\right)^{0.3}$

4.16 The average friction coefficient for flow over a 0.6 m-long plate is 0.01. What is the value of the drag force in N per m width of the plate for the following fluids: (a) air at 15°C, (b) steam at 100°C and 1 N/m^2, (c) water at 40°C, (d) mercury at 100°C, and (e) ethyl alcohol at 100°C?

4.17 Hydrogen at 15°C and at a pressure of 1 atm is flowing along a flat plate at a velocity of 3 m/s. If the plate is 0.3 m wide and at 71°C, calculate the following quantities at $x = 0.3$ m and at the distance corresponding to the transition point, i.e., $Re_x = 5 \times 10^5$. (Take properties at 43°C.)
(a) Hydrodynamic boundary layer thickness, in cm.
(b) Local friction coefficient, dimensionless.
(c) Average friction coefficient, dimensionless.
(d) Drag force, in N$_f$.
(e) Thickness of thermal boundary layer, in cm.
(f) Local convective-heat-transfer coefficient, in W/m^2°C.
(g) Average convective-heat-transfer coefficient, in W/m^2°C.
(h) Rate of heat transfer, in W.

4.18 Repeat Prob. 4.17 for $x = 0.9$ m and $U_\infty = 61$ m/s, (a) taking the laminar boundary layer into account and (b) assuming that the turbulent boundary layer starts at the leading edge.

4.19 Determine the rate of heat loss in Btu/hr from the wall of a building in a 10-mph wind blowing parallel to its surface. The wall is 80 ft long, 20 ft high its surface temperature is 80°F, and the temperature of the ambient air is 40°F.

4.20 A spacecraft heat exchanger is to operate in a nitrogen atmosphere at a pressure of about 10^4 N/m^2 and 38°C. For a flat-plate heat exchanger designed to operate on earth in air at one atmosphere and 38°C in turbulent flow estimate the ratio of heat-transfer coefficients on the earth to that in nitrogen assuming forced circulation cooling of the plate surface at the same velocity in both cases.

4.21 A thin flat plate 6 in. square is suspended from a balance into a uniformly flowing stream of glycerin in such a way that the glycerin flows parallel to and along the top and bottom surfaces of the plate. The total drag on the plate measured and found to be 13 lb$_f$. If the glycerin flows at the rate of 50 fps and at a temperature of 112°F, what is the heat-transfer coefficient $\bar{h}_c$ in Btu/hr sq ft °F.

4.22 Mercury at 15°C flows over and parallel to a flat surface at a velocity of 3 m/s. Calculate the thickness of the hydrodynamic boundary layer at a distance 0.3 m from the leading edge of the surface.

4.23 A thin flat plate 6 in. square is tested for drag in a wind tunnel with air at 100 fps, 14.7 psia, and 60°F flowing across and parallel to the top and bottom surfaces. The observed total drag force is 0.00150 lb. Calculate the rate of heat transfer from this plate when the surface temperature is maintained at 250°F. Neglect radiation.

Ans. 370 Btu/hr

4.24 Mercury at 60°F flows parallel to the short side of a thin flat smooth plate with a velocity of 1 ft/sec. The plate is 6 in. wide and 1 ft long and its surface temperature is 160°F. Find:

(a) the local friction coefficient at the middle point of the plate, and the total drag force on the plate

(b) the temperature of the mercury at a point 4 in. from the leading edge and 0.05 in. from the surface of the plate

(c) the Nusselt number at the end of the plate.

4.25 Water at a velocity of 8 ft/sec flows parallel to the surface of a 3-ft-long horizontal, smooth and thin flat plate. Determine the local thermal and hydrodynamic boundary-layer thicknesses, and the local friction coefficient, at the midpoint of the plate. What is the rate of heat transfer from the plate to the water per ft width of the plate, if the surface temperature is kep uniformly at 300 F, and the temperature of the main water stream is 60°F?

4.26 A thin flat plate is placed in an atmospheric pressure air stream flowing parallel to it at a velocity of 15 ft/sec. The temperature at the surface of the plate is maintained uniformly at 400°F, and that of the main air stream is 70°F. Calculate the temperature and horizontal velocity at a point 1 ft from the leading edge and 0.03 in. above the surface of the plate.   *Ans.* 340.5°F

4.27 The surface temperature of a thin flat plate located parallel to an air stream is 196°F. The free stream velocity is 200 ft/sec and the temperature of the air is 32°F. The plate is 24 in. wide and 18 in. long in the direction of the air stream. Neglecting the end effect of the plate and assuming that the flow in the boundary layer changes abruptly from laminar to turbulent at a transition Reynolds number of $Re_{tr} = 4 \times 10^5$, find:

(a) the average heat transfer coefficient in the laminar and turbulent regions

(b) the rate of heat transfer for the entire plate, considering both sides

(c) the average friction coefficient in the laminar and turbulent regions

(d) the total drag force.

Also, plot the heat transfer coefficient and local friction coefficient as a function of the distance from the leading edge of the plate.

[*Ans.* (a) 18.3, 30.9 Btu/ft^2 hr°F; (b) 28,340 Btu/hr; (c) 0.0021, 0.036; (d) 0.934 lb$_f$]

4.28 Assuming a linear velocity distribution and a linear temperature distribution in the boundary layer over a flat plate, derive a relation between the thermal and hydrodynamic boundary-layer thicknesses and the Prandtl number.

4.29 Derive the integral momentum boundary-layer equation for steady incompressible two-dimensional flow over a flat porous wall through which fluid is injected with a velocity v_o normal to the surface.

4.30 A fluid at temperature T_∞ is flowing at a velocity U_∞ over a flat plate which is at the same temperature as the fluid for a distance x_o from the leading edge, but at a temperature T_s beyond this point. Show by means of the integral boundary-layer equations that ζ, the ratio of the thermal boundary-layer thickness to the hydrodynamic boundary-layer thickness, over the heated portion of the plate is approximately

$$\zeta \sim Pr^{-1/3}\left[1 - \left(\frac{x_o}{x}\right)^{3/4}\right]^{1/3}$$

if the flow is laminar.

Hint: Assume that the temperature distribution is a cubic parabola and use T_s as your datum to simplify the boundary conditions, i.e., let

$$T - T_s = ay + cy^3$$

4.31 Air at 30 m/s flows between two parallel flat plates spaced 5 cm apart. Estimate the distance from the entrance where the boundary layers meet.

4.32 A preliminary design study for a nuclear moon-rocket reactor is to be made. The reactor consideration consists of a stack of parallel flat plates which are 2 ft square, 2 in. apart, and heated to 3000°F. Gaseous hydrogen at 20-atm pressure and a temperature of 0°F centers at one end at a velocity of 200 fps and is heated as it passes between the plates. (*a*) Calculate the average heat-transfer coefficient assuming that transition occurs at a Reynolds number of 5×10^5. (*b*) Estimate the average heat-transfer coefficient assuming that a turbulence screen is placed at the entrance so that the entire boundary layer is turbulent. (*c*) For the turbulent flow conditions, determine the temperature of the hydrogen at the exit assuming that the surface temperature of the plate is uniformly at 3000°F. State all your assumptions clearly. (*d*) If the plates are $\frac{1}{10}$-in. thick and have a thermal conductivity of 10 Btu/hr ft °F, determine the required rate of heat generation within the plate and the maximum temperature at the center of the plate assuming that uniform heat generation by nuclear fission occurs within the plates. (*e*) How many times would the hydrogen have to pass between the two plates in order to heat it to an average temperature of 2000°F?

4.33 A refrigeration truck is traveling at 80 mph on a desert highway where the air temperature is 50°C. The body of the truck may be idealized as a rectangular box, 3 m wide, 2.1 m high, and 6 m long, at a surface temperature of 10°C. Assume that the heat transfer from the front and back of the truck may be neglected, that the stream does not separate from the surface, and that the boundary layer is turbulent over the whole surface. If, for every 3600 W of heat loss one ton capacity of the refrigerating unit is necessary, calculate the required tonnage of the refrigeration unit.

4.34 The wing of an airplane has a polished chromium skin. At a 3000 m altitude it receives 650 W/m² sq ft by solar radiation. Assuming that the interior surface of the wing's skin is well insulated and the wing has a chord of 6 m length, i.e., $L \simeq 6$ m, estimate the equilibrium temperature of the wing at a flight speed of 150 m/s.

4.35 A cooling fin for a heat exchanger, situated parallel to an atmospheric pressure air stream, measures 0.075 m along the leading edge and 0.45 m in the flow direction. Its base temperature is 88°C, and the air is at 10°C. The velocity of the air is 27 m/s. Determine the total drag force and the total rate of heat transfer from the fin to the air.

4.36 A 1-in.-diam, 6-in.-long transite rod ($k = 0.56$ Btu/hr ft °F, $\rho = 100$ lb/cu ft, $c = 0.20$ Btu/lb °F) on the end of a 1-in.-diam wood rod at a uniform temperature of 212°F is suddenly placed into a 60°F, 100 ft/sec air stream flowing parallel to the axis of the rod. Estimate the center line temperature of the transite rod 8 min after cooling starts. Assume radial heat conduction, but include radiation losses, based on an emissivity of 0.90, to black surroundings at air temperature.

4.37 Glycerol flows over a sheet of ice for which the average Nusselt number is given by

$$\overline{Nu}_L = 0.2 \, Re_L^{0.7} Pr^{0.3}$$

The ice melts and dissolves in the glycerol under conditions for which the average convection heat transfer coefficient, $\bar{h}_L$, is 80 W/m² K. What is the average convection mass transfer coefficient under these conditions?

4.38 Dry atmospheric air at 320 K flows at a velocity of 20 m/s over a smooth porous plate saturated with liquid water at 320 K. If the plate is 1 m by 1 m, placed with its surface parallel to the air stream, estimate the mass rate of evaporation from the upper side in kilograms per meter.

4.39 If the plate in Problem 4.38 was saturated with water at 350 K heat and mass transfer would occur simultaneously. How would you analyze this case?

4.40 Air at 100°C flows over a streamlined naphthalene body. Naphthalene sublimes into air and its vapor pressure at 100°C is 20 mm Hg. The heat transfer coefficient for this system was previously found to be 3 Btu/h ft²°F. The mass diffusivity of naphthalene vapor in air at 100°C is 0.32 ft²/h. The concentration of naphthalene in the bulk air stream is negligibly small. Calculate the mass transfer coefficient and the mass flux for the system.

REFERENCES

1. H. Schlichting, *Boundary Layer Theory*, 6th ed., J. Kestin, transl., McGraw-Hill, New York, 1968.
2. E. R. Van Driest, "Calculation of the Stability of the Laminar Boundary Layer in a Compressible Fluid on a Flat Plate with Heat Transfer," *J. Aero. Sci.*, vol. 19, pp. 801–813, 1952.
3. H. L. Langhaar, *Dimensional Analysis and Theory of Models*, Wiley, New York, 1951.
4. E. R. Van Driest, "On Dimensional Analysis and the Presentation of Data in Fluid Flow Problems," *J. Appl. Mech.*, vol. 13, p. A-34, 1940.
5. *Handbook of Heat Transfer*, 2d ed., W. M. Rohsenow, J. P. Hartnett, and E. M. Ganic, eds., McGraw-Hill, New York, 1985.
6. S. V. Patankar and D. B. Spalding, *Heat and Mass Transfer in Boundary Layers*, 2d ed., International Textbook Co., London, 1970.
7. M. Blasius, "Grenzschichten in Flüssigkeiten mit Kleiner Reibung," *Z. Math. Phys.*, vol. 56, no. 1, 1908.
8. M. Hansen, "Velocity Distribution in the Boundary Layer of a Submerged Plate," NACA TM 585, 1930.
9. E. Pohlhausen, "Der Wärmeaustausch zwischen festen Körpern und Flüssigkeiten mit kleiner Reibung und kleiner Wärmeleitung," *Z. Angew. Math. Mech.*, vol. 1, p. 115, 1921.
10. B. Gebhart, *Heat Transfer*, 2d ed., McGraw-Hill, New York, 1971.
11. A. P. Colburn, "A Method of Correlating Forced Convection Heat Transfer Data and a Comparison with Fluid Friction," *Trans. AIChE*, vol. 29, pp. 174–210, 1933.
12. E. R. G. Eckert and R. M. Drake, *Heat and Mass Transfer*, 2d ed., McGraw-Hill, New York, 1959.
13. L. Prandtl, "Bemerkungen über den Wärmeübergang im Rohr," *Phys. Zeit.*, vol. 29, p. 487, 1928.
14. L. Prandtl, "Eine Beziehung zwischen Wärmeaustauch und Ströhmungswiederstand der Flüssigkeiten," *Phys. Zeit.*, vol. 10, p. 1072, 1910.

15. S. E. Isakoff and T. B. Drew, "Heat and Momentum Transfer in Turbulent Flow of Mercury," *Institute of Mechanical Engineers and ASME, Proceedings, General Discussion on Heat Transfer*, pp. 405–409, 1951.

16. W. Forstall, Jr., and A. H. Shapiro, "Momentum and Mass Transfer in Co-axial Gas Jets," *J. Appl. Mech.*, vol. 17, p. 399, 1950.

17. R. C. Martinelli, "Heat Transfer to Molten Metals," *Trans. ASME*, vol. 69, pp. 947–959, 1947.

18. F. P. Incopera and D. P. DeWitt, *Fundamentals of Heat Transfer*, Wiley, New York, 1981.

19. E. L. Diret, W. James, and M. Stracy, "Heat Transmission from Fine Wires to Water," *Ind. Eng. Chem.*, vol. 39, pp. 1098–1103, 1947.

20. A. H. Davis, "Convective Cooling of Wires in Streams of Viscous Liquids," *Philos. Mag.*, vol. 47, pp. 1057–1091, 1924.

21. R. Hilpert, "Wärmeabgabe von geheizten Drähten und Rohren," *Forsch. Geb. Ingenieurwes.*, vol. 4, pp. 215–224, 1933.

22. W. J. King, "The Basic Laws and Data of Heat Transmission," *Mech. Eng.*, vol. 54, pp. 410–415, 1932.

23. A. Bejan, *Convection Heat Transfer*, Wiley, New York, 1984.

Chapter 5

Natural Convection

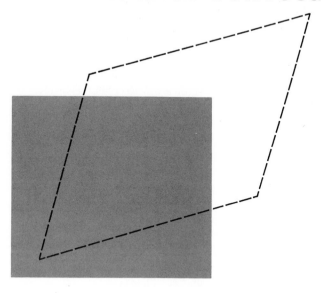

5.1 INTRODUCTION

Natural-convection heat transfer occurs whenever a body is placed in a fluid at a higher or a lower temperature than that of the body. As a result of the temperature difference, heat flows between the fluid and the body and causes a change in the density of the fluid in the vicinity of the surface. The difference in density leads to downward flow of the heavier fluid and upward flow of the lighter one. If the motion of the fluid is caused solely by differences in density resulting from temperature gradients, without the aid of a pump or a fan, the associated heat transfer mechanism is called natural convection. Natural-convection currents transfer internal energy stored in the fluid in essentially the same manner as forced-convection currents. However, the intensity of the mixing motion is generally less in natural convection, and consequently the heat transfer coefficients are lower than in forced convection.

Although natural-convection heat transfer coefficients are relatively small many devices depend largely on this mode of heat transfer for cooling. In the electrical engineering field, transmission lines, transformers, rectifiers, and electrically heated wires such as the heating elements of an electric furnace are cooled in part by natural convection. As a result of the heat generated internally, the temperature of these bodies rises above that of the surroundings. As the temperature difference increases, the rate of heat flow also increases until a state of equilibrium is reached where the rate of heat generation is equal to the rate of heat dissipation.

Natural convection is the dominant heat flow mechanism from steam radiators, walls of a building, or the stationary human body in a quiescent atmosphere. The determination of the heat load on heating and air-conditioning equipment and computers requires, therefore, a knowledge of natural-convection heat transfer coefficients. Natural convection is also responsible for heat losses from pipes carrying steam or other heated fluids. Natural convection has been proposed in nuclear power applications to cool the surfaces of bodies in which heat is generated by fission (1). The importance of natural-convection heat transfer has led to the recent publication (2) of a textbook devoted entirely to the subject.

In all of the aforementioned examples the body force responsible for the convection currents is the gravitational attraction. Gravity, however, is not the only body force that can produce natural convection. In certain aircraft applications there are components such as the blades of gas turbines and helicopter ramjets which rotate at high speeds. Associated with these rotative speeds are large centrifugal forces whose magnitudes, like the gravitational force, are also proportional to the fluid density and hence can generate strong natural-convection currents. Cooling of rotating components by natural convection is therefore feasible even at high heat fluxes.

The fluid velocities in natural-convection currents, especially those generated by gravity, are generally low, but the characteristics of the flow in the vicinity of the heat transfer surface are similar to those in forced convection. A boundary layer forms near the surface and the fluid velocity at the interface is zero. Figure 5.1 shows the velocity and temperature distributions near a heated flat plate placed in a vertical position in air (3). At a given distance from the bottom of the plate, the local upward velocity increases with increasing distance from the surface to reach a maximum value at a distance between 3 and 4 mm, then decreases and approaches zero again about 30 to 50 mm from the surface. Although the velocity profile is different from that observed in forced convection over a flat plate, where the velocity approaches the free-stream velocity asymptotically, in the vicinity of the surface the characteristics of both types of boundary layers are similar. In natural convection, as in forced convection, the flow may be laminar or turbulent, depending on the distance from the leading edge, the fluid properties, the body force, and the temperature difference between the surface and the fluid.

The temperature field in natural convection (Fig. 5.1) is similar to that observed in forced convection. Hence, the physical interpretation of the Nusselt number presented in Section 4.6 applies. For practical application however, Newton's equation, Eq. (1.11)

$$dq = h_c \, dA(T_s - T_\infty)$$

is generally used. The reason for writing the equation for a differential area dA is that, in natural convection, the heat transfer coefficient h_c is not uniform over a surface. As in forced convection over a flat plate, we shall therefore distinguish

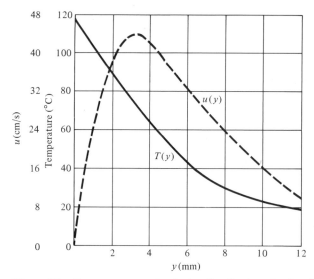

Figure 5.1 Temperature and velocity distributions in the vicinity of a heated flat plate placed vertically in still air. [After E. Schmidt and W. Beckmann (3).]

between a local value of h_c and an average value $\bar{h}_c$ obtained by averaging h_c over the entire surface. The temperature T_∞ refers to a point in the fluid sufficiently removed from the body that the temperature of the fluid is not affected by the presence of a heating (or cooling) source in the body.

Exact evaluation of the heat transfer coefficient for natural convection from the boundary layer is very difficult. The problem has been solved only for simple geometries, such as a vertical flat plate and a horizontal cylinder (3,4). We shall not discuss these specialized solutions here. Instead, we shall set up the differential equations for natural convection from a vertical flat plate by using only fundamental physical principles. From these equations, without actually solving them, we shall determine the similarity conditions and associated dimensionless parameters that correlate experimental data. In Section 5.3 pertinent experimental data for various shapes of practical interest will be presented in terms of these dimensionless parameters, and their physical significance will be discussed. Section 5.4 treats natural convection from rotating objects, in which the body force due to centrifugal acceleration may be more important than the gravitational body force. Section 5.5 deals with problems in which natural convection and forced convection act at the same time—that is, mixed convection. Section 5.6 treats mass transfer problems in which flow is generated solely by species concentration gradients.

5.2 SIMILARITY PARAMETERS FOR NATURAL CONVECTION

In the analysis of natural convection we shall make use of a phenomenon observed by the Greeks over 2000 years ago and phrased by Archimedes somewhat as follows: A body immersed in a fluid experiences a buoyant or lifting

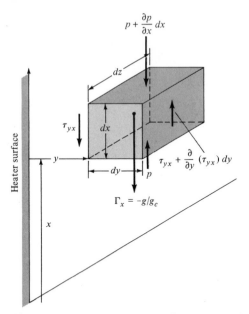

Figure 5.2 Forces acting on a fluid element in natural convection flow.

force equal to the mass of the displaced fluid. Hence, a submerged body rises when its density is less than that of the surrounding fluid and sinks when its density is greater. The buoyant effect is the driving force in natural convection.

For the purpose of analysis, consider a domestic heating panel, which can be idealized as a vertical flat plate, very long and wide in the plane perpendicular to the floor so that the flow is two-dimensional (Fig. 5.2). When the heater is turned off, the panel is at the same temperature as the surrounding air. The gravitational or body force acting on each fluid element is in equilibrium with the hydrostatic pressure gradient, and the air is motionless. When the heater is turned on, the fluid in the vicinity of the panel will be heated and its density will decrease. Hence, the body force (defined as the force per unit mass) on a unit volume in the heated portion of the fluid is less than in the unheated fluid. This imbalance causes the heated fluid to rise, a phenomenon that is well known from experience. In addition to the buoyant force, there are pressure forces and also frictional forces that act when the air is in motion. Once steady-state conditions have been established, the total force on a volume element $dx\,dy\,dz$ in the positive x direction perpendicular to the floor consists of the following:

1. The force due to the pressure gradient

$$p\,dy\,dz - \left(p + \frac{\partial p}{\partial x}\,dx\right)dy\,dz = -\frac{\partial p}{\partial x}(dx\,dy\,dz)$$

2. The body force $\Gamma_x \rho(dx\,dy\,dz)$, where $\Gamma_x = -g/g_c$, since gravity alone is active.*

* g_c is the gravitational constant, equal to $1\ \text{kg}\,\text{m/N}\,\text{s}^2$ in the SI system.

3. The frictional shearing forces due to the velocity gradient

$$(-\tau_{yx})\,dx\,dz + \left(\tau_{yx} + \frac{\partial \tau_{yx}}{\partial y}\,dy\right)dx\,dz$$

Since $\tau_{yx} = \mu(\partial u/\partial y)/g_c$ in laminar flow, the net frictional force is

$$\left(\frac{\mu}{g_c}\frac{\partial^2 u}{\partial y^2}\right)dx\,dy\,dz$$

Forces due to the deformation of the fluid element will be neglected in view of the low velocity. Ostrach (1) has shown that the effects of compression work and frictional heat may be important in natural-convection problems when very large temperature differences exist, very large length scales are involved, or very high body forces occur, such as in high-speed rotating machinery.

The rate of change of momentum of the fluid element is $\rho\,dx\,dy\,dz$ $\times\,[u(\partial u/\partial x) + v(\partial u/\partial y)]$, as shown in Section 4.5. Applying Newton's second law to the elemental volume yields

$$\rho\left(u\frac{\partial u}{\partial x} + v\frac{\partial u}{\partial y}\right) = -g_c\frac{\partial p}{\partial x} - \rho g + \mu\frac{\partial^2 u}{\partial y^2} \tag{5.1}$$

after canceling $dx\,dy\,dz$. The unheated fluid far removed from the plate is in hydrostatic equilibrium, or $g_c(\partial p_e/\partial x) = -\rho_e g$, where the subscript e denotes equilibrium conditions. At any elevation the pressure is uniform and therefore $\partial p/\partial x = \partial p_e/\partial x$. Substituting $\rho_e g$ for $-(\partial p/\partial x)$ in Eq. (5.1) gives

$$\rho\left(u\frac{\partial u}{\partial x} + v\frac{\partial u}{\partial y}\right) = (\rho_e - \rho)g + \mu\frac{\partial^2 u}{\partial y^2} \tag{5.2}$$

A further simplification can be made by assuming that the density ρ depends only on the temperature, and not on the pressure. For an incompressible fluid this is self-evident, but for a gas it implies that the vertical dimension of the body is small enough that the hydrostatic density ρ_e is constant. This is referred to as the Boussinesq approximation. With these assumptions, the buoyant term can be written

$$g(\rho_e - \rho) = g(\rho_\infty - \rho) = -g\rho\beta(T_\infty - T) \tag{5.3}$$

where β is the coefficient of thermal expansion, defined as

$$\beta = -\frac{1}{\rho}\frac{\partial \rho}{\partial T}\bigg|_p \cong \frac{\rho_\infty - \rho}{\rho(T - T_\infty)} \tag{5.4}$$

for an ideal gas (i.e., $\rho = p/RT$) the coefficient of thermal expansion is

$$\beta = \frac{1}{T_\infty} \tag{5.5}$$

where the temperature T_∞ is absolute temperature far from the plate.

The equation of motion for natural convection is obtained by substituting the buoyant term Eq. (5.3) into Eq. (5.2), yielding

$$u \frac{\partial u}{\partial x} + v \frac{\partial u}{\partial y} = g\beta(T - T_\infty) + v \frac{\partial^2 u}{\partial y^2} \tag{5.6}$$

In deriving the conservation of energy equation for the flow near the plate, we follow the same steps used in Chapter 4 to derive the conservation of energy equation for the forced flow near a flat plate. This leads to Eq. (4.13), which also describes the temperature field for the natural-convection problem:

$$u \frac{\partial T}{\partial x} + v \frac{\partial T}{\partial y} = \alpha \frac{\partial^2 T}{\partial y^2}$$

The dimensionless parameters may be determined from the Buckingham π theorem, Section 4.8. We have seven physical quantities:

U_∞—characteristic velocity

L—characteristic length

g—acceleration due to gravity

β—coefficient of expansion

$(T - T_\infty)$—temperature difference

v—kinematic viscosity

α—thermal diffusivity

which may be expressed in four primary dimensions: mass, length, time, and temperature. We should therefore be able to express the dimensionless heat transfer coefficient (Nusselt number) in terms of $7 - 4 = 3$ dimensionless groups

$$\text{Nu} = \text{Nu}(\pi_1, \pi_2, \pi_3) \tag{5.7}$$

Using the method described in Section 4.8, we find

$$\pi_1 = \frac{U_\infty L}{v}$$

$$\pi_2 = \frac{v}{\alpha} \tag{5.8}$$

$$\pi_3 = \frac{\rho^2 g \beta (T - T_\infty) L^3}{\mu^2}$$

We recognize π_1 as the Reynolds number and π_2 as the Prandtl number. The third dimensionless group is called the Grashof number Gr and represents the ratio of buoyant forces to viscous forces. Consistent units are:

α, v (m²/s)	g (m/s²)
L (m)	β (1/K)
U_∞ (m/s)	$(T - T_\infty)$(K)

Since the flow velocity is determined by the temperature field, π_1 is not an independent parameter. Therefore, we eliminate the dependence of Nusselt number on π_1. Experimental results for natural-convection heat transfer can therefore be correlated by an equation of the type

$$\text{Nu} = \phi(\text{Gr})\psi(\text{Pr}) \tag{5.9}$$

Often, the Grashof number and the Prandtl number are grouped together as a product GrPr, called the Rayleigh number Ra. Then the Nusselt number relation becomes

$$\text{Nu} = \phi(\text{Ra, Pr}) \tag{5.10}$$

Using an equation of this type, experimental data from various sources for natural convection from horizontal wires and tubes of diameter D are correlated in Fig. 5.3 by plotting $\bar{h}_c D/k$, the average Nusselt number, against $c_p \rho^2 g \beta \, \Delta T \, D^3/\mu k$, the Rayleigh number. The physical properties are evaluated at the arithmetic mean temperature. We observe that data for fluids as different as air, glycerin, and water are well correlated over a range of Rayleigh numbers from 10^{-5} to 10^7 for cylinders ranging from small wires to large pipes.

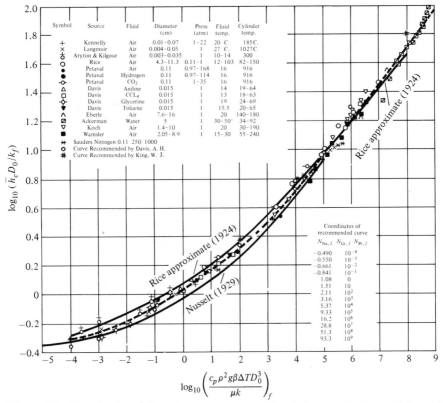

Figure 5.3 Correlation of data for natural convection heat transfer from horizontal cylinders in gases and liquids. (By permission from W. H. McAdams, *Heat Transmission*, 3d ed., McGraw-Hill, New York, 1954.)

EXAMPLE 5.1 ─────────────────────────────────

Calculate the rate of convective heat loss per unit length from a horizontal wire, 1 mm in diameter, held at 127°C in quiescent air at 27°C. Repeat for the case where the wire is held at 127°C in a carbon dioxide atmosphere.

Solution.   Using the average temperature of 77°C to calculate properties from Appendix 2, Table 27, the Rayleigh number is

$$\mathrm{Ra}_D = \frac{g\beta \, \Delta T D^3}{\nu^2} \, \mathrm{Pr}$$

$$= \frac{(9.8 \text{ m/s}^2)(350 \text{ K})^{-1}(100 \text{ K})(0.001 \text{ m})^3}{(2.12 \times 10^{-5} \text{ m}^2/\text{s})^2}(0.71)$$

$$= 4.43, \log_{10} \mathrm{Ra}_D = 0.646$$

From Fig. 5.3, $\log_{10} \mathrm{Nu}_D = 0.12$, $\mathrm{Nu}_D = 1.32$

$$\bar{h}_c = \frac{(1.32)(0.0291 \text{ W/m K})}{0.001 \text{ m}}$$

$$= 38.4 \text{ W/m}^2 \text{ K}$$

The rate of heat loss per meter length in air is:

$$q = (38.4 \text{ W/m}^2 \text{ K})(100 \text{ K})(\pi \, 0.001 \text{ m}^2/\text{m})$$

$$= 12.1 \text{ W/m}$$

Using Table 28 for properties of carbon dioxide

$$\mathrm{Ra}_D = 16.90$$
$$\log_{10} \mathrm{Ra}_D = 1.23$$
$$\log_{10} \mathrm{Nu}_D = 0.21$$
$$\mathrm{Nu}_D = 1.62$$
$$\bar{h}_c = 33.2 \text{ W/m}^2 \text{ K}$$
$$q = 10.4 \text{ W/m}$$

───

It has been claimed (6) that the correlation in Fig. 5.3 gives approximate results also for three-dimensional shapes such as short cylinders and blocks if the characteristic length dimension is determined by

$$\frac{1}{L} = \frac{1}{L_{\mathrm{hor}}} + \frac{1}{L_{\mathrm{vert}}}$$

where L_{vert} is the height and L_{hor} the average horizontal dimension of the body. Sparrow and Ansari (7), however, have shown that the characteristic length given by this equation may lead to large errors in predicting $\overline{\mathrm{Nu}}_L$ for some three-dimensional bodies. Their data suggest, in fact, that it is likely that no such simple characteristic length will collapse data for a wide range of geometric shapes and that a separate correlation equation may be required for each shape.

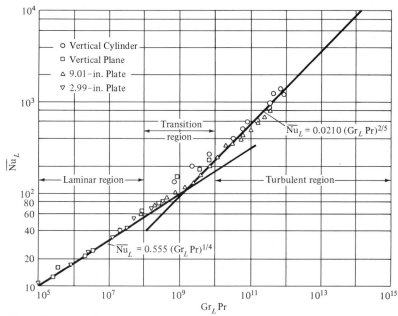

Figure 5.4 Correlation of data for natural-convection heat transfer from vertical plates and cylinders (49).

A correlation for natural convection from vertical plates and vertical cylinders is shown in Fig. 5.4.* The ordinate is $\bar{h}_c L/k$, the average Nusselt number based on the height of the body, and the abscissa is $c_p \rho^2 \beta g\, \Delta T L^3/\mu k$, the Rayleigh number. We note that there is a change in the slope of the line correlating the experimental data at a Rayleigh number of 10^9. The reason for the change in slope is that the flow is laminar up to a Rayleigh number of about 10^8, passes through a transition regime between 10^8 and 10^{10}, and becomes fully turbulent at Rayleigh numbers above 10^{10}. This is illustrated in the photographs of Fig. 5.5. These pictures show lines of constant density in natural convection from a vertical flat plate to air at atmospheric pressure obtained with a Mach-Zehnder (9, 10) optical interferometer. This instrument produces interference fringes, which are recorded by a camera. The fringes are the result of density gradients caused by temperature gradients in gases. The spacing of the fringes is a direct measure of the density distribution, which is related to the temperature distribution. Figure 5.5 shows the fringe pattern observed near a heated vertical flat plate, 0.91 m high and 0.46 m wide, in air. The flow is laminar for about 51 cm from the bottom of the plate. Transition to turbulent flow begins at 53 cm, corresponding to a critical Rayleigh number about 4×10^8. Near the top of the plate, turbulent flow is approached. This type of behavior is typical of natural convection on vertical surfaces, and under normal conditions the

---

* According to Gebhart (8), a vertical cylinder of diameter D may be treated as a flat plate of height L when $D/L > 35\, Gr_L^{-1/4}$.

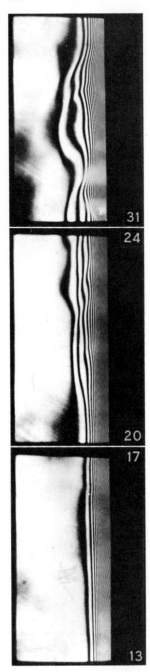

Figure 5.5 Interference photograph illustrating laminar and turbulent natural-convection flow of air along a vertical flat plate. (Courtesy of Professor E. R. G. Eckert.)

critical value of the Rayleigh number is usually taken as 10^9 for air. An extensive treatment of transition and stability in natural-convection systems is presented in (2) and (8).

EXAMPLE 5.2 ————————————————————————————

Estimate the electrical power required to maintain a vertical-plate resistance heater at 130°C in ambient air at 20°C. The plate is 15 cm high and 10 cm wide. Compare with results for a plate 450 cm high. The unit-surface conductance for radiation $\bar{h}_r$ is 8.5 W/m² K for the specified surface temperature.

Solution. The arithmetic mean temperature is 75°C and the corresponding value of Gr_L is found to be 65 $L^3(T_s - T_\infty)$, where L is in centimeters, from the last column in Appendix 2, Table 27 by interpolation. For the specified conditions we get

$$\mathrm{Gr}_L = (65)(15)^3(110) = 2.41 \times 10^7$$

for the smaller plate. Since the Grashof number is less than 10^9, the flow is laminar. For air at 75°C the Prandtl number is 0.71 and GrPr is therefore 1.17×10^7. From Fig. 5.4 the average Nusselt number is 35.7 at $\mathrm{GrPr} = 1.71 \times 10^7$ and therefore

$$\bar{h}_c = 35.7\,\frac{k}{L} = (35.7)\,\frac{2.9 \times 10^{-2}\ \mathrm{W/m\ K}}{0.15}\,\frac{}{\mathrm{m}} = 6.90\ \mathrm{W/m^2\ K}$$

Combining the effects of convection and radiation as shown in Chapt. 1, the total heat dissipation rate from both sides of the plate is therefore

$$q = A(\bar{h}_c + \bar{h}_r)(T_s - T_\infty)$$
$$= [(2)(0.15)(0.10)\ \mathrm{m^2}][(6.9 + 8.5)\ \mathrm{W/m^2\ K}](110\ \mathrm{K}) = 50.8\ \mathrm{W}$$

For the large plate the Rayleigh number is $(450/15)^3$ times larger or Ra = 4.62×10^{11}. This indicates that the flow is turbulent. From Fig. 5.4, the average Nusselt number is 973 and $\bar{h}_c = 6.3$ W/m² K. The total heat dissipation rate from both sides of the plate is therefore

$$q = [(2)(4.5)(0.10)\ \mathrm{m^2}][(6.3 + 8.5)\ \mathrm{W/m^2\ K}](110\ \mathrm{K}) = 1465\ \mathrm{W}$$

When the physical properties of the fluid vary considerably with temperature and the temperature difference between the body surface T_s and the surrounding medium T_∞ is large, satisfactory results can be obtained by evaluating the physical properties in Eq. (5.10) at the mean temperature $(T_s + T_\infty)/2$. However, when the surface temperature is not known, a value must be assumed initially. It can then be used to calculate the unit-surface conductance to a first approximation. The surface temperature is then recalculated with this value of the surface conductance, and if there is a significant discrepancy between the assumed and the calculated value of T_s the latter is used to recalculate the heat transfer coefficient for the second approximation. Correlations that specifically include the effect of variable properties are given by Clausing (11).

5.3 EMPIRICAL CORRELATION FOR VARIOUS SHAPES

After experimental data have been correlated by dimensional analysis, it is general practice to write an equation for the line that best fits the data. It is also useful to compare the experimental results with those obtained by analytic means, if they are available. This comparison allows one to determine whether the analytic method adequately describes the experimental results. If the two agree, one can describe the physical mechanisms that are important for the problem with confidence.

In this section the results of some experimental studies on natural convection for a number of geometric shapes of practical interest are presented. Each shape has been associated with a characteristic dimension, such as its distance from the leading edge x, length L, diameter D, and so on. The characteristic dimension is attached as a subscript to the dimensionless parameters Nu and Gr. Average values of the Nusselt number for a given surface are identified by a bar, that is, $\overline{\text{Nu}}$; local values are without a bar. All physical properties are to be evaluated at the arithmetic mean between the surface temperature T_s and the temperature of the undisturbed fluid T_∞. The temperature difference in the Grashof number ΔT represents the absolute value of the difference between the temperatures T_s and T_∞. The accuracy with which in practice the unit-surface conductance can be predicted from any of the equations is generally no better than 20 percent, because most experimental data scatter by as much as ± 15 percent or more, and in a majority of engineering applications stray currents due to some interaction with surfaces other than the one transferring the heat are unavoidable.

In the following subsections we present correlation equations for several important geometries. That information is also contained in condensed form in the Summary, Section 5.7, where a brief description and simple illustration of the geometry is given along with the appropriate correlation equation.

5.3.1 Vertical Plates and Cylinders

For a flat vertical surface, it is possible to find analytical and approximate solutions to the momentum and energy equations, Eqs. (5.6) and (4.13), by the integral boundary-layer analysis introduced in Section 4.10. Details of the method for natural convection may be found in (2). The results indicate that the local value of the heat transfer coefficient for laminar natural convection from an isothermal vertical plate or cylinder at a distance x from the leading edge is

$$h_{cx} = 0.508 \text{Pr}^{1/2} \frac{\text{Gr}_x^{1/4}}{(0.952 + \text{Pr})^{1/4}} \frac{k}{x} \tag{5.11}$$

Since $Gr_x \sim x^3$, Eq. (5.11) shows that the heat transfer coefficient decreases with the distance from the leading edge to the 1/4 power. The leading edge is the lower edge for a heated surface and the upper edge for a surface cooler than the surrounding fluid. The average value of the heat transfer coefficient for a height L is obtained by integrating Eq. (5.11) and dividing by L, or

$$\bar{h}_c = \frac{1}{L} \int_0^L h_{cx} \, dx = 0.68 Pr^{1/2} \frac{Gr_L^{1/4}}{(0.952 + Pr)^{1/4}} \frac{k}{L} \qquad (5.12a)$$

In dimensionless form, the average Nusselt number is

$$\overline{Nu}_L = \frac{\bar{h}_c L}{k} = 0.68 Pr^{1/2} \frac{Gr_L^{1/4}}{(0.952 + Pr)^{1/4}} \qquad (5.12b)$$

Gryzagoridis (12) has shown experimentally that Eq. (5.12b) adequately represents the data in the regime $10 < Gr_L Pr < 10^8$.

For a vertical plane submerged in a liquid metal ($Pr < 0.03$), the average Nusselt number in laminar flow is (13)

$$\overline{Nu}_L = \frac{\bar{h}_c L}{k} = 0.68(Gr_L Pr^2)^{1/4} \qquad (5.12c)$$

In the turbulent region, the value of h_{cx}, the local heat transfer coefficient, is nearly constant over the surface. In fact, McAdams (6) recommends for $Gr_L > 10^9$ the equation

$$\overline{Nu}_L = \frac{\bar{h}_c L}{k} = 0.13(Gr_L Pr)^{1/3} \qquad (5.13)$$

according to which the heat transfer coefficient is independent of the length L.

A theoretical analysis by Sparrow and Gregg (14), supported by experimental data from Dotson (15), indicates that the equations for laminar natural convection from a vertical flat plate apply to a constant surface temperature as well as to a uniform heat flux over the surface. In the latter case the surface temperature T_s is to be taken at one-half of the total height of the plate. Other types of correlations for constant heat flux are presented in (16) and (17).

For a vertical plate, or a plate tilted at an angle θ from the vertical, with the heated surface facing downward (or cooled surface facing upward), Fujii and Imura (18) found that the equation

$$\overline{Nu}_L = 0.56(Gr_L Pr \cos \theta)^{1/4} \qquad (5.14)$$

applies in the range

$$10^5 < Gr_L Pr \cos \theta < 10^{11} \text{ and } 0 \le \theta \le 89°$$

In Eq. (5.14) the plate length L is the dimension which rotates in a vertical plane as θ increases. If the heated surface is facing upward (or cooled surface facing downward), Eq. (5.14) is only recommended up to the critical Rayleigh number, because at higher Ra the data deviate significantly from Eq. (5.14) and boundary-layer separation was observed. The critical Rayleigh number depends on θ and is given by Figure 12 in (18). Beyond this critical Rayleigh number, Fujii and Imura (18) give correlation equations for predicting Nu_L; however, the data for two plates of different size differ slightly, resulting in two correlation equations.

5.3.2 Horizontal Plates

For a two-dimensional horizontal plate of length L with the heated surface facing downward, Fujii and Imura (18) recommend the relation

$$\overline{\mathrm{Nu}}_L = 0.58(\mathrm{Gr}_L\mathrm{Pr})^{1/5}$$

in the range

$$10^6 < \mathrm{Gr}_L\mathrm{Pr} < 10^{11} \tag{5.15}$$

When the heated surface faces upward, Fujii and Imura (18) recommend the relations

$$\overline{\mathrm{Nu}}_L = 0.16(\mathrm{Gr}_L\mathrm{Pr})^{1/3} \quad \text{for} \quad 7 \times 10^6 < \mathrm{Gr}_L\mathrm{Pr} < 2 \times 10^8 \tag{5.16a}$$
$$\overline{\mathrm{Nu}}_L = 0.13(\mathrm{Gr}_L\mathrm{Pr})^{1/3} \quad \text{for} \quad 5 \times 10^8 < \mathrm{Gr}_L\mathrm{Pr} \tag{5.16b}$$

For circular plates of diameter D with the heated surface facing downward, Kadambi and Drake (19) recommend

$$\overline{\mathrm{Nu}}_D = 0.82(\mathrm{Gr}_D\mathrm{Pr})^{1/5}\mathrm{Pr}^{0.034} \tag{5.17}$$

Experimental data for a cooled circular horizontal plate facing down in a liquid metal are correlated by the relation (20)

$$\overline{\mathrm{Nu}}_D = \frac{\bar{h}_cD}{k} = 0.26(\mathrm{Gr}_D\mathrm{Pr}^2)^{0.35} \tag{5.18}$$

EXAMPLE 5.3 ───

Calculate the rate of convective heat loss from a 1-m-square, horizontal plate heated to 227°C in ambient air at 27°C.

Solution. Using properties of air at the mean temperature, we find

$$\mathrm{Ra}_L = \frac{(9.8 \text{ m/s}^2)(396 \text{ K})^{-1}(200 \text{ K})(1 \text{ m})^3(0.71)}{(2.7 \times 10^{-5} \text{ m}^2/\text{s})^2}$$
$$= 4.82 \times 10^9$$

From Eq. (5.15) the Nusselt number for heat transfer from the bottom of the plate is

$$\overline{\mathrm{Nu}}_L = 0.58(4.82 \times 10^9)^{0.2}$$
$$= 50.1$$

and from Eq. (5.16) the Nusselt number from the top surface is

$$\overline{\mathrm{Nu}}_L = 0.13(4.82 \times 10^9)^{1/3}$$
$$= 220$$

The corresponding heat transfer coefficients are

$$\text{bottom: } \overline{h_c} = (50.1)(0.032 \text{ W/m K})/(1 \text{ m}) = 1.60 \text{ W/m}^2 \text{ K}$$
$$\text{top: } \overline{h_c} = (221)(0.032 \text{ W/m K})/(1 \text{ m}) = 7.07 \text{ W/m}^2 \text{ K}$$

and the total convective heat loss is therefore

$$Q = (1 \text{ m}^2)[(1.61 + 7.07) \text{ W/m}^2 \text{ K}](200 \text{ K})$$
$$= 1740 \text{ W}$$

Note that the heat dissipated by the upward-facing surface is nearly 82 percent of the total.

5.3.3 Cylinders, Spheres, and Cones

The temperature field around a horizontal cylinder heated in air is illustrated in Fig. 5.6, which shows interference fringes photographed by Eckert and Soehnghen (10). The flow is laminar over the entire surface. The closer spacing of the interference fringes over the lower portion of the cylinder indicates a steeper temperature gradient and consequently a larger local unit-surface conductance than over the top portion. The variation of the surface conductance with angular position α is shown in Fig. 5.7 for two Grashof numbers. The

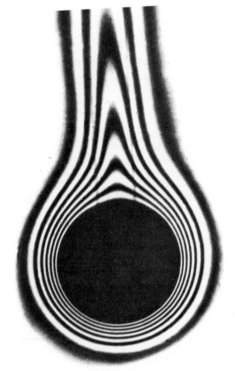

Figure 5.6 Interference photograph illustrating temperature field around a horizontal cylinder in laminar flow. (Courtesy of Professor E. R. G. Eckert.)

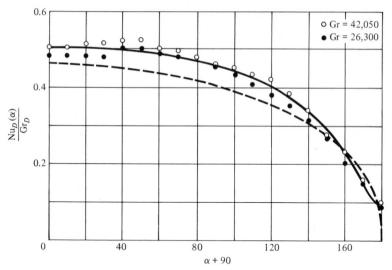

Figure 5.7 Local dimensionless unit-surface conductance along the circumference of a horizontal cylinder in laminar natural convection [dashed line according to (4)]. [Courtesy of the U.S. Air Force; from "Studies on Heat Transfer in Laminar Free Convection with the Zehnder-Mach Interferometer," by E. R. G. Eckert and E. E. Soehnghen (10).]

experimental results do not differ appreciably from the theoretical calculations of Hermann (4), who derived the equation

$$\text{Nu}_{D\alpha} = 0.604 \text{Gr}_D{}^{1/4} \phi(\alpha) \tag{5.19}$$

for air, that is, $\text{Pr} = 0.71$. The angle α is measured from the horizontal position and numerical values of the function $\phi(\alpha)$ are as follows:

	Bottom half				Top half			
α	-90	-60	-30	0	30	60	75	90
$\phi(\alpha)$	0.76	0.75	0.72	0.66	0.58	0.46	0.36	0

An equation for the average heat transfer coefficient from single horizontal wires or pipes in natural convection, recommended by McAdams (6) on the basis of the experimental data in Fig. 5.3, is

$$\overline{\text{Nu}}_D = 0.53(\text{Gr}_D \text{Pr})^{1/4} \tag{5.20}$$

This equation is valid for Prandtl numbers larger than 0.5 and Grashof numbers ranging from 10^3 to 10^9. For very small diameters, Langmuir showed that the rate of heat dissipation per unit length is nearly independent of the wire diameter, a phenomenon he applied in his invention of the coiled filaments in gas-filled incandescent lamps. The average unit-surface conductance for Gr_D less than 10^3 is most conveniently evaluated from the curve $A-A$ drawn through the experimental points in Fig. 5.3 in the low-Grashof-number range.

In turbulent flow it has been observed (21) that the heat flux can be increased substantially without a corresponding increase in the surface temperature. It appears that in natural convection the turbulent-exchange mechanism increases in intensity as the rate of heat flow is increased and thereby reduces the thermal resistance.

EXAMPLE 5.4

At what temperature will a heated, horizontal steel drum, 1 m in diameter, produce turbulent flow in air at 27°C? Repeat for the case where the drum is placed in a water bath at 27°C. Use property values at 27°C.

Solution. The criterion for transition is $Ra_D = 10^9$. For air at 27°C this gives

$$Ra_D = \frac{(9.8 \text{ m/s}^2)(300 \text{ K})^{-1}(\Delta T)(1 \text{ m})^3(0.71)}{(1.64 \times 10^{-5} \text{ m}^2/\text{s})^2} = 10^9$$

Therefore,
$$\Delta T = 12°C$$
$$T_{\text{drum}} = 12 + 27 = 39°C$$

For water (Table 13, Appendix 2) we get

$$Ra_D = \frac{(9.8 \text{ m/s}^2)(2.73 \times 10^{-4} \text{ K}^{-1})(\Delta T)(1 \text{ m})^3(5.9)}{(0.861 \times 10^{-6} \text{ m}^2/\text{s})^2} = 10^9$$

Solving for ΔT we find $\Delta T = 0.05°C$. Note that in water even a small temperature difference will induce turbulence.

For liquid metals in laminar flow the equation

$$\overline{Nu}_D = 0.53(Gr_D Pr^2)^{1/4} \tag{5.21}$$

correlates the available data (21) for horizontal cylinders.

Al-Arabi and Khamis (22) have correlated heat transfer data for cylinders of various lengths, diameters, and angles of inclination from the vertical. Their results are of the form $\overline{Nu}_L = m(Gr_L Pr)^n$, where m and n are functions of the cylinder diameter and angle of inclination from the vertical, θ. Transition to turbulent flow occurred near

$$(Gr_L Pr)_{cr} = 2.6 \times 10^9 + 1.1 \times 10^9 \tan \theta \tag{5.22}$$

In the laminar regime, $9.88 \times 10^7 \leq Gr_L Pr \leq (Gr_L Pr)_{cr}$, they found

$$\overline{Nu}_L = [2.9 - 2.32(\sin \theta)^{0.8}](Gr_D)^{-1/12}(Gr_L Pr)^{[1/4 + (1/12)(\sin \theta)1.2]} \tag{5.23}$$

and in the turbulent regime, $(Gr_L Pr)_{cr} \leq Gr_L Pr \leq 2.95 \times 10^{10}$, they found

$$\overline{Nu}_L = [0.47 + 0.11(\sin \theta)^{0.8}](Gr_D)^{-1/12}(Gr_L Pr)^{1/3} \tag{5.24}$$

In both regimes the Grashof number based in cylinder diameter is restricted to the range $1.08 \times 10^4 \leq Gr_D \leq 6.9 \times 10^5$.

For natural convection to or from spheres of diameter D the empirical equation

$$\overline{\mathrm{Nu}}_D = 2 + 0.392(\mathrm{Gr}_D)^{1/4} \text{ for } 1 < \mathrm{Gr}_D < 10^5 \tag{5.25}$$

is recommended (23). For very small spheres, as the Grashof number approaches zero, the Nusselt number approaches a value of 2, that is, $\bar{h}_c D/k \to 2$. This condition corresponds to pure conduction through a stagnant layer of fluid surrounding the sphere.

Experimental data for natural convection from vertical cones pointing downward with vertex angles between 3 and 12 degrees have been correlated (24) by

$$\overline{\mathrm{Nu}}_L = 0.63(1 + 0.72\epsilon)\mathrm{Gr}_L^{1/4} \tag{5.26}$$

where $3° < \phi < 12°$, $7.5 < \log \mathrm{Gr}_L < 8.7$, $0.2 \leq \epsilon \leq 0.8$, and

$$\epsilon = \frac{2}{\mathrm{Gr}_L^{1/4} \tan(\phi/2)}$$

ϕ = vertex angle

L = slant height of the cone

5.3.4 Enclosed Spaces

Natural-convection heat transfer across enclosures such as shown in Fig. 5.8 is important for determining heat loss through windows, from flat-plate solar collectors, through building walls, and in many other applications. The enclosure consists of two isothermal vertical surfaces at temperatures T_1 and T_2 spaced a distance δ apart and of height L. The top and bottom of the enclosure are insulated. The Grashof number is defined by

$$\mathrm{Gr}_\delta = \frac{g\beta(T_1 - T_2)\delta^3}{\nu^2} \tag{5.27}$$

Any temperature difference will produce flow in the enclosure. Hollands and Konicek (25) found that for $\mathrm{Gr}_\delta \gtrsim 8000$ the flow consists of one large cell rotating in the enclosure. The heat transfer mechanism is equivalent to con-

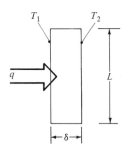

Figure 5.8 Nomenclature for natural convection in enclosed vertical spaces.

duction across the enclosure for $Gr_\delta < 8000$. As the Grashof number is increased beyond this value, the flow becomes more of a boundary-layer type with fluid rising in a layer near the heated surface, turning the corner at the top, and flowing downward in a layer near the cooled surface. The boundary-layer thickness decreases with $Gr_\delta^{1/4}$ and the core region is more or less inactive and thermally stratified.

For the geometry in Fig. 5.8 Catton (26) recommends the correlations of Berkovsky and Polevikov (27):

$$\overline{Nu}_\delta = 0.22 \left(\frac{L}{\delta}\right)^{-1/4} \left(\frac{Pr}{0.2 + Pr} Ra_\delta\right)^{0.28} \tag{5.28}$$

in the range

$$2 < L/\delta < 10, \ Pr < 10, \text{ and } Ra_\delta < 10^{10}$$

and

$$\overline{Nu}_\delta = 0.18 \left(\frac{Pr}{0.2 + Pr} Ra_\delta\right)^{0.29} \tag{5.29}$$

in the range

$$1 < L/\delta < 2, \ 10^{-3} < Pr < 10^5, \text{ and } 10^3 < \frac{Ra_\delta Pr}{0.2 + Pr}$$

Data are lacking for aspect ratios L/δ less than one. Imberger (28) found that as $Ra_\delta \to \infty$, $Nu_\delta \to (L/\delta)Ra_\delta^{1/4}$ for $L/\delta = 0.01$ and 0.02. Bejan et al. (29) found that $Nu_\delta = 0.014Ra_\delta^{0.38}$ for $L/\delta = 0.0625$ and $2 \times 10^8 < Ra_\delta < 2 \times 10^9$. Nansteel and Greif (30) found $Nu_\delta = 0.748Ra_\delta^{0.226}$ for $L/\delta = 0.5, 2 \times 10^{10} < Ra_\delta \le 10^{11}$, and $3.0 \le PR \le 4.3$.

In a horizontal fluid layer with heating from above, heat transfer is by conduction only. Heating from below results in conduction heat transfer only if $Ra_\delta < 1700$, where the length scale is the spacing enclosing the layer. Above this value of Ra_δ the fluid motion is in the form of multiple cells rotating with a horizontal axis, which are known as Benard cells. The flow begins to become turbulent for $Ra_\delta \sim 5500$ for $Pr = 0.7$ and for $Ra_\delta \sim 55,000$ for $Pr = 8500$ (31) and becomes fully turbulent for $Ra_\delta \sim 10^6$.

Hollands et al. (32) correlated data for horizontal air layers heated from below over a very wide range of Rayleigh numbers with

$$\overline{Nu}_\delta = 1 + 1.44 \left[1 - \frac{1708}{Ra_\delta}\right]^\cdot + \left[\left(\frac{Ra_\delta}{5830}\right)^{1/3} - 1\right]^\cdot \tag{5.30}$$

where the notation $[\ ]^\cdot$ indicates that if the quantity inside the bracket is negative the quantity is to be taken as zero. This equation closely represented data for air from the critical Rayleigh number ($Ra_\delta = 1700$) to $Ra_\delta = 10^8$. To closely match data for water, it was necessary to add a term to the above

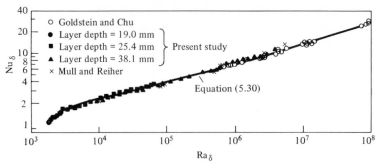

Figure 5.9 Correlation of data for natural-convection heat transfer across a horizontal layer of air heated from below. [Reprinted from (32) with permission from Pergamon Press, Ltd.]

equation:

$$\overline{Nu}_\delta = 1 + 1.44\left[1 - \frac{1708}{Ra_\delta}\right] + \left[\left(\frac{Ra_\delta}{5830}\right)^{1/3} - 1\right]$$
$$+ 2.0\left[\frac{Ra_\delta^{1/3}}{140}\right]^{[1 - \ln(Ra_\delta^{1/3}/140)]} \tag{5.31}$$

which is then valid from the critical Rayleigh number (~ 1700) to $Ra_\delta = 3.5 \times 10^9$. These two correlation equations with experimental data are shown in Figs. 5.9 and 5.10.

EXAMPLE 5.5 ———————————————————————

A pan of water 8 cm deep is placed on a stove-top burner. The burner element is thermostatically controlled and maintains the bottom of the pan at 100°C. Assuming the top surface of the water is initially at room

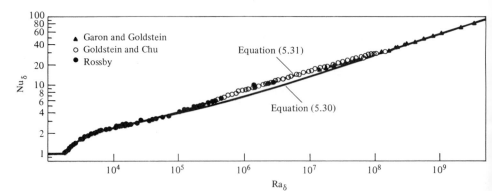

Figure 5.10 Correlation of data for natural-convection heat transfer across a horizontal layer of water heated from below. [Reprinted from (32) with permission from Pergamon Press, Ltd.]

temperature, 20°C, what is the initial rate of heat transfer from the burner to the water? The pan is circular and is 15 cm in diameter.

Solution.   For the properties of water at 60°C

$$Ra_\delta = \frac{(9.8 \text{ m/s}^2)(5.18 \times 10^{-4} \text{ K}^{-1})(80 \text{ K})(0.08 \text{ m})^3(3.02)}{(0.478 \times 10^{-6} \text{ m}^2/\text{s})^2}$$

$$Ra_\delta = 2.75 \times 10^9$$

From Eq. (5.31) we find

$$Nu_\delta = 1 + 1.44 + 76.8 + 0.1$$
$$= 79.3$$

$$\bar{h}_c = \overline{Nu_\delta} \, \frac{k}{\delta} = \frac{(79.3)(0.657 \text{ W/m K})}{0.08 \text{ m}}$$

$$= 651 \text{ W/m}^2 \text{ K}$$

The initial rate of heat transfer is therefore

$$Q = (651 \text{ W/m}^2 \text{ K})\left(\frac{\pi \, 0.15^2 \text{ m}^2}{4}\right)(80 \text{ K})$$

$$= 920 \text{ W}$$

For natural convection inside spherical cavities of diameter D the relation

$$\frac{D\bar{h}_c}{k} = C(Gr_D Pr)^n \tag{5.32}$$

is recommended (33) with the constants C and n selected from the tabulation below

$Gr_D Pr$	C	n
10^4–10^9	0.59	1/4
10^9–10^{12}	0.13	1/3

For natural-convection heat transfer across the gap between two horizontal concentric cylinders, Raithby and Hollands (34) suggest the correlation equation

$$\frac{k_{eff}}{k} = 0.386\left[\frac{\ln(D_o/D_i)}{b^{3/4}(1/D_i^{3/5} + 1/D_o^{3/5})^{5/4}}\right]\left(\frac{Pr}{0.861 + Pr}\right)^{1/4} Ra_b^{1/4} \tag{5.33}$$

Here, D_o is the diameter of the outer cylinder, D_i is the diameter of the inner cylinder, $2b = D_o - D_i$, and the Rayleigh number Ra_b is based on the temperature difference across the gap. The effective thermal conductivity k_{eff} is the thermal conductivity that a motionless fluid (with conductivity k) in the gap must have to transfer the same amount of heat as the moving fluid.

The correlation Eq. (5.33) is valid over the following range of parameters:

$$0.70 \leq \text{Pr} \leq 6000$$

$$10 \leq \left[\frac{\ln(D_o/D_i)}{b^{3/4}(1/D_i^{3/5} + 1/D_o^{3/5})^{5/4}}\right]^4 \text{Ra}_b \leq 10^7$$

For concentric spheres, Raithby and Hollands (34) recommend

$$\frac{k_{\text{eff}}}{k} = 0.74\left[\frac{b^{1/4}}{D_o D_i(D_i^{-7/5} + D_o^{-7/5})^{5/4}}\right]\text{Ra}_b^{1/4}\left(\frac{\text{Pr}}{0.861 + \text{Pr}}\right)^{1/4} \tag{5.34}$$

Eq. 5.34 is valid for

$$0.70 \leq \text{Pr} \leq 4200$$

and

$$10 \leq \left[\frac{b}{(D_o D_i)^4(D_i^{-7/5} + D_o^{-7/5})^5}\right]\text{Ra}_b \leq 10^7$$

5.4* ROTATING CYLINDERS, DISKS, AND SPHERES

Heat transfer by convection between a rotating body and a surrounding fluid is of importance in the thermal analysis of shafting, flywheels, turbine rotors, and other rotating components of various machines. Convection from a heated rotating horizontal cylinder to ambient air has been studied by Anderson and Saunders (35). Turbulence begins to appear at a critical peripheral-speed Reynolds number $\text{Re}_\omega = \omega\pi D^2/\nu$ of about 50, where ω is the rotational speed in radians per second. With heat transfer, the critical speed is reached when the circumferential speed of the cylinder surface becomes approximately equal to the upward natural-convection velocity at the side of a heated stationary cylinder.

Below the critical velocity simple natural convection, characterized by the conventional Grashof number $\beta g(T_s - T_\infty)D^3/\nu^2$, controls the rate of heat transfer. At speeds greater than critical ($\text{Re}_\omega > 8000$ in air) the peripheral-speed Reynolds number $\pi D^2\omega/\nu$ becomes the controlling parameter. The combined effects of the Reynolds, Prandtl, and Grashof numbers on the average Nusselt number for a horizontal cylinder rotating in air above the critical velocity can be expressed by the empirical equation (36)

$$\overline{\text{Nu}}_D = \frac{\bar{h}_c D}{k} = 0.11(0.5\text{Re}_\omega^2 + \text{Gr}_D\text{Pr})^{0.35} \tag{5.35}$$

Heat transfer from a rotating disk has been investigated experimentally by Cobb and Saunders (37) and theoretically by Millsap and Pohlhausen (38) and Kreith and Taylor (5), among others. The boundary layer on the disk is laminar and of uniform thickness at rotational Reynolds numbers $\omega D^2/\nu$ below about 10^6. At higher Reynolds numbers the flow becomes turbulent and the bound-

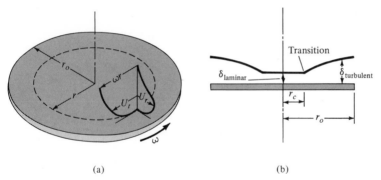

Figure 5.11 Velocity and boundary-layer profiles for a disk rotating in an infinite environment.

ary layer thickens with increasing radius (see Fig. 5.11). The average Nusselt number for a disk rotating in air is (37, 39)

$$\overline{Nu}_D = \frac{\bar{h}_c D}{k} = 0.35 \left(\frac{\omega r_o^2}{\nu} \right)^{1/2} \tag{5.36}$$

for $\omega D^2/\nu < 5 \times 10^5$. The radius of the disk, $r_o = D/2$, in Eq. (5.36).

In the turbulent flow regime of a disk rotating in air (37), the local value of the Nusselt number at a radius r is given approximately by

$$Nu_r = \frac{h_c r}{k} = 0.0195 \left(\frac{\omega r^2}{\nu} \right)^{0.8} \tag{5.37}$$

and the average value of the Nusselt number for laminar flow between $r = 0$ and r_c, and turbulent flow in the outer ring between $r = r_c$ and r_o, is approximately

$$\overline{Nu}_r = \frac{\bar{h}_c r_o}{k} = 0.015 \left(\frac{\omega r_o^2}{\nu} \right)^{0.8} - 100 \left(\frac{r_c}{r_o} \right)^2 \tag{5.38}$$

EXAMPLE 5.6 _____

A 20-cm-diameter steel shaft is heated to 400°C for heat-treating. The shaft is then allowed to cool in air (at 20°C) while rotating about its own (horizontal) axis at 3 revolutions per minute. Compute the rate of convective heat transfer from the shaft when it has cooled to 100°C.

Solution. The rotation speed of the shaft is

$$\omega = \frac{3 \text{ rev/min } (2\pi \text{ rad/rev})}{(60 \text{ s/min})} = 0.31 \text{ rad}$$

From the properties of air at 60°C, the Reynolds number is

$$Re_\omega = \frac{\pi (0.2 \text{ m})^2 (0.31 \text{ s}^{-1})}{1.94 \times 10^{-5} \text{ m}^2/\text{s}} = 2008$$

and the Rayleigh number is

$$\text{Ra} = \frac{(9.8 \text{ m/s}^2)(333 \text{ K})^{-1}(80 \text{ K})(0.2 \text{ m})^3(0.71)}{(1.94 \times 10^{-5} \text{ m}^2/\text{s})^2} = 3.55 \times 10^7$$

From Eq. (5.35)

$$\overline{\text{Nu}}_D = 0.11[0.5(2008)^2 + 3.55 \times 10^7]^{0.35} = 49.2$$

$$\bar{h}_c = \frac{(49.2)(0.0279 \text{ W/m K})}{0.20 \text{ m}} = 6.86 \text{ W/m}^2\text{K}$$

$$q = (6.86 \text{ W/m}^2 \text{ K})[\pi(0.2)(1) \text{ m}^2](80 \text{ K}) = 345 \text{ W/m}$$

Note that the effect of gravity-induced natural convection is large relative to that induced by the rotation of the shaft.

For a disk rotating in a fluid having a Prandtl number larger than unity, the local Nusselt number can be obtained, according to (40), from the equation

$$\overline{\text{Nu}}_r = \frac{\text{Re}_r \text{Pr} \sqrt{(C_{Dr}/2)}}{5\text{Pr} + 5 \ln(5\text{Pr} + 1) + \sqrt{(2/C_{Dr})} - 14} \tag{5.39}$$

where C_{Dr} is the local drag coefficient at radius r, which, according to (41), is given by

$$\frac{1}{\sqrt{C_{Dr}}} = -2.05 + 4.07 \log_{10} \text{Re}_r \sqrt{C_{Dr}} \tag{5.40}$$

For a sphere of diameter D rotating in an infinite environment with $\text{Pr} > 0.7$ in the laminar-flow regime ($\text{Re}_\omega = \omega D^2/\nu < 5 \times 10^4$) the average Nusselt number ($\bar{h}_c D/k$) can be obtained from

$$\overline{\text{Nu}}_D = 0.43\text{Re}_\omega^{0.5}\text{Pr}^{0.4} \tag{5.41}$$

while in the Reynolds number range between 5×10^4 and 7×10^5 the equation

$$\overline{\text{Nu}}_D = 0.066\text{Re}_\omega^{0.67}\text{Pr}^{0.4} \tag{5.42}$$

correlates the available experimental data (42).

5.5 COMBINED FORCED AND NATURAL CONVECTION

In Chapter 4 forced convection in flow over a flat surface was treated, and the preceding sections of this chapter dealt with heat transfer in natural-convection systems. In this section the interaction between natural- and forced-convection processes will be considered.

In any heat transfer process density gradients occur, and in the presence of a force field natural-convection currents arise. If the forced-convection effects are very large, the influence of natural-convection currents may be negligible and, similarly, when the natural-convection forces are very strong, the forced-convection effects may be negligible. The questions we wish to consider now are: Under what circumstances can either forced or natural convection be neglected, and what are the conditions when both effects are of the same order of magnitude?

To obtain an indication of the relative magnitudes of natural- and forced-convection effects, we consider the differential equation describing the uniform flow over a vertical flat plate with the buoyancy effect and the free-stream velocity U_∞ in the same direction. This would be the case when the plate is heated and the forced flow is upward, or when the plate is cooled and the forced flow is downward. Taking the flow direction as x and assuming that the physical properties are uniform except for the temperature effect on the density, the Navier-Stokes boundary-layer equation including natural-convection forces is

$$u \frac{\partial u}{\partial x} + v \frac{\partial u}{\partial y} = -\frac{1}{\rho} \frac{\partial p}{\partial x} + \frac{\mu}{\rho} \frac{\partial^2 u}{\partial y^2} + g\beta(T - T_\infty) \tag{5.43}$$

This equation can be generalized as follows. Substituting X for x/L, Y for y/L, θ for $(T - T_\infty)/(T_0 - T_\infty)$, P for $(p - p_\infty)/(\rho U_\infty^2/2g_c)$, U for u/U_∞, and V for v/U_∞ in Eq. (5.43) gives

$$U \frac{\partial U}{\partial X} + V \frac{\partial U}{\partial Y} = -\frac{1}{2} \frac{\partial P}{\partial X} + \left(\frac{\mu}{\rho U_\infty L}\right) \frac{\partial^2 U}{\partial Y^2} + \left[\frac{g\beta L^3(T_0 - T_\infty)}{\nu^2}\right] \frac{\nu^2}{U_\infty^2 L^2} \theta \tag{5.44}$$

In the region near the surface—that is, in the boundary layer—$\partial U/\partial X$ and U are of the order of unity. Since U changes from 1 at $x = 0$ to a very small value at $x = 1$, and u is of the same order of magnitude as U_∞, the left-hand side of Eq. (5.44) is of the order of unity. Similar reasoning indicates that the first two terms on the right-hand side as well as θ are of the order of unity. Consequently, the buoyancy effect will influence the velocity distribution, on which, in turn, the temperature distribution depends, if the coefficient of θ is of the order of 1 or larger, that is, if

$$\frac{[g\beta L^3(T_0 - T_\infty)]/\nu^2}{(U_\infty L/\nu)^2} = \frac{\mathrm{Gr}_L}{\mathrm{Re}_L^2} \cong 1 \tag{5.45}$$

In other words, the ratio $\mathrm{Gr}/\mathrm{Re}^2$ gives a qualitative indication of the influence of buoyancy on forced convection. When the Grashof number is of the same order of magnitude as or larger than the square of the Reynolds number, natural-convection effects cannot be ignored, compared to forced convection. Similarly, in a natural-convection process the influence of forced convection becomes significant when the square of the Reynolds number is of the same order of magnitude as the Grashof number.

Several special cases have been treated in the literature (43–45). For example, for laminar forced convection over a vertical flat plate, Sparrow and Gregg (43) showed that for Prandtl numbers between 0.01 and 10 the effect of buoyancy on the local heat transfer coefficient for pure forced convection will be less than 10 percent if

$$\mathrm{Gr}_x \leq 0.150\mathrm{Re}_x^2 \tag{5.46}$$

Eckert and Diaguila (45) studied mixed convection in a vertical tube with air, primarily in the turbulent regime. When the buoyancy-induced flow was in the same direction as the forced flow, they found that the local heat transfer coefficient differed from that for pure natural-convection behavior by less than 10 percent if

$$Gr_x > 0.007 Re_x^{2.5} \tag{5.47}$$

and from pure forced-convection behavior by less than 10 percent if

$$Gr_x < 0.0016 Re_x^{2.5} \tag{5.48}$$

In Fig. 5.12 we have plotted Eqs. (5.46)–(5.48) to delineate the regimes of pure natural convection in boundary-layer flow, mixed-convection boundary-layer flow, and pure forced convection in boundary-layer flow for geometries in which the buoyancy and forced-flow effects are parallel.

Siebers et al. (46) measured mixed-convection heat transfer from a large (3.03 m high × 2.954 m wide) vertical flat plate. The plate was heated electrically to produce natural-convection flow up the plate, and it was located in a wind tunnel to simultaneously expose it to a horizontal forced flow parallel to the plate. Therefore, this study was concerned with vertical buoyancy flow and horizontal forced flow. They based the magnitude of the natural convection on Gr_H, where H is the plate height, and the magnitude of the forced convection on Re_L, where L is the plate width. Their results indicate that if $Gr_H/Re_L^2 < 0.7$ then the heat transfer is essentially due to forced convection, and if $Gr_H/Re_L^2 > 10$ then natural convection dominates. For intermediate values, that is, mixed convection, they provide correlating equations for local heat transfer coefficients. Figure 5.13 shows the different flow regimes for this geometry for laminar flow (Fig. 5.13a) and turbulent flow (Fig. 5.13b).

The influence of natural convection on forced flow in tubes and ducts is discussed in Chapter 6, Section 6.3.

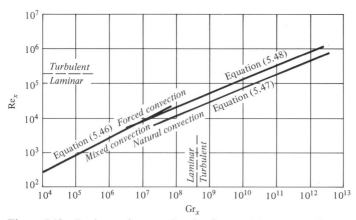

Figure 5.12 Regimes of convection for flow and bouyancy effects parallel; boundary-layer processes. (By permission from B. Gebhart, *Heat Transfer*, 2d ed., McGraw-Hill, New York, 1971.)

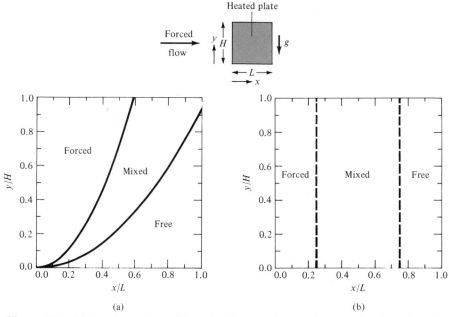

Figure 5.13 (*a*) Laminar zones of forced, mixed, and natural convection for a heated vertical plate in horizontal forced flow (46). (*b*) Turbulent zones of forced, mixed, and natural convection for a heated vertical plate in horizontal forced flow (46).

5.6* NATURAL-CONVECTION MASS TRANSFER

Density differences required to produce buoyancy forces may result from concentration gradients in a fluid body just as they can arise from temperature gradients. Thus, even an isothermal body of fluid may exhibit naturally driven circulation if there exists a concentration gradient of one or more species. For example, consider a vertical surface placed in an extensive body of liquid. If the material comprising the vertical surface is soluble in the liquid, the liquid near the surface will exhibit a finite concentration of the solute, while far away from the surface this concentration will be zero. If the pure liquid is less dense than the mixture, then the liquid near the surface will begin to fall just as if the surface had been cooled.

Natural-convection mass transfer is found often in nature and in industrial processes. Evaporation from the surface of ponds or from the human body and industrial drying operations are examples. In many practical applications one finds both heat and mass transfer occurring simultaneously and interacting with one another, resulting in a very complex problem. Problems of that complexity are beyond the scope of this text. However, it is instructive to demonstrate the similarity between heat and mass transfer for a simple natural-convection problem just as was done for forced convection in Chapter 4.

We return now to the vertical-plate problem. If the concentration of the solute, say species *A*, in the liquid, say species *B*, is dilute and if the system is isothermal, the governing equations which replace the momentum equation Eq.

(5.6) and the energy equation Eq. (4.13) are

$$u \frac{\partial u}{\partial x} + v \frac{\partial u}{\partial y} = g\beta^*(C_A - C_{A,\infty}) + v \frac{\partial^2 u}{\partial y^2} \tag{5.49}$$

$$u \frac{\partial C_A}{\partial x} + v \frac{\partial C_A}{\partial y} = D_{AB} \frac{\partial^2 C_A}{\partial y^2} \tag{5.50}$$

respectively. Here, $\beta^* = -(1/\rho)(\partial \rho / \partial C_A)|_{T,p}$, C_A is the local concentration of species A, $C_{A,\infty}$ is the concentration of species A far from the plate, and D_{AB} is the binary diffusion coefficient. Thus, the buoyancy term in the momentum equation represents the buoyancy force due to a local change in density which, in turn, has been brought about by a local concentration difference $C_A - C_{A,\infty}$ rather than a local temperature difference. The energy equation has been replaced by an equation expressing the local balance between the convection and diffusion of species A rather than of thermal energy. Repeating the procedure used to nondimensionalize Eqs. (5.6) and (4.13), we find that the important dimensionless groups for this problem are

$$Gr_{CL} = \frac{g\beta^*(C_{A,0} - C_{A,\infty})L^3}{v^2} \tag{5.51}$$

$$Sc = \frac{v}{D_{AB}} \tag{5.52}$$

that is, the mass diffusion Grashof number Gr_{CL} and the Schmidt number Sc. The concentration of species A at the surface is $C_{A,0}$. Recall from Chapter 4 that the Sherwood number is a dimensionless mass transfer coefficient and is the mass transfer equivalent of the Nusselt number. By analogy with Eq. (5.10) it would seem that, in general, the Sherwood number could be expressed as a function of the mass diffusion Grashof number and the Schmidt number

$$Sh_L = \phi(Gr_{CL}, Sc) \tag{5.53}$$

The dimensionless equations describing thermally driven natural convection are identical to these describing concentration-driven natural convection if Gr in the former is replaced by Gr_C, and Pr in the former is replaced by Sc. Thus, the value of the Sherwood number can often be determined by analogy with heat transfer. That is, the values of Gr_C and Sc for the mass transfer problem are calculated and inserted into a correlation equation for Nu = ϕ(Gr, Pr) by replacing Gr with Gr_C and Pr with Sc. The resulting value of Nu is equal to the desired value of the Sherwood number. The correlation equation must represent heat transfer data from a geometry and boundary conditions similar to those in the mass transfer problem for which Sh is desired. For example, constant surface temperature is similar to a uniform surface concentration.

An inverse procedure, which is obviously just as valid, is often used to obtain heat transfer correlations by performing a mass transfer experiment, which, in some cases, may be simpler to perform.

For example, Goldstein and Lau (47) measured the natural-convection mass transfer from a horizontal square plate facing downward and equated the Sherwood number to the Nusselt number for the equivalent heat transfer problem—an isothermal heated square plate facing upward. By measuring the rate of mass loss from square plates of naphthalene, which sublimes in air, ranging in size from 2.6 to 20.3 cm, they were able to determine the Sherwood number for the range of Rayleigh numbers from 10 to 6000, that is,

$$\overline{Sh}_L = 0.746 Ra_{CL}{}^{1/5} \tag{5.54}$$

where $Ra_{CL} = Gr_{CL} Sc$. One important reason for measuring the mass transfer instead of heat transfer is that edge conduction errors and radiation errors are eliminated. In these experiments the condition requiring low concentration of naphthalene in the air was met, but the Schmidt number was approximately 2.5, whereas the Prandtl number for air is 0.7.

EXAMPLE 5.7 ————————————————————————————————

A horizontal cylinder, 9.2 cm in diameter and 1 m long, has been fabricated out of naphthalene and is supported in a quiescent room at 53°C. Determine the weight loss of the cylinder after 45 min.

Solution. To determine the weight loss from the cylinder we will relate the mass transfer problem to a similar heat transfer problem. Equation (5.20) gives for the heat transfer problem

$$\overline{Nu}_D = 0.53 (Gr_D Pr)^{1/4}$$

in the range

$$Pr > 0.5 \quad \text{and} \quad 10^3 < Gr_D < 10^9$$

We must first determine whether the Grashof number for mass transfer falls in this range. Writing Eq. (5.51) in terms of density

$$Gr_{CD} = \frac{g\beta^*(C_{A,0} - C_{A,\infty})D^3}{\nu^2} = \frac{g(\rho_w - \rho_\infty)D^3}{\rho_\infty \nu^2}$$

To evaluate the overall density difference $\rho_w - \rho_\infty$ we must determine the density at the cylinder surface ρ_w. (The density far from the cylinder ρ_∞ is that of air.) The density near the cylinder depends on the vapor pressure of naphthalene in air at 53°C, which is 1 mm Hg [e.g., see (48)]. The average molecular weight of the gas near the cylinder is

$$\overline{\mathscr{M}}_w = \frac{\mathscr{M}_N}{760} + \frac{759\mathscr{M}_a}{760} = \frac{128}{760} + \frac{(759)(28.8)}{760} = 28.93 \frac{g}{g\ mole}$$

and from the ideal gas law

$$\rho_w - \rho_\infty = \frac{P}{RT}(\overline{\mathscr{M}}_w - \mathscr{M}_\infty) = \frac{(1\ atm)(28.93 - 28.8)(g/g\ mole)}{(0.0821\ l\ atm/g\ mole\ K)(326\ K)}$$

$$= 0.0049\ g/l$$

Using the density and kinematic viscosity of air at 53°C and 1 atm for ρ_∞ and ν, we have

$$Gr_{CD} = \frac{(9.8\ m/s^2)(0.0049\ kg/m^3)(0.092\ m)^3}{(1.049\ kg/m^3)(18.8 \times 10^{-6}\ m^2/s)^2} = 10^5$$

which falls in the range for the correlation equation.

Since Sc = 2.5, this is also in the range Pr > 0.5 and

$$\overline{\text{Nu}} = \overline{\text{Sh}} = 0.53(2.5 \times 10^5)^{0.25} = 11.9$$

From Eq. (4.33) the Sherwood number is

$$\overline{\text{Sh}} = \frac{\overline{h}_c D}{D_{AB}}$$

The average mass transfer coefficient, by analogy to the average heat transfer coefficient, is the mass flux divided by the driving force

$$\overline{h}_c = \frac{\dot{m}}{\rho_w - \rho_\infty}$$

so

$$\frac{\dot{m}}{\rho_w - \rho_\infty} \frac{D}{D_{AB}} = 11.8 \quad \text{or} \quad \dot{m} = 11.8 \frac{D_{AB}}{D} (\rho_w - \rho_\infty)$$

from Table 38, for air at 25°C, $D_{AB} = 0.611 \times 10^{-5}$ m²/s. (No data are readily available for higher temperatures,

$$\dot{m} = \frac{(11.8)(0.611 \times 10^{-5} \text{ m}^2/\text{s})}{0.092 \text{ m}} (0.0049 \text{ kg/m}^3) = 3.8 \times 10^{-6} \text{ kg/s m}^2$$

The surface area of the cylinder is

$$A = \pi(0.092 \text{ m})(1 \text{ m}) = 0.289 \text{ m}^2$$

Therefore the mass loss is

$$\dot{m} \Delta t A = \left(3.8 \times 10^{-6} \frac{\text{kg}}{\text{s m}^2}\right)(0.289 \text{ m}^2)(0.75 \times 3600)\left(\frac{1000 \text{ g}}{\text{kg}}\right) = 2.97 \text{ g}$$

When heat and mass transfer occur simultaneously, as they do in many important practical problems such as evaporation, one must consider the simultaneous solution of the momentum, energy, and mass diffusion equations. Since both temperature differences and concentration differences affect the buoyancy, both $\beta(T - T_\infty)$ and $\beta^*(C_A - C_{A,\infty})$ terms appear in the momentum equation. When both terms are small relative to unity, a sum of the two terms determines the buoyancy force in the momentum equation. For a further discussion of the simultaneous natural convection of thermal energy and species the reader is referred to (2).

5.7 SUMMARY

For the convenience of the reader, useful correlation equations for the determination of the average value of the natural-convection heat transfer coefficients for several important geometries are presented in Table 5.1 below.

TABLE 5.1 Natural Convection Heat Transfer Correlations[a]

Geometry	Correlation equation	Restrictions
Long vertical or tilted plate with heated surface facing downward	$\overline{\mathrm{Nu}}_L = 0.56(\mathrm{Gr}_L \mathrm{Pr} \cos \theta)^{1/4}$	$10^5 < \mathrm{Gr}_L \mathrm{Pr} \cos \theta < 10^{11}$ $0 \le \theta \le 89°$
Long horizontal plate with heated surface facing downward	$\overline{\mathrm{Nu}}_L = 0.58(\mathrm{Gr}_L \mathrm{Pr})^{1/5}$	$10^6 < \mathrm{Gr}_L \mathrm{Pr} < 10^{11}$
Long horizontal plate with heated surface facing upward	$\overline{\mathrm{Nu}}_L = 0.16(\mathrm{Gr}_L \mathrm{Pr})^{1/3}$ $\overline{\mathrm{Nu}}_L = 0.13(\mathrm{Gr}_L \mathrm{Pr})^{1/3}$	$7 \times 10^6 < \mathrm{Gr}_L \mathrm{Pr} < 2 \times 10^8$ $5 \times 10^8 < \mathrm{Gr}_L \mathrm{Pr}$
Horizontal circular plate with heated surface facing downward	$\overline{\mathrm{Nu}}_D = 0.82(\mathrm{Gr}_D \mathrm{Pr})^{1/5} \mathrm{Pr}^{0.034}$	

(*Continued*)

TABLE 5.1 (Continued)

Geometry	Correlation equation	Restrictions
Single long horizontal cylinder	$\overline{\mathrm{Nu}}_D = 0.53(\mathrm{Gr}_D\mathrm{Pr})^{1/4}$ $\overline{\mathrm{Nu}}_D = 0.53(\mathrm{Gr}_D\mathrm{Pr}^2)^{1/4}$	$\mathrm{Pr} > 0.5;\ 10^3 < \mathrm{Gr}_D < 10^9$ Liquid metals, laminar flow
Inclined cylinder, length L	$\overline{\mathrm{Nu}}_L = [2.9 - 2.32(\sin\theta)^{0.8}](\mathrm{Gr}_D)^{-1/12}[\mathrm{Gr}_L\mathrm{Pr}]^{1/4 + 1/12\,(\sin\theta)1.2}$ $\overline{\mathrm{Nu}}_L = [0.47 + 0.11(\sin\theta)^{0.8}](\mathrm{Gr}_D)^{-1/12}(\mathrm{Gr}_L\mathrm{Pr})^{1/3}$	Laminar: $9.88 \times 10^7 \le \mathrm{Gr}_L\mathrm{Pr} \le (\mathrm{Gr}_L\mathrm{Pr})_{cr}$ $1.08 \times 10^4 \le \mathrm{Gr}_D \le 6.9 \times 10^5$ Turbulent: $(\mathrm{Gr}_L\mathrm{Pr})_{cr} \le \mathrm{Gr}_L\mathrm{Pr} \le 2.95 \times 10^{10}$ $1.08 \times 10^4 \le \mathrm{Gr}_D \le 6.9 \times 10^5$ where $(\mathrm{Gr}_L\mathrm{Pr})_{cr} = 2.6 \times 10^9 + 1.1 \times 10^9 \tan\theta$
Sphere Diameter D	$\overline{\mathrm{Nu}}_D = 2 + 0.392(\mathrm{Gr}_D)^{1/4}$	$1 < \mathrm{Gr}_0 < 10^5$
Vertical cone	$\overline{\mathrm{Nu}}_L = 0.63(1 + 0.73\epsilon)\mathrm{Gr}_{L}^{1/4}$	$3° < \phi < 12°$ $7.5 < \log \mathrm{Gr}_L < 8.7$ $0.2 \le \epsilon \le 0.8$

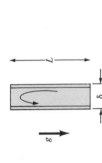

Vertical enclosed space heated from the side

$$\overline{\mathrm{Nu}}_\delta = 0.22\left(\frac{L}{\delta}\right)^{-1/4}\left(\frac{\mathrm{Pr}}{0.2+\mathrm{Pr}}\,\mathrm{Ra}_\delta\right)^{0.28}$$

$$\left\{\begin{array}{l} 2 < \dfrac{L}{\delta} < 10,\ \mathrm{Pr} < 10 \\[1mm] \mathrm{Ra}_\delta < 10^{10} \end{array}\right.$$

$$\overline{\mathrm{Nu}}_\delta = 0.18\left(\frac{\mathrm{Pr}}{0.2+\mathrm{Pr}}\,\mathrm{Ra}_\delta\right)^{0.29}$$

$$\left\{\begin{array}{l} 1 < \dfrac{L}{\delta} < 2,\ 10^{-3} < \mathrm{Pr} < 10^{5} \\[1mm] 10^{3} < \dfrac{\mathrm{Ra}_\delta\,\mathrm{Pr}}{0.2+\mathrm{Pr}} \end{array}\right.$$

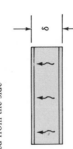

Horizontal enclosed space heated from below

$$\overline{\mathrm{Nu}}_\delta = 1 + 1.44\left[1 - \frac{1708}{\mathrm{Ra}_\delta}\right]^{\bullet} + \left[\left(\frac{\mathrm{Ra}_\delta}{5830}\right)^{1/3} - 1\right]^{\bullet}$$

Air, $1700 < \mathrm{Ra}_\delta < 10^{8}$

$$\overline{\mathrm{Nu}}_\delta = 1 + 1.44\left[1 - \frac{1708[1-\ln(\mathrm{Ra}_\delta^{1/3}/140)]}{\mathrm{Ra}_\delta}\right]^{\bullet} + \left[\left(\frac{\mathrm{Ra}_\delta}{5830}\right)^{1/3} - 1\right]^{\bullet}$$
$$+\, 2.0\left[\frac{\mathrm{Ra}_\delta^{1/3}}{140}\right]$$

Water, $1700 < \mathrm{Ra}_\delta < 3.5 \times 10^{9}$

Diameter D

$$\overline{\mathrm{Nu}}_D = C(\mathrm{Gr}_D\mathrm{Pr})^{n}$$

See table following Eq. (5.32)

Spherical cavity interior

$$\frac{k_{\mathrm{eff}}}{k} = 0.386\left[\frac{\ln(D_o/D_i)}{b^{3/4}(1/D_i^{3/5} + 1/D_o^{3/5})^{5/4}}\right]\left(\frac{\mathrm{Pr}}{0.861+\mathrm{Pr}}\right)^{1/4}\mathrm{Ra}_b^{1/4}$$

$$0.70 \le \mathrm{Pr} \le 6000$$
$$10 \le \left[\frac{\ln(D_o/D_i)}{b^{3/4}(1/D_i^{3/5} + 1/D_o^{3/5})^{5/4}}\right]^{4}\mathrm{Ra}_b \le 10^{7}$$

Long concentric cylinders

(Continued)

TABLE 5.1 (Continued)

Geometry	Correlation equation	Restrictions
Concentric spheres	$$\frac{k_{\text{eff}}}{k} = 0.74 \left[\frac{b^{1/4}}{D_o D_i (D_i^{-7/5} + D_o^{-7/5})^{5/4}} \right] \text{Ra}_b^{1/4} \left(\frac{\text{Pr}}{0.861 + \text{Pr}} \right)^{1/4}$$	$0.70 \leq \text{Pr} \leq 4200$ $10 \leq \left[\frac{b}{(D_o D_i)^4 (D_i^{-7/5} + D_o^{-7/5})^5} \right] \text{Ra}_b \leq 10^9$
Long rotating cylinder	$$\overline{\text{Nu}}_D = \frac{\overline{h_c} D}{k} = 0.11(0.5\text{Re}_\omega{}^2 + \text{Gr}_D \text{Pr})^{0.35}$$	$\text{Re}_\omega = \frac{\pi D^2 \omega}{\nu} > 8000$
Rotating disk	$$\overline{\text{Nu}}_D = \frac{\overline{h_c} D}{k} = 0.35 \left(\text{Re}_\omega \right)^{1/2}$$	$\text{Re}_\omega = \frac{\omega D^2}{4\nu} < 5 \times 10^5$
Rotating sphere	$$\overline{\text{Nu}}_D = 0.43 \text{Re}_\omega{}^{0.5} \text{Pr}^{0.4}$$ $$\overline{\text{Nu}}_D = 0.066 \text{Re}_\omega{}^{0.67} \text{Pr}^{0.4}$$	$\text{Re}_\omega = \frac{\omega D^2}{\nu} < 5 \times 10^4$ $\text{Pr} > 0.7$ $5 \times 10^4 < \text{Re}_\omega < 7 \times 10^5$

[a] To obtain the equivalent correlation for mass transfer replace Nu by Sh and Pr by Sc.

The problems for this chapter are related to the subject matter as shown below

Topic	Section	Problem number
Correlations	5.3	5.1
Correlations	5.3	5.2
Correlations	5.3	5.3
Correlations	5.3	5.4
Correlations	5.3	5.5
Correlations	5.3	5.6
Correlations	5.3	5.7
Dimensionless parameters	5.2	5.8
Dimensionless parameters	5.2	5.9
Correlations	5.3	5.10
Correlations	5.3	5.11
Correlations	5.3	5.12
Properties	5.2	5.13
Correlations	5.3	5.14
Correlations	5.3	5.15
Correlations	5.3	5.16
Correlations	5.3	5.17
Mixed convection	5.5	5.18
Correlations	5.3	5.19
Rotating parts	5.4	5.20
Rotating parts	5.4	5.21
Correlations	5.3	5.22
Correlations	5.3	5.23
Correlations	5.3	5.24
Correlations	5.3	5.25
Correlations	5.3	5.26

PROBLEMS

5.1 A 2 cm OD bare aluminum electric power transmission line carries 5000 amps at 400 kV. The wire has an electrical resistivity of 1.72 micro-ohms cm^2/cm at 20°C and is suspended horizontally between two towers separated by 1 km. Determine the surface temperature of the transmission line. What fraction of the dissipated power is due to radiation heat transfer?

5.2 A pot of coffee has been allowed to cool to room temperature, 17°C. If the electrical coffee maker is turned back on, the hot plate on which the pot rests is brought up to 70°C immediately and held at that temperature by a thermostat. Consider the pot to be a vertical cylinder 130 mm in diameter and the depth of coffee in the pot to be 100 mm. Neglect losses from the pot. How long will it take before the coffee is drinkable (50°C)? How much did it cost to heat the coffee if electricity costs $0.05 per kilowatt-hour?

5.3 In petroleum processing plants it is often necessary to pump highly viscous liquids such as asphalt through pipes. In order to keep pumping costs within reason, the pipelines are electrically heated to reduce the viscosity of the asphalt. Consider a

6 inch OD uninsulated pipe. How much power per foot of pipe length is necessary to maintain the pipe surface at 120°F? If the pipe is insulated with 2 in. of fiberglass insulation, what is the power requirement? If the pipe is uninsulated, write a general expression for the pipe surface temperature which gives minimum operating costs including pumping power and electrical heating power.

5.4 Compare the rate of heat loss from a human body with the typical energy intake from consumption of food (1300 kcal/day). Model the body as a vertical cylinder 30 cm in diameter and 1.8 m high in still air. Assume the skin temperature is 2°C below normal body temperature. Neglect radiation, transpiration cooling (sweating), and the effects of clothing.

5.5 A long, 2 cm OD horizontal copper pipe carries wet saturated steam at 1.2 atm absolute pressure. The pipe is contained within an environmental testing chamber in which the ambient air pressure can be adjusted from 0.5 to 2.0 atm, absolute. What is the effect of this pressure change on the rate of condensate flow per meter length of pipe? Assume that the pressure change does not affect the absolute viscosity, thermal conductivity, or specific heat of the air.

5.6 Compare the rate of condensate flow from the pipe in the preceding problem (air pressure = 2.0 atm) with that for a 3.89 cm OD pipe and 0.5 atm air pressure. What is the rate of condensate flow if the 2-cm pipe is submerged in a 20°C constant-temperature water bath?

5.7 A gas-fired industrial furnace is used to generate steam. The furnace is a 3-m cubic structure and the interior surfaces are completely covered with boiler tubes transporting wet steam at 5 atm. It is desired to keep the furnaces' losses to 1 percent of the total heat input of 1 MW. The outside of the furnace can be insulated with a blanket-type mineral wool insulation ($k = 0.13$ W/m°C), which is protected by a polished metal sheet outer shell. Assume that the floor of the furnace is insulated. What thickness of insulation is required? What is the temperature of the metal shell sides?

5.8 An empirical equation proposed by Heilman (*Trans. ASME*, vol. 51, p. 287, 1929) for the unit-surface conductance in natural convection from long horizontal cylinders to air is

$$\bar{h}_c = \frac{4.709(T_s - T_\infty)^{0.266}}{D^{0.2} T^{0.181}}$$

The corresponding equation in dimensionless form is

$$\frac{\bar{h}_c D}{k} = C \mathrm{Gr}_D{}^m \mathrm{Pr}_D{}^n$$

By comparing the two equations, determine those values of C, m, and n in the second equation that will give the same results as the first equation.

5.9 A laboratory experiment has been performed to determine the natural-convection heat transfer correlation for a horizontal cylinder of elliptical cross section in air. The cylinder is 1 m long, has a hydraulic diameter of 1 cm, and is heated internally by electrical resistance heating. Recorded data include power dissipation, cylinder surface temperature, and ambient air temperature. The power dissipation has been corrected for radiation effects:

$T_s - T_\infty$ (°C)	Q (W)
15.2	4.60
40.7	15.76
75.8	34.29
92.1	43.74
127.4	65.62

Assume that all air properties may be evaluated at 27°C and determine the constants in the correlation equation $\overline{Nu} = CGr^mPr^n$.

5.10 An 8 in. ID horizontal steam pipe carries 220 lbm/h dry saturated steam at 2 atm absolute pressure. If ambient air temperature is 70°F, determine the rate of condensate flow at the end of the pipe. Use an emissivity of 0.85 for the pipe surface. If it is desired to keep heat losses below 1 percent of the rate of energy transport by the steam, what thickness of fiberglass insulation is required? The rate of energy transport by the steam is the heat of condensation of the steam flow.

5.11 Only 10 percent of the energy dissipated by the tungsten filament of an incandescent lamp is in the form of useful visible light. Consider a 1000 W lamp with a 10-cm spherical glass bulb. Assuming an emissivity of 0.85 for the glass and ambient air temperature of 27°C, what is the temperature of the glass bulb?

5.12 A vertical steel tank 2 m tall and 30 cm OD is to be charged with nitrogen at 20°C. If the final tank pressure is to be 15 atm, what should the flow rate of nitrogen be to ensure that the tank does not exceed 35°C?

5.13 Show that the coefficient of thermal expansion for an ideal gas is $1/T$, where T is the absolute temperature.

5.14 Consider a design for a nuclear reactor using natural-convection heating of liquid bismuth. The reactor is to be constructed of parallel vertical plates 6 ft tall and 4 ft wide, in which heat is generated uniformly. Estimate the maximum possible heat dissipation rate from each plate if the surface temperature of the plate is not to exceed 1600°F and the lowest allowable bismuth temperature is 600°F.

5.15 A sphere 20 cm in diameter containing liquid air (-140°C) is covered with 5-cm-thick glass wool. Estimate the rate of heat transfer to the liquid air from the surrounding air at 20°C by convection and radiation. How would you reduce the heat transfer?

5.16 Estimate the rate of heat transfer across a double-pane window assembly in which the outside pane is at 0°C and the inside pane is at 20°C. The panes are spaced 1 cm apart. What is the thermal resistance ("R" value) of the window?

5.17 Repeat the above problem if the inner surfaces of the glass panes are coated with a very thin metallic layer giving an emissivity of 0.05.

5.18 A vertical isothermal plate 30 cm high is suspended in an atmospheric air stream flowing at 2 m/s in a vertical direction. If the air is at 16°C, estimate the plate temperature for which the natural-convection effect on the heat transfer coefficient will be less than 5 percent.

5.19 A thermocouple (1/32 in. OD) is located horizontally in a large enclosure whose walls are at 100°F. The enclosure is filled with a transparent quiescent gas which has the same properties as air. The electromotive force (emf) of the thermocouple indicates a temperature of 230°C. Estimate the true gas temperature if the emissivity of the thermocouple is 0.8.

5.20 A mild steel, 2 cm OD shaft, rotating in 20°C air at 20,000 rev/min, is attached to two bearings 0.7 m apart. If the temperature at the bearings is 90°C, determine the temperature distribution along the shaft. Hint: Show that for high rotational speeds Eq. (5.35) approaches $\overline{\mathrm{Nu}}_D = 0.086(\pi D^2 \omega/\nu)^{0.7}$.

5.21 Estimate the rate of heat transfer from one side of a 2 m diameter disk rotating at 600 rev/min in 20°C air, if its surface temperature is 50°C.

5.22 A 1-m-square copper plate is placed horizontally on 2-m-high legs. The plate has been coated with a material that provides a solar absorptance of 0.9 and an infrared emittance of 0.25. If the air temperature is 30°C, determine the equilibrium temperature on an average clear day in which the solar radiation incident on a horizontal surface is 850 W/m².

5.23 If cooling coils are attached to the back of the plate in the preceding problem, determine the cooling-water flow required such that 60 percent of the solar energy is absorbed by the water. Assume that cooling water is available at 15°C. What is the outlet temperature?

5.24 An 8 × 8 ft steel sheet 1/16 in. thick is removed from an annealing oven at a uniform temperature of 800°F and placed in a large room at 70°F in a horizontal position. (*a*) Calculate the rate of heat transfer from the steel sheet immediately after its removal from the furnace, considering both radiation and convection. (*b*) Determine the time required for the steel sheet to cool to a temperature of 100°F. Hint: This will require numerical integration.

5.25 A long steel rod (2 cm in diameter, 2 m long) has been heat-treated and quenched to a temperature of 100°C in an oil bath. In order to cool it further it is necessary to remove it from the bath and expose it to room air. Will the faster cool-down result from cooling the cylinder in the vertical or the horizontal position? How long will the two methods require to allow the rod to cool to 40°C in 20°C air?

5.26 Calculate the rate of heat transfer between a pair of concentric cylinders 20 mm and 126 mm in diameter. The inner cylinder is maintained at 37°C and the outer cylinder is maintained at 17°C.

REFERENCES

1. S. Ostrach, "New Aspects of Natural-Convection Heat Transfer," *Trans. ASME*, vol. 75, pp. 1287–1290, 1953.
2. Y. Jaluria, *Natural Convection Heat and Mass Transfer*, Pergamon, New York, 1980.
3. E. Schmidt and W. Beckmann, "Das Temperatur und Geschwindigkeitsfeld vor einer wärmeabgebenden senkrechten Platte bei natürlicher Konvektion," *Tech. Mech. Thermodyn.*, vol. 1, no. 10, pp. 341–349, October 1930; cont. vol. 1, no. 11, pp. 391–406, November 1930.
4. R. Hermann, "Wärmeübergang bei frier Ströhmung am wagrechten Zylinder in

zwei-atomic Gasen," *VDI Forschungsh.*, no. 379, 1936; translated in NACA Tech. Memo. 1366, November 1954.

5. F. Kreith and J. H. Taylor, Jr., "Heat Transfer from a Rotating Disk in Turbulent Flow," ASME paper 56-A-146, 1956.

6. W. H. McAdams, *Heat Transmission*, 3d ed., McGraw-Hill, New York, 1954.

7. E. M. Sparrow and M. A. Ansari, "A Refutation of King's Rule for Multi-Dimensional External Natural Convection," *Int. J. Heat Mass Transfer*, vol. 26, pp. 1357–1364, 1983.

8. B. Gebhart, *Heat Transfer*, 2d ed., chap. 8, McGraw-Hill, New York, 1970.

9. E. R. G. Eckert and E. Soehnghen, "Interferometric Studies on the Stability and Transition to Turbulence of a Free-Convection Boundary Layer," Proceedings of the General Discussion on Heat Transfer, pp. 321–323, ASME-IME, London, 1951.

10. E. R. G. Eckert and E. Soehnghen, "Studies on Heat Transfer in Laminar Free Convection with the Zehnder-Mach Interferometer," USAF Tech. Rept. 5747, December 1948.

11. A. M. Clausing, "Natural Convection Correlations for Vertical Surfaces Including Influences of Variable Properties," *J. Heat Transfer*, vol. 105, no. 1, pp. 138–143, 1983.

12. J. Gryzagoridis, "Natural Convection from a Vertical Flat Plate in the Low Grashof Number Range," *Int. J. Heat Mass Transfer*, vol. 14, pp. 162–164, 1971.

13. O. E. Dwyer, "Liquid-Metal Heat Transfer," chapter 5 in Sodium and NaK Supplement to *Liquid Metals Handbook*, Atomic Energy Commission, Washington, D.C., 1970.

14. E. M. Sparrow and J. L. Gregg, "Laminar Free Convection from a Vertical Flat Plate," *Trans. ASME*, vol. 78, pp. 435–440, 1956.

15. J. P. Dotson, *Heat Transfer from a Vertical Flat Plate by Free Convection*, M.S. thesis, Purdue University, May 1954.

16. G. C. Vliet, "Natural Convection Local Heat Transfer on Constant Heat Flux Inclined Surfaces," *Trans. ASME, Ser. C, J. Heat Transfer*, vol. 91, pp. 511–516, 1969.

17. G. C. Vliet and C. K. Liu, "An Experimental Study of Turbulent Natural Convection Boundary Layers," *Trans. ASME, Ser. C, J. Heat Transfer*, vol. 91, pp. 517–531, 1969.

18. T. Fujii and H. Imura, "Natural Convection Heat Transfer from a Plate with Arbitrary Inclination," *Int. J. Heat Mass Transfer*, vol. 15, pp. 755–767, 1972.

19. V. Kadambi and R. M. Drake, Jr., "Free Convection Heat Transfer from Horizontal Surfaces for Prescribed Variations in Surface Temperature and Mass Flow through the Surface," Tech. Rept. MECH Eng. HT-1, Princeton University, 1959.

20. J. S. McDonald and T. J. Connally, "Investigation of Natural Convection Heat Transfer in Liquid Sodium," *Nucl. Sci. Eng.*, vol. 8, pp. 369–377, 1960.

21. S. C. Hyman, C. F. Bonilla, and S. W. Ehrlich, "Heat Transfer to Liquid Metals and Non-Metals at Horizontal Cylinders," in *AIChE Symposium on Heat Transfer*, Atlantic City, pp. 21–23, 1953.

22. M. Al-Arabi and M. Khamis, "Natural Convection Heat Transfer from Inclined Cylinders," *Int. J. Heat Mass Transfer*, vol. 25, pp. 3–15, 1982.

23. T. Yuge, "Experiments on Heat Transfer from Spheres Including Combined Natural and Forced Convection," *Trans. ASME, Ser. C, J. Heat Transfer*, vol. 82, pp. 214–220, 1960.

24. P. H. Oosthuizen and E. Donaldson, "Free Convection Heat Transfer from Vertical Cones," *Trans. ASME, Ser. C, J. Heat Transfer*, vol. 94, pp. 330–331, 1972.

25. K. G. T. Hollands and L. Konicek, "Experimental Study of the Stability of Differentially Heated Inclined Air Layers," *Int. J. Heat Mass Transfer*, vol. 16, pp. 1467–1476, 1973.

26. I. Catton, "Natural Convection in Enclosures," in *Proceedings, Sixth International Heat Transfer Conference, Toronto*, vol. 6, pp. 13–31, Hemisphere, Washington, D.C., 1978.

27. D. B. Spalding and H. Afgan, eds., *Heat Transfer and Turbulent Buoyant Convection*, vols. 1 and 2, Hemisphere, Washington, D.C., 1977.

28. J. Imberger, "Natural Convection in a Shallow Cavity with Differentially Heated End Walls, Part 3. Experimental Results," *J. Fluid Mech.*, vol. 65, pp. 247–260, 1974.

29. A. Bejan, A. A. Al-Homoud, and J. Imberger, "Experimental Study of High Rayleigh Number Convection in a Horizontal Cavity with Different End Temperatures," *J. Heat Transfer*, vol. 109, pp. 283–299, 1981.

30. M. Nansteel and R. Greif, "Natural Convection in Undivided and Partially Divided Rectangular Enclosures," *J. Heat Transfer*, vol. 103, pp. 623–629, 1981.

31. R. Krisnamurti, "On the Transition to Turbulent Convection, Part 2, The Transition to Time-Dependent Flow," *J. Fluid Mech.*, vol. 42, pp. 309–320, 1970.

32. K. G. T. Hollands, G. D. Raithby, and L. Konicek, "Correlation Equations for Free Convection Heat Transfer in Horizontal Layers of Air and Water," *Int. J. Heat Mass Transfer*, vol. 18, pp. 879–884, 1975.

33. F. Kreith, "Thermal Design of High Altitude Balloons and Instrument Packages," *J. Heat Transfer*, vol. 92, pp. 307–332, 1970.

34. G. D. Raithby and K. G. T. Hollands, "A General Method of Obtaining Approximate Solutions to Laminar and Turbulent Free Convection Problems," in *Advances in Heat Transfer*, Academic, New York, 1974.

35. J. T. Anderson and O. A. Saunders, "Convection from an Isolated Heated Horizontal Cylinder Rotating about Its Axis," *Proc. R. Soc. London Ser. A*, vol. 217, pp. 555–562, 1953.

36. W. M. Kays and I. S. Bjorklund, "Heat Transfer from a Rotating Cylinder with and without Cross Flow," *Trans. ASME, Ser. C*, vol. 80, pp. 70–78, 1958.

37. E. C. Cobb and O. A. Saunders, "Heat Transfer from a Rotating Disk," *Proc. R. Soc. London Ser. A*, vol. 220, pp. 343–351, 1956.

38. K. Millsap and K. Pohlhausen, "Heat Transfer by Laminar Flow from a Rotating Plate," *J. Aero-sp. Sci.*, vol. 19, pp. 120–126, 1952.

39. C. Wagner, "Heat Transfer from a Rotating Disk to Ambient Air," *J. Appl. Phys.*, vol. 19, pp. 837–841, 1948.

40. F. Kreith, J. H. Taylor, and J. P. Chang, "Heat and Mass Transfer from a Rotating Disk," *Trans. ASME, Ser. C*, vol. 81, pp. 95–105, 1959.

41. T. Theodorsen and A. Regier, "Experiments on Drag of Revolving Disks, Cylinders, and Streamlined Rods at High Speeds," NACA Rept. 793, Washington, D.C., 1944.

42. F. Kreith, L. G. Roberts, J. A. Sullivan, and S. N. Sinha, "Convection Heat Transfer and Flow Phenomena of Rotating Spheres," *Int. J. Heat Mass Transfer*, vol. 6, pp. 881–895, 1963.

43. E. M. Sparrow and J. L. Gregg, "Buoyancy Effects in Forced Convection Flow and Heat Transfer," *Trans. ASME, J. Appl. Mech.*, sect. E, vol. 81, pp. 133–135, 1959.

44. Y. Mori, "Buoyancy Effects in Forced Laminar Convection Flow over a Horizontal Flat Plate," *Trans. ASME, J. Heat Transfer*, sect. C, vol. 83, pp. 479–482, 1961.

45. E. Eckert and A. J. Diaguila, "Convective Heat Transfer for Mixed, Free, and Forced Flow through Tubes," *Trans. ASME*, vol. 76, pp. 497–504, 1954.

46. D. L. Siebers, R. G. Schwind, and R. J. Moffat, "Experimental Mixed Convection Heat Transfer from a Large, Vertical Surface in Horizontal Flow," Sandia Rept. SAND 83-8225, Sandia National Laboratories, Albuquerque, N.M., 1983.

47. R. J. Goldstein and K. S. Lau, "Laminar Natural Convection from a Horizontal Plate and the Influence of Plate-Edge Extensions," *Int. J. Heat Mass Transfer*, vol. 129, pp. 55–75, 1983.

48. *Chemical Engineers Handbook*, 6th ed., R. H. Perry and D. W. Green, eds., McGraw-Hill, New York, pp. 3–58.

49. E. R. G. Eckert and T. W. Jackson, "Analysis of Turbulent Free Convection Boundary Layer on Flat Plate," NACA Rept. 1015, July 1950.

Chapter 6

Forced Convection Inside Tubes and Ducts

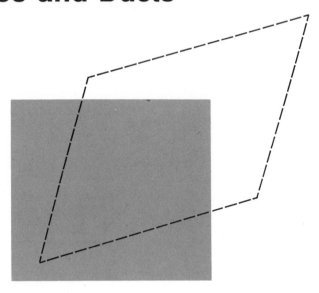

6.1 INTRODUCTION

Heating and cooling of fluids flowing inside conduits are among the most important heat transfer processes in engineering. The design and analysis of heat exchangers require a knowledge of the heat transfer coefficient between the wall of the conduit and the fluid flowing inside it. The sizes of boilers, economizers, superheaters, and preheaters depend largely on the unit-convective conductance between the inner surface of the tubes and the fluid. Also, in the design of air-conditioning and refrigeration equipment, it is necessary to evaluate heat transfer coefficients for fluids flowing inside ducts. Once the heat transfer coefficient for a given geometry and specified flow conditions is known, the rate of heat transfer at the prevailing temperature difference can be calculated from the equation

$$q_c = \bar{h}_c A(T_{\text{surface}} - T_{\text{fluid}}) \tag{6.1}$$

The same relation can also be used to determine the area required to transfer heat at a specified rate for a given temperature potential. But when heat is transferred to a fluid inside a conduit, the fluid temperature varies along the conduit and at any cross section. The fluid temperature must therefore be defined with care and precision for flow inside a duct.

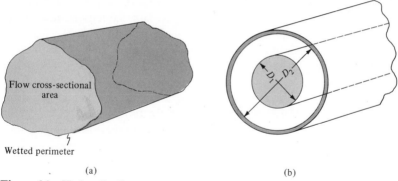

Figure 6.1 Hydraulic diameter for (a) irregular cross section and (b) annulus.

The heat transfer coefficient $\bar{h}_c$ can be calculated from the Nusselt number $\bar{h}_c D_H/k$, as shown in Section 4.6. For flow in long tubes or conduits (Fig. 6.1a) the significant length in the Nusselt number is the *hydraulic diameter* D_H, defined as

$$D_H = 4 \frac{\text{flow cross-sectional area}}{\text{wetted perimeter}} \tag{6.2}$$

For a circular tube or a pipe the flow cross-sectional area is $\pi D^2/4$, the wetted perimeter is πD, and therefore the inside diameter of the tube equals the hydraulic diameter. For an annulus formed between two concentric tubes (Fig. 6.1b) we have

$$D_H = 4 \frac{(\pi/4)(D_2{}^2 - D_1{}^2)}{\pi(D_1 + D_2)} = D_2 - D_1 \tag{6.3}$$

In engineering practice the Nusselt number for flow in conduits is usually evaluated from empirical equations based on experimental results. From a dimensional analysis, as shown in Section 4.6, the experimental results obtained in forced-convection heat transfer experiments in long ducts and conduits can be correlated by an equation of the form

$$\text{Nu} = \phi(\text{Re})\psi(\text{Pr}) \tag{6.4}$$

where the symbols ϕ and ψ denote functions of the Reynolds number and Prandtl number, respectively. For short ducts, particularly in laminar flow, the right-hand side of Eq. (6.3) must be modified by including the aspect ratio x/D_H, or

$$\text{Nu} = \phi(\text{Re})\psi(\text{Pr})f\left(\frac{x}{D_H}\right)$$

where $f(x/D_H)$ denotes the functional dependence on the aspect ratio.

6.1.1 Reference Fluid Temperature

The convective heat transfer coefficient used to build the Nusselt number for heat transfer to a fluid flowing in a conduit is defined by Eq. (6.1). The numerical value of $\bar{h}_c$, as mentioned previously, depends on the choice of the reference temperature in the fluid. For flow over a plane surface the temperature of the fluid far away from the heat source is generally constant, and its value is a natural choice for the fluid temperature in Eq. (6.1). In heat transfer to or from a fluid flowing in a conduit, the temperature of the fluid does not level out but varies both along the direction of mass flow and in the direction of heat flow. At a given cross section of the conduit, the temperature of the fluid at the center could be selected as the reference temperature in Eq. (6.1). However, the center temperature is difficult to measure in practice; furthermore, it is not a measure of the change in internal energy of all the fluid flowing in the conduit. It is therefore a common practice, and one we shall follow here, to use the *average fluid bulk temperature* T_b as the reference fluid temperature in Eq. (6.1). The average fluid temperature at a station of the conduit is often called the *mixing cup temperature* because it is the temperature which the fluid passing a cross-sectional area of the conduit during a given time interval would assume if the fluid were collected and mixed in a cup.

Use of the fluid bulk temperature as the reference temperature in Eq. (6.1) allows us to make heat balances readily because, in the steady state, the difference in average bulk temperature between two sections of a conduit is a direct measure of the rate of heat transfer, or

$$q_c = mc_p \Delta T_b \qquad (6.5)$$

where q_c = rate of heat transfer to fluid, W

   m = flow rate, kg/s

   c_p = specific heat at constant pressure, kJ/kg K

   ΔT_b = difference in average fluid bulk temperature between cross
        sections in question, K or °C

The problems associated with variations of the bulk temperature in the direction of flow will be considered in detail in Chapter 8, where the analysis of heat exchangers is taken up. For preliminary calculations, it is common practice to use the bulk temperature halfway between the inlet and the outlet section of a duct as the reference temperature in Eq. (6.1). This procedure is satisfactory when the wall temperature of the duct is constant, but may require some modification when the heat is transferred between two fluids separated by a wall as, for example, in a heat exchanger where one fluid flows inside a pipe while another passes over the outside of the pipe. Although this type of problem is of considerable practical importance, it will not concern us in this chapter, where the emphasis is placed on the evaluation of convection heat transfer coefficients, which can be determined in a given flow system when the pertinent bulk and wall temperatures are specified.

6.1.2 Effect of Reynolds Number on Heat Transfer and
Pressure Drop in Fully Established Flow

For a given fluid the Nusselt number depends primarily on the flow conditions, which can be characterized by the Reynolds number Re. For flow in long conduits the characteristic length in the Reynolds number, as in the Nusselt number, is the hydraulic diameter and the velocity to be used is the average over the flow cross-sectional area, $\bar{U}$, or

$$\text{Re}_{D_H} = \frac{\bar{U}D_H\rho}{\mu} = \frac{\bar{U}D_H}{\nu} \tag{6.6}$$

In long ducts, where the entrance effects are not important, the flow is laminar when the Reynolds number is below about 2100. In the range of Reynolds numbers between 2100 and 10,000 a transition from laminar to turbulent flow takes place. The flow in this regime is called transitional. At a Reynolds number of about 10,000 the flow becomes fully turbulent.

In laminar flow through a duct, just as in laminar flow over a plate, there is no mixing of warmer and colder fluid particles by eddy motion and the heat transfer takes place solely by conduction. Since all fluids with the exception of liquid metals have small thermal conductivities, the heat transfer coefficients in laminar flow are relatively small. In transitional flow a certain amount of mixing occurs by means of eddies, which carry warmer fluid into cooler regions, and vice versa. Since the mixing motion, even if it is only on a small scale, accelerates the transfer of heat considerably, a marked increase in the heat transfer coefficient occurs above Re = 2100. This is illustrated in Fig. 6.2, where experimentally measured values of the average Nusselt number for atmospheric air flowing through a long heated tube are plotted as a function of Reynolds number.

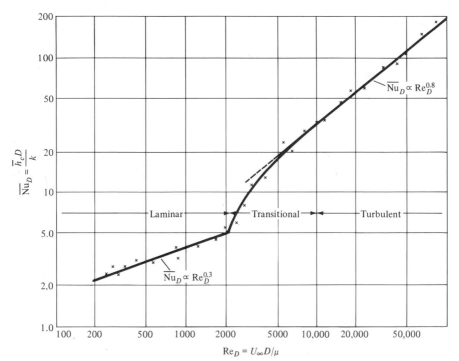

Figure 6.2 Nusselt number versus Reynolds number for air flowing in a long pipe.

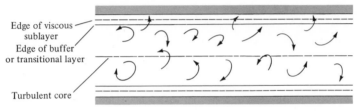

Edge of viscous
sublayer

Edge of buffer
or transitional layer

Turbulent core

Figure 6.3   Flow structure for a fluid in turbulent flow through a pipe.

Since the Prandtl number for air does not vary appreciably, Eq. (6.4) reduces to Nu = ϕ(Re), and the curve drawn through the experimental points shows the dependence of Nu on the flow conditions. We note that, in the laminar regime, the Nusselt number remains small, increasing from about 3.5 at Re = 300 to 5.0 at Re = 2100. Above a Reynolds number of 2100, the Nusselt number begins to increase rapidly until the Reynolds number reaches about 8000. As the Reynolds number is further increased, the Nusselt number continues to increase, but at a slower rate.

A qualitative explanation for this behavior can be given by observing the fluid flow field shown schematically in Fig. 6.3. At Reynolds numbers above 8000, the flow inside the conduit is fully turbulent except for a very thin layer of fluid adjacent to the wall. In this layer turbulent eddies are damped out as a result of the viscous forces that predominate near the surface, and therefore heat flows through it mainly by conduction.* The edge of this sublayer is indicated by a dashed line in Fig. 6.3. The flow beyond it is turbulent and the circular arrows in the turbulent-flow regime represent the eddies which sweep the edge of the layer, probably penetrate it, and carry along with them fluid at the temperature prevailing there. The eddies mix the warmer and cooler fluids so effectively that heat is transferred very rapidly between the edge of the viscous sublayer and the turbulent bulk of the fluid. It is thus apparent that, except for fluids of high thermal conductivity (e.g., liquid metals), the thermal resistance of the sublayer controls the rate of heat transfer, and most of the temperature drop between the bulk of the fluid and the surface of the conduit occurs in this layer. The turbulent portion of the flow field, on the other hand, offers little resistance to the flow of heat. The only effective method of increasing the heat transfer coefficient is therefore to decrease the thermal resistance of the sublayer. This can be accomplished by increasing the turbulence in the main stream so that the turbulent eddies can penetrate deeper into the layer. An increase in turbulence, however, is accompanied by large energy losses which increase the frictional pressure drop in the conduit. In the design and selection of industrial heat exchangers, where not only the initial cost but also the operating expenses must be considered, the pressure drop is an important factor. An increase of the flow velocity yields higher heat transfer coefficients which, in accordance

---

* Although recent studies have shown that turbulent transport exists to some extent also near the wall (1), especially when the Prandtl number is larger than 5, the layer near the wall is commonly referred to as the "viscous sublayer."

with Eq. (6.1), decrease the size and consequently also the initial cost of the equipment for a specified heat transfer rate. At the same time, however, the pumping cost increases. The optimum design therefore requires a compromise between the initial and operating costs. In practice, it has been found that increases in pumping costs and operating expenses often outweigh the saving in the initial cost of heat transfer equipment under continuous operating conditions. As a result, the velocities used in a majority of commercial heat exchange equipment are relatively low, corresponding to Reynolds numbers of no more than 50,000. Laminar flow is usually avoided in heat exchange equipment because of the low heat transfer coefficients obtained. However, in the chemical industry, where frequently very viscous liquids must be handled, laminar flow sometimes cannot be avoided without producing undesirably large pressure losses.

It was shown in Section 4.11 that for turbulent flow of liquids and gases over a flat plate, the Nusselt number is proportional to the Reynolds number raised to the 0.8 power. Since in turbulent forced convection the viscous sublayer generally controls the rate of heat flow irrespective of the geometry of the system, it is not surprising that also for turbulent forced convection in conduits the Nusselt number is related to the Reynolds number by the same type of power law. For the case of air flowing in a pipe, this relation is illustrated in the graph of Fig. 6.2.

6.1.3 Effect of Prandtl Number

The Prandtl number Pr is a function of the fluid properties alone. It has been defined as the ratio of the kinematic viscosity of the fluid to the thermal diffusivity of the fluid, that is,

$$\mathrm{Pr} = \frac{\nu}{\alpha} = \frac{c_p \mu}{k}$$

The kinematic viscosity ν, or μ/ρ, is often referred to as the molecular diffusivity of momentum because it is a measure of the rate of momentum transfer between the molecules. The thermal diffusivity of a fluid $k/c_p\rho$ is often called the molecular diffusivity of heat. It is a measure of the ratio of the heat transmission and energy storage capacities of the molecules.

The Prandtl number relates the temperature distribution to the velocity distribution, as shown in Section 4.6 for flow over a flat plate. For flow in a pipe, just as over a flat plate, the velocity and temperature profiles are similar for fluids having a Prandtl number of unity. When the Prandtl number is smaller, the temperature gradient near a surface is less steep than the velocity gradient, and for fluids whose Prandtl number is larger than one, the temperature gradient is steeper than the velocity gradient. The effect of Prandtl number on the temperature gradient in turbulent flow at a given Reynolds number in tubes is illustrated schematically in Fig. 6.4, where temperature profiles at different Prandtl numbers are shown at $\mathrm{Re}_D = 10,000$. These curves reveal that, at a specified Reynolds number, the temperature gradient at the wall is steeper

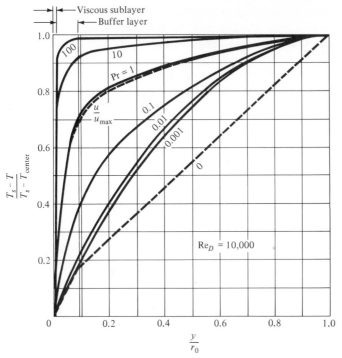

Figure 6.4 Effect of Prandtl number on temperature profile for turbulent flow in a long pipe. [From R. C. Martinelli (35), with permission of the publishers, the American Society of Mechanical Engineers.]

in a fluid having a large Prandtl number than in a fluid having a small Prandtl number. Consequently, at a given Reynolds number, fluids with larger Prandtl numbers have larger Nusselt numbers.

Liquid metals generally have a high thermal conductivity and a small specific heat; their Prandtl numbers are therefore small, ranging from 0.005 to 0.01. The Prandtl numbers of gases range from 0.6 to 0.9. Most oils, on the other hand, have large Prandtl numbers, some up to 5000 or more, because their viscosity is large at low temperatures and their thermal conductivity is small.

6.1.4 Entrance Effects

In addition to the Reynolds number and the Prandtl number, several other factors can influence heat transfer by forced convection in a duct. For example, when the conduit is short, entrance effects are important. As a fluid enters a duct with a uniform velocity, the fluid immediately adjacent to the tube wall is brought to rest. For a short distance from the entrance a laminar boundary layer is formed along the tube wall. If the turbulence in the entering fluid stream is high, the boundary layer will quickly become turbulent. Irrespective of whether the boundary layer remains laminar or becomes turbulent, it will increase in thickness until it fills the entire duct. From this point on, the velocity profile across the duct remains essentially unchanged.

The development of the thermal boundary layer in a fluid that is heated or cooled in a duct is qualitatively similar to that of the hydrodynamic boundary layer. At the entrance, the temperature is generally uniform transversely, but as the fluid flows along the duct, the heated or cooled layer increases in thickness until heat is transferred to or from the fluid in the center of the duct. Beyond this point the temperature profile remains essentially constant if the velocity profile is fully established.

The final shapes of the velocity and temperature profiles depend on whether the fully developed flow is laminar or turbulent. Figures 6.5 and 6.6 illustrate qualitatively the growth of the boundary layers as well as the variations in the local unit-convective conductance near the entrance of a tube for laminar and turbulent conditions, respectively. Inspection of these figures shows that the unit-thermal conductance varies considerably near the entrance. If the entrance is square-edged, as in most heat exchangers, the initial development

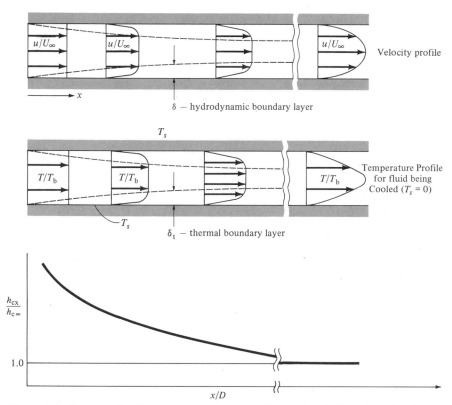

Figure 6.5 Velocity distribution, temperature profiles, and variation of the local heat transfer coefficient near the inlet of a tube for air being cooled in laminar flow (surface temperature, T_s, uniform).

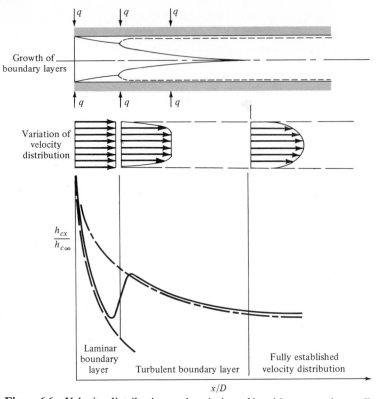

Figure 6.6 Velocity distribution and variation of local heat transfer coefficient near the entrance of a tube for a fluid in turbulent flow.

of the hydrodynamic and thermal boundary layers along the walls of the tube is quite similar to that along a flat plate. Consequently, the conductance is largest near the entrance and decreases along the duct until both the velocity and the temperature profiles for the fully developed flow have been established. If the pipe Reynolds number for the fully developed flow $\bar{U}D\rho/\mu$ is below 2100, the entrance effects may be appreciable for a length as much as 100 hydraulic diameters from the entrance. For laminar flow in a circular tube the hydraulic entry length at which the velocity profile approaches its fully developed shape can be obtained from the relation (2)

$$\left(\frac{x_{\text{fully developed}}}{D}\right)_{\text{lam}} = 0.05 \, \text{Re}_D \tag{6.7}$$

whereas the distance from the inlet at which the temperature profile approaches its fully developed shape is given by the relation (3)

$$\left(\frac{x_{\text{fully developed}}}{D}\right)_{\text{lam},T} = 0.05 \, \text{Re}_D \text{Pr} \tag{6.8}$$

In turbulent flow, conditions are essentially independent of Prandtl number, and for average pipe velocities corresponding to turbulent-flow Reynolds numbers entrance effects disappear about 10 or 20 diameters from the inlet.

6.1.5 Variation of Physical Properties

Another factor that can influence the heat transfer and friction considerably is the variation of physical properties with temperature. When a fluid flowing in a duct is heated or cooled, its temperature and consequently its physical properties vary along the duct as well as over any given cross section. For liquids, only the temperature dependence of the viscosity is of major importance. For gases, on the other hand, the temperature effect on the physical properties is more complicated than for liquids because the thermal conductivity and the density, in addition to the viscosity, vary significantly with temperature. In either case, the numerical value of the Reynolds number depends on the location at which the properties are evaluated. It is believed that the Reynolds number based on the average bulk temperature is the significant parameter to describe the flow conditions. However, considerable success in the empirical correlation of experimental heat transfer data has been achieved by evaluating the viscosity at an *average film* temperature, defined as a temperature approximately halfway between the wall and the average bulk temperatures. Another method of taking account of the variation of physical properties with temperature is to evaluate all properties at the average bulk temperature and to correct for the thermal effects by multiplying the right-hand side of Eq. (6.4) by a function proportional to the ratio of bulk to wall temperatures or bulk to wall viscosities.

6.1.6 Thermal Boundary Conditions and Compressibility Effects

For fluids having a Prandtl number of unity or less, the heat transfer coefficient also depends on the thermal boundary condition. For example, in geometrically similar liquid metal or gas heat transfer systems a uniform wall temperature yields smaller convective conductances than a uniform heat input at the same Reynolds and Prandtl numbers (4, 5, 6). When heat is transferred to or from gases flowing at very high velocities, compressibility effects influence the flow and the heat transfer. Problems associated with heat transfer to or from fluids at high Mach numbers are discussed in (7–9).

6.1.7 Limits of Accuracy in Predicted Values of Convective Heat Transfer Coefficients

In the application of any empirical equation for forced convection to practical problems it is important to bear in mind that the predicted values of the heat transfer coefficient are not exact. The results obtained by various experimenters, even under carefully controlled conditions, differ appreciably. In turbulent flow the accuracy of a heat transfer coefficient predicted from any available equation or graph is no better than ± 20 percent, whereas in laminar flow the accuracy may be of the order of ± 30 percent. In the transition region, where experimental data are scant, the accuracy of the Nusselt number predicted from available information may be even lower.

6.2* ANALYSIS OF LAMINAR FORCED CONVECTION IN A LONG TUBE

To illustrate some of the most important concepts in forced convection, we will analyze a simple case and calculate the heat transfer coefficient for laminar flow through a tube under fully developed conditions with a constant heat flux at the wall. We begin by deriving the velocity distribution. Consider a fluid element as shown in Fig. 6.7. The pressure is uniform over the cross section and the pressure forces are balanced by the viscous shear forces acting over the surface, or

$$\pi r^2 [p - (p + dp)] = \tau 2\pi r \, dx = -\left(\mu \frac{du}{dr}\right) 2\pi r \, dx$$

From this relation we obtain

$$du = \frac{1}{2\mu}\left(\frac{dp}{dx}\right) r \, dr$$

where dp/dx is the axial pressure gradient. The radial distribution of the axial velocity is then

$$u(r) = \frac{1}{4\mu}\left(\frac{dp}{dx}\right) r^2 + C$$

where C is a constant of integration, whose value is determined by the boundary condition that $u = 0$ at $r = r_s$. Using this condition to evaluate C gives the

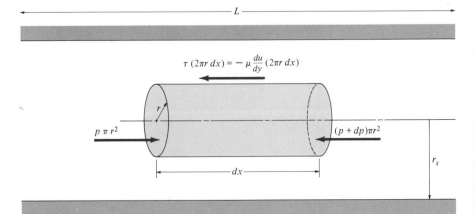

Figure 6.7 Force balance on a cylindrical fluid element inside a tube of radius r_s.

velocity distribution

$$u(r) = \frac{r^2 - r_s^2}{4\mu} \frac{dp}{dx} \tag{6.9}$$

The maximum velocity u_{max} at the center ($r = 0$) is

$$u_{max} = -\frac{r_s^2}{4\mu} \frac{dp}{dx} \tag{6.10}$$

so that the velocity distribution can be written in dimensionless form as

$$\frac{u}{u_{max}} = 1 - \left(\frac{r}{r_s}\right)^2 \tag{6.11}$$

The above relation shows that the velocity distribution in fully developed laminar flow is parabolic.

In addition to the heat transfer characteristics, engineering design requires consideration of the pressure loss and pumping power required to sustain the convection flow through the conduit. The pressure loss in a tube of length L is obtained from a force balance on a fluid element inside the tube (see Fig. 6.7)

$$\Delta p \pi r_s^2 = 2\pi r_s \tau_s L \tag{6.12}$$

where Δp = pressure drop in length L ($\Delta p = -(dp/dx)L$) and
τ_s = wall shear stress ($\tau_s = -\mu \, du/dr|_{r=r_s}$)

The pressure drop can also be related to a so-called *Darcy friction factor* f according to

$$\Delta p = f \frac{L}{D} \frac{\rho \bar{U}^2}{2g_c} \tag{6.13}$$

where $\bar{U}$ is the average velocity in the tube.

It is important to note that f, the friction factor in Eq. (6.14), is not the same quantity as the friction coefficient C_f, which was defined in Chapter 4 as

$$C_f = \frac{\tau_s}{\rho \bar{U}^2 / 2g_c} \tag{6.14}$$

C_f is often referred to as the *Fanning friction coefficient*. Since $\tau_s = -\mu(du/dr)_{r=r_s}$ it is apparent from Eqs. (6.12), (6.13), and (6.14) that

$$C_f = \frac{f}{4}$$

For flow through a pipe the mass flow rate is obtained from Eq. (6.9)

$$\dot{m} = \rho \int_0^{r_s} u 2\pi r \, dr = \frac{\Delta p \pi}{2L\mu} \int_0^{r_s} (r^2 - r_s^2) r \, dr = -\frac{\Delta p \pi r_s^4}{8L\mu} \tag{6.15}$$

and the average velocity $\bar{U}$ is

$$\bar{U} = \frac{\dot{m}}{\rho \pi r_s^2} = -\frac{\Delta p r_s^2}{\rho 8 L\mu} \tag{6.16}$$

equal to one-half of the maximum velocity in the center. Equation (6.13) can be rearranged into the form

$$\Delta p = \frac{64 L \mu}{\rho \bar{U}^2 D} \frac{\bar{U}^2}{2} = \frac{64}{\mathrm{Re}_D} \frac{L}{D} \frac{\rho \bar{U}^2}{2 g_c} \tag{6.17}$$

Comparing Eq. (6.17) with Eq. (6.13), we see that for fully developed laminar flow in a tube the friction factor in a pipe is a simple function of Reynolds number

$$f = \frac{64}{\mathrm{Re}_D} \tag{6.18}$$

The pumping power, P_p, is equal to the product of the pressure drop and the volumetric flow rate of the fluid divided by the pump efficiency.

The analysis above is limited to laminar flow with a parabolic velocity distribution in pipes or circular tubes, known as Poiseuille flow; but the approach taken to derive this relation is more general. If we know the shear stress as a function of the velocity and its derivative, the friction factor could also be obtained for turbulent flow. However, for turbulent flow the relationship between the shear and the average velocity is not well understood. Moreover, while in laminar flow the friction factor is independent of surface roughness, in turbulent flow the quality of the pipe surface influences the pressure loss. Therefore, friction factors for turbulent flow cannot be derived analytically, but must be measured and correlated empirically.

6.2.1 Uniform Heat Flux

For the energy analysis, consider the control volume shown in Fig. 6.8. In laminar flow heat is transferred by conduction into and out of the element in a radial direction, whereas in the axial direction the energy transport is by convection. Thus, the rate of heat conduction into the element is

$$dq_{k,r} = -k 2\pi r\, dx\, \frac{\partial T}{\partial r}$$

while the rate of heat conduction out of the element is

$$dq_{k,r+dr} = -k 2\pi (r + dr)\, dx \left[\frac{\partial T}{\partial r} + \frac{\partial^2 T}{\partial r^2}\, dx \right]$$

The net rate of convection out of the element is

$$dq_c = 2\pi r\, dr\, \rho c_p u(r)\, \frac{\partial T}{\partial x}\, dx$$

Writing a net energy balance in the form

Net rate of conduction = net rate of convection
into the element   out of the element

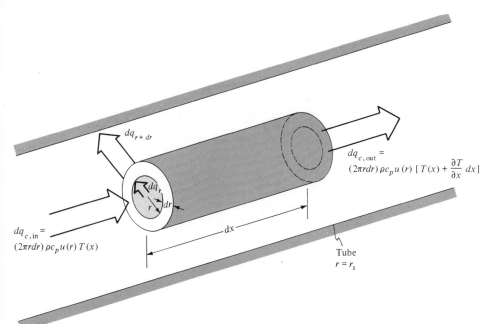

Figure 6.8 Schematic sketch of control volume for energy analysis in flow through a pipe.

we get, neglecting second-order terms,

$$k\left(\frac{\partial T}{\partial r} + r\frac{\partial^2 T}{\partial r^2}\right) dx\, dr = r\rho c_p u \frac{\partial T}{\partial x} dx\, dr \qquad (6.19)$$

which can be recast in the form

$$\frac{1}{ur}\frac{\partial}{\partial r}\left(r\frac{\partial T}{\partial r}\right) = \frac{\rho c_p}{k}\frac{\partial T}{\partial x} \qquad (6.20)$$

The fluid temperature must increase linearly with distance x since the heat flux over the surface is specified to be uniform, or

$$\frac{\partial T}{\partial x} = \text{constant} \qquad (6.21)$$

When the axial temperature gradient $\partial T/\partial x$ is constant, Eq. (6.20) reduces from a partial to an ordinary differential equation with r as the only space coordinate.

The boundary conditions for the temperature distribution in Eq. (6.20) are

$$\frac{\partial T}{\partial r} = 0 \qquad\qquad \text{at } r = 0$$

$$k\frac{\partial T}{dr}\bigg|_{r=r_s} = q_s'' = \text{constant} \qquad \text{at } r = r_s$$

To solve Eq. (6.20) we substitute the velocity distribution from Eq. (6.11). Assuming that the temperature gradient does not affect the velocity profile, that is, the properties do not change with temperature, we get

$$\frac{\partial}{\partial r}\left(r\frac{\partial T}{\partial r}\right) = \frac{1}{\alpha}\frac{\partial T}{\partial x}u_{max}\left(1 - \frac{r^2}{r_s^2}\right)r \tag{6.22}$$

The first integration with respect to r gives

$$r\frac{\partial T}{\partial r} = \frac{1}{\alpha}\frac{\partial T}{\partial x}\frac{u_{max}r^2}{2}\left(1 - \frac{r^2}{2r_s^2}\right) + C_1 \tag{6.23}$$

A second integration with respect to r gives

$$T(r, x) = \frac{1}{\alpha}\frac{\partial T}{\partial x}\frac{u_{max}}{4}r^2\left(1 - \frac{r^2}{4r_s^2}\right) + C_1 \ln r + C_2 \tag{6.24}$$

But note that $C_1 = 0$ since $(\partial T/\partial r)_{r=0} = 0$ and that the second boundary condition is satisfied by the requirement that the axial temperature gradient $\partial T/\partial x$ is constant. If we let the temperature at the center ($r = 0$) be T_c, then $C_2 = T_c$ and the temperature distribution becomes

$$T - T_c = \frac{1}{\alpha}\frac{\partial T}{\partial x}\frac{u_{max}r_s^2}{4}\left[\left(\frac{r}{r_s}\right)^2 - \frac{1}{4}\left(\frac{r}{r_s}\right)^4\right] \tag{6.25}$$

The average bulk temperature T_b that was used in defining the heat transfer coefficient can be calculated from

$$T_b = \frac{\int_0^{r_s}(\rho u c_p T)(2\pi r\, dr)}{\int_0^{r_s}(\rho u c_p)2\pi r\, dr} = \frac{\int_0^{r_s}(\rho u c_p T)2\pi r\, dr}{\rho c_p \dot{m}} \tag{6.26}$$

Since the heat flux from the tube wall is uniform, the enthalpy of the fluid in the tube must increase linearly with x and thus $\partial T_b/\partial x$ = constant. We can calculate the bulk temperature by substituting Eqs. (6.25) and (6.11) for T and u, respectively, in Eq. (6.26). This yields

$$T_b - T_c = \frac{7}{96}\frac{u_{max}r_s^2}{\alpha}\frac{\partial T}{\partial x} \tag{6.27}$$

while the wall temperature is

$$T_s - T_c = \frac{3}{16}\frac{u_{max}r_s^2}{x}\frac{\partial T}{\partial x} \tag{6.28}$$

In deriving the temperature distribution we used a parabolic velocity distribution, which exists in fully developed flow in a long tube. Hence, with $\partial T/\partial x$ equal to a constant, the average heat transfer coefficient is

$$\bar{h}_c = \frac{q_c}{A(T_s - T_b)} = \frac{k(\partial T/\partial r)_{r=r_s}}{T_s - T_b} \tag{6.29}$$

Evaluating the radial temperature gradient at $r = r_s$ from Eq. (6.23) and substituting it with Eqs. (6.27) and (6.28) in the above definition yields

$$\bar{h}_c = \frac{24k}{11r_s} = \frac{48k}{11D} \tag{6.30}$$

or

$$\overline{Nu}_D = \frac{\bar{h}_c D}{k} = 4.364 \qquad \text{for } q_s'' = \text{const} \tag{6.31}$$

EXAMPLE 6.1

Water entering at 10°C is to be heated to 40°C in a tube of 0.02 m ID at a mass flow rate of 0.01 kg/s. The outside of the tube is wrapped with an insulated electric heating element that produces a uniform flux of 15,000 W/m² over the surface. Neglecting any entrance effects, determine

(a) The Reynolds number
(b) The heat transfer coefficient
(c) The length of pipe needed for a 30°C increase in average temperature
(d) The inner tube surface temperature at the outlet
(e) The friction factor
(f) The pressure drop in the pipe
(g) The pumping power required if the pump is 50 percent efficient.

Solution. From Table 13 in Appendix 2, the appropriate properties of water at an average temperature between inlet and outlet of 25°C are obtained by interpolation:

$$\rho = 997 \text{ kg/m}^3$$
$$c_p = 4180 \text{ J/kg K}$$
$$k = 0.608 \text{ W/m K}$$
$$\mu = 910 \times 10^{-6} \text{ N s/m}^2$$

(a) The Reynolds number is

$$\text{Re}_D = \frac{\rho \bar{U} D}{\mu} = \frac{4\dot{m}}{\pi D \mu} = \frac{(4)(0.01 \text{ kg/s})}{(\pi)(0.02 \text{ m})(910 \times 10^{-6} \text{ N s/m}^2)} = 670$$

This establishes that the flow is laminar.

(b) Since the thermal boundary is one of uniform heat flux, $\text{Nu}_D = 4.36$ from Eq. (6.31) and

$$\bar{h}_c = 4.36 \frac{k}{D} = 4.36 \frac{0.608 \text{ W/m K}}{0.02 \text{ m}} = 132.5 \text{ W/m}^2 \text{ K}$$

(c) The length of pipe needed for a 30°C temperature rise is obtained from a heat balance

$$q'' \pi D L = \dot{m} c_p (T_{\text{out}} - T_{\text{in}})$$

Solving for L when $T_{out} - T_{in} = 30$ K gives

$$L = \frac{\dot{m}c_p 10}{\pi D q''} = \frac{(0.01 \text{ kg/s})(4180 \text{ J/kg K})(30 \text{ K})}{(\pi)(0.02 \text{ } D)(15{,}000 \text{ W/m}^2)} = 1.33 \text{ m}$$

Since $L/D = 66.5$ and $0.05\text{Re}_D = 33.5$, entrance effects are negligible according to Eq. (6.7). Note that if L/D had been significantly less than 33.5, the calculations would have to be repeated with entrance effects taken into account, using relations to be presented.

(d) From Eq. (6.1)

$$q'' = \frac{q_c}{A} = \bar{h}_c(T_s - T_b)$$

and

$$T_s = \frac{q_c}{A\bar{h}_c} + T_b = \frac{15{,}000 \text{ W/m}^2}{132.5 \text{ W/m}^2 \text{ °C}} + 40°\text{C} = 153.6°\text{C}$$

(e) The friction factor is found from Eq. (6.18):

$$f = \frac{64}{\text{Re}_D} = \frac{64}{670} = 0.0955$$

(f) The pressure drop in the pipe is, from Eq. (6.17),

$$\Delta p = f\left(\frac{L}{D}\right)\left(\frac{\rho \bar{U}^2}{2g_c}\right) = (0.0955)(66.5)\frac{(997 \text{ kg/m}^3)(0.0032 \text{ m/s}^2)}{2g_c(\text{m/s}^2)}$$

$$= 18.67 \text{ kg/m}^2$$

(g) The pumping power P_p is equal to the pressure loss times the volumetric mass flow rate divided by the pump efficiency, or

$$P_p = \dot{m}\frac{\Delta P}{\rho\eta_p} = \frac{(0.01 \text{ kg/s})(18.67 \text{ kg/m}^2)}{(0.5)(997 \text{ kg/m}^3)} = 0.0375 \text{ W}$$

6.2.2* Uniform Surface Temperature

When the tube surface temperature is uniform rather than the heat flux, the analysis is more complicated because the temperature difference between the wall and the bulk varies along the tube, that is, $\partial T_b/\partial x = f(x)$. Equation (6.20) can be solved subject to the second boundary condition that at $r = r_s$, $T(x, r_s) = $ constant, but an iterative procedure is necessary. The result is not a simple algebraic expression, but the Nusselt number is found [for example, see (10)] to be a constant, or

$$\overline{\text{Nu}}_D = \frac{\bar{h}_c D}{k} = 3.66 \qquad (T_s = \text{const}) \qquad (6.32)$$

In addition to the value of the Nusselt number, the constant-temperature boundary condition requires a different temperature difference to evaluate the rate of heat transfer to or from a fluid flowing through a duct. Except for the entrance region, in which the boundary layer develops and the heat transfer coefficient decreases, the temperature difference between the surface of the duct and the bulk remains constant along the duct when the heat flux is uniform. This is apparent from an examination of Eq. (6.20) and is illustrated graphically in Fig. 6.9. For a constant wall temperature, on the other hand, only the bulk temperature increases along the duct and the temperature potential decreases (see Fig. 6.9). Write the heat balance equation

$$dq_c = \dot{m}c_p\, dT_b = q_s''P\, dx$$

where P is the perimeter of the duct and q_s'' is the surface heat flux. From the above we can obtain a relation for the bulk temperature gradient in the x-direction

$$\frac{dT_b}{dx} = \frac{q_s''P}{\dot{m}c_p} = \frac{P}{\dot{m}c_p}h_c(T_s - T_b) \tag{6.33}$$

Since $dT_b/dx = d(T_b - T_s)/dx$ for a constant surface temperature, we have after separating variables

$$\int_{\Delta T_{in}}^{\Delta T_{out}} \frac{d(\Delta T)}{\Delta T} = -\frac{P}{\dot{m}c_p}\int_0^L h_c\, dx \tag{6.34}$$

where $\Delta T = T_b - T_s$ and the subscripts "in" and "out" denote conditions at the inlet $(x = 0)$ and the outlet $(x = L)$ of the duct, respectively. Integrating Eq. (6.34) yields

$$\frac{\Delta T_{out}}{\Delta T_{in}} = -\frac{PL}{\dot{m}c_p}\bar{h}_c \tag{6.35}$$

where

$$\bar{h}_c = \frac{1}{L}\int_0^L h_c\, dx$$

Rearranging Eq. (6.35) gives

$$\frac{\Delta T_{out}}{\Delta T_{in}} = \exp\left(\frac{-\bar{h}_c PL}{\dot{m}c_p}\right) \tag{6.36}$$

The rate of heat transfer by convection to or from a fluid flowing through a duct with $T_s = $ constant can be expressed in the form

$$q_c = \dot{m}c_p[(T_s - T_{b,in}) - (T_s - T_{b,out})] = \dot{m}c_p(\Delta T_{in} - \Delta T_{out})$$

and substituting from $\dot{m}c_p$ from Eq. (6.35), we get

$$q_c = \bar{h}_c A_s\left[\frac{\Delta T_{out} - \Delta T_{in}}{\ln(\Delta T_{out}/\Delta T_{in})}\right] \tag{6.37}$$

The expression in the square bracket is called the *log mean temperature difference* (*LMTD*).

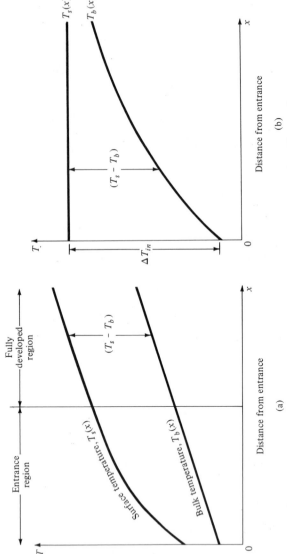

Figure 6.9 Variation of average bulk temperature with constant heat flux and constant wall temperature. (*a*) Constant heat flux, $q_s(x) =$ constant; (*b*) constant surface temperature, $T_s(x) =$ constant.

EXAMPLE 6.2 ────────────────────────────────────

An engine oil is flowing through a 1-cm-ID, 0.02-cm-wall copper tube at the rate of 0.05 kg/s. The oil enters at 35°C and is to be heated to 45°C by atmospheric-pressure steam condensing on the outside. Calculate the length of tube required.

Solution. We shall assume that the tube is long and that its temperature is uniform at 100°C. The first assumption must be checked; the second assumption is an engineering approximation justified by the high thermal conductivity of copper and the large heat transfer coefficient for a condensing vapor (see Table 1.4). From the appendix we get the following properties for oil at 40°C:

$$c_p = 1964 \text{ J/kg K}$$
$$\rho = 876 \text{ kg/m}^3$$
$$k = 0.144 \text{ W/m K}$$
$$\mu = 0.210 \text{ N s/m}^2$$
$$\text{Pr} = 28.7$$

The Reynolds number is

$$\text{Re}_D = \frac{4\dot{m}}{\mu\pi D} = \frac{(4)(0.05 \text{ k/s})}{(\pi)(0.210 \text{ N s/m}^2)(0.01 \text{ m})} = 30.3$$

The flow is therefore laminar and the Nusselt number for a constant surface temperature is 3.66. The average heat transfer coefficient is

$$\bar{h}_c = \overline{\text{Nu}}_D \frac{k}{D} = 3.66 \frac{0.144 \text{ W/m K}}{0.01 \text{ m}} = 52.7 \text{ W/m}^2 \text{ K}$$

The rate of heat transfer is

$$q_c = c_p\dot{m}(T_{b,\text{out}} - T_{b,\text{in}})$$
$$= (1966 \text{ J/kg K})(0.05 \text{ kg/s})(45 - 35) \text{ K} = 983 \text{ W/s}$$

Recalling that $\ln(1/x) = -\ln x$, the LMTD is

$$\text{LMTD} = \frac{\Delta T_{\text{out}} - \Delta T_{\text{in}}}{\ln(\Delta T_{\text{out}}/\Delta T_{\text{in}})} = \frac{55 - 65}{\ln(55/65)} = \frac{10}{0.167} = 59.86$$

Substituting the above information in Eq. (6.37), where $A_s = L\pi D_i$, gives

$$L = \frac{q_c}{\pi D_i\bar{h}_c\text{LMTD}} = \frac{983 \text{ W/s}}{(\pi)(0.01 \text{ m})(52.7 \text{ W/m}^2 \text{ K})(59.86)} = 9.92 \text{ m}$$

Checking our first assumption, we find $L/D \sim 1000$, justifying neglect of entrance effects. Note also that LMTD is very nearly equal to the difference between the surface temperature and the average bulk fluid temperature halfway between the inlet and the outlet. The required length is not suitable for a practical design with a straight pipe. To achieve the desired thermal performance in a more convenient shape, one could route the tube back and forth several times or use a coiled tube. The first approach will be discussed in Chapter 8 on heat exchanger design, and the coiled-tube design is illustrated in an example in the next section.

6.3 CORRELATIONS FOR LAMINAR FORCED CONVECTION

This section presents empirical correlations and analytic results that can be used in thermal design of heat transfer systems which contain tubes and ducts with gaseous or liquid fluids in laminar flow. Although heat transfer coefficients in laminar flow are considerably smaller than in turbulent flow, in the design of heat exchange equipment for viscous liquids it is often necessary to accept a smaller heat transfer coefficient in order to reduce the pumping-power requirements. Laminar gas flow is encountered in high-temperature, compact heat exchangers, where tube diameters are very small and gas densities low. Other applications of laminar-flow forced convection occur in chemical processes and food industries, as well as solar and nuclear power plants where liquid metals are used as heat transfer media. Since liquid metals have a high thermal conductivity, their heat transfer coefficients are relatively large even in laminar flow.

6.3.1 Short Circular and Rectangular Ducts

The details of the mathematical solutions for laminar flow in short ducts with entrance effects are beyond the scope of this text. References listed at the end of this chapter, especially (3) and (10), contain the mathematical background for the engineering equations and graphs that are presented and discussed in this section.

For engineering applications it is most convenient to present the results of analytic and experimental investigations in terms of a Nusselt number defined in the conventional manner as $h_c D/k$. But the heat transfer coefficient h_c can vary along the tube, and for practical applications the average value of the conductance is most important. Consequently, for the equations and charts presented in this section we shall use a mean Nusselt number $\overline{Nu}_D = \bar{h}_c D/k$, averaged with respect to the circumference and length of the duct L, or

$$\overline{Nu}_D = \frac{1}{L} \int_0^L Nu_x \, dx$$

where the subscript x refers to local conditions at x. This Nusselt number is often termed the *log mean Nusselt number* because it can be used directly in the log mean rate equations presented in the preceding section and applied to heat exchangers in Chapter 8.

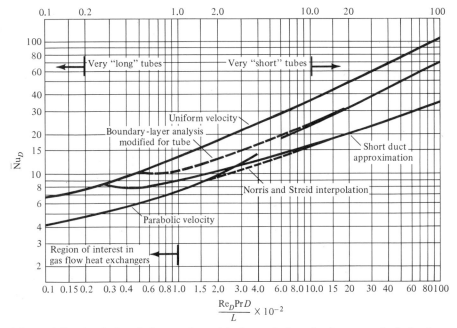

Figure 6.10 Analytic solutions and empirical correlations for heat transfer in laminar flow through circular tubes at constant wall temperature, $\overline{Nu}_D$ versus $Re_D Pr D/L$. [From W. M. Kays (13), with permission of the publishers, the American Society of Mechanical Engineers.]

Mean Nusselt numbers for laminar flow in tubes at a uniform wall temperature have been calculated analytically by various investigators. Their results are shown in Fig. 6.10 for several velocity distributions. All of these solutions are based on the idealizations of a constant tube-wall temperature and a uniform temperature distribution at the tube inlet, and they apply strictly only when the physical properties are independent of temperature. The abscissa is the dimensionless quantity $Re_D Pr D/L$.* To determine the mean value of the Nusselt number for a given tube of length L and diameter D, one evaluates the Reynolds number Re_D and the Prandtl number Pr, forms the dimensionless parameter $Re_D Pr D/L$, and enters the appropriate curve of Fig. 6.10. The selection of the curve representing the conditions that most nearly correspond to the physical conditions depends on the nature of the fluid and the geometry of the system. For high Prandtl number fluids, such as oils, the velocity profile is established much more rapidly than the temperature profile. Consequently, application of the curve labeled "parabolic velocity" does not lead to a serious error in long tubes when $Re_D Pr D/L$ is less than 100. For very long tubes the Nusselt number approaches a limiting minimum value of 3.66 when the tube

---

* Instead of the dimensionless ratio $Re_D Pr D/L$ some authors use the Graetz number Gz, which is $\pi/4$ times this ratio (14).

temperature is uniform. When the heat transfer rate instead of the tube temperature is uniform, the limiting value of $\overline{\mathrm{Nu}}_D$ is 4.36.

For very low Prandtl number fluids, such as liquid metals, the temperature profile is established much more rapidly than the velocity profile. For typical applications the assumption of a uniform velocity profile, called "slug flow," may give satisfactory results, although experimental evidence is insufficient for a quantitative evaluation of the possible deviation from the analytic solution for slug flow. For very short tubes or rectangular ducts with initially uniform velocity and temperature distribution, the flow conditions along the wall approximate those along a flat plate, and the boundary-layer analysis presented in Chapter 4 is expected to yield satisfactory results for liquids having Prandtl numbers between 1.0 and 15.0. The boundary-layer solution applies (11, 12) when L/D is less than $0.0048\mathrm{Re}_D$ for tubes and when L/D_H is less than $0.0021\mathrm{Re}_{D_H}$ for flat ducts of rectangular cross section. For these conditions the equation for convection over a flat plate can be converted to the coordinates of Fig. 6.10, leading to

$$\overline{\mathrm{Nu}}_{D_H} = \frac{\mathrm{Re}_{D_H}\mathrm{Pr}D_H}{4L}\ln\left[\frac{1}{1-(2.654/\mathrm{Pr}^{0.167})(\mathrm{Re}_{D_H}\mathrm{Pr}D_H/L)^{0.5}}\right] \quad (6.38)$$

An analysis for longer tubes is presented in (13), and the results are shown in Fig. 6.10 for $\mathrm{Pr} = 0.73$ in the $\mathrm{Re}_D\mathrm{Pr}D/L$ range of 100 to 1500, where this approximation is applicable.

6.3.2 Ducts of Noncircular Cross Section

Heat transfer and friction in fully developed laminar flow through ducts with a variety of cross sections has been treated analytically (14). The results are summarized in Table 6.1, using the following nomenclature:

$\overline{\mathrm{Nu}}_{H1}$—average Nusselt number for uniform heat flux in flow direction and uniform wall temperature at any cross section

$\overline{\mathrm{Nu}}_{H2}$—average Nusselt number for uniform heat flux both axially and circumferentially

$\overline{\mathrm{Nu}}_T$—average Nusselt number for uniform wall temperature

$f\,\mathrm{Re}_{D_H}$—product of friction factor and Reynolds number

A duct geometry encountered quite often is the concentric tube annulus, shown schematically in Fig. 6.1b. Heat transfer to or from the fluid flowing through the space formed between the two concentric tubes may occur at the inner surface, the outer surface, or both surfaces simultaneously. Moreover, the heat transfer surface may be at constant temperature or constant heat flux. An extensive treatment of this topic has been presented by Kays and Perkins (10), including entrance effects and the impact of eccentricity. Here we shall consider only the most commonly encountered case of an annulus with one side insulated and the other at constant temperature.

TABLE 6.1 NUSSELT NUMBER AND FRICTION FACTOR FOR FULLY DEVELOPED LAMINAR FLOW OF A NEWTONIAN FLUID THROUGH SPECIFIED DUCTS[a]

Geometry $\left(\dfrac{L}{D_H} > 100\right)$	$\overline{Nu}_{H1}$	$\overline{Nu}_{H2}$	$\overline{Nu}_T$	$f Re_D$	$\dfrac{\overline{Nu}_{H1}}{\overline{Nu}_T}$
$2a$ 60° $\dfrac{2b}{2a} = \dfrac{\sqrt{3}}{2}$ $2b$	3.111	1.892	2.47	13.333	1.26
$2b$ $\dfrac{2b}{2a} = 1$ $2a$	3.608	3.091	2.976	14.227	1.21
(hexagon)	4.002	3.862	3.34[b]	15.054	1.20
$2b$ $\dfrac{2b}{2a} = \dfrac{1}{2}$ $2a$	4.123	3.017	3.391	15.548	1.22
(circle)	4.364	4.364	3.657	16.000	1.19
$2b$ $\dfrac{2b}{2a} = \dfrac{1}{4}$ $2a$	5.331	2.930	4.439	18.233	1.20
$2b$ $\dfrac{2b}{2a} = .9$ $2a$	5.099	4.35[b]	3.66	18.700	1.39
$2b$ $\dfrac{2b}{2a} = \dfrac{1}{8}$ $2a$	6.490	2.904	5.597	20.585	1.16
$\dfrac{2b}{2a} = 0$	8.235	8.235	7.541	24.000	1.09
$\dfrac{b}{a} = 0$ insulated	5.385	—	4.861	24.000	1.11

[a] Abstracted from Shah and London (14).
[b] Interpolated values.

TABLE 6.2 NUSSELT NUMBER AND
              FRICTION FACTOR FOR FULLY
              DEVELOPED LAMINAR FLOW IN
              AN ANNULUS[a]

$\dfrac{D_i}{D_o}$	$\overline{Nu}_i$	$\overline{Nu}_o$	$f\,Re_{D_H}$
0.00	—	3.66	16.00
0.05	17.46	4.06	21.56
0.10	11.56	4.11	22.34
0.25	7.37	4.23	23.27
0.50	5.74	4.43	23.78
1.00	4.86	4.86	24.00

[a] One surface at constant temperature and the other insulated (14).

Denoting the inner surface by the subscript i and the outer surface by o, the rate of heat transfer and the corresponding Nusselt numbers are

$$q_{c,i} = \bar{h}_{c,i}\pi D_i L\,(T_{s,i} - T_b)$$

$$q_{c,o} = \bar{h}_{c,o}\pi D_o L\,(T_{s,o} - T_b)$$

$$\overline{Nu}_i = \frac{\bar{h}_{c,i}D_H}{k}$$

$$\overline{Nu}_o = \frac{\bar{h}_{c,o}D_H}{k}$$

where $D_H = D_o - D_i$.

The Nusselt numbers for heat flow at the inner surface only with the outer surface insulated, $\overline{Nu}_i$, and heat flow at the outer surface with the inner surface insulated, $\overline{Nu}_o$, as well as the product of the friction factor and the Reynolds number for fully developed laminar flow are presented in Table 6.2. For other conditions, such as constant heat flux and short annuli, the reader is referred to (14).

EXAMPLE 6.3 _____

Calculate the average heat transfer coefficient and the friction factor for flow of *n*-butyl alcohol at a bulk temperature of 293 K through a 0.1 m × 0.1 m duct, 5 m long, with walls at 300 K, if the average velocity is 0.03 m/s.

Solution. The hydraulic diameter is

$$D_H = 4\left(\frac{0.1 \times 0.1}{4 \times 0.1}\right) = 0.1 \text{ m}$$

Physical properties at 293 K from Table 18 in Appendix 2 are:

$$\rho = 810 \text{ kg/m}^3 \qquad\qquad c_p = 2366 \text{ J/kg K}$$
$$\mu = 29.5 \times 10^{-4} \text{ N s/m}^2 \qquad \nu = 3.64 \times 10^{-6} \text{ m}^2/\text{s}$$
$$k = 0.167 \text{ W/m K} \qquad\qquad Pr = 41.8$$

The Reynolds number is

$$\text{Re}_{D_H} = \frac{\bar{U}D_H\rho}{\mu} = \frac{(0.03 \text{ m/s})(0.1 \text{ m})(810 \text{ kg/m}^3)}{29.5 \times 10^{-4} \text{ N s/m}^2} = 824$$

Hence, the flow is laminar. Assuming fully developed flow, we get from Table 6.1 for a uniform wall temperature the Nusselt number:

$$\overline{\text{Nu}}_{D_H} = \frac{\bar{h}_c D_H}{k} = 2.976$$

This yields for the average heat transfer coefficient

$$\bar{h}_c = 2.976 \frac{0.167 \text{ W/m K}}{0.1 \text{ m}} = 4.97 \text{ W/m}^2 \text{ K}$$

Similarly, from Table 6.1 the product $\text{Re}_{D_H}f = 14.227$ and

$$f = \frac{14.227}{42.6} = 0.0172$$

Recall that for a fully developed velocity profile the duct length must be at least $0.05\text{Re} \times D_H = 4.1$ m, but for a fully developed temperature profile the duct must be 172 m long. Thus *fully developed flow will not exist.*

If we use Fig. 6.10 with $\text{Re}_{D_H}\text{Pr}D/L = (824)(41.8)(0.1/5) = 6.88$, the average Nusselt number is about 15 and $\bar{h}_c = (15)(0.167 \text{ W/m K})/0.1 \text{ m} = 25 \text{ W/m}^2$ K. This value is five times larger than that for fully developed flow.

Note that for this problem the difference between bulk and wall temperature is small. Hence, property variations are not significant in this case.

6.3.3 Effect of Property Variations

Since the microscopic heat flow mechanism in laminar flow is conduction, as shown in Section 6.2 the rate of heat flow between the walls of a conduit and the fluid flowing in it can be obtained analytically by solving the equations of motion and of conduction heat flow simultaneously. But to obtain a solution it is necessary to know or assume the velocity distribution in the duct. In fully developed laminar flow through a tube without heat transfer, the velocity distribution at any cross section is parabolic. But when appreciable heat transfer occurs, temperature differences are present and the fluid properties of the wall and the bulk may be quite different. These property variations distort the velocity profile.

In liquids the viscosity decreases with increasing temperature, while in gases the reverse trend is observed. When a liquid is heated, the fluid near the wall is less viscous than the fluid in the center. Consequently, the velocity of the heated fluid is larger than that of an unheated fluid near the wall, but less in the center. The distortion of the parabolic velocity profile for liquids on heating

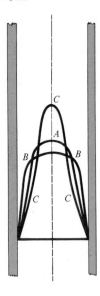

Figure 6.11 Effect of heat transfer on velocity profiles in fully developed laminar flow through a pipe. Curve A, isothermal flow; curve B, heating of liquid or cooling of gas; curve C, cooling of liquid or heating of gas.

or cooling is shown in Fig. 6.11. For gases, the conditions are reversed, but the variation of density with temperature introduces additional complications.

Empirical viscosity correction factors are merely approximate rules and recent data indicate that they may not be satisfactory when very large temperature gradients exist. As an approximation in the absence of a more satisfactory method, it is suggested (15) that, for liquids, the Nusselt number obtained from analytic solutions presented in Fig. 6.10 be multiplied by the ratio of the viscosity at the bulk temperature μ_b to the viscosity at the surface temperature μ_s, raised to the 0.14 power, that is, $(\mu_s/\mu_b)^{0.14}$, to correct for the variation of properties due to the temperature gradient. For gases, Kays and London (16) suggest that the Nusselt number be multiplied by a temperature correction factor. If all fluid properties are evaluated at the average bulk temperature, the corrected Nusselt number is

$$\overline{\mathrm{Nu}}_D = \overline{\mathrm{Nu}}_{D,\,\mathrm{Fig.\ 6.10}} \left(\frac{T_b}{T_s}\right)^n$$

where $n = 0.25$ for a gas heating in a tube and 0.08 for a gas cooling in a tube. Hausen (17) recommended the following relation for the average convection coefficient in laminar flow through ducts with uniform surface temperature:

$$\overline{\mathrm{Nu}}_{D_H} = 3.66 + \frac{0.0668\ \mathrm{Re}_{D_H}\mathrm{Pr}D/L}{1 + 0.045(\mathrm{Re}_{D_H}\mathrm{Pr}D/L)^{0.66}} \left(\frac{\mu_b}{\mu_s}\right)^{0.14} \tag{6.39}$$

A relatively simple empirical equation suggested by Sieder and Tate (15) which has been widely used to correlate experimental results for liquids in tubes can be written in the form

$$\overline{\mathrm{Nu}}_{D_H} = 1.86 \left(\frac{\mathrm{Re}_D\mathrm{Pr}D}{L}\right)^{0.33} \left(\frac{\mu_b}{\mu_s}\right)^{0.14} \tag{6.40}$$

where all the properties in Eqs. (6.39) and (6.40) are based on the bulk temperature and the empirical correction factor $(\mu_b/\mu_s)^{0.14}$ is introduced to account for the effect of the temperature variation on the physical properties. Equation (6.40) can be applied when the surface temperature is uniform in the range $0.48 < \text{Pr} < 16{,}700$ and $0.0044 < (\mu_b/\mu_s) < 9.75$. Whitaker (18) recommends use of Eq. (6.40) only when $(\text{Re}_D \text{Pr} D/L)^{0.33}(\mu_b/\mu_s)^{0.14}$ is larger than 2.

For laminar flow of gases between two parallel, uniformly heated plates a distance $2y_0$ apart, Swearingen and McEligot (19) showed that gas property variations can be taken into account by the relation

$$\overline{\text{Nu}} = \text{Nu}_{\text{constant properties}} + 0.024Q^{+0.3}\text{Gz}_b^{0.75} \tag{6.41}$$

where $\quad Q^+ = q_s'' y_0/(kT)_{\text{entrance}}$

$\qquad q_s'' = \text{surface heat flux at the walls}$

The variation in physical properties also affects the friction factor. To evaluate the friction factor of fluids being heated or cooled it is suggested that, for liquids, the isothermal friction factor be modified by

$$f_{\text{heat transfer}} = f_{\text{isothermal}} \left(\frac{\mu_s}{\mu_b}\right)^{0.14} \tag{6.42}$$

and, for gases, by

$$f_{\text{heat transfer}} = f_{\text{isothermal}} \left(\frac{T_s}{T_b}\right)^{0.14} \tag{6.43}$$

EXAMPLE 6.4 ───────────────────────────────────

Water at an inlet temperature of 333 K flows at a velocity of 0.2 m/s through a 0.3-m-long capillary tube, 2.54×10^{-3} m ID. Assuming that the tube temperature is maintained at 353 K, calculate the outlet temperature of the water.

Solution. The properties of water at 333 K are, from Table 13 in Appendix 2,

$$\rho = 983 \text{ kg/m}^3$$
$$c_p = 4181 \text{ J/kg K}$$
$$\mu = 4.72 \times 10^{-4} \text{ N s/m}^2$$
$$k = 0.658 \text{ W/m K}$$
$$\text{Pr} = 3.00$$

To ascertain whether the flow is laminar, evaluate the Reynolds number at the inlet bulk temperature,

$$\text{Re}_D = \frac{\rho \bar{U} D}{\mu} = \frac{(983)(0.2)(0.00254)}{4.72 \times 10^{-4}} = 1058$$

The flow is laminar and because

$$\text{Re}_D \text{Pr} \frac{D}{L} = \frac{(1058)(3.00)(0.00254)}{0.3} = 26.9 > 10$$

Eq. (6.40) can be used to evaluate the heat transfer coefficient. But since the mean bulk temperature is not known, we shall evaluate all the properties first at the inlet bulk temperature T_{b1}, then determine an exit bulk temperature, and make a second iteration to obtain a more precise value. Designating inlet and outlet conditions with the subscripts 1 and 2, respectively, the energy balance becomes

$$q_c = \bar{h}_c \pi D L \left(T_s - \frac{T_{b1} + T_{b2}}{2} \right) = \dot{m} c_p (T_{b2} - T_{b1}) \qquad (a)$$

At the wall temperature of 353 K, $\mu_s = 3.52 \times 10^{-4}$ N s/m² from Table 13. From Eq. (6.40) we can calculate the average Nusselt number:

$$\overline{Nu}_D = 1.86 \left[\frac{(1058)(3.00)(0.00254)}{0.3} \right]^{0.33} \left(\frac{4.72}{3.52} \right)^{0.14} = 5.74$$

and thus

$$\bar{h}_c = \frac{k \overline{Nu}_D}{D} = \frac{(0.658)(5.74)}{0.00254} = 1487 \text{ W/m}^2 \text{ K}$$

The mass flow rate is

$$\dot{m} = \rho \frac{\pi D^2}{4} \bar{U} = \frac{(983)\pi(0.00254)^2(0.2)}{4} = 0.996 \times 10^{-3} \text{ kg/s}$$

Inserting the calculated values for $\bar{h}_c$ and $\dot{m}$ into Eq. (a), along with $T_{b1} = 333$ K and $T_s = 353$ K, gives

$$(1487)\pi(0.00254)(0.3)\left(353 - \frac{333 + T_{b2}}{2} \right)$$

$$= (0.996 \times 10^{-3})(4181)(T_{b2} - 333) \qquad (b)$$

Solving for T_{b2} gives

$$T_{b2} = 345 \text{ K}$$

For the second iteration we shall evaluate all properties at the new average bulk temperature

$$\bar{T}_b = \frac{345 + 333}{2} = 339 \text{ K}$$

At this temperature we get, from Table 13,

$$\rho = 980 \text{ kg/m}^3$$
$$c_p = 4185 \text{ J/kg K}$$
$$\mu = 4.36 \times 10^{-4} \text{ N s/m}^2$$
$$k = 0.662 \text{ W/m K}$$
$$Pr = 2.78$$

Recalculating the Reynolds number with properties based on the new mean bulk temperature gives

$$\mathrm{Re}_D = \frac{\rho \bar{U} D}{\mu} = \frac{(980)(0.2)(2.54 \times 10^{-3})}{4.36 \times 10^{-4}} = 1142$$

With this value of Re_D, the heat transfer coefficient can now be calculated. One obtains on the second iteration $\mathrm{Re}_D \mathrm{Pr}(D/L) = 25.9$, $\overline{\mathrm{Nu}}_D = 5.61$, and $\bar{h}_c = 1490$ W/m² K. Substituting the new value of $\bar{h}_c$ in Eq. (b) gives $T_{b2} = 345$ K. Further iterations will not affect the results appreciably in this example because of the small difference between bulk and wall temperature. In cases where the temperature difference is large, a second iteration may be necessary.

It is recommended that the reader verify the results using the LMTD method with Eq. (6.37).

6.3.4 Effect of Natural Convection

An additional complication in the determination of a heat transfer coefficient in laminar flow arises when the buoyancy forces are of the same order of magnitude as the external forces due to the forced circulation. Such a condition may arise in oil coolers when low flow velocities are employed. Also, in the cooling of rotating parts, such as rotor blades of gas turbines and ramjets attached to the propellers of helicopters, the natural-convection forces may be so large that their effect on the velocity pattern cannot be neglected even in high-velocity flow. When the buoyancy forces are in the same direction as the external forces, such as the gravitational forces superimposed on upward flow, they increase the rate of heat transfer. When the external and buoyancy forces act in opposite directions, the heat transfer is reduced. Eckert (11, 12) studied heat transfer in mixed flow, and his results are shown qualitatively in Fig. 6.12a and b. In the darkly shaded area, the contribution of free convection to the total heat transfer is less than 10 percent, whereas in the lightly shaded area, forced-convection effects are less than 10 percent and free convection predominates. In the unshaded area, natural and forced convection are of the same order of magnitude. In practice, free-convection effects are hardly ever significant in turbulent flow (20). In cases where it is doubtful whether forced- or free-convection flow applies, the heat transfer coefficient is generally calculated by using forced- and free-convection relations separately, and the larger one is used (21). The accuracy of this rule is estimated to be about 25 percent.

The influence of natural convection on the heat transfer to fluids in horizontal isothermal tubes has been investigated by Depew and August (22). They found that their own as well as previously available data for tubes with $L/D > 50$ could be correlated by the equation

$$\overline{\mathrm{Nu}}_D = 1.75 \left(\frac{\mu_b}{\mu_s} \right)^{0.14} [\mathrm{Gz} + 0.12(\mathrm{Gz Gr}_D{}^{1/3} \mathrm{Pr}^{0.36})^{0.88}]^{1/3} \qquad (6.44)$$

Correlations for vertical tubes and ducts are considerably more complicated because they depend on the relative direction of the heat flow and the natural convection. A summary of available information is given in (23) and (24).

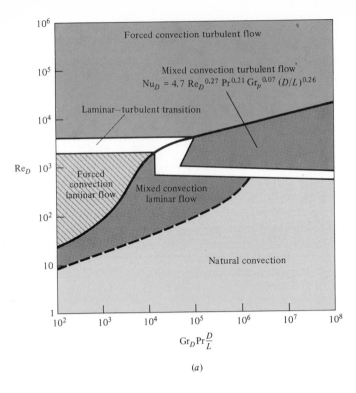

(a)

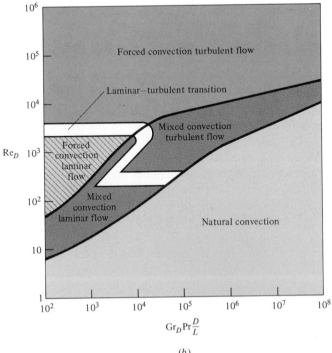

(b)

Figure 6.12 Forced, natural, and mixed convection regimes for (a) horizontal pipe flow and (b) vertical pipe flow.

6.3.5 Coiled Tubes

Coiled tubes are used in heat exchange equipment to obtain a large heat transfer area per unit volume and to enhance the heat transfer coefficient on the inside surface. The basic configuration is shown in Fig. 6.13. As a result of the centrifugal forces, a secondary flow pattern consisting of two vortices perpendicular to the axial flow direction is set up and heat transport will occur not only by diffusion in the radial direction, but also by convection. The contribution of this secondary convective transport dominates the overall process and enhances the rate of heat transfer per unit length of tube compared to a straight tube of equal length.

The flow characterization and the associated convection heat transfer coefficient in coiled tubes are governed by the flow Reynolds number and the ratio of tube diameter to coil diameter, D/d_c. The product of these two dimensionless numbers is called the *Dean number*, $\mathrm{Dn} \equiv \mathrm{Re}_D(D/d_c)$. Three regions can be distinguished (25):

1. The region of small Dean numbers, $\mathrm{Dn} < 20$, in which inertial forces due to secondary flow are negligible. In this region the peripherally averaged Nusselt number is given by the relation (25)

$$\overline{\mathrm{Nu}} = 1.7(\mathrm{Dn}^2\mathrm{Pr})^{1/5} \tag{6.45}$$

if $\mathrm{Dn}^2\mathrm{Pr} > 10{,}000$.

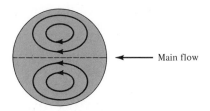

Double vortex flow in a curved tube

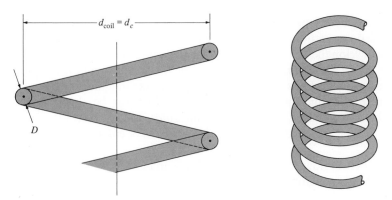

Figure 6.13 Schematic diagram illustrating flow and nomenclature for heat transfer in helically coiled tubes.

2. The region of intermediate Dean numbers, $20 < \mathrm{Dn} < 100$, in which the inertial forces due to secondary flow balance the viscous forces. In this region the recommended correlation is (25)

$$\overline{\mathrm{Nu}} = 0.9(\mathrm{Re}^2\mathrm{Pr})^{1/6} \tag{6.46}$$

3. The region of large Dean numbers, $\mathrm{Dn} > 100$. In this region viscous forces are significant only in the boundary near the tube wall. In the range $100 < \mathrm{Dn} < 1000$, the peripherally average Nusselt number is

$$\overline{\mathrm{Nu}} = 0.7\,\mathrm{Re}_D{}^{0.43}\mathrm{Pr}^{1/5}\left(\frac{D}{d_c}\right)^{0.07} \tag{6.47}$$

There is no appreciable difference in the value of the average Nusselt number for a uniform heat flux and a uniform surface temperature in coiled tubes. The friction factor in a coiled tube f_c in laminar flow, according to Ito (26), can be obtained from

$$f_c = \left(\frac{64}{\mathrm{Re}_D}\right)\frac{21.5\,\mathrm{Dn}}{(1.56 + \log_{10}\mathrm{Dn})}\,5.73 \tag{6.48}$$

in the range $2000 > \mathrm{Dn} > 13.5$. Ito also gives the relation

$$\mathrm{Re}_{D,\,\text{transition}} = 2\left(\frac{D}{d_c}\right)^{0.32} \times 10^4 \tag{6.49}$$

for the value of the Reynolds number at which the flow becomes turbulent. Equation (6.49) is in agreement with experimental results in the range $15 < D/d < 860$, while at larger values the critical Reynolds number for a curved pipe is the same as for a straight pipe.

6.4* ANALOGY BETWEEN HEAT AND MOMENTUM TRANSFER IN TURBULENT FLOW

To illustrate the most important physical variables affecting heat transfer by turbulent forced convection to or from fluids flowing in a long tube or duct, we shall now develop the so-called Reynolds analogy between heat and momentum transfer (27). The assumptions necessary for the simple analogy are valid only for fluids having a Prandtl number of unity, but the fundamental relation between heat transfer and fluid friction for flow in ducts can be illustrated for this case without introducing mathematical difficulties. The results of the simple analysis can also be extended to other fluids by means of empirical correction factors.

The rate of heat flow per unit area in a fluid can be related to the temperature gradient by the equation developed previously:

$$\frac{q_c}{A\rho c_p} = -\left(\frac{k}{\rho c_p} + \epsilon_H\right)\frac{dT}{dy} \tag{6.50}$$

Similarly, the shearing stress caused by the combined action of the viscous forces and the turbulent momentum transfer is given by

$$\frac{\tau}{\rho} = \left(\frac{\mu}{\rho} + \epsilon_M\right)\frac{du}{dy} \tag{6.51}$$

According to the Reynolds analogy, heat and momentum are transferred by analogous processes in turbulent flow. Consequently, both q and τ vary with y, the distance from the surface, in the same manner. For fully developed turbulent flow in a pipe, the local shearing stress decreases linearly with the radial distance r. Hence we can write

$$\frac{\tau}{\tau_s} = \frac{r}{r_s} = 1 - \frac{y}{r_s} \tag{6.52}$$

and

$$\frac{q_c/A}{(q_c/A)_s} = \frac{r}{r_s} = 1 - \frac{y}{r_s} \tag{6.53}$$

where the subscript s denotes conditions at the inner surface of the pipe. Introducing Eqs. (6.52) and (6.53) into Eqs. (6.50) and (6.51), respectively, yields

$$\frac{\tau_s}{\rho}\left(1 - \frac{y}{r_s}\right) = \left(\frac{\mu}{\rho} + \epsilon_M\right)\frac{du}{dy} \tag{6.54}$$

and

$$\frac{q_{c,s}}{A_s\rho c_p}\left(1 - \frac{y}{r_s}\right) = -\left(\frac{k}{\rho c_p} + \epsilon_H\right)\frac{dT}{dy} \tag{6.55}$$

If $\epsilon_H = \epsilon_M$, expressions in parentheses on the right-hand side of Eqs. (6.54) and (6.55) are equal, provided the molecular diffusivity of momentum μ/ρ equals the molecular diffusivity of heat $k/\rho c_p$, that is, the Prandtl number is unity. Dividing Eq (6.55) by Eq. (6.54) yields, under these restrictions,

$$\frac{q_{c,s}}{A_s c_p \tau_s}\,du = -dT \tag{6.56}$$

Integration of Eq. (6.56) between the wall, where $u = 0$ and $T = T_s$, and the bulk of the fluid, where $u = \bar{U}$ and $T = T_b$ yields

$$\frac{q_s\bar{U}}{A_s c_p \tau_s} = T_s - T_b$$

which can also be written in the form

$$\frac{\tau_s}{\rho\bar{U}^2} = \frac{q_s}{A_s(T_s - T_b)}\frac{1}{c_p\rho\bar{U}} = \frac{\bar{h}_c}{c_p\rho\bar{U}} \tag{6.57}$$

since $\bar{h}_c$ is by definition equal to $q_s/A_s(T_s - T_b)$. Multiplying the numerator and the denominator of the right-hand side by $D_H \mu k$ and regrouping yields

$$\frac{\bar{h}_c}{c_p \rho \bar{U}} \frac{D_H \mu k}{D_H \mu k} = \left(\frac{\bar{h}_c D_H}{k}\right)\left(\frac{k}{c_p \mu}\right)\left(\frac{\mu}{\bar{U} D_H \rho}\right) = \frac{\overline{Nu}}{RePr} = \overline{St}$$

where $\overline{St}$ is the *Stanton number*.

To bring the left-hand side of Eq. (6.57) into a more convenient form, we use Eq. (6.14):

$$\tau_s = f\frac{\rho \bar{U}^2}{8}$$

Substituting Eq. (6.14) for τ_s in Eq. (6.57) finally yields a relation between the Stanton number $\overline{St}$ and the friction factor

$$\overline{St} = \frac{\overline{Nu}}{RePr} = \frac{f}{8} \tag{6.58}$$

which is known as the *Reynolds analogy* for flow in a tube. It agrees fairly well with experimental data for heat tranfer in gases whose Prandtl number is nearly unity.

According to experimental data for fluids flowing in smooth tubes in the range of Reynolds numbers from 10,000 to 1,000,000, the friction factor is given by the empirical relation (16)

$$f = 0.184 \, Re_D^{-0.2} \tag{6.59}$$

Using this relation, Eq. (6.58) can be written as

$$\overline{St} = \frac{\overline{Nu}}{RePr} = 0.023 \, Re_D^{-0.2} \tag{6.60}$$

Since Pr was assumed unity,

$$\overline{Nu} = 0.023 \, Re_D^{0.8} \tag{6.61}$$

or

$$\bar{h}_c = 0.023 \bar{U}^{0.8} D^{-0.2} k \left(\frac{\mu}{\rho}\right)^{-0.8} \tag{6.62}$$

Note that, in fully established turbulent flow, the unit-convective conductance is directly proportional to the velocity raised to the 0.8 power, but inversely proportional to the tube diameter raised to the 0.2 power. For a given flow rate, an increase in the tube diameter reduces the velocity and thereby causes a decrease in $\bar{h}_c$ proportional to $1/D^{1.8}$. The use of small tubes and high velocities is therefore conducive to large heat transfer coefficients, but at the same time the power required to overcome the frictional resistance is increased. In the design of heat exchange equipment it is therefore necessary to strike a balance between the gain in heat transfer rates achieved by the use of ducts having small cross-sectional areas and the accompanying increase in pumping requirements.

Figure 6.14 shows the effect of surface roughness on the friction coefficient. We observe that the friction coefficient increases appreciably with the relative

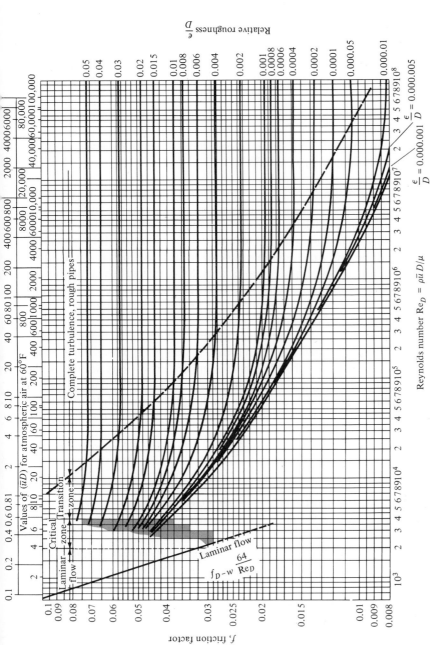

Figure 6.14 Friction factor versus Reynolds number for laminar and turbulent flow in tubes with various surface roughnesses. [From L. F. Moody (56), with permission of the publishers, the American Society of Mechanical Engineers.]

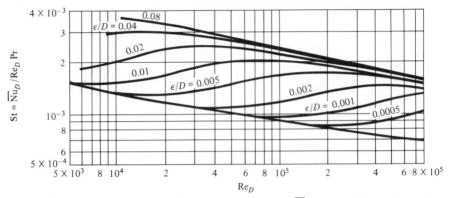

Figure 6.15 Heat transfer in artificially roughened tubes, $\overline{St}$ versus Re for various values of ϵ/D according to Dipprey and Sabersky (30).

roughness, defined as ratio of the average asperity height ϵ to the diameter D. According to Eq. (6.58), one would expect that roughening the surface, which increases the friction coefficient, also increases the convective conductance. Experiments performed by Cope (28) and Nunner (29) are qualitatively in agreement with this prediction, but a considerable increase in surface roughness is required to improve the rate of heat transfer appreciably. Since an increase in the surface roughness causes a substantial increase in the frictional resistance, for the same pressure drop, the rate of heat transfer obtained from a smooth tube is larger than from a rough one in turbulent flow.

Measurements by Dipprey and Sabersky (30) in tubes artificially roughened with sand grains are summarized in Fig. 6.15, where the Stanton number is plotted against the Reynolds number for various values of the roughness ratio ϵ/D. The lower straight line is for smooth tubes. At small Reynolds numbers St has the same value for rough and smooth tube surfaces. The larger the value ϵ/D, the smaller the value of Re at which the heat transfer begins to improve with increase in Reynolds number. But for each value of ϵ/D the Stanton number reaches a maximum and, with a further increase in Reynolds number, begins to decrease.

6.5 EMPIRICAL CORRELATIONS FOR TURBULENT FORCED CONVECTION

The Reynolds analogy presented in the preceding section was extended semi-analytically to fluids with Prandtl numbers larger than unity in (31)–(34) and to liquid metals with very small Prandtl numbers in (35). But the phenomena of turbulent forced convection are so complex that for engineering design empirical correlations are used in practice.

6.5.1 Circular Ducts and Tubes

The Dittus-Boelter equation (36) extends the Reynolds analogy to fluids with Prandtl numbers between 0.7 and 160 by multiplying the right-hand side of

Eq. (6.61) by a correction factor of the form Pr^n:

$$\overline{Nu}_D = \frac{\bar{h}_c D}{k} = 0.023\ Re_D^{0.8} Pr^n \qquad (6.63)$$

where

$$n = \begin{cases} 0.4 & \text{for heating} \quad (T_s > T_b) \\ 0.3 & \text{for cooling} \quad (T_s < T_b) \end{cases}$$

With all properties in this correlation evaluated at the bulk temperature T_b, Eq. (6.63) has been confirmed experimentally to within ± 25 percent for uniform wall temperature as well as uniform heat flux conditions within the following ranges of parameters:

$$0.7 < Pr < 160$$
$$Re_D > 6000$$
$$(L/D) > 60$$

Since this correlation does not take into account variations in physical properties due to the temperature gradient at a given cross section, it should only be used for situations with moderate temperature differences $(T_b - T_s)$.

For situations in which significant property variations due to a large temperature difference $(T_s - T_b)$ exist, a correlation developed by Sieder and Tate (15) is recommended:

$$\overline{Nu}_D = 0.027\ Re_D^{0.8} Pr^{0.3} \left(\frac{\mu_b}{\mu_s}\right)^{0.14} \qquad (6.64)$$

In Eq. (6.64) all properties except μ_s are evaluated at the bulk temperature. The viscosity μ_s is evaluated at the surface temperature. Equation (6.64) is appropriate for uniform wall temperature and uniform heat flux in the following range of conditions:

$$0.7 < Pr < 16{,}000$$
$$Re_D > 6000$$
$$(L/D) > 60$$

To account for the variation in physical properties due to the temperature gradient in the flow direction, the surface and bulk temperatures should be the values half-way between the inlet and the outlet of the duct. For ducts of other than circular cross-sectional shapes Eqs. (6.63) and (6.64) can be used if the diameter D is replaced by the hydraulic diameter D_H.

A correlation similar to Eq. (6.64), but restricted to gases, was proposed by Kays and London (16) for long ducts:

$$\overline{Nu}_{D_H} = C\ Re_{D_H}^{0.8} Pr^{0.3} \left(\frac{T_b}{T_s}\right)^n \qquad (6.65)$$

where all properties are based on the bulk temperature T_b. The constant C and

TABLE 6.3 HEAT TRANSFER CORRELATIONS FOR LIQUIDS AND GASES IN INCOMPRESSIBLE FLOW THROUGH TUBES

Name (reference)	Formula[a]	Conditions	Equation
Dittus-Boelter (36)	$\overline{Nu}_D = 0.023\,Re_D^{0.8}Pr^{0.4}$	$0.5 < Pr < 120$ $2300 < Re_D < 10^7$	(6.63)
Sieder-Tate (15)	$\overline{Nu}_D = 0.027\,Re_D^{0.8}Pr^{1/3}\left(\dfrac{\mu_b}{\mu_s}\right)^{0.14}$	$2300 < Re_D < 10^7$	(6.64)
Petukhov-Popov (37)	$\overline{Nu}_D = \dfrac{(f/8)Re_D Pr}{K_1 + K_2(f/8)^{1/2}(Pr^{2/3} - 1)}$ where $f = (1.82\log_{10}Re_D - 1.64)^{-2}$ $K_1 = 1 + 3.4f$ $K_2 = 11.7 + \dfrac{1.8}{Pr^{1/3}}$	$0.5 < Pr < 2000$ $10^4 < Re_D < 5 \times 10^6$	(6.66)
Sleicher-Rouse (38)	$\overline{Nu}_D = 5 + 0.015\,Re_D^{a}Pr_s^{b}$ where $a = 0.88 - \dfrac{0.24}{4 + Pr_s}$ $b = 1/3 + 0.5e^{-0.6Pr_s}$	$0.1 < Pr < 10^5$ $10^4 < Re_D < 10^6$	(6.67)

[a] All properties are evaluated at the bulk fluid temperature except where noted. Subscripts b and s indicate film and surface temperatures, respectively.

the exponent n are:

$$C = \begin{cases} 0.020 & \text{for uniform surface temperature } T_s \\ 0.021 & \text{for uniform heat flux } q_s'' \end{cases}$$

$$n = \begin{cases} 0.575 & \text{for heating} \\ 0.150 & \text{for cooling} \end{cases}$$

More complex empirical correlations have been proposed by Petukhov and Popov (37) and by Sleicher and Rouse (38). Their results are shown in Table 6.3, which presents four empirical correlation equations widely used by engineers to predict the heat transfer coefficient for turbulent forced convection in long, smooth, circular tubes. A careful experimental study with water heated in smooth tubes at Prandtl numbers of 6.0 and 11.6 showed that the Petukhov-Popov and the Sleicher-Rouse correlations agreed with the data over a Reynolds number range between 10,000 and 100,000 to within ± 5 percent, while the Dittus-Boelter and Sieder-Tate correlations, popular with heat transfer engineers, underpredicted the data by 5 to 15 percent (39). Figure 6.16 shows a comparison of these equations with experimental data at $Pr = 6.0$ (water at 26.7°C). The following example illustrates the use of some of these empirical correlations.

EXAMPLE 6.5

Determine the unit thermal convective conductance for water flowing at a velocity of 10 ft/s in an annulus formed between a 1-in.-OD tube and a

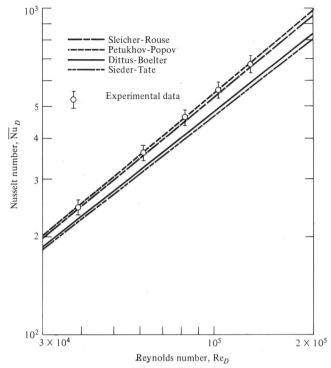

Figure 6.16 Comparison of predicted and measured Nusselt number for turbulent flow of water in a tube (26.7°C; Pr = 6.0).

$1\frac{1}{2}$-in.-ID tube. The water is at 180°F and is being cooled. The temperature of the inner wall is 100°F, and the outer wall of the annulus is insulated. Neglect entrance effects and compare the results obtained from all four equations in Table 6.3. The properties of water are given below in engineering units.

T (°F)	μ (lb$_m$/h ft)	k (Btu/h ft °F)	ρ (lb$_m$/ft^3)	c (Btu/lb$_m$ °F)
100	1.67	0.36	62.0	1.0
140	1.14	0.38	61.3	1.0
180	0.75	0.39	60.8	1.0

Solution. The hydraulic diameter D_H for this geometry is 0.5 in. The Reynolds number based on the hydraulic diameter and the bulk temperature properties is

$$\text{Re}_{D_H} = \frac{\rho \bar{U} D_H}{\mu} = \frac{(10\ \text{ft/s})(0.5/12\ \text{ft})(60.8\ \text{lb}_m/\text{ft}^3)(3600\ \text{s/h})}{0.75\ \text{lb}_m/\text{h ft}}$$

$$= 125{,}000$$

The Prandtl number is

$$Pr = \frac{c_p \mu}{k} = \frac{(1.0 \text{ Btu/lb}_m \text{ °F})(0.75 \text{ lb}_m/\text{h ft})}{0.39 \text{ Btu/h ft °F}} = 1.92$$

The Nusselt number according to the Dittus-Boelter correlation (6.63) is

$$\overline{Nu} = 0.023 \, Re_{D_H}^{0.8} Pr^{0.3} = (0.023)(11,954)(1.22) = 334$$

Using the Sieder-Tate correlation (6.64), we get

$$\overline{Nu} = 0.027 \, Re_{D_H}^{0.8} Pr^{0.33} \left(\frac{\mu_b}{\mu_s}\right)^{0.14}$$

$$= (0.027)(11,954)(1.24)\left(\frac{0.75}{1.67}\right)^{0.14} = 358$$

The Petukhov-Popov correlation (6.66) gives

$$f = (1.82 \log_{10} Re_{D_H} - 1.64)^{-2} = (9.276 - 1.64)^{-2} = 0.01715$$

$$K_1 = 1 + 3.4f = 1.0583$$

$$K_2 = 11.7 + \frac{1.8}{Pr^{0.33}} = 13.15$$

$$\overline{Nu} = \frac{f \, Re_{D_H} Pr/8}{K_1 + K_2(f/8)^{1/2}(Pr^{0.67} - 1)}$$

$$= \frac{(0.01715)(125,000)(1.92/8)}{1.0583 + (13.15)(0.01715/8)^{1/2}(0.548)} = 370$$

The Sleicher-Rouse correlation (6.67) yields

$$\overline{Nu} = 5 + 0.015 \, Re_f^a Pr_s^b$$

$$a = 0.88 - \frac{0.24}{4 + 4.64} = 0.88 - 0.0278 = 0.852$$

$$b = \frac{1}{3} + \frac{0.5}{e^{0.6Pr_s}} = 0.333 + \frac{0.5}{16.17} = 0.364$$

$$Re_{D_{H,f}} = 82,237$$

$$\overline{Nu} = 5 + (0.015)(82,237^{0.852})(4.64^{0.364})$$

$$= 5 + (0.015)(15,404)(1.748) = 409$$

Assuming that the correct answer is Nu = 370, the first two correlations underpredict $\overline{Nu}$ by about 10 percent and 3.5 percent, respectively, while the Sleicher-Rouse method overpredicts by about 10.5 percent.

It should be noted that in general the surface and film temperatures are not known and therefore the use of Eq. (6.67) requires iteration for large temperature differences. The main difficulty in applying Eq. (6.66) for conditions with varying properties is that the friction factor f may be affected by heating or cooling to an unknown extent.

For gases and liquids flowing in short circular tubes ($2 < L/D < 60$) with abrupt contraction entrances, the entrance configuration of greatest interest in heat exchanger design, the entrance effect for Reynolds numbers corresponding to turbulent flow (40) can be represented approximately by the equation

$$\frac{\bar{h}_{c,L}}{\bar{h}_c} = 1 + \left(\frac{D}{L}\right)^{0.7} \tag{6.68}$$

when L/D is less than 20 but larger than 2, and by the equation

$$\frac{\bar{h}_{c,L}}{\bar{h}_c} = 1 + \frac{6D}{L} \tag{6.69}$$

when L/D is larger than 20. In both of the above equations, $\bar{h}_{c,L}$ is the average unit conductance for the tube of finite length L and $\bar{h}_c$ is the conductance for an infinitely long tube.

An extensive theoretical analysis of the heat transfer and the friction drop in the entrance regions of smooth passages is given in (41) and a complete survey of experimental results for various types of inlet condition is given in (42) and (43).

6.5.2 Ducts of Noncircular Shape and Coiled Tubes

Heat transfer performance for cooling of air in turbulent flow with 21 different tubes having integral internal spiral and longitudinal fins has been studied by Carnavos (44). For the 21 tube profiles shown in Fig. 6.17, the heat transfer data were correlated within ± 6 percent at Reynolds numbers between 10^4 and 10^5 by the equation

$$\frac{\overline{\text{Nu}}_{D_H}}{\text{Pr}^{0.4}} = 0.023 \, \text{Re}_{D_H}^{0.8} \left(\frac{A_{fa}}{A_{fc}}\right)^{0.1} \left(\frac{A_n}{A_a}\right)^{0.5} (\sec \alpha^3) \tag{6.70}$$

Figure 6.17 Profiles of internally finned tubes. [Reprinted from T. C. Carnavos (44), with permission from the publishers, Hemisphere Publishing Corp.]

The Fanning friction factor f was correlated within ± 7 percent for all configurations except 11, 12, and 28 by the relation

$$f = \frac{0.046}{\mathrm{Re}_{D_H}^{0.20}} \left(\frac{A_{fa}}{A_{fn}}\right)^{0.5} \cos \alpha^{0.5} \tag{6.71}$$

where A_{fa} = actual free-flow cross-sectional area

A_{fc} = open core flow area inside fins

A_a = actual heat transfer area

A_n = nominal heat transfer area based on tube ID without fins

α = helix angle for spiral fins

A_{fn} = nominal flow area based on tube ID without fins

To apply these correlations, all physical properties should be based on the average bulk temperature.

It is claimed that the capacity of an existing heat exchanger can be increased at constant pumping power between 12 and 66 percent by substitution of tubes with inner fins for smooth tubes.

For turbulent flow through a long and smooth annulus, Kays and Perkins (10) recommend that the friction factor be increased by 10 percent and propose the relation

$$f = 0.085(\mathrm{Re}_{D_H})^{-0.25} \tag{6.72}$$

valid for $6000 < \mathrm{Re}_{D_H} < 300,000$ and $0.0625 < D_i/D_o < 0.562$. To calculate the heat transfer coefficient in turbulent flow through an annulus, one generally uses one of the equations from Table 6.3 with the hydraulic diameter as the pertinent length dimension. Equation (6.72) can be used for f in Eq. (6.65). Another approach, based on data with gases, has been proposed by Nemira et al. (45).

For turbulent flow and forced-convection heat transfer in rectangular and triangular ducts, experimental data have been obtained with air. The correlation proposed (46, 47) is

$$\overline{\mathrm{Nu}}_{D_H} = 0.021 \, \mathrm{Re}_{D_H}^{0.8} \mathrm{Pr}^{0.4} \left(\frac{T_s}{T_b}\right)^{-0.7} \left[1 + \left(\frac{L}{D_H}\right)^{-0.7} \left(\frac{T_s}{T_b}\right)^{0.7}\right] \tag{6.73}$$

In the turbulent regime the flow and heat transfer in helically coiled tubes have been investigated by several authors, including Seban and McLaughlin (48) at low Reynolds numbers and by Jeschke (49) at Reynolds numbers above 10^5. The data are somewhat scattered but, in general, coiling the tube enhances the heat transfer in turbulent flow less than in laminar flow. For example, when the ratio of coil to tube diameter d_c/D is 50 the heat transfer coefficient in the helically coiled tube is only 7 percent higher than in a straight tube. To calculate the Nusselt number for convection heat transfer in a helically coiled tube, Hausen (17) has proposed the following correlation for Reynolds numbers larger than the critical value:

$$\frac{\overline{\mathrm{Nu}}_{D,\,\mathrm{helical}}}{\overline{\mathrm{Nu}}_{D,\,\mathrm{straight}}} = 1 + \left(\frac{21}{\mathrm{Re}_D^{0.14}}\right)\left(\frac{D}{d_c}\right) \tag{6.74}$$

6.5.3 Liquid Metals

Liquid metals have in recent years been employed as heat transfer media because they have certain advantages over other common liquids used for heat transfer purposes. Liquid metals, such as sodium, mercury, lead, and lead-bismuth alloys, have relatively low melting points and combine high densities with low vapor pressures at high temperatures as well as with large thermal conductivities, ranging from 5 to 50 Btu/h ft °F. These metals can be used over wide ranges of temperatures, they have a large heat capacity per unit volume, and they have large unit thermal convective conductances. They are especially suitable for use in nuclear power plants, where large amounts of heat are liberated and must be removed in a small volume. Liquid metals pose some difficulties in handling and pumping, but the development of electromagnetic pumps has eliminated most of these problems.

In liquid metals even in a highly turbulent stream the effect of eddying is of secondary importance compared to conduction. As a result, the empirical equations for gases and liquids do not apply. Several theoretical analyses for the evaluation of the Nusselt number are available, but there still exist some unexplained discrepancies between many of the experimental data and the analytic results. This is illustrated in Fig. 6.18, where experimentally measured Nusselt numbers for heating of mercury in long tubes are compared with the analysis of Martinelli (35).

Lubarsky and Kaufman (50) found that the relation

$$\overline{Nu}_D = 0.625(Re_D Pr)^{0.4} \tag{6.75}$$

empirically correlated most of the data in Fig. 6.18, but the error band is seen to be substantial. Those points in Fig. 6.18 that fall far below the average are believed to have been obtained in systems where the liquid metal did not wet the surface. However, no final conclusions regarding the effect of wetting have been reached to date.

According to Skupinski et al. (51), the Nusselt number for liquid metals flowing in smooth tubes can be obtained from

$$\overline{Nu}_D = 4.82 + 0.0185(Re_D Pr)^{0.827} \tag{6.76}$$

if the heat flux is uniform in the range $Re_D Pr > 100$ and $L/D > 30$ with all properties evaluated at the bulk temperature.

According to an investigation of the thermal entry region for turbulent flow of a liquid metal in a pipe with uniform heat flux, the Nusselt number depends only on the Reynolds number when $Re_D Pr < 100$. For these conditions, Lee (52) found that the equation

$$\overline{Nu}_D = 3.0\, Re_D^{0.0833} \tag{6.77}$$

fits data and analysis well. Convection in the entrance regions for fluids with small Prandtl numbers has also been investigated analytically by Deissler (41) and experimental data supporting the analysis are summarized in (53) and (54). In turbulent flow the thermal entry length $(L/D_H)_{entry}$ is approximately 10 equivalent diameters when the velocity profile is already developed, and 30 equivalent diameters when it is developing simultaneously with the temperature profile.

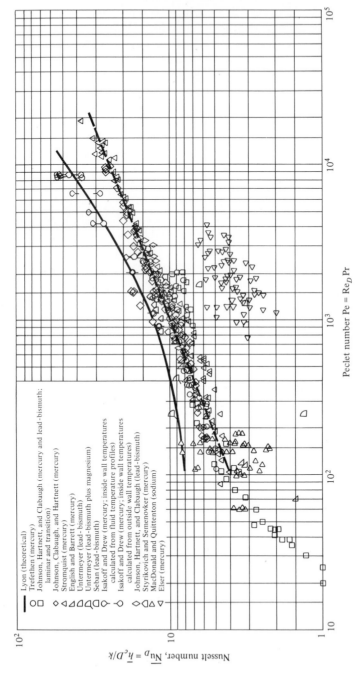

Figure 6.18 Comparison of measured and predicted Nusselt numbers for liquid metals heated in long tubes with uniform heat flux. (Courtesy of the National Advisory Committee for Aeronautics, NACA TN 3336.)

For a constant surface temperature the data are correlated, according to Seban and Shimazaki (55), by the equation

$$\overline{\mathrm{Nu}}_D = 5.0 + 0.025(\mathrm{Re}_D\mathrm{Pr})^{0.8} \tag{6.78}$$

in the range $\mathrm{Re}\mathrm{Pr} > 100$, $L/D > 30$.

EXAMPLE 6.6 ───

A liquid metal flows at a mass rate of 3 kg/s through a constant-heat-flux 5-cm-ID tube in a nuclear reactor. The fluid at 473 K is to be heated with the tube wall 30 K above the fluid temperature. Determine the length of the tube required for a 1 K rise in bulk fluid temperature, using the following properties:

$$\rho = 7.7 \times 10^3 \text{ kg/m}^3$$
$$v = 8.0 \times 10^{-8} \text{ m}^2/\text{s}$$
$$c_p = 130 \text{ J/kg K}$$
$$k = 12 \text{ W/m K}$$
$$\mathrm{Pr} = 0.011$$

Solution. The rate of heat transfer per unit temperature rise is

$$q = \dot{m}c_p \Delta T = (3.0)(130)(10) = 3900 \text{ W}$$

The Reynolds number is

$$\mathrm{Re}_D = \frac{\dot{m}D}{\rho A v} = \frac{(3)(0.06)}{(7.7 \times 10^3)[\pi(0.05)^2/4](8.0 \times 10^{-8})} = 1.24 \times 10^5$$

The heat transfer coefficient is obtained from Eq. (6.75):

$$\bar{h}_c = \left(\frac{k}{D}\right)0.625(\mathrm{Re}_D\mathrm{Pr})^{0.4}$$

$$= \left(\frac{12}{0.05}\right)0.625(1.24 \times 10^5)(0.011)^{0.4}$$

$$= 2692 \text{ W/m}^2 \text{ K}$$

The surface area required is

$$A = \pi DL = \frac{q}{\bar{h}_c(T_s - T_b)}$$

$$= \frac{3900}{(2692)(30)} = 0.483 \text{ m}^2$$

Finally, the required length is

$$L = \frac{A}{\pi D} = \frac{0.0483}{(\pi)(0.05)} = 0.307 \text{ m}$$

6.6* MASS TRANSFER BY CONVECTION IN FLOW THROUGH DUCTS

Transfer of mass by convection for flow inside conduits is encountered in numerous engineering systems—for example, in vaporization of a liquid that wets the surface into a gas stream flowing inside a circular column, or sublimation from the surface to a fluid passing through a duct. A concentration boundary layer similar to the temperature boundary layer will develop under these conditions and the mass transfer analog of the mean bulk temperature T_b is the mean species concentration $\rho_{A,b}$, defined for flow through a pipe or tube of inner radius r_s by the relation

$$\rho_{A,b} = \frac{1}{\pi \bar{u}_b r_s^2} \int_0^{r_s} 2\pi r u(r) \rho_A(r)\, dr \tag{6.79}$$

In mass transfer, as in heat transfer, there exist an entrance region, in which the boundary layer develops, and a fully developed region, in which the shape of the concentration profile remains constant with distance. Using the analogy between heat and mass transfer, in laminar flow, the concentration profile will approach its fully developed shape when $x/D \sim 0.05\,(\mathrm{Re}_D \mathrm{Sc})$, while in turbulent flow, where conditions are nearly independent of Sc, Eq. (6.8) holds.

The rate of mass transfer of species A can be calculated from

$$n_{A,s} = \bar{h}_m A (\rho_{A,s} - \rho_{A,b}) \tag{6.80}$$

The convection mass transfer coefficient h_m can be obtained from the Sherwood number Sh_{D_H}, defined as

$$\overline{\mathrm{Sh}}_{D_H} = \frac{\bar{h}_m D_H}{D_{AB}} \tag{6.81}$$

for flow through a duct of hydraulic diameter D_H. The Sherwood number in turn can be obtained from the same type of correlations used for heat transfer by replacing Nu_{D_H} by Sh_{D_H} and Pr by Sc.

EXAMPLE 6.7

Dry air at 25°C is passing at a flow rate of 6×10^{-4} kg/s through a vertical pipe of inner diameter $D = 2$ cm. A thin film of benzene is flowing very slowly down over the inner surface, as shown in Fig. 6.19. If the tube is 2 m long, calculate the average mass transfer coefficient, assuming that the film is smooth.

Solution. Assume that evaporative cooling effects are negligible and that the film of benzene does not occupy an appreciable part of the inner flow area. From property tables in the appendices we get:

air: $\nu = 15.7 \times 10^{-6}$ m^2/s $\mu = 183.6 \times 10^{-7}$ N s/m^2

benzene-air: $D_{AB} = 0.88 \times 10^{-5}$ m^2/s $\mathrm{Sc} = \dfrac{\nu}{D_{AB}} = 1.78$

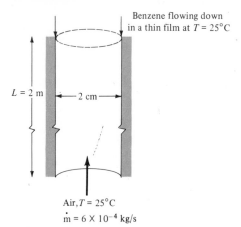

Benzene flowing down in a thin film at $T = 25°C$

$L = 2$ m

2 cm

Air, $T = 25°C$
$\dot{m} = 6 \times 10^{-4}$ kg/s

Figure 6.19 Schematic diagram of mass transfer system for Example 6.8.

The Reynolds number for the air is

$$Re_D = \frac{4\dot{m}}{\pi D \mu} = \frac{(4)(6 \times 10^{-4} \text{ kg/s})}{(\pi)(0.02 \text{ m})(1.836 \times 10^{-5} \text{ N s/m}^2)} = 2080$$

Hence, the flow is laminar.

A constant benzene concentration over the tube surface corresponds to a constant temperature. The ratio $[Re_D Sc/(L/D)]^{1/3}$ is about 3.3 and an appropriate equation for calculating the Sherwood number is Eq. (6.40), with Sc replacing Pr, or

$$\overline{Sh}_D = 1.86\left(\frac{Re_D Sc D}{L}\right)^{1/3} = (1.86)(3.3) = 6.14$$

Hence,

$$\bar{h}_m = \frac{\overline{Sh}_D D_{AB}}{D} = \frac{(6.14)(0.88 \times 10^{-5} \text{ m}^2/\text{s})}{0.02 \text{ m}} = 0.0028 \text{ m/s}$$

If we had assumed that the flow was fully developed, $\overline{Sh}_D = 3.66$, and the predicted mass transfer coefficient would have been about 40 percent less than the value calculated by taking the entrance effects into accounts.

6.7 CLOSURE

In this chapter we have presented theoretical and empirical correlations that may be used to calculate the Nusselt or Sherwood number, from which the heat or mass transfer coefficient for convection heat transfer to or from a fluid flowing through a duct can be obtained. It cannot be overemphasized that empirical equations derived from experimental data by means of dimensional analysis are applicable only over the range of parameters for which data exist to verify the relation within a specified error band. Serious errors can result if an empirical relation is applied beyond the parameter range over which it has been verified.

When applying an empirical relationship to calculate a convection heat or mass transfer coefficient, the following sequence of steps should be followed:

1. Collect appropriate physical properties for the fluid in the temperature range of interest.
2. Establish the appropriate geometry for the system and the correct significant length for the Reynolds and Nusselt (or Sherwood) numbers.
3. Determine whether the flow is laminar, turbulent, or transitional by calculating the Reynolds number.
4. Determine whether free-convection effects may be appreciable by calculating the Grashof number and comparing it with the square of the Reynolds number.
5. Select an appropriate equation that applies to the geometry and flow required. If necessary, iterate initial calculations of dimensionless parameters in accordance with the stipulations of the equation selected.
6. Make an order-of-magnitude estimate of the heat transfer coefficient (see Table 1.4).
7. Calculate the value of the heat transfer coefficient from the equation in step 5 and compare with the estimate in step 6 to spot possible errors in the decimal point or units.

It should be noted that experimental data on which empirical relations are based have generally been obtained under controlled conditions in a laboratory, whereas most practical applications occur under conditions that deviate from laboratory conditions in one way or another. Consequently, the predicted value of a heat transfer coefficient may deviate from the actual value, and since such uncertainties are unavoidable it is often satisfactory to use a simple correlation, especially for preliminary designs.

A special note of caution is in order for the transition regime. The mechanisms of heat transfer and fluid flow in the transition region (Re_D between 2100 and 6000) vary considerably from system to system. In this region the flow may be unstable, and fluctuations in pressure drop and heat transfer have been observed. There exists a large uncertainty in the basic heat transfer and flow-friction performance, and consequently the designer is advised to design equipment to operate outside this region, if possible; the curves of Fig. 6.20 may be used, but the actual performance may deviate considerably from that predicted on the basis of these curves.

To aid in the rapid selection of an appropriate relation to obtain the heat transfer coefficient for flow in a duct, some of the most commonly used empirical equations are summarized in Table 6.4. A more complete summary of additional equations can be found in the *Handbook of Heat Transfer* (24).

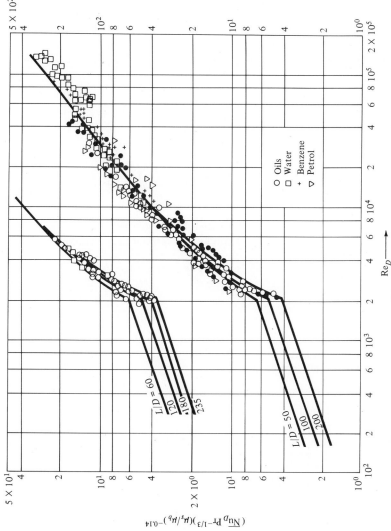

Figure 6.20 Recommended correlation curves for heat transfer coefficients in the transition regime. [From E. N. Sieder and C. E. Tate (15), with permission of the copyright owner, the American Chemical Society.]

TABLE 6.4 SUMMARY OF FORCED CONVECTION CORRELATION FOR INCOMPRESSIBLE FLOW INSIDE TUBES AND DUCTS[a,b,c,d]

System description	Recommended correlation	Equation in text
Friction factor for laminar flow in long tubes and ducts	Liquids: $f = (64/\mathrm{Re}_D)(\mu_s/\mu_b)^{0.14}$ Gases: $f = (64/\mathrm{Re}_D)(T_s/T_b)^{0.14}$	(6.42) (6.43)
Nusselt number for fully developed laminar flow in long tubes with uniform heat flux, $\mathrm{Pr} > 0.6$	$\overline{\mathrm{Nu}}_D = 4.36$	(6.31)
Nusselt number for fully developed laminar flow in long tubes with uniform wall temperature, $\mathrm{Pr} > 0.6$	$\overline{\mathrm{Nu}}_D = 3.66$	(6.32)
Average Nusselt number for laminar flow in tubes and ducts of intermediate length with uniform wall temperature, $(\mathrm{Re}_{D_H}\mathrm{Pr}D_H/L)^{0.33}(\mu_b/\mu_s)^{0.14} > 2$, $0.004 < (\mu_b/\mu_s) < 10$, $0.5 < \mathrm{Pr} < 16{,}000$	$\overline{\mathrm{Nu}}_{D_H} = 1.86(\mathrm{Re}_{D_H}\mathrm{Pr}D_H/L)^{0.33}(\mu_b/\mu_s)^{0.14}$	(6.40)
Average Nusselt number for laminar flow in short tubes and ducts with uniform wall temperature, $100 < (\mathrm{Re}_{D_H}\mathrm{Pr}D_H/L) < 1500$, $\mathrm{Pr} > 0.7$	$\overline{\mathrm{Nu}}_{D_H}$ $= 3.66 + \left(\dfrac{0.0668\, L/D_H\mathrm{Re}_{D_H}\mathrm{Pr}}{1 + 0.04\, L/D_H\mathrm{Re}_{D_H}\mathrm{Pr}}\right)\left(\dfrac{\mu_b}{\mu_s}\right)^{0.14}$	(6.39)
Friction factor for fully developed turbulent flow through smooth long tubes and ducts,	$f = 0.184/\mathrm{Re}_{D_H}^{0.2}$ ($30{,}000 < \mathrm{Re}_{D_H} < 10^6$) $f = 0.316/\mathrm{Re}_{D_H}^{0.25}$ ($5{,}000 < \mathrm{Re}_{D_H} < 30{,}000$)	(6.59)
Average Nusselt number for fully developed turbulent flow through smooth, long tubes and ducts, $\mathrm{Re}_{D_H} > 6000$, $0.7 < \mathrm{Pr} < 16{,}000$, $L/D_H > 60$	$\overline{\mathrm{Nu}}_{D_H} = 0.027\,\mathrm{Re}_{D_H}^{0.8}\mathrm{Pr}^{0.33}(\mu_b/\mu_s)^{0.14}$ or Table 6.3	(6.64) (6.66)
Average Nusselt number for liquid metals in turbulent, fully developed flow through smooth tubes with uniform heat flux, $10^4 > \mathrm{Re}_D\mathrm{Pr} > 100$, $L/D > 30$	$\overline{\mathrm{Nu}}_D = 4.82 + 0.0185(\mathrm{Re}_D\mathrm{Pr})^{0.827}$	(6.75)
Same as above, but in thermal entry region when $\mathrm{Re}_D\mathrm{Pr} < 100$	$\overline{\mathrm{Nu}}_D = 3.0\,\mathrm{Re}_D^{0.0833}$	(6.76)
Average Nusselt number for liquid metals in turbulent fully developed flow through smooth tubes with uniform surface temperature, $\mathrm{Re}_D\mathrm{Pr} > 100$, $L/D > 30$	$\overline{\mathrm{Nu}}_D = 5.0 + 0.025(\mathrm{Re}_D\mathrm{Pr})^{0.8}$	(6.77)

[a] Mass transfer correlations corresponding to the heat transfer correlations can be obtained by replacing Nu by Sh and Pr by Sc (see Section 6.6).

[b] All physical properties in the correlation are evaluated at the bulk temperature T_b except μ_s, which is evaluated at the surface temperature T_s.

[c] $\mathrm{Re}_{D_H} = D_H\bar{U}\rho/\mu$, $D_H = 4A_c/P$, $\bar{U} = \dot{m}/\rho A_c$.

[d] Incompressible flow correlations apply when average velocity is less than half the speed of sound (Mach number < 0.5) to gases and vapors.

PROBLEMS

The problems for this chapter are organized by subject matter as shown below.

Topic	Section	Problem number
Evaluation of heat transfer coefficient and pressure drop	6.3 and 6.5	6.1–6.13
Determination of temperature rise in flow through a duct with convection	All	6.14–6.17
Derivation of theoretical relations for convection in laminar or turbulent flow	6.2 and 6.4	6.18–6.26
Engineering analysis of a thermal system in which convection plays a major role	All	6.27–6.38

6.1 Water at an average temperature of 27°C is flowing through a smooth 5.08-cm-ID pipe at a velocity of 0.91 m/s. If the temperature at the inner surface of the pipe is 49°C, determine (*a*) the unit-surface conductance, (*b*) the rate of heat flow per foot of pipe, (*c*) the bulk temperature rise per foot, and (*d*) the pressure drop per meter.

6.2 An aniline-alcohol solution is flowing at a velocity of 10 fps through a long, 1-in.-ID thin-wall tube. On the outer surface of the tube, steam is condensing at atmospheric pressure, and the tube-wall temperature is 212°F. The tube is clean, and there is no thermal resistance due to a scale deposit on the inner surface. Using the physical properties tabulated below, estimate the unit-surface conductance between the fluid and the pipe by means of Eqs. (6.20) and (6.22) and Fig. 6.9, and compare the results. Assume that the bulk temperature of the aniline solution is 68°F and neglect entrance effects.

Physical properties of the aniline solution:

Temperature (°F)	Viscosity (centipoise)	Thermal conductivity (Btu/h ft °F)	Specific gravity	Specific heat (Btu/lb °F)
68	5.1	0.100	1.03	0.50
140	1.4	0.098	0.98	0.53
212	0.6	0.095		0.56

6.3 For water at a bulk temperature of 32°C flowing at a velocity of 1.5 m/s through a 2.54-cm.-ID duct with a wall temperature of 43°C, calculate the Nusselt number and the convection heat transfer coefficient by three different methods and compare the results.   *Ans.* $Nu = 260 \, (\pm 10\%)$.

6.4 Atmospheric air at a velocity of 61 m/s and a temperature of 16°C enters a 0.61-m-long square metal duct of 20 × 20 cm cross section. If the duct wall is at 149°C, determine the average unit-surface conductance. Comment briefly on the L/D_H effect.   *Ans.* $\bar{h}_c = 172 \, W/m^2 \, K \, (\pm 15\%)$.

6.5 Mercury flows inside a copper tube 9 m long with a 5.1-cm inside diameter at an average velocity of 7 m/s. The temperature at the inside surface of the tube is 38°C uniformly throughout the tube, and the arithmetic mean bulk temperature of the mercury is 66°C. Assuming the velocity and temperature profiles are fully developed, calculate the rate of heat transfer by convection for the 9-m length by considering the mercury as (*a*) an ordinary liquid and (*b*) liquid metal. Compare the results.

6.6 Determine the heat transfer coefficient for liquid bismuth flowing through an annulus (5 cm ID, 6.1 cm OD) at a velocity of 4.5 m/s. The wall temperature of the inner surface is 427°C and the bismuth is at 316°C. It may be assumed that heat losses from the outer surface are negligible.

6.7 Engine oil flows at a rate of 0.5 k/s through a 2.5-cm-ID tube. The oil enters at 25°C while the tube wall is at 100°C. (a) If the tube is 4 m long, determine whether the flow is fully developed. (b) Calculate the heat transfer coefficient.

6.8 Water at a bulk inlet temperature of 93°C is flowing with a velocity of 0.015 m/s through a 0.015-m-diameter tube, 0.3 m long. If the tube wall temperature is 204°C, determine the average heat transfer coefficient and estimate the bulk temperature rise of the water.   *Ans.* 368 W/m² K; 22°C.

6.9 If air is used instead of water in the tube of Problem 6.8, but the velocity of the air is increased until the heat transfer coefficient with the air equals that obtained with water at 1.5 cm/s, determine the velocity required and the pressure drop.
                                                                        Ans. 109 m/s.

6.10 In a long annulus (25 cm ID, 38 cm OD), atmospheric air is heated by steam condensing at 149°C on the inner surface. If the velocity of the air is 6 m/s and its bulk temperature is 38°C, calculate the heat transfer coefficient.

6.11 The equation

$$\overline{Nu} = 0.116(Re^{2/3} - 125)Pr^{1/3}\left[1 + \left(\frac{D}{L}\right)^{2/3}\right]\left(\frac{\mu_b}{\mu_s}\right)^{0.14}$$

has been proposed by Hausen for the transition range (2300 < Re < 8000) as well as for higher Reynolds numbers. Compare the values of $\overline{Nu}$ predicted by Hausen's equation for Re = 3000 and Re = 20,000 at $D/L = 0.1$ and 0.01 with those obtained from appropriate equations or charts in the text. Assume the fluid is water at 15°C flowing through a pipe at 100°C.

6.12 Compute the average unit-surface conductance $\bar{h}_c$ for 10°C water flowing at 3 m/s in a long, 2.5-cm-ID pipe (surface temperature 39°C) by three different equations and compare your results. Also determine the pressure drop per meter length of pipe.

6.13 Air at an average temperature of 149°C flows through a short square duct (10 × 10 × 2.25 cm) at a rate of 53 kg/h. The duct wall temperature is 430°C. Determine the average heat transfer coefficient, using the duct equation with appropriate L/D correction. Compare your results with flow-over-flat-plate relations.

6.14 Air at 16°C and atmospheric pressure enters a 1.25-cm-ID tube at 30 m/s. For an average wall temperature of 100°C, determine the discharge temperature of the air and the pressure drop in inches of water if the pipe is (a) 10 cm long and (b) 102 cm long. Use the average bulk temperature of the air between the inlet and the outlet to evaluate the rate of heat transfer between the wall and the air.

$$\text{Ans. (a) } T_{out} = 27.8°C, \Delta p = 9.65 \text{ cm } H_2O; (b) T_{out} = 77.8°C,$$
$$\Delta p = 23.1 \text{ cm } H_2O$$

6.15 In a refrigeration system, brine (10 percent NaCl by weight) having a viscosity of 16.5 N s/m² and a thermal conductivity of 0.85 W/m² K is flowing through a long 2.5-cm-ID pipe at 6.1 m/s. Under these conditions the heat transfer coefficient was found to be 1135 W/m² K. For a brine temperature of −1°C and a pipe temperature of 18.3°C, determine the temperature rise of the brine per foot length of pipe if the velocity of the brine is doubled. Assume that the specific heat of the brine is 3768 J/kg K and that its density is equal to that of water.

6.16 Water at 82.2°C is flowing through a thin copper tube (15.2 cm ID) at a velocity of 7.6 m/s. The duct is located in a room at 15.6°C and the unit-surface conductance at the outer surface of the duct is 14.1 W/m² K. (a) Determine the heat transfer coefficient at the inner surface. (b) Estimate the length of duct in which the water temperature drops (5/9)°C   *Ans.* (a) $\bar{h}_c = 21{,}200$ W/m² K

6.17 Determine the rate of heat transfer per foot length to a light oil flowing through a 1-in.-ID, 2-ft-long copper tube at a velocity of 6 fpm. The oil enters the tube at 60°F and the tube is heated by steam condensing on its outer surface at atmospheric pressure with a unit-surface conductance of 2000 Btu/h ft² °F. The properties of the oil at various temperatures are listed in accompanying tabulation:

$T(°F)$	60	80	100	150	212
ρ (lb/ft³)	57	57	56	55	54
c (Btu/lb °F)	0.43	0.44	0.46	0.48	0.51
k (Btu/h ft °F)	0.077	0.077	0.076	0.075	0.074
μ (lb/h ft)	215	100	55	19	8
Pr	1210	577	330	116	55

6.18 The equation

$$\overline{Nu} = \frac{\bar{h}_c D}{k} = \left[3.65 + \frac{0.0668(D/L)RePr}{1 + 0.04[(D/L)RePr]^{2/3}}\right]\left(\frac{\mu_b}{\mu_s}\right)^{0.14}$$

was recommended by H. Hausen (*Zeitschr. Ver. Deut. Ing., Beiheft* No. 4, 1943) for forced-convection heat transfer in fully developed laminar flow through tubes. Compare the values of the Nusselt number predicted by Hausen's equation for Re = 1000, Pr = 1, and $D/L = 2$, 10, and 100, respectively, with those obtained from two other appropriate equations or graphs in the text.

6.19 Derive an equation of the form $\bar{h}_c = f(T, D, V)$ for turbulent flow of water through a long tube in the temperature range between 20° and 100°C.

6.20 Show that for fully developed laminar flow between two flat plates spaced 2a apart, the Nusselt number based on the "bulk mean" temperature is 4.12 if the temperature of both walls varies linearly with the distance x, i.e., $\partial T/\partial x = C$. The "bulk mean" temperature is defined as

$$\bar{T} = \frac{\int_0^{2a} u(y)T(y)\,dy}{\int_0^{2a} u(y)\,dy}$$

Hint: Assume a solution of the form $T(x, y) = Cx + \bar{\Theta}(y)$ where $\bar{\Theta}$ is the local temperature difference between the fluid at y and the wall.

6.21 Repeat Problem 6.20, but assume that one wall is insulated while the temperature of the other walls increases linearly with x.   *Ans.* Nu = 2.70.

6.22 Show that for fully developed laminar flow in a tube with a parabolic velocity profile, i.e.,

$$u(r) = \frac{2w}{\rho\pi r_o^2}\left[1 - \left(\frac{r}{r_o}\right)^2\right]$$

the Nusselt number $2h_c r_o/k$ is 48/11 if the wall temperature increases linearly with x, i.e., $\partial T/\partial x = C$. *Hint:* Assume a solution of the form $T = Cx + \bar{\theta}(r)$, where $\bar{\theta}$ represents the difference between the local fluid temperature and the wall temperature at the same location.

6.23 For fully turbulent flow in a long tube of diameter D, develop a relation between the ratio $L\Delta T/D$ in terms of flow and heat transfer parameters, where $L\Delta T$ is the tube length required to raise the bulk temperature of the fluid by ΔT. Use Eq. 6.63 for fluids with Prandtl number of the order of unity or larger and Eq. 6.75 for liquid metals.

6.24 Water in *turbulent* flow is to be heated in a single-pass tubular heat exchanger by steam condensing on the outside of the tubes. The flow rate of the water, its pressure drop, its inlet and outlet temperatures, and the steam pressure are fixed. Assuming that the tube wall temperature remains constant, determine the dependence of the total required heat exchanger area on the inside diameter of the tubes.

6.25 An incompressible fluid is flowing in steady laminar flow between two parallel infinite plates whose wetted surfaces are a distance $2a$ apart and are at uniform temperature T_w. Owing to a peculiar variation of viscosity with temperature, the velocity of the fluid is uniform, i.e., $u(y) = U$. Assuming that the physical properties c_p and k are constant, that the fluid enters the system $(x = 0)$ at a uniform temperature T_o, and that conduction in the direction flow is negligible, develop an expression for the temperature distribution in the fluid in terms of the spacing, $2a$; the fluid velocity, U; and the fluid properties, c_p, μ, and k. After having derived an expression for the temperature $T(x, y)$, find an expression for the local rate of heat transfer and the local Nusselt number $h(x)2a/k$.

6.26 The energy conservation equation for steady laminar flow through a tube as shown in the accompanying sketch is

$$u \frac{\partial T}{\partial x} = a \frac{1}{r} \frac{\partial}{\partial r}\left(r \frac{\partial T}{\partial r}\right)$$

if the fluid properties are uniform. Under certain conditions it is quite reasonable to assume that the velocity profile is flat, i.e., $u(r) = \bar{U}$, so that "rod" or "slug" flow prevails. For this type of flow, assuming that the fluid enters at $x = 0$ at uniform temperature T_o and the wall of the tube is kept at temperature T_w, determine the temperature $T(r, x)$ as a function of relevant dimensionless parameters. (a) Letting $x' = x/L$ and $r' = r/r_o$, show that $aL/\bar{U}r_o^2$ is an appropriate dimensionless parameter for the independent variable. (b) Starting with the continuity equation in the form $w = \rho\bar{U}\pi r_o^2$ show that $\pi kL/wc_p$ may also be used as a dimensionless parameter. (c) Derive an expression for the dimensionless temperature distribution $[T(r, x) - T_o]/(T_w - T_o)$ for slug flow. (d) Defining the "bulk mean" temperature at a given cross section as

$$\bar{T} = \frac{\displaystyle\int_0^{r_o} 2\pi u T(r)r\, dr}{\displaystyle\int_0^{r_o} 2\pi u r\, dr}$$

verify that for slug flow at a distance $x = L$ from the inlet

$$\frac{\bar{T} - T_o}{T_w - T_o} = 1 - 0.692e^{-5.78z} + 0.131e^{-30.4z}$$

where $z = aL/\bar{U}r_o^2$. (e) Derive an expression for the local Nusselt number based on the difference between T_w and the "bulk mean" temperature defined under part (d) and show that Nu $\rightarrow 5.78$ as $x \rightarrow \infty$

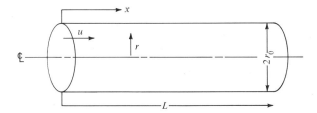

6.27 It is proposed to heat dry sand by passing it steadily downward through a vertical pipe that is heated by a vapor condensing on the outside. The sand is assumed to flow through the pipe with a uniform velocity profile. At the exit of the pipe the sand flows into a mixer and its temperature becomes uniform. The inside wall of the pipe is maintained at 104.4°C, and the thermal resistance between the pipe wall and the sand is assumed to be negligible. The sand, initially at a temperature of 48.9°C, is fed to the pipe at a rate of 0.03 m³/h. The pipe is 5.49 m long and has an inside diameter of 3.05 cm. Estimate the temperature of the mixed sand leaving the heater, using the following properties of sand: density = 1601 kg/m³; thermal conductivity = 0.35 W/m K; specific heat = 1000 J/kg. *Hint:* See Problem 6.26 for an approach.

6.28 If the total resistance between the steam and the air (including the pipe wall and scale on the steam side) in Problem 6.10 is 0.05 m² K/W, calculate the temperature difference between the outer surface of the inner pipe and the air. Show the thermal circuit.

6.29 A plastic tube of 7.6-cm ID and 1.27 cm thick having a thermal conductivity of 1.7 W/m² K, a density of 2400 kg/m³, and a specific heat of 1675 J/kg K is cooled from an initial temperature of 77°C by passing air at 21°C inside and outside the tube parallel to its axis. The velocities of the two air streams are such that the coefficients of heat transfer are the same on the interior and exterior surfaces. Measurements show that at the end of 26 min, the temperature difference between the tube surfaces and the air is 10 percent of the initial temperature difference. It is proposed to cool a tube of a similar material having an inside diameter of 15 cm and a wall thickness of 2.5 cm from the same initial temperature, also using air at 21°C and feeding to the inside of the tube the same number of kilograms of air per hour that was used in the first experiment. The air-flow rate over the exterior surfaces will be adjusted to give the same heat transfer coefficient on the outside as on the inside of the tube. It may be assumed that the air-flow rate is so high that the temperature rise along the axis of the tube may be neglected. Using the experience gained initially with the 4.5-cm tube, estimate how long it will take to cool the surface of the larger tube to 27°C under the conditions described. Indicate all assumptions and approximations in your solution.   *Ans.* 164.5 min.

6.30 In a pipe within a pipe heat exchanger, water is flowing in the annulus and an aniline-alcohol solution having the properties listed in Problem 6.2 is flowing in the central pipe. The inner pipe is 0.527 in. ID, 0.625 in. OD, and the ID of the outer pipe is 0.750 in. For a water bulk temperature of 80°F and an aniline bulk temperature of 175°F, determine the overall heat transfer coefficient based on the outer diameter of the central pipe and the frictional pressure drop per unit length of the water and the aniline for the following velocities: (a) water rate 1 gpm, oil rate 1 gpm; (b) water rate 10 gpm, oil rate 1 gpm; (c) water rate 1 gpm, oil rate 10 gpm; and (d) water rate 10 gpm, oil rate 10 gpm ($L/D = 400$).

6.31 Evaluate the rate of heat loss per foot from superheated steam flowing at 315°C and 1.724×10^6 N/m² pressure through a 10-cm-ID pipe at a velocity of 30 m/s. The pipe is lagged with a 5-cm-thick layer of asbestos. Heat is transferred to the surroundings by free convection and radiation. Draw the thermal circuit and state all assumptions.

6.32 Assume that the inner cylinder in Problem 6.6 is a heat source of an aluminum-clad rod of uranium, 5 cm OD and 2 m long. Estimate the heat flux that will raise the temperature of the bismuth 40°C and the maximum center and surface temperatures necessary to transfer heat at this rate.

6.33 The following thermal-resistance data were obtained on a 50,000-ft² condenser constructed with 1-in.-OD brass tubes, $23\frac{3}{4}$ ft long, 0.049 in. wall thickness, at various water velocities inside the tubes [*Trans. ASME*, vol. 58, p. 672, 1936].

$\dfrac{1}{U_o} \times 10^3$ (h ft² °F/Btu)	Water velocity (fps)	$\dfrac{1}{U_o} \times 10^3$ (h ft² °F/Btu)	Water velocity (fps)
2.060	6.91	3.076	2.95
2.113	6.35	2.743	4.12
2.212	5.68	2.498	6.76
2.374	4.90	3.356	2.86
3.001	2.93	2.209	6.27
2.081	7.01		

Assuming that the unit-surface conductance on the steam side is 2000 Btu/h ft² °F, determine the scale resistance. *Hint:* Plot $(1/U_o)$ against $(1/\bar{U}^{0.8})$. (This method is called the *Wilson plot*.)

6.34 A double-pipe heat exchanger is used to condense steam at 6894 N/m². Water at an average bulk temperature of 10°C flows at 3.05 m/s through the inner pipe (copper, 2.54 cm ID, 3.05 cm OD). Steam at its saturation temperature flows in the annulus formed between the outer surface of the inner pipe and an outer pipe of 5.08 cm ID. The average unit-surface conductance of the condensing steam is 5680 W/m² K, and the thermal resistance of a surface scale on the outer surface of the copper pipe is 0.000176 m² K/W. (a) Determine the overall heat transfer coefficient between the steam and the water based on the outer area of the copper pipe. Also sketch the thermal circuit and (b) evaluate the temperature at the inner surface of the pipe. (c) Estimate the length required to condense 0.45 kg of steam.
Ans. $U_o = 1986$ W/m² K; 18.3°C.

6.35 A 2.54-cm-OD, 1.9-cm-ID steel pipe carries dry air at a velocity of 7.6 m/s and a temperature of -7°C. Ambient air is at 21°C and has a dew point of 10°C. How much insulation with a conductivity of 0.0018 W/m K is needed to prevent condensation on the exterior of the insulation if $\bar{h} = 0.024$ W/m K on the outside.

6.36 A light oil with a 18°C inlet temperature flows at a rate of 1453.5 kg/min through a 5.1-cm-ID pipe that is enclosed by a jacket containing condensing steam at 149°C. If the pipe is 9 m long, determine the outlet temperature of the oil.

6.37 Mercury at an inlet bulk temperature of 90°C flows through a 1.2-cm-ID tube at a flow rate of 4535 kg/h. This tube is a part of a nuclear reactor in which heat can be generated uniformly at any desired rate by adjusting the neutron flux level. Determine the length of tube required to raise the bulk temperature of the mercury to 230°C without generating any mercury vapor, and determine the corresponding heat flux.

6.38 Power generation in a nuclear reactor is limited principally by the ability of the coolant to absorb the heat generated. Compare the relative merits of water, liquid sodium, and carbon dioxide as coolants of a nuclear reactor, assuming that either a gas or a steam turbine may be used to generate electrical energy.

REFERENCES

1. R. H. Notter and C. A. Sleicher, "The Eddy Diffusivity in the Turbulent Boundary Layer near a Wall," *Eng. Sci.*, vol. 26, pp. 161–171, 1971.
2. H. L. Langhaar, "Steady Flow in the Transition Length of a Straight Tube," *J. Appl. Mech.*, vol. 9, pp. 55–58, 1942.
3. W. M. Kays and M. E. Crawford, *Convective Heat and Mass Transfer*, 2d ed., McGraw-Hill, New York, 1980.
4. O. E. Dwyer, "Liquid-Metal Heat Transfer," chapter 5 in Sodium and NaK Supplement to *Liquid Metals Handbook*, Atomic Energy Commission, Washington, D.C., 1970.
5. J. R. Sellars, M. Tribus, and J. S. Klein, "Heat Transfer to Laminar Flow in a Round Tube or Flat Conduit—the Graetz Problem Extended," *Trans. ASME*, vol. 78, pp. 441–448, 1956.
6. C. A. Schleicher and M. Tribus, "Heat Transfer in a Pipe with Turbulent Flow and Arbitrary Wall-Temperature Distribution," *Trans. ASME*, vol. 79, pp. 789–797, 1957.
7. E. R. A. Eckert, "Engineering Relations for Heat Transfer and Friction in High Velocity Laminar and Turbulent Boundary Layer Flow over Surfaces with Constant Pressure and Temperature," *Trans. ASME*, vol. 78, pp. 1273–1284, 1956.
8. W. D. Hayes and R. F. Probstein, *Hypersonic Flow Theory*, Academic Press, New York, 1959.
9. F. Kreith, *Principles of Heat Transfer*, 2d ed., chap. 12, International Textbook Co., Scranton, Pa., 1965.
10. W. M. Kays and K. R. Perkins, "Forced Convection, Internal Flow in Ducts," in *Handbook of Heat Transfer Applications*, W. R. Rohsenow, J. P. Hartnett, and E. N. Garric, eds., vol. 1, chap. 7, McGraw-Hill, New York, 1985.
11. R. G. Eckert and A. J. Diaguila, "Convective Heat Transfer for Mixed Free and Forced Flow through Tubes," *Trans. ASME*, vol. 76, pp. 497–504, 1954.
12. B. Metais and E. R. G. Eckert, "Forced, Free, and Mixed Convection Regimes," *Trans. ASME, Ser. C, J. Heat Transfer*, vol. 86, pp. 295–296, 1964.
13. W. M. Kays, "Numerical Solution for Laminar Flow Heat Transfer in Circular Tubes," *Trans. ASME*, vol. 77, pp. 1265–1274, 1955.
14. R. K. Shah and A. L. London, *Laminar Flow Forced Convection in Ducts*, Academic Press, New York, 1978.
15. E. N. Sieder and C. E. Tate, "Heat Transfer and Pressure Drop of Liquids in Tubes," *Ind. Eng. Chem.*, vol. 28, p. 1429, 1936.

16. W. M. Kays and A. L. London, *Compact Heat Exchangers*, 3rd ed., McGraw-Hill, New York, 1984.

17. H. Hausen, *Heat Transfer in Counter Flow, Parallel Flow and Cross Flow*, McGraw-Hill, New York, 1983.

18. S. Whitaker, "Forced Convection Heat Transfer Correlations for Flow in Pipes, Past Flat Plates, Single Cylinders, and for Flow in Packed Beds and Tube Bundles," *AIChE J.*, vol. 18, pp. 361–371, 1972.

19. T. W. Swearingen and D. M. McEligot, "Internal Laminar Heat Transfer with Gas-Property Variation," *Trans. ASME, Ser. C, J. Heat Transfer*, vol. 93, pp. 432–440, 1971.

20. "Engineering Sciences Data," Heat Transfer Subsciences, Technical Editing and Production Ltd., London, 1970.

21. W. M. McAdams, *Heat Transmission*, 3d ed., McGraw-Hill, New York, 1954.

22. C. A. Depew and S. E. August, "Heat Transfer due to Combined Free and Forced Convection in a Horizontal and Isothermal Tube," *Trans. ASME, Ser. C, J. Heat Transfer*, vol. 93, pp. 380–384, 1971.

23. B. Metais and E. R. G. Eckert, "Forced, Free, and Mixed Convection Regimes," *Trans. ASME, Ser. C, J. Heat Transfer*, vol. 86, pp. 295–296, 1964.

24. W. M. Rohsenow, J. P. Hartnett, and E. N. Ganic, eds., *Handbook of Heat Transfer*, 2nd ed., McGraw-Hill, New York, 1985.

25. L. A. M. Janssen and C. J. Hoogendoorn, "Laminar Convective Heat Transfer in Helically Coiled Tubes," *Int. J. Heat Mass Transfer*, vol. 21, pp. 1197–1206, 1978.

26. H. Ito, "Friction Factors for Turbulent Flow in Curved Pipes," *J. Basic Eng., Trans. ASME*, vol. 81, pp. 123–134, 1959.

27. O. Reynolds, "On the Extent and Action of the Heating Surface for Steam Boilers," *Proc. Manchester Lit. Philos. Soc.*, vol. 8, 1874.

28. W. F. Cope, "The Friction and Heat Transmission Coefficients of Rough Pipes," *Proc. Inst. Mech. Eng.*, vol. 145, p. 99, 1941.

29. W. Nunner, "Wärmeübergang and Druckabfall in Rauhen Rohren," *VDI Forschungsh.*, no. 455, 1956.

30. D. F. Dipprey and R. H. Sabersky, "Heat and Momentum Transfer in Smooth and Rough Tubes at Various Prandtl Numbers," *Int. J. Heat Mass Transfer*, vol. 5, pp. 329–353, 1963.

31. L. Prandtl, "Eine Beziehung zwischen Wärmeaustausch und Strömungswiederstand der Flüssigkeiten," *Phys. Z.*, vol. 11, p. 1072, 1910.

32. T. von Karman, "The Analogy between Fluid Friction and Heat Transfer," *Trans. ASME*, vol. 61, p. 705, 1939.

33. L. M. K. Boelter, R. C. Martinelli, and F. Jonassen, "Remarks on the Analogy between Heat and Momentum Transfer," *Trans. ASME*, vol. 63, pp. 447–455, 1941.

34. R. G. Deissler, "Investigation of Turbulent Flow and Heat Transfer in Smooth Tubes Including the Effect of Variable Properties," *Trans. ASME*, vol. 73, p. 101, 1951.

35. R. C. Martinelli, "Heat Transfer to Molten Metals," *Trans. ASME*, vol. 69, p. 947, 1947.

36. F. W. Dittus and L. M. K. Boelter, *Univ. Calif. Berkeley Publ. Eng.*, vol. 2, p. 433, 1930.

37. B. S. Petukhov, "Heat Transfer and Friction in Turbulent Pipe Flow with Variable Properties," *Adv. Heat Transfer*, vol. 6, Academic Press, New York, pp. 503–564, 1970.

38. C. A. Sleicher and M. W. Rouse, "A Convenient Correlation for Heat Transfer to Constant and Variable Property Fluids in Turbulent Pipe Flow," *Int. J. of Heat Mass Transfer*, vol. 18, pp. 677–683, 1975.

39. J. J. Lorentz, D. T. Yung, C. B. Parchal, and G. E. Layton, "An Assessment of Heat Transfer Correlations for Turbulent Water Flow through a Pipe at Prandtl Numbers of 6.0 and 11.6," ANL/OTEC-PS-11, Argonne Natl. Lab., Argonne, Ill., January 1982.

40. W. M. McAdams, *Heat Transmission*, 3d ed., McGraw-Hill, New York, 1954.

41. R. G. Deissler, "Turbulent Heat Transfer and Friction in the Entrance Regions of Smooth Passages," *Trans. ASME*, vol. 77, pp. 1221–1234, 1955.

42. J. P. Hartnett, "Experimental Determination of the Thermal Entrance Length for the Flow of Water and of Oil in Circular Pipes," *Trans. ASME*, vol. 77, pp. 1211–1234, 1955.

43. L. M. K. Boelter, D. Young, and H. W. Iverson, "An Investigation of Aircraft Heaters—XXVII. Distribution of Heat Transfer Rate in the Entrance Section of a Circular Tube," NACA TN 1451, 1948.

44. T. C. Carnavos, "Cooling Air in Turbulent Flow with Internally Finned Tubes," *Heat Transfer Eng.*, vol. 1, pp. 43–46, 1979.

45. M. A. Nemira, J. V. Vilemes, and V. M. Simonis, "Heat Transfer in Turbulent Flows of Gases through Annuli with Variable Physical Properties," *Heat Transfer—Sov. Res.*, vol. 12, pp. 104–112, 1980.

46. D. A. Campbell and H. C. Perkins, "Variable Property Turbulent Heat and Momentum Transfer for Air in a Vertical Round-Corner, Triangular Duct," *Int. J. Heat Mass Transfer*, vol. 11, pp. 1003–1013, 1968.

47. K. R. Perkins, K. W. Shade, and D. M. McEligot, "Heated Laminarizing Gas Flow in a Square Duct," *Int. J. Heat Mass Transfer*, vol. 16, pp. 897–916, 1973.

48. R. A. Seban and E. F. McLaughlin, "Heat Transfer in Tube Coils with Laminar and Turbulent Flow," *Int. J. Heat Mass Transfer*, vol. 7, pp. 387–395, 1963.

49. H. Jeschke, "Wärmeübergang und Druckverlust in Rohrschlangen," Suppl. Technische Mechanik, in *VDI Z.*, vol. 69, pp. 24–28, 1925.

50. B. Lubarsky and S. J. Kaufman, "Review of Experimental Investigations of Liquid-Metal Heat Transfer," NACA TN 3336, 1955.

51. E. Skupinski, J. Tortel, and L. Vautrey, "Determination des Coefficients de Convection d'un Alliage Sodium-Potassium dans un Tube Circulative," *Int. J. Heat Mass Transfer*, vol. 8, pp. 937–951, 1965.

52. S. Lee, "Liquid Metal Heat Transfer in Turbulent Pipe Flow with Uniform Wall Flux," *Int. J. Heat Mass Transfer*, vol. 26, pp. 349–356, 1983.

53. R. P. Stein, "Heat Transfer in Liquid Metals," in *Advances in Heat Transfer*, J. P. Hartnett and T. F. Irvine, eds., vol. 3, Academic Press, New York, 1966.

54. N. Z. Azer, "Thermal Entry Length for Turbulent Flow of Liquid Metals in Pipes with Constant Wall Heat Flux," *Trans. ASME, Ser. C, J. Heat Transfer*, vol. 90. pp. 483–485, 1968.

55. R. A. Seban and T. T. Shimazaki, "Heat Transfer to Fluid Flowing Turbulently in a Smooth Pipe with Walls at Constant Temperature," *Trans. ASME*, vol. 73, pp. 803–807, 1951.

56. L. F. Moody, "Friction Factor for Pipe Flow," *Trans. ASME*, vol. 66, 1944.

Chapter 7

Forced Convection over Exterior Surfaces

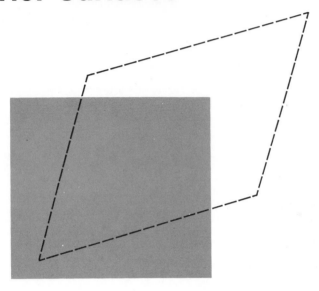

7.1 FLOW OVER BLUFF BODIES

In this chapter we shall consider heat transfer by forced convection between the exterior surface of bluff bodies, such as spheres, wires, tubes, and tube bundles, and fluids flowing perpendicularly to the axes of these bodies. The heat transfer phenomena for these systems, as for those in which a fluid flows inside a duct or along a flat plate, are closely related to the nature of the flow. The most important difference between the flow over a bluff body and the flow over a flat plate or a streamlined body lies in the behavior of the boundary layer. We recall that the boundary layer of a fluid flowing over the surface of a streamlined body will separate when the pressure rise along the surface becomes too large. On a streamlined body the separation, if it takes place at all, occurs near the rear. On a bluff body, on the other hand, the point of separation often lies not far from the leading edge. Beyond the point of separation of the boundary layer, the fluid in a region near the surface flows in a direction opposite to the main stream, as shown in Fig. 7.1. The local reversal in the flow results in disturbances which produce turbulent eddies. This is illustrated in Fig. 7.2, which is a photograph of the flow pattern of a stream flowing at a right angle to a cylinder. We can

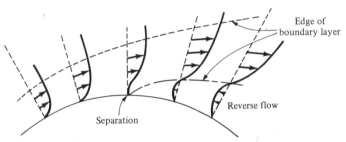

Figure 7.1 Schematic sketch of boundary layer on a circular cylinder near separation point.

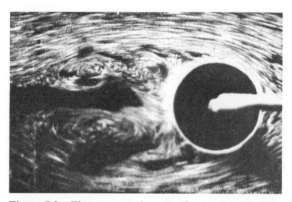

Figure 7.2 Flow pattern in cross-flow over a single horizontal cylinder. (Photograph by H. L. Rubach, *Mitt. Forschungsarb.*, vol. 185, 1916.)

see that eddies from both sides of the cylinder extend downstream, so that a turbulent wake is formed at the rear of the cylinder.

Associated with the separation of the flow are large pressure losses, since the kinetic energy of the eddies that pass off into the wake cannot be regained. In flow over a streamlined body, the pressure loss is caused mainly by the skin friction drag. For a bluff body, on the other hand, the skin friction drag is small compared to the form drag in the Reynolds number range of commercial interest. The form or pressure drag arises from the separation of the flow, which prevents closing of the streamlines and thereby induces a low-pressure region in the rear of the body. When the pressure over the rear of the body is lower than over the front, there exists a pressure difference which produces a drag force over and above that of the skin friction. The magnitude of the form drag decreases as the separation moves farther toward the rear.

The geometric shapes that are most important for engineering work are the long circular cylinder and the sphere. The heat transfer phenomena for these two shapes in cross-flow have been studied by a number of investigators, and representative data are summarized in Section 7.2. In addition to the average surface conductance over a cylinder, the variation of conductance around the circumference will be considered. A knowledge of the peripheral variation of the heat transfer associated with flow over a cylinder is important for many practical problems such as heat transfer calculations for airplane wings, whose leading-edge contours are approximately cylindrical. The interrelation between heat transfer and flow phenomena will also be stressed because it can be applied to the measurement of the velocity and its fluctuations in a turbulent stream by means of a hot-wire anemometer.

Section 7.3 treats heat transfer in packed beds, that is, systems in which heat transfer to or from spherical or other particles is important.

Section 7.4 deals with heat transfer to or from bundles of tubes in cross-flow, a configuration that is widely used in boilers, air-preheaters, and conventional shell-and-tube heat exchangers. Representative experimental data are presented and applied to typical engineering problems in Section 7.5.

7.2 CYLINDERS, SPHERES, AND OTHER BLUFF SHAPES

Photographs of typical flow patterns for flow over a single cylinder and a sphere are shown in Figs. 7.2 and 7.3, respectively. The most forward points of these bodies are called stagnation points. Fluid particles striking there are brought to rest, and the pressure at the stagnation point p_0 rises approximately one velocity head, that is, $(\rho U_\infty^2/2g)$, above the pressure in the oncoming free stream p_∞. The flow divides at the stagnation point of the cylinder, and a boundary layer builds up along the surface. The fluid accelerates when it flows past the surface of the cylinder, as can be seen by the crowding of the streamlines shown in Fig. 7.4. This flow pattern for a nonviscous fluid in irrotational flow, a highly idealized case, is called *potential flow*. The velocity reaches a maximum at both sides of the cylinder, then falls again to zero at the stagnation point in the rear. The pressure distribution around the cylinder corresponding to this idealized flow pattern is shown by the solid line in Fig. 7.5. Since the pressure distribution is symmetric about the vertical center plane of the cylinder, it is clear that there will be no pressure drag in irrotational flow. However, unless the Reynolds number is very low, a real fluid will not adhere to the entire surface of the cylinder but, as mentioned previously, the boundary layer in which the flow is not irrotational will separate from the sides of the cylinder as a result of the adverse pressure gradient. The separation of the boundary layer and the resultant wake in the rear of the cylinder give rise to pressure distributions which are shown for different Reynolds numbers by the dashed lines in Fig. 7.5. It can be seen that there is fair agreement between the ideal and the actual pressure distribution in the neighborhood of the forward stagnation point. In the rear of the cylinder, however, the actual and ideal distributions differ considerably. The characteristics of the flow pattern and of the boundary layer depend on the Reynolds number, $\rho U_\infty D/\mu$, which for flow over a cylinder or a sphere is based on the velocity of the oncoming free stream U_∞ and the outside diameter of the body D. Properties are evaluated at free-stream conditions. The flow pattern around the cylinder undergoes a series of changes as the Reynolds number is increased, and since the heat transfer depends largely on the flow, we shall consider first the effect of Reynolds number of the flow and then interpret the heat transfer data in the light of this information.

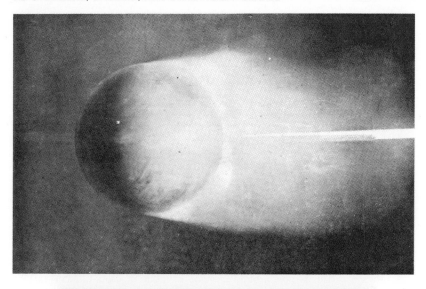

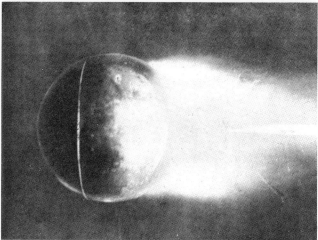

Figure 7.3 Photographs of air flowing over a sphere. In the lower picture a "tripping" wire induced early transition and delayed separation. (Courtesy of L. Prandtl and the *Journal of the Royal Aeronautical Society*.)

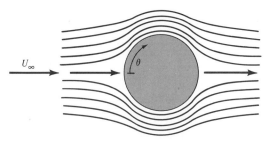

Figure 7.4 Streamlines for potential flow over a circular cylinder.

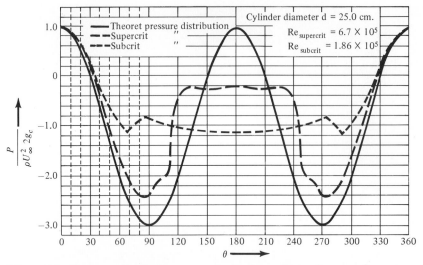

Figure 7.5 Pressure distribution around circular cylinder in cross-flow at various Reynolds numbers; p, local pressure; $\rho V_\infty^2/2$, free-stream impact pressure; θ, angle measured from stagnation point. (By permission from L. Flachsbart, *Handbuch der Experimental Physik*, vol. 4, part 2.)

The sketches in Fig. 7.6 illustrate flow patterns typical of the characteristic ranges of Reynolds numbers. The letter symbols of the sketches in Fig. 7.6 correspond to the flow regimes indicated in the curve of Fig. 7.7, where the total drag coefficients of a cylinder and a sphere, C_D, are plotted as a function of the Reynolds number. The force term in the total drag coefficient is the sum of the pressure and frictional forces; it is defined by the equation

$$C_D = \frac{\text{drag force/unit length}}{(\rho\, U_\infty^2/2g_c)D}$$

where ρ = free-stream density

U_∞ = free-stream velocity

D = outside diameter

The following discussion strictly applies only to long cylinders, but it also gives a qualitative picture of the flow past a sphere. The letters (*a*) to (*e*) refer to Figs. 7.6 and 7.7.

(*a*) At Reynolds numbers of the order of unity or less, the flow adheres to the surface and the streamlines follow those predicted from potential-flow theory. The inertia forces are negligibly small and the drag is caused only by viscous forces, since there is no flow separation. Heat is transferred by conduction alone.

(*b*) At Reynolds numbers of the order of 10, the inertia forces become appreciable and two weak eddies stand in the rear of the cylinder. The pressure drag accounts now for about one-half of the total drag.

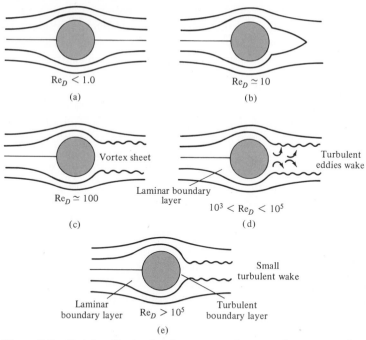

Figure 7.6 Sketches illustrating flow pattern for cross-flow over a circular cylinder at various Reynolds numbers.

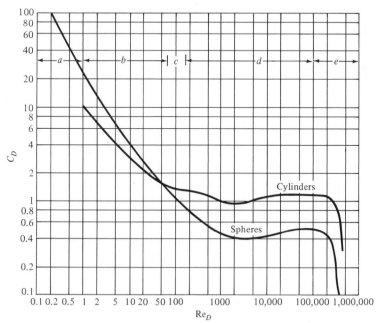

Figure 7.7 Drag coefficient versus Reynolds number for long circular cylinders and spheres in cross-flow.

(c) At a Reynolds number of the order of 100, vortices separate alternately from both sides of the cylinder and stretch a considerable distance downstream. These vortices are referred to as *von Karman vortex-streets* in honor of the scientist Theodore von Karman, who studied the shedding of vortices from bluff objects. The pressure drag now predominates.

(d) In the Reynolds number range between 10^3 and 10^5, the skin friction drag becomes negligible compared to the pressure drag caused by turbulent eddies in the wake. The drag coefficient remains approximately constant because the boundary layer remains laminar from the leading edge to the point of separation, which lies throughout this Reynolds number range at an angular position θ between 80 and 85 degrees measured from the direction of the flow.

(e) At Reynolds numbers larger than about 10^5 (the exact value depends on the turbulence level of the stream) the kinetic energy of the fluid in the laminar boundary layer over the forward part of the cylinder is sufficient to overcome the unfavorable pressure gradient without separating. The flow in the boundary layer becomes turbulent while it is still attached, and the separation point moves toward the rear. The closing of the streamlines reduces the size of the wake, and the pressure drag is therefore also substantially reduced. Experiments by Fage and Falkner (1, 2) indicate that, once the boundary layer has become turbulent, it will not separate before it has reached an angular position corresponding to a θ of about 130 degrees.

Analyses of the boundary-layer growth and the variation of the local unit-surface conductances with angular position around circular cylinders and spheres have been only partially successful. Squire (3) has solved the equations of motion and energy for a cylinder at constant temperature in cross-flow over that portion of the surface to which a laminar boundary layer adheres. He showed that, at the stagnation point and in its immediate neighborhood, the convective unit-surface conductance can be calculated from the equation

$$\text{Nu}_D = \frac{h_c D}{k} = C \sqrt{\frac{\rho U_\infty D}{\mu}} \tag{7.1}$$

where C is a constant whose numerical value at various Prandtl numbers is tabulated below:

Pr	0.7	0.8	1.0	5.0	10.0
C	1.0	1.05	1.14	2.1	1.7

Over the forward portion of the cylinder ($0 < \theta < 80$ deg), the empirical equation for $h_{c\theta}$, the local value of the unit-surface conductance at θ,

$$\frac{h_{c\theta} D}{k} = 1.14 \left(\frac{\rho U_\infty D}{\mu}\right)^{0.5} \text{Pr}^{0.4} \left[1 - \left(\frac{\theta}{90}\right)^3\right] \tag{7.2}$$

has been found to agree satisfactorily (4) with experimental data.

Giedt (5) has measured the local pressures and the local unit-convective conductances over the entire circumference of a long, 10.2-cm-OD cylinder in an air stream over a Reynolds number range from 90,000 to 220,000. Giedt's results are shown in Fig. 7.8, and similar data for lower Reynolds numbers are shown in Fig. 7.9. If the data shown in Figs. 7.8 and 7.9 are compared at corresponding Reynolds numbers with the flow patterns and the boundary-layer characteristics described earlier, some important observations can be made.

At Reynolds numbers below 100,000, separation of the laminar boundary layer occurs at an angular position of about 80 degrees. The heat transfer and the flow characteristics over the forward portion of the cylinder resemble those for laminar flow over a flat plate, which were discussed earlier. The local conductance is largest at the stagnation point and decreases with distance along the surface as the boundary-layer thickness increases. The conductance reaches a minimum on the sides of the cylinder near the separation point. Beyond the separation point the local conductance increases because considerable turbulence exists over the rear portion of the cylinder, where the eddies of the wake

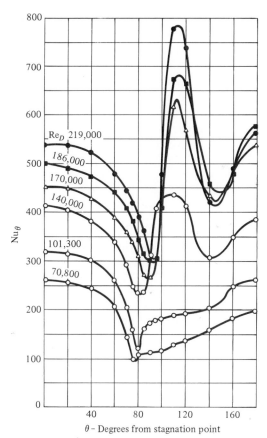

Figure 7.8 Circumferential variation of the unit-surface conductance at high Reynolds numbers for a circular cylinder in cross-flow. [Extracted from W. H. Giedt (5), with permission of the publishers, the American Society of Mechanical Engineers.]

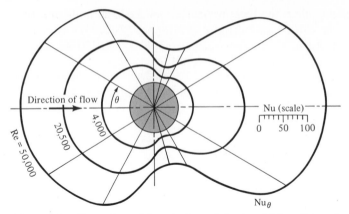

Figure 7.9 Circumferential variation of the local Nusselt number $Nu_\theta = h_{c\theta}D_o/k_f$ at low Reynolds numbers for a circular cylinder in cross-flow. (According to W. Lorisch, from M. ten Bosch, *Die Wärmeübertragung*, 3d ed., Springer Verlag, Berlin, 1936.)

sweep the surface. However, the conductance over the rear is no larger than over the front, because the eddies recirculate part of the fluid and, despite their high turbulence, are not as effective in mixing the fluid in the vicinity of the surface with the fluid in the main stream as a turbulent boundary layer.

At Reynolds numbers large enough to permit transition from laminar to turbulent flow in the boundary layer without separation of the laminar boundary layer, the unit-surface conductance has two minima around the cylinder. The first minimum occurs at the point of transition. As the transition from laminar to turbulent flow progresses, the unit conductance increases and reaches a maximum approximately at the point where the boundary layer becomes fully turbulent. Then the unit-surface conductance begins to decrease again and reaches a second minimum at about 130°, the point at which the turbulent boundary layer separates from the cylinder. Over the rear of the cylinder the unit conductance increases to another maximum at the rear stagnation point.

EXAMPLE 7.1

To design a heating system for the purpose of preventing ice formation on an aircraft wing it is necessary to know the unit-surface conductance over the outer surface of the leading edge. The leading-edge contour may be approximated by a half-cylinder of 30 cm diameter. The ambient air is at $-34°C$ and the surface temperature is to be no less than 0°C. The plane is designed to fly at 7500 m altitude at a speed of 150 m/s. Calculate the distribution of the convective unit-surface conductance over the forward portion of the wing.

Solution. At an altitude of 7500 m the standard atmospheric air pressure is 38.9 kPa and the density of the air is 0.566 kg/m³ (see Table 37 in Appendix 2).

The unit-surface conductance at the stagnation point ($\theta = 0$) is, according to Eq. (7.2),

$$h_c(\theta = 0) = 1.14 \left(\frac{\rho U_\infty D}{\mu} \right)^{0.5} Pr^{0.4} \frac{k}{D}$$

$$= (1.14) \left(\frac{0.566 \times 150 \times 0.30}{1.74 \times 10^{-5}} \right)^{0.5} (0.72)^{0.4} \left(\frac{0.024}{0.30} \right)$$

$$= 96.7 \text{ W/m}^2 \text{ °C}$$

The variation of h_c with θ is obtained by multiplying the value of the unit-surface conductance at the stagnation point by $1 - (\theta/90)^3$. The results are tabulated below.

θ (deg)	0	15	30	45	60	75
$h_{c\theta}$ (W/m² °C)	96.7	96.3	93.1	84.6	68.0	40.7

It is apparent from the foregoing discussion that the variation of the unit-surface conductance around a cylinder or a sphere is a very complex problem. For many practical applications it is fortunately not necessary to know the local value $h_{c\theta}$, but is sufficient to evaluate the average value of the conductance around the body. A number of observers have measured mean conductances for flow over single cylinders and spheres. Hilpert (6) accurately measured the average conductances for air flowing over cylinders of diameters ranging from 19 μm to 15 cm. His results are shown in Fig. 7.10, where the average Nusselt $\bar{h}_c D/k$ is plotted as a function of the Reynolds number $U_\infty D/v$.

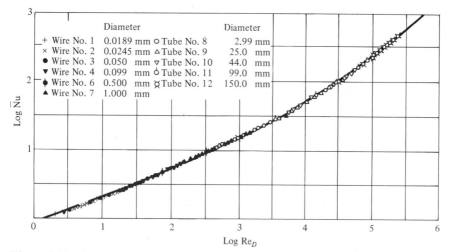

Figure 7.10 Average heat transfer coefficient versus Reynolds number for a circular cylinder in cross-flow with air. [After R. Hilpert (6, p. 220).]

TABLE 7.1 COEFFICIENTS FOR EQ. (7.3).

Re_D	C	m
1–40	0.75	0.4
40–1 × 10³	0.51	0.5
1 × 10⁴–2 × 10⁵	0.26	0.6
2 × 10⁵–1 × 10⁶	0.076	0.7

A correlation for a cylinder in cross-flow of liquids and gases has been proposed by Zukauskas (7)

$$\overline{Nu}_D = \frac{\overline{h}_c D}{k} = C \left(\frac{U_\infty D}{\nu} \right)^m Pr^n \left(\frac{Pr}{Pr_s} \right)^{0.25} \tag{7.3}$$

where all fluid properties are evaluated at the bulk fluid temperature except for Pr_s, which is evaluated at the tube wall temperature. The constants in Eq. (7.3) are given in Table 7.1. For $Pr < 10$, $n = 0.36$, and for $Pr > 10$, $n = 0.37$.

For cylinders with noncircular cross sections in gases Jakob (8) compiled data from two sources and presented coefficients of the correlation equation

$$Nu_D = B\, Re_D{}^n \tag{7.4}$$

in Table 7.2.

TABLE 7.2 CONSTANTS OF EQ. (7.4) FOR FORCED CONVECTION PERPENDICULAR TO NONCIRCULAR TUBES

Flow direction and profile	Re_D From	Re_D To	n	B
→ ◇ D	5,000	100,000	0.588	0.222
→ ⬭ D	2,500	15,000	0.612	0.224
→ ◇ D	2,500	7,500	0.624	0.261
→ ⬡ D	5,000	100,000	0.638	0.138
→ ⬡ D	5,000	19,500	0.638	0.144
→ ▢ D	5,000	100,000	0.675	0.092
→ ▢ D	2,500	8,000	0.699	0.160
→ ❘ D	4,000	15,000	0.731	0.205
→ ⬡ D	19,500	100,000	0.782	0.035
→ ◖ D	3,000	15,000	0.804	0.085

For heat transfer from a cylinder in cross-flow of liquid metals, Ishiguro et al. (9) recommend the correlation equation

$$\overline{Nu}_D = 1.125(Re_D Pr)^{0.413} \tag{7.5}$$

in the range $1 \leq Re_D Pr \leq 100$. Equation (7.5) predicts a somewhat lower Nu_D than that of analytic studies for either constant temperature $[\overline{Nu}_D = 1.015(Re_D Pr)^{0.5}]$ or constant flux $[\overline{Nu}_D = 1.145(Re_D Pr)^{0.5}]$. As pointed out in (9), neither boundary condition was achieved for the experimental effort. The difference between Eq. (7.5) and correlation equations for the two analytic studies is apparently due to the assumption of inviscid flow in the analytic studies. Such an assumption cannot allow for a separated region at large values of $Re_D Pr$, which is where Eq. (7.5) deviates from the analytic results.

Quarmby and Al-Fakhri (10) found experimentally that the effect of tube aspect ratio (length-to-diameter ratio) is negligible for aspect ratio values greater than four. The forced air flow over the cylinder was essentially that of an infinite cylinder in cross-flow. They examined the effect of heated-length variations, and thus aspect ratio, by independently heating five longitudinal sections of the cylinder. Their data for infinite aspect ratios compared favorably with the data of Zukauskas (7) for cylinders in cross-flow. For aspect ratios less than four they recommend

$$\overline{Nu}_D = 0.123 \, Re_D^{0.651} + \left(\frac{D}{L}\right)^{0.85} Re_D^{0.792} \tag{7.6}$$

in the range $\qquad 7 \times 10^4 < Re_D < 1.1 \times 10^5$

Equation (7.6) agrees well with data of Zukauskas (7) in the limit $L/D \rightarrow \infty$ for this relatively small Reynolds number range.

Turbulence in the free stream approaching the cylinder can have a relatively strong influence on the average heat transfer. Yardi and Sukhatme (11) determined experimentally an increase in the average heat transfer coefficient of 16 percent as the free-stream turbulence intensity was increased from 1 to 8 percent in the Reynolds number range 6000 to 60,000. On the other hand, the length scale of the turbulence did not affect the average heat transfer coefficient. Their local heat transfer measurements showed that the effect of free-stream turbulence was largest at the front stagnation point and diminished to an insignificant effect at the rear stagnation point.

7.2.1 Hot-Wire Anemometer

The relationship between the velocity and the rate of heat transfer from a single cylinder in cross-flow is used to measure velocity and velocity fluctuations in turbulent flow and in combustion processes by means of a hot-wire anemometer. This instrument consists basically of a thin (3 to 30 μm diameter) electrically heated wire stretched across the ends of two prongs. When the wire is exposed to a cooler fluid stream, it loses heat by convection. The temperature of the wire, and consequently its electrical resistance, depends on the temperature and the velocity of the fluid and the heating current. To determine the fluid velocity, either the wire is maintained at a constant temperature by adjusting the current

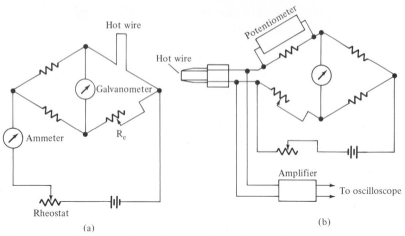

Figure 7.11   Schematic circuits for hot-wire probes and associated equipment.

and the fluid speed determined from the measured value of the current, or the wire is heated by a constant current and the speed deduced from a measurement of the electrical resistance or the voltage drop in the wire. In the first method the hot wire forms one arm in the circuit of a Wheatstone bridge, as shown in Fig. 7.11a. The resistance of the rheostat arm R_e is adjusted to balance the bridge when the temperature, and consequently the resistance, of the wire has reached some desired value. When the fluid velocity increases, the current required to maintain the temperature and resistance of the wire constant also increases. This change in the current is accomplished by adjusting the rheostat in series with the voltage supply. When the galvanometer indicates that the bridge is in balance again, the change in current, read on the ammeter, indicates the change in speed. In the other method the fluctuations in voltage drop caused by variations in the fluid velocity are impressed across the input of an amplifier, the output of which is connected to an oscilloscope. Figure 7.11b illustrates schematically an arrangement for the voltage measurement. Additional information on the hot-wire method is given in (12) and (13).

EXAMPLE 7.2 ————————————————————————————————

A 25-μm-diameter polished-platinum wire 6 mm long is to be used for a hot-wire anemometer to measure the velocity of 20°C air in the range between 2 and 10 m/s. The wire is to be placed into the circuit of the Wheatstone bridge shown in Fig. 7.11a. Its temperature is to be maintained at 230°C by adjusting the current by means of the rheostat. To design the electric circuit it is necessary to know the required current as a function of air velocity. The electrical resistivity of platinum at 230°C is 17.1 microohms cm.

Solution.   Since the wire is very thin, conduction along it can be neglected; also, the temperature gradient in the wire at any cross section may be disregarded. At the mean film temperature of 125°C, the air has a thermal conductivity of 0.032 W/m °C and a kinematic viscosity of 2.66×10^{-5} m²/s. At a velocity of 2 m/s the Reynolds number is

$$\mathrm{Re}_D = \frac{(2 \text{ m/s})(25 \times 10^{-6} \text{ m})}{2.66 \times 10^{-5} \text{ m}^2/\text{s}} = 1.88$$

The Reynolds number range of interest is therefore 1 to 40, so the correlation equation from Eq. (7.3) and Table 7.1 is

$$\frac{\bar{h}_c D}{k} = 0.75 \, \mathrm{Re}_D{}^{0.4} \mathrm{Pr}^{0.36} \left(\frac{\mathrm{Pr}}{\mathrm{Pr}_s}\right)^{0.25}$$

Neglecting the small variation in Prandtl number from 20° to 230°C, the average convective unit-surface conductance as a function of velocity is

$$\bar{h}_c = (0.75)(1.88)^{0.4} \left(\frac{U_\infty}{2}\right)^{0.4} (0.71)^{0.36} \left(\frac{0.032}{25 \times 10^{-6}}\right)$$

$$= 828 U_\infty{}^{0.4} \text{ W/m}^2 \text{ °C}$$

At this point it is necessary to estimate the unit-surface conductance for radiant heat flow. According to Eq. (1.21) we have

$$\bar{h}_r = \frac{q_r}{A(T_s - T_\infty)} = \frac{\sigma \epsilon (T_s{}^4 - T_\infty{}^4)}{T_s - T_\infty} = \sigma \epsilon (T_s{}^2 + T_\infty{}^2)(T_s + T_\infty)$$

or, since

$$(T_s{}^2 + T_\infty{}^2)(T_s + T_\infty) \approx 4 \left(\frac{T_s + T_\infty}{2}\right)^3$$

we have approximately

$$\bar{h}_r = \sigma \epsilon 4 \left(\frac{T_s + T_\infty}{2}\right)^3$$

The emissivity of polished platinum from Appendix 2, Table 7 is about 0.05, so $\bar{h}_r$ is about 0.05 W/m² °C. This shows that the amount of heat transferred by radiation is negligible compared to the heat transferred by forced convection. The rate at which heat is transferred from the wire is therefore

$$q_c = \bar{h}_c A(T_s - T_\infty) = (828 U_\infty{}^{0.4})(\pi)(25 \times 10^{-6})(6 \times 10^{-3})(210)$$
$$= 0.0819 U_\infty{}^{0.4} \text{ W}$$

which must equal the rate of dissipation of electrical energy to maintain the wire at 230°C. The electrical resistance of the wire is

$$R_e = (17.1 \times 10^{-6} \text{ ohm cm}) \frac{0.6 \text{ cm}}{\pi (25 \times 10^{-4} \text{ cm})^2/4}$$

$$= 2.09 \text{ ohm}$$

A heat balance with the current i in amperes gives

$$i^2 R_e = 0.0819 V_\infty^{0.4}$$

Solving for the current as a function of velocity, we get

$$i = \left(\frac{0.0819}{2.09}\right)^{1/2} U_\infty^{0.2} = 0.20 U_\infty^{0.20} \text{ amp}$$

7.2.2 Spheres

A knowledge of heat transfer characteristics to or from spherical bodies is important for predicting the thermal performance of systems where clouds of particles are heated or cooled in a stream of fluid. An understanding of the heat transfer from isolated particles is generally needed before attempting to correlate data for packed beds, clouds of particles, or other situations where the particles may interact. When the particles have an irregular shape, the data for spheres will yield satisfactory results if the sphere diameter is replaced by an equivalent diameter, that is, if D is taken as the diameter of a spherical particle having the same surface area as the irregular particle.

The total drag coefficient of a sphere is shown as a function of the free-stream Reynolds number in Fig. 7.7*, and corresponding data for heat transfer between a sphere and air are shown in Fig. 7.12. In the Reynolds number range from about 25 to 100,000, the equation recommended by McAdams (14) for calculating the average unit-surface conductance for spheres heated or cooled by a gas is

$$\frac{\bar{h}_c D}{k} = 0.37 \left(\frac{\rho D U_\infty}{\mu}\right)^{0.6} = 0.37 \, \text{Re}_D^{0.6} \tag{7.7}$$

For Reynolds numbers between 1.0 and 25, the equation

$$\bar{h}_c = c_p U_\infty \rho \left(\frac{2.2}{\text{Re}_D} + \frac{0.48}{\text{Re}_D^{0.5}}\right) \tag{7.8}$$

may be used for heat transfer in a gas. For heat transfer in liquids as well as gases, the equation

$$\frac{\bar{h}_c D}{k} = 2 + (0.4 \, \text{Re}_D^{0.5} + 0.06 \, \text{Re}_D^{0.67}) \text{Pr}^{0.4} \left(\frac{\mu_\infty}{\mu_s}\right)^{0.25} \tag{7.9}$$

correlates available data in the ranges of Reynolds numbers between 3.5 and 7.6×10^4 and Prandtl numbers between 0.7 and 380 (15).

Achenbach (16) has measured the average heat transfer from a constant-surface-temperature sphere in air for Reynolds numbers beyond the critical

---

* When the sphere is dragged along by a stream, as, for example, a liquid droplet in a gas stream, the pertinent velocity for the Reynolds number is the velocity difference between the stream and the body.

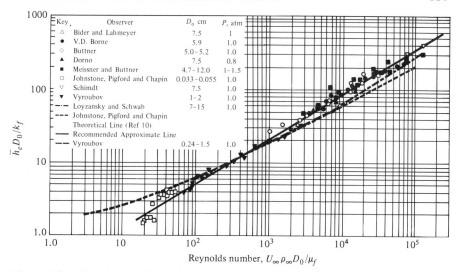

Figure 7.12 Correlation of experimental average heat transfer coefficients for flow over a sphere. [By permission from W. H. McAdams (14).]

value. For Reynolds numbers below the critical value $100 < Re_D < 2 \times 10^5$ he found

$$\overline{Nu}_D = 2 + \left(\frac{Re_D}{4} + 3 \times 10^{-4} Re_D^{1.6}\right)^{1/2} \tag{7.10}$$

which may be compared with the data from several sources presented in Fig. 7.12. In the limiting case when the Reynolds number is less than unity, Johnston et al. (17) have shown from theoretical considerations that the Nusselt number approaches a constant value of two for a Prandtl number of unity unless the spheres have diameters of the order of the mean free path of the molecules in the gas. Beyond the critical point, $4 \times 10^5 < Re_D < 5 \times 10^6$, Achenbach recommends

$$\overline{Nu}_D = 430 + 5 \times 10^{-3} Re_D + 0.25 \times 10^{-9} Re_D^2 - 3.1 \times 10^{-17} Re_D^3 \tag{7.11}$$

In the case of heat transfer from a sphere to a liquid metal, Witte (18) used a transient measurement technique to determine the correlation equation

$$\overline{Nu}_D = \frac{\bar{h}_c D}{k} = 2 + 0.386 \, (Re_D Pr)^{1/2} \tag{7.12}$$

in the range $3.6 \times 10^4 < Re_D < 2 \times 10^5$. The only liquid metal they tested was sodium. The data fell somewhat below those for previous results for air or water, but gave close agreement with previous analyses that assumed potential flow of liquid sodium around a sphere.

7.2.3 Bluff Objects

Sogin (19) determined experimentally the heat transfer coefficient in the separated wake regions behind a flat plate of width D placed perpendicular to the flow and a half-round cylinder of diameter D over Reynolds numbers between 1 and 4×10^5, and found that the following equations correlated the mean heat transfer results in air:

Normal flat plate:

$$\overline{\mathrm{Nu}}_D = \frac{\overline{h}_c D}{k} = 0.20\ \mathrm{Re}_D{}^{2/3} \qquad\qquad (7.13)$$

Half-round cylinder with flat rear surface:

$$\mathrm{Nu}_D = \frac{\overline{h}_c D}{k} = 0.16\ \mathrm{Re}_D{}^{2/3} \qquad\qquad (7.14)$$

These results are in agreement with an analysis by Mitchell (20).

Tien and Sparrow (21) used the naphthalene sublimation technique to measure mass transfer coefficients from square plates to air at various angles to a free stream. They studied the range $2 \times 10^4 < \mathrm{Re}_L < 10^5$ for angles of attack and pitch of 25°, 45°, 65°, and 90° and yaw angles of 0°, 22.5°, and 45°. They found the rather unexpected result that all the data could be correlated accurately (± 5 percent) with a single equation

$$(\overline{h}_c/c_p \rho U_\infty)\mathrm{Pr}^{2/3} = 0.930\ \mathrm{Re}_L{}^{-1/2} \qquad\qquad (7.15)$$

where the length scale L is the length of the plate edge.

The insensitivity to flow approach angle was attributed to a relocation of the stagnation point as the angle was changed and the flow adjusted to minimize the drag force on the plate. Because the plate was square, this movement of the stagnation point did not appear to alter the mean flow path length. For shapes other than squares, this insensitivity to flow approach angle may not hold.

EXAMPLE 7.3 ————————————————————————————————

Determine the rate of convective heat loss from a solar collector panel array attached to a roof and exposed to an air velocity of 0.5 m/s. The array is 2.5 m square, the surface of the collectors is 70°C, and the ambient air temperature is 20°C.

Solution. At the film temperature of 45°C the kinematic viscosity of air is 1.81×10^{-5} m²/s, the density is 1.08 kg/m³, and the specific heat is 1015 W s/kg °C. The Reynolds number is then

$$\mathrm{Re}_L = \frac{U_\infty L}{v} = \frac{(0.5\ \mathrm{m/s})(2.5\ \mathrm{m})}{(1.81 \times 10^{-5}\ \mathrm{m^2/s})} = 69{,}061$$

Equation (7.15) gives

$$(\bar{h}_c/c_p\rho U_\infty)Pr^{2/3} = 0.930(69061)^{-1/2} = 0.0035$$

The average heat transfer coefficient is

$$\bar{h}_c = (0.0035)(0.71)^{-2/3}(1.08)(1015)(0.5) = 2.41 \text{ W/m}^2 \text{ °C}$$

and the rate of heat loss from the array is

$$q = (2.41)(70 - 20)(2.5)(2.5) = 753 \text{ W}$$

7.3* PACKED BEDS

Many important processes require contact between a gas or a liquid stream and solid particles. These processes include catalytic reactors, grain dryers, beds for storage of solar thermal energy, gas chromatography, regenerators, and desiccant beds. By contacting the fluid with the surface of the particle one may transfer heat and/or mass between the fluid and the particle. The device may consist of a pipe, vessel, or some other containment for the particle bed through which the gas or liquid flows. Figure 7.13a depicts a packed bed which could be used for sensible storage of solar energy. The bed would be heated during the charging cycle by pumping hot air or another heated working fluid through the bed. The particles, which comprise the packed bed, heat up to the air temperature thereby storing heat sensibly. During the discharge cycle, cooler air

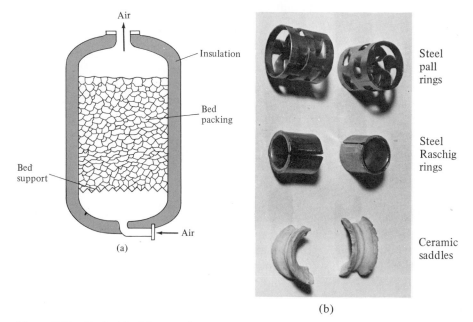

Figure 7.13 Packed-bed heat exchanger.

would be pumped through the bed, cooling the particles and removing the stored heat. The particles, sometimes called the bed packing, may take one of several forms including rocks, catalyst pellets, or commercially manufactured shapes as shown in Fig. 7.13b, depending on the intended use of the packed bed.

Depending on the use of the packed bed, it may be necessary to transfer heat or mass between the particle and the fluid, or it may be necessary to transfer heat through the wall of the containment vessel. For example, in the packed bed in Fig. 7.3a one needs to predict the rate of heat transfer between the air and the particles. On the other hand, a catalytic reactor may need to reject the heat of reaction (which occurs on the particle surface) through the walls of the reactor vessel. The presence of the catalyst particles modifies the wall heat transfer to the extent that correlations for flow through an empty tube are not applicable.

Correlations for heat or mass transfer in packed beds utilize a Reynolds number based on the superficial fluid velocity U_s, that is, the fluid velocity that would exist if the bed were empty. The length scale used in the Reynolds number and Nusselt or Sherwood number is generally the equivalent diameter of the packing D_p. Since spheres are only one possible type of packing, an equivalent particle diameter must be defined which is based in some way on the particle volume and surface area. Such a definition may vary from one correlation to another, so some care is needed before attempting to apply the correlation. Another important parameter in packed beds is the void fraction ϵ, which is the fraction of the bed volume that is empty. The void fraction sometimes appears explicitly in correlations and is sometimes used in the Reynolds number. In addition, the Prandtl number may appear explicitly in the correlation even though the original data may have been for gases only. In such a case, the correlation is probably not reliable for liquids.

Whitaker (15) correlated data for heat transfer from gases to different kinds of packing from several sources. The types of packing included cylinders with diameter equal to height, spheres, and several types of commercial packings such as Raschig rings, partition rings, and Berl saddles. The data are correlated within ± 25 percent by the equation

$$\frac{\bar{h}_c D_p}{k} = \frac{1 - \epsilon}{\epsilon} (0.5 \, \mathrm{Re}_{D_p}^{1/2} + 0.2 \, \mathrm{Re}_{D_p}^{2/3}) \mathrm{Pr}^{1/3} \tag{7.16}$$

in the range $20 < \mathrm{Re}_{D_p} < 10^4$, $0.34 < \epsilon < 0.78$.

The packing diameter D_p is defined as six times the volume of the particle divided by the particle surface area, which for a sphere reduces to the diameter. All fluid properties are to be evaluated at the bulk fluid temperature. If the bulk fluid temperature varies significantly through the heat exchanger one may use the average of the inlet and outlet values. Whitaker defined the Reynolds number as

$$\mathrm{Re}_{D_p} = \frac{D_p U_s}{v(1 - \epsilon)}$$

Equation (7.16) does not correlate data for cubes as well because a significant reduction in surface area can occur when the cubes stack against each other. Also, data for a regular arrangement (body-centered cubic) of spheres lie well above the correlation Eq. (7.16).

Upadhyay (22) used the mass transfer analogy to study heat and mass transfer in packed beds at very low Reynolds numbers. Upadhyay recommends the correlation

$$(\bar{h}_c/c_p\rho U_s)\mathrm{Pr}^{2/3} = \frac{1}{\epsilon} 1.075 \, \mathrm{Re}_{D_p}^{-0.826} \tag{7.17}$$

in the range $0.01 < \mathrm{Re}_{D_p} < 10$ and

$$(\bar{h}_c/c_p\rho U_s)\mathrm{Pr}^{2/3} = \frac{1}{\epsilon} 0.455 \, \mathrm{Re}_{D_p}^{-0.4} \tag{7.18}$$

in the range $10 < \mathrm{Re}_{D_p} < 200$.
The Reynolds number in Eqs. (7.17) and (7.18) is defined as

$$\mathrm{Re}_{D_p} = \frac{D_p U_s}{\nu},$$

where the particle diameter is

$$D_p = \sqrt{\frac{A_p}{\pi}}$$

and A_p is the particle surface area.

The range of void fraction tested by Upadhyay was fairly narrow, $0.371 < \epsilon < 0.451$, and data were for cylindrical pellets only. The actual data were for a mass transfer operation, dissolution of the solid particles in water. Since the Schmidt number ranged from 770 to 42,400, use of this correlation for air, $\mathrm{Pr} = 0.71$, may be suspect.

For computing heat transfer from the wall of the packed bed to a gas, Beek (23) recommends

$$\frac{\bar{h}_c D_p}{k} = 2.58 \, \mathrm{Re}_{D_p}^{1/3}\mathrm{Pr}^{1/3} + 0.094 \, \mathrm{Re}_{D_p}^{0.8}\mathrm{Pr}^{0.4} \tag{7.19}$$

for particles like cylinders which can pack next to the wall, and

$$\frac{\bar{h}_c D_p}{k} = 0.203 \, \mathrm{Re}_{D_p}^{1/3}\mathrm{Pr}^{1/3} + 0.220 \, \mathrm{Re}_{D_p}^{0.8}\mathrm{Pr}^{0.4} \tag{7.20}$$

for particles like spheres which contact the wall at one point. In Eqs. (7.19) and (7.20), the Reynolds number is

$$40 < \mathrm{Re}_{D_p} = \frac{U_s D_p}{v} < 2000$$

where D_p is defined by Beek as the diameter of the sphere or cylinder. For other types of packings a definition such as that used by Whitaker should suffice.

Beek also gives a correlation equation for the friction factor

$$f = \frac{D_p}{l} \frac{\Delta p}{\rho U_s^2} = \frac{1 - \epsilon}{\epsilon^3} \left(1.75 + 150 \frac{1 - \epsilon}{Re_{D_p}} \right) \qquad (7.21)$$

In Eq. (7.21) Δp is the pressure drop over a length l of the packed bed.

EXAMPLE 7.4 ————————————————————————————

It is desired to preheat a flow of carbon monoxide at atmospheric pressure from 50° to 350°C in a packed bed. The bed is a pipe with 7.62 cm ID, filled with a random arrangement of solid cylinders 1 cm in diameter and 1 cm long. The flow rate of carbon monoxide is 5 kg/h and the inside surface of the pipe is held at 400°C. Determine the required length of the bed.

Solution. The average film temperature at the preheater inlet is 225°C and at the preheater outlet is 375°C. Evaluating properties of carbon monoxide (Table 29, Appendix 2) at the average of these, or 300°C, we find a kinematic viscosity of 4.82×10^{-5} m²/s, a thermal conductivity of 0.042 W/m °C, a density of 0.60 kg/m³, a specific heat of 1081 J/kg °C, and Prandtl number of 0.71. The superficial velocity is

$$U_s = \frac{5 \text{ kg/h}}{(0.6 \text{ kg/m}^3)(\pi \ 0.0762^2/4 \text{ m}^2)} = 1827 \text{ m/h}$$

The cylindrical packing volume is $(\pi \ 1^2/4)(1) = 0.785$ cm³ and the surface area is $(2)(\pi \ 1^2/4) + \pi(1)(1) = 4.712$ cm². Therefore, the equivalent packing diameter is

$$D_p = \frac{(6)(0.785)}{4.712} = 1 \text{ cm}$$

giving a Reynolds number of

$$Re_{D_p} = \frac{(1827/3600)(0.01)}{4.82 \times 10^{-5}} = 105$$

From Eq. (7.19) we find

$$\frac{\bar{h}_c D_p}{k} = 2.58(105)^{1/3}(0.71)^{1/3} + 0.094(105)^{0.8}(0.71)^{0.4}$$

$$= 14.3$$

or

$$\bar{h}_c = \frac{(14.3)(0.042)}{0.01} = 60.1 \text{ W/m}^2 \text{ °C}$$

The required rate of heat transfer to the carbon monoxide is

$$q = \frac{(5 \text{ kg/h})(1081 \text{ J/kg °C})(300°C)}{3600 \text{ s/h}} = 450 \text{ W}$$

which must equal the rate of heat transfer by convection

$$450 \text{ W} = \bar{h}_c A \text{ (LMTD)}$$

The log mean temperature difference is

$$\text{LMTD} = \frac{(400 - 50) - (400 - 350)}{\ln\left[(400 - 50)/(400 - 350)\right]} = 154°\text{C}$$

We may now calculate the length of preheater:

$$q = (\bar{h}_c \pi D L) \text{ LMTD}$$
$$450 = (60.1)(\pi \times 0.0762 \times L)(154)$$

giving

$$L = 20.3 \text{ cm}$$

7.4 TUBE BUNDLES IN CROSS-FLOW

The evaluation of the convective conductance between a bank of tubes and a fluid flowing at right angles to the tubes is an important step in the design and performance analysis of many types of commercial heat exchangers. There are, for example, a large number of gas heaters in which a hot fluid inside the tubes heats a gas passing over the outside of the tubes. Figure 7.14 shows several arrangements of tubular air heaters in which the products of combustion, after they leave a boiler, economizer, or superheater, are used to preheat the air going to the steam-generating units. The shells of these gas heaters are usually rectangular and the shell-side gas flows in the space between the outside of the tubes and the shell. Since the flow cross-sectional area is continuously changing along the path, the shell-side gas speeds up and slows down periodically. A similar situation exists in some unbaffled short-tube liquid-to-liquid heat exchangers in which the shell-side fluid flows over the tubes. In these units the tube arrangement is similar to that in a gas heater except that the shell cross-sectional area varies where a cylindrical shell is used.

Heat transfer and pressure drop data for a large number of these heat exchanger cores have been compiled by Kays and London (24). Included are data on banks of bare tubes as well as tubes with plate fins, strip fins, wavy plate fins, pin fins, and so on.

In this section, we discuss some of the flow and heat transfer characteristics of bare tube bundles. Rather than concern ourselves with detailed information on a specific heat exchanger core or tube arrangement, or a particular type of tube fin, we shall focus on the common element of most heat exchangers, the tube bundle in cross-flow. This information is directly applicable to one of the most common heat exchangers, shell-and-tube, and will provide a basis for understanding the engineering data on specific heat exchangers presented in (24).

The heat transfer in flow over tube bundles depends largely on the flow pattern and the degree of turbulence, which in turn are functions of the velocity

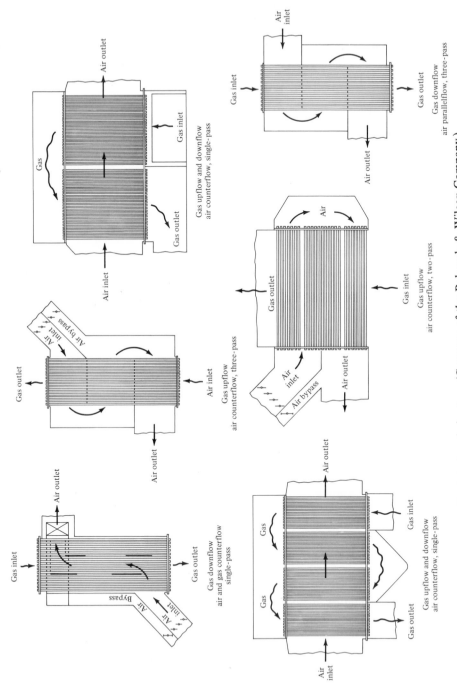

Figure 7.14 Some arrangements for tubular air heaters. (Courtesy of the Babcock & Wilcox Company.)

of the fluid and the size and arrangement of the tubes. The photographs of Figs. 7.15 and 7.16 illustrate the flow patterns for water flowing in the low turbulent range over tubes arranged *in line* and *staggered*, respectively. The photographs were obtained (25) by sprinkling fine aluminum powder on the surface of water flowing perpendicularly to the axis of vertically placed tubes. We observe that the flow patterns around tubes in the first transverse rows are similar to those for flow around single tubes. Focusing our attention on a tube in the first row of the in-line arrangement, we see that the boundary layer

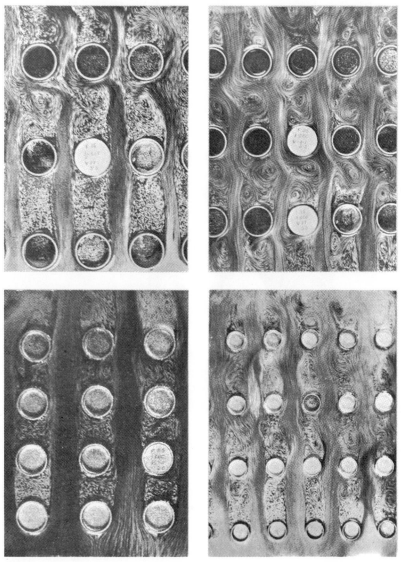

Figure 7.15 Flow patterns for in-line tube bundles. [By permission from R. D. Wallis (25).]

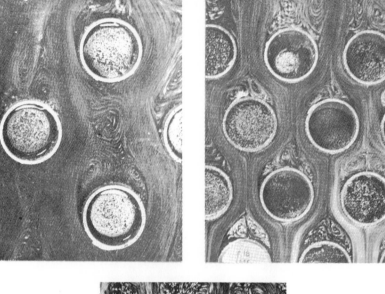

Figure 7.16   Flow patterns for staggered tube bundles. [By permission from R. D. Wallis (25).]

separates from both sides of the tube and a wake forms behind it. The turbulent wake extends to the tube located in the second transverse row. As a result of the high turbulence in the wakes, the boundary layers around tubes in the second and subsequent rows become progressively thinner. It is therefore not unexpected that, in turbulent flow, the heat transfer coefficients of tubes in the first row are smaller than the heat transfer coefficients of tubes in subsequent rows. In laminar flow, on the other hand, the opposite trend has been observed (26) due to a shading effect by the upstream tubes.

For a closely spaced staggered-tube arrangement (Fig. 7.16), the turbulent wake behind each tube is somewhat smaller than for similar in-line arrangements, but there is no appreciable reduction in the overall energy dissipation. Experiments on various types of tube arrangements (7) have shown that, for practical units, the relation between heat transfer and energy dissipation depends primarily on the velocity of the fluid, the size of the tubes, and the distance between the tubes. However, in the transition zone the performance of a closely spaced, staggered tube arrangement is somewhat superior to that of a similar in-line tube arrangement. In the laminar regime, the first row of tubes exhibits lower heat transfer than the downstream rows, just the opposite behavior of the in-line arrangement.

The equations available for the calculation of heat transfer coefficients in flow over tube banks are based entirely on experimental data because the flow pattern is too complex to be treated analytically. Experiments have shown that, in flow over staggered-tube banks, the transition from laminar to turbulent flow is more gradual than in flow through a pipe, whereas for in-line tube bundles the transition phenomena resemble those observed in pipe flow. In either case the transition from laminar to turbulent flow begins at a Reynolds number based on the velocity at the minimum flow area, about 200, and the flow becomes fully turbulent at a Reynolds number of about 6000.

For engineering calculations the average heat transfer coefficient for the entire tube bundle is of primary interest. The experimental data for heat transfer in flow over banks of tubes are usually correlated by an equation of the form $\overline{Nu}_D = \text{const} (Re_D)^m (Pr)^n$, which was previously used to correlate the data for flow over a single tube. To apply this equation to flow over tube bundles it is necessary to select a reference velocity, since the speed of the fluid varies along its path. The velocity used to build the Reynolds number for flow over tube bundles is based on the *minimum free area* available for fluid flow, regardless of whether the minimum area occurs in the transverse or diagonal openings. For in-line tube arrangements (Fig. 7.17), the minimum free-flow area per unit length of tube A_{min} is always $A_{min} = S_T - D$, where S_T is the distance between centers

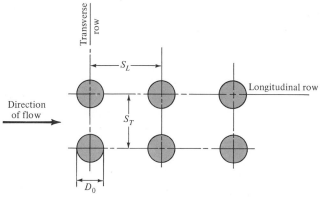

Figure 7.17 Sketch illustrating nomenclature for in-line tube arrangements.

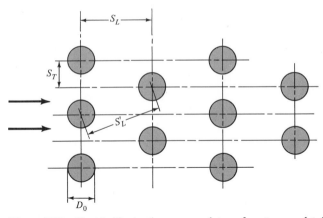

Figure 7.18 Sketch illustrating nomenclature for staggered tube arrangements.

of the tubes in adjacent longitudinal rows (measured perpendicularly to the direction of flow), or the *transverse pitch*.

For staggered arrangements (Fig. 7.18) the minimum free-flow area may occur, as in the previous case, either between adjacent tubes in a row or if S_L/S_T is so small that $\sqrt{S_T^2 + S_L^2} < S_T + (D/2)$, between diagonally opposed tubes. In the latter case, the maximum velocity U_{max} is $S_T/(\sqrt{S_L^2 + S_T^2} - D)$ times the free-flow velocity based on the shell area without tubes. The symbol S_L denotes the center-to-center distance between adjacent transverse rows of tubes or pipes (measured in the direction of flow) and is called the *longitudinal pitch*, and $S_L'^2 = S_L^2 + S_T^2$.

Zukauskas (7) has developed correlation equations for predicting the mean heat transfer from tube banks. The equations are primarily for tubes in the inner rows of the tube bank. However, the mean heat transfer coefficients for rows 3, 4, 5, . . . are indistinguishable from one another; the second row will exhibit 10 to 25 percent lower heat transfer than the internal rows for $\mathrm{Re} < 10^4$ and equal heat transfer for $\mathrm{Re} > 10^4$; and the heat transfer of the first row may be 60 to 75 percent of that of the internal rows, depending on longitudinal pitch. Therefore, the correlation equations will predict tube-bank heat transfer within 6 percent for 10 or more rows. The correlations are valid for $0.7 < \mathrm{Pr} < 500$.

The correlation equations are of the form

$$\overline{\mathrm{Nu}}_D = C \, \mathrm{Re}_D{}^m \mathrm{Pr}^{0.36} \left(\frac{\mathrm{Pr}}{\mathrm{Pr}_s}\right)^{0.25} \tag{7.22}$$

where the subscript s means that the fluid property value is to be evaluated at the tube wall temperature. Other fluid properties are to be evaluated at the bulk fluid temperature.

For laminar flow in the range $10 < \mathrm{Re}_D < 100$

$$\overline{\mathrm{Nu}}_D = 0.8 \, \mathrm{Re}_D{}^{0.4} \mathrm{Pr}^{0.36} \left(\frac{\mathrm{Pr}}{\mathrm{Pr}_s}\right)^{0.25} \tag{7.23}$$

for in-line tubes and

$$\overline{Nu}_D = 0.9 \, Re_D{}^{0.4} Pr^{0.36} \left(\frac{Pr}{Pr_s}\right)^{0.25} \tag{7.24}$$

for staggered tubes.

In the transition regime $10^3 < Re_D < 2 \times 10^5$ the exponent on Re_D, m, varies from 0.55 to 0.73 for in-line banks, depending on the tube pitch. A mean value of 0.63 is recommended for in-line banks with $S_T/S_L \geq 0.7$:

$$\overline{Nu}_D = 0.27 \, Re_D{}^{0.63} Pr^{0.36} \left(\frac{Pr}{Pr_s}\right)^{0.25} \tag{7.25}$$

[For $S_T/S_L < 0.7$ Eq. (7.25) will significantly overpredict $\overline{Nu}_D$; however, this tube arrangement yields an ineffective heat exchanger.] For staggered banks with $S_T/S_L < 2$

$$\overline{Nu}_D = 0.35 \left(\frac{S_T}{S_L}\right)^{0.2} Re_D{}^{0.60} Pr^{0.36} \left(\frac{Pr}{Pr_s}\right)^{0.25} \tag{7.26}$$

and for $S_T/S_L \geq 2$

$$\overline{Nu}_D = 0.40 \, Re_D{}^{0.60} Pr^{0.36} \left(\frac{Pr}{Pr_s}\right)^{0.25} \tag{7.27}$$

In the turbulent regime, $Re_D > 2 \times 10^5$, heat transfer for the inner tubes increases rapidly due to turbulence generated by the upstream tubes. In some cases, the Reynolds number exponent m exceeds 0.8, which corresponds to the exponent on Reynolds numbers for the turbulent boundary layer on the front of the tube. This means that the heat transfer on the rear portion of the tube must increase even more rapidly. Therefore, the value of m depends on tube arrangement, tube roughness, fluid properties, and free-stream turbulence. An average value $m = 0.84$ is recommended.

For in-line tube banks

$$\overline{Nu}_D = 0.021 \, Re_D{}^{0.84} Pr^{0.36} \left(\frac{Pr}{Pr_s}\right)^{0.25} \tag{7.28}$$

For staggered rows with $Pr > 1$

$$\overline{Nu}_D = 0.022 \, Re_D{}^{0.84} Pr^{0.36} \left(\frac{Pr}{Pr_s}\right)^{0.25} \tag{7.29}$$

and if $Pr = 0.7$

$$\overline{Nu}_D = 0.019 \, Re_D{}^{0.84} \tag{7.30}$$

The preceding correlation equations, Eqs. (7.23)–(7.30), are compared with experimental data from several sources in Fig. 7.19 for in-line arrangements and in Fig. 7.20 for staggered arrangements. Solid lines in the figures represent the correlation equations.

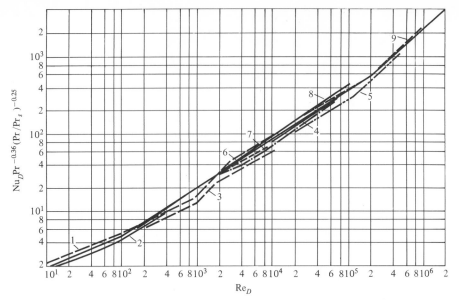

Figure 7.19 Comparison of heat transfer of in-line banks. Curve 1, 1.25 × 1.25, and curve 2, 1.5 × 1.5, after Bergelin et al.; curve 3, 1.25 × 1.25, after Kays and London; curve 4, 1.45 × 1.45, after Kuznetsov and Turilin; curve 5, 1.3 × 1.5, after Lyapin; curve 6, 2.0 × 2.0, after Isachenko; curve 7, 1.9 × 1.9, after Grimison; curve 8, 2.4 × 2.4, after Kuznetsov and Turilin; curve 9, 2.1 × 1.4, after Hammecke et al. [From (7) by permission, Academic Press.]

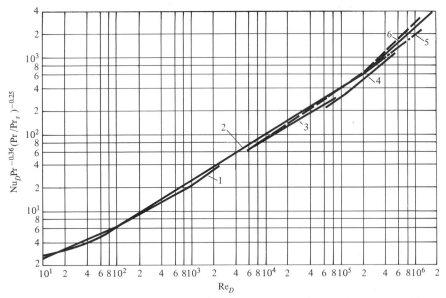

Figure 7.20 Comparison of heat transfer of staggered banks. Curve 1, 1.5 × 1.3, after Bergelin et al.; curve 2, 1.5 × 1.5 and 2.0 × 2.0, after Grimson and Isachenko; curve 3, 2.0 × 2.0, after Antuf'yev and Beletsky, Kuznetsov and Turilin, and Kazakevich; curve 4, 1.3 × 1.5, after Lyapin; curve 5, 1.6 × 1.4, after Dwyer and Sheeman; curve 6, 2.1 × 1.4, after Hammecke et al. [From (7) by permission, Academic Press.]

For closely spaced in-line banks it is necessary to base the Reynolds number on the average velocity integrated over the perimeter of the tube so that the results for various spacings will collapse to a single correlation line. Such results, presented in (7), indicate that this procedure correlates data for $2 \times 10^3 < \mathrm{Re}_D < 2 \times 10^5$ and for spacings $1.01 \leq S_T/D = S_L/D \leq 1.05$. However, Aiba et al. (27) show that for a single row of closely spaced tubes a critical Reynolds number, Re_{D_c}, exists. Below Re_{D_c} a stagnant region forms behind the first cylinder, reducing heat transfer to the remaining (three) cylinders below that for a single cylinder. Above Re_{D_c} the stagnant region rolls up into a vortex and significantly increases the heat transfer from the downstream cylinders.

In the range $1.15 \leq S_L/D \leq 3.4$, Re_{D_c} may be calculated from

$$\mathrm{Re}_{D_c} = 1.14 \times 10^5 \left(\frac{S_L}{D}\right)^{-5.84} \tag{7.31}$$

From data (7) on closely spaced tube banks ($1.01 \leq S_T/D = S_L/D \leq 1.05$) one would conclude that the discontinuous behavior does not occur when the single row of tubes is placed in a bank consisting of several such tube rows.

The pressure drop for a bank of tubes in cross-flow may be calculated from

$$\Delta p = f \frac{\rho U_{\max}}{2} N \tag{7.32}$$

where the velocity is that in the minimum free-flow area and the friction coefficient f depends on the Re_D (also based on velocity in the minimum free-flow area) according to Fig. 7.21 for in-line banks and Fig. 7.22 for staggered banks

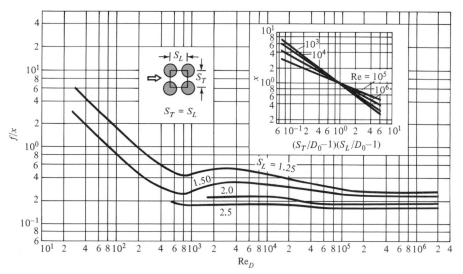

Figure 7.21 Pressure drop coefficient of in-line banks as referred to the relative longitudinal pitch S_L. [From (7) by permission, Academic Press.]

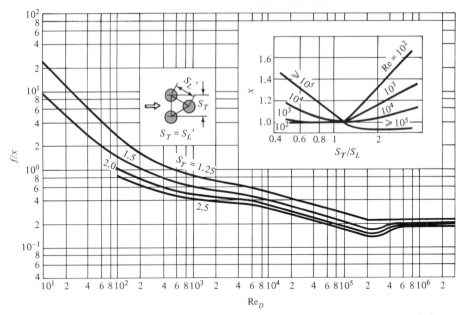

Figure 7.22 Pressure drop coefficients of staggered banks as referred to the relative transverse pitch S_T. [From (7) by permission, Academic Press.]

(7). The correlation factor X defined in those figures accounts for nonsquare in-line arrangements and nonequilateral-triangle staggered arrangements.

The variation of the average heat transfer coefficient of a tube bank with the number of transverse rows is shown in Table 7.3 for turbulent flow. To calculate the average heat transfer coefficient for tube banks with less than 10 rows, the $\bar{h}_c$ obtained from Eqs. (7.28)–(7.30) should be multiplied by the appropriate ratio $\bar{h}_{cN}/\bar{h}_c$.

7.4.1 Liquid Metals

Experimental data for the heat transfer characteristics of liquid metals in cross-flow over a tube bank have been obtained at Brookhaven National Laboratory (28, 29). In these tests mercury [Pr = 0.022 (28)] and NaK [Pr = 0.017 (29)]

TABLE 7.3 RATIO OF $\bar{h}_c$ FOR N TRANSVERSE ROWS TO $\bar{h}_c$ FOR 10 TRANSVERSE ROWS IN TURBULENT FLOW[a]

Ratio $\bar{h}_{cN}/\bar{h}_c$	N									
	1	2	3	4	5	6	7	8	9	10
Staggered tubes	0.68	0.75	0.83	0.89	0.92	0.95	0.97	0.98	0.99	1.0
In-line tubes	0.64	0.80	0.87	0.90	0.92	0.94	0.96	0.98	0.99	1.0

[a] From W. M. Kays and R. K. Lo (36).

were heated while flowing normal to a staggered tube bank consisting of 60 to 70 1.2 cm tubes, 10 rows deep, arranged in an equilateral triangular array with a 1.375 pitch-to-diameter ratio. Both local and average heat transfer coefficents were measured in turbulent flow. The average heat transfer coefficients in the interior of the tube bank are well correlated by the equation

$$\overline{Nu}_D = 4.03 + 0.228(Re_D Pr)^{0.67} \tag{7.33}$$

in the Reynolds number range 20,000 to 80,000. Additional data are presented in (30).

The measurements of the distribution of the local unit-surface conductance around the circumference of a tube indicate that for a liquid metal the turbulent effects in the wake upon heat transfer are small compared to the heat transfer by conduction within the fluid. Whereas with air and water a marked increase in the local heat transfer coefficient occurs in the wake region of the tube (see Fig. 7.8), with mercury the unit-surface conductance decreases continuously with increasing θ. At a Reynolds number of 83,000 the ratio $h_{c\theta}/\overline{h}_c$ was found to be 1.8 at the stagnation point, 1.0 at $\theta = 90°$, 0.5 at $\theta = 145°$, and 0.3 at $\theta = 180°$.

7.5 APPLICATION TO HEAT EXCHANGER DESIGN

In the design and selection of a stationary commercial heat exchanger, the power requirement and the initial cost of the unit must be considered. Results obtained by Pierson (31) show that the smallest possible pitch in each direction results in the lowest power requirement for a specified rate of heat transfer. Since smaller values of pitch also permit the use of a smaller shell, the cost of the unit is reduced when the tubes are closely packed. There is little difference in performance between in-line and staggered arrangements, but the former are easier to clean. The Tubular Exchanger Manufacturers Association recommends that tubes be spaced with a minimum center-to-center distance of 1.25 times the outside diameter of the tube and, when tubes are on a square pitch, that a minimum clearance lane of 0.65 cm be provided.

EXAMPLE 7.5 _____

Atmospheric air at 58°F is to be heated to 86°F by passing it over a bank of brass tubes inside which steam at 212°F is condensing. The unit-surface conductance on the inside of the tubes is about 1000 Btu/h ft² °F. The tubes are 2 ft long, $\frac{1}{2}$ in. OD, BWG No. 18 (0.049 in. wall thickness). They are to be arranged in-line in a square pattern with a pitch of $\frac{3}{4}$ in. inside a rectangular shell 2 ft wide and 15 in. high. If the total mass rate of flow of the air to be heated is 32,000 lb$_m$/h, estimate (a) the number of transverse rows required and (b) the pressure drop.

Solution. (a) Since the thermal resistance on the air side will be much larger than the combined resistance of the pipe wall and the steam, we shall first assume that the outside surface of the pipe is at the steam temperature. The mean film temperature of the air T_f will then be approximately equal to

$$\frac{1}{2}\left(\frac{58+86}{2}+212\right)=142°F$$

The mass velocity at the minimum cross-sectional area, which is between adjacent tubes, is calculated next. The shell is 15 in. high and consequently holds 19 longitudinal rows of tubes. The minimum free area is

$$A_{min}=(20)(2)\left(\frac{0.75-0.50}{12}\right)=0.833\text{ ft}^2$$

and the maximum mass velocity ρU_{max} is

$$G_{max}=\frac{32,000}{0.833}=38,400\text{ lb}_m/\text{h ft}^2$$

Hence, the Reynolds number is

$$\text{Re}_{max}=\frac{G_{max}D_0}{\mu}=\frac{(38,400\text{ lb/h ft}^2)(0.5/12\text{ ft})}{0.0462\text{ lb/h ft}}=34,600$$

Assuming that more than 10 rows will be required, the unit-surface conductance is calculated from Eq. (7.25). We get

$$\bar{h}_c=\left(\frac{0.016\text{ Btu/h ft °F}}{0.5/12\text{ ft}}\right)(0.27)(34,600)^{0.63}(0.72)^{0.36}$$
$$=66.7\text{ Btu/h ft}^2\text{ °F}$$

We can now determine the temperature at the outer tube wall, which was originally assumed equal to the steam temperature. There are three thermal resistances in series between the steam and the air. The resistance at the steam side per tube is

$$R_1=\frac{1/h_i}{\pi D_i L}=\frac{1/1000}{3.14(0.402/12)2}=0.00474\text{ h °F/Btu}$$

The resistance of the pipe wall ($k=60$ Btu/h ft °F) is approximately

$$R_2=\frac{0.049/k}{\pi[(D_0+D_i)/2]L}=\frac{0.049/60}{(3.14)(0.451)(2)}=0.000287\text{ h °F/Btu}$$

The resistance at the outside of the tube is

$$R_3=\frac{1/h_0}{\pi D_0 L}=\frac{1/66.7}{3.14(0.5/12)2}=0.0573\text{ h °F/Btu}$$

The total resistance is then

$$R_1+R_2+R_3=0.0623\text{ h °F/Btu}$$

Since the sum of the resistance at the steam side and the resistance of the tube wall is about 8 percent of the total resistance, about 8 percent of the total temperature drop occurs between the steam and the outer tube wall. The mean film temperature can now be corrected, and we get

$$T_f = 137°F$$

This will not change the values of the physical properties appreciably, and no adjustment in the previously calculated value of $\bar{h}_c$ is necessary.

The mean temperature difference between the steam and the air can now be calculated. Using the arithmetic average, we get

$$T_{steam} - T_{air} = 212 - \left(\frac{58 + 86}{2}\right) = 140°F$$

The specific heat of air at constant pressure is 0.241 Btu/lb$_m$ °F. Equating the rate of heat flow from the steam to the air to the rate of enthalpy rise of the air gives

$$\frac{19N \, \Delta T_{avg}}{R_1 + R_2 + R_3} = \dot{m}_{air} c_p (T_{out} - T_{in})_{air}$$

Solving for N, the number of transverse rows, yields

$$N = \frac{(32,000)(0.24)(86 - 58)(0.0623)}{(19)(140)} = 5.04, \text{ i.e., 5 rows}$$

Since the number of tubes is less than 10, it is necessary to correct $\bar{h}_c$ in accordance with Table 7.3, or

$$\bar{h}_{c6\ rows} = 0.92\bar{h}_{c10\ rows} = (0.92)(66.7) = 61.4 \text{ Btu/h ft}^2 \text{ °F}$$

Repeating the calculations with the corrected value of the average unit-surface conductance on the air side, we find that six transverse rows are sufficient for heating the air according to the specifications.

(b) The pressure drop is obtained from Eq. (7.32) and Fig. 7.20. Since $S_T = S_L = 1.5D_0$ we have

$$\left(\frac{S_T}{D_0} - 1\right)\left(\frac{S_L}{D_0} - 1\right) = 0.5^2 = 0.25$$

For $Re_{D_0} = 34,600$ and $(S_T/D_0 - 1)(S_L/D_0 - 1) = 0.25$ the correction factor is $X = 2.5$ and the friction factor from Fig. 7.21 is

$$f = (2.5)(0.3) = 0.75$$

The velocity is

$$U_{max} = \frac{G_{max}}{\rho} = \frac{(40,400 \text{ lb}_m/\text{h ft}^2)}{(0.0664 \text{ lb}_m/\text{ft}^3) \, 3600 \text{ s/h}}$$
$$= 169 \text{ ft/s}$$

The pressure drop is therefore

$$\Delta p = \frac{(0.75)(0.066)(169)^2}{2} \frac{6}{32.2} = 132 \text{ lb}_f/\text{ft}^2$$

Figure 7.23   Heat exchanger tube bundle with baffles. (Courtesy of the Aluminum Company of America.)

In many commercial shell-and-tube heat exchangers, baffles are used to increase the velocity and consequently the heat transfer coefficient on the shell side. Figure 7.23 is a photograph of a large baffled exchanger for vegetable-oil service. The flow of the shell-side fluid in baffled heat exchangers is partly perpendicular and partly parallel to the tubes. The heat transfer coefficient on the shell side in this type of unit depends not only on the size and spacing of the tubes and the velocity and physical properties of the fluid, but also on the spacing and shape of the baffles. In addition, there is always leakage through the tube holes in the baffle and between the baffle and the inside of the shell, and there is bypassing between the tube bundle and the shell. Because of these complications, the heat transfer coefficient can be estimated only by approximate methods or from experience with similar units. According to one approximate method, which is widely used for design calculations (32), the average heat transfer coefficient calculated for the corresponding tube arrangement in simple cross-flow is multiplied by 0.6 to allow for leakage and other deviations from the simplified model. For additional information the reader is referred to (32–35).

7.6 SUMMARY

For the convenience of the reader useful correlation equations for the determination of the average value of the convection heat transfer coefficients in cross-flow over exterior surfaces are tabulated in Table 7.4.

TABLE 7.4 HEAT TRANSFER CORRELATIONS FOR EXTERNAL FLOW

Geometry	Correlation equation	Restrictions
Long cylinder in gas or liquid	$\overline{Nu}_D = C\,Re_D^m\,Pr^n(Pr/Pr_s)^{1/4}$ (see Table 7.1)	$1 < Re_D < 10^6$
Noncircular cylinder in gas	$\overline{Nu}_D = B\,Re_D^n$ (see Table 7.2)	$2500 < Re_D < 10^5$
Cylinder in a liquid metal	$\overline{Nu}_D = 1.125(Re_D Pr)^{0.413}$	$1 < Re_D Pr < 100$
Short cylinder in gas	$\overline{Nu}_D = 0.123\,Re_D^{0.651} + (D/L)^{0.85}\,Re_D^{0.792}$	$7\times10^4 < Re_D < 1.1\times10^5$ $L/D < 4$
Sphere in a gas	$\dfrac{\overline{h}_c}{c_p\rho U_\infty} = (2.2/Re_D + 0.48/Re_D^{0.5})$	$1 < Re_D < 25$
	$\overline{Nu}_D = 0.37\,Re_D^{0.6}$	$25 < Re_D < 10^5$
	$\overline{Nu}_D = 430 + 5\times10^{-3}\,Re_D + 0.25\times10^{-9}\,Re_D^2 - 3.1\times10^{-17}\,Re_D^3$	$4\times10^5 < Re_D < 5\times10^6$
Sphere in a gas or a liquid	$\overline{Nu}_D = 2 + (0.4\,Re_D^{1/2} + 0.06\,Re_D^{2/3})Pr^{0.4}(\mu/\mu_s)^{1/4}$	$3.5 < Re_D < 7.6\times10^4$ $0.7 < Pr < 380$
Sphere in a liquid metal	$\overline{Nu}_D = 2 + 0.386(Re_D Pr)^{1/2}$	$3.6\times10^4 < Re_D < 2\times10^5$
Long flat plate, width D, perpendicular to flow in gas	$\overline{Nu}_D = 0.20\,Re_D^{2/3}$	$1 < Re_D < 4\times10^5$
Half-round cylinder with flat rear surface in gas	$\overline{Nu}_D = 0.16\,Re_D^{2/3}$	$1 < Re_D < 4\times10^5$
Square plate, dimension, L, perpendicular to flow of gas or liquid	$(\overline{h}_c/c_p\rho U_\infty)Pr^{2/3} = 0.930\,Re_L^{-1/2}$	$2\times10^4 < Re_L < 10^5$
Packed bed— heat transfer to or from packing, gas	$\overline{Nu}_{D_p} = \dfrac{1-\epsilon}{\epsilon}(0.5\,Re_{D_p}^{1/2} + 0.2\,Re_{D_p}^{2/3})Pr^{1/3}$	$20 < Re_{D_p} < 10^4$ $0.34 < \epsilon < 0.78$

(*Continued*)

TABLE 7.4 (Continued)

Geometry	Correlation equation	Restrictions
(ϵ = void fraction) (D_p = equivalent packing diameter, see Eq. 7.16)	$(\bar{h}_c/c_p \rho U_s)\text{Pr}^{2/3} = \dfrac{1.075}{\epsilon} \, \text{Re}_{D_p}^{-0.826}$	$0.01 < \text{Re}_{D_p} < 10$
Packed bed— heat transfer to or from containment wall, gas	$\overline{\text{Nu}}_{D_p} = 2.58\,\text{Re}_{D_p}^{1/3}\text{Pr}^{1/3} + 0.094\,\text{Re}_{D_p}^{0.8}\text{Pr}^{0.4}$	$40 < \text{Re}_{D_p} < 2000$ cylinder-like packing
	$\overline{\text{Nu}}_{D_p} = 0.203\,\text{Re}_{D_p}^{1/3}\text{Pr}^{1/3} + 0.220\,\text{Re}_{D_p}^{0.8}\text{Pr}^{0.4}$	$40 < \text{Re}_{D_p} < 2000$ sphere-like packing
Tube bundle in cross-flow (see Figs. 7.17 and 7.18)	$\overline{\text{Nu}}_D \text{Pr}^{-0.36}(\text{Pr}/\text{Pr}_s)^{-0.25} = C(S_T/S_L)^n \, \text{Re}_D^m$	

C	m	n	Restrictions
0.8	0.4	0	$10 < \text{Re}_D < 100$, in-line
0.9	0.4	0	$10 < \text{Re}_D < 100$, staggered
0.27	0.63	0	$1000 < \text{Re}_D < 2 \times 10^5$, in-line $S_T/S_L \geq 0.7$
0.35	0.60	0.2	$1000 < \text{Re}_D < 2 \times 10^5$, staggered $S_T/S_L < 2$
0.40	0.60	0	$1000 < \text{Re}_D < 2 \times 10^5$, staggered $S_T/S_L \geq 2$
0.021	0.84	0	$\text{Re}_D > 2 \times 10^5$, in-line
0.022	0.84	0	$\text{Re}_D > 2 \times 10^5$, staggered $\text{Pr} > 1$

Correlation equation	Restrictions
$\overline{\text{Nu}}_D = 0.019\,\text{Re}_D^{0.84}$	$\text{Re}_D > 2 \times 10^5$, staggered $\text{Pr} = 0.7$
$\overline{\text{Nu}}_D = 4.03 + 0.228(\text{Re}_D\text{Pr})^{2/3}$	$2 \times 10^4 < \text{Re}_D < 8 \times 10^4$, staggered liquid metals

PROBLEMS

The problems for this chapter are related to
the subject matter as shown below.

7.1 Determine the unit-surface conductance at the stagnation point and average value
of the conductance for a single 5-cm-OD, 60-cm-long tube in cross-flow. The
temperature of the tube surface is 260°C, the velocity of the fluid flowing perpendicularly to the tube axis is 6 m/s, and its temperature is 38°C. The following
fluids are to be considered: (a) air, (b) hydrogen, and (c) water.

7.2 A spherical water droplet of 1.5 mm diameter is freely falling in atmospheric air.
Calculate the average convection heat transfer coefficient when the droplet has
reached its terminal velocity. Assume that the water is at 50°C and the air is at
20°C and neglect mass transfer and radiation.

7.3 A mercury-in-glass thermometer at 100°F (OD = 0.35 in.) is inserted through the
duct wall into a 100 ft/s air stream at 150°F. Estimate the unit-convective conductance between the air and the thermometer.

7.4 Steam at 1 atm and 100°C is flowing across a 5-cm-OD tube at a velocity of 6 m/s.
Estimate the Nusselt number, the heat transfer coefficient, and the rate of heat
transfer per meter length of pipe if the pipe is at 200°C.

7.5 Repeat Problem 7.4 for a tube bank in which all of the tubes are spaced with their
centerlines 7.5 cm apart.

7.6 Determine the average unit-surface conductance for air at 60°C flowing at a velocity of 1 m/s over a bank of 6-cm-OD tubes arranged as shown in the accompanying sketch. The tube-wall temperature is 117°C.

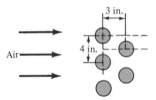

7.7 A stainless steel pin fin, 2 in. long, $\frac{1}{4}$ in. OD, extends from a flat plate into a 400 mph air stream as shown in the accompanying sketch. (*a*) Estimate the average heat transfer coefficient between the air and the fin. (*b*) Estimate the temperature at the end of the fin. (*c*) Estimate the rate of heat flow from the fin.

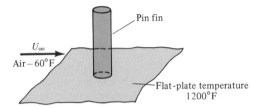

7.8 Repeat Problem 7.7 with glycerin flowing over the fin at 7 ft/s.

7.9 An inventor claims that pumping power can be reduced if the tubes in a bank in cross-flow are replaced by hollow streamlined bodies whose cross sections have the shape of an ellipse. He claims that energy losses in the wake would be reduced

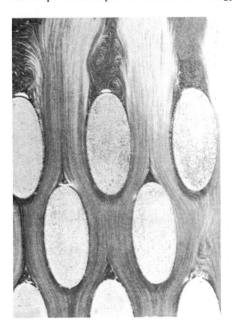

without affecting the rate of heat transfer adversely. See the accompanying sketch. Present your evaluation of the inventor's claim in the form of a short report, and substantiate your conclusions by order-of-magnitude calculations. State all of your assumptions.

7.10 The instruction manual for a hot-wire anemometer states that "roughly speaking, the current varies as the fourth power of the average velocity at a fixed wire resistance." Check this statement, using the heat transfer characteristics of a thin wire in air and water.

7.11 In a lead-shot tower, spherical 0.95-cm-diameter BB shots are formed by drops of molten lead which solidify as they descend in cooler air. At the terminal velocity, i.e., when the drag equals the gravitational force, estimate the total unit-surface conductance if the lead surface is at 171°C ($\epsilon = 0.63$) and the air temperature is 16°C. Assume $C_D = 0.75$ for the first trial calculation.

7.12 Water at 180°C and at 3 m/s enters a bare, 15-m-long, 2.5-cm wrought-iron pipe. If air at 10°C flows perpendicular to the pipe at 12 m/s, determine the outlet temperature of the water. (Note that the temperature difference between the air and the water varies along the pipe.)

7.13 Estimate the unit-surface conductance for liquid sodium at 1000°F flowing over a 10-row staggered-tube bank arranged in an equilateral-triangular arrow with a 1.5 pitch-to-diameter ratio. The entering velocity is 10 ft/s, based on the area of the shell, and the tube surface temperature is 400°F. What is the outlet temperature of the sodium?

7.14 Estimate (a) the unit-surface conductance for a spherical fuel droplet injected into a diesel engine at 80°C and 90 m/s. The oil droplet is 0.025 mm in diameter, the cylinder pressure is 4800 kPa, and the gas temperature is 944 K. (b) What is the time required to heat the droplet to its self-ignition temperature of 300°C?

7.15 An electrical transmission line of 1.2 cm OD carries 200 A and has a resistance of 3×10^{-4} ohm per meter of length. If the air around this line is at 16°C, determine the surface temperature on a windy day, assuming a wind blows across the line at 33 km/h.

7.16 Derive an equation in the form $\bar{h}_c = f(T, D, U_\infty)$ for flow of air over a long horizontal cylinder for the temperature range 0° to 200°F, using Eq. (7.3) as a basis.

7.17 Repeat Problem 7.16 for water in the temperature range 50° to 100°F.

7.18 A copper sphere 2.5 cm in diameter is suspended by a fine wire in the center of an experimental hollow cylindrical furnace whose inside wall is maintained uniformly at 430°C. Dry air at a temperature of 90°C and a pressure of 1.2 atm is blown steadily through the furnace at a velocity of 14 m/s. The inside diameter of the furnace is 20 cm, the furnace is 80 cm long, and the emissivity of the interior surface of the furnace wall is 0.9. The copper is slightly oxidized, and its emissivity is 0.4. Assuming that the air is completely transparent to radiation, calculate for the steady state: (a) the overall heat transfer coefficient between the copper sphere and the air, and (b) the temperature of the sphere.

7.19 The temperature of air flowing through a 25-cm-diameter duct whose inner walls are at 320°C is to be measured with a thermocouple soldered in a cylindrical steel well of 1.2 cm OD, whose exterior is oxidized as shown in the accompanying sketch. The air flows normal to the cylinder at a mass velocity of 17,600 kg/h m². If the temperature indicated by the thermocouple is 200°C, estimate the actual temperature of the air.

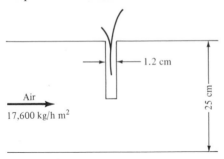

7.20 Develop an expression for the ratio of the rate of heat transfer to water at 38°C from a thin flat strip of width πD and length L at zero angle of attack and a tube of the same length and diameter D in cross-flow with its axis normal to the air stream in the Reynolds number range between 50 and 4000. Assume both surfaces are at 90°C.

7.21 Repeat Problem 7.20 for air flowing over the same two surfaces in the Reynolds number range between 40,000 and 400,000, assuming the surface temperature is 38°C. Neglect radiation.

7.22 Liquid mercury at a temperature 315°C flows at a velocity of 3 m/s over a staggered bank of $\frac{5}{8}$-in. 16 BWG stainless steel tubes, arranged in an equilateral triangular array with a pitch-to-diameter ratio of 1.375. If water at 2 atm pressure is being evaporated inside the tubes, estimate the average rate of heat transfer to the water per meter length of the bank, if the bank is 10 rows deep and has 60 tubes in it.

7.23 Compare the rate of heat transfer and the pressure drop for an in-line and a staggered arrangement of a tube bank consisting of 300 tubes, 6 ft long and 1 in. OD. The tubes are to be arranged in 15 rows with normal and parallel spacing of 2 in. The tube surface temperature is 200°F and water at 100°F is flowing at a mass rate of 12,000 lb/s over the tubes.

7.24 One method of storing solar energy for use during cloudy days, or at night, is to store it in the form of sensible heat in a rock bed. Suppose such a rock bed has been heated to 70°C and it is desired to heat a stream of air by blowing it through the bed. If the air inlet temperature is 10°C and the mass velocity of the air in the bed is 0.5 kg/s m², how long must the bed be in order for the initial outlet air temperature to be 65°C? Assume that the rocks are spherical, 2 cm in diameter, and that the bed void fraction is 0.5. *Hint:* The surface area of the rocks per unit volume of the bed is $(6/D_p)(1 - \epsilon)$.

REFERENCES

1. A. Fage, "The Air Flow around a Circular Cylinder in the Region Where the Boundary Layer Separates from the Surface," Brit. Aero. Res. Comm. R and M 1179, 1929.

2. A. Fage and V. M. Falkner, "The Flow around a Circular Cylinder," Brit. Aero Res. Comm. R and M 1369, 1931.

3. H. B. Squire, *Modern Developments in Fluid Dynamics*, 3d ed., vol. 2, Clarendon, Oxford, 1950.

4. R. C. Martinelli, A. G. Guibert, E. H. Morin, and L. M. K. Boelter, "An Investigation of Aircraft Heaters VIII—a Simplified Method for Calculating the Unit-Surface Conductance over Wings," NACA ARR, March 1943.

5. W. H. Giedt, "Investigation of Variation of Point Unit-Heat-Transfer Coefficient around a Cylinder Normal to an Air Stream," *Trans. ASME*, vol. 71, pp. 375–381, 1949.

6. R. Hilpert, "Wärmeabgabe von geheizten Drähten und Rohren," *Forsch. Geb. Ingenieurwes.*, vol. 4, p. 215, 1933.

7. A. A. Zukauskas, "Heat Transfer from Tubes in Cross Flow," *Adv. Heat Transfer*, Academic, vol. 8, pp. 93–106, 1972.

8. M. Jakob, *Heat Transfer*, vol. 1, Wiley, New York, 1949.

9. R. Ishiguro, K. Sugiyama, and T. Kumada, "Heat Transfer around a Circular Cylinder in a Liquid-Sodium Crossflow," *Int. J. Heat Mass Transfer*, vol. 22, pp. 1041–1048, 1979.

10. A. Quarmby and A. A. M. Al-Fakhri, "Effect of Finite Length on Forced Convection Heat Transfer from Cylinders," *Int. J. Heat Mass Transfer*, vol. 23, pp. 463–469, 1980.

11. N. R. Yardi and S. P. Sukhatme, "Effects of Turbulence Intensity and Integral Length Scale of a Turbulent Free Stream on Forced Convection Heat Transfer from a Circular Cylinder in Cross Flow," *Proc. 6th Int. Heat Transfer Conf.*, Hemisphere, Washington, D.C., 1978.

12. H. Dryden and A. N. Kuethe, "The Measurement of Fluctuations of Air Speed by the Hot-Wire Anemometer," NACA Rept. 320, 1929.

13. C. E. Pearson, "Measurement of Instantaneous Vector Air Velocity by Hot-Wire Methods," *J. Aerosp. Sci.*, vol. 19, pp. 73–82, 1952.

14. W. H. McAdams, *Heat Transmission*, 3d ed., McGraw-Hill, New York, 1953.

15. S. Whitaker, "Forced Convection Heat Transfer Correlations for Flow in Pipes, Past Flat Plates, Single Cylinders, Single Spheres, and for Flow in Packed Beds and Tube Bundles," *AIChE J.*, vol. 18, pp. 361–371, 1972.

16. E. Achenbach, "Heat Transfer from Spheres up to $Re = 6 \times 10^6$," *Proc. 6th Int. Heat Transfer Conf.*, vol. 5, Hemisphere, Washington, D.C., 1978.

17. H. F. Johnston, R. L. Pigford, and J. H. Chapin, "Heat Transfer to Clouds of Falling Particles," *Univ. of Ill. Bull.*, vol. 38, no. 43, 1941.

18. L. C. Witte, "An Experimental Study of Forced-Convection Heat Transfer from a Sphere to Liquid Sodium," *J. Heat Transfer*, vol. 90, pp. 9–12, 1968.

19. H. H. Sogin, "A Summary of Experiments on Local Heat Transfer from the Rear of Bluff Obstacles to a Lowspeed Airstream," *Trans. ASME, Ser. C, J. Heat Transfer*, vol. 86, pp. 200–202, 1964.

20. J. W. Mitchell, "Base Heat Transfer in Two-Dimensional Subsonic Fully Separated Flows," *Trans. ASME, Ser. C, J. Heat Transfer*, vol. 93, pp. 342–348, 1971.

21. K. K. Tien and E. M. Sparrow, "Local Heat Transfer and Fluid Flow Characteristics for Airflow Oblique or Normal to a Square Plate," *Int. J. Heat Mass Transfer*, vol. 22, pp. 349–360, 1979.

22. S. N. Upadhyay, B. K. D. Agarwal, and D. R. Singh, "On the Low Reynolds Number Mass Transfer in Packed Beds," *J. Chem. Eng. Jpn.*, vol. 8, pp. 413–415, 1975.

23. J. Beek, "Design of Packed Catalytic Reactors," *Adv. Chem. Eng.*, vol. 3, pp. 203–271, 1962.
24. W. M. Kays and A. L. London, *Compact Heat Exchangers*, 2d ed., McGraw-Hill, New York, 1964.
25. R. D. Wallis, "Photographic Study of Fluid Flow between Banks of Tubes," *Engineering*, vol. 148, pp. 423–425, 1934.
26. W. E. Meece, "The Effect of the Number of Tube Rows upon Heat Transfer and Pressure Drop during Viscous Flow across In-Line Tube Banks," M.S. thesis, Univ. of Delaware, 1949.
27. S. Aiba, T. Ota, and H. Tsuchida, "Heat Transfer of Tubes Closely Spaced in an In-Line Bank," *Int. J. Heat Mass Transfer*, vol. 23, pp. 311–319, 1980.
28. R. J. Hoe, D. Dropkin, and O. E. Dwyer, "Heat Transfer Rates to Crossflowing Mercury in a Staggered Tube Bank—I," *Trans. ASME*, vol. 79, pp. 899–908, 1957.
29. C. L. Richards, O. E. Dwyer, and D. Dropkin, "Heat Transfer Rates to Crossflowing Mercury in a Staggered Tube Bank—II," *ASME—AIChE Heat Transfer Conf.*, paper 57-HT-11, 1957.
30. S. Kalish and O. E. Dwyer, "Heat Transfer to NaK Flowing through Unbaffled Rod Bundles," *Int. J. Heat Mass Transfer*, vol. 10, pp. 1533–1558, 1967.
31. O. L. Pierson, "Experimental Investigation of Influence of Tube Arrangement on Convection Heat Transfer and Flow Resistance in Cross Flow of Gases over Tube Banks," *Trans. ASME*, vol. 59, pp. 563–572, 1937.
32. T. Tinker, "Analysis of the Fluid Flow Pattern in Shell-and-Tube Heat Exchangers and the Effect Distribution on the Heat Exchanger Performance," *Inst. Mech. Eng., ASME Proc. General Discuss. Heat Transfer*, pp. 89–115, September 1951.
33. B. E. Short, "Heat Transfer and Pressure Drop in Heat Exchangers," Bull. 3819, Univ. of Texas, 1938. (See also revision, Bull. 4324, June 1943.)
34. D. A. Donohue, "Heat Transfer and Pressure Drop in Heat Exchangers," *Ind. Eng. Chem.*, vol. 41, pp. 2499–2511, 1949.
35. A. C. Mueller, "Thermal Design of Heat Exchangers," Eng. Bull. 121, Res. Ser., Purdue Univ., 1954.
36. W. M. Kays and R. K. Lo, "Basic Heat Transfer and Flow Friction Design Data for Gas Flow Normal to Banks of Staggered Tubes—Use of a Transient Technique," Tech. Rept. 15, Navy Contract N6-ONR-251 T. O. 6, Stanford Univ., 1952.

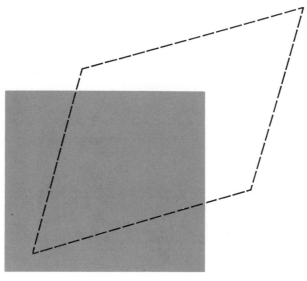

Chapter 8

Heat Exchangers

8.1 INTRODUCTION

This chapter deals with the thermal analysis of various types of heat exchangers that transfer heat between two fluids. Two methods of predicting the performance of conventional industrial heat exchangers will be outlined and techniques of estimating the required size and the most suitable type of heat exchanger to accomplish a specified task will be presented. Heat exchangers for solar thermal energy conversion will be considered in Chapter 11.

When a heat exchanger is placed into a thermal transfer system, a temperature drop is required to transfer the heat. The size of this temperature drop can be decreased by utilizing a larger heat exchanger, but this will require an increase in size and economic investment. Considerations along these lines are important in engineering design and in a complete engineering design of heat exchange equipment, not only the thermal performance characteristics, but also the pumping power requirements and the economics of the system are important. The role of heat exchangers has taken on increasing importance recently as engineers are becoming energy-conscious and want to optimize designs not only in terms of a thermal analysis and economic return on the investment, but also in terms of the energy payback of a system. Thus economics, as well as considerations such as the availability and amount of energy and raw materials necessary to accomplish a given task, should be considered.

8.2 BASIC TYPES OF HEAT EXCHANGERS

A heat exchanger is a device in which heat is transferred between a warmer and a colder substance, usually fluids. There are three basic types of heat exchangers:

> *Recuperators.* In this type of heat exchanger the hot and cold fluids are separated by a wall and heat is transferred by a combination of convection to and from the wall and conduction through the wall. The wall can include extended surfaces, such as fins (see Chapter 2) or enhancement devices (see Chapter 11).
>
> *Regenerators.* In a regenerator the hot and the cold fluids occupy alternately the same space in the exchanger core. The exchanger core, or "matrix," serves as a heat storage device that is periodically heated by the warmer of the two fluids and then transfers heat to the colder fluid. In a *fixed matrix* configuration the hot and cold fluids pass alternately through a stationary exchanger, and for continuous operation two or more matrices are necessary as shown in Fig. 8.1a. One commonly used arrangement for the matrix is the "packed bed" discussed in Chapter 6. Another approach is the *rotary regenerator* in which a circular matrix rotates and alternately exposes a portion of its surface to the hot and then to the cold fluid, as shown in Fig. 8.1b.
>
> *Direct Contact Heat Exchangers.* In this type of heat exchanger the hot and cold fluids contact each other directly. An example of such a device is a cooling tower in which a spray of water falling from the top of the tower is contacted directly and cooled by a stream of air flowing upward. Other direct contact systems use imiscible liquids or solid to gas exchange. The direct contact approach is still in the research and development stage and the reader is referred to (20) for further information.

The coverage in this chapter is confined to the first type of heat exchanger and will emphasize the so-called "shell-and-tube" design. The simplest arrangement of this type of heat exchanger consists of a tube within a tube, as shown in Fig. 8.2a. Such an arrangement can be operated either in counterflow or in parallel flow, with either the hot or the cold fluid passing through the annular space and the other fluid passing through the inside of the inner pipe.

A more common type of heat exchanger, widely used in the chemical and process industry, is the shell-and-tube arrangement shown in Fig. 8.2b. In this type of heat exchanger one fluid flows inside the tubes while the other fluid is forced through the shell and over the outside of the tubes. The reason for forcing the fluid to flow over the tubes rather than along the tubes is that a higher heat transfer coefficient can be achieved in cross-flow than in flow parallel to the tubes. To achieve cross-flow on the shell side, baffles are placed inside the

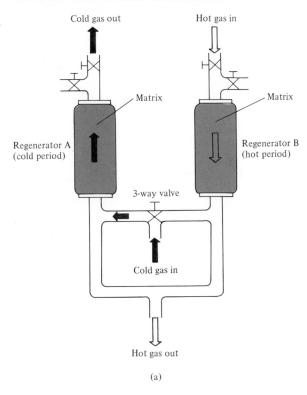

(a)

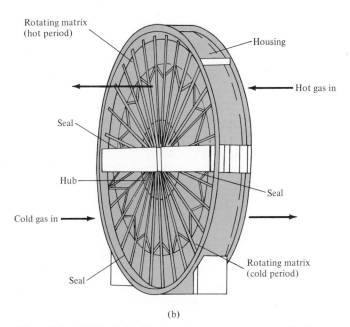

(b)

Figure 8.1 (a) Fixed dual-bed regenerator or system. (b) Rotary regenerator.

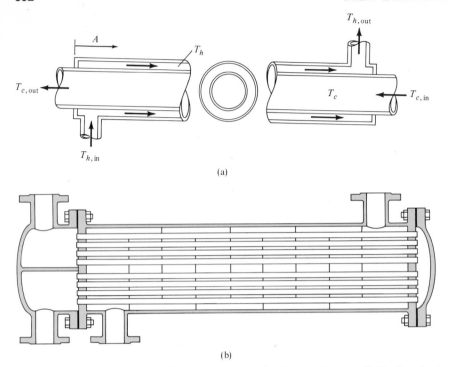

Figure 8.2 (a) Simple tube-within-a-tube counterflow heat exchanger. (b) Shell-and-tube heat exchanger with segmental baffles: two tube passes, one shell pass.

shell as shown in Fig. 8.2b. These baffles ensure that in each section the flow passes across the tubes flowing downward in the first, upward in the second section, and so on. Depending on the header arrangements at the two ends of the heat exchanger, one or more tube passes can be achieved. For a two-tube-pass arrangement, the inlet header is split so that the fluid flowing into the tubes passes through half of the tubes in one direction, then turns around and returns through the other half of the tubes to where it started, as shown in Fig. 8.2b. Three and four tube passes can be achieved by rearrangement of the header space. A variety of baffles have been used in industry, but the most common kind is the disk-and-doughnut baffle shown in Fig. 8.3b.

In gas heating or cooling it is often convenient to use a cross-flow heat exchanger such as shown in Fig. 8.4. In such a heat exchanger, one of the fluids passes through the tubes, whereas the gaseous fluid is forced across the tube bundle. The flow of the exterior fluid may be by forced or by free convection. In this type of exchanger the gas flowing across the tube is considered to be *mixed*, whereas the fluid in the tube is considered to be *unmixed*. The exterior gas flow is mixed because it can move about freely between the tubes as it exchanges heat, whereas the fluid within the tubes is confined and cannot mix with any other stream during the heat exchange process.

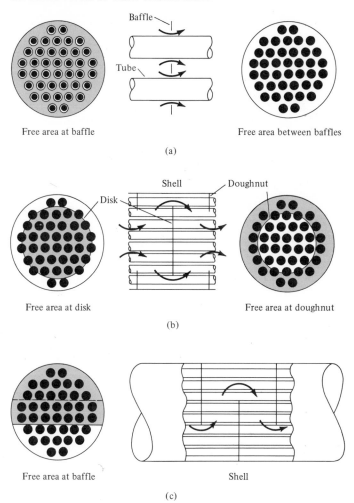

Free area at baffle   Free area between baffles

(a)

Free area at disk   Free area at doughnut

(b)

Free area at baffle   Shell

(c)

Figure 8.3 Three types of baffles used in shell-and-tube heat exchangers: (*a*) orifice baffle; (*b*) disk-and-doughnut baffle; (*c*) segmental baffle. [From C. B. Cramer (17).]

Another type of cross-flow heat exchanger, which is widely used in the space and comfort heating industry for homes, is shown in Fig. 8.5. In this arrangement the gas flows across a finned tube bundle and is unmixed because it is confined to separate flow passages. As shown in the temperature profile above the cross-flow heat exchanger, when the fluid is unmixed it will have a temperature gradient both parallel and normal to the flow direction. When the fluid is well mixed, there will be a tendency for the fluid temperature to equalize in the direction normal to the flow, and there will only be a temperature gradient in the direction of the flow. In the design of heat exchangers it is important to specify whether the fluids are mixed or unmixed, and which of the fluids is mixed. It is also important to balance the temperature drop by obtaining approximately equal heat transfer coefficients on the exterior and interior of the

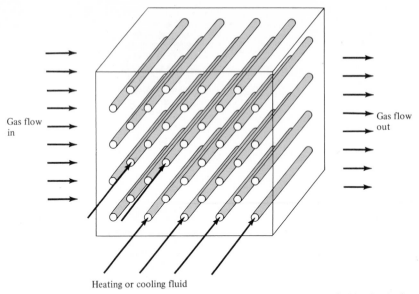

Figure 8.4 Cross-flow air heater illustrating cross-flow with one fluid mixed, the other unmixed.

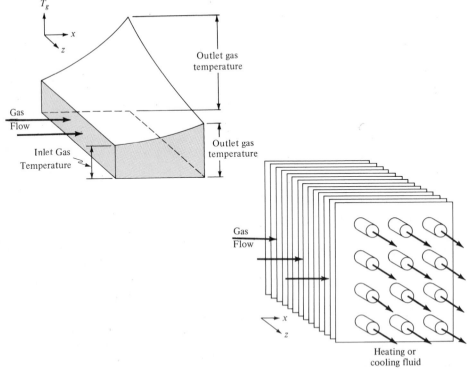

Figure 8.5 Type of cross-flow heat exchanger, widely used in space and comfort heating industry for homes.

tubes. If this is not done, one of the thermal resistances may be unduly large and cause an unnecessarily high overall temperature drop for a given rate of heat transfer, which in turn demands larger equipment and results in poor economics.

The shell-and-tube heat exchanger illustrated in Fig. 8.2*b* has fixed tube plates at each end and the tubes are welded or expanded into the plates. This type of construction has the lowest initial cost but can be used only for small temperature differences between the hot and the cold fluid because no provision is made to prevent thermal stresses due to the differential expansion between the tubes and the shell. Another disadvantage is that the tube bundle cannot be removed for cleaning. These drawbacks can be overcome by modification of the basic design as shown in Fig. 8.6. In this arrangement one tube plate is fixed but the other is bolted to a floating-head cover which permits the tube bundle to move relative to the shell. The floating tube sheet is clamped between the floating head and a flange so that it is possible to remove the tube bundle for cleaning. The heat exchanger shown in Fig. 8.6 has one shell pass and two tube passes.

For certain special applications such as aircraft gas turbines, the rate of heat transfer per unit weight and unit volume is the prime consideration. Compact, lightweight heat exchangers for this type of service have been investigated

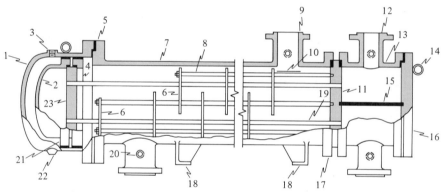

Figure 8.6 Shell-and-tube heat exchanger with floating head. (Courtesy of the Tubular Exchange Manufacturers Association.)

Key:

1. Shell cover
2. Floating head
3. Vent connection
4. Floating-head backing device
5. Shell cover—end flange
6. Transverse baffles or support plates
7. Shell

8. Tie rods and spacers
9. Shell nozzle
10. Impingement baffle
11. Stationary tube sheet
12. Channel nozzle
13. Channel
14. Lifting ring
15. Pass partition
16. Channel cover

17. Shell channel— end flange
18. Support saddles
19. Heat transfer tube
20. Test connection
21. Floating-head flange
22. Drain connection
23. Floating tube sheet

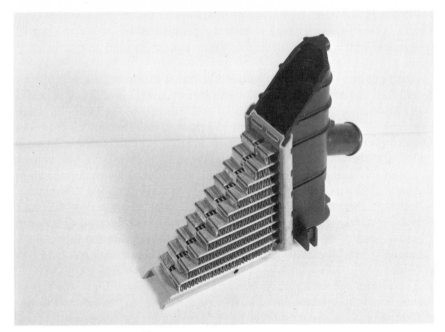

Figure 8.7 Vacuum brazed aluminum radiator. (Courtesy of Climate Control Division, Ford Motor Company.)

by Kays and London (1). A typical design is shown in Fig. 8.7. For a complete description and analysis of compact heat exchangers, especially for the application of fins to increase the effectiveness of such units, the reader is referred to (1–4).

8.3 OVERALL HEAT TRANSFER COEFFICIENT

One of the first tasks in a thermal analysis of a shell-and-tube heat exchanger is to evaluate the overall heat transfer coefficient between the two fluid streams. It was shown in Chapter 1 that the overall heat transfer coefficient between a hot fluid at temperature T_h and a cold fluid at temperature T_c separated by a solid plane wall is defined by

$$q = UA(T_h - T_c) \tag{8.1}$$

where

$$UA = \frac{1}{\sum\limits_{n=1}^{n=3} R_n} = \frac{1}{(1/h_1 A_1) + (L/kA_k) + (1/h_2 A_2)}$$

For a tube-within-a-tube heat exchanger, as shown in Fig. 8.2a, the area at the inner heat transfer surface is $2\pi r_i L$ and the area at the outer surface is $2\pi r_o L$.

Thus, if the overall heat transfer coefficient is *based on the outer area* A_0,

$$U_o = \frac{1}{(A_o/A_i h_i) + [A_o \ln(r_o/r_i)/2\pi kL] + (1/h_o)} \tag{8.2}$$

while *on the basis of the inner area* A_i we get

$$U_i = \frac{1}{(1/h_i) + [A_i \ln(r_o/r_i)/2\pi kL] + (A_i/A_o h_o)} \tag{8.3}$$

Although for a careful and precise design it is always necessary to calculate the individual heat transfer coefficients, for preliminary estimates it is often useful to have an approximate value of U, typical of conditions encountered in practice. Table 8.1 lists a few typical values of U for various applications (5). It should be noted that in many cases the value of U is almost completely determined by the thermal resistance at one of the fluid/solid interfaces, as when one of the fluids is a gas and the other a liquid, or when one of the fluids is a boiling liquid with a very large heat transfer coefficient.

8.3.1 Fouling Factors

The overall heat transfer coefficient of a heat exchanger under operating conditions, especially in the process industry, can often not be predicted from thermal analysis alone. During operation with most liquids and some gases, a deposit gradually builds up on the heat transfer surface. This deposit may be rust, boiler scale, silt, coke, or any number of things. Its effect, which is referred to as *fouling*, is to increase the thermal resistance. The manufacturer cannot usually predict the nature of the dirt deposit or the rate of fouling. Therefore, only the performance of clean exchangers can be guaranteed. The thermal resistance of the deposit can generally be obtained only from actual tests or from experience. If performance tests are made on a clean exchanger and repeated later after the unit has been in service for some time, the thermal resistance of the deposit (or *fouling factor*) R_d can be determined from the relation

$$R_d = \frac{1}{U_d} - \frac{1}{U} \tag{8.4}$$

where U = overall heat transfer coefficient of clean exchanger

         U_d = overall heat transfer coefficient after fouling has occurred

         R_d = unit thermal resistance of deposit

A convenient working form of Eq. (8.4) is

$$U_d = \frac{1}{R_d + 1/U}$$

Fouling factors for various applications have been compiled by the Tubular Exchanger Manufacturers Association (TEMA) and are available in their publication (6). A few samples are given in Table 8.2. The fouling factors should be

TABLE 8.1 OVERALL HEAT TRANSFER COEFFICIENTS FOR VARIOUS APPLICATIONS (W/m² K)ᵃ

Heat flow → to: from:	Gas (stagnant) $\bar{h}_c = 5-15$	Gas (flowing) $\bar{h}_c = 10-100$	Liquid (stagnant) $\bar{h}_c = 50-1,000$	Liquid (flowing) Water $\bar{h}_c = 3,000-1,000$ Other liquids $\bar{h}_c = 500-2,000$	Boiling liquid Water $\bar{h}_c = 3,500-60,000$ Other liquids $\bar{h}_c = 1,000-20,000$
Gas (free convection) $\bar{h}_c = 5-15$	Room/outside air through glass $U = 1-2$	Superheaters $U = 3-10$		Combustion chamber $U = 10-40 +$ radiation	Steam boiler $U = 10-40 +$ radiation
Gas (flowing) $\bar{h}_c = 10-100$		Heat exchangers for gases $U = 10-30$	Gas boiler $U = 10-50$		
Liquid (free convection) $\bar{h}_c = 50-10,000$			Oil bath for heating $U = 25-500$	Cooling coil $U = 500-1,500$ with stirring	
Liquid (flowing) water $\bar{h}_c = 3,000-10,000$ other liquids $\bar{h}_c = 500-3,000$	Radiator central heating $U = 5-15$	Gas coolers $U = 10-50$	Heating coil in vessel water/water without stirring $U = 50-250$ with stirring $U = 500-2,000$	Heat exchanger water/water $U = 900-2,500$ water/other liquids $U = 200-1,000$	Evaporators of refrigerators or brine coolers $U = 300-1,000$
Condensing vapor water $\bar{h}_c = 5,000-30,000$ other liquids $\bar{h}_c = 1,000-4,000$	Steam radiators $U = 5-20$	Air heaters $U = 10-50$	Steam jackets around vessels with stirrers, water $U = 300-1,000$ other liquids $U = 150-500$	Condensers steam/water $U = 1,000-4,000$ other vapor/ water $U = 300-1,000$	Evaporators steam/water $U = 1,500-6,000$ steam/other liquids $U = 300-2,000$

ᵃ Adapted from W. J. Beek and K. M. K. Muttall (5).

TABLE 8.2 NORMAL FOULING FACTORS[a]

Type of fluid	Fouling factor, R_d (m² K/W)
Seawater	
Below 325 K	0.00009
Above 325 K	0.0002
Treated boiler feedwater above 325 K	0.0002
Fuel oil	0.0009
Quenching oil	0.0007
Alcohol vapors	0.00009
Steam, non-oil-bearing	0.00009
Industrial air	0.0004
Refrigerating liquid	0.0002

[a] From *Standards of Tubular Exchanger Manufacturers Association* (6).

applied as indicated in the following equation for the overall design heat transfer coefficient U_d of *unfinned* tubes with deposits:

$$U_d = \frac{1}{(1/\bar{h}_o) + R_o + R_k + (R_i A_o/A_i) + (A_o/\bar{h}_i A_i)} \tag{8.5}$$

where U_d = design overall coefficient of heat transfer, W/m² K, based on unit area of outside tube surface

$\bar{h}_o$ = average heat transfer coefficient of fluid on outside of tubing, W/m² K

$\bar{h}_i$ = average heat transfer coefficient of fluid inside tubing, W/m² K

R_o = unit fouling resistance on outside of tubing, m² K/W

R_i = unit fouling resistance on inside of tubing, m² K/W

R_k = unit thermal resistance of tubing, m² K/W, based on outside tube surface area

$\dfrac{A_o}{A_i}$ = ratio of outside tube surface to inside tube surface area

8.4 LOG MEAN TEMPERATURE DIFFERENCE

The temperatures of fluids in a heat exchanger are generally not constant, but vary from point to point as heat flows from the hotter to the colder fluid. Even for a constant thermal resistance, the rate of heat flow will therefore vary along the path of the exchangers because its value depends on the temperature difference between the hot and the cold fluid at the section. Figures 8.8–8.11 illustrate the changes in temperature that may occur in either or both fluids in a simple shell-and-tube exchanger (Fig. 8.2b). The distances between the solid lines are proportional to the temperature differences ΔT between the two fluids.

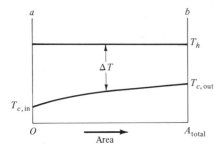

Figure 8.8 Temperature distribution in single-pass condenser.

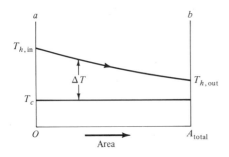

Figure 8.9 Temperature distribution in single-pass evaporator.

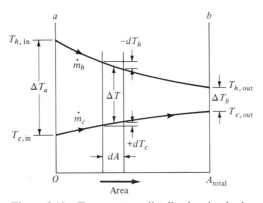

Figure 8.10 Temperature distribution in single-pass parallel-flow heat exchanger.

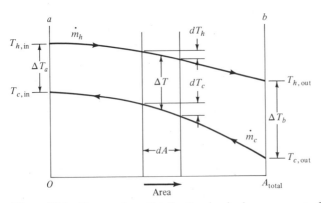

Figure 8.11 Temperature distribution in single-pass counterflow heat exchanger.

Figure 8.8 illustrates the case where a vapor is condensing at a constant temperature while the other fluid is being heated. Figure 8.9 is representative of a case where a liquid is evaporated at constant temperature while heat is flowing from a warmer fluid whose temperature decreases as it passes through the heat exchanger. For both of these cases the direction of flow of either fluid is immaterial and the constant-temperature medium may also be at rest. Figure 8.10 represents conditions in a parallel-flow exchanger, and Fig. 8.11 applies to counterflow. No change of phase occurs in the latter two cases. Inspection of Fig. 8.10 shows that, no matter how long the exchanger is, the final temperature of the colder fluid can never reach the exit temperature of the hotter fluid in parallel flow. For counterflow, on the other hand, the final temperature of the cooler fluid may exceed the outlet temperature of the hotter fluid, since a favorable temperature gradient exists all along the heat exchanger. An additional advantage of the counterflow arrangement is that, for a given rate of heat flow, less surface area is required than in parallel flow. In fact, the counterflow arrangement is the most effective of all heat exchanger arrangements.

To determine the rate of heat transfer in any of the aforementioned cases the equation

$$dq = U \, dA \, \Delta T \tag{8.6}$$

must be integrated over the heat transfer area A along the length of the exchanger. If the overall unit conductance U is constant, if changes in kinetic energy are neglected, and if the shell of the exchanger is perfectly insulated, Eq. (8.6) can be easily integrated analytically for parallel flow or counterflow. An energy balance over a differential area dA yields

$$dq = -\dot{m}_h c_{ph} \, dT_h = \pm \dot{m}_c c_{pc} \, dT_c = U \, dA(T_h - T_c) \tag{8.7}$$

where $\dot{m}$ is the mass rate of flow in kg/s, c_p is the specific heat at constant pressure in J/kg K, and T is the average bulk temperature of the fluid in K. The subscripts h and c refer to the hot and cold fluid, respectively; the plus sign in the third term applies to parallel flow and the minus sign to counterflow. If the specific heats of the fluids do not vary with temperature, we can write a heat balance from the inlet to an arbitrary cross section in the exchanger, or

$$-C_h(T_h - T_{h,\,in}) = C_c(T_c - T_{c,\,in}) \tag{8.8}$$

where $C_h = \dot{m}_h c_{ph}$, heat capacity rate of hotter fluid, W/K
$C_c = \dot{m}_c c_{pc}$, heat capacity rate of colder fluid, W/K

Solving Eq. (8.8) for T_h gives

$$T_h = T_{h,\,in} - \frac{C_c}{C_h}(T_c - T_{c,\,in}) \tag{8.9}$$

from which we obtain

$$T_h - T_c = -\left(1 + \frac{C_c}{C_h}\right)T_c + \frac{C_c}{C_h}T_{c,\,in} + T_{h,\,in} \tag{8.10}$$

Substituting Eq. (8.10) for $T_h - T_c$ in Eq. (8.7) yields, after some rearrangement,

$$\frac{dT_c}{-[1 + (C_c/C_h)]T_c + (C_c/C_h)T_{c,\,in} + T_{h,\,in}} = \frac{U\,dA}{C_c} \tag{8.11}$$

Integrating Eq. (8.11) over the entire length of the exchanger (i.e., from $A = 0$ to $A = A_{total}$) yields

$$\ln\left\{\frac{-[1 + (C_c/C_h)]T_{c,\,out} + (C_c/C_h)T_{c,\,in} + T_{h,\,in}}{-[1 + (C_c/C_h)]T_{c,\,in} + (C_c/C_h)T_{c,\,in} + T_{h,\,in}}\right\} = -\left(\frac{1}{C_c} + \frac{1}{C_h}\right)UA$$

which can be simplified to

$$\ln\left[\frac{(1 + C_c/C_h)(T_{c,\,in} - T_{c,\,out}) + T_{h,\,in} - T_{c,\,in}}{T_{h,\,in} - T_{c,\,in}}\right] = -\left(\frac{1}{C_c} + \frac{1}{C_h}\right)UA \tag{8.12}$$

From Eq. (8.8) we obtain

$$\frac{C_c}{C_h} = -\frac{T_{h,\,out} - T_{h,\,in}}{T_{c,\,out} - T_{c,\,in}} \tag{8.13}$$

which can be used to eliminate the heat capacity rates in Eq. (8.12). After some rearrangement we get

$$\ln\left(\frac{T_{h,\,out} - T_{c,\,out}}{T_{h,\,in} - T_{c,\,in}}\right) = [(T_{h,\,out} - T_{c,\,out}) - (T_{h,\,in} - T_{c,\,in})]\frac{UA}{q} \tag{8.14}$$

since

$$q = C_c(T_{c,\,out} - T_{c,\,in}) = C_h(T_{h,\,in} - T_{h,\,out})$$

Letting $T_h - T_c = \Delta T$, Eq. (8.14) can be written

$$q = UA\frac{\Delta T_a - \Delta T_b}{\ln(\Delta T_a/\Delta T_b)} \tag{8.15}$$

where the subscripts a and b refer to the respective ends of the exchanger and ΔT_a is the temperature difference between the hot and cold fluid streams at the inlet while ΔT_b is the temperature difference at the outlet end as shown in Figs. 8.10 and 8.11. In practice, it is convenient to use an average effective temperature difference $\overline{\Delta T}$ for the entire heat exchanger, defined by

$$q = UA\,\overline{\Delta T} \tag{8.16}$$

Comparing Eqs. (8.15) and (8.16), one finds that, for parallel flow or counterflow,

$$\overline{\Delta T} = \frac{\Delta T_a - \Delta T_b}{\ln(\Delta T_a/\Delta T_b)} \tag{8.17}$$

As mentioned previously in Chapter 6, the average temperature difference, $\overline{\Delta T}$, is called the *logarithmic mean temperature difference* often designated by LMTD. The LMTD also applies when the temperature of one of the fluids is constant, as shown in Figs. 8.8 and 8.9. When $\dot{m}_h c_{ph} = \dot{m}_c c_{pc}$, the temperature difference

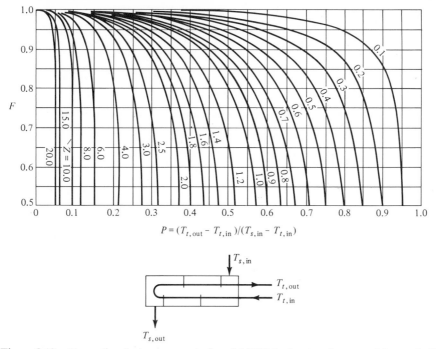

$$P = (T_{t,\text{out}} - T_{t,\text{in}})/(T_{s,\text{in}} - T_{t,\text{in}})$$

Figure 8.12 Correction factor to counterflow LMTD for heat exchanger with one shell pass and two, or a multiple of two, tube passes. (Courtesy of the Tubular Exchange Manufacturers Association.)

is constant in counterflow and $\overline{\Delta T} = \Delta T_a = \Delta T_b$. If the temperature difference ΔT_a is not more than 50 percent greater than ΔT_b, the arithmetic mean temperature difference will be within 1 percent of the LMTD and may be used to simplify calculations.

The use of the logarithmic mean temperature is only an approximation in practice because U is generally not constant. In design work, however, the overall conductance is usually evaluated at a mean section, halfway between ends, and treated as constant. If U varies considerably, numerical step-by-step integration of Eq. (8.6) may be necessary.

For more complex heat exchangers such as the shell-and-tube arrangements with several tube or shell passes and with cross-flow exchangers having mixed and unmixed flow, the mathematical derivation of an expression for the mean temperature difference becomes quite complex. The usual procedure is to modify the simple LMTD by correction factors, which have been published in chart form by Bowman et al. (7) and by TEMA (6). Four of these graphs* are shown in Figs. 8.12–8.15. The ordinate of each is the correction factor F. To obtain the true mean temperature for any of these arrangements, the LMTD calculated for *counterflow* must be multiplied by the appropriate correction

* Correction factors for several other arrangements are presented in (6).

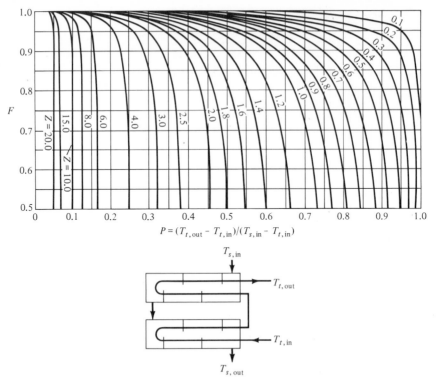

Figure 8.13 Correction factor to counterflow LMTD for heat exchanger with two shell passes and a multiple of two tube passes. (Courtesy of the Tubular Exchange Manufacturers Association.)

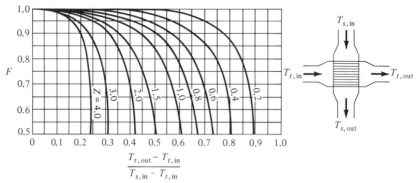

Figure 8.14 Correction factor to counterflow LMTD for cross-flow heat exchangers with the fluid on the shell side mixed, the other fluid unmixed, and one tube pass. [Extracted from R. A. Bowman, A. C. Mueller, and W. M. Nagel (7), with permission of the publishers, the American Society of Mechanical Engineers.]

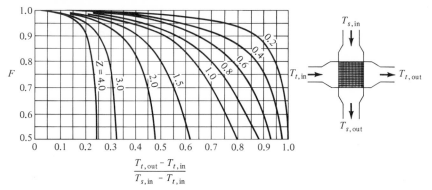

Figure 8.15 shown with axes: F (vertical, from 0.5 to 1.0) versus $\dfrac{T_{t,\text{out}} - T_{t,\text{in}}}{T_{s,\text{in}} - T_{t,\text{in}}}$ (horizontal, from 0 to 1.0). Curves labeled $Z = 4.0$, 3.0, 2.0, 1.5, 1.0, 0.8, 0.6, 0.4, 0.2. Schematic shows cross-flow exchanger with $T_{s,\text{in}}$ entering top, $T_{s,\text{out}}$ leaving bottom, $T_{t,\text{in}}$ entering left, $T_{t,\text{out}}$ leaving right.

Figure 8.15 Correction factor to counterflow LMTD for a cross-flow heat exchanger with both fluids unmixed and one tube pass. [Extracted from R. A. Bowman, A. C. Mueller, and W. M. Nagel (7); with permission of the publishers, the American Society of Mechanical Engineers.)

factor, that is,

$$\Delta T_{\text{mean}} = (\text{LMTD})(F) \tag{8.18}$$

The values shown on the abscissa are for the dimensionless temperature-difference ratio

$$P = \frac{T_{t,\text{out}} - T_{t,\text{in}}}{T_{s,\text{in}} - T_{t,\text{in}}} \tag{8.19}$$

where the subscripts t and s refer to the tube and shell fluid, respectively, and the subscripts "in" and "out" refer to the inlet and outlet conditions, respectively. The ratio P is an indication of the heating or cooling effectiveness and can vary from zero for a constant temperature of one of the fluids to unity for the case when the inlet temperature of the hotter fluid equals the outlet temperature of the colder fluid. The parameter for each of the curves Z is equal to the ratio of the products of the mass flow rate times the heat capacity of the two fluids $\dot{m}_t c_{pt}/\dot{m}_s c_{ps}$. This ratio is also equal to the temperature change of the shell fluid divided by the temperature change of the fluid in the tubes:

$$Z = \frac{\dot{m}_t c_{pt}}{\dot{m}_s c_{ps}} = \frac{T_{s,\text{in}} - T_{s,\text{out}}}{T_{t,\text{out}} - T_{t,\text{in}}} \tag{8.20}$$

In applying the correction factors it is immaterial whether the warmer fluid flows through shell or tubes. If the temperature of either of the fluids remains constant, the direction of flow is also immaterial, since F equals 1 and the LMTD applies directly.

EXAMPLE 8.1 ————————————————————————————————————

　　　Determine the heat transfer surface area required for a heat exchanger constructed from a 0.0254-m-OD tube to cool 6.93 kg/s of a 95 percent ethyl alcohol solution ($c_p = 3810$ J/kg K) from 65.6° to 39.4°C, using 6.30

kg/s of water available at 10°C. Assume that the overall coefficient of heat transfer based on the outer-tube area is 568 W/m² K and consider each of the following arrangements:

> (a) Parallel-flow tube and shell
> (b) Counterflow tube and shell
> (c) Reversed-current exchanger with 2 shell passes and 72 tube passes, the alcohol flowing through the shell and the water flowing through the tubes
> (d) Cross-flow, with one tube pass and one shell pass, shell-side fluid mixed

Solution.

(a) The outlet temperature of the water for any of the four arrangements can be obtained from an overall energy balance, assuming that the heat loss to the atmosphere is negligible. Writing the energy balance as

$$\dot{m}_h c_{ph}(T_{h,\,in} - T_{h,\,out}) = \dot{m}_c c_{pc}(T_{c,\,out} - T_{c,\,in})$$

and substituting the data in the equation above, we obtain

$$(6.93)(3810)(65.6 - 39.4) = (6.30)(4187)(T_{c,\,out} - 10)$$

from which the outlet temperature of the water is found to be 36.2°C. The rate of heat flow from the alcohol to the water is

$$q = \dot{m}_h c_{ph}(T_{h,\,in} - T_{h,\,out}) = (6.93)(3810)(65.6 - 39.4)$$
$$= 691{,}800 \text{ W}$$

From Eq. (8.17) the LMTD for parallel flow is

$$\text{LMTD} = \frac{\Delta T_a - \Delta T_b}{\ln(\Delta T_a/\Delta T_b)} = \frac{55.6 - 3.2}{\ln(55.6/3.2)} = 18.4°C$$

From Eq. (8.12) the heat transfer surface area is

$$A = \frac{q}{(U)(\text{LMTD})} = \frac{691{,}800}{(568)(18.4)} = 66.2 \text{ m}^2$$

The 830-m length of the exchanger for a 0.0254-m-OD tube would be too great to be practical.

(b) For the counterflow arrangement, the appropriate mean temperature difference is $65.6 - 36.2 = 29.4°C$, because $\dot{m}_c c_{pc} = \dot{m}_h c_{ph}$. The required area is

$$A = \frac{q}{(U)(\text{LMTD})} = \frac{691{,}800}{(568)(29.4)} = 41.4 \text{ m}^2$$

which is about 40 percent less than the area necessary for parallel flow.

(c) For the reversed-current arrangement, we determine the appropriate mean temperature difference by applying the correction factor found from Fig. 8.13 to the mean temperature for counterflow

$$P = \frac{T_{c,\,out} - T_{c,\,in}}{T_{h,\,in} - T_{c,\,in}} = \frac{36.2 - 10}{65.6 - 10} = 0.47$$

and the heat capacity rate ratio is

$$Z = \frac{\dot{m}_c c_{pc}}{\dot{m}_h c_{ph}} = 1$$

From the chart of Fig. 8.13, $F = 0.97$ and the heat transfer area is

$$A = \frac{41.4}{0.97} = 42.7 \text{ m}^2$$

The length of the exchanger for seventy-two 0.0254-m-OD tubes in parallel would be

$$L = \frac{A/72}{\pi D} = \frac{42.7/72}{\pi(0.0254)} = 7.4 \text{ m}$$

This length is not unreasonable, but if it is desirable to shorten the exchanger, more tubes could be used.

(d) For the cross-flow arrangement (Fig. 8.3), the correction factor is found from the chart of Fig. 8.14 to be 0.88. The required surface area is thus 47.0 m², about 10 percent larger than that for the reversed-current exchanger.

8.5 HEAT EXCHANGER EFFECTIVENESS

In the thermal analysis of the various types of heat exchangers presented in the preceding section, an equation [Eq. (8.16)] of the type

$$q = U A \Delta T_{\text{mean}}$$

was used. This form will be found convenient when all the terminal temperatures necessary for the evaluation of the appropriate mean temperature are known, and Eq. (8.16) is widely employed in the design of heat exchangers to given specifications. There are, however, numerous occasions when the performance of a heat exchanger (i.e., U) is known, or can at least be estimated, but the temperatures of the fluids leaving the exchanger are not known. This type of problem is encountered in the selection of a heat exchanger or when the unit has been tested at one flow rate, but service conditions require different flow rates for one or both fluids. The outlet temperatures and the rate of heat flow can only be found by a rather tedious trial-and-error procedure if the charts presented in the preceding section are used. In such cases it is desirable to circumvent entirely any reference to the logarithmic or any other mean temperature difference. A method that accomplishes this has been proposed by Nusselt (8) and Ten Broeck (9).

To obtain an equation for the rate of heat transfer which does not involve any of the outlet temperatures, we introduce the *heat exchanger effectiveness* $\mathcal{E}$. The heat exchanger effectiveness is defined as the ratio of the actual rate of heat transfer in a given heat exchanger to the maximum possible rate of heat exchange. The latter would be obtained in a counterflow heat exchanger of infinite heat transfer area. In this type of unit, if there are no external heat losses, the outlet temperature of the colder fluid equals the inlet temperature of the hotter fluid when $\dot{m}_c c_{pc} < \dot{m}_h c_{ph}$; when $\dot{m}_h c_{ph} < \dot{m}_c c_{pc}$, the outlet temperature of the warmer fluid equals the inlet temperature of the colder one. In other words, the effectiveness compares the actual heat transfer rate to the maximum rate whose only limit is the second law of thermodynamics. Depending on which of the heat capacity rates is smaller, the effectiveness is

$$\mathcal{E} = \frac{C_h(T_{h,\,in} - T_{h,\,out})}{C_{min}(T_{h,\,in} - T_{c,\,in})} \tag{8.21a}$$

or

$$\mathcal{E} = \frac{C_c(T_{c,\,out} - T_{c,\,in})}{C_{min}(T_{h,\,in} - T_{c,\,in})} \tag{8.21b}$$

where C_{min} is the smaller of the $\dot{m}_h c_{ph}$ and $\dot{m}_c c_{pc}$ magnitudes.

Once the effectiveness of a heat exchanger is known, the rate of heat transfer can be determined directly from the equation

$$q = \mathcal{E}C_{min}(T_{h,\,in} - T_{c,\,in}) \tag{8.22}$$

since

$$\mathcal{E}C_{min}(T_{h,in} - T_{c,\,in}) = C_h(T_{h,\,in} - T_{h,\,out}) = C_c(T_{c,\,out} - T_{c,\,in})$$

Equation (8.22) is the basic relation in this analysis because it expresses the rate of heat transfer in terms of the effectiveness, the smaller heat capacity rate, and the difference between the inlet temperatures. It replaces Eq. (8.16) in the LMTD analysis but does not involve the outlet temperatures. Equation (8.22) is, of course, also suitable for design purposes instead of Eq. (8.16).

We shall illustrate the method of deriving an expression for the effectiveness of a heat exchanger by applying it to a parallel-flow arrangement. The effectiveness can be introduced into Eq. (8.13) by replacing $(T_{c,\,in} - T_{c,\,out})/(T_{h,\,in} - T_{c,\,in})$ by the effectiveness relation from Eq. (8.21b). We obtain

$$\ln\left[1 - \mathcal{E}\left(\frac{C_{min}}{C_h} + \frac{C_{min}}{C_c}\right)\right] = -\left(\frac{1}{C_c} + \frac{1}{C_h}\right)UA$$

or

$$1 - \mathcal{E}\left(\frac{C_{min}}{C_h} + \frac{C_{min}}{C_c}\right) = e^{-(1/C_c + 1/C_h)UA}$$

Solving for $\mathcal{E}$ yields

$$\mathcal{E} = \frac{1 - e^{-[1 + (C_h/C_c)]UA/C_h}}{(C_{min}/C_h) + (C_{min}/C_c)} \tag{8.23}$$

When C_h is less than C_c, the effectiveness becomes

$$\mathscr{E} = \frac{1 - e^{-[1 + (C_h/C_c)]UA/C_h}}{1 + (C_h/C_c)} \tag{8.24a}$$

and when $C_c < C_h$, we obtain

$$\mathscr{E} = \frac{1 - e^{-[1 + (C_c/C_h)]UA/C_c}}{1 + (C_c/C_h)} \tag{8.24b}$$

The effectiveness for both cases can therefore be written in the form

$$\mathscr{E} = \frac{1 - e^{-[1 + (C_{min}/C_{max})]UA/C_{min}}}{1 + (C_{min}/C_{max})} \tag{8.25}$$

The foregoing derivation illustrates how the effectiveness for a given flow arrangement can be expressed in terms of two dimensionless parameters, the heat capacity rate ratio C_{min}/C_{max} and the ratio of the overall conductance to the smaller heat capacity rate, UA/C_{min}. The latter of the two parameters is called the *number of heat transfer units*, or NTU. The number of heat transfer units is a measure of the heat transfer size of the exchanger. The larger the value of NTU, the closer the heat exchanger approaches its thermodynamic limit. By analyses which, in principle, are similar to the one presented here for parallel flow, effectiveness may be evaluated for most flow arrangements of practical interest. The results have been put by Kays and London (1) into convenient graphs from which the effectiveness can be determined for given values of NTU and C_{min}/C_{max}. The effectiveness curves for some common flow arrangements are shown in Figs. 8.16–8.20. The abscissas of these figures are the NTUs of the

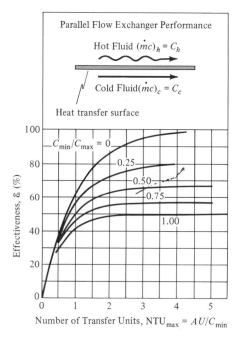

Figure 8.16 Heat exchanger effectiveness for parallel flow. [With permission from W. M. Kays and A. L. London (1).]

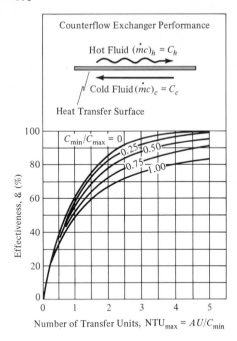

Figure 8.17 Heat exchanger effectiveness for counterflow. [With permission from W. M. Kays and A. L. London (1).]

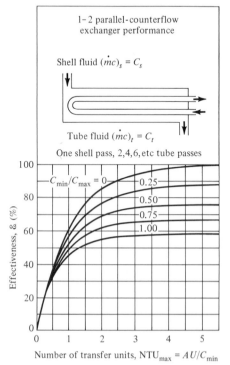

Figure 8.18 Heat exchanger effectiveness for shell-and-tube heat exchanger with one well-baffled shell pass and two, or a multiple of two, tube passes. [With permission from W. M. Kays and A. L. London (1).]

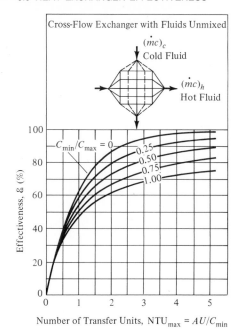

Figure 8.19 Heat exchanger effectiveness for cross-flow with both fluids unmixed. [With permission from W. M. Kays and A. L. London (1).]

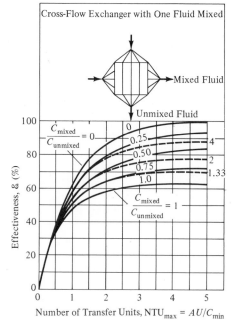

Figure 8.20 Heat exchanger effectiveness for cross-flow with one fluid mixed and the other unmixed. When $C_{\text{mixed}}/C_{\text{unmixed}} > 1$, NTU_{max} is based on C_{unmixed}. [With permission from W. M. Kays and A. L. London (1).]

heat exchangers. The constant parameter for each curve is the heat capacity rate ratio C_{min}/C_{max}, and the effectiveness is read on the ordinate. Note that, for an evaporator or condenser, $C_{min}/C_{max} = 0$, because if one fluid remains at constant temperature throughout the exchanger, its effective specific heat, and thus its heat capacity rate, is by definition equal to infinity.

EXAMPLE 8.2 ──

From a performance test on a well-baffled single-shell, two-tube-pass heat exchanger, the following data are available: oil ($c_p = 2100$ J/kg K) in turbulent flow inside the tubes entered at 340 K at the rate of 1.00 kg/s and left at 310 K; water flowing on the shell side entered at 290 K and left at 300 K. A change in service conditions requires the cooling of a similar oil from an initial temperature of 370 K but at three-fourths of the flow rate used in the performance test. Estimate the outlet temperature of the oil for the same water rate and inlet temperature as before.

Solution. The test data may be used to determine the heat capacity rate of the water and the overall conductance of the exchanger. The heat capacity rate of the water is, from Eq. (8.13),

$$C_c = C_h \frac{T_{h,in} - T_{h,out}}{T_{c,out} - T_{c,in}} = (1.00)(2100)\frac{340 - 310}{300 - 290}$$

$$= 6300 \text{ W/K}$$

and the temperature ratio P is, from Eq. (8.19),

$$P = \frac{T_{t,out} - T_{t,in}}{T_{s,in} - T_{t,in}} = \frac{340 - 310}{340 - 290} = 0.6$$

$$Z = \frac{300 - 290}{340 - 310} = 0.33$$

From Fig. 8.12, $F = 0.94$ and the mean temperature difference is

$$\overline{\Delta T} = (F)(\text{LMTD}) = (0.94)\frac{(340 - 300) - (310 - 290)}{\ln[(340 - 300)/(310 - 290)]} = 27.1 \text{ K}$$

From Eq. (8.16) the overall conductance is

$$UA = \frac{q}{\overline{\Delta T}} = \frac{(1.00)(2100)(340 - 310)}{27.1} = 2325 \text{ W/K}$$

Since the thermal resistance on the oil side is controlling, a decrease in velocity to 75 percent of the original value will increase the thermal resistance roughly by the velocity ratio raised to the 0.8 power. This can be verified by reference to Eq. (6.62). Under the new conditions the conductance, the

NTU, and the heat capacity rate ratio will therefore be approximately

$$UA \simeq (2325)(0.75)^{0.8} = 1850 \text{ W/K}$$

$$NTU = \frac{UA}{C_{oil}} = \frac{1850}{(0.75)(1.00)(2100)} = 1.17$$

and

$$\frac{C_{oil}}{C_{water}} = \frac{C_{min}}{C_{max}} = \frac{(0.75)(1.00)(2100)}{6300} = 0.25$$

From Fig. 8.18 the effectiveness is equal to 0.61. Hence from the definition of $\mathscr{E}$ in Eq. (8.21a), the oil outlet temperature is

$$T_{oil\ out} = T_{oil\ in} - \mathscr{E}\,\Delta T_{max} = 370 - [0.61(370 - 290)] = 321.2 \text{ K}$$

EXAMPLE 8.3

A flat-plate-type heater (Fig. 8.21) is to be used to heat air with the hot exhaust gases from a turbine. The required air-flow rate is 0.75 kg/s, entering at 290 K; the hot gases are available at a temperature of 1150 K and at a mass flow rate of 0.60 kg/s. Determine the temperature of the air leaving the heat exchanger for the parameters listed below.

P_a = wetted perimeter on air side, 0.703 m

P_g = wetted perimeter on gas side, 0.416 m

A_a = cross-sectional area of air passage (per passage), $2.275 \times 10^{-3} \text{ m}^2$

Solution. Inspection of Fig. 8.21 shows that the unit is of the cross-flow type, both fluids unmixed. As a first approximation the end effects will be

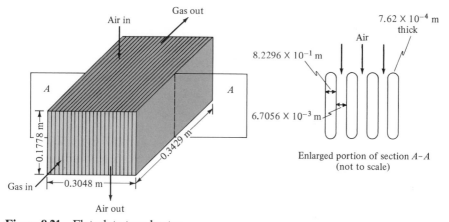

Figure 8.21 Flat-plate-type heater.

neglected. The flow systems for the air and gas streams are similar to flow in straight ducts having the following dimensions:

$$\text{Length of air duct } L_a = 0.20 \text{ m}$$

$$\text{Hydraulic diameter of air duct } D_{Ha} = \frac{4A_a}{P_a} = 0.0128 \text{ m}$$

$$\text{Length of gas duct } L_g = 0.35 \text{ m}$$

$$\text{Hydraulic diameter of gas duct } D_{Hg} = \frac{4A_g}{P_g} = 0.0154 \text{ m}$$

$$\text{Heat transfer surface area } A = 2.52 \text{ m}^2$$

The heat transfer coefficients may be evaluated from Eq. (6.63) for flow in ducts ($L_a/D_{Ha} = 15.6$, $L_g/D_{Hg} = 22.7$). A difficulty arises, however, because the temperatures of both fluids vary along the duct. It is therefore necessary to estimate an average film temperature and refine the calculations after the outlet and wall temperatures have been found. Selecting the average air-side film temperature at 573 K and the average gas-side film temperature at 973 K, the properties at those temperatures are, from Appendix 2, Table 27 (assuming that the properties of the gas are approximated by those of air):

$$\mu_{\text{air}} = 2.93 \times 10^{-5} \text{ N s/m}^2$$
$$\text{Pr}_{\text{air}} = 0.69$$
$$k_{\text{air}} = 0.0446 \text{ W/m K}$$
$$\mu_{\text{gas}} = 4.085 \times 10^{-5} \text{ N s/m}^2$$
$$\text{Pr}_{\text{gas}} = 0.73$$
$$k_{\text{gas}} = 0.0623 \text{ W/m K}$$
$$c_{p\text{air}} = 1047 \text{ J/kg K}$$
$$c_{p\text{gas}} = 1101 \text{ J/kg K}$$

The mass flow rates per unit area are

$$\left(\frac{\dot{m}}{A}\right)_{\text{air}} = \frac{0.75}{(19)(2.275 \times 10^{-3})} = 17.35 \text{ kg/m}^2 \text{ s}$$

$$\left(\frac{\dot{m}}{A}\right)_{\text{gas}} = \frac{0.60}{(18)(1.600 \times 10^{-3})} = 20.83 \text{ kg/m}^2 \text{ s}$$

The Reynolds numbers are

$$\text{Re}_{\text{air}} = \frac{(\dot{m}/A)_{\text{air}} D_{Ha}}{\mu_a} = \frac{(17.35)(0.0128)}{2.932 \times 10^{-5}} = 7570$$

$$\text{Re}_{\text{gas}} = \frac{(\dot{m}/A)_{\text{gas}} D_{Hg}}{\mu_g} = \frac{(20.83)(0.0154)}{4.085 \times 10^{-5}} = 7850$$

Using Eq. (6.63), the average heat transfer coefficients are

$$\bar{h}_{air} = \left(0.023 \frac{k_a}{D_H} Re_{D_H}{}^{0.8} Pr^{0.3}\right)\left(\frac{D_H}{1+L}\right)^{0.7}$$

$$= \left[(0.023)\frac{0.0446}{0.0128}(7570)^{0.8}(0.69)^{0.3}\right]\left(\frac{1}{1+15.6}\right)^{0.7}$$

$$= 95.0 \text{ W/m}^2 \text{ K}$$

$$\bar{h}_{gas} = \left[(0.023)\frac{0.0623}{0.0154}(7850)^{0.8}(0.73)^{0.3}\right]\left(\frac{1}{1+22.7}\right)^{0.7}$$

$$= 114 \text{ W/m}^2 \text{ K}$$

The above values show that approximately 60 percent of the overall temperature drop occurs on the air side. If the thermal resistance of the metal wall is neglected, the overall conductance is

$$UA = \frac{1}{\dfrac{1}{\bar{h}_a A} + \dfrac{1}{\bar{h}_g A}} = \frac{1}{\dfrac{1}{(95.0)(2.52)} + \dfrac{1}{(114)(2.52)}}$$

$$= 131 \text{ W/K}$$

The number of transfer units, based on the warmer fluid, which has the smaller heat capacity rate, is

$$NTU = \frac{UA}{C_{min}} = \frac{142}{(0.60)(1101)} = 0.198$$

The heat capacity-rate ratio is

$$\frac{C_g}{C_a} = \frac{(0.60)(1101)}{(0.75)(1047)} = 0.841$$

and from Fig. 8.19 the effectiveness is 0.13. Finally, the average outlet temperatures of the gas and air are

$$T_{gas\,out} = T_{gas\,in} - \mathscr{E}\,\Delta T_{max}$$
$$= 1150 - 0.13(1150 - 290) = 1038 \text{ K}$$

$$T_{air\,out} = T_{air\,in} + \frac{C_g}{C_a}\mathscr{E}\,\Delta T_{max} = 290 + (0.841)(0.13)(1150 - 290)$$

$$= 384 \text{ K}$$

A check on the average air-side and gas-side film temperatures gives values (567 K, 946 K) sufficiently close to the assumed values (573 K, 973 K) to make a second approximation unnecessary. To appreciate the usefulness of the approach based on the concept of heat exchanger effectiveness, it is suggested that this same problem be worked out by trial and error, using Eq. (8.16) and the chart of Fig. 8.15.

The effectiveness of the heat exchanger in Example 8.3 is very low (13 percent) because the heat transfer area is too small to utilize the available energy efficiently. The relative gain in heat transfer performance which can be achieved by increasing the heat transfer area is well represented on the effectiveness curves. A fivefold increase in area would raise the effectiveness to 60 percent. If, however, a particular design falls near or above the knee of these curves, increasing the surface area will not improve the performance appreciably but may cause an undue increase in the frictional pressure drop.

EXAMPLE 8.4 ──

A heat exchanger (condenser) using steam from the exhaust of a turbine at a pressure of 4.0-in. Hg abs. is to be used to heat 25,000 lb/h of seawater ($c = 0.95$ Btu/lb °F) from 60 to 110°F. The exchanger is to be sized for one shell pass and four tube passes with 60 parallel tube circuits of 0.995-in. ID and 1.125-in. OD brass tubing ($k = 60$ Btu/h ft °F). For the clean exchanger the average heat transfer coefficients at the steam and water sides are estimated to be 600 and 300 Btu/h ft² °F, respectively. Calculate the tube length required for long-term service.

Solution. At 4.0-in. Hg abs. the temperature of condensing steam will be 125.4°F, so that the required effectiveness of the exchanger is

$$\epsilon = \frac{T_{h,\,out} - T_{c,\,in}}{T_{h,\,in} - T_{c,\,in}} = \frac{110 - 60}{125.4 - 60} = 0.765$$

For a condenser $C_{min}/C_{max} = 0$ and from Figure 8.18, NTU = 1.4. The fouling factors from Table 8.2 are 0.0005h ft² °F/Btu for both sides of the tubes. The overall design heat-transfer coefficient per unit outside area of tube is from Eq. 1.5

$$U_d = \cfrac{1}{\cfrac{1}{600} + 0.0005 + \cfrac{1.125}{2 \times 12 \times 60} \ln \cfrac{1.125}{0.995} + \cfrac{0.0005x \times 1.125}{0.995} + \cfrac{1.125}{300 \times 0.995}}$$

$$= 152 \text{ Btu/h ft}^2 \text{ °F}$$

The total area A_o is $20\pi D_o L$ and, since $U_d A_o / C_{min} = 1.4$, the length of the tube is

$$L = \frac{1.4 \times 25,000 \times 0.95 \times 12}{60 \times \pi \times 1.125 \times 152} = 12.3 \text{ ft}$$

8.6 CLOSURE

In this chapter we have studied the thermal design of heat exchangers in which two fluids at different temperatures flow in spaces separated by a wall and exchange heat by convection to and from, and conduction through the wall. Such heat exchangers, sometimes called *recuperators*, are by far the most common and industrially important heat transfer devices. The most common configuration is the shell-and-tube heat exchanger for which two methods of thermal analysis have been presented: the LMTD (log mean temperature difference) and the NTU or effectiveness method. The former is most convenient when all the terminal temperatures are specified and the heat exchanger area is to be determined, while the latter is preferred when the thermal performance or the area is known, specified, or can be estimated. Both of these methods are useful, but it is important to reemphasize the rather stringent assumptions on which they are based:

1. The overall heat transfer coefficient U is uniform over the entire heat exchanger surface.
2. The physical properties of the fluids do not vary with temperature.
3. Available correlations are satisfactory to predict the individual heat transfer coefficients required to determine U.

Current design methodology is based usually on suitably chosen average values. When the spacial variation of U can be predicted, the appropriate value is an area average, $\bar{U}$, given by

$$\bar{U} = \frac{1}{A} \int_A U \, dA \tag{8.26}$$

The integration can be carried out numerically by methods presented in Chapter 3. But even this approach leaves the final result with a margin of error that is difficult to pin down quantitatively. It is probable that in the future increased emphasis will be placed on computer-aided design (CAD), and the reader is encouraged to follow developments in this area. These tools will be particularly important in the design of condensers, and some preliminary information on this topic will be presented on Section 10.5 in the chapter on heat transfer with change in phase.

In addition to recuperators there are two other *generic* types of heat exchangers in use. In both of these types the hot and cold fluid streams occupy the same space, a channel with or without solid inserts. In one type, the *regenerator*, the hot and the cold fluid pass alternately over the same heat transfer surface. In the other type, exemplified by the *cooling tower*, both fluids flow through the same passage simultaneously and contact each other directly. These types of exchangers are therefore often called *direct contact devices*. In many of the latter type the transfer of heat is accompanied by simultaneous transfer of mass. The transfer mechanism is complex and the reader is referred to (10), (11) (12), and (20) for more information.

Periodic flow regenerators have been used in practice only with gases. The regenerator consists of one or more flow passages which are partially filled either with solid pellets or with metal matrix inserts. During one part of the cycle the inserts store internal energy as the warmer fluid flows over their surfaces. During the other part of the cycle internal energy is released as the colder fluid passes through the regenerator and is heated. Thus, heat is transferred in a cyclic process. The principal advantage of the regenerator is a high heat transfer effectiveness per unit weight and space. The major problem is to prevent

TABLE 8.3 APPROXIMATE OVERALL HEAT TRANSFER
              COEFFICIENTS FOR PRELIMINARY ESTIMATES

	Overall Coefficient, U	
Duty	(Btu/h ft^2 °F)	(W/m^2 K)
Steam to water		
Instantaneous heater	400–600	2,270–3,400
Storage-tank heater	175–300	990–1,700
Steam to oil		
Heavy fuel	10–30	57–170
Light fuel	30–60	170–340
Light petroleum distillate	50–200	280–1,130
Steam to aqueous solutions	100–600	570–3,400
Steam to gases	5–50	28–280
Water to compressed air	10–30	57–170
Water to water, jacket water coolers	150–275	850–1,560
Water to lubricating oil	20–60	110–340
Water to condensing oil vapors	40–100	220–570
Water to condensing alcohol	45–120	255–680
Water to condensing Freon-12	80–150	450–850
Water to condensing ammonia	150–250	850–1,400
Water to organic solvents, alcohol	50–150	280–850
Water to boiling Freon-12	50–150	280–850
Water to gasoline	60–90	340–510
Water to gas oil or distillate	35–60	200–340
Water to brine	100–200	570–1,130
Light organics to light organics	40–75	220–425
Medium organics to medium organics	20–60	110–340
Heavy organics to heavy organics	10–40	57–220
Heavy organics to light organics	10–60	57–340
Crude oil to gas oil	30–55	170–310

leakage between the warmer and cooler fluids at elevated pressures. Regenerators have been used successfully as air preheaters in open-hearth and blast furnaces, in gas liquefication processes, and gas turbines.

The analysis of regenerators is quite involved and the reader interested in the design and operation of these units is referred to (1), (14), (15), and (16) for detailed information. Reference (14) contains a summary of the design theory with particular emphasis on the exhaust-gas thermal-energy regenerator in gas turbine power plants. Reference (15) presents calculated values for the effectiveness of regenerators and (16) gives a complete and detailed treatment of regenerator theory and practice.

For preliminary estimates of shell-and-tube heat exchanger sizes and performance parameters, it is often sufficient to know the order of magnitude of the overall heat transfer coefficient under average service conditions. Typical values of overall heat transfer coefficients recommended for preliminary estimates by Mueller (17) are given in Table 8.3.

For an up-to-date summary of specialized topics on the design and performance of heat exchangers, including evaporation and condensation, heat exchanger vibration, compact heat exchangers, fouling of heat exchangers, and heat exchange enhancement methods, the reader is referred to Taborek et al. (18) and Schlünder (19).

PROBLEMS

The problems for this chapter are organized as shown in the table below. Problems marked with an asterisk* require the direct or indirect evaluation of heat transfer coefficients before the heat exchanger can be analyzed.

Problem number	Section	Subject
8.1 to 8.8	8.3	Overall Heat Transfer Coefficient
8.9 to 8.18	8.4	Log Mean Temperature Difference
8.19 to 8.28	8.5	Heat Exchanger Effectiveness
8.29 to 8.38	All	Design Problems

8.1   In a heat exchanger air flows over brass tubes of 1.8 cm ID and 2.1 cm OD that contain steam. The convective heat-transfer coefficients on the air and steam sides of the tubes are 70 W/m^2 K and 210 W/m^2 K, respectively. Calculate the overall heat transfer coefficient for the heat exchanger (a) based on the inner tube area, (b) based on the outer tube area.

8.2   Repeat Problem 8.1 but assume that a fouling factor on the inside of the tube of 0.0018 m^2 K/W has developed during operation.

8.3   A light oil flows through a copper tube of 2.6 cm ID and 3.2 cm OD. Air is flowing over the exterior of the tube. The convective heat transfer coefficient for the oil is 120 W/m^2 K and for the air is 35 W/m^2 K. Calculate the overall heat transfer coefficient based on the outside area of the tube (a) considering the thermal resistance of the tube, (b) neglecting the resistance of the tube.

8.4   Repeat Problem 8.3, but assume that fouling factors of 0.0009 m^2 K/W on the inside and 0.0004 m^2 K/W on the outside respectively have developed.

*8.5   Water flowing in a long aluminum tube is to be heated by air flowing perpendicular to the exterior of the tube. The ID of the tube is 1.85 cm and its OD is 2.3 cm. The mass flow rate of the water through the tube is 0.65 kg/s and the temperature of the water in the tube averages 30°C. The free stream velocity and ambient temperature of the air are 10 m/s and 120°C, respectively. Estimate the overall heat-transfer coefficient for the heat exchanger using appropriate correlations from previous chapters. State all your assumptions.

8.6   Hot water is used to heat air in a double pipe heat exchanger. If the heat transfer coefficients on the water side and on the air side are 100 Btu/hr ft^2 °F and 10 Btu/hr ft^2 °F, respectively, calculate the overall heat transfer coefficient per unit length based on the outer diameter. The heat exchanger pipe is 2 inch, schedule 40, made of steel (k = 54 W/m K), with water inside. Express your answer in Btu/hr ft °F and W/m °C.

8.7   Repeat Problem 8.6, but assume that a fouling factor of 0.001 hr °F ft^2/Btu has developed over time.

8.8 The heat transfer coefficient on the inside of a copper tube (1.9 cm ID and 2.3 cm OD) is 500 W/m² K and 120 W/m² K on the outside, but a deposit with a fouling factor of 0.009 m² K/W has built up over time. Estimate the percent increase in the overall heat transfer coefficient if the deposit were removed.

8.9 A shell-and-tube heat exchanger has one shell pass and four tube passes. The fluid in the tubes enters at 200°C and leaves at 100°C. The temperature of the fluid entering the shell is 20°C and is 90°C as it leaves the shell. The overall heat transfer coefficient based on a surface area of 12 m² is 300 W/m² K. Calculate the heat transfer rate between the fluids.

8.10 Exhaust gases from a power plant are used to preheat air in a cross-flow heat exchanger. The exhaust gases enter the heat exchanger at 450°C and leave at 200°C. The air enters the heat exchanger at 70°C, leaves at 250°C and has a mass flow rate of 10 kg/s. Assume the properties of the exhaust gases can be approximated by those of air. The overall heat transfer coefficient of the heat exchanger is 154 W/m² K. Calculate the heat exchanger surface area required if (a) the air is unmixed and the exhaust gases are mixed and (b) both fluids are unmixed.

8.11 A shell-and-tube heat exchanger with two tube passes and a single shell pass is used to heat water by condensing steam in the shell. The flow rate of the water is 15 kg/s and it is heated from 60 to 80°C. The steam condenses at 140°C and the overall heat transfer coefficient of the heat exchanger is 820 W/m² K. If there are 45 tubes with an OD of 2.75 cm, calculate the length of tubes required.

8.12 Water entering a shell-and-tube heat exchanger at 35°C is to be heated to 75°C by an oil. The oil enters at 110°C and leaves at 75°C. The heat exchanger is arranged for counterflow with the water making one shell pass and the oil two tube passes. If the water flow rate is 68 kg per minute and the overall heat transfer coefficient is estimated from Table 8.1 to be 320 W/m² K, calculate the required heat exchanger area.

8.13 In a tubular heat exchanger with two shell passes and eight tube passes, 100,000 lb/hr of water are heated in the shell from 180 to 300°F. Hot exhaust gases having roughly the same physical properties as air enter the tubes at 650°F and leave at 350°F. The total surface, based on the outer tube surface, is 10,000 sq ft. Determine (a) the log-mean temperature if the heat exchanger were a simple counterflow type, (b) the correction factor F for the actual arrangement, (c) the effectiveness of the heat exchanger, (d) the average overall heat transfer coefficient.

8.14 In a single-pass counterflow heat exchanger, 4536 kg/hr of water enter at 15°C and cool 9071 kg/hr of an oil having a specific heat of 2093 J/kg °C lb F from 93 to 65°C. If the overall heat transfer coefficient is 284 W/m² °C, determine the surface area required.

8.15 Carbon dioxide at 27°C is to be used to heat 100,000 lb/hr of water from 37°C to 148 C while the gas temperature drops 204°C. For an overall heat transfer coefficient of 57 W/m² K, compute the required area of the exchanger in square feet for (a) parallel flow, (b) counterflow, (c) a 2–4 reversed current exchanger, and (d) crossflow, gas mixed.

8.16 An economizer is to be purchased for a power plant. The unit is to be large enough to heat 7.5 kg/s of water from 71 to 182°C. There are 26 kg/s of flue gases ($c_p =$

1000 J/kg K) available at 426°C. Estimate (a) the outlet temperature of the flue gases, (b) the heat transfer area required for a counterflow arrangement if the overall heat transfer coefficient is 57 W/m² K.

8.17 It is proposed to preheat the water for a boiler with flue gases from the stack ($c_p = 0.24$ Btu/lb$_m$ F). The flue gases are available at 300°F, at the rate of 2000 lb$_m$/hr. The water entering the exchanger at 60°F at the rate of 400 lb$_m$/hr is to be heated to 200°F. The heat exchanger is to be of the reversed current type, one shell pass and 4 tube passes. The water flows inside the tubes which are made of copper (1-in.-ID, 1.25-in.-OD). The heat transfer coefficient at the gas side is 20 Btu/hr sq ft °F, while the heat transfer coefficient on the water side is 200 Btu/hr sq ft °F. A scale on the water side offers an additional thermal resistance of 0.01 hr sq ft °F/Btu. (a) Determine the overall heat transfer coefficient based on the *outer* tube diameter. (b) Determine the appropriate mean temperature difference for the heat exchanger. (c) Estimate the required tube length. (d) What would be the improvement in the effectiveness if the water flow rate would be doubled, giving an average-unit conductance of 320 Btu/hr sq ft °F?

8.18 Water is to be heated from 50 to 90°F at the rate of 300,000 gal/hr in a single-pass shell-and-tube heat exchanger consisting of 1-in. schedule 40 steel pipe. The surface coefficient on the steam side is estimated to be 2000 Btu/hr sq ft °F. A pump is available which can deliver the desired quantity of water provided the pressure drop through the pipes does not exceed 15 psi. Calculate the number of tubes in parallel and the length of each tube necessary to operate the heat exchanger with the available pump.

8.19 Water is heated by hot air in a heat exchanger. The flow rate of the water is 12 kg/s and that of the air is 2 kg/s. The water enters at 40°C and the air enters at 460°C. The overall heat transfer coefficient of the heat exchanger is 275 W/m² K based on a surface area of 14 m². Determine the effectiveness of the heat exchanger if it is (a) parallel-flow type or a (b) cross-flow type. Then calculate the heat transfer rate for the two types of heat exchangers described and the outlet temperatures of the hot and cold fluids for the conditions given.

8.20 Oil ($c_p = 2.1$ kJ/kg K) is used to heat water in a shell and tube heat exchanger with a single shell and two tube passes. The overall heat transfer coefficient is 525 W/m² K. The mass flow rates are 7 kg/s for the oil and 10 kg/s for the water. The oil and water enter the heat exchanger at 240°C and 20°C, respectively. The heat exchanger is to be designed so that the water leaves the heat exchanger with a minimum temperature of 80°C. Calculate the heat transfer surface area required to achieve this temperature.

8.21 12.5 kg/s of benzene are to be cooled continuously from 82°C to 54°C by 10 kg/s of water available at 15.5°C. Using Table 8.2 estimate the surface area required for (a) crossflow, six tube passes, one-shell pass, neither of the fluids mixed; (b) reversed current exchanger, two-shell passes and eight tube passes, colder fluid inside of tubes.

8.22 Starting with a heat balance, show that the effectiveness for a counterflow arrangement is

$$\mathscr{E} = \frac{1 - e^{-[1-(C_{min}/C_{max})]NTU_{max}}}{1 - (C_{min}/C_{max})e^{-[1-(C_{min}/C_{max})]NTU_{max}}}$$

8.23 In gas turbine recuperators the exhaust gases are used to heat the incoming air and C_{min}/C_{max} is therefore approximately equal to unity. Show that for this case $\mathscr{E} = 1 - e^{-NTU}$ for counterflow and $\mathscr{E} = \frac{1}{2}(1 - e^{-2NTU})$ for parallel flow.

8.24 A steam-heated single-pass tubular preheater is designed to raise 45,000 lb/hr of air from 70 to 170°F, using saturated steam at 20 pisa. It is proposed to double the flow rate of air and, in order to be able to use the same heat exchanger and achieve the desired temperature rise, it is proposed to increase the steam pressure. Calculate the steam pressure necessary for the new conditions and comment on the design characteristics of the new arrangement. *Ans.* 2.54 psia

8.25 Water is heated while flowing through a pipe by steam condensing on the outside of the pipe. (a) Assuming a uniform overall conductance along the pipe, derive an expression for the water temperature as a function of distance from the entrance. (b) For an overall conductance of 570 W/m² K, based on the inside diameter of 5 cm, a steam temperature of 104°C, and a water-flow rate of 0.063 kg/s, calculate the length required to raise the water temperature from 15.5 to 65.5°C.

8.26 At a rate of 5.43 gpm, water at 80°F enters a No. 18 BWG $\frac{5}{8}$-in. condenser tube made of nickel chromium steel ($k = 15$ Btu/hr ft °F). The tube is 10 ft long and its outside is heated by steam condensing at 120°F. Under these conditions the average heat transfer coefficient on the water side is 1750 Btu/hr sq ft °F, and the heat transfer coefficient on the steam side may be taken as 2000 Btu/hr sq ft °F. On the interior of the tube, however, there is a scale having a thermal conductance equivalent to 1000 Btu/hr sq ft °F. (a) Calculate the overall heat transfer coefficient U per square foot of exterior surface area. (b) Calculate the exit temperature of the water. *Ans.* (a) 376 Btu/hr ft² °F; (b) 96°F

8.27 Water flowing at a rate of 100,000 lb/hr is to be cooled from 200 to 150°F by means of an equal flow rate of cold water entering at 100°F. The water velocity will be such that the overall coefficient of heat transfer U is 400 Btu/hr sq ft °F. Calculate the square feet of heat-exchanger surface needed for each of the following arrangements: (a) parallel flow, (b) counterflow, (c) a mutli-pass heat exchanger with the hot water making one pass through a well-baffled shell and the cold water making two passes through the tubes, and (d) a crossflow heat exchanger with the hot water making one pass through the shell and the cold water making one pass through the tubes. *Ans.* (b) 250; (c) 310; (d) 285

8.28 Water flowing at a rate of 10 kg/s through 50 double-pass tubes in a shell and tube heat exchanger heats air that flows through the shell side. The length of the brass tubes is 6.7 m and they have an outside diameter of 2.6 cm. The heat transfer coefficients of the water and air are 470 W/m² K and 210 W/m² K, respectively. The air enters the shell at a temperature of 15°C and a flow rate of 1.6 kg/s. The temperature of the water as it enters the tubes is 75°C. Calculate (a) the heat exchanger effectiveness, (b) the heat transfer rate to the air, and (c) the outlet temperature of the air and water.

8.29 A small space heater is constructed of $\frac{1}{2}$-in., 18-gauge brass tubes, 2 ft long. The tubes are arranged in equilateral, staggered triangles on $1\frac{1}{2}$-in. centers, four rows of 15 tubes each. A fan blows 2000 cfm of atmospheric air at 70°F uniformly over the tubes (see sketch). Estimate: (a) heat transfer rate; (b) exit temperature of the air; (c) rate of steam condensation, assuming that saturated steam at 2 psig inside the tubes as the heat source. State your assumptions. NOTE. Work parts a, b, and c of this problem by two methods. First use the LMTD, which requires a trial-and-error or graphical solution; then use the effectiveness method. (d) Also, estimate pressure drop of the air, in inches of water; (e) size motor required to drive the fan.

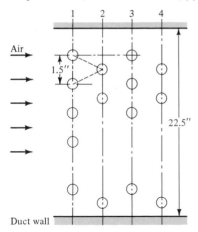

8.30 A one-tube pass cross-flow heat exchanger is considered for recovering energy from the exhaust gases of a turbine-driven engine. The heat exchanger is constructed of flat plates, forming an egg-crate pattern as shown in the sketch below. The velocities of the entering air (10°C) and exhaust gases (425°C) are both equal to 61 m s. Assuming that the properties of the exhaust gases are the same as those of the air, estimate for a path length of 1.2 m the overall heat transfer coefficient U, neglecting the thermal resistance of the intermediate metal wall. Then determine the outlet temperature of the air, comment on the suitability of the proposed design, and if possible, suggest improvements. State your assumptions.

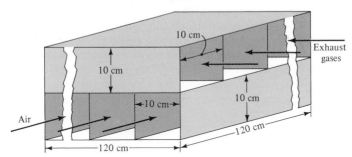

*8.31 A shell-and-tube counterflow heat exchanger is to be designed for heating an oil from 80 to 180°F. The oil is to pass through $1\frac{1}{2}$-in. schedule 40 pipes at a velocity of 200 fpm and steam is to condense at 215°F on the outside of the pipes. The specific heat of the oil is 0.43 Btu/lb °F and its mass density is 58 lb/cu ft. The steam-side heat transfer coefficient is approximately 1800 Btu/hr sq ft, and the thermal conductivity of the metal of the tubes is 17 Btu/hr ft °F. The results of previous experiments giving the oil-side heat transfer coefficients for the same pipe size at the same oil velocity as those to be used in the exchanger are shown below:

ΔT (°F)	135	115	95	75	35	
T_{oil} (°F)	80	100	120	140	160	180
h_{ci} (Btu/hr sq ft °F)	14	15	18	25	45	96

(a) Find the overall heat transfer coefficient U, based on the outer surface area at the point where the oil is 100°F. (b) Find the temperature of the inside surface of the pipe when the oil temperature is 100°F. (c) Find the required length of the tube bundle.   *Ans.* (a) 12.5 Btu/hr sq ft °F; (b) 213°F; (c) 520 ft

*8.32 A shell-and-tube heat exchanger in an ammonia plant is preheating 1132 m³/hr standard cubic feet of nitrogen per hour from 21 to 65°C using steam condensing at 138,000 N/m². The tubes in the heat exchanger have an inside diameter of 1 in. and the mass velocity of the nitrogen through the tubes is 13.6 kg/s m². In order to change from ammonia synthesis to methanol synthesis, the same heater is to be used to preheat carbon dioxide from 21 to 77°C, using steam condensing at 241,000 N/m². Calculate the flow rate which can be anticipated from this heat exchanger in pounds of carbon dioxide per hour.   *Ans.* About 3.429 kg/s.

*8.33 In an industrial plant a shell-and-tube heat exchanger is heating dirty water at the rate of 300,000 lb/hr from 140 to 235°F by means of steam condensing at 240°F on the outside of the tubes. The heat exchanger has 500 steel tubes (ID = 0.052 ft, OD = 0.700 ft) in a tube bundle which is 30 ft long. The water flows through the tubes while the steam condenses in the shell. If it may be assumed that the thermal resistance of the scale on the inside pipe wall is unaltered when the mass rate of flow is increased and that changes in water properties with temperature are negligible, estimate (a) the heat transfer coefficient on the water side and (b) the exit temperature of the dirty water if its mass rate of flow is doubled.
                     Ans. (a) About 600 Btu/hr sq ft °F; (b) 222°F

*8.34 Liquid benzene (specific gravity = 0.86) is to be heated in a counterflow concentric-pipe heat exchanger from 90 to 190°F. For a tentative design, the velocity of the benzene through the inside pipe (1-in., schedule 40) can be taken as 8 fps. *Saturated process steam at 200 psia* is available for heating. Two methods of using this steam are proposed: (a) Pass the process steam directly through the annular pipe of the exchanger; this would require that the latter be designed for the high pressure. (b) Throttle the steam adiabatically to 20 psia before passing it through the heater. In both cases the operation would be controlled so that *saturated water leaves the heater.* As an approximation, assume that for both cases the film coefficient for *condensing* steam remains constant at 2250 Btu/hr sq ft °F, that the thermal resistance of the pipe wall is negligible, and that the pressure drop for the steam is negligible. If the inside diameter of the outer pipe is 2 in., calculate the mass rate of flow of steam (lb/hr per pipe) and the length of heater required for each arrangement.
                     Ans. (a) 485 lb/hr per pipe, 14.7 ft; (b) 407 lb/hr per pipe, 45 ft

8.35 Calculate the overall conductance and the rate of heat flow from the hot gases to the cold air in the crossflow tube-bank type of heat exchanger shown in the accompanying illustration for the following operating conditions:

Air flow rate = 3000 lb/hr.

Hot gas flow rate = 5000 lb/hr.

Temperature of hot gases entering exchanger = 1600°F.

Temperature of cold air entering exchanger = 100°F.

Both gases are approximately at atmospheric pressure.

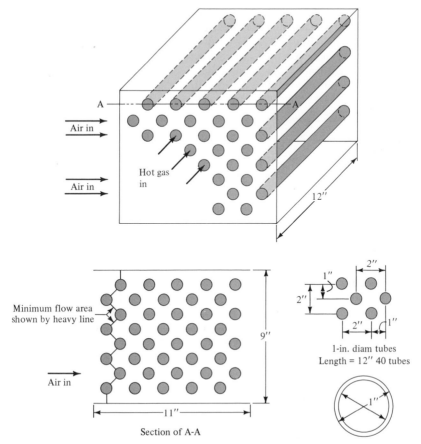

Section of A-A

8.36 An oil having a specific heat of 2100 J/kg K enters an oil cooler at 82°C at the rate of 2.5 kg/s. The cooler is a counterflow unit with water as the coolant, the transfer area being 28 m² and the overall heat transfer coefficient being 570 W/m² K. The water enters the exchanger at 27°C. Determine the water rate required if the oil is to leave the cooler at 38°C.

8.37 Dry air is cooled from 66 to 38°C, while flowing at the rate of 1.25 kg/s in a simple adiabatic counterflow heat exchanger, by means of cold air which enters at 15.5°C and flows at a rate of 1.6 kg/s. It is planned to lengthen the heat exchanger so that 1.25 kg/s of air can be cooled from 65 to 26.5°C with a counterflow current of air at 1.6 kg/s entering at 15.5°C. Assuming that the specific heat of the air is constant, calculate the ratio of the length of the new heat exchanger to the length of the original.
Ans. Approx. 2.36

8.38 Saturated steam at 5 psi condenses on the outside of an 8.5-ft length of copper tubing heating 0.6 gpm of water flowing in the tube. The water temperatures, measured at 10 equally spaced stations along the tube length are:

Station	1	2	3	4	5	6	7	8	9	10	11
Temperature (°F)	65	109	135	152	163	172	179	186	190	195	198

Calculate (a) average overall heat transfer coefficient U_0 based on the outside tube area; (b) average water-side heat transfer coefficient $\bar{h}_w$ (assume steam-side coefficient at $\bar{h}_s = 2000$ Btu/sq ft hr °F), (c) local overall coefficient U_x based on the outside tube area for each of the 10 sections between temperature stations, and (d) local waterside coefficients h_{wx} for each of the 10 sections.

Plot all items vs. tube length. Tube dimensions: ID = 0.790 in., OD = 0.985 in., length = 8.5 ft. Temperature station 1 is at tube entrance and station 11 at tube exit.

REFERENCES

1. W. M. Kays and A. L. London, *Compact Heat Exchangers*, 3rd ed., McGraw-Hill, New York, 1984.
2. W. M. Kays and A. L. London, "Remarks on the Behavior and Application of Compact High-Performance Heat Transfer Surfaces," Inst. Mech. Eng. and ASME, *Proc. General Discussion on Heat Transfer*, pp. 127–132, 1951.
3. R. K. Shah, "What's New is Heat Exchanger Design," *Mech. Eng.*, May 1984, pp. 50–59.
4. A. L. London and W. M. Kays, "The Gas Turbine Regenerator—the Use of Compact Heat Transfer Surfaces," *Trans. ASME*, vol. 72, p. 611, 1950.
5. W. J. Beek and K. M. K. Muttzall, *Transport Phenomena*, Wiley, New York, 1975.
6. *Standards of Tubular Exchanger Manufacturers Association*, 6th ed., 1978.
7. R. A. Bowman, A. C. Mueller, and W. M. Nagle, "Mean Temperature Difference in Design," *Trans. ASME*, vol. 62, pp. 283–294, 1940.
8. W. Nusselt, "A New Heat Transfer Formula for Cross-Flow," *Technische Mechanik and Thermodynamik*, vol. 12, 1930.
9. H. Ten Broeck, "Multipass Exchanger Calculations," *Ind. Eng. Chem.*, vol. 30, pp. 1041–1042, 1938
10. N. Afgan and E. U. Schluender, eds., *Heat Exchangers: Design and Theory Sourcebook*, McGraw-Hill, New York, 1974.
11. W. M. Rohsenow, J. P. Hartnett, and E. N. Ganic, eds., *Heat Transfer Handbook*, (*Applications*), 2nd ed., McGraw-Hill, New York, 1985.
12. D. W. Green and J. O. Maloney, eds., *Perry's Chemical Engineers' Handbook*, 6th ed., McGraw-Hill, New York, 1984.
13. A. H. P. Skelland, *Diffusional Mass Transfer*, Wiley, New York, 1974.
14. J. E. Coppage and A. L. London, "The Periodic-Flow Regenerator—A Summary of Design Theory," *Trans. ASME*, vol. 75, pp. 779–787, 1953.
15. T. J. Lambertson, "Performance Factors of a Periodic-Flow Heat Exchanger," M.S. thesis, USN Postgraduate School, Monterey, Calif., 1957; ASME Paper 57-SA-13, 1957.
16. H. Hausen, *Heat Transfer in Counterflow, Parallel Flow and Cross Flow*, McGraw-Hill, New York, 1983.
17. A. C. Mueller, "Thermal Design of Shell-and-Tube Heat Exchangers for Liquid-to-Liquid Heat Transfer," *Eng Bull., Res. Ser. 121*, Purdue Univ. Eng. Exp. Stn., 1954.
18. J. Taborek, J. F. Hewitt, A. Afgan, eds., *Heat Exchangers—Theory and Practice*, McGraw-Hill, New York, 1983.
19. E. U. Schlünder (ed.), *Heat Exchanger Design Handbook*, 5 vols., Hemisphere Pub., New York, 1982.
20. R. F. Boehm and F. Kreith, eds., *Direct Contact Heat Transfer*, Hemisphere Pub. Co., New York, 1986.

Chapter 9

Heat Transfer by Radiation

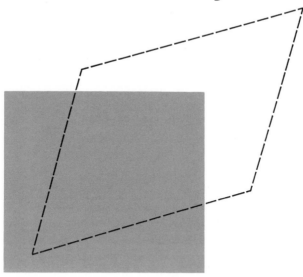

9.1 THERMAL RADIATION

When a body is placed in an enclosure whose walls are at a temperature below that of the body, the temperature of the body will decrease even if the enclosure is evacuated. The process by which heat is transferred from a body by virtue of its temperature, without the aid of any intervening medium, is called *thermal radiation*. This chapter deals with the characteristics of thermal radiation and radiation exchange, that is, heat transfer by radiation.

The physical mechanism of radiation is not yet completely understood. Radiant energy is envisioned sometimes as transported by electromagnetic waves, at other times as transported by photons. Neither viewpoint completely describes the nature of all observed phenomena. It is known, however, that radiation travels with the speed of light c, equal to about 3×10^8 m/s in a vacuum. This speed is equal to the product of the frequency and the wavelength of the radiation, or

$$c = \lambda v$$

where λ = wavelength, m

   v = frequency, s^{-1}

The unit of wavelength is the meter, but it is usually more convenient to use the micrometer (μm), equal to 10^{-6} m [1 μm = 10^4 Å (angstroms) or 3.94 × 10^{-5} in. (inches)]. In the older engineering literature the micron, equal to a micrometer, is also used and is denoted by the symbol μ.

From the viewpoint of electromagnetic theory, the waves travel at the speed of light, while from the quantum point of view, energy is transported by photons which travel at that speed. Although all the photons have the same velocity, there is always a distribution of energy among them. The energy associated with a photon, e_p, is given by $e_p = h\nu$, where h is Planck's constant, equal to 6.625 × 10^{-34} J s, and ν is the frequency of the radiation in s^{-1}. The energy spectrum can also be described in terms of the wavelength of radiation, λ, which is related to the propagation velocity and the frequency by $\lambda = c/\nu$.

Radiation phenomena are usually classified by their characteristic wavelength (Fig. 9.1). Electromagnetic phenomena encompass many types of radiation, from short-wavelength γ-rays and X-rays to long-wavelength radio waves. The wavelength of radiation depends on how the radiation is produced. For example, a metal bombarded by high-frequency electrons emits X-rays, while

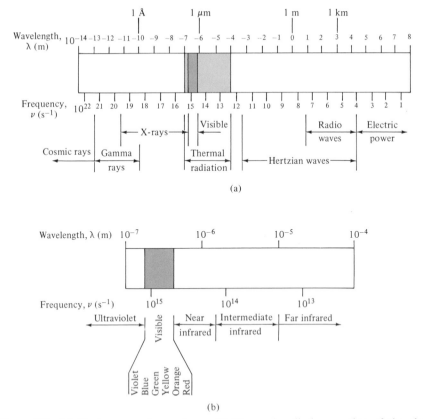

Figure 9.1 (a) Electromagnetic spectrum. (b) Thermal radiation portion of the electromagnetic spectrum.

certain crystals can be excited to emit long-wavelength radio waves. *Thermal radiation* is defined as radiant energy emitted by a medium by virtue of its temperature. In other words, the emission of thermal radiation is governed by the temperature of the emitting body. The wavelength range encompassed by thermal radiation falls approximately between 0.1 and 100 μm. This range is usually subdivided into the ultraviolet, the visible, and the infrared, as shown in Fig. 9.1.

Thermal radiation always encompasses a range of wavelengths. The amount of radiation emitted per unit wavelength is called *monochromatic radiation*; it varies with wavelength, and the word *spectral* is used to denote this dependence. The spectral distribution depends on the temperature and the surface characteristics of the emitting body. The sun, with an effective surface temperature of about 5800 K (10,400 R), emits most of its energy below 3 μm, whereas the earth, at a temperature of about 290 K (520 R), emits over 99 percent of its radiation at wavelengths longer than 3 μm. This difference in the spectral ranges gives rise to the "greenhouse effect." A greenhouse is warm inside even when the outside air is cool, because glass permits radiation at the wavelength of the sun to pass, but is almost opaque to radiation in the wavelength range emitted by the interior of the greenhouse. Thus, solar radiation may enter, but once it has been absorbed it cannot leave the interior of the greenhouse.

9.2 BLACKBODY RADIATION

A *blackbody*, or ideal radiator, is a body that emits and absorbs at any temperature the maximum possible amount of radiation at any given wavelength. The ideal radiator is a theoretical concept which sets an upper limit to the emission of radiation in accordance with the second law of thermodynamics. It is a standard with which the radiation characteristics of other media are compared.

For laboratory purposes, a blackbody can be approximated by a cavity, such as a hollow sphere, whose interior walls are maintained at a uniform temperature T. If a small hole is provided in the wall, any radiation entering through it is partly absorbed and partly reflected at the interior surfaces. The reflected radiation, as shown schematically in Fig. 9.2, will not immediately escape from the cavity but will first strike repeatedly the interior surface. Each time it strikes, a part of it is absorbed; when the original radiation beam finally reaches the hole again and escapes, it has been so weakened by repeated reflection that the energy leaving the cavity is negligible. This is true regardless of the surface and composition of the wall of the cavity. Thus, a small hole in the walls surrounding a large cavity acts like a blackbody because practically all the radiation incident upon it is absorbed.

In a similar manner, the radiation emitted by the interior surface of a cavity is absorbed and reflected many times and eventually fills the cavity uniformly. If a blackbody at the same temperature as the interior surface is

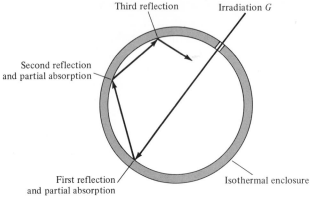

Figure 9.2 Schematic diagram of blackbody cavity.

placed in the cavity, it receives radiation uniformly; that is, it is irradiated isotropically. The blackbody absorbs all of the incident radiation and, since the system consisting of the blackbody and the cavity is at a uniform temperature, the rate of emission of radiation by the body must equal its rate of irradiation. Otherwise there would be a net transfer of energy as heat between two bodies at the same temperature in an isolated system, an obvious violation of the second law of thermodynamics. Denoting the rate at which radiant energy from the walls of the cavity is incident on the blackbody, that is, the *blackbody irradiation*, by G_b and the rate at which the blackbody emits energy by E_b, we thus obtain $G_b = E_b$. This means that the irradiation in a cavity whose walls are at a temperature T is equal to the emissive power of a blackbody at the same temperature. A small hole in the wall of a cavity will not disturb this condition appreciably, and the radiation escaping from it will therefore have blackbody characteristics. Since this radiation is independent of the nature of the surface, it follows that the emissive power of a blackbody depends only·on its temperature.

9.2.1 Blackbody Laws

The spectral radiant energy emission per unit time and per unit area from a blackbody at wavelength λ in the wavelength range $d\lambda$ will be denoted by $E_{b\lambda}\, d\lambda$. The quantity $E_{b\lambda}$ is usually called the *monochromatic blackbody emissive power*. A relationship which shows how the emissive power of a blackbody is distributed among the different wavelengths was derived by Max Planck, in 1900, by means of his quantum theory. According to *Planck's law*, an ideal radiator at temperature T emits radiation according to the relation (1)

$$E_{b\lambda}(T) = \frac{C_1}{\lambda^5(e^{C_2/\lambda T} - 1)} \tag{9.1}$$

where $E_{b\lambda}$ = monochromatic emissive power of a blackbody at absolute temperature T, W/m³ (Btu/h ft² μ)

         λ = wavelength, m (μ)

T = absolute temperature of the body, K (degrees R = 460 + °F)

C_1 = first radiation constant
= 3.7415 × 10⁻¹⁶ W m² (1.1870 × 10⁸ Btu/μ^4/ft² h)

C_2 = second radiation constant
= 1.4388 × 10⁻² m K (2.5896 × 10⁴ μ R)

The monochromatic emissive power for a blackbody at various tempera-ture is plotted in Fig. 9.3 as a function of wavelength. Observe that at tem-peratures below 5800 K the emission of radiation energy is appreciable between 0.2 and about 50 μm. The wavelength at which the monochromatic emissive power is a maximum, $E_{b\lambda}(\lambda_{max}, T)$, decreases with increasing temperature.

The relationship between the wavelength λ_{max} at which $E_{b\lambda}$ is a maximum and the absolute temperature is called *Wien's displacement law* (1). It can be derived from Planck's law by satisfying the condition for a maximum of $E_{b\lambda}$, or

$$\frac{dE_{b\lambda}}{d\lambda} = \frac{d}{d\lambda}\left[\frac{C_1}{\lambda^5(e^{C_2/\lambda T} - 1)}\right]_{T=\text{const}} = 0$$

The result of this operation is

$$\lambda_{max}T = 2.898 \times 10^{-3} \text{ m K} \quad (5216.4 \ \mu R) \tag{9.2}$$

The visible range of wavelengths, shown as a shaded band in Fig. 9.3, extends over a narrow region from about 0.4 to 0.7 μm. Only a very small amount

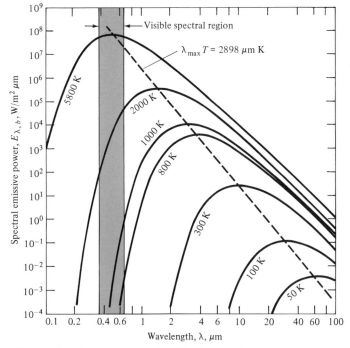

Figure 9.3 Monochromatic blackbody emissive power.

of the total energy falls in this range of wavelengths at temperatures below 800 K. At higher temperatures, however, the amount of radiant energy within the visible range increases and the human eye begins to detect the radiation. The sensation produced on the retina and transmitted to the optic nerve depends on the temperature, a phenomenon which is still used to estimate the temperatures of metals during heat treatment. At about 800 K an amount of radiant energy sufficient to be observed is emitted at wavelengths between 0.6 and 0.7 μm, and an object at that temperature glows with a dull-red color. As the temperature is further increased, the color changes to bright red and yellow, becoming nearly white at about 1500 K. At the same time, the brightness also increases because more and more of the total radiation falls within the visible range.

Recall from Chapter 1 that the total emission of radiation per unit surface area per unit time from a blackbody is related to the fourth power of the absolute temperature according to the *Stefan-Boltzmann law*

$$E_b(T) = \frac{q_r}{A} = \sigma T^4 \tag{9.3}$$

where A = area of the blackbody emitting the radiation, m^2 (ft^2)

T = absolute temperature of the area A in K (R)

σ = Stefan-Boltzmann constant
= 5.670 × 10^{-8} W/m^2 K^4 (0.1714 × 10^{-8} Btu/h ft^2 R^4)

The total emissive power given by Eq. (9.3) represents the total thermal radiation emitted over the entire wavelength spectrum. At a given temperature T the area under a curve such as that shown in Fig. 9.4 is E_b. The total emissive power and the monochromatic emissive power are related by

$$\int_0^\infty E_{b\lambda}\, d\lambda = \sigma T^4 = E_b \tag{9.4}$$

Substituting Eq. (9.1) for $E_{b\lambda}$ and performing the integration indicated above shows that the Stefan-Boltzmann constant σ and the constants C_1 and C_2 in Planck's law are related by

$$\sigma = \left(\frac{\pi}{C_2}\right)^4 \frac{C_1}{15} = 5.670 \times 10^{-8} \text{ W/m}^2 \text{ K}^4 \tag{9.5}$$

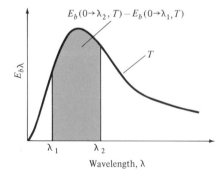

Figure 9.4 Radiation band and radiation function.

The Stefan-Boltzmann law shows that under most circumstances the effects of radiation are insignificant at low temperatures, owing to the small value for σ. At room temperature (~ 300 K) the total emissive power of a black surface is approximately 460 W/m². This value is only about one-tenth of the heat flux transferred from a surface to a fluid by convection, even when the convective heat transfer coefficient and temperature difference are reasonably low values of 100 W/m² K and 50 K, respectively. At low temperatures we can, therefore, often neglect radiative effects. However, we must include radiation effects at high temperatures because the emissive power increases with the fourth power of the absolute temperature.

9.2.2 Radiation Functions and Band Emission

For engineering calculations involving real surfaces it is often important to know the energy radiated at a specified wavelength or in a finite band between specific wavelengths λ_1 and λ_2, that is, $\int_{\lambda_1}^{\lambda_2} E_{b\lambda}(T) \, d\lambda$. Numerical calculations for such cases are facilitated by the use of the *radiation functions* (2). The derivation of these functions and their application are illustrated below.

At any given temperature the monochromatic emissive power is a maximum at the wavelength $\lambda_{max} = 2.898 \times 10^{-3}/T$, according to Eq. (9.2). Substituting λ_{max} into Eq. (9.1) gives the maximum monochromatic emissive power at temperature T, $E_{b\lambda max}(T)$, or

$$E_{b\lambda max}(T) = \frac{C_1 T^5}{(0.002898)^5 (e^{C_2/0.002898} - 1)} = 12.87 \times 10^{-6} \, T^5 \text{ W/m}^3 \quad (9.6)$$

If we divide the monochromatic emissive power of a blackbody, $E_{b\lambda}(T)$, by its maximum emissive power at the same temperature, $E_{b\lambda max}(T)$, we obtain the dimensionless ratio

$$\frac{E_{b\lambda}(T)}{E_{b\lambda max}(T)} = \left(\frac{2.898 \times 10^{-3}}{\lambda T} \right)^5 \left(\frac{e^{4.965} - 1}{e^{0.014388/\lambda T} - 1} \right) \quad (9.7)$$

where λ is in micrometers and T is in Kelvin.

Observe that the right-hand side of Eq. (9.7) is a unique function of the product λT. To determine the monochromatic emissive power $E_{b\lambda}$ for a blackbody at given values of λ and T, evaluate $E_{b\lambda}/E_{b\lambda max}$ from Eq. (9.7) and $E_{b\lambda max}$ from Eq. (9.6) and multiply.

EXAMPLE 9.1 ———————————————————————————

Determine (a) the wavelength at which the monochromatic emissive power of a tungsten filament at 1400 K is a maximum, (b) the monochromatic emissive power at that wavelength, and (c) the monochromatic emissive power at 5 μm.

Solution. From Eq. (9.2) the wavelength at which the emissive power is a maximum is

$$\lambda_{max} = 2.898 \times 10^{-3}/1400 = 2.07 \times 10^{-6} \text{ m}$$

From Eq. (9.6) at $T = 1400$ K,

$$E_{b\lambda max} = 12.87 \times 10^{-6} \times (1400)^5 = 6.92 \times 10^{10} \text{ W/m}^3$$

At $\lambda = 5 \mu m$, $\lambda T = 5 \times 1400 = 7.0 \times 10^{-3}$ mK; substituting this value into Eq. (9.7) we get

$$\frac{E_{b\lambda}(1400)}{E_{b\lambda max}(1400)} = \left(\frac{2.848 \times 10^{-3}}{7.0 \times 10^{-3}}\right)^5 \left(\frac{e^{4.965} - 1}{e^{0.014388/\lambda T} - 1}\right)$$

$$= (0.01216)\left(\frac{e^{4.965} - 1}{e^{2.055} - 1}\right) = 0.254$$

Thus, $E_{b\lambda}$ at 5 μm is 25.4 percent of the maximum value $E_{b\lambda max}$, or 1.758×10^{10} W/m^3.

It is often necessary to determine the fraction of the total blackbody emission in a spectral band between wavelengths λ_1 and λ_2. To obtain the emis-

TABLE 9.1   BLACKBODY RADIATION FUNCTIONS

λT (mK × 10³)	$\dfrac{E_b(0 \to \lambda T)}{\sigma T^4}$	λT (mK × 10³)	$\dfrac{E_b(0 \to \lambda T)}{\sigma T^4}$
0.2	0.341796×10^{-26}	6.2	0.754187
0.4	0.186468×10^{-11}	6.4	0.754182
0.6	0.929299×10^{-7}	6.6	0.783248
0.8	0.164351×10^{-4}	6.8	0.796180
1.0	0.320780×10^{-3}	7.0	0.808160
1.2	0.213431×10^{-2}	7.2	0.819270
1.4	0.779084×10^{-2}	7.4	0.829580
1.6	0.197204×10^{-1}	7.6	0.839157
1.8	0.393449×10^{-1}	7.8	0.848060
2.0	0.667347×10^{-1}	8.0	0.856344
2.2	0.100897	8.5	0.874666
2.4	0.140268	9.0	0.890090
2.6	0.183135	9.5	0.903147
2.8	0.227908	10.0	0.914263
3.0	0.273252	10.5	0.923775
3.2	0.318124	11.0	0.931956
3.4	0.361760	11.5	0.939027
3.6	0.403633	12	0.945167
3.8	0.443411	13	0.955210
4.0	0.480907	14	0.962970
4.2	0.516046	15	0.969056
4.4	0.548830	16	0.973890
4.6	0.579316	18	0.980939
4.8	0.607597	20	0.985683
5.0	0.633786	25	0.992299
5.2	0.658011	30	0.995427
5.4	0.680402	40	0.998057
5.6	0.701090	50	0.999045
5.8	0.720203	75	0.999807
6.0	0.737864	100	1.000000

sion in a band, shown in Fig. 9.4 by the shaded area, we must first calculate $E_b(0 - \lambda_1 T)$, the blackbody emission in the interval from 0 to λ_1 at T, or

$$\int_0^{\lambda_1} E_{b\lambda}(T)\,d\lambda = E_b(0 - \lambda_1 T) \tag{9.8}$$

This expression can be recast in dimensionless form as a function of λT, the product of wavelength and temperature

$$\frac{E_b(0 - \lambda_1 T)}{\sigma T^4} = \int_0^{\lambda_1 T} \frac{E_{b\lambda}}{\sigma T^5}\,d(\lambda T) \tag{9.9}$$

The fraction of the total blackbody emission between 0 and a given value of λ is presented in Table 9.1 and Fig. 9.5 as a universal function of λT.

To determine the amount of radiation emitted in the band between λ_1 and λ_2 for a black surface at temperature T we evaluate the difference between the two integrals below

$$\int_0^{\lambda_2} E_{b\lambda}(T)\,d\lambda - \int_0^{\lambda_1} E_{b\lambda}(T)\,d\lambda = E_b(0 - \lambda_2 T) - E_b(0 - \lambda_1 T) \tag{9.10}$$

The procedure is illustrated in the following example.

EXAMPLE 9.2 ──

Silica glass transmits 92 percent of the incident radiation in the wavelength range between 0.35 and 2.7 μm and is opaque at longer and shorter wavelengths. Estimate the percentage of solar radiation which the glass will transmit. The sun may be assumed to radiate as a blackbody at 5800 K.

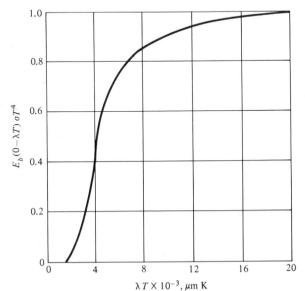

Figure 9.5 Ratio of blackbody emission between 0 and λ to the total emission, $E_b(0 - \lambda T)/\sigma T^4$ versus λT.

Solution.   For the wavelength range within which the glass is transparent, $\lambda T = 2{,}030$ μm K at the lower limit and $15{,}660$ μm K at the upper limit. From Table 9.1 we find

$$\frac{\int_0^{2{,}030} E_{b\lambda}\, d\lambda}{\int_0^\infty E_{b\lambda}\, d\lambda} = 6.7 \text{ percent}$$

and

$$\frac{\int_0^{15{,}660} E_{b\lambda}\, d\lambda}{\int_0^\infty E_{b\lambda}\, d\lambda} = 97.0 \text{ percent}$$

Thus 90.3 percent of the total radiant energy incident upon the glass from the sun is in the wavelength range between 0.35 and 2.7 μm and 83.1 percent of the solar radiation is transmitted through the glass.

9.2.3 Intensity of Radiation

In our discussion so far we have only considered the total amount of radiation leaving a surface, that is, the emissive power. This concept, however, is inadequate for a heat transfer analysis when the amount of radiation passing in a given direction and intercepted by some other body is sought. The amount of radiation passing in a given direction is described in terms of the intensity of radiation, I. Before defining the intensity of radiation, we must have measures of the direction and the space into which a body radiates. As shown in Fig. 9.6a, a differential plane angle $d\alpha$ is defined as the ratio of an element of arc length dl on a circle to the radius r of that circle. Similarly, a *differential solid angle* $d\omega$ is defined (see Fig. 9.6b) as the ratio of the element of area dA_n on a sphere to the square of the radius of sphere, or

$$d\omega = \frac{dA_n}{r^2} \tag{9.11}$$

The unit of the solid angle is the steradian (sr).

The rate of radiation heat flow per unit surface area from a body which passes in a given direction can be measured by determining the radiation through

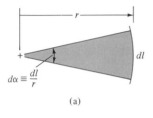

$$d\alpha \equiv \frac{dl}{r}$$

(a)

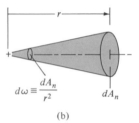

$$d\omega \equiv \frac{dA_n}{r^2}$$

(b)

Figure 9.6   (a) Differential plane angle and (b) differential solid angle.

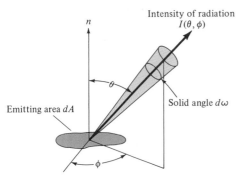

Figure 9.7 Schematic diagram illustrating intensity of radiation.

an element on the surface of a hemisphere constructed around the radiating surface. If the radius of this hemisphere equals unity, the hemisphere has a surface area of 2π and subtends a solid angle of 2π steradians about a point at the center of its base. The surface area on such a hemisphere with a radius of unity has the same numerical value as the so-called *solid angle* ω measured from the radiating surface element. The solid angle can be used to define simultaneously the direction and the space into which radiation from a body propagates.

The *intensity of radiation* $I(\theta, \phi)$ is the energy emitted per unit area per unit time into a solid angle $d\omega$ centered around a direction which can be defined in terms of the zenith angle θ and the azimuthal angle ϕ in the spherical coordinate system of Fig. 9.7. The differential area dA_n in Fig. 9.7 is perpendicular to the (θ, ϕ) direction. But for a spherical surface $dA_n = r\, d\theta\, r \sin\theta\, d\phi$ and therefore

$$d\omega = \sin\theta\, d\theta\, d\phi \qquad (9.12)$$

With the above definitions the intensity of radiation $I(\theta, \phi)$ is the rate at which radiation is emitted in the direction (θ, ϕ) per unit area of the emitting surface normal to this direction, per unit solid angle centered about (θ, ϕ).

Since the projected area of emission from Fig. 9.7 is $dA_1 \cos\theta$, we obtain for the intensity of a black surface, $I_b(\theta, \phi)$

$$I_b(\theta, \phi) = \frac{dq_r}{dA_1 \cos\theta\, d\omega} \;(\text{W/m}^2\ \text{sr}) \qquad (9.13)$$

where dq_r is the rate at which radiation emitted from dA_1 passes through dA_n.

EXAMPLE 9.3 _____

A flat black surface of area $A_1 = 10\ \text{cm}^2$ emits 1000 W/m^2 sr in the normal direction. A small surface A_2 having the same area as A_1 is placed as shown in Fig. 9.8 relative to A_1 at a distance of 0.5 m^2. Determine the solid angle subtended by A_2 and the rate at which A_2 is irradiated by A_1.

Solution. Since A_1 is black, it is a diffuse emitter and its intensity I_b is independent of direction. Moreover, since both areas are quite small,

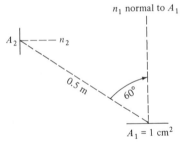

they may be approximated as differential surface areas and the solid angle can be calculated from Eq. (9.11) or $d\omega_{2-1} = dA_{n,2}/r^2$.

The area $dA_{n,2}$ is the projection of A_2 in the direction *normal* to the incident radiation for dA_1, or $dA_{n2} = dA_2 \cos\theta_2$, where θ is the angle between the normal $\bar{n}_2$ and the radiation ray connecting dA_1 and dA_2, that is, $\theta = 30°$. Thus

$$d\omega_{2-1} = \frac{A_2 \cos\theta_2}{r^2} = \frac{10^{-3}\ \text{m}^2 \cos 30°}{(0.5)^2} = 0.00346\ \text{sr}$$

The irradiation of A_2 by A_1, $q_{r,1\rightarrow 2}$, is

$$q_{r,1\rightarrow 2} = I_1 A_1 \cos\theta_1\, d\omega_{2-1}$$

$$= \left(1000\ \frac{\text{W}}{\text{m}^2\ \text{sr}}\right)(10^{-3}\ \text{m}^2)(\cos 60°)(0.00346\ \text{sr}) = 0.00173\ \text{W}$$

9.2.4 Relation between Intensity and Emissive Power

To relate the intensity of radiation to the emissive power, one simply determines the energy from a surface radiating into a hemispherical enclosure placed above it, as shown in Fig. 9.9. Since the hemisphere will intercept all the radiant rays emanating from the surface, the total amount of radiation passing through the hemispherical surface equals the emissive power. From Eq. (9.13), the rate of radiation emitted from dA_1 passing through dA_n is

$$\frac{dq_r}{dA_1} = I_b(\theta, \phi) \cos\theta\, d\omega \tag{9.14}$$

Substituting Eq. (9.12) for the solid angle $d\omega$ and integrating over the entire hemisphere yields the total rate of radiant emission per unit area, called the emissive power:

$$\left(\frac{q}{A}\right)_r = \int_0^{2\pi} \int_0^{\pi/2} I_b(\theta, \phi) \cos\theta \sin\theta\, d\theta\, d\phi \tag{9.15}$$

In order to integrate Eq. (9.15), the variation of the intensity with θ and ϕ must be known. As will be discussed more fully in the next section, the intensity of real surfaces exhibits no appreciable variation with ϕ but does vary with θ.

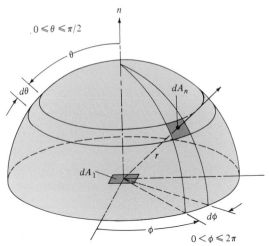

$0 \leqslant \theta \leqslant \pi/2$

$0 < \phi \leqslant 2\pi$

Figure 9.9 Radiation from a differential area dA into surrounding hemisphere centered at dA.

Although this variation can be taken into account, for most engineering calculations it may be assumed that the surface is diffuse and the intensity is uniform in all angular directions. Blackbody radiation is actually perfectly diffuse and radiation from industrial rough surfaces approaches diffuse characteristics. If the intensity from a surface is independent of direction, it is said to conform to *Lambert's cosine law*. For a black surface, integration of Eq. (9.15) yields the *blackbody emissive power* E_b:

$$\left(\frac{q}{A}\right)_r = E_b = \pi I_b \tag{9.16}$$

Thus for a black surface, the emissive power equals π times the intensity. The same relation between emissive power and intensity obtains for any surface that conforms to Lambert's cosine law.

The concept of intensity can be applied to the total radiation over the entire wavelength spectrum, as well as to monochromatic radiation. The relation between the total and the monochromatic intensity I_λ is simply

$$I(\phi, \theta) = \int_0^\infty I_\lambda(\phi, \theta) \, d\lambda \tag{9.17}$$

If a surface radiates diffusely, it is apparent that also

$$E_\lambda = \pi I_\lambda \tag{9.18}$$

since I_λ is uniform in all directions.

9.2.5 Irradiation

To make a heat balance on a body we need to know not only the radiation leaving but also the radiation incident on the surface. This radiation originates

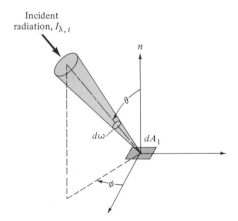

Figure 9.10   Radiation incident on a differential area dA_1 in a spherical coordinate system.

from emission and reflection occurring at other surfaces and will in general have a specific directional and spectral distribution. As shown in Fig. 9.10, the incident radiation can be characterized in terms of the incident spectral intensity, $I_{\lambda,i}$, defined as the rate at which radiant energy at wavelength λ from direction (θ, ϕ) per unit area of the intercepting surface normal to this direction, per unit solid angle about the direction (θ, ϕ), and per unit wavelength interval $d\lambda$ at λ. The term irradiation denotes radiation incident from all directions on a surface. *Spectral irradiation*, G_λ (W/m^2 m) is defined as the rate at which monochromatic radiation at wavelength λ is incident on a surface per unit area of that surface, or

$$G_\lambda = \int_0^{2\pi} \int_0^{\pi/2} I_{\lambda,i}(\lambda, \theta, \phi)\cos\theta \sin\theta \, d\theta \, d\phi \qquad (9.19a)$$

where $\sin\theta \, d\theta \, d\phi$ is the unit solid angle. Observe that the factor $\cos\theta$ originates from the fact that G_λ is a flux based on the actual surface area, whereas $I_{\lambda,i}$ is defined in terms of the projected area. The total irradiation represents the rate of radiation incident per unit area from all direction over all wave lengths and is given by

$$G = \int_0^\infty G_\lambda(\lambda) \, d\lambda = \int_0^\infty \int_0^{2\pi} \int_0^{\pi/2} I_{\lambda,i}(\lambda, \theta, \phi)\cos\theta \sin\theta \, d\theta \, d\phi \, d\lambda \quad (9.19b)$$

If the incident radiation is diffuse—that is the intercepting area is diffusely irradiated and $I_{\lambda,i}$ is independent of direction—it follows that

$$G = \pi I_i \qquad (9.20)$$

9.3 RADIATION PROPERTIES

Most surfaces encountered in engineering practice do not behave like blackbodies. To characterize the radiation properties of nonblack surfaces, dimensionless quantities such as the emittance, absorptance, and transmittance are used to relate the emitting, absorbing, and transmitting capabilities of a real surface to those of a blackbody. Radiation properties of real surfaces are functions of wavelength, temperature, and direction. The properties that describe how a surface behaves as a function of wavelength are called monochromatic or spectral properties, and the properties that describe the distribution of radiation with angular direction are called directional properties. For precise heat transfer calculation, we must know the radiative properties of the emitting surface as well as those of all other surfaces with which radiation exchange occurs.

Taking into account the spectral and directional properties of all surfaces, even if they are known, results in complex and involved analyses which can only be carried out by computer. However, engineering calculations of acceptable accuracy can usually be carried out by a simplified approach, using a single radiation property value averaged over the wavelength range of interest and directions. Radiation properties that are averaged over all wavelengths and directions are called total properties. Although we will almost exclusively use total radiative properties here, it is important to be aware of the spectral and directional characteristics of surfaces in order to account for them in problems in which these variations are significant. In the following we will discuss radiation properties in order of increasing complexity, beginning with total properties, followed by spectral properties, and finally directional properties.

9.3.1 Total Radiation Properties

For most engineering calculations total radiation properties as defined in this subsection are sufficiently accurate. The definition of total radiation properties is illustrated in Fig. 9.11. When radiation is incident on a surface at rate G, a portion of the total irradiation is absorbed in the material, a portion is reflected

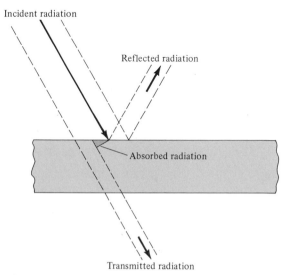

Incident radiation

Reflected radiation

Absorbed radiation

Transmitted radiation

Figure 9.11 Schematic diagram illustrating incident, reflected, and absorbed radiation in terms of total radiation properties.

from the surface, and the remainder is transmitted through the body. The absorptance, reflectance, and transmittance* describe how the total irradiation is distributed.

The *absorptance* α of a surface is the fraction of the total irradiation absorbed by the body. The *reflectance* ρ of a surface is defined as the fraction of the irradiation that is reflected from the surface. The *transmittance* τ of a body is the fraction of the incident radiation that is transmitted. If an energy balance is made on a surface as illustrated in Fig. 9.11 one obtains

$$\alpha G + \rho G + \tau G = G \qquad (9.21)$$

From Eq. (9.21), it is apparent that the sum of the absorptance, reflectance, and transmittance must equal unity:

$$\alpha + \rho + \tau = 1 \qquad (9.22)$$

When a body is opaque it will not transmit any of the incident radiation, that is, $\tau = 0$. For an opaque body Eq. (9.22) reduces to

$$\alpha + \rho = 1 \qquad (9.23)$$

If a surface is also a perfect reflector, from which all irradiation is reflected, ρ is unity and the transmittance as well as the absorptance is zero. A good mirror approaches a reflectivity of 1. As mentioned previously, a blackbody absorbs all of the irradiation and therefore has an absorptance equal to unity and a reflectance equal to zero.

Another important total radiation property of real surfaces is the emittance. The *emittance* of a surface, ϵ, is defined as the total radiation emitted divided by the total radiation that would be emitted by a blackbody at the same temperature, or

$$\epsilon = \frac{E(T)}{E_b(T)} = \frac{E(T)}{\sigma T^4} \qquad (9.24)$$

Since a blackbody emits the maximum possible radiation at a given temperature, the emittance of a surface is always between zero and unity. But when a surface is black, $E(T) = E_b(T)$ and $\epsilon_b = \alpha_b = 1.0$.

9.3.2 Monochromatic Radiation Properties and Kirchhoff's Law

Total radiation properties can be obtained from monochromatic properties, which apply only at a single wavelength. Designating E_λ as the monochromatic emissive power of an arbitrary surface, the monochromatic hemispherical emittance of the surface, ϵ_λ, is given by

$$\epsilon_\lambda = \frac{E_\lambda(T)}{E_{b\lambda}(T)} \qquad (9.25)$$

---

* Some authors prefer to use absorptivity, reflectivity, and transmissivity. But the suffix "tivity" should only be used for pure substances, e.g., gold.

In other words, ϵ_λ is the fraction of the blackbody radiation emitted by the surface at wavelength λ. Similarly, the hemispherical monochromatic absorptance of a surface, α_λ, is defined as the fraction of the total irradiation at wavelength λ which is absorbed by the surface,

$$\alpha_\lambda = \frac{G_\lambda(T)}{G_{b\lambda}(T)} \tag{9.26}$$

An energy balance on a monochromatic basis, similar to Eq. (9.22), yields

$$\alpha_\lambda + \epsilon_\lambda + \tau_\lambda = 1 \tag{9.27}$$

An important relation between ϵ_λ and α_λ can be obtained from *Kirchhoff's radiation law*, which states in essence that the monochromatic emittance is equal to the monochromatic absorptance for any surface. A rigorous derivation of this law has been presented by Planck (1), but the essential features can be illustrated more simply from the following consideration. Suppose we place a small body inside a black enclosure whose walls are fixed at temperature T (see Fig. 9.12). After thermal equilibrium is established, the body must attain the temperature of the walls. In accordance with the second law of thermodynamics the body must, under these conditions, emit at every wavelength as much radiation as it absorbs. If the monochromatic radiation per unit time per unit area incident on the body is $G_{b\lambda}$, the equilibrium condition is expressed by

$$E_\lambda = \alpha_\lambda G_{b\lambda} \tag{9.28}$$

or

$$\frac{E_\lambda}{\alpha_\lambda} = G_{b\lambda} \tag{9.29}$$

But since the incident radiation depends only on the temperature of the enclosure, it would be the same on any other body in thermal equilibrium with the enclosure, irrespective of the absorptance of the body's surface. One can therefore conclude that the ratio of the monochromatic emissive power to the absorptance at any given wavelength is the same for all bodies at thermal

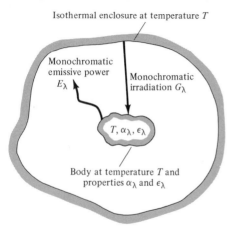

Isothermal enclosure at temperature T

Monochromatic emissive power E_λ

Monochromatic irradiation G_λ

$T, \alpha_\lambda, \epsilon_\lambda$

Body at temperature T and properties α_λ and ϵ_λ

Figure 9.12 Radiation emitted and received at wavelength λ by a body in an isothermal enclosure at temperature T.

equilibrium. Since the absorptance must always be less than unity and can be equal to one only for a perfect absorber, that is, a blackbody, Eq. (9.29) shows also that at any given temperature the emissive power is a maximum for a blackbody. Thus, when $\alpha_\lambda = 1$, $E_\lambda = E_{b\lambda}$, and $G_{b\lambda} = E_{b\lambda}$ in Eq. (9.29). Replacing E_λ by $\epsilon_\lambda E_{b\lambda}$ in Eq. (9.28) gives

$$\epsilon_\lambda E_{b\lambda} = \alpha_\lambda G_{b\lambda} = \alpha_\lambda E_{b\lambda}$$

which shows that at any wavelength λ at temperature T,

$$\epsilon_\lambda(\lambda, T) = \alpha_\lambda(\lambda, T) \tag{9.30}$$

as stated at the outset.

Although the above relation was derived under the condition that the body is in equilibrium with its surroundings, it is actually a general relation which applies under any conditions. The reason for this is that both α_λ and ϵ_λ are surface properties which depend solely on the condition of the surface and its temperature. We can therefore conclude that unless changes in temperature cause physical alteration in the surface characteristics, the hemispherical monochromatic absorptance equals the monochromatic emittance of a surface.

The *total hemispherical emittance* for a nonblack surface is obtained from Eqs. (9.8) and (9.25). Combining these two relations, we find that at a given temperature T the total hemispherical emittance is

$$\epsilon(T) = \frac{E(T)}{E_b(T)} = \frac{\int_0^\infty \epsilon_\lambda(\lambda)E_{b\lambda}(\lambda, T)\,d\lambda}{\int_0^\infty E_{b\lambda}(\lambda, T)\,d\lambda} \tag{9.31}$$

This relation shows that when the monochromatic emittance of a surface is a function of wavelength, it will vary with the temperature of the surface, even though the monochromatic emittance is solely a surface property. The reason for this variation is that the percentage of the total radiation that falls within a given wavelength band depends on the temperature of the emitting surface.

EXAMPLE 9.4 ————————————————————————

The hemispherical emittance of an aluminum paint is approximately 0.4 at wavelengths below 3 μm and 0.8 at longer wavelengths, as shown in Fig. 9.13. For this surface determine the total emittance at a room temperature of 27°C and at a temperature of 527°C. Why are the two values different?

Solution. At room temperature the product λT at which the emittance changes is equal to 3 μm $\times$ (27 + 273) K = 900 μm K, while at the elevated temperature $\lambda T = 2700$ μm K. From Table 9.1 we obtain

$$\frac{E_b(0 \to \lambda T)}{\sigma T^4} \cong 0.0001 \quad \text{for} \quad \lambda T = 900 \ \mu\text{m K}$$

$$\frac{E_b(0 \to \lambda T)}{\sigma T^4} = 0.140 \quad \text{for} \quad \lambda T = 2400 \ \mu\text{m K}$$

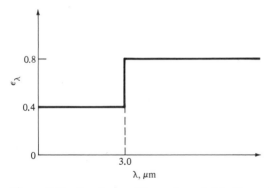

Figure 9.13 Spectral emittance for paint in Example 9.4.

Thus, the emittance at 27°C is essentially equal to 0.8, while at 527°C we get from Eq. (9.31)

$$
\epsilon = \frac{\displaystyle\int_0^{\lambda_1} \epsilon_\lambda(\lambda) E_{b\lambda}(\lambda T)\, d\lambda}{\displaystyle\int_0^{\lambda_1} E_{b\lambda}(\lambda T)\, d\lambda} + \frac{\displaystyle\int_{\lambda_1}^{\infty} \epsilon_\lambda(\lambda) E_{b\lambda}(\lambda T)\, d\lambda}{\displaystyle\int_{\lambda_1}^{\infty} E_{b\lambda}(\lambda T)\, d\lambda}
$$

$$
= (0.4)(0.14) + (0.8)(0.86) = 0.744
$$

The reason for the difference in the total emittance is that at the higher temperature the percentage of the total emissive power in the low-emittance region of the paint is appreciable, while at the lower temperature practically all the radiation is emitted at wavelengths above 3 μm.

Similarly, the *total absorptance* of a surface can be obtained from basic definitions. Consider a surface at temperature T subject to incident radiation from a source at T^* given by

$$
G = \int_0^\infty G_\lambda(\lambda^*, T^*)\, d\lambda \tag{9.32}
$$

where the asterisk is used to denote the conditions of the source. If the variation of the monochromatic absorptance with wavelength of the receiving surface is given by $\alpha_\lambda(\lambda)$, the total absorptance is

$$
\alpha(\lambda^*, T^*) = \frac{\displaystyle\int_0^\infty \alpha_\lambda(\lambda) G_\lambda(\lambda^*, T^*)\, d\lambda}{\displaystyle\int_0^\infty G_\lambda(\lambda^*, T^*)\, d\lambda} \tag{9.33}
$$

Note that the total absorptance of a surface depends on the temperature and on the spectral characteristics of the incident radiation. Therefore, although the relation $\epsilon_\lambda = \alpha_\lambda$ is always valid, the total values of absorptance and emittance are, in general, different for real surfaces.

9.3.3 Gray Bodies

Gray bodies are surfaces with monochromatic emittances and absorptances whose values are independent of wavelength. Even though real surfaces do not meet this specification exactly, it is often possible to choose suitable average values for the emittance and absorptance $\bar{\epsilon}$ and $\bar{\alpha}$ to make the gray-body assumption acceptable for engineering analysis. For a completely gray body, with subscript g denoting gray,

$$\epsilon_\lambda = \bar{\epsilon} = \bar{\alpha} = \alpha_\lambda = \epsilon_g = \alpha_g$$

The emissive power E_g is given by

$$E_g = \epsilon_g \sigma T^4 \tag{9.34}$$

Thus, if the emittance of a gray body is known at one wavelength, the total emittance and the total absorptance are also known. Moreover, the total values of absorptance and emittance are equal even if the body is not in thermal equilibrium with its surroundings. In practice, however, the choice of suitable average values should reflect the conditions of the source for the average absorptance and the temperature of the surface of the body that receives and emits radiation for the choice of the average emittance. A surface that is idealized as having uniform properties, but whose average emittance is not equal to the average absorptance, is called a selectively gray body.

EXAMPLE 9.5

The aluminum paint from Example 9.4 is used to cover the surface of a body that is maintained at 27°C. In one installation this body is irradiated by the sun, in another by a source at 527°C. Calculate the effective absorptance of the surface for both conditions, assuming the sun is a blackbody at 5800 K.

Solution. For the case of solar irradiation we find from Table 9.1 for $\lambda T = 3 \ \mu m \times 5800 \ K = 17,400 \ \mu m \ K = 17.4 \times 10^{-3} \ mK$, that

$$\frac{E_b(0 \to \lambda T)}{\sigma T^4} = 0.98$$

This means that 98 percent of the solar radiation falls below 3 μm and the effective absorptance is, from Eq. (9.33),

$$\alpha(\lambda_{sun}, T_{sun}) = \int_0^{3 \ \mu m} \alpha(\lambda) G_\lambda(\lambda_s, T_s) \, d\lambda + \int_{3 \ \mu m}^\infty \alpha(\lambda) G_\lambda(\lambda_s, T_s) \, d\lambda$$
$$= (0.4)(0.98) + (0.8)(0.02) = 0.408$$

For the second condition with the source at 527°C (800 K) the absorptance can be calculated in a similar manner. However, the calculation is the same as for the emittance at 800 K in Example 9.3 since $\epsilon_\lambda = \alpha_\lambda$ and $\bar{\epsilon} = \bar{\alpha}$ in equilibrium. Hence, $\bar{\alpha} = 0.744$ for a source at 800 K.

The preceding two examples illustrate the limits of gray-body assumptions. Whereas it may be acceptable to treat the aluminum-painted surface as totally gray with an average $\bar{\alpha} = \bar{\epsilon} = (0.8 + 0.744)/2 = 0.77$ for radiation exchange between it and a source at 800 K or less, for radiation exchange between the aluminum-painted surface and the sun such an approximation would lead to a serious error. The surface in the latter case would have to be treated as selectively gray with the averaged values for $\bar{\alpha}$ and $\bar{\epsilon}$ equal to 0.408 and 0.80, respectively.

9.3.4 Real Surface Characteristics

Radiation from real surfaces differs in several aspects from black- or gray-body radiation. Any real surface radiates less than a blackbody at the same temperature. Gray surfaces radiate a constant fraction ϵ_g of the monochromatic emissive power of a black surface at the same temperature T over the entire spectrum; real surfaces radiate a fraction ϵ_λ at any wavelength, but this fraction is not constant and varies with wavelength. Figure 9.14 shows a comparison of spectral emission from black, gray, and real surfaces. Both gray and black surfaces radiate diffusely, and the shape of the spectroradiometric curve for a gray surface is similar to that for a black surface at the same temperature, with the height reduced proportionally by the numerical value of the emittance.

The spectral emission from the real surface, shown by the solid line in Fig. 9.14, differs in detail from the gray-body spectral emission, but for the purpose of analysis the two may be sufficiently similar on the average to characterize the surface as approximately gray with $\epsilon_g = 0.6$, and the emissive power is given by Eq. (9.34)

$$E_{\text{real}} \cong \epsilon_g \sigma T^4$$

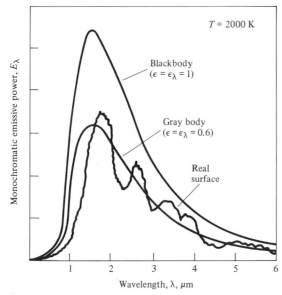

Figure 9.14 Comparison of hemispherical monochromatic emission for a black, a gray ($\varepsilon = 0.6$), and a real surface.

Observe, however, that Fig. 9.14 compares the emissive power of the real surface
with that of a gray surface with $\epsilon_g = 0.6$ at a temperature of 2000 K. At wavelengths above 1.5 μm the fit is fairly good, but at wavelengths below 1.5 μm
the emittance of the real surface is only about 50 percent of that of the gray
body. For temperatures below 2000 K this difference will not introduce a serious error because most of the radiant emission occurs at wavelengths above
1.5 μm. At higher temperatures, however, it may be necessary to approximate
the real surface with an emittance value less than 0.6 for $\lambda < 1.5$ μm. For the
absorptance to solar radiation, which falls largely below 2.0 μm, a value closer
to 0.3 would be a good approximation.

EXAMPLE 9.6 _____

The spectral hemispherical emittance of a painted surface is shown in
Fig. 9.15. Using a selective gray approximation, calculate (a) the effective
emittance over the entire spectrum, (b) the emissive power at 1000 K, and
(c) the percentage of solar radiation that this surface would absorb. Assume
that solar radiation corresponds to a blackbody source at 5800 K.

Solution. We shall approximate the real surface characteristics by a
three-band gray model. Below 2.0 μm the emittance is 0.3, between 2.0
and 4.0 μm the emittance is about 0.9, and above 4.0 μm the emittance
is about 0.5.

(a) The effective emittance over the entire spectrum is

$$
\bar{\epsilon} = \frac{\int_0^\infty \epsilon_\lambda E_{b\lambda}\, d\lambda}{\int_0^\infty E_{b\lambda}\, d\lambda}
$$

$$
= \epsilon_1 \left[\frac{E_b(0 \rightarrow \lambda_1 T)}{\sigma T^4} \right] + \epsilon_2 \left[\frac{E_b(0 \rightarrow \lambda_2 T) - E_b(0 \rightarrow \lambda_1 T)}{\sigma T^4} \right]
$$

$$
+ \epsilon_3 \left[\frac{E_b(0 \rightarrow \infty) - E_b(0 \rightarrow \lambda_2 T)}{\sigma T^4} \right]
$$

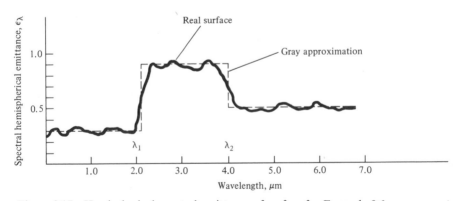

Figure 9.15 Hemispherical spectral emittance of surface for Example 9.6.

From the data, $\lambda_1 T = 2 \times 10^{-3}$ mK and $\lambda_2 T = 4 \times 10^{-3}$ mK. Evaluating the blackbody emission in the three bands from Table 9.1 yields

$$\bar\epsilon = (0.3)(0.0667) + 0.9(0.4809 - 0.0667) + 0.5(1.0 - 0.4809)$$
$$= 0.0200 + 0.373 + 0.255 = 0.6485$$

(b) The emissive power is then

$$E = \bar\epsilon \sigma T^4 = (0.6485)(5.67 \times 10^{-8})(1,000)^4$$
$$= 3.67 \times 10^4 \text{ W/m}^2$$

The emissive power of a blackbody at 1000 K is, for comparison, 5.67×10^4 W/m^2.

(c) To calculate the average solar absorptance we use Eq. (9.33), or

$$\bar\alpha_s = \frac{\displaystyle\int_0^\infty \alpha_\lambda G_\lambda^* \, d\lambda}{\displaystyle\int_0^\infty G_\lambda^* \, d\lambda}$$

According to Kirchhoff's law $\alpha_\lambda = \epsilon_\lambda$ and therefore

$$\bar\alpha_s = \frac{\epsilon_1 \displaystyle\int_0^{2\,\mu m} G_\lambda^* \, d\lambda}{\sigma T^4} + \frac{\epsilon_2 \displaystyle\int_{2\,\mu m}^{4\,\mu m} G_\lambda^* \, d\lambda}{\sigma T^4} + \frac{\epsilon_3 \displaystyle\int_{4\,\mu m}^\infty G_\lambda^* \, d\lambda}{\sigma T^4}$$

Assuming the sun radiates as a blackbody at 5800 K, we get from Table 9.1

$$\bar\alpha_s = (0.3)(0.941) + 0.9(0.990 - 0.94) + 0.5(1.0 - 0.990)$$
$$= 0.332$$

Thus, about 33 percent of the solar radiation would be absorbed. Note that the ratio of emittance at 1000 K to absorptance from a 5800 K source is almost 2.

For convenience, the total hemispherical emittances of a selected group of industrially important surfaces at different temperatures are presented in Table 9.2. A more extensive tabulation of experimentally measured radiation properties for many surfaces has been prepared by Gubareff et al. (8). Some general features and trends of their results are discussed below.

Figure 9.16 shows the measured monochromatic emittance (or absorptance) of some electrical conductors as a function of wavelength (9). Polished surfaces of metals have low emittances but, as shown in Fig. 9.17, the presence of an oxide layer may increase the emittance appreciably. The monochromatic emittance of an electrical conductor (e.g., see the curves for Al or Cu in Fig. 9.16) increases with decreasing wavelength. Consequently, in accordance with Eq. (9.31), the total emittance of electrical conductors increases with increasing temperature, as illustrated in Fig. 9.18 for several metals and one dielectric.

TABLE 9.2 HEMISPHERICAL EMITTANCES OF VARIOUS SURFACES[a]

Material	Wavelength and average temperature				
	9.3 μm 310 K	5.4 μm 530 K	3.6 μm 800 K	1.8 μm 1700 K	0.6 μm Solar 6000 K
Metals					
Aluminum					
Polished	~0.04	0.05	0.08	~0.19	~0.3
Oxidized	0.11	~0.12	0.18		
24-ST weathered	0.4	0.32	0.27		
Surface roofing	0.22				
Anodized (at 1000°F)	0.94	0.42	0.60	0.34	
Brass					
Polished	0.10	0.10			
Oxidized	0.61				
Chromium					
Polished	~0.08	~0.17	0.26	~0.40	0.49
Copper					
Polished	0.04	0.05	~0.18	~0.17	
Oxidized	0.87	0.83	0.77		
Iron					
Polished	0.06	0.08	0.13	0.25	0.45
Cast, oxidized	0.63	0.66	0.76		
Galvanized, new	0.23			0.42	0.66
Galvanized, dirty	0.28			0.90	0.89
Steel plate, rough	0.94	0.97	0.98		
Oxide	0.96		0.85		0.74
Molten				0.3–0.4	
Magnesium	0.07	0.13	0.18	0.24	0.30
Molybdenum filament			~0.09	~0.15	~0.2[b]
Silver					
Polished	0.01	0.02	0.03		0.11
Stainless steel					
18–8, polished	0.15	0.18	0.22		
18–8, weathered	0.85	0.85	0.85		
Steel tube					
Oxidized		0.80			
Tungsten filament	0.03			~0.18	0.35[c]
Zinc					
Polished	0.02	0.03	0.04	0.06	0.46
Galvanized sheet	~0.25				
Building and Insulating Materials					
Asbestos paper	0.93	0.93			
Asphalt	0.93		0.9		0.93
Brick					
Red	0.93				0.7
Fire clay	0.9		~0.7	~0.75	
Silica	0.9		~0.75	0.84	
Magnesite refractory	0.9			~0.4	
Enamel, white	0.9				
Marble, white	0.95		0.93		0.47
Paper, white	0.95		0.82	0.25	0.28

TABLE 9.2 Continued

| | Wavelength and average temperature | | | | |
Material	9.3 μm 310 K	5.4 μm 530 K	3.6 μm 800 K	1.8 μm 1700 K	0.6 μm Solar 6000 K
Plaster	0.91				
Roofing board	0.93				
Enameled steel, white				0.65	0.47
Asbestos cement, red				0.67	0.66
Paints					
Aluminized lacquer	0.65	0.65			
Cream paints	0.95	0.88	0.70	0.42	0.35
Lacquer, black	0.96	0.98			
Lampblack paint	0.96	0.97		0.97	0.97
Red paint	0.96				0.74
Yellow paint	0.95		0.5		0.30
Oil paints (all colors)	~0.94	~0.9			
White (ZnO)	0.95		0.91		0.18
Miscellaneous					
Ice	~0.97[d]				
Water	~0.96				
Carbon					
T-carbon, 0.9 percent ash	0.82	0.80	0.79		
Filament	~0.72			0.53	
Wood	~0.93				
Glass	0.90				(Low)

[a] Since the emittance at a given wavelength equals the absorptance at that wavelength, the values in this table can be used to approximate the absorptance to radiation from a source at the temperature listed. For example, polished aluminum will absorb 30 percent of incident solar radiation.
[b] At 3000 K.
[c] At 3600 K.
[d] At 273 K.
Source: Refs. (3–8).

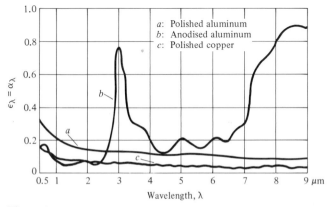

Figure 9.16 Variation of monochromatic absorptance (or emittance) with wavelength for three electrical conductors at room temperature.

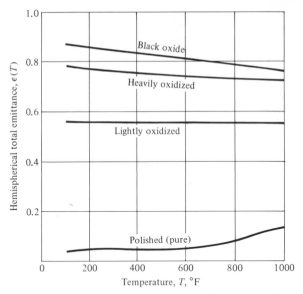

Figure 9.17 Effect of oxide coating on hemispherical total emittance of copper. [Data from Gubareff et al. (8).]

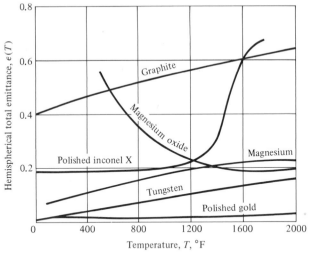

Figure 9.18 Effect of temperature on hemispherical total emittance of several metals and one dielectric. [Data from Gubareff et al. (8).]

As a group, electrical nonconductors exhibit the opposite trend and have generally high values of infrared emittance. Figure 9.19 illustrates the variation of the monochromatic emittance of several electrical nonconductors with wavelength.

For heat transfer calculations an average emittance or absorptance for the wavelength band in which the bulk of the radiation is emitted or absorbed is

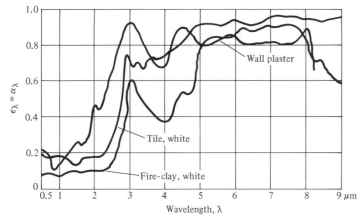

Figure 9.19 Variation of monochromatic absorptance (or emittance) with wavelength for three electrical nonconductors. [According to W. Sieber (9).]

desired. The wavelength band of interest depends on the temperature of the body from which the radiation originates, as pointed out in Section 9.1. If the distribution of the monochromatic emittance is known, the total emittance can be evaluated from Eq. (9.31) and the total absorptance can be calculated from Eq. (9.33), if the temperature and the spectral characteristics of the source are also specified. Sieber (9) evaluated the total absorptance of the surfaces of several materials as a function of source temperature, with the receiving surfaces at room temperature and the emitter a blackbody. His results are shown in Fig. 9.20, where the ordinate is the total absorptance for radiation normal to

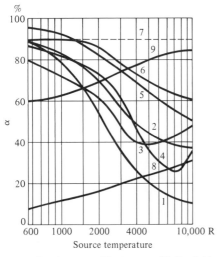

(1) White fire clay	(4) Wood	(7) Roof shingles
(2) Asbestos	(5) Porcelain	(8) Aluminum
(3) Cork	(6) Concrete	(9) Graphite

Figure 9.20 Variation of total absorptance with source temperature for several materials at room temperature. [According to W. Sieber (9).]

the surface and the abscissa is the source temperature. We observe that the absorptance of aluminum, typical of good conductors, increases with increasing source temperature, whereas the absorptance of nonconductors exhibits the opposite trend.

Figure 9.21 illustrates that the emittance of real surfaces is also a function of direction. The directional emittance $\epsilon(\theta, \phi)$ is defined as the intensity of radiation emitted from a surface in the direction θ, ϕ divided by the blackbody intensity, or

$$\epsilon(\theta, \phi) = \frac{I(\theta, \phi)}{I_b} \tag{9.35}$$

Referring to Fig. 9.7, the monochromatic hemispherical emittance is defined by the relation

$$\epsilon_\lambda = \frac{E_\lambda}{E_{b\lambda}} = \frac{\int_{\phi=0}^{2\pi} \int_{\theta=0}^{\pi/2} I_\lambda(\theta, \phi) \sin\theta \cos\theta \, d\theta \, d\phi}{\pi I_{b\lambda}} \tag{9.36}$$

But, as mentioned previously, the variation of the emittance with the azimuthal angle ϕ is usually negligible. If the emittance is only a function of the elevation angle θ, Eq. (9.36) can be integrated over the angle ϕ and simplified to

$$\epsilon_\lambda = \frac{2\pi \int_{\theta=0}^{\pi/2} I_\lambda(\theta) \sin\theta \cos\theta \, d\theta}{\pi I_b} \tag{9.37}$$

Substituting Eq. (9.35) for I_λ/I_b, we get

$$\epsilon_\lambda = 2 \int_{\theta=0}^{\pi/2} \epsilon_\lambda(\theta) \sin\theta \cos\theta \, d\theta \tag{9.38}$$

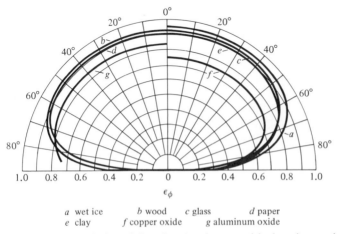

a wet ice	b wood	c glass	d paper
e clay	f copper oxide	g aluminum oxide	

Figure 9.21 Variation of directional emittance with elevation angle for several electrical nonconductors. [From E. Schmidt and E. Eckert (10) with permission.]

EXAMPLE 9.7 ————————————————————————————————————

The directional emittance of an oxidized surface at 800 K can be approximated by

$$\epsilon(\theta) = 0.70 \cos \theta$$

Determine (*a*) the emittance perpendicular to the surface, (*b*) the hemispherical emittance, and (*c*) the radiant emissive power if the surface is 5 cm × 10 cm.

Solution

(*a*) $\epsilon(0)$, the emittance for $\theta = 0°$ or $\cos \theta = 1$, is 0.70.

(*b*) The hemispherical emittance is obtained by performing the integration indicated by Eq. (9.38), or

$$\bar{\epsilon} = 2 \int_0^{\pi/2} 0.70 \cos^2 \theta \sin \theta \, d\theta = -0.7 \cos^3 \theta \Big|_0^{\pi/2}$$

Substituting the above limits gives 0.467. Note that the ratio $\epsilon(0)/\bar{\epsilon} = 1.5$.

(*c*) The emissive power is

$$E = \bar{\epsilon} A \sigma T^4 = (0.467)(5 \times 10^{-3} \text{ m}^2)(5.37 \times 10^{-8} \text{ W/m}^2 \text{ K}^4)(1800 \text{ K}^4)$$

$$= 54.2 \text{ W}$$

The polar plots in Figs. 9.21 and 9.22, from (10), illustrate the directional emittances for some electrical nonconductors and conductors, respectively. In these plots θ is the angle between the normal to the surface and the direction of the radiant beam emitted from the surface. For surfaces whose radiation intensity follows Lambert's cosine law and depends only on the projected area, the emittance curves would be semicircles. Figure 9.21 shows that, for nonconductors such as wood, paper, and oxide films, the emittance decreases at large values of the emission angle θ, whereas for polished metals the opposite trend is observed (see Fig. 9.22). For example, the emittance of polished chromium, which is widely used as a radiation shield, is as low as 0.06 in the normal direction but increases to 0.14 when viewed from an angle θ of 80°. Experimental data on the directional variation of emittance are scant, and until more information becomes available a satisfactory approximation for

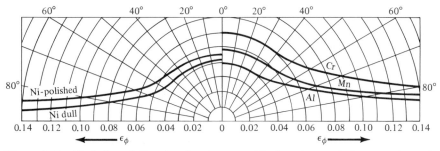

Figure 9.22 Variation of directional emittance with elevation angle for several metals. [From E. Schmidt and E. Eckert (10) with permission.]

engineering calculations is to assume for polished metallic surfaces a mean value of $\epsilon/\epsilon_n = 1.2$ and for nonmetallic surfaces $\epsilon/\epsilon_n = 0.96$, where ϵ is the average emittance through a hemispherical solid angle of 2π steradians and ϵ_n is the emittance in the direction of the normal to the surface.

Reflectance and Transmittance   When a surface does not absorb all of the incident radiation, the portion not absorbed will either be transmitted or reflected. Most solids are opaque and do not transmit radiation. The portion of the radiation which is not absorbed is, therefore, reflected back into hemispherical space. This may be characterized by the monochromatic hemispherical reflectance ρ_λ, defined as

$$\rho_\lambda = \frac{\text{radiant energy reflected/time-area-wavelength}}{G_\lambda} \tag{9.39}$$

or by the total reflectance ρ, defined as

$$\rho = \frac{\text{radiant energy reflected/time-area}}{\int_0^\infty G_\lambda \, d\lambda} \tag{9.40}$$

For nontransmitting materials the relations

$$\rho_\lambda = 1 - \alpha_\lambda \tag{9.41}$$

and

$$\rho = 1 - \alpha$$

must obviously hold at every wavelength and over the entire spectrum, respectively.

For the most general case of a material which partly absorbs, partly reflects, and partly transmits radiation incident on its surface, we define τ_λ as the fraction transmitted at wavelength λ and τ as the fraction of the total incident radiation which is transmitted. Referring to Fig. 9.11, the monochromatic relation is

$$\rho_\lambda + \alpha_\lambda + \tau_\lambda = 1 \tag{9.42}$$

whereas the total relation between reflectance, absorptance, and transmittance is given by Eq. (9.22). Glass, rock salt, and other inorganic crystals are examples of the few solids which, unless very thick, are to a certain degree transparent to radiation of certain wavelengths. Many liquids and all gases are also transparent.

There are two basic types of radiation reflections: *specular* and *diffuse*. If the angle of reflection is equal to the angle of incidence, the reflection is called specular. On the other hand, when an incident beam is reflected uniformly in all directions, the reflection is called diffuse. No real surface is either specular or diffuse. In general, reflection from highly polished and smooth surfaces approaches specular characteristics, while reflection from industrial "rough" surfaces approaches diffuse characteristics. An ordinary mirror reflects specularly in the visible wavelength range, but not necessarily over the longer-wavelength range of thermal radiation.

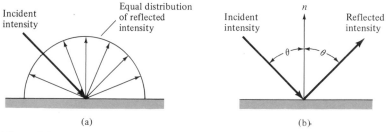

Figure 9.23 Schematic diagram illustrating (a) diffuse and (b) specular reflection.

Figure 9.23 illustrates schematically the behavior of diffuse and specular reflectors. For engineering calculations, industrially plated, machined, or painted surfaces may be treated as though they were diffuse, according to experiments by Schonhorst and Viskanta (11). Methods for treating problems with surfaces that are partly specular and partly diffuse are presented in (12–14).

9.4 THE RADIATION SHAPE FACTOR

In most practical problems involving radiation, the intensity of thermal radiation passing between surfaces is not appreciably affected by the presence of intervening media because, unless the temperature is so high as to cause ionization or dissociation, monatomic and most diatomic gases as well as air are transparent. Moreover, since most industrial surfaces can be treated as diffuse emitters and reflectors of radiation in a heat transfer analysis, a key problem in calculating radiation heat transfer between surfaces is to determine the fraction of the total diffuse radiation leaving one surface which is intercepted by another surface and vice versa. *The fraction of diffusely distributed radiation leaving a surface A_i that reaches surface A_j is called the radiation shape factor F_{i-j}.* The first subscript appended to the radiation shape factor denotes the surface from which the radiation emanates, while the second subscript denotes the surface receiving the radiation.

Consider two black surfaces A_1 and A_2, as shown in Fig. 9.24. The radiation leaving A_1 and arriving at A_2 is

$$q_{1 \to 2} = E_{b1}A_1F_{1-2} \tag{9.43}$$

and the radiation leaving A_2 and arriving at A_1 is

$$q_{2 \to 1} = E_{b2}A_2F_{2-1} \tag{9.44}$$

Since both surfaces are black, all the incident radiation will be absorbed and the net rate of energy exchange, $q_{1 \rightleftharpoons 2}$, is

$$q_{1 \rightleftharpoons 2} = E_{b1}A_1F_{1-2} - E_{b2}A_2F_{2-1} \tag{9.45}$$

If both surfaces are at the same temperature $E_{b1} = E_{b2}$ and there can be no net heat flow between them. Therefore, $q_{1 \rightleftharpoons 2} = 0$, and since neither areas nor shape

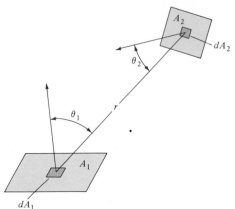

Figure 9.24 Nomenclature for geometric shape-factor derivation.

factors are functions of temperature

$$A_1 F_{1-2} = A_2 F_{2-1} \qquad (9.46)$$

Equation (9.46) is known as the *reciprocity theorem*. The net rate of transfer between any two black surfaces, A_1 and A_2, can thus be written in two forms

$$q_{1 \rightleftharpoons 2} = A_1 F_{1-2}(E_{b1} - E_{b2}) = A_2 F_{2-1}(E_{b1} - E_{b2}) \qquad (9.47)$$

Inspection of Eq. (9.47) reveals that the net rate of heat flow between two black bodies can be determined by evaluating the radiation from either one of the surfaces to the other surface and replacing its emissive power by the difference between the emissive powers of the two surfaces. Since the end result is independent of the choice of the emitting surface, one selects that surface whose shape factor can be determined more easily. For example, the shape factor F_{1-2} for any surface A_1 completely enclosed by another surface is unity. In general, however, the determination of a shape factor for any but the most simple geometric configuration is rather complex.

To determine the fraction of the energy leaving surface A_1 that strikes surface A_2, consider first the two differential surfaces dA_1 and dA_2. If the distance between them is r, then $dq_{1 \rightarrow 2}$, the rate at which radiation from dA_1 is received by dA_2, is, from Eq. (9.13), given by

$$dq_{1 \rightarrow 2} = I_1 \cos \theta_1 \, dA_1 \, d\omega_{1-2} \qquad (9.48)$$

where I_1 = intensity of radiation from dA_1

$dA_1 \cos \theta_1$ = projection of area element dA_1 as seen from dA_2

$\quad d\omega_{1-2}$ = solid angle subtended by receiving area dA_2 with respect
to center point of dA_1

The subtended angle $d\omega_{1-2}$ is equal to the projected area of the receiving surface in the direction of the incident radiation divided by the square of the distance between dA_1 and dA_2, or, using the nomenclature of Fig. 9.24,

$$d\omega_{1-2} = \cos \theta_2 \frac{dA_2}{r^2} \qquad (9.49)$$

Substituting Eqs. (9.49) and (9.13) for $d\omega_{1-2}$ and I_1, respectively, into Eq. (9.48) yields

$$dq_{1\to2} = E_{b1}\,dA_1\left(\frac{\cos\theta_1\cos\theta_2\,dA_2}{\pi r^2}\right) \tag{9.50}$$

where the term in parentheses is equal to the fraction of the total radiation emitted from dA_1 that is intercepted by dA_2. By analogy, the fraction of the total radiation emitted from dA_2 that strikes dA_1 is

$$dq_{2\to1} = E_{b2}\,dA_2\left(\frac{\cos\theta_2\cos\theta_1\,dA_1}{\pi r^2}\right) \tag{9.51}$$

so that the net rate of radiant heat transfer between dA_1 and dA_2 is

$$dq_{1\rightleftharpoons2} = (E_{b1} - E_{b2})\frac{\cos\theta_1\cos\theta_2\,dA_1\,dA_2}{\pi r^2} \tag{9.52}$$

To determine $q_{1\rightleftharpoons2}$, the net rate of radiation between the entire surfaces A_1 and A_2, we simply integrate the fraction in the preceding equation over both surfaces and obtain

$$q_{1\rightleftharpoons2} = (E_{b1} - E_{b2})\int_{A_1}\int_{A_2}\frac{\cos\theta_1\cos\theta_2\,dA_1\,dA_2}{\pi r^2} \tag{9.53}$$

The double integral is conveniently written in shorthand notation as either A_1F_{1-2} or A_2F_{2-1}, where F_{1-2} is called the shape factor evaluated on the basis of area A_1 and F_{2-1} is called the shape factor evaluated on the basis of A_2. The method of evaluation of the double integral is illustrated below.

EXAMPLE 9.8 _____

Determine the geometric shape factor for a very small disk A_1 and a large parallel disk A_2 located a distance L directly above the smaller one, as shown in Fig. 9.25.

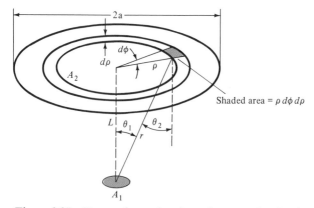

Figure 9.25 Nomenclature for shape-factor evaluation between two disks in Example 9.8.

Solution. From Eq. (9.53) the geometric shape factor is

$$A_1 F_{1-2} = \int_{A_1} \int_{A_2} \frac{\cos \theta_1 \cos \theta_2}{\pi r^2} dA_1 dA_2$$

but since A_1 is very small, the shape factor is given by

$$A_1 F_{1-2} = \frac{A_1}{\pi} \int_{A_2} \frac{\cos \theta_1 \cos \theta_2}{r^2} dA_2$$

From Fig. 9.25, $\cos \theta_1 = \cos \theta_2 = L/r$, $r = \sqrt{\rho^2 + L^2}$, and $dA_2 = \rho \, d\phi \, d\rho$. Substituting these relations, we obtain

$$A_1 F_{1-2} = \frac{A_1}{\pi} \int_0^a \int_0^{2\pi} \frac{L^2}{(\rho^2 + L^2)^2} \rho \, d\rho \, d\phi$$

which can be integrated directly to yield

$$A_1 F_{1-2} = \frac{A_1 a^2}{a^2 + L^2} = A_2 F_{2-1}$$

TABLE 9.3   GEOMETRIC SHAPE FACTORS FOR USE IN EQS. (9.47) AND (9.55)

Surfaces between which radiation is being interchanged	Shape factor, F_{1-2}
1. Infinite parallel planes.	1
2. Body A_1 completely enclosed by another body, A_2. Body A_1 cannot see any part of itself.	1
3. Surface element dA (A_1) and rectangular surface (A_2) above and parallel to it, with one corner of rectangle contained in normal to dA.	See Fig. 9.26
4. Element dA (A_1) and parallel circular disk (A_2) with its center directly above dA. (See Example 9.3.)	$\dfrac{a^2}{(a^2 + L^2)}$
5. Two parallel and equal squares, rectangles, or disks of width or diameter D, a distance L apart.	See Fig. 9.27
6. Two parallel disks of unequal diameter, distance L apart with centers on same normal to their planes, smaller disk A_1 of radius a, larger disk of radius b.	$\dfrac{1}{2a^2}\left[L^2 + a^2 + b^2 - \sqrt{(L^2 + a^2 + b^2)^2 - 4a^2 b^2} \right]$
7. Two rectangles in perpendicular planes with a common side.	See Fig. 9.28
8. Radiation between an infinite plane A_1 and one or two rows of infinite parallel tubes in a parallel plane A_2 if the only other surface is a refractory surface behind the tubes.	See Fig. 9.30

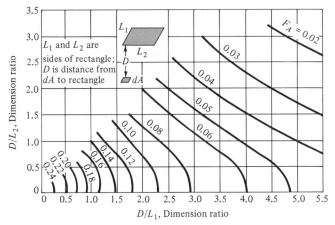

Figure 9.26 Shape factor for a surface element and a rectangular surface parallel to it. [From H. C. Hottel (25) with permission.]

The preceding example shows that the determination of a shape factor by evaluating the double integral of Eq. (9.53) is generally very tedious. Fortunately, the shape factors for a large number of geometric arrangements have been evaluated and a majority of them can be found in (3–7). A selected group of practical interest is summarized in Table 9.3 and Figs. 9.26 to 9.30.

The shape factors for surfaces that are two-dimensional, infinitely long in one direction, and characterized by identical cross sections normal to the

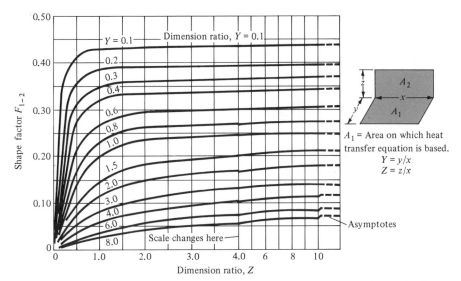

Figure 9.27 Shape factor for adjacent rectangles in perpendicular planes sharing a common edge. [From H. C. Hottel (25) with permission.]

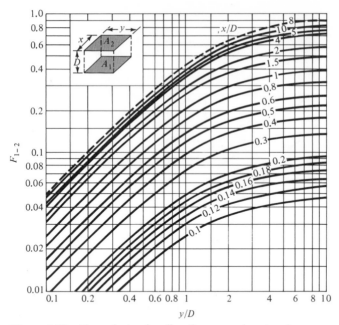

Figure 9.28   Shape factor for directly opposed rectangles.

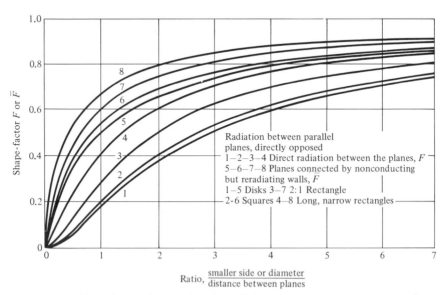

Figure 9.29   Shape factors for equal and parallel squares, rectangles, and disks. [From H. C. Hottel (25) with permission.]

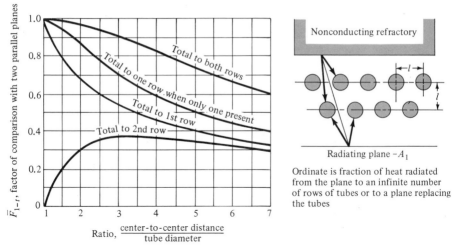

Figure 9.30 Shape factor for a plane and one or two rows of tubes parallel to it. [From H. C. Hottal (25) with permission.]

infinite direction can be determined by a simple procedure called the crossed-string method. Figure 9.31 shows two surfaces that satisfy the geometric restrictions for the crossed-string method. Hottel and Sarofim (15) have shown that the shape factor F_{1-2} is equal to the sum of the lengths of the crossed strings stretched between the ends of the two surfaces minus the sum of the uncrossed strings divided by twice the length L_1. In the form of an equation,

$$F_{1-2} = \frac{(\overline{ad} + \overline{eb}) - (\overline{ab} + \overline{cd})}{2L_1} \tag{9.54}$$

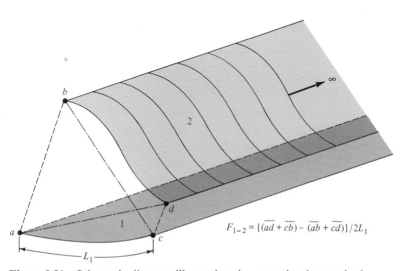

Figure 9.31 Schematic diagram illustrating the crossed-string method.

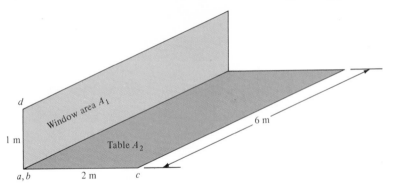

Figure 9.32   Window and table for Example 9.9.

EXAMPLE 9.9 _____

A window arrangement consists of a long opening 1 m high and 5 m long. Under this window, as shown in Fig. 9.32, is a working table 2 m wide. Determine the shape factor between the window and the table.

Solution.   Assume that the window and the table are sufficiently long to be approximated as infinitely long surfaces. Then we can use the crossed-string method, and since for this case points *a* and *b* are the same, we have

$$ab = 0$$
$$cb = L_1 = 2 \text{ m}$$
$$ad = L_2 = 1 \text{ m}$$
$$cd = L_3 = \sqrt{5} \text{ m}$$

and

$$F_{1-2} = \tfrac{1}{2}(1 + 2 - \sqrt{5}) = 0.382$$

9.4.1 Shape-Factor Algebra

The basic shape factors from the charts in Figs. 9.26 to 9.30 can be used to obtain shape factors for a larger class of geometries that can be built up from the elementary curves. This process is known as shape factor algebra. Shape-factor algebra is based on the principle of conservation of energy. Suppose we want to determine the shape factor from surface A_1 to the combined areas $A_2 + A_3$ as shown in Fig. 9.33. We can write

$$F_{1 \to (2+3)} = F_{1-2} + F_{1-3} \qquad (9.55)$$

That is, the total shape factor is equal to the sum of its parts. Rewriting Eq. (9.55) as

$$A_1 F_{1-2,3} = A_1 F_{1-2} + A_1 F_{1-3}$$

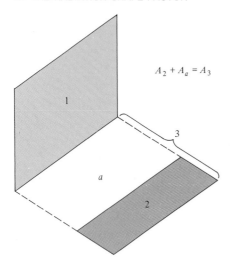

$A_2 + A_a = A_3$

Figure 9.33 Schematic sketch illustrating shape-factor algebra.

and using the reciprocity relations

$$A_1 F_{1-2,3} = (A_2 + A_3)F_{2,3-1}$$
$$A_1 F_{1-2} = A_2 F_{2-1}$$
$$A_1 F_{1-3} = A_3 F_{3-1}$$

yields

$$(A_2 + A_3)F_{2,3-1} = A_2 F_{2-1} + A_3 F_{3-1} \qquad (9.56)$$

Thus, the total radiation received by A_1 is the sum of the radiant energy fractions from A_2 and A_3. This simple relation can be used to evaluate the shape factor F_{1-3} in terms of the shape factors for perpendicular rectangles with a common edge given in Fig. 9.27. Other combinations can be obtained in a similar manner. The following example illustrates the numerical evaluation procedure.

EXAMPLE 9.10 ────────────────────────────────────

A room 12 ft on one side by 24 ft on the other has a ceiling height of 12 ft. A small skylight of area A_1 is located in the ceiling, as shown in Fig. 9.34, 6 ft from two walls. Assuming that the incoming light is diffuse, evaluate the shape factor of the floor with respect to the skylight.

Solution. Let $A_3 = A_a + A_1$ and $A_4 = A_b + A_2$. Applying Eqs. (9.55) and (9.56) gives

$$A_3 F_{3-4} = A_a F_{a-b} + A_a F_{a-2} + A_1 F_{1-b} + A_1 F_{1-2}$$
$$A_3 F_{3-b} = A_a F_{a-b} + A_1 F_{1-b}$$
$$F_{a-4} = F_{a-b} + F_{a-2}$$

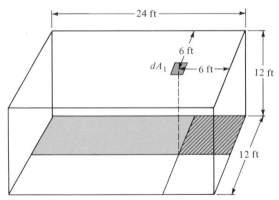

Figure 9.34   Sketch for Example 9.10.

Combining these three equations and solving for F_{1-2} gives

$$F_{1-2} = \frac{1}{A_1}(A_3 F_{3-4} - A_a F_{a-4} - A_3 F_{3-b} + A_a F_{a-b})$$

The shape factors on the right-hand side of this equation are plotted in Fig. 9.27. The values are

$$F_{3-4} = 0.19$$
$$F_{a-4} = 0.32$$
$$F_{3-b} = 0.08$$
$$F_{a-b} = 0.19$$

Therefore, F_{1-2} is

$$F_{1-2} = \frac{1}{30}[50(0.19) - 20(0.32) - 50(0.08) + 20(0.19)]$$

$$F_{1-2} = 0.097$$

Thus, 9.7 percent of the radiation emitted by surface 1 will directly impinge on surface 2.

9.5 ENCLOSURES WITH BLACK SURFACES

To determine the net radiation heat transfer to or from a surface, it is necessary to account for radiation coming from all directions. This procedure is facilitated by figuratively constructing an enclosure around the surface and specifying the radiation characteristics of each surface. The surfaces comprising the enclosure for a given surface i are all the surfaces that can be seen by an observer standing on surface i in the surrounding space. The enclosure need not necessarily consist only of solid surfaces, but may include open spaces denoted as "windows." Each such open window may be assigned an equivalent blackbody temperature corresponding to the entering radiation. If no radiation enters, a window acts like a blackbody at zero temperature which absorbs all outgoing radiation and emits and reflects none.

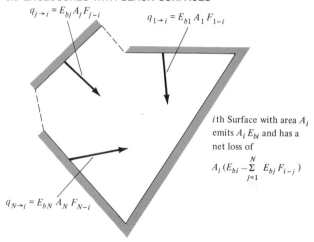

Figure 9.35 Schematic diagram of enclosure of N black surfaces with energy quantities incident upon and leaving surface i.

The net rate of radiation loss from a typical surface A_i in an enclosure (see Fig. 9.35) consisting of N black surfaces is equal to the difference between the emitted radiation and the absorbed radiation, or

$$q_{i\rightleftarrows\text{enclosure}} = A_i(E_{bi} - G_i) \tag{9.57}$$

where G_i is the radiation incident on surface i per unit time and unit area, called the irradiation.

The radiation incident on A_i comes from the other N surfaces in the enclosure. From a typical surface j, the radiation incident on i is $E_{bj}A_jF_{j-i}$. Summing the contributions from all N surfaces gives

$$A_iG_i = E_{b1}A_1F_{1-i} + E_{b2}A_2F_{2-i} + \cdots + E_{bN}A_NF_{N-i}$$

which can be written compactly in the form

$$A_iG_i = \sum_{j=1}^{N} E_{bj}A_jF_{j-i} \tag{9.58}$$

Using the reciprocity law $A_iF_{i-j} = A_jF_{j-i}$ and substituting Eq. (9.58) for G_i in Eq. (9.57) yields for the net rate of radiation heat loss from *any* surface in an enclosure of black surfaces

$$q_{i\rightleftarrows\text{enclosure}} = A_i \left(E_{bi} - \sum_{j=1}^{N} E_{bj}F_{i-j} \right) \tag{9.59}$$

An alternative approach to the problem is by extension of Eqs. (9.43) and (9.44). Since the radiant energy leaving any surface i must impinge on the N surfaces forming the enclosure,

$$\sum_{j=1}^{N} F_{i-j} = 1.0 \tag{9.60}$$

Equation (9.60) includes a term F_{i-i} which is not zero when a surface is concave so that some radiation leaving surface i will be directly incident on it. The total emissive power of A_i is therefore distributed between the N surfaces according to

$$A_i E_{bi} = \sum_{j=1}^{N} E_{bi} A_i F_{i-j} \qquad (9.61)$$

Introducing Eq. (9.61) for $A_i E_{bi}$ in Eq. (9.59) gives the net rate of heat loss from surface i in the form

$$q_{i \rightleftharpoons \text{enclosure}} = \sum_{j=1}^{N} (E_{bi} - E_{bj}) A_i F_{i-j} \qquad (9.62)$$

Thus, the net heat loss may be calculated by summing the differences in emissive power and multiplying each by the appropriate area-shape factor.

An inspection of Eq. (9.62) shows that there is also an analogy between heat flow by radiation and the flow of electric current. If the blackbody emissive power E_b is considered to act as a potential and the area-shape factor $A_i F_{i-j}$ as the conductance between two nodes at potentials E_{bi} and E_{bj}, then the resulting net flow of heat is analogous to the flow of electric current in an analogous network. Examples of networks for blackbody enclosures consisting of three and four heat transfer surfaces at given temperatures are shown in Fig. 9.36, a and b, respectively.

In engineering problems there are situations when, for one or more surfaces in an enclosure, not the temperature but the heat flux is prescribed. In such cases the temperatures of these surfaces are unknown. For the case when the net radiation heat transfer rate $q_{r,k}$ from one surface A_k is prescribed, while for all the other surfaces of the enclosure the temperature is specified, Eq. (9.59) can be rearranged to solve for T_k. Since $E_{bk} = \sigma T_k^4$ one obtains

$$T_k = \left[\frac{\displaystyle\sum_{j \neq k}^{N} \sigma T_j^4 F_{k-j} + (q_r/A)_k}{\sigma(1 - F_{k-k})} \right]^{1/4} \qquad (9.63)$$

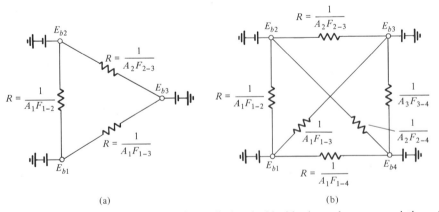

(a)   (b)

Figure 9.36 Equivalent networks for radiation in blackbody enclosures consisting of (a) three and (b) four surfaces.

where $j = k$ is specifically excluded from the summation. Once T_k is known, the heat transfer rates at all other surfaces can be obtained from Eq. (9.62).

Of special interest is the case of a *no-flux or adiabatic surface* which diffusely reflects and emits radiation at the same rate at which it receives it. Under steady-state conditions the interior surfaces of refractory walls in industrial furnaces can be treated as adiabatic surfaces. The interior walls of these surfaces receive heat by convection as well as radiation and lose heat to the outside by conduction. In practice, however, the heat flow by radiation is so much larger than the difference between the heat flow by convection to and the heat flow by conduction from the surface that the walls act essentially as reradiators, that is, no-flux surfaces.

A simplified sketch of a pulverized-fuel furnace is shown in Fig. 9.37a. The floor is assumed to be at a uniform temperature T_1 radiating to a nest of oxidized-steel tubes at T_2 which fill the ceiling of the furnace. The side walls and the ceiling are assumed to act as reradiators at a *uniform temperature* T_R. If we let A_R denote the reradiating area and assume that the floor and the tubes are black, the equivalent network representing the radiation exchange between the floor and the tubes in the presence of the reradiating walls is that

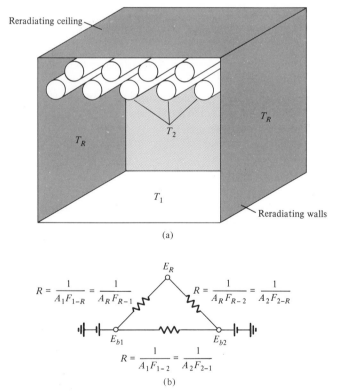

(a)

$$R = \frac{1}{A_1 F_{1-R}} = \frac{1}{A_R F_{R-1}} \qquad E_R \qquad R = \frac{1}{A_R F_{R-2}} = \frac{1}{A_2 F_{2-R}}$$

$$E_{b1} \qquad E_{b2}$$

$$R = \frac{1}{A_1 F_{1-2}} = \frac{1}{A_2 F_{2-1}}$$

(b)

Figure 9.37 Simplified sketch of a furnace and equivalent network for radiation in an enclosure consisting of two black surfaces and an adiabatic surface.

shown in Fig. 9.37*b*. A part of the radiation emitted from A_1 goes directly to A_2, while the rest strikes A_R and is reflected from there. Of the reflected radiation, a part is returned to A_1, a part to A_2, and the rest to A_R for further reflection. However, since the refractory walls must get rid of all the incident radiation by either reflection or radiation, their emissive power will act in the steady state like a floating potential whose actual value, that is, its emissive power and temperature, depends only on the relative values of the conductances between E_R and E_{b1} and E_R and E_{b2}. Thus, the net effect of this rather complicated radiation pattern can be represented in the equivalent network by two parallel heat flow paths between A_1 and A_2, one having an effective conductance of $A_1 F_{1-2}$, the other having an effective conductance equal to

$$\frac{1}{1/A_1 F_{1-R} + 1/A_2 F_{2-R}}$$

The net heat flow by radiation between a black heat source and a black heat sink in such a simple furnace is then equal to

$$q_{1 \rightleftarrows 2} = A_1(E_{b1} - E_{b2})\left(F_{1-2} + \frac{1}{1/F_{1-R} + A_1/A_2 F_{2-R}}\right) \qquad (9.64)$$

If neither of the surfaces can see any part of itself, F_{1-R} and F_{2-R} can be eliminated by using Eqs. (9.46) and (9.60). This yields, after some simplification,

$$q_{1 \rightleftarrows 2} = A_1 \sigma(T_1^4 - T_2^4) \frac{A_2 - A_1 F_{1-2}^2}{A_1 + A_2 - 2A_1 F_{1-2}} = A_1 \bar{F}_{1-2}(E_{b1} - E_{b2}) \qquad (9.65)$$

where $\bar{F}_{1-2}$ is the effective shape factor for the configuration shown in Fig. 9.37. The same result would, of course, be obtained from Eqs. (9.62) and (9.63). The details of this derivation are left as an exercise.

9.6 ENCLOSURES WITH GRAY SURFACES

In the preceding section, radiation between black surfaces was considered. The assumption that a surface is black simplifies heat transfer calculations because all of the incident radiation is absorbed. In practice, one may generally neglect reflections without introducing serious errors if the absorptivity of the radiating surfaces is larger than 0.9. There are, however, numerous problems involving surfaces of low absorptance and emittance, especially in installations where radiation is undesirable. For example, the inner walls of a thermos bottle are silvered in order to reduce the heat flow by radiation. Also, thermocouples for high-temperature work are frequently surrounded by radiation shields to reduce the difference between the indicated temperature and the temperature of the medium to be measured.

     If the radiating surfaces are not black, the analysis becomes exceedingly difficult unless the surfaces are considered to be gray. The analysis in this section is limited to gray surfaces which follow Lambert's cosine law and also reflect diffusely. The radiation from such surfaces can be treated conveniently in terms of the *radiosity J*, which is defined as the rate at which radiation leaves a given surface per unit area. The radiosity is the sum of radiation emitted, reflected, and transmitted. For opaque bodies which transmit no radiation, the radiosity from a typical surface i can be defined (16)

$$J_i = \rho_i G_i + \epsilon_i E_{bi} \tag{9.66}$$

where J_i = radiosity, W/m^2

G_i = irradiation or radiation per unit time incident on a unit surface area, W/m^2

E_{bi} = blackbody emissive power, W/m^2

ρ_i = reflectance

ϵ_i = emittance

Consider the ith surface having area A_i in an enclosure consisting of N surfaces as shown in Fig. 9.35. To maintain surface i at temperature T_i a certain amount of heat, q_i, must be supplied from some external source to make up for the net radiative loss in a steady-state condition. The net rate of heat transfer from a surface i by radiation is equal to the difference between the outgoing and the incoming radiation. Using the terminology of Eq. (9.66), the net rate of heat loss is the difference between the radiosity and the irradiation, or

$$q_i = A_i(J_i - G_i) \tag{9.67}$$

It should be noted that Eq. (9.67) is strictly valid only when the temperature as well as the irradiation over A_i is uniform. To satisfy both of these conditions simultaneously, it is sometimes necessary to subdivide a physical surface into smaller sections for the purpose of analysis.

If the surfaces exchanging radiation are gray, $\epsilon_i = \alpha_i$ and $\rho_i = (1 - \epsilon_i)$ for each of them. The irradiation G_i can then be eliminated from Eq. (9.67) by combining it with Eq. (9.66). This yields

$$q_i = \frac{A_i \epsilon_i}{\rho_i} (E_{bi} - J_i) = \frac{A_i \epsilon_i}{1 - \epsilon_i} (E_{bi} - J_i) \tag{9.68}$$

Another relation for the *net rate of heat loss* by radiation from A_i can be obtained by evaluating the irradiation in terms of the radiosity of all the other surfaces which can be seen from it. The incident radiation G_i can be evaluated by the same approach used previously in a blackbody enclosure. The incident radiation consists of the portions of radiation from the other $N - 1$ surfaces which impinge on A_i. If the surface A_i can see a part of itself, also a portion of the radiation emitted by A_i will contribute to the irradiation. The shape factors for diffusely reflecting gray surfaces are obviously the same as for black surfaces since they depend only on geometric relations defined by Eq. (9.53). We can, therefore, write in symbolic form

$$A_i G_i = J_1 A_1 F_{1-i} + J_2 A_2 F_{2-i} + \cdots + J_i A_i F_{i-i}$$
$$+ \cdots + J_j A_j F_{j-i} + \cdots + J_N A_N F_{N-i} \tag{9.69}$$

Using the reciprocity relations

$$A_1F_{1-i} = A_iF_{i-1}$$
$$A_2F_{2-i} = A_iF_{i-2}$$
$$A_NF_{N-i} = A_iF_{i-N}$$

Eq. (9.69) can be written so that the only area appearing is A_i:

$$A_iG_i = J_1A_iF_{i-1} + J_2A_iF_{i-2} + \cdots + J_iA_iF_{i-i}$$
$$+ \cdots + J_jA_iF_{i-j} + \cdots + J_NA_iF_{i-N}$$

This can be expressed compactly as

$$G_i = \sum_{j=1}^{N} J_jF_{i-j} \tag{9.70}$$

Equation (9.70) is identical to Eq. (9.61) for a black enclosure, except that the blackbody emissive power has been replaced by the radiosity. Substituting the summation of Eq. (9.70) for G_i in Eq. (9.67) yields

$$q_i = A_i\left(J_i - \sum_{j=1}^{N} J_jF_{i-j}\right) \tag{9.71}$$

Equations (9.68) and (9.71) can be written for each of the N surfaces of the enclosure, giving $2N$ equations for $2N$ unknowns. There will always be N unknown J's, while the remaining unknowns will consist of q's or T's, depending on what boundary conditions are specified. The J's can always be eliminated, giving N equations relating the N unknown temperatures and net rates of radiation transfer.

In terms of an equivalent electrical circuit we could write Eq. (9.68) in the form

$$q_i = \frac{E_{bi} - J_i}{(1 - \epsilon_i)/A_i\epsilon_i} \tag{9.72}$$

and consider the rate of radiation heat transfer q_i as the current in a network between potentials E_{bi} and J_i with a resistance of $(1 - \epsilon_i)/A_i\epsilon_i$ between them. Since the effect of the system geometry on the net radiation between any two gray surfaces A_i and A_k emitting radiation at the rates J_i and J_k, respectively, is the same as for geometrically similar black surfaces, it can be expressed in terms of the geometric shape factor defined by Eq. (9.53). The direct radiation exchange between any two opaque and diffuse surfaces A_i and A_j is given by

$$q_{i\rightleftharpoons j} = (J_i - J_j)A_iF_{i-j} = (J_i - J_j)A_jF_{j-i} \tag{9.73}$$

Equations (9.68) and (9.73) provide the basis for determining the net rate of radiant heat transfer between gray bodies in a gray enclosure by means of an equivalent network. The effect of the reflectance and emittance can be taken into account by connecting a *blackbody potential node* E_b to each of the nodal points in the network by means of a *finite resistance* $(1 - \epsilon)/A\epsilon$. In the case of a blackbody this resistance is zero since $\epsilon = 1$. In Fig. 9.38 the equivalent networks for radiation in an enclosure consisting of two and four gray bodies are shown. For two-component gray enclosures, such as two parallel and infinite

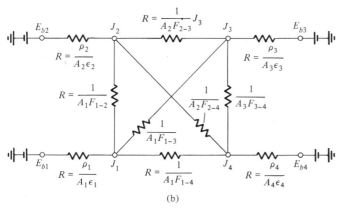

(a)

(b)

Figure 9.38 Equivalent networks for radiation in gray enclosures consisting of two and four surfaces: (a) two gray-body surfaces and (b) four gray-body surfaces.

plates, concentric cylinders of infinite height, and concentric spheres, the network reduces to a single line of resistances in series, as shown in Fig. 9.37a.

To illustrate the procedure for calculating radiation heat transfer between gray surfaces we will derive an expression for the rate of radiation heat transfer between two long concentric cylinders of area A_1 and A_2 and temperatures T_1 and T_2, respectively, and compare the result with the network in Fig. 9.38a.

Referring to Fig. 9.39, the shape factor for the smaller cylinder of area A_1 relative to the larger cylinder which encloses it, F_{1-2}, is 1.0. From Eq. (9.37),

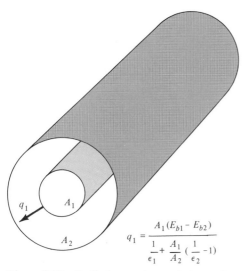

$$q_1 = \frac{A_1(E_{b1} - E_{b2})}{\dfrac{1}{\epsilon_1} + \dfrac{A_1}{A_2}\left(\dfrac{1}{\epsilon_2} - 1\right)}$$

Figure 9.39 Radiation exchange between two gray cylindrical surfaces.

$A_1F_{1-2} = A_2F_{2-1}$ and $F_{2-1} = A_1/A_2$. Since surface 2 can partly view itself, from Eq. (9.60) we have also $F_{2-2} = 1 - (A_1/A_2)$. From Eqs. (9.68) and (9.71) the net rates of heat loss from A_1 and A_2 are

$$q_1 = \frac{A_1\epsilon_1}{1 - \epsilon_1}(E_{b1} - J_1) = A_1(J_1 - J_2)$$

and

$$q_2 = \frac{A_2\epsilon_2}{1 - \epsilon_2}(E_{b2} - J_2) = A_2(J_2 - J_1F_{2-1} - J_2F_{2-2})$$

Substituting the appropriate expressions for F_{2-1} and F_{2-2} yields the relation $q_2 = A_1(-J_1 + J_2) = -q_1$, as expected from an overall heat balance. Eliminating J_2 and substituting for J_1 in the heat loss equation for A_1 gives

$$q_1 = \frac{A_1(E_{b1} - E_{b2})}{1/\epsilon_1 + (A_1/A_2)[(1 - \epsilon_2)/\epsilon_2]} \tag{9.74}$$

From the equivalent network in Fig. 9.38a the sum of the three resistances is

$$\frac{1 - \epsilon_1}{\epsilon_1 A_1} + \frac{1}{A_1F_{1-2}} + \frac{1 - \epsilon_2}{\epsilon_2 A_2} = \frac{1}{A_1}\left[\frac{1}{\epsilon_1} + \frac{A_1}{A_2}\left(\frac{1 - \epsilon_2}{\epsilon_2}\right)\right]$$

which gives the identical result for the net rate of heat loss from A_1, as expected.

The net rate of heat transfer in simple systems where radiation is transferred only between two gray surfaces can also be written in terms of an equivalent conductance $A_1\mathscr{F}_{1-2}$ in the form

$$q_{1\rightleftharpoons 2} = A_1\mathscr{F}_{1-2}(E_{b1} - E_{b2}) \tag{9.75}$$

where A_1 is the smaller of the two surfaces and $\mathscr{F}_{1-2}$ is given below. For two infinitely long concentric cylinders or two concentric spheres,

$$\mathscr{F}_{1-2} = \frac{1}{(1 - \epsilon_1)/\epsilon_1 + 1 + A_1(1 - \epsilon_2)/A_2\epsilon_2} \tag{9.76}$$

For two equal parallel plates of the same emittance ϵ spaced a finite distance apart,

$$\mathscr{F}_{1-2} = \frac{\epsilon[1 + (1 - \epsilon)F_{1-2}]}{1 + [(1 - \epsilon)F_{1-2}]^2} \tag{9.77}$$

where the shape factor F_{1-2} can be obtained from Fig. 9.29. For two infinitely large parallel plates,

$$\mathscr{F}_{1-2} = \frac{1}{1/\epsilon_1 + 1/\epsilon_2 - 1} \tag{9.78}$$

For a small gray body of area A_1 inside a large enclosure of area A_2 ($A_1 \ll A_2$),

$$\mathscr{F}_{1-2} = \epsilon_1$$

In many real problems radiation heat transfer will cause the internal energy and the temperature of a body to change. The heat transfer rate should then be interpreted as a quasi-steady-state result. Under those circumstances the solution will require a transient analysis similar to that presented in Chapter 2, with the surface temperature of the body a function of time.

EXAMPLE 9.11 _____

Liquefied oxygen (boiling temperature, $-297°F$) is to be stored in a spherical container of 1 ft diameter. The system is insulated by an evacuated space between the inner sphere and a surrounding 1.5-ft-ID concentric sphere. Both spheres are made of polished aluminum ($\epsilon = 0.03$), and the temperature of the outer sphere is $30°F$. Estimate the rate of heat flow by radiation to the oxygen in the container.

Solution. Although the internal energy of the oxygen will change, its temperature will remain constant since it is undergoing a change in phase. The absolute temperatures of the surfaces are

$$T_1 = 460 - 297 = 163 \text{ R}$$
$$T_2 = 460 + 30 \ = 490 \text{ R}$$

From Eq. (9.75) the rate of heat loss from the inner sphere is

$$q_1 = \frac{A_1\sigma(T_1^4 - T_2^4)}{1/\epsilon_1 + (A_1/A_2)[(1-\epsilon_2)/\epsilon_2]} = \frac{\pi \times 0.1714(1.63^4 - 4.9^4)}{1/0.03 + (1/2.25)(0.97/0.03)}$$
$$= -6.5 \text{ Btu/h}$$

Since the radiation loss from A_1 is negative, the heat is actually transferred to the oxygen, as expected.

_____

The radiant heat flow in an enclosure consisting of two gray surfaces connected by reradiating surfaces can also be solved without difficulty by means of the equivalent circuit. According to Eqs. (9.72) and (9.73), it is only necessary to replace E_{b1} and E_{b2}, the potentials used in Section 9.5 for black surfaces, by J_1 and J_2 and connect the new potentials with the resistances $\rho_1/\epsilon_1 A_1$ and $\rho_2/\epsilon_2 A_2$ to their respective blackbody potentials E_{b1} and E_{b2}. The resulting network is shown in Fig. 9.40, and from it we see that the total conductance

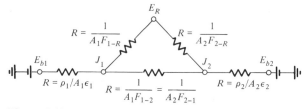

Figure 9.40 Equivalent circuit for radiation in an enclosure consisting of two gray surfaces connected by a reradiating surface.

between E_1 and E_2 is now

$$A_1 \mathscr{F}_{1-2} = \cfrac{1}{\cfrac{\rho_1}{\epsilon_1 A_1} + \cfrac{\rho_2}{\epsilon_2 A_2} + \cfrac{1}{A_1[F_{1-2} + 1/(1/F_{1-R} + A_1/A_2 F_{2-R})]}}$$

where the denominator of the last term is the conductance for the blackbody network given by Eq. (9.65). The expression for the conductance can be recast into the more convenient form

$$A_1 \mathscr{F}_{1-2} = \cfrac{1}{\cfrac{1}{A_1}\left(\cfrac{1}{\epsilon_1} - 1\right) + \cfrac{1}{A_2}\left(\cfrac{1}{\epsilon_2} - 1\right) + \cfrac{1}{A_1 \bar{F}_{1-2}}} \qquad (9.79)$$

where $A_1 \bar{F}_{1-2}$ is the effective conductance for the blackbody network, equal to the last term in the denominator of the original expression. The equation for the net radiant heat transfer per unit time between two gray surfaces at uniform temperatures in the presence of reradiating surfaces can then be written

$$q_{1 \rightleftharpoons 2} = A_1 \mathscr{F}_{1-2} \sigma (T_1{}^4 - T_2{}^4) \qquad (9.80)$$

For enclosures consisting of several surfaces the radiation heat transfer from any one of them can be calculated by drawing the equivalent circuit and performing a circuit analysis. This analysis can be made by applying Kirchhoff's current law, which states that the sum of the currents entering a given node is zero. When a computer is available, the same result can be obtained by a matrix method outlined in Sec. 9.7.

9.7* MATRIX INVERSION

Matrix methods have been used in Chapter 3 to solve conduction problems numerically. In this section we will consider the matrix inversion method, which is a powerful tool for solving radiation problems, although it requires certain assumptions and simplifications in practice. The method can be applied only when the radiation over each surface is uniform and each surface is isothermal. Any surface in the enclosure that does not meet these two requirements must be subdivided into smaller segments until the temperature and radiation flux over each are approximately uniform. However, with a computer the addition of surfaces does not significantly increase the amount of work required to obtain a numerical solution (5, 13).

9.7.1 Enclosures with Gray Surfaces

The problem at hand is solving N linear algebraic equations in N unknowns. The equations are obtained by evaluating the emittances of the surfaces and the shape factors between them and writing Eqs. (9.68) and (9.71) for each nodal point:

$$(q_i)''_{net} = \frac{\epsilon_i}{\rho_i}(E_{bi} - J_i) = \frac{\epsilon_i}{1 - \epsilon_i}(E_{bi} - J_i) \tag{9.68}$$

and

$$(q_i)''_{net} = J_i - \sum_{j=1}^{j=N} J_j F_{i-j} \tag{9.71}$$

For a gray enclosure consisting of *three surfaces at specified temperatures* this procedure yields

$$(q_1)''_{net} = \frac{\epsilon_1}{1 - \epsilon_1}(E_{b1} - J_1) = J_1 - J_1 F_{1-1} - J_2 F_{1-2} - J_3 F_{1-3} \tag{9.81a}$$

$$(q_2)''_{net} = \frac{\epsilon_2}{1 - \epsilon_2}(E_{b2} - J_2) = J_2 - J_1 F_{2-1} - J_2 F_{2-2} - J_3 F_{2-3} \tag{9.81b}$$

$$(q_3)''_{net} = \frac{\epsilon_3}{1 - \epsilon_3}(E_{b3} - J_3) = J_3 - J_1 F_{3-1} - J_2 F_{3-2} - J_3 F_{3-3} \tag{9.81c}$$

In this set of equations $N = 3$ and the three unknowns are the radiosities J_1, J_2, and J_3. The above set of equations can be recast into the more convenient form

$$\left(1 - F_{1-1} + \frac{\epsilon_1}{1 - \epsilon_1}\right)J_1 + (-F_{1-2})J_2 + (-F_{1-3})J_3 = \frac{\epsilon_1}{1 - \epsilon_1}E_{b1} \tag{9.82a}$$

$$(-F_{2-1})J_1 + \left(1 - F_{2-2} + \frac{\epsilon_2}{1 - \epsilon_2}\right)J_2 + (-F_{1-3})J_3 = \frac{\epsilon_2}{1 - \epsilon_2}E_{b2} \tag{9.82b}$$

$$(-F_{3-1})J_1 + (-F_{3-2})J_2 + \left(1 - F_{3-3} + \frac{\epsilon_3}{1 - \epsilon_3}\right)J_3 = \frac{\epsilon_3}{1 - \epsilon_3}E_{b3} \tag{9.82c}$$

Using matrix notation, we get

$$a_{11}J_1 + a_{12}J_2 + a_{13}J_3 = C_1 \tag{9.83a}$$

$$a_{21}J_1 + a_{22}J_2 + a_{23}J_3 = C_2 \tag{9.83b}$$

$$a_{31}J_1 + a_{32}J_2 + a_{33}J_3 = C_3 \tag{9.83c}$$

These equations can be written in the condensed matrix form as presented in Chapter 3:

$$\mathbf{AJ} = \mathbf{C} \tag{9.84}$$

where $\mathbf{A}$ is the 3×3 matrix

$$\mathbf{A} = \begin{bmatrix} a_{11} & a_{12} & a_{13} \\ a_{21} & a_{22} & a_{23} \\ a_{31} & a_{32} & a_{33} \end{bmatrix} \tag{9.85}$$

and **J** and **C** are column matrices consisting of three elements each:

$$\mathbf{J} = \begin{bmatrix} J_1 \\ J_2 \\ J_3 \end{bmatrix} \tag{9.86}$$

$$\mathbf{C} = \begin{bmatrix} \dfrac{\epsilon_1}{1 - \epsilon_1} E_{b1} = C_1 \\ \dfrac{\epsilon_2}{1 - \epsilon_2} E_{b2} = C_2 \\ \dfrac{\epsilon_3}{1 - \epsilon_3} E_{b3} = C_3 \end{bmatrix} \tag{9.87}$$

For the general case of an enclosure with N surfaces the matrix will have the same form as Eq. (9.84), but

$$\mathbf{A} = \begin{bmatrix} a_{11} & a_{12} & \cdots & a_{1N} \\ a_{21} & a_{22} & \cdots & \\ a_{31} & & & \\ \vdots & & & \\ a_{N1} & a_{N2} & \cdots & a_{NN} \end{bmatrix}, \qquad \mathbf{C} = \begin{bmatrix} C_1 \\ C_2 \\ \vdots \\ C_N \end{bmatrix}, \qquad \mathbf{J} = \begin{bmatrix} J_1 \\ J_2 \\ \vdots \\ J_N \end{bmatrix}$$

The off-diagonal elements of **A** are

$$a_{ij} = -F_{i-j} \qquad (i \neq j) \tag{9.88}$$

and the diagonal terms are

$$a_{ii} = \left(1 - F_{ii} + \frac{\epsilon_i}{1 - \epsilon_1} \right) \tag{9.89}$$

The elements of **C** are

$$C_i = \frac{\epsilon_i}{1 - \epsilon_i} E_{bi} \tag{9.90}$$

When a surface in the enclosure is black and its temperature, T_i, is specified, the radiosity J_i is equal to E_{bi}. Hence, it is no longer unknown and the terms in the matrix for a black element are

$$a_{ij} = 0 \qquad (i \neq j) \tag{9.91}$$
$$a_{ii} = 1.0 \tag{9.92}$$
$$C_i = E_{bi} = \sigma T^4 \tag{9.93}$$

When the heat flux instead of the temperature is specified for a surface A_i the off-diagonal elements of **A** remain the same as in Eq. (9.88). However, the diagonal elements, a_{ii}, become -

$$a_{ii} = 1 - F_{ii} \tag{9.94}$$

and the elements in the **C** matrix are

$$C_i = (q_i)''_{net} \tag{9.95}$$

This can easily be verified for a three-surface enclosure by inspection of Eqs. (9.81). For example, if the heat flux for surface 1 were specified Eq. (9.82a) becomes, upon eliminating the unknown E_{b1},

$$(q_1)''_{net} = (1 - F_{1-1})J_1 + (-F_{1-2}) J_2 + (-F_{1-3})J_3 \qquad (9.96)$$

To obtain a numerical solution we must invert the matrix $\mathbf{A}$. If $\mathbf{A}^{-1}$ denotes the inverse of $\mathbf{A}$ the solution for the radiosities is given by

$$\mathbf{J} = \mathbf{A}^{-1}\mathbf{C} \qquad (9.97)$$

where

$$\mathbf{A}^{-1} = \begin{bmatrix} b_{11} & b_{12} & \cdots & b_{1N} \\ b_{21} & \cdots & & \\ \vdots & & & \\ b_{N1} & b_{N2} & \cdots & b_{NN} \end{bmatrix} \qquad (9.98)$$

The solution for each radiosity can then be written in the form of a series:

$$
\begin{aligned}
J_1 &= b_{11}C_1 + b_{12}C_2 + b_{13}C_3 + \cdots + b_{1N}C_N \\
J_2 &= b_{21}C_1 + \cdots \\
&\vdots \\
J_N &= b_{N1}C_1 + b_{N2}C_2 + \cdots + b_{NN}C_N
\end{aligned}
\qquad (9.99)
$$

In practical terms the problem of solving the simultaneous linear algebraic equations for the radiosities reduces to the inversion of a matrix. Once the radiosities are known, the rate of heat flow can be obtained from Eq. (9.71) for each surface. When the heat flux is specified, Eq. (9.68) can be solved for the temperature T_i,

$$T_i = \left[\frac{1 - \epsilon_i}{\sigma\epsilon_i} (q_i)''_{net} + J_i/\sigma \right]^{1/4} \qquad (9.100)$$

The following examples illustrate this procedure.

EXAMPLE 9.12 ────────────────────────────────────

The temperatures of the top and bottom surfaces of the frustum of the cone shown in Fig. 9.41 are maintained at 600 K and 1200 K, respectively, while side A_2 is perfectly insulated ($q_2 = 0$). If all surfaces are gray and diffuse, determine the net radiative exchange between the top and bottom surfaces, i.e., A_3 and A_1.

Solution. From Table 9.3 we find that $F_{31} = 0.333$ and from Eq. 9.60 we obtain $F_{32} = 1 - F_{31} = .667$.

According to the reciprocity theorem $A_1F_{13} = A_3F_{31}$ and $A_2F_{23} = A_3F_{32}$. Therefore, $F_{13} = 0.147$ and $F_{23} = 0.130$. From Eq. 9.60 we get $F_{12} = 1 - F_{13} = 0.853$ and by reciprocity, $F_{21} = F_{12}A_1/A_2 = 0.372$. Finally, $F_{22} = 1 - F_{21} - F_{23} = 0.498$.

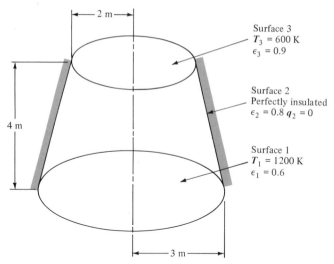

Figure 9.41 Schematic sketch of cone for Example 9.12.

According to the general relations given by Equations 9.68 and 9.71 the system of equations to be solved for this problem may be written:

$$E_{b1} \cdot \frac{\epsilon_1}{1 - \epsilon_1} = J_1\left(1 - F_{11} + \frac{\epsilon_1}{1 - \epsilon_1}\right) + J_2(-F_{12}) + J_3(-F_{13})$$

$$0 = J_1(-F_{21}) + J_2(1 - F_{22}) + J_3(-F_{23})$$

$$E_{b3} \cdot \frac{\epsilon_3}{1 - \epsilon_3} = J_1(-F_{31}) + J_2(-F_{32}) + J_3\left(1 - F_{33} + \frac{\epsilon_3}{1 - \epsilon_3}\right)$$

or in matrix notation $\mathbf{A} \cdot \mathbf{J} = \mathbf{C}$.

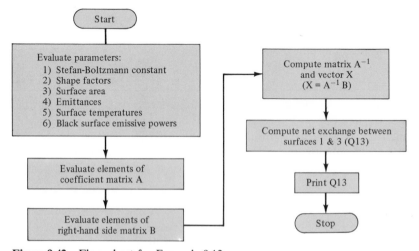

Figure 9.42 Flow chart for Example 9.12.

A FORTRAN subroutine called LINEQZ which solves linear algebraic systems of equations of the form $\mathbf{A} \cdot \mathbf{X} = \mathbf{B}$ will be used to evaluate all the J's. The net rate of heat transfer between top and bottom, i.e., the value of $q_{3 \rightleftharpoons 1}$, can then be determined from Equation 9.73. Fig. 9.42 gives the flow diagram for the computer operations to solve this problem. The FORTRAN program and the solution are presented in Table 9.4. The

TABLE 9.4 FORTRAN PROGRAM FOR EXAMPLE 9.12

```
DIMENSION A(3,3),B(3,1),X(3,1),DIGITS(1)
DIMENSION SCRA1(3,3),SCRA2(3),SCRA3(3),SCRA4(3)
DIMENSION F(3,3),AR(3),EPS(3),T(3),EB(3)
```

The arrays required for the program are declared. See Table 9.5 for the precise definition of each. The arrays SCRA1, SCRA2, etc. are peculiar to the matrix inversion subroutine LINEQZ.

```
PI=4*ATAN(1.)
SIGMA=.567E-07
F(1,1)=0.
F(1,2)=.853
F(1,3)=.147
F(2,1)=.372
F(2,2)=.498
F(2,3)=.130
F(3,1)=.333
F(3,2)=.667
F(3,3)=0.
AR(1)=9.*PI
EPS(1)=.6
EPS(3)=.9
T(1)=1200.
T(3)=600.
EB(1)=SIGMA*T(1)**4.
EB(3)=SIGMA*T(3)**4.
```

The physical parameters, e.g., shape factors and emittance, are evaluated.

```
A(1,1)=1.-F(1,1)+EPS(1)/(1.-EPS(1))
A(1,2)=-F(1,2)
A(1,3)=-F(1,3)
A(2,1)=-F(2,1)
A(2,2)=1.-F(2,2)
A(2,3)=-F(2,3)
A(3,1)=-F(3,1)
A(3,2)=-F(3,2)
A(3,3)=1.-F(3,3)+EPS(3)/(1.-EPS(3))
```

The values of the elements of the coefficient matrix A in the equation $[A][X] = [B]$ are specified.

```
B(1) = EB(1)*EPS(1)/(1. - EPS(1))
B(2)=0.
B(3) = EB(3)*EPS(3)/(1. - EPS(3))
```

The values of the right hand side vector **B** are specified.

```
CALL LINEQZ(3,3,1,A,B,X,D,DIGITS,SCRA1,SCRA2,SCRA3,SCRA4)
```

The matrix inversion subroutine LINEQZ is called to compute the solution vector $X = A^{-1} \cdot B$. This is one of many matrix inversion routines. The reader may use any routine available and specify the arguments accordingly.

TABLE 9.5 SYMBOLS USED IN FORTRAN PROGRAM FOR EXAMPLE 9.12

FORTRAN Symbol	Heat Balance Equation Notation	Description	Units
A(I,J)	a_{ij}	coefficient matrix elements	—
AR(1), AR(3)	A_1, A_3	lower and upper surface areas	m^2
B(I)	C_i	right-hand side matrix elements	W/m^2
EB(1), EB(3)	E_{b1}, E_{b3}	blackbody emissive powers	W/m^2
EPS(1), etc.	ϵ_1, etc.	total hemispheric emittance	—
F(1,1), F(1,2), etc.	F_{11}, F_{12}, etc.	shape factors	—
PI	π	3.14159 ...	—
Q31	$q_{3 \rightleftharpoons 1}$	net exchange between surfaces 3 and 1	W
SIGMA	σ	Stefan-Boltzmann constant (0.567×10^{-7})	W/m^2K^4
T(1), T(3)	T_1, T_3	surface temperatures	K
X(I)	J_i	radiosities (elements of solution vector)	W/m^2

symbols used in the FORTRAN program are defined in Table 9.5. The symbols which appear in the program but are not listed in Table 9.5 are associated only with the subroutine LINEQZ and do not have significance in the solution of the heat transfer problem.

EXAMPLE 9.13 ─────────────────────────────

Determine the temperature of surface 1 for the cone shown in Fig. 9.41 if $q_1 = 3 \times 10^5$ W/m² and $\epsilon_3 = 1$. Assume that all other parameters are the same as in Example 9.12.

Solution. From equations 9.94, 9.95, and 9.97, the following system of equations must be solved for J_1, J_2, and J_3.

$$q_1/A_1 = J_1(1 - F_{11}) + J_2(-F_{12}) + J_3(-F_{13})$$
$$0 = J_1(-F_{21}) + J_2(1 - F_{22}) + J_3(-F_{23})$$
$$E_{b3} = J_3$$

Once the J_i's are known, Equation 9.100 gives T_1. The FORTRAN program for the solution of this problem is shown in Table 9.6. Since it is very similar to the preceding program the flow diagram is essentially the same as that used in Example 9.12.

9.7.2 Enclosure with Nongray Surfaces

The method of approach used to calculate heat transfer in gray surface enclosures can easily be adapted to nongray surfaces. If the surface properties are functions of wavelength, they can be approximated by gray "bands" within which an average value of emittance and absorptance is used. Then the same calculation method used previously for gray enclosures can be used to determine the ra-

TABLE 9.6 FORTRAN PROGRAM FOR EXAMPLE 9.12

```
DIMENSION A(3,3),B(3,1),X(3,1),DIGITS(1)
DIMENSION SCRA1(3,3),SCRA2(3),SCRA3(3),SCRA4(3)
DIMENSION F(3,3),AR(3),EPS(3),T(3),EB(3)
COMMENT EVALUATE CONSTANTS AND PARAMETERS
PI = 4*ATANI(1.)
SIGMA = .567E-07
F(1,1) = 0.
F(1,2) = .853
F(1,3) = .147
F(2,1) = .372
F(2,2) = .498
F(2,3) = .130
F(3,1) = .333
F(3,2) = .667
F(3,3) = 0.
AR(1) = 9*PI
EPS(1) = 0.6
Q1 = 300000.
T(3) = 600.
EB(3) = SIGMA*T(3)**4.
COMMENT EVALUATE ELEMENTS OF COEFFICIENT MATRIX
A(1,1) = 1.-F(1,1)
A(1,2) = -F(1,2)
A(1,3) = -F(1,3)
A(2,1) = -F(2,1)
A(2,2) = 1.-F(2,2)
A(2,3) = -F(2,3)
A(3,1) = 0.
A(3,2) = 0.
A(3,3) = 1.
COMMENT EVALUATE ELEMENTS OF RIGHT HAND SIDE MATRIX
B(1) = Q1/AR(1)
B(2) = 0.
B(3) = EB(3)
COMMENT CALL SUBROUTINE TO SOLVE SYSTEM OF EQUATIONS
CALL LINEQZ(3,3,1,A,B,X,D,DIGITS,SCRA1,SCRA2,SCRA3,SCRA4)
COMMENT COMPUTE T(1) FROM EQUATION 5-49
T(1) = ((X(1) + Q*(1. - EPS(1))/(AR(1)*EPS(1)))/SIGMA)**.25
PRINT 2,T(1)
FORMAT (1H1.*TEMPERATURE OF SURFACE 1, T(1), = *,F7.1,*K*)
END
TEMPERATURE OF SURFACE 1, T(1), =934.6 K
```

diation heat transfer within each band. The following example illustrates the procedure.

EXAMPLE 9.14 ─────────────────────────────────

Determine the rate of heat transfer between two large parallel flat plates placed 2 in. apart, if one plate (*A*) is at 2040°F and the other (*B*) at 540°F.

Plate A has an emittance of 0.1 between 0 and 2.5 μm and an emittance of 0.9 for wavelengths longer than 2.5 μm. The emittance of plate B is 0.9 between 0 and 4.0 μm and 0.1 at longer wavelengths.

Solution. The shape factor F_{A-B} for two large parallel rectangular plates is 1.0 if end effects are negligible. The radiosity of A is given by

$$\int_0^\infty J_{\lambda A}\, d\lambda = \int_0^\infty \epsilon_{\lambda A} E_{b\lambda A}\, d\lambda + \int_0^\infty \rho_{\lambda A} G_{\lambda A}\, d\lambda$$

and the radiosity of B by

$$\int_0^\infty J_{\lambda B}\, d\lambda = \int_0^\infty \epsilon_{\lambda B} E_{b\lambda B}\, d\lambda + \int_0^\infty \rho_{\lambda B} G_{\lambda B}\, d\lambda$$

However, using spectral bands between 0 and 2.5 μm, 2.5 and 4.0 μm, and 4.0 μm or larger, the system obeys gray surface radiation laws within each band and the rate of heat transfer can be calculated from Eq. (9.75) in three steps as shown below:

Band 1:

$$q_{A\rightleftharpoons B}\Big|_0^{2.5\ \mu m} = \mathscr{F}_{A-B}(\epsilon_A = 0.1, \epsilon_B = 0.9)$$
$$\times \left[\frac{E_{b,0-2.5}(T_A)}{E_{b,0-\infty}(T_A)}\, \sigma T_A^4 - \frac{E_{b,0-2.5}(T_B)}{E_{b,0-\infty}(T_B)}\, \sigma T_B^4 \right]$$

Band 2:

$$q_{A\rightleftharpoons B}\Big|_{2.5\ \mu m}^{4.0\ \mu m} = \mathscr{F}_{A-B}(\epsilon_A = 0.9, \epsilon_B = 0.9)$$
$$\times \left[\frac{E_{b,2.5-4.0}(T_A)}{E_{b,0-\infty}(T_A)}\, \sigma T_A^4 - \frac{E_{b,2.5-4.0}(T_B)}{E_{b,0-\infty}(T_B)}\, \sigma T_B^4 \right]$$

Band 3:

$$q_{A\rightleftharpoons B}\Big|_{4.0\ \mu m}^{\infty} = \mathscr{F}_{A-B}(\epsilon_A = 0.9, \epsilon_B = 0.1)$$
$$\times \left[\frac{E_{b,4.0-\infty}(T_A)}{E_{b,0-\infty}(T_A)}\, \sigma T_A^4 - \frac{E_{b,4.0-\infty}(T_B)}{E_{b,0-\infty}(T_B)}\, \sigma T_B^4 \right]$$

where

$$\mathscr{F}_{A-B} = \frac{1}{1/\epsilon_A + 1/\epsilon_B - 1}$$

from Fig. 9.38.

The percentage of the total radiation within a given band is obtained from Table 9.1. For example, $(E_{b,0-2.5}/E_{b,0-\infty})$ for a temperature of $T_A = 2500$ R is 0.375 and for a temperature of $T_B = 1000$ R it is about 0.004. Thus, for the first band,

$$q_{A\rightleftharpoons B1}\Big|_0^{2.5\ \mu m} = 0.10 \times 0.1714(0.375 \times 25^4 - 0.004 \times 10^4)$$

$$= 2530 \text{ Btu/h ft}^2$$

Similarly, for the second band,

$$q_{A \rightleftharpoons B2} \Big|_{2.5 \, \mu m}^{4.0 \, \mu m} = 23,000 \text{ Btu/h ft}^2$$

and for the third band,

$$q_{A \rightleftharpoons B3} \Big|_{4.0 \, \mu m}^{\infty} = 1240 \text{ Btu/h ft}^2$$

Finally, summing over all three bands, the total rate of radiation heat transfer is

$$q_{A \rightleftharpoons B} \Big|_{0}^{\infty} = \sum_{N=1}^{N=3} q_{A \rightleftharpoons B^N} = 2530 + 23,000 + 1240 + 26,770 \text{ Btu/h ft}^2$$

It should be noted that most of the radiation is transferred within the second band, where both surfaces are nearly black.

Enclosures consisting of several nongray surfaces can be treated in a similar manner by dividing the radiation spectrum into finite bands within which the radiation properties can be approximated by constant values. This procedure can become particularly useful when the enclosure is filled with a gas that absorbs and emits radiation only at certain wavelengths.

9.7.3* Enclosures with Absorbing and Transmitting Media

The method of analysis outlined in the preceding sections can be extended to solve problems in which heat is transferred by radiation in an enclosure containing a medium that is both absorbing and transmitting. Various glasses, plastics, and gases are examples of such media. To illustrate the method of approach, consider first radiation between two plates when the space between them is filled with a "gray" gas that does not reflect any of the incident radiation. The geometry is shown in Fig. 9.43a. The two solid surfaces are at temperatures T_1 and T_2; the properties of the transmitting gas are denoted by the subscript m.

Kirchhoff's law applied to the transmitting gray gas requires that $\alpha_m = \epsilon_m$ and since the reflectivity of the medium is zero,

$$\tau_m = 1 - \alpha_m = 1 - \epsilon_m \tag{9.101}$$

We will derive the equations for the heat transfer rate between the surfaces by developing the thermal circuit for the problem. The portion of the total radiation leaving surface 1 that arrives after passing through the gas at surface 2 is

$$J_1 A_1 F_{12} \tau_m$$

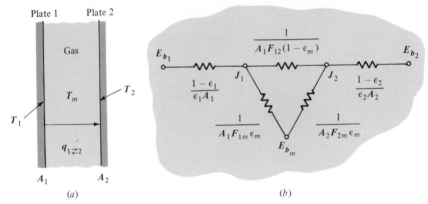

Figure 9.43 Electric analog for radiation between finite planes separated by a gas.

and that from surface 2 which reaches 1 is

$$J_2 A_2 F_{21} \tau_m$$

The net rate of heat transfer between the two surfaces is therefore

$$q_{1 \rightleftharpoons 2} = A_1 F_{12} \tau_m (J_1 - J_2) = \frac{J_1 - J_2}{1/A_1 F_{12}(1 - \epsilon_m)} \tag{9.102}$$

Thus, the equivalent resistance between nodal points J_1 and J_2 for this case in a network will be $1/A_1 F_{12}(1 - \epsilon_m)$.

Radiation heat transfer occurs also between each of the surfaces and the gas. If the gas is at temperature T_m it will emit radiation at a rate

$$J_m = \epsilon_m E_{bm} \tag{9.103}$$

The fraction of the energy emitted by the gaseous medium which reaches surface 1 is

$$A_m F_{m-1} J_m = A_m F_{m-1} \epsilon_m E_{bm} \tag{9.104}$$

Similarly, the fraction of the radiation leaving A_1 which is absorbed by the transparent medium is

$$J_1 A_1 F_{1m} \alpha_m = J_1 A_1 F_{1m} \epsilon_m \tag{9.105}$$

The net rate of heat transfer by radiation between the gas and surface 1 is the difference between the radiation emitted by the gas toward A_1 and the radiation emanating from A_1 which is absorbed by the gas. Thus

$$q_{m \rightleftharpoons 1} = A_m F_{m1} \epsilon_m E_{bm} - J_1 A_1 F_{1m} \epsilon_m \tag{9.106}$$

Using the reciprocity theorem, $A_1 F_{1m} = A_m F_{m1}$, the net exchange can be written in the form

$$q_{m \rightleftharpoons 1} = \frac{E_{bm} - J_1}{1/A_1 F_{1m} \epsilon_m} \tag{9.107}$$

Similarly, the net exchange between the gas and A_2 is

$$q_{m \rightleftharpoons 2} = \frac{E_{bm} - J_2}{1/A_2 F_{2m}\epsilon_m} \tag{9.108}$$

Using the above relations to construct an equivalent circuit, radiation between two surfaces at T_1 and T_2 respectively, separated by an absorbing medium at T_m, can be represented as shown in Fig. 9.43b. If the gas is not maintained at a specified temperature, but reaches an equilibrium temperature at which it emits radiation at the same rate at which it absorbs it, E_{bm} becomes a floating node in the network. For this case the net rate of heat transfer between A_1 and A_2 is

$$q_{1 \rightleftharpoons 2} = \frac{\sigma(T_1^4 - T_2^4)}{\dfrac{1 - \epsilon_1}{\epsilon_1 A_1} + \dfrac{1 - \epsilon_2}{\epsilon_2 A_2} + \dfrac{1}{A_1[F_{1-2}(\tau_m) + 1/(F_{1-m}\epsilon_m + A_1/A_2 F_{2-m}\epsilon_m)]}} \tag{9.109}$$

When A_1 and A_2 are very large so that F_{1-2}, F_{1-m}, and F_{2-m} approach unity the last factor is the denominator approaches $(2/\epsilon_m + 2\tau_m)$.

More complex enclosures with several surfaces can be treated by the matrix method once the appropriate thermal network has been drawn. Details for method of solution of such cases can be found in advanced radiation texts (12, 13).

9.8 RADIATION PROPERTIES OF GASES AND VAPORS

In this section we shall consider some basic concepts of gaseous radiation. A comprehensive treatment of this subject is beyond the scope of this text, and the reader should consult (13, 18, 20–28) for details of the theoretical background and complete derivations of the calculation techniques.

Elementary gases such as O_2, N_2, H_2, and dry air have a symmetrical molecular structure and neither emit nor absorb radiation unless they are heated to such extremely high temperatures that they become ionized plasmas and electronic energy transformations occur. On the other hand, gases which have polar molecules with an electronic moment such as a dipole or quadrupole absorb and emit radiation in limited spectral ranges, called bands. In practice, the most important of these gases are H_2O, CO_2, CO, SO_2, NH_3, and the hydrocarbons. These gases are asymmetric in one or more of their modes of vibration. During molecular collisions, rotation and vibrations of individual atoms in a molecule can be excited so that atoms which possess free electrical charges can emit electromagnetic waves. Similarly, when radiation of the appropriate wavelength impinges on such a gas, it can be absorbed in the process. We shall restrict our consideration here to the evaluation of the radiation properties of H_2O and CO_2. They are the most important gases in thermal radiation calculations and also illustrate the basic principles of gaseous radiation.

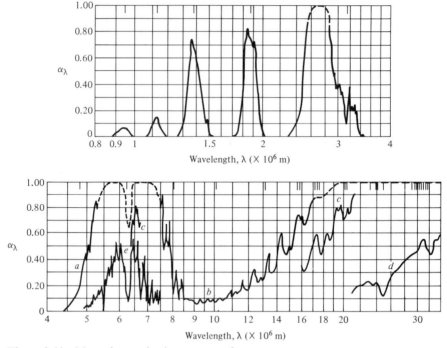

Figure 9.44   Monochromatic absorptance of water vapor.

Typical changes in energy level due to changes in vibrational frequency or rotation manifest themselves in a strong peak at the wavelength corresponding to the vibrational transformation, with multiple rotational energy changes slightly above or below the peak. This process results in absorption or emission bands. The shape and width of these bands depend on the temperature and on the pressure of the gas while the magnitude of the monochromatic absorptance is primarily a function of the thickness of the gas layer. The absorption spectrum of steam shown in Fig. 9.44 illustrates the complexity of the process. The most important absorption bands for steam lie between 1.7 and 2.0μm, 2.2 and 3.0μm, 4.8 and 8.5μm, and 11 and 25μm.

Experimental measurements generally yield the absorptance of a gas layer over a band width corresponding to the width of the spectrometer slit used. Thus, experimental data are usually presented in terms of the monochromatic absorptance, as shown in Fig. 9.44. For most engineering calculations, however, the quantity of primary interest is the effective total absorptance or emittance. This quantity assumes that the gas is gray and, as shown below, its value depends not only on the pressure, temperature, and composition, but also on the geometry of the radiating gas.

Whereas for opaque solids the emission and absorption of radiation are surface phenomena, in calculating the radiation emitted or absorbed by a gas layer its thickness, pressure, and shape as well as its surface area must be taken into account. When monochromatic radiation at an intensity $I_{\lambda 0}$ passes through a gas layer of thickness L, the radiant-energy absorption in a differential distance dx is governed by the relation

$$dI_{\lambda x} = -k'_\lambda I_{\lambda x} \, dx \tag{9.110}$$

where $I_{\lambda x}$ = intensity at a distance x;

k'_λ = monochromatic absorption coefficient, a proportionality constant whose value depends on the pressure and temperature of the gas.

Integration between the limits $x = 0$ and $x = L$ yields

$$I_{\lambda L} = I_{\lambda 0} e^{-k'_\lambda L} \tag{9.111}$$

where $I_{\lambda L}$ is the intensity of radiation at L. The difference between the intensity of radiation entering the gas at $x = 0$ and the intensity of radiation leaving the gas layer at $x = L$ is the amount of energy absorbed by the gas

$$I_{\lambda 0} - I_{\lambda L} = I_{\lambda 0}(1 - e^{-K'_\lambda L}) = \alpha_{G\lambda} \tag{9.112}$$

The quantity in the parentheses represents the *monochromatic absorptance of the gas* $\alpha_{G\lambda}$, or, according to Kirchhoff's law, also the emittance at the wavelength λ, $\epsilon_{G\lambda}$. To obtain effective values of the emittance or absorptance, a summation over all of the radiation bands is necessary. We observe that, for large values of L, i.e., for thick layers, gas radiation approaches blackbody conditions within the wavelengths of its absorption bands.

For gas bodies of finite dimensions, however, the effective absorptance or emittance depends on the shape and the size of the gas body, since radiation is not confined to one direction. The precise method of calculating the effective absorptance or emittance is quite complex (24–27), but for engineering calculations an approximate method developed by Hottel and Egbert (17, 18) yields results of satisfactory accuracy. Hottel evaluated the effective total emittances of a number of gases at various temperatures and pressures and presented his results in graphs similar to those shown in Figs. 9.45 and 9.46. The graphs apply strictly only to a system in which a hemispherical gas mass of radius L radiates to an element of surface located at the center of the base of a hemisphere. However, for shapes other than hemispheres, an effective beam length can be calculated. Table 9.7 lists the constants by which the characteristic dimensions of several simple shapes are to be multiplied to obtain an equivalent mean hemispherical beam length L for use in Figs. 9.45 and 9.46. For approximate calculations and for shapes other than those listed in Table 9.7, L can be taken as 3.4 × volume/surface area.

In Figs. 9.45 and 9.46 the symbols P_{H_2O} and P_{CO_2} represent the partial pressures of the gases. The total pressure for both figures is 1 atm. When the

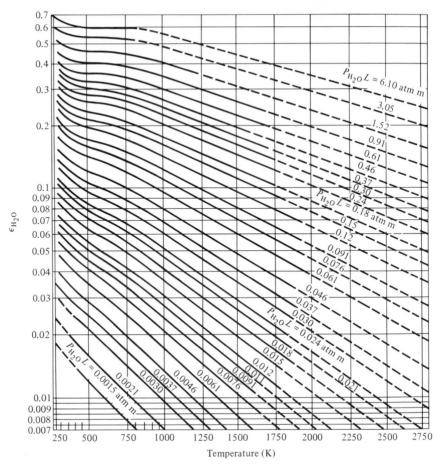

Figure 9.45 Emittance of water vapor at a total pressure of 1 atm. [With permission from H. C. Hottel, in W. H. McAdams (7, chapter 4).]

total gas pressure differs from 1 atm the values from Figs. 9.45 and 9.46 must be multiplied by a correction factor. The emittance of H_2O and CO_2 at a total pressure P_T other than 1 atm are then given by the expressions (24)

$$(\epsilon_{H_2O})_{P_T} = C_{H_2O}(\epsilon_{H_2O})_{P_T=1} \tag{9.113a}$$

$$(\epsilon_{CO_2})_{P_T} = C_{CO_2}(\epsilon_{CO_2})_{P_T=1} \tag{9.113b}$$

and the correction factors C_{H_2O} and C_{CO_2} are plotted in Figs. 9.47 and 9.48, respectively. When both H_2O and CO_2 exist in a mixture, the emittance of the mixture may be calculated by adding the emittance of the gases determined by assuming that each gas exists alone and then subtracting a factor $\Delta\epsilon$, which accounts for emission in overlapping wavelength bands. The factor $\Delta\epsilon$ for H_2O and CO_2 is plotted in Fig. 9.49. The emittance of a mixture of H_2O and CO_2

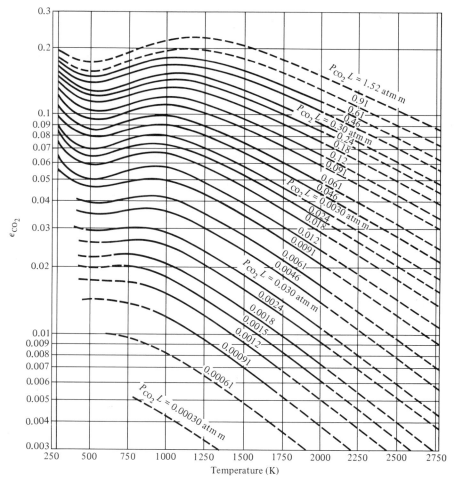

Figure 9.46 Emittance of carbon dioxide at a total pressure of 1 atm. With permission from H. C. Hottel, in W. H. McAdams (7, chapter 4).]

is therefore given by the expression

$$\epsilon_{mix} = C_{H_2O}(\epsilon_{H_2O})_{P_T=1} + C_{CO_2}(\epsilon_{CO_2})_{P_T=1} - \Delta\epsilon \tag{9.114}$$

EXAMPLE 9.15

Determine the emittance of a gas mixture consisting of N_2, H_2O, and CO_2 at a temperature of 800 K. The gas mixture is in a sphere with diameter of 0.4 m and the partial pressures of the gases are $P_{N_2} = 1$ atm, $P_{H_2O} = 0.4$ atm, and $P_{CO_2} = 0.6$ atm.

Solution. The mean beam length for a spherical mass of gas is obtained from Table 9.6:

$$L = (2/3)D = 0.27 \text{ m}$$

TABLE 9.7   MEAN BEAM LENGTH OF VARIOUS GAS SHAPES

Geometry	L
Sphere	$\frac{2}{3}$ (diameter)
Infinite cylinder	Diameter
Infinite parallel planes	2 (distance between planes)
Semi-infinite cylinder, radiating to center of base	Diameter
Right circular cylinder, height equal to diameter:	
Radiating to center of base	Diameter
Radiating to whole surface	$\frac{2}{3}$ (diameter)
Infinite cylinder of half-circular cross section radiating to spot in middle of flat side	Radius
Rectangular parallelepipeds:	
Cube	$\frac{2}{3}$ (edge)
1:1:4 radiating to 1 × 4 face	0.9 (shortest edge)
Radiating to 1 × 1 face	0.86 (shortest edge)
Radiating to all faces	0.891 (shortest edge)
Space outside infinite bank of tubes with centers on equilateral triangles:	
Tube diameter = clearance	3.4 (clearance)
Tube diameter = $\frac{1}{2}$ (clearance)	4.44 (clearance)

Source: W. M. Rohsenow, J. P. Hartnett and E. Ganic (27).

The emittances are given in Figs. 9.45 and 9.46 and appropriate values for the parameters to be used are

$$T = 800 \text{ K}$$

$$P_{H_2O}L = 0.107 \text{ atm m}$$

$$P_{CO_2}L = 0.160 \text{ atm m}$$

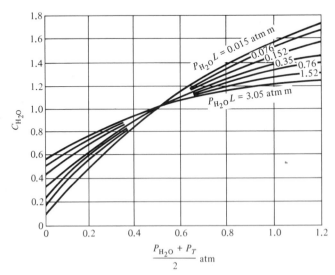

Figure 9.47   Correction factor for the emittance of water vapor at pressures other than 1 atm. [From H. C. Hottel and R. B. Egbert (18) and R. B. Egbert (26).]

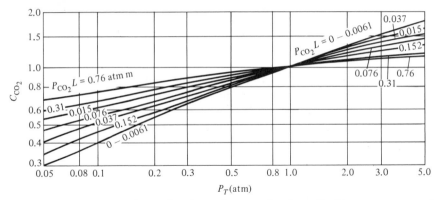

Figure 9.48 Correction factor for the emittance of carbon dioxide at pressures other than 1 atm. [From H. C. Hottel and R. B. Egbert (18).]

The emittances for water vapor and carbon dioxide at 1 atm total pressure are, from Figs. 9.45 and 9.46, respectively

$$(\epsilon_{H_2O})_{P_T=1} = 0.15,$$
$$(\epsilon_{CO_2})_{P_T=1} = 0.125,$$

N_2 does not radiate appreciably at 800 K, but since the total gas pressure is 2 atm, we must correct the 1-atm values for ϵ. From Figs. 9.47 and 9.48 the pressure correction factors are

$$C_{H_2O} = 1.62$$
$$C_{CO_2} = 1.12$$

The value for $\Delta\epsilon$ used to correct for emission in overlapping wavelength bands is determined from Fig. 9.49:

$$\Delta\epsilon = 0.014$$

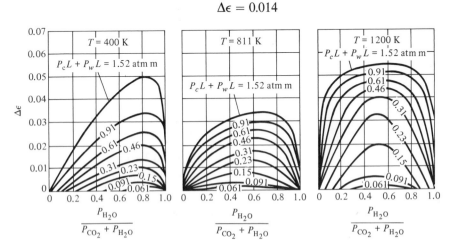

Figure 9.49 Factor $\Delta\epsilon$ to correct the emittance of a mixture of water vapor and CO_2. [From H. C. Hottel and R. B. Egbert (18).]

Finally, the emittance of the mixture can be obtained from Eq. (9.114):

$$\epsilon_{mix} = 1.62 \times 0.15 + 1.12 \times 0.125 - 0.014 = 0.369$$

Gas absorptance can be obtained from the emittance charts shown previously by modifying the parameters in the charts. As an example, consider water vapor at a temperature of T_{H_2O} with incident radiation from a source surface at a temperature of T_s. The absorptance of the H_2O vapor is approximately given by the relation

$$\alpha_{H_2O} = C_{H_2O}\epsilon'_{H_2O}\left(\frac{T_{H_2O}}{T_s}\right)^{0.45} \tag{9.115}$$

if C_{H_2O} is taken from Fig. 9.47 and the value for the emittance of water vapor ϵ'_{H_2O} from Fig. 9.45 is evaluated at temperature T_s and a pressure–mean beam length product equal to $P_{H_2O}L(T_s/T_{H_2O})$. Similarly, the absorptance of CO_2 can be obtained from

$$\alpha_{CO_2} = C_{CO_2}\epsilon'_{CO_2}\left(\frac{T_{CO_2}}{T_s}\right)^{0.65} \tag{9.116}$$

where the value for C_{CO_2} is taken from Fig. 9.48 and the value for ϵ'_{CO_2} is evaluated from Fig. 9.46 at a pressure–mean beam length product equal to $P_{CO_2}L(T_s/T_{CO_2})$.

EXAMPLE 9.16

Determine the absorptance of a mixture of H_2O vapor and N_2 gas at a total pressure of 2.0 atm and a temperature of 500 K if the mean beam length is 0.75 m. Assume that the radiation passing through the gas is emitted by a source at 1000 K and the partial pressure of the water vapor is 0.4 atm.

Solution. Since nitrogen is transparent, the absorption in the mixture is due to the water vapor alone. From Eq. (9.115) the absorptivity of H_2O is

$$\alpha_{H_2O} = C_{H_2O}\epsilon'_{H_2O}(T_{H_2O}/T_s)^{0.45}$$

The values of the parameters needed to evaluate the absorptivity of the gas are obtained from the data:

$$P_{H_2O} \cdot L = 0.4 \times 0.75 = 0.3 \text{ atm m}$$
$$\tfrac{1}{2}(P_T + P_{H_2O}) = \tfrac{1}{2}(2 + 0.4) = 1.2 \text{ atm}$$

From Figs. 9.45 and 9.47 we find

$$\epsilon_{H_2O} = 0.29$$
$$C_{H_2O} = 1.40$$

Substituting the above values in Eq. (9.115) gives the absorptance of the mixture

$$\alpha = 1.4 \times 0.29(500/1000)^{0.45} = 0.30$$

To calculate the rate of heat flow by radiation between a nonluminous gas at T_G and the walls of a blackbody container at T_w, the absorptance α_G of the gas should be evaluated at the temperature T_w and the emittance ϵ_G at the temperature T_G. The net rate of radiant heat flow is then equal to the difference between the emitted and absorbed radiation, or

$$q_r = \sigma A_G(\epsilon_G T_G{}^4 - \alpha_G T_w{}^4) \qquad (9.117)$$

EXAMPLE 9.17 ───────────────────────────────────

Flue gas at 2000°F containing 5 percent water vapor flows at atmospheric pressure through a 2-ft-square flue made of refractory brick. Estimate the rate of heat flow per foot length from the gas to the wall if the inner-wall surface temperature is 1850°F and the average unit-surface convective conductance is 1 Btu/h ft^2 °F.

Solution The rate of heat flow from the gas to the wall by convection per unit length is

$$q_c = \bar{h}_c A(T_{\text{gas}} - T_{\text{wall}})$$
$$= (1)(8)(150) = 1200 \text{ Btu/h ft length of flue}$$

To determine the rate of heat flow by radiation, we calculate first the effective beam length, or

$$L = \frac{3.4 \times \text{volume}}{\text{surface area}} = \frac{(3.4)(4)}{8} = 1.7 \text{ ft } (0.52 \text{ m})$$

The product of partial pressure and L is

$$pL = (0.05)(0.52) = 0.026 \text{ atm m}$$

From Fig. 9.45, for $pL = 0.026$ and $T_G = 1367$ K (2000°F), we find $\epsilon_G = 0.035$. Similarly, we find $\alpha_G = 0.039$ at $T_w = 1283$ K (1850°F). The pressure correction is negligible since $\bar{C}_p \cong 1$ according to Fig. 9.47. Assuming that the brick surface is black, the net rate of heat flow from the gas to the wall by radiation is, according to Eq. (9.117),

$$q_r = 0.171 \times 8[0.035(24.6)^4 - 0.039(23.1)^4] = 2340 \text{ Btu/h}$$

The total heat flow from the gas to the duct is therefore 3540 Btu/h. It is interesting to note that the small amount of moisture in the gas contributes about one-half of the total heat flow.

A recent review of radiation properties of gases showed that when the radiation properties of H_2O and CO_2 are evaluated from the graphs in Figs. 9.45–9.48, they can be used for industrial heat transfer calculations with satisfactory accuracy as long as the enclosure surface is not highly reflecting. But calculation of the radiant heat transfer in a gas-filled enclosure becomes considerably more complicated when the enclosure surfaces are not black and reflect a part of the incident radiation. When the emittance of the enclosure is larger than 0.7, an approximate answer may be obtained by multiplying the rate of heat flow calculated from Eq. (9.117) by $(\epsilon_s + 1)/2$, where ϵ_s is the emittance of the enclosure surface. When the enclosure walls have smaller emittances, the procedure outlined in Section 9.5 can be used, provided the assumption that all surfaces as well as the gas are "gray" is acceptable. If one or more of the surfaces are not gray or if the gas cannot be treated as a gray body, a band approximation procedure similar to that in Example 9.14 must be used. Details for such refinement in the calculation procedures are presented in (12, 13, 19, 20). Scaling rules that extend the application of one-atmosphere total spectrum emissivity data to determine gas emissivities at higher and lower pressures are available (28).

9.9 RADIATION COMBINED WITH CONVECTION AND CONDUCTION

In the preceding sections of this chapter we have considered radiation as an isolated phenomenon. Energy exchange by radiation is the predominant heat flow mechanism at high temperatures because the rate of heat flow depends on the fourth power of the absolute temperature. In many practical problems, however, convection and conduction cannot be neglected, and in this section we shall consider problems which involve two or all three modes of heat flow simultaneously.

To include radiation in a thermal network involving convection and conduction it is often convenient to define a unit thermal radiative conductance, or radiant heat transfer coefficient, $\bar{h}_r$, as

$$\bar{h}_r = \frac{q_r}{A_1(T_1 - T_2')} = \mathscr{F}_{1-2}\left[\frac{\sigma(T_1^4 - T_2^4)}{T_1 - T_2'}\right] \qquad (9.118)$$

where A_1 = area upon which $\mathscr{F}_{1-2}$ is based, m^2

$T_1 - T_2'$ = a reference temperature difference, in K, in which T_2' may be chosen equal to T_2 or any other convenient temperature in the system

$\bar{h}_r$ = radiant heat transfer coefficient, $W/m^2\ K$

Once a radiant heat transfer coefficient has been calculated, it can be treated similarly to the convective heat transfer coefficient, because the rate of heat flow becomes linearly dependent on the temperature difference and radiation can be incorporated directly in a thermal network for which the temperature is the driving potential. Knowledge of the value of $\bar{h}_r$ is also essential in determining the overall conductance $\bar{h}$ for a surface to or from which heat flows by convection and radiation, since according to Chapter 1,

$$\bar{h} = \bar{h}_c + \bar{h}_r$$

If $T_2 = T'_2$, the bracketed expression in Eq. (9.118) is called the *temperature factor* F_T, and

$$\bar{h}_r = \mathscr{F}_{1-2}F_T \qquad (9.119)$$

EXAMPLE 9.18

A butt-welded thermocouple (Fig. 9.50) having an emittance of 0.8 is used to measure the temperature of a transparent gas flowing in a large duct whose walls are at a temperature of 440°F. The temperature indicated by the thermocouple is 940°F. If the convective heat transfer coefficient between the surface of the couple and the gas $\bar{h}_c$ is 25 Btu/h ft² °F, estimate the *true* gas temperature.

Solution. The temperature of the thermocouple is below the gas temperature because the couple loses heat by radiation to the wall. Under steady-state conditions the rate of heat flow by radiation from the thermocouple junction to the wall equals the rate of heat flow by convection from the gas to the couple. We can write this heat balance as

$$q = \bar{h}_c A_T(T_G - T_T) = A_T \epsilon \sigma(T_T{}^4 - T_{\text{wall}}{}^4)$$

where A_T is the surface area, T_T the temperature of the thermocouple, and T_G the temperature of the gas. Substituting the data of the problem we obtain

$$\frac{q}{A_T} = 0.8 \times 0.1714\left[\left(\frac{1400}{100}\right)^4 - \left(\frac{900}{100}\right)^4\right] = 4410 \text{ Btu/h ft}^2$$

and the true gas temperature is

$$T_G = \frac{q}{\bar{h}_c A_T} + T_T = \frac{4410}{25} + 940 = 1116°\text{F}$$

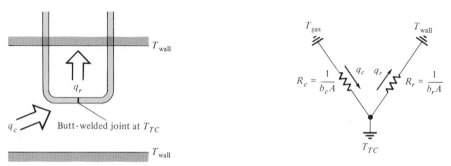

Figure 9.50 Physical system and thermal network for butt-welded thermocouple without radiation shield.

In systems where heat is transferred simultaneously by convection and radiation, it is frequently not possible to determine the radiant heat transfer coefficient directly. Since the temperature factor F_T contains the temperatures of the radiation emitter and the receiver, it can be evaluated only when both of these temperatures are known. If one of the temperatures depends on the rate of heat flow, that is, if one of the potentials in the network is "floating," one must assume a value for the floating potential and then determine whether that value will satisfy continuity of heat flow in the steady state. If the rate of heat flow to the potential node is not equal to the rate of heat flow from the node, another temperature must be assumed. The trial-and-error process is continued until the energy balance is satisfied. The general technique is illustrated in the next example.

EXAMPLE 9.19

Determine the correct gas temperature in Example 9.18 if the thermocouple was shielded by a thin cylindrical radiation shield having an inside diameter four times as large as the outer diameter of the thermocouple. Assume that the convective heat transfer coefficient of the shield is 20 Btu/h ft^2 °F on both sides and that the emittance of the shield, made of stainless steel, is 0.3 at 1000°F.

Solution. A sketch of the physical system is shown in Fig. 9.51. Heat flows by convection from the gas to the thermocouple and its shield. At the same time, heat flows by radiation from the thermocouple to the inside

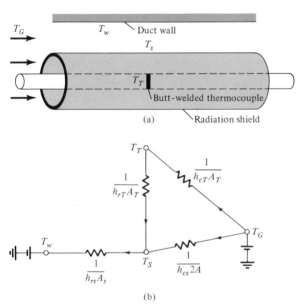

Figure 9.51 Physical system and thermal network for butt-welded thermocouple with radiation shield.

surface of the shield, is conducted through the shield, and flows by radiation from the outer surface of the shield to the walls of the duct. If we assume that the temperature of the shield is uniform (that is, if we neglect the thermal resistance of the conduction path because the shield is very thin), the thermal network is as shown in Fig. 9.51. The temperature of the duct wall T_w and the temperature of the thermocouple T_T are known, while the temperatures of the shield T_s and of the gas T_G must be determined. The latter two temperatures are floating potentials. A heat balance on the shield can be written as

$$\begin{matrix} \text{Rate of heat flow from} \\ T_G \text{ and } T_T \text{ to } T_s \end{matrix} = \begin{matrix} \text{rate of heat flow} \\ \text{from } T_s \text{ to } T_w \end{matrix}$$

or

$$\bar{h}_{cs} 2A_s(T_G - T_s) + h_{rT}A_T(T_T - T_s) = \bar{h}_{rs}A_s(T_s - T_w)$$

A heat balance on the thermocouple yields

$$\bar{h}_{cT}A_T(T_G - T_T) = \bar{h}_{rT}A_T(T_T - T_s)$$

where the nomenclature is given in Fig. 9.50. Taking A_T as unity, A_s equals 4 and we obtain from Eq. (9.75)

$$A_T \mathscr{F}_{T-s} = \cfrac{1}{\cfrac{1 - \epsilon_T}{A_T \epsilon_T} + \cfrac{1}{A_T} + \cfrac{1 - \epsilon_s}{A_s \epsilon_s}} = \cfrac{1}{\cfrac{0.2}{0.8} + 1 + \cfrac{0.7}{(4)(0.3)}} = 0.547$$

and

$$A_s \mathscr{F}_{s-w} = A_s \epsilon_s = (4)(0.3) = 1.2$$

Assuming a shield temperature of $900°F$, we have, according to Eq. (9.118),

$$\bar{h}_{rT}A_T = A_T \mathscr{F}_{T-s}F_T = (0.547)(18.1) = 9.85$$

and

$$\bar{h}_{rs}A_s = A_s \mathscr{F}_{s-w}F_T = (1.2)(11.4) = 12.5$$

Substituting these values into the first heat balance permits the evaluation of the gas temperature, and we get

$$T_G = \frac{h_{rs}A_s(T_s - T_w) - h_{rT}A_T(T_T - T_s)}{(\bar{h}_{cs})(2A_s)} + T_s$$

$$= \frac{5750 - 581}{(20)(2)(4)} + 900 = 932°F$$

Since the temperature of the gas cannot be less than that of the thermocouple, the assumed shield temperature was too low. Repeating the calculations with a new shield temperature of $930°F$ yields $T_G = 970°F$. We now substitute this value to see if it satisfies the second heat balance and get:

$$\begin{array}{ll}
\text{Heat flow rate by convection} \\
\textit{to} \ \text{ thermocouple}
\end{array} = 25 \, A_T(970 - 940) = 750 \text{ Btu/h}$$

$$\begin{array}{ll}
\text{Net heat flow rate by radiation} \\
\textit{from} \ \text{ thermocouple}
\end{array} = \bar{h}_{rT} A_T(T_T - T_s) = 203 \text{ Btu/h}$$

Since the rate of heat flow to the thermocouple exceeds the rate of heat flow from the thermocouple, our assumed shield temperature was too high. Repeating the calculations with an assumed shield temperature of 923°F yields a gas temperature of 966°F, which satisfies the heat balance on the thermocouple. The details of this calculation are left as an exercise.

A comparison of the results in Examples 9.19 and 9.20 shows that the indicated temperature of the unshielded thermocouple differs from the true gas temperature by 176°F, while the shielded couple reads only 26°F less than the true gas temperature. A double shield would reduce the temperature error to less than 10°F for the conditions specified in the example.

9.10 CLOSURE

In this chapter the basic characteristics of thermal radiation and methods for calculating heat exchange by radiation have been presented. Radiant energy emission is proportional to the absolute temperature raised to the fourth power and radiation heat transfer therefore becomes increasingly important at higher temperatures. The ideal radiator or "blackbody" is a concept convenient in the analysis of radiation heat transfer because it provides an upper limit to emission, absorption, and heat exchange by radiation. Blackbody radiation has geometric and spectral characteristics that can be treated analytically or numerically.

Real surfaces differ from black surfaces by virtue of their surface characteristics. Real surfaces always absorb and emit less radiation than black surfaces at the same temperature. Their surface characteristics can often be approximated by "gray" bodies that emit and absorb a given fraction of blackbody radiation over the entire wavelength spectrum. Radiation heat transfer between real surfaces can be analyzed by assuming that the surfaces are gray or by using gray band approximations.

The geometric relation between bodies is characterized by the shape factor, which determines the amount of radiation leaving a given surface that is intercepted by another. Using the shape factors and surface characteristics, it is possible to construct equivalent networks for radiation between surfaces in an enclosure which result in a series of linear relations that can be formulated in the form of a matrix. The temperatures and rate of radiation heat transfer for each of the surfaces in an enclosure can be determined by a matrix inversion, which can be most conveniently performed with a computer. When radiation and convection occur simultaneously, the analysis requires solution of nonlinear equations, which can become complex, especially in systems with gaseous radiation. These types of problems usually require trial-and-error solutions.

PROBLEMS

The problems for this chapter are organized by subject matter as shown below.

Problem number	Section	Subject
9.1 to 9.8	9.2 and 9.3	Spectral Characteristics of Radiation
9.9 to 9.15	9.4	Shape Factors
9.16 to 9.23	9.5	Radiation in Black Body Enclosures
9.24 to 9.31	9.6	Radiation in Gray Surface Enclosures
9.32 to 9.35	9.7 and 9.8	Gaseous Radiation
9.36 to 9.43	9.9	Radiation Combined with Convection and/or Conduction
9.44 to 9.52	All	Design Problems requiring engineering analysis and judgement

9.1 For an ideal radiator (hohlraum) with a 4-in.-diam opening, located in black surroundings at 60°F, calculate for hohlraum temperatures of 212°F and 1040°F (a) the net radiant-heat-transfer rate, in Btu/hr; (b) the wavelength at which the emission is a maximum, in microns; (c) the monochromatic emission at λ_{max}, in Btu/hr sq ft μ; (d) the wavelengths at which the monochromatic emission is 1 percent of the maximum value.
Ans. (a) 19.6, 746; (b) 7.76, 3.48; (c) 30, 1641; and (d) 2.45 and 56.6, 1.10 and 25.4

9.2 A tungsten filament is heated to 2700 K. At what wavelength is the maximum amount of radiation emitted? What fraction of the total energy is in the visible range (0.4 to 0.75 μm)? Assume that the filament radiates as a gray body.

9.3 Determine the total average hemispherical emittance and the emissive power of a surface which has a spectral hemispherical emittance of 0.8 at wavelengths less than 1.5 μm, 0.6 from 1.5 to 2.5 μm, and 0.4 at wavelengths longer than 2.5 μm. The surface temperature is 1111 K. *Ans.* 0.446

9.4 Show that (a) $(E_{b\lambda})_1/(E_{b\lambda})_2 = T_2^5/T_1^5$, and (b) $E_{b\lambda}/T^5 = f(\lambda T)$. Also, for $\lambda T = 5000$ μm K (c) calculate $E_{b\lambda}/T^5$.

9.5 Compute the average emittance of anodised aluminum at 100°C and 650°C from the spectral curve in Fig. 9.16. Assume $\epsilon_\lambda = 0.8$ for $\lambda > 9$ μm.

9.6 A large body of nonluminous gas at a temperature of 1100°C has emission bands between 2.5 and 3.5 μm and between 5 and 8 μm. At 1100°C the effective emittance in the first band is 0.8 and in the second 0.6. Determine the emissive power of this gas in W/m^2.

9.7 A flat plate is in a solar orbit 93,000,000 miles from the sun. It is always oriented normal to the rays of the sun and both sides of the plate have a finish which has a spectral absorptance of 0.95 at wavelengths shorter than 3 μ and a spectral absorptance of 0.06 at wavelengths longer than 3 μ. Assuming that the sun is a 10,000 R blackbody source, determine the equilibrium temperature of the plate.
Ans. 1040 R

9.8 By substituting Eq. 9.1 for $E_{b\lambda}(T)$ in Eq. 9.4 and performing the integration over the entire spectrum, derive a relationship between σ and the constants C_1 and C_2 in Eq. 9.1.

9.9 Derive an expression for the geometric shape factor F_{1-2} for a rectangular surface A_1. 1 by 20 m placed parallel to and centered 5 m above a 20-m-square surface A_2.

9.10 Determine the shape factor F_{1-4} for the geometrical configuration shown below.

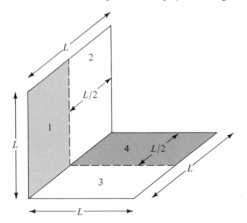

9.11 Determine the shape factor F_{1-2} for the geometrical configuration shown below.

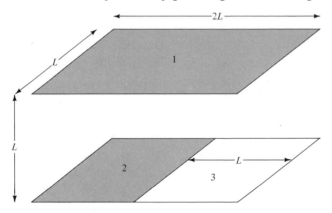

9.12 Determine the ratio of the total hemispherical emittance to the normal emittance for a nondiffuse surface if the intensity of emission varies as the cosine of the angle measured from the normal.

9.13 Using basic shape-factor definitions, estimate the equilibrium temperature of the planet Mars which has a diameter of 4150 miles and revolves around the sun at a distance of 141×10^6 miles. Assume that both the planet Mars and the sun act as blackbodies with the sun having an equivalent blackbody temperature of 10,000 R. Then repeat your calculations assuming that the albedo of Mars (the fraction of the incoming radiation returned to space) is 0.15.

9.14 Show that the moon would appear as a disk if its surface were perfectly diffuse.

9.15 A 4-cm-diam cylindrical enclosure of black surfaces, as shown in the accompanying sketch, has a 2-cm hole in the top cover. Assuming the walls of the enclosure are at the same temperature, determine the percentage of the total radiation emitted from the walls which will escape through the hole in the cover.

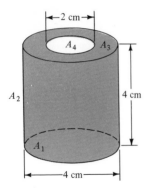

9.16 Show that the temperature of the re-radiating surface T_R in Fig. 9.37 is

$$T_R = \left(\frac{A_1 F_{1R} T_1^4 + A_2 F_{2R} T_2^4}{A_1 F_{1R} + A_2 F_{2R}} \right)^{1/4}$$

9.17 A radiation source is to be built, as shown in the diagram, for an experimental study of radiation. The base of the hemisphere is to be covered by a circular plate having a centered hole of radius $R/2$. The underside of the plate is to be held at 555 K by heaters embedded in its surface. The heater surface is black. The hemispherical surface is well-insulated on the outside. Assume gray diffuse processes and uniform distribution of radiation. (a) Find the ratio of the radiant intensity at the opening to the intensity of emission at the surface of the heated plate. (b) Find the radiant energy loss through the opening in watts for $R = 0.3$ m. (c) Find the temperature of the hemispherical surface.

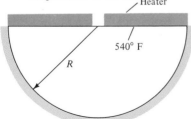

9.18 The radiant-heating ceiling of a 12- by 20-ft room is 8 ft from the floor and is main-tained at 110°F while the room air is 50°F. Assuming both surfaces are black, es-timate the net rate of heat transfer per square foot of floor surface at 80°F, located (a) in the center of the room, (b) in the corner of the room.

$Ans.$ 18.3; 7.4 Btu/hr sq ft

9.19 A large slab of steel 0.1 m thick has in it a 0.1 m-diam hole, with axis normal to the surface. Considering the sides of the hole to be black, specify the rate of heat loss from the hole in W and Btu/hr. The plate is at 811 K, the surroundings at 300 K.

9.20 A 0.5-ft-diam black disk is placed halfway between two black 10-ft-diam disks which are 20 ft apart with all disk surfaces parallel to each other. If the surroundings are at 0 R, determine the temperature of the two larger disks required to maintain the smaller disk at 1000°F. What would be the effect of replacing the small disk by a black sphere of the same diameter? $Ans.$ 2180 R, 2350 R

9.21 Show that $A_1\bar{F}_{1-2}$ for two black parallel planes of equal area connected by re-radiating walls at a constant temperature is:

$$A_1\bar{F}_{1-2} = A_1\left(\frac{1 + F_{1-2}}{2}\right)$$

Compare the results from this problem with the curves in the text where allowance is made for a continuous variation in the temperature of the re-radiating walls.

9.22 Calculate the net radiant-heat-transfer rate if the two surfaces in Prob. 9.9 are black and connected by a refractory surface of 500-sq-m area. A_1 is at 555 K and A_2 is at 278 K. What is the refractory surface temperature?

9.23 A black sphere (1 in. diam) is placed in a large infrared heating oven whose walls are maintained at 700°F. The temperature of the air in the oven is 200°F and the heat-transfer coefficient for convection between the surface of the sphere and the air is 5 Btu/hr sq ft °F. Estimate the *net rate of heat flow* to the sphere when its surface temperature is 100°F.

9.24 A hollow cylindrical container is shown in the accompanying sketch. Designating the top, bottom, and walls, respectively, by subscripts 1, 2, and 3 and assuming that all surfaces are gray and that areas A_1 and A_2 are at the same temperature T while A_3 is at temperature T_3, which is lower than T, derive an equation for the rate of radiant heat transfer from A_3 to the top and bottom surfaces of the container.

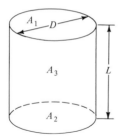

9.25 The wedge-shaped cavity shown in the accompanying sketch consists of two long strips joined along one edge. Surface 1 is 1 m wide and has an emittance of 0.4 and a temperature of 1000 K. The other wall has a temperature of 600 K and is black. The enclosure may be assumed to be at a temperature of 0 K. Assuming gray diffuse processes and uniform flux distribution, calculate the rate of energy loss from surface 1 and 2 per meter length.

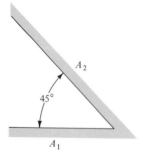

9.26 Derive an equation for the net rate of radiant heat transfer from surface 1 in the system shown in the accompanying sketch. Assume that each surface is at a uniform temperature and that the geometrical shape factor F_{1-2} is 0.1.

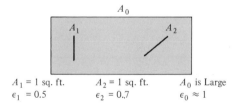

A_0

A_1 $\quad\quad A_2$

$A_1 = 1$ sq. ft. $\quad\quad A_2 = 1$ sq. ft. $\quad\quad A_0$ is Large
$\epsilon_1 = 0.5$ $\quad\quad\quad \epsilon_2 = 0.7$ $\quad\quad\quad \epsilon_0 \approx 1$

9.27. Two 5-ft-square and parallel flat plates are 1 ft apart. Plate A_1 is maintained at a temperature of $1540°$F and A_2 at $460°$F. The emittances are 0.5 and 0.8. Considering the surroundings black at 0 R and including multiple inter-reflections, determine (a) the net radiant exchange and (b) the heat input required by surface A_1 to maintain its temperature.

9.28 Two concentric spheres 0.2 m and 0.3 m in diam, with the space between them evacuated, are to be used to store liquid air (-133 K) in a room at 293 K. If the surfaces of the spheres have been flashed with aluminum and the liquid air has a latent heat of vaporization of 209 kJ/kg determine the number of kilograms of liquid air evaporated per hour.

9.29 Determine the steady-state temperatures of two radiation shields placed in the evacuated space between two infinite planes at temperatures of 555 K and 278 K. The emittances of all surfaces are 0.8.

9.30 Three thin sheets of polished aluminum are placed parallel to each other so that the distance between them is very small compared to the size of the sheets. If one of the outer sheets is at $540°$F, whereas the other outer sheet is at $140°$F, calculate the net rate of heat flow by radiation and the temperature of the intermediate sheet. Convection may be ignored. $\qquad$ *Ans.* Approx. 15 Btu/hr sq ft, $407°$F

9.31 Determine the rate of heat transfer between two 1 by 1 m parallel flat plates placed 0.2 m apart and connected by re-radiating walls. Assume that plate A is maintained at 1500 K and plate B at 500 K. (a) Plate A has an emittance of 0.9 over the entire spectrum and plate B has an emittance of 0.1. (b) Plate A has an emittance of 0.1 between 0 and 2.5 μm and an emittance of 0.9 at wavelengths longer than 2.5 μm, while plate B has an emittance of 0.1 over the entire spectrum. (c) The emittance of plate A is the same as in part (b), and plate B has an emittance of 0.9 between 0 and 4.0 μm and an emittance of 0.1 at wavelengths larger than 4.0 μm.

9.32 A small sphere (1 in. diam) is placed in a heating oven whose cavity is a 1-ft cube filled with air at 14.7 psia, contains 3 percent water vapor at $1000°$F, and whose walls are at $2000°$F. The emittance of the sphere is equal to $0.4 - 0.0001T$, where T is the surface temperature in $°$F. When the surface temperature of the sphere is $1000°$F, determine (a) the total irradiation received *by* the walls of the oven *from* the sphere, (b) the net heat transfer by radiation between the sphere and the walls of the oven, and (c) the radiant heat transfer coefficient.

9.33 The radiant section of a small thermal-cracking combustion chamber has a volume of 3200 cu ft and a total surface of 2000 sq ft. A single-row tube-curtain of 4 in.

schedule 40 pipes on 7-in. centers covers 760 sq ft of the wall. When the furnace is fired with gas of composition $(CH_2)_x$ at a rate of 2000 lb/hr, an Orsat analysis shows that 13 percent CO_2 are present. Estimate the rate of heat transfer to the tubes under these conditions, using the simplest furnace model that can be justified.

> Additional data:
>
> Tube emittance = 0.9
>
> Fuel heating value = 250,000 Btu/lb-mole of C.

Mean molar heat capacity of combustion products = 8.2 Btu/lb-mole °F.

Mean radiating gas temperature is 200°F above the bridge-wall gas-temperature.

Air and fuel enter at 60°F.

9.34 A 0.61 m radius hemisphere (811 K surface temperature) is filled with a gas mixture at 533 K and 2-atm pressure containing 6.67 percent CO_2 and water vapor at 0.5 percent relative humidity. Determine the emittance and absorptance of the gas, and the net rate of radiant heat flow to the gas.

9.35 Two infinitely large black plane surface are 0.3 m apart and the space between them is filled by an isothermal gas mixture at 833 K and atmospheric pressure consisting of 25% CO_2, 25% H_2O, and 50% N_2 by volume. If one of the surface is maintained at 278 K and the other at 1390 K respectively, calculate

(a) the effective emittance of the gas at its temperature (*ans.* 0.27)
(b) the effective absorptance of the gas to radiation from the 1390 K surface (*ans.* 0.17)
(c) the effective absorptance of the gas to radiation from the 278 K surface
(d) the net rate of heat transfer to the gas per square foot of surface area
(e) the net rate of heat transfer from the hotter surface

9.36 A manned spacecraft capsule has a shape of a cylinder 8 ft in diameter and 30 ft long. The air inside the capsule is maintained at 55°F and the convection-heat-transfer coefficient on the interior surface is 3 Btu/hr sq ft °F. Between the outer skin and the inner surface is a 6-in. layer of glass-wool insulation having a thermal conductivity of 0.01 Btu/hr ft °F. If the emittance of the skin is 0.05 and there is no aerodynamic heating or irradiation from astronomical bodies, calculate the total heat transfer rate into space at 0 R.

9.37 A package of electronic equipment is enclosed in a sheet-metal box which has a 0.093 m^2 base and is 0.15 m high. The equipment uses 1200 W of electrical power and is placed on the floor of a large room. The emittance of the walls of the box is 0.80 and the room air and the surrounding temperature is 21°C. Assuming that the average temperature of the container wall is uniform, estimate that temperature.

9.38 An 0.2 m O.D. oxidized steel pipe at a surface temperature of 756 K passes through a large room in which the air and the walls are at 38°C. If the heat transfer co-efficient by convection from the surface of the pipe to the air in the room is 28.4 W/m^2 K °F, estimate the total heat loss per foot length of pipe in W and Btu/hr.

9.39 A $\frac{1}{4}$-in.-thick sheet of polished stainless steel is suspended in a comparatively large vacuum-drying oven with black walls. The dimensions of the sheet are 12-by-12 in. and its specific heat is 0.135 Btu/lb °F. If the walls of the oven are uniformly at 300°F and the metal is to be heated from 50 to 250 °F, estimate how long the sheet should be left in the oven if (a) heat transfer by convection may be neglected and (b) the units-surface conductance for convection is 0.5 Btu/hr sq ft°F.

Ans. (a) 2.8 hr; (b) 0.4 hr

9.40 Calculate the equilibrium temperature of a thermocouple in a large air duct if the air temperature is 1367 K, the duct-wall temperature 533 K, the emittance of the couple 0.5, and the convective heat transfer coefficient, $\bar{h}_c$, is 114 W/m² K.

9.41 Repeat Prob. 9.40 with the addition of a radiation shield ($\epsilon = 0.1$, $\bar{h}_c = 114$ W/m² K) using a FORTRAN computer program.

9.42 A thermocouple is used to measure the temperature of a flame in a combustion chamber. If the thermocouple temperature is 1033 K and the walls of the chamber areat 700 K, what is the error in the thermocouple reading due to radiation to the walls? Assume all surfaces are black and the convection coefficient is 568 W/m² °C on the thermocouple. *Ans.* 93 K

9.43 A metal plate is placed in the sunlight. The incident radiant energy G is 250 Btu/hr sq ft. The air and the surroundings are at 50°F. The heat transfer coefficient by free convection from the upper surface of the plate is 3 Btu/hr sq ft °F. The plate has an average emittance of 0.9 at solar wavelengths and 0.1 at long wavelengths. Neglecting conduction losses on the lower surface, determine the equilibrium temperature of the plate. *Ans.* $\simeq 122°F$

9.44 A 2-ft-square section of panel heater is installed in the corner of the ceiling of a room having a 9-by-12-ft floor area with an 8-ft ceiling. If the surface of the heater, made from oxidized iron, is at 300°F and the walls and the air of the room are at 68°F in the steady state, determine (a) the rate of heat transfer to the room by radiation: (b) the rate of heat transfer to the room by convection ($\bar{h}_c \simeq 2$ Btu/hr sq ft °F); (c) the cost of heating the room at 1 cent per kwhr in cents per hour.

9.45 Calculate the equilibrium temperature of the earth if the concentration of CO_2 in the atmosphere were twice as large as it is today.

9.46 One hundred pounds of carbon dioxide are stored in a high-pressure cylinder 10 in. in diam (OD), 4 ft long and $\frac{1}{2}$ in. thick. The cylinder is fitted with a safety rupture diaphragm designed to fail at 2000 psia (with the specified charge, this pressure will be reached when the temperature increases to 120°F). During a fire, the cylinder is completely exposed to the irradiation from flames at 2000°F ($\epsilon = 1.0$). For the specified conditions, $c_v = 0.60$ Btu/lb °F for CO_2. Neglecting the convective heat transfer, determine the time the cylinder may be exposed to this irradiation before the diaphragm will fail if the initial temperature is 70°F and (a) the cylinder is bare oxidized steel ($\epsilon = 0.79$), (b) the cylinder is painted with aluminum paint ($\epsilon = 0.30$).

Ans. (a) 0.47 min, (b) 1.24 min

9.47 A hydrogen bomb may be approximated by a fireball at a temperature of 7200 K according to a report published in 1950 by the Atomic Energy Commission. (a) Calculate the total rate of radiant-energy emission in watts, assuming that the gas radiates as a blackbody and has a diameter of 1 mile. (b) If the surrounding atmosphere absorbs radiation below 0.3 μm, determine the percent of the total radiation emitted by the bomb which is absorbed by the atmosphere. (c) Calculate the rate of irradiation on a 1 m² area of the wall of a house 25 miles from the center of the blast if the blast occurs at an altitude of 10 miles and the wall faces in the direction of the blast. (d) Estimate the total amount of radiation absorbed assuming that the blast lasts approximately 10 sec and that the wall is covered by a coat of red paint. (e) If the wall were made of oak whose inflammability limit is estimated to be 650 K and which had a thickness of 1 cm., determine whether or not the wood would catch on fire. Justify you answer by an engineering analysis stating carefully all assumptions.

9.48 An electric furnace is to be used for batch heating a certain material (specific heat of 0.16 Btu/lb °F) from 70 to 1400°F. The material is placed on the furnace floor which is 6-by-12 ft in area as shown in the accompanying sketch. The side walls of the furnace are made of a refractory material. Parallel to the plane of the roof, but several inches below it, a grid of round resistor rods is installed. The resistors are $\frac{1}{2}$ in. in diameter and are spaced 2 in. center to center. The resistor temperature is to be maintained at 2000°F, under which conditions the emissivity of the resistor surface is 0.6. If the top surface of the stock may be assumed to have an emissivity of 0.9, estimate the time required for heating a 6-ton batch. External heat losses from the furnace may be neglected, the temperature gradient through the stock may be considered negligibly small, and steady-state conditions may be assumed.   *Ans.* Approx. 2 hr

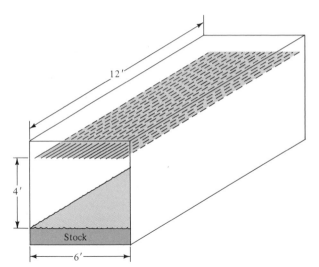

9.49 A rectangular flat water tank is placed on the roof of a house with its lower portion perfectly insulated. A sheet of glass whose transmission characteristics are tabulated below is placed 1 cm above the water surface. Assuming that the average incident solar radiation is 630 W/m², calculate the equilibrium water temperature for a water depth of 12 cm if the unit-convective conductance at the top of the glass is 8.5 W/m² K and the surrounding air temperature is 20°C. Disregard interflections.

$$\tau_\lambda \text{ of glass} = 0 \quad \text{for wavelength from 0 to 0.35 } \mu m$$

$$= 0.92 \text{ for wavelength from 0.35 to 2.7 } \mu m$$

$$= 0 \quad \text{for wavelength larger than 2.7 } \mu m$$

$$\tau_\lambda \text{ of glass} = 0.08 \text{ for all wavelengths}$$

9.50 Mercury is to be evaporated at 605°F in a furnace. The mercury flows through a 1-in. BWG No. 18 gauge stainless-steel tube, which is placed in the center of the furnace whose cross section, perpendicular to the tube axis, is a square 8 by 8 in. The furnace is made of brick having an emissivity of 0.85, with the walls maintained uniformly at 1800°F. If the surface heat transfer coefficient on the inside of the tube is 500 Btu/hr sq ft °F and the emittance of the outer surface of the tube is 0.60, calculute the rate of the heat transfer per foot of tube, neglecting convection within the furnace. *Ans.* 5550 Btu/hr ft

9.51 A 1-in.-diam cylindrical refractory crucible for melting lead is to be built for thermocouple calibration. An electrical heater immersed in the metal is shut off at some temperature above the melting point. The fusion-cooling curve is obtained by observing the thermocouple emf as a function of time. Neglecting heat losses through the wall of the crucible, estimate the cooling rate (Btu/hr) for the molten lead surface (melting point 621.2°F, surface emissivity 0.8) if the crucible depth above the lead surface is (a) 1 in., (b) 5 in. Assume that the emittance of the refractory surface is unity and the surroundings are at 70°F. (c) Noting that the crucible would hold about 0.2 lb of lead for which the heat of fusion is 10 Btu/lb, comment on the suitability of the crucible for the purpose intended.

9.52 A spherical satellite circling the sun is to be maintained at a temperature of 25°C. The satellite rotates continuously and is covered partly with solar cells having a gray surface with an absorptance of 0.1. The rest of the sphere is to be covered by a special coating which has an absorption of 0.8 for solar radiation and an emittance of 0.2 for the emitted radiation. Estimate the portion of the surface of the sphere which can be covered by solar cells. The solar irradiation may be assumed to be 1,420 W/m^2 of surface perpendicular to the rays of the sun.

REFERENCES

1. M. Planck, *The Theory of Heat Radiation*, Dover, New York, 1959.
2. R. V. Dunkle, "Thermal-Radiation Tables and Applications," *Trans. ASME*, vol. 65, pp. 549–552, 1954.
3. M. Fischenden and O. A. Saunders, *The Calculation of Heat Transmission*. His Majesty's Stationery Office, London, 1932.
4. D. C. Hamilton and W. R. Morgan, "Radiant Interchange Configuration Factors," NACA TN2836, Washington, D.C., 1962.
5. F. Kreith and W. Z. Black, *Basic Heat Transfer*, Harper & Row, New York, 1980.
6. H. Schmidt and E. Furthman, "Üeber die Gesamtstrahlung faester Koerper," *Mitt. Kaiser-Wilhelm-Inst. Eisenforsch.*, Abh. 109, Dusseldorf, 1928.
7. W. H. McAdams, *Heat Transmission*, 3d ed., McGraw-Hill, New York, 1954.
8. G. G. Gubareff, J. E. Janssen, and R. H. Torborg, Thermal Radiation Properties Survey, Honeywell Research Center, Minneapolis, Minn., 1960.
9. W. Sieber, "Zusammensetzung der von Werk-und Baustoffen Zurückgeworfenen Wärmestrahlung," *Z. Tech. Phys.*, vol. 22, pp. 130–135, 1941.
10. E. Schmidt and E. Eckert, "Über die Richtungsverteilung der Wärmestrahlung von Oberflächen," *Forsch. Geb. Ingenieurwes.*, vol. 6, pp. 175–183, 1935.
11. J. R. Schonhorst and R. Viskanta, "An Experimental Examination of the Validity of the Commonly Used Methods of Radiant-Heat Transfer Analysis," *Trans. ASME, Ser. C., J. Heat Transfer*, vol. 90, pp. 429–436, 1968.
12. E. M. Sparrow and R. D. Cess, *Radiation Heat Transfer*, Wadsworth Publishing Company, Belmont, Calif., 1966.

13. R. Siegel and J. R. Howell, *Thermal Radiation Heat Transfer*, 2nd ed., McGraw-Hill, New York, 1980.

14. R. G. Hering and T. F. Smith, "Surface Roughness Effects on Radiant Energy Interchange," *Trans. ASME, Ser. C., J. Heat Transfer*, vol. 93, pp. 88–96, 1971.

15. H. C. Hottel and A. F. Sarofim, *Radiative Heat Transfer*, pp. 31–39, McGraw-Hill, New York, 1967.

16. A. K. Oppenheim, "The Network Method of Radiation Analysis," *Trans. ASME*, vol. 78, pp. 725–735, 1956.

17. H. C. Hottel, "Heat Transmission by Radiation from Nonluminous Gases," *AIChE Trans.*, vol. 19, pp. 173–205, 1927.

18. H. C. Hottel and R. B. Egbert, "Radiant Heat Transmission from Water Vapor," *AIChE Trans.*, vol. 38, pp. 531–565, 1942.

19. J. A. Wibelt, *Engineering Radiation Heat Transfer*, Holt, Rinehart & Winston, New York, 1966.

20. C. L. Tien, "Thermal Radiation Properties of Gases," *Adv. Heat Transfer*, vol. 5, pp. 254–321, 1968.

21. R. Goldstein, "Measurements of Infrared Absorption by Water Vapor at Temperatures to 1000 K," *J. Quant. Spectrosc. Radiat. Transfer*, vol. 4, pp. 343–352, 1964.

22. D. K. Edwards and W. Sun, "Correlations for Absorption by the 9.4- and 10.4-micron CO_2 Bands," *Appl. Opt.*, vol. 3, p. 1501, 1964.

23. D. K. Edwards, B. J. Flornes, L. K. Glassen, and W. Sun, "Correlations of Absorption by Water Vapor at Temperatures from 300 to 1100 K," *Appl. Opt.*, vol. 4, pp. 715–722, 1965.

24. H. C. Hottel, in *Heat Transmission*, by W. C. McAdams, 3d ed., Chapter 4, McGraw-Hill, New York, 1954.

25. H. C. Hottel, "Radiant Heat Transmission," *Mech. Eng.*, vol. 52, pp. 699–704, 1930.

26. R. B. Egbert, Sc.D. thesis, Massachusetts Institute of Technology, 1941.

27. W. M. Rohsenow, J. P. Hartnett, and E. Ganic, eds., *Handbook of Heat Transfer*, 2nd ed. McGraw-Hill, New York, 1985.

28. D. K. Edwards and R. Matovosian, "Scaling Rules for Total Absorptivity and Emissivity of Gases," *J. Heat Transfer*, vol. 106, pp. 685–689, 1984.

Chapter 10

Heat Transfer with Change in Phase

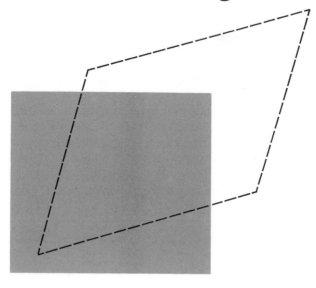

10.1 INTRODUCTION TO BOILING

Heat transfer to boiling liquids is a convection process involving a change in phase from liquid to vapor. The phenomena of boiling heat transfer are considerably more complex than those of convection without phase change because, in addition to all of the variables associated with convection, those associated with the change in phase are also relevant. Whereas in liquid-phase convection, the geometry of the system, the viscosity, the density, the thermal conductivity, the expansion coefficient, and the specific heat of the fluid are sufficient to describe the process, in boiling heat transfer, the surface characteristics, the surface tension, the latent heat of evaporation, the pressure, the density, and possibly other properties of the vapor play an important part. As a result of the large number of variables involved, neither general equations describing the boiling process nor general correlations of boiling heat transfer data are available to date. Considerable progress has been made in gaining a physical understanding of the boiling mechanism (1–4). By observing the boiling phenomena with the aid of high-speed photography, it has been found that there are various distinct regimes of boiling in which the heat transfer mechanisms differ radically. To correlate the experimental data it is therefore best to describe and analyze each of the boiling regimes separately.

10.2 POOL BOILING

10.2.1 Pool Boiling Regimes

To acquire a physical understanding of the phenomena which are characteristic of the various boiling regimes we shall first consider a simple system consisting of a heating surface, such as a flat plate or a wire, submerged in a pool of water at saturation temperature without external agitation. This is called *pool boiling*. A familiar example of such a system is the boiling of water in a kettle on a stove. As long as the temperature of the surface does not exceed the boiling point of the liquid by more than a few degrees, heat is transferred to liquid near the heating surface by natural convection. The convection currents circulate the superheated liquid, and evaporation takes place at the free surface of the liquid. The heat transfer mechanism in this process, although some evaporation occurs, is simply natural convection, because only liquid is in contact with the heating surface.

As the temperature of the heating surface is increased, a point is reached where, in certain places known as nucleation sites, vapor bubbles are formed and escape from the heated surface. Nucleation sites are very small inclusions in the surface which result from the process used to manufacture the surface. The inclusions are too small to admit liquid because of the liquid surface tension and the resulting vapor pocket acts as a site for bubble growth and release. As a bubble is released, liquid flows over the inclusion trapping vapor, thus providing a start for the next bubble. This process occurs simultaneously at a number of nucleation sites on the heating surface. The vapor bubbles are at first small and condense before reaching the surface, but as the temperature is raised further, they become more numerous and larger until they finally rise to the free surface. These phenomena may be observed when boiling water in a kettle.

The boiling regimes are illustrated in Fig. 10.1 for a horizontal wire heated electrically in a pool of distilled water at atmospheric pressure with corresponding saturation temperature of 100°C (5, 6). In this curve the heat flux is plotted as a function of the temperature difference between the surface and the saturation temperature. This temperature difference, ΔT_x, is called the excess temperature above the boiling point, or *excess temperature* for short. We observe that, in regimes 2 and 3, the heat flux increases rapidly with increasing surface temperature. The process in these two regimes is called *nucleate boiling*. In the individual bubble regime most of the heat is transferred from the heating surface to the surrounding liquid by a vapor-liquid exchange action (7). As vapor bubbles form and grow on the heating surface, they push hot liquid from the vicinity

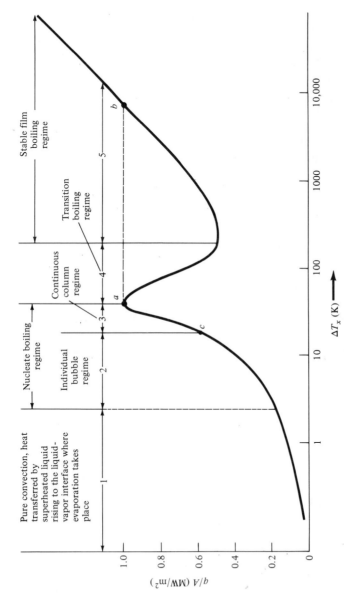

Figure 10.1 Typical boiling curves for a wire, tube, or horizontal surface in a pool of water at atmospheric pressure.

of the surface into the colder bulk of the liquid. In addition, intense microconvection currents are set up as vapor bubbles are emitted and colder liquid from the bulk rushes toward the surface to fill the void. As the surface heat flux is raised and the number of bubbles increases to the point where they begin to coalesce, heat transfer by evaporation becomes more important and eventually predominates at very large heat fluxes in regime 3 (8).

If the excess temperature in a temperature-controlled system is raised to about 35°C, we observe that the heat flux reaches a maximum (about 10^6 W/m^2 in a pool of water), and a further increase of the temperature causes a decrease in the rate of heat flow. This maximum heat flux, called the *critical heat flux*, is said to occur at the *critical excess temperature.*

The reason for the inflection point near c in the curve may be found by examining the heat transfer mechanism during boiling. At the onset of boiling, bubbles grow at nucleation sites on the surface until the buoyant force or currents of the surrounding liquid carry them away. But as the heat flux or the surface temperature is increased in nucleate boiling, the number of sites at which bubbles grow increases. Simultaneously, the rate of growth of the bubbles increases and so does the frequency of formation. As the rate of bubble emission from a site increases, bubbles collide and coalesce with their predecessors (9). This point is indicated as a transition from regime 2 to regime 3 in Fig. 10.1. Eventually, successive bubbles merge into more or less continuous vapor columns (2, 4, 8).

As the maximum heat flux is approached, the number of vapor columns increases. But since each new column occupies space formerly occupied by liquid, there is a limit to the number of vapor columns that can be emitted from the surface. This limit is reached when the space between these columns is no longer sufficient to accommodate the streams of liquid which must move toward the hot surface to replace the liquid which evaporated to form the vapor columns.

If the surface temperature is raised further so that the ΔT_x at the maximum heat flux is exceeded, any of three things can occur, depending on the method of heat control and the material of the heating surface (10):

1. If the heater surface temperature is the independent variable and the heat flux is controlled by it, the mechanism will change to transition boiling and the heat flux will decrease. This corresponds to operation in regime 4 in Fig. 10.1.
2. If the heat flux is controlled, as in an electrically heated wire, the surface temperature is dependent on it. Provided the melting point of the heater material is sufficiently high, a transition from nucleate to film boiling will take place, and the heater will operate at a very much higher temperature. This case corresponds to a transition from point a to point b in Fig. 10.1.
3. If the heat flux is independent, but the heater material has a low melting point, burnout occurs. For a very short time the heat supplied to the heater exceeds the amount of heat removed because when the peak heat flux is reached, an increase in heat generation is accom-

panied by a decrease in the rate of heat flow from the heater surface. Consequently, the temperature of the heater material will rise to the melting point, and the heater will burn out.

In the stable film-boiling regime a vapor film blankets the entire heater surface, whereas in the transition film-boiling regime nucleate and stable film boiling occur alternately at a given location on the heater surface (11). The photographs in Figs. 10.2 and 10.3 illustrate the nucleate- and film-boiling mechanisms on a wire submerged in water at atmospheric pressure. Note the

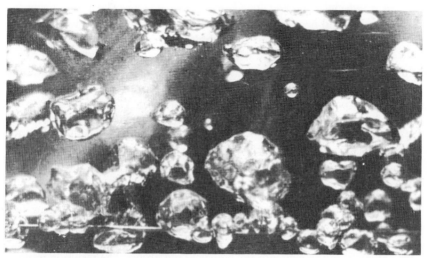

Figure 10.2   Photograph showing nucleate boiling on a wire in water. (Courtesy of J. T. Castles.)

Figure 10.3   Photograph showing film boiling on a wire in water. (Courtesy of J. T. Castles.)

film of vapor which completely covers the wire in Fig. 10.3. A phenomenon which closely resembles this condition is also observed when a drop of water falls on a red-hot stove. The drop does not evaporate immediately but dances on the stove because a steam film forms at the interface between the hot surface and the liquid and insulates the droplet from the surface.

When the surface temperature exceeds the saturation temperature, local boiling in the vicinity of the surface may take place even if the bulk temperature is below the boiling point. The boiling process in a liquid whose bulk temperature is below the saturation temperature but whose boundary layer is sufficiently superheated that bubbles form next to the heating surface is usually called *heat transfer to a subcooled boiling liquid* or *surface boiling*. The mechanisms of bubble formation and heat transfer are similar to those described for liquids at saturation temperature. However, the bubbles increase in number while their size and average lifetime decrease with decreasing bulk temperature at a given heat flux (12). As a result of the increase in the bubble population, the agitation of the liquid caused by the motion of the bubbles is more intense in a subcooled liquid than in a pool of saturated liquid, and much larger heat fluxes can be attained before the critical temperature is achieved. The mechanism by which a typical bubble transfers heat in subcooled and degassed water is illustrated by the sketches in Fig. 10.4 (13). The letters for the sequence of events described in the following paragraphs correspond to the designation of the sketches.

(a) The liquid next to the wall is superheated.

(b) A vapor nucleus of sufficient size to permit a bubble to grow has formed at a pit or scratch in the surface.

(c) The bubble grows and pushes the layer of superheated liquid above it away from the wall into the cooler liquid above. The resulting motion of the liquid is indicated by arrows.

(d) The top of the bubble surface extends into cooler liquid. The temperature in the bubble has dropped. The bubble continues to grow by

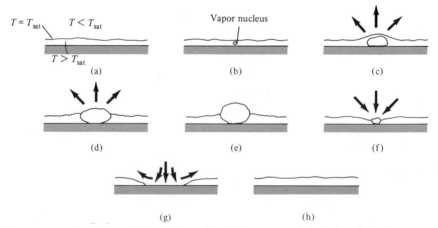

Figure 10.4 Flow pattern induced by a bubble in a subcooled boiling liquid.

virtue of the inertia of the liquid, but at a slower rate than during stage c because it receives less heat per unit volume.

(e) The inertia of the liquid has caused the bubble to grow so large that its upper surface extends far into cooler liquid. It loses more heat by evaporation and convection than it received by conduction from the heating surface.

(f) The inertia forces have been dissipated and the bubble begins to collapse. Cold liquid from above follows in its wake.

(g) The vapor phase has been condensed, the bubble has disappeared, and the hot wall is splashed by a stream of cold liquid at high velocity.

(h) The film of superheated liquid has settled and the cycle repeats.

The foregoing description of the life cycle of a typical bubble also applies qualitatively through stage e to liquids containing dissolved gases, to solutions of more than one liquid, and to saturated liquids. In these cases, however, the bubble does not collapse but is carried away from the surface by buoyant forces or convection currents. A void is created all the same and the surface is swept by cooler fluid rushing in from above. What eventually happens to the bubbles, whether they collapse on the surface or are swept away, has little influence on the heat transfer mechanism, which depends mostly on the pumping action and liquid agitation.

The primary variable controlling the bubble mechanism is the excess temperature. It should be noted, however, that in the nucleate boiling regime the total variation of the excess temperature irrespective of the fluid bulk temperature is relatively small for a very large range of heat flux. For design purposes the conventional heat transfer coefficient, which is based on the difference in temperature between the bulk of the fluid and the surface, is therefore only of secondary interest compared to the maximum heat flux attainable in nucleate boiling and the wall temperature at which boiling begins.

Generation of steam in the tubes of a boiler, vaporization of fluids such as gasoline in the chemical industry, and boiling of a refrigerant in the cooling coils of a refrigerator are processes which closely resemble those described above, except that in these industrial applications of boiling the fluid generally flows past the heating surface by forced convection. The heating surface is frequently the inside of a tube or a duct, and the fluid at the discharge end is a mixture of liquid and vapor. The foregoing descriptions of bubble formation and behavior also apply qualitatively to forced convection, but the heat transfer mechanism is further complicated by the motion of the bulk of the fluid. Boiling in forced convection will be discussed in Section 10.3.

10.2.2 Nucleate Pool Boiling

The dominant mechanism by which heat is transferred in single phase forced convection is the turbulent mixing of hot and cold fluid particles. As shown in

Chapter 4 experimental data for forced convection without boiling can be correlated by a relation of the type

$$Nu = \phi(Re, Pr)$$

where the Reynolds number Re is a measure of the turbulence and mixing motion associated with the flow. The increased heat transfer rates attained with nucleate boiling are the result of the intense agitation of the fluid produced by the motion of vapor bubbles. To correlate experimental data in the nucleate-boiling regime the conventional Reynolds number in Eq. (4.31) is modified so that it is significant of the turbulence and mixing motion for the boiling process. A type of Reynolds number Re_b, which is a measure of the agitation of the liquid in nucleate-boiling heat transfer, is obtained by combining the average bubble diameter D_b, the mass velocity of the bubbles per unit area G_b, and the liquid viscosity μ_l to form the dimensionless modulus

$$Re_b = \frac{D_b G_b}{\mu_l}$$

This parameter, often called the bubble Reynolds number, takes the place of the conventional Reynolds in nucleate boiling. If we use the bubble diameter D_b as the significant length in the Nusselt number, we have

$$Nu_b = \frac{h_b D_b}{k_l} = \phi(Re_b, Pr_l) \tag{10.1}$$

where Pr_l is the Prandtl number of the saturated liquid and h_b is the *nucleate-boiling heat transfer coefficient*, defined as

$$h_b = \frac{q''}{\Delta T_x}$$

In nucleate boiling the excess temperature ΔT_x is the physically significant temperature potential. It replaces the temperature difference between the surface and the bulk of the fluid, ΔT, used in single-phase convection. Numerous experiments have shown the validity of this method, which obviates the need to know the exact temperature of the liquid and can therefore be applied to saturated as well as subcooled liquids.

Using experimental data on pool boiling as a guide, Rohsenow (14) modified Eq. (10.1) by means of simplifying assumptions. An equation found convenient for the reduction and correlation of experimental data (15) for many different fluids is

$$\frac{c_l \Delta T_x}{h_{fg} Pr_l^n} = C_{sf} \left[\frac{q''}{\mu_l h_{fg}} \sqrt{\frac{g_c \sigma}{g(\rho_l - \rho_v)}} \right]^{0.33} \tag{10.2}$$

where c_l = specific heat of saturated liquid, J/kg K

q'' = heat flux, W/m^2

h_{fg} = latent heat of vaporization, J/kg

g = gravitational acceleration, m/s^2

ρ_l = density of the saturated liquid, kg/m^3

ρ_v = density of the saturated vapor, kg/m^3

σ = surface tension of the liquid-to-vapor interface, N/m

Pr_l = Prandtl number of the saturated liquid

μ_l = viscosity of the liquid, kg/ms

n = 1.0 for water, 1.7 for other fluids

C_{sf} = empirical constant which depends on the nature of the heating surface-fluid combination and whose numerical value varies from system to system

Usage of Eq. (10.2) requires that property values be known very accurately. In particular, note the sensitivity of the Prandtl number effect on heat flux.

The most important variables affecting C_{sf} are the surface roughness of the heater, which determines the number of nucleation sites at a given temperature (11), and the angle of contact between the bubble and the heating surface, which is a measure of the wettability of a surface with a particular fluid. The sketches of Fig. 10.5 show that the contact angle θ decreases with greater wetting. A totally wetted surface has the smallest area covered by vapor at a given excess temperature and consequently represents the most favorable condition for efficient heat transfer. In the absence of quantitative information on the effect of wettability and surface conditions on the constant C_{sf}, its value must be determined empirically for each fluid-surface combination.

Figure 10.6 shows experimental data obtained by Addoms (16) for pool boiling of water on a 0.61-mm-diameter platinum wire at various saturation pressures. These data, plotted as heat flux versus excess temperature in Fig. 10.6, are replotted in Fig. 10.7, using

$$\frac{q''}{\mu_l h_{fg}} \sqrt{\frac{g_c \sigma}{g(\rho_l - \rho_v)}}$$

as the ordinate and $c_l \Delta T_x / h_{fg} Pr_l$ as the abscissa. The slope of the straight line

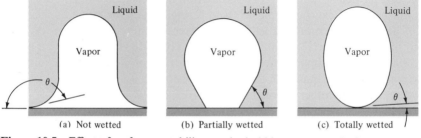

(a) Not wetted   (b) Partially wetted   (c) Totally wetted

Figure 10.5   Effect of surface wettability on the bubble contact angle θ.

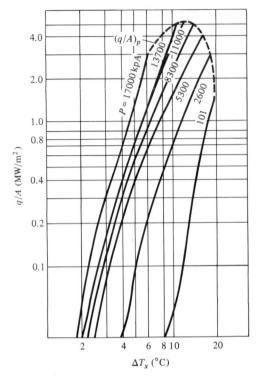

Figure 10.6 Heat flux versus excess temperature for nucleate boiling of water on a 0.61-mm-diameter electrically heated platinum wire. [From W. M. Rohsenow (14), with permission of the publishers, the American Society of Mechanical Engineers.]

faired through the experimental points is 0.33; for water boiling on platinum, the value of C_{sf} is 0.013. For comparison, the experimental values of C_{sf} for a number of other fluid-surface combinations are listed in Table 10.1.

Selected values of the vapor-liquid surface tension for water at various temperatures are shown in Table 10.2 for use in Eq. (10.2).

The principal advantage of the Rohsenow correlation is that the performance of a particular fluid-surface combination in nucleate boiling at any pressure and heat flux can be predicted from a single test. One value of the heat flux q'' and the corresponding value of the excess temperature difference ΔT_x are all that are required to evaluate C_{sf} in Eq. (10.2). It should be noted, however, that Eq. (10.2) applies only to clean surfaces. For contaminated surfaces the exponent of Pr_l, n, has been found to vary between 0.8 and 2.0. Contamination also affects the other exponent in Eq. (10.2) and C_{sf}.

The geometric shape of the heating surface has no appreciable effect on the nucleate-boiling mechanism (17, 18). This is not unexpected, since the influence of the bubble motion on the fluid conditions is limited to a region very near the surface.

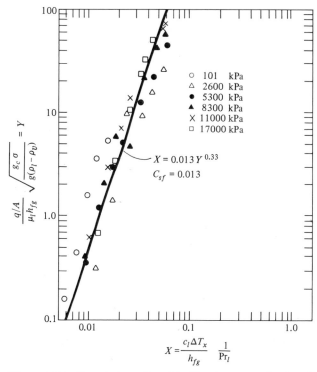

Figure 10.7 Correlation of pool-boiling heat transfer data for water by the method of Rohsenow. [From W. M. Rohsenow (14), with permission of the publishers, the American Society of Mechanical Engineers; data of Addoms (16).]

For calculating the heat flux, Collier (19) recommends the following correlation equation as being much simpler to use than Eq. (10.2).

$$q'' = 0.000481 \, \Delta T_x^{3.33} p_{cr}^{2.3} \left[1.8 \left(\frac{p}{p_{cr}} \right)^{0.17} + 4 \left(\frac{p}{p_{cr}} \right)^{1.2} + 10 \left(\frac{p}{p_{cr}} \right)^{10} \right]^{3.33}$$

(10.3)

In Eq. (10.3) ΔT_x is the excess temperature in °C, p is the operating pressure in atm, p_{cr} is the critical pressure in atm, and q'' is in W/m².

10.2.3 Critical Heat Flux in Nucleate Pool Boiling

The Rohsenow method correlates data for all types of nucleate-boiling processes, including pool boiling of saturated or subcooled liquids and boiling of subcooled or saturated liquids flowing by forced or natural convection in tubes or ducts. Specifically, the correlation equation, Eq. (10.2), relates the boiling heat flux to the excess temperature, provided the relevant fluid properties and the pertinent coefficient C_{sf} are available. The correlation is restricted to nucleate boiling and does not reveal the excess temperature at which the heat flux reaches

TABLE 10.1 VALUES OF THE COEFFICIENT C_{sf} IN EQ. (10.2) FOR VARIOUS LIQUID-SURFACE COMBINATIONS

Fluid-heating surface combination	C_{sf}
Water–copper (35)[a]	0.0130
Carbon tetrachloride–copper (35)	0.0130
35% K_2CO_3–copper (35)	0.0054
50% K_2CO_3–copper (35)	0.00275
n-Butyl alcohol–copper (35)	0.00305
Isopropyl alcohol–copper (35)	0.00225
n-Pentane–chromium (68)	0.0150
Water–platinum (16)	0.0130
Benzene–chromium (68)	0.0100
Water–brass (69)	0.0060
Ethyl alcohol–chromium (68)	0.0027
n-Pentane on emery-polished copper (15)	0.0154
n-Pentane on emery-polished nickel (15)	0.0127
Water on emery-polished copper (15)	0.0128
Carbon tetrachloride on emery-polished copper (15)	0.0070
Water on emery-polished, paraffin-treated copper (15)	0.0147
n-Pentane on lapped copper (15)	0.0049
n-Pentane on emery-rubbed copper (15)	0.0074
Water on scored copper (15)	0.0068
Water on ground and polished stainless steel (15)	0.0080
Water on Teflon pitted stainless steel (15)	0.0058
Water on chemically etched stainless steel (15)	0.0133
Water on mechanically polished stainless steel (15)	0.0132

[a] Numbers in parentheses are those of references listed at the end of the chapter.

TABLE 10.2 VAPOR-LIQUID SURFACE TENSION FOR WATER

Surface tension $\sigma(\times 10^3$ N/m)	Saturation temperature °C
75.5	0
72.9	20
69.5	40
66.1	60
62.7	80
58.9	100
48.7	150
37.8	200
26.1	250
14.3	300
3.6	350

Source: N. B. Vargaftik, *Tables on the Thermophysical Properties of Liquids and Gases*, 2nd ed., Hemisphere, Washington, D.C., 1975, p. 53.

a maximum, or what the value of this flux is at which nucleate boiling breaks down and an insulating vapor film forms. As mentioned earlier, the maximum heat flux attainable with nucleate boiling is sometimes of greater interest to the designer than the exact surface temperature, because for efficient heat transfer (20) and operating safety (1, 21), particularly in high-performance constant-heat-input systems, operation in the film-boiling regime must be avoided.

Although there exists no satisfactory theory for predicting boiling heat transfer coefficients, the maximum heat flux condition in nucleate pool boiling, that is, the critical heat flux, can be predicted with reasonable accuracy.

Close inspection of the nucleate-boiling regime (Fig. 10.1) shows (9) that it consists of at least two major subregimes. In the first region, which corresponds to low heat flux densities, the bubbles behave as isolated entities and do not interfere with one another. But as the heat flux is increased, the process of vapor removal from the heating surface changes from an intermittent to a continuous one, and as the frequency of bubble emission from the surface increases, the isolated bubbles merge into continuous vapor columns.

The stages of the transition process from isolated bubbles to continuous vapor columns are shown schematically in Fig. 10.8a. The photographs in Fig. 10.8b and c, show the two regimes for water boiling on a horizontal surface at atmospheric pressure (9). At the transition from the region of isolated bubbles to that of vapor columns, only a small portion of the heating surface is covered by vapor. But as the heat flux is increased, the column diameter increases and additional vapor columns form. As the fraction of a cross-sectional area parallel to the heating surface occupied by vapor increases, neighboring vapor columns and the enclosed liquid begin to interact. Eventually a vapor generation rate is attained where the close spacing between adjacent vapor columns leads to high relative velocities between the vapor moving away from the surface and the liquid streams flowing toward the surface to maintain continuity. The point of maximum heat flux occurs when the velocity of the liquid relative to the velocity of the vapor is so great that a further increase would either cause the vapor columns to drag the liquid away from the heating surface or cause the liquid streams to drag the vapor back toward the heating surface. Either case is obviously physically impossible without a decrease in the heat flux.

With this type of flow model as a guide, Zuber and Tribus (22) and Moissis and Berenson (9) derived analytical relations for the maximum heat flux which are in essential agreement with an equation proposed earlier by Kutateladze (23) by empirical means. The recommended expression (24) for the peak flux (in W/m^2) in saturated pool boiling is

$$q'' = \frac{\pi}{24} \rho_v^{1/2} h_{fg} [\sigma g(\rho_l - \rho_v) g_c]^{1/4} \tag{10.4}$$

Lienhard and Dhir (25) recommend replacing the constant $\pi/24$ with 0.149.

Equation (10.4) predicts that water will sustain a larger peak heat flux than any of the common liquids because water has such a large heat of vaporization. Further inspection of Eq. (10.4) suggests ways and means for increasing

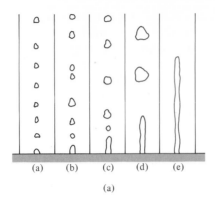

(a)

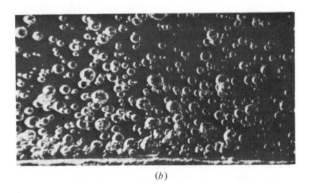

(b)

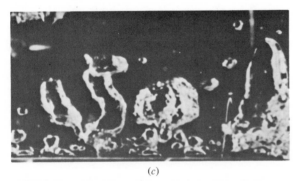

(c)

Figure 10.8 Transition from isolated-bubble regime to continuous-column regime in nucleate boiling. (a) Schematic sketch of transition. (b) Photograph of isolated-bubble regime for water at atmospheric pressure and a heat flux of 121,100 W/m². (c) Photograph of continuous-column regime for water at atmospheric pressure and a heat flux of 366,000 W/m². [From R. Moissis and P. J. Berenson (9), with permission of the publishers, the American Society of Mechanical Engineers.]

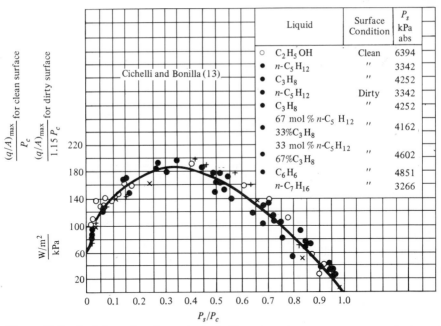

Figure 10.9 Peak heat flux in nucleate boiling at various pressures—correlation of Cichelli and Bonilla. [By permission from M. T. Cichelli and C. F. Bonilla (68).]

maximum heat flux. Pressure affects peak heat flux because it changes the vapor density and also the boiling point. Changes in the boiling point affect the heat of vaporization and the surface tension. For each liquid there exists, therefore, a certain pressure which yields the highest heat flux. This is illustrated in Fig. 10.9, where the peak nucleate-boiling heat flux is plotted as a function of the ratio of system pressure to critical pressure. For water the optimum pressure is about 10,300 kPa and the peak heat flux is about 3.8 MW/m². The quantity in brackets in Eq. (10.4) also shows that the gravitational field affects the peak heat flux. The reason for this behavior is that in a given field the liquid phase, by virtue of its higher density, is subject to a larger force per unit volume than the vapor phase. Since this difference in forces acting on the two phases brings about a separation of the two phases, an increase in the field strength, as in a large centrifugal force field, increases the separating tendency and will also increase the peak flux. Conversely, experiments by Usiskin and Siegel (26) indicate that a reduced gravitational field decreases the peak heat flux in accordance with Eq. (10.4); in a field of zero gravity, vapor does not leave the heated solid and the heat flux tends toward zero.

When the liquid bulk is subcooled, the maximum heat flux can be estimated (24) from the equation

$$q''_{max} = q''_{max \, sat} \left\{ 1 + \left[\frac{2k_l(T_{sat} - T_{liquid})}{\sqrt{\pi \alpha_l \tau}} \right] \frac{24}{\pi h_{fg} \rho_v} \left[\frac{\rho_v^2}{g_c \sigma g(\rho_l - \rho_v)} \right]^{1/4} \right\} \qquad (10.5)$$

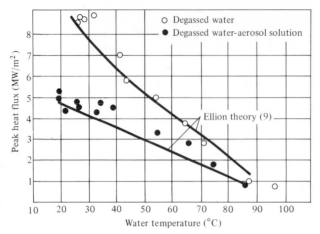

Figure 10.10 Effect of bulk temperature on peak heat flux in pool boiling. [By permission from M. E. Ellion (13).]

where

$$\tau = \frac{\pi}{3}\sqrt{2\pi}\left[\frac{g_c\sigma}{g(\rho_l - \rho_v)}\right]^{1/2}\left[\frac{\rho_v^{\,2}}{g_c\sigma g(\rho_l - \rho_v)}\right]^{1/4}$$

and $q''_{\text{max sat}}$ can be determined from Eq. (10.4). Figure 10.10 illustrates the influence of the bulk temperature on the peak heat flux for distilled water and a 1 percent aqueous solution of a surface-active agent boiling on a stainless steel heater. The addition of the surface-active agent decreased the surface tension of water from 72 to 34 dyne/cm. This caused an appreciable decrease in the peak heat flux, an effect which is in agreement with Eq. (10.4). Noncondensable gases and nonwetting surfaces also reduce the peak heat flux at a given bulk temperature.

Westwater (10), Huber and Hoehne (27), and others found that certain additives (e.g., small amounts of Hyamine 1622) can increase the peak heat flux. Also, the presence of an ultrasonic or electrostatic field can increase the peak heat flux attainable in nucleate boiling. The peak heat flux is affected by orientation of the boiling surface. Bernath (28) has shown that the peak heat flux for a vertical flat surface may be as much as 25 percent less than that for a horizontal surface.

EXAMPLE 10.1 —————————————————————————————

Water at atmospheric pressure is boiling on a mechanically polished stainless steel surface which is heated electrically from below. Determine the heat flux from the surface to the water when the surface temperature is 106°C, and compare with the critical heat flux for nucleate boiling. Repeat for the case of water boiling on a teflon coated stainless steel surface.

Solution. From Table 10.1, C_{sf} is 0.0132 for the mechanically polished surface. From steam tables, $h_{fg} = 2250$ J/g, $\rho_l = 962$ kg/m³, and $\rho_v = 0.60$ kg/m³. From Appendix 2, Table 13, $c_l = 4211$ J/kg °C, $Pr_l = 1.75$, $\mu_l = 2.77 \times 10^{-4}$ kg/m s. From Table 10.2, the surface tension at 100°C is 58.8×10^{-3} N/m. Substituting these properties in Eq. (10.2) with $\Delta T_x = 106 - 100 = 6$°C gives

$$q'' = \left(\frac{c_l \Delta T_x}{C_{sf} h_{fg} Pr_l}\right)^3 \mu_l h_{fg} \sqrt{\frac{g(\rho_l - \rho_v)}{g_c \sigma}}$$

$$= \left[\frac{(4211 \text{ J/kg °C})(6°C)}{(0.0132)(2.25 \times 10^6 \text{ J/kg})(1.75)}\right]^3 (2.83 \times 10^{-4} \text{ kg/m s})$$

$$\times (2.25 \times 10^6 \text{ J/kg})\left[\sqrt{\frac{(962 \text{ kg/m}^3)(9.8 \text{ m/s}^2)}{58.8 \times 10^{-3} \text{ N/m}}}\right]$$

$$= 29,289 \text{ W/m}^2$$

Note that we have neglected the vapor density relative to the liquid density.

To determine the critical heat flux use Eq. (10.4):

$$q'' = \frac{\pi}{24} \rho_v^{1/2} h_{fg} [\sigma g(\rho_l - \rho_v)g_c]^{1/4}$$

$$= \frac{\pi}{24}(0.60)^{1/2}(2.25 \times 10^6)[(58.8 \times 10^{-3})(9.8)(962)]^{0.25}$$

$$= 1.107 \times 10^6 \text{ W/m}^2$$

At 6°C excess temperature the heat flux is less than the critical value, therefore nucleate pool boiling exists. Had the calculated critical heat flux been less than the heat flux calculated from Eq. (10.2), film boiling would exist and the assumptions underlying the application of Eq. (10.2) would not be satisfied.

Since $q'' \sim C_{sf}^{-3}$, we have for the teflon coated stainless steel surface

$$q'' = 29,289 \left(\frac{0.0132}{0.0058}\right)^3 = 345,300 \text{ W/m}^2$$

a remarkable increase in heat flux, which is, however, still below the critical value.

In applying the theoretical equations for the critical heat flux in practice, a few words of caution are in order. Data have been presented in the literature indicating lower critical heat fluxes than those predicted from Eq. (10.4) or (10.5). Berenson (11) explains this as follows. Boiling is a local phenomenon, while in most experiments and industrial installations an average heat flux is measured or specified. Therefore, if different locations of a heating surface have different heat fluxes or different nucleate-boiling curves, the measured result will represent an average. But the largest local heat flux at a given temperature difference will always be higher than the average measured value, and if the heat flux is not uniform—for instance, if considerable difference in subcooling or in surface condition exists or if gravity variations occur as around the periphery of a horizontal tube—a burnout may occur locally even if the average value of heat flux is below the critical value.

10.2.4 Pool Film Boiling

This boiling regime has less industrial significance because of the very high surface temperature encountered. As shown in Fig. 10.3, the surface is blanketed by a film of vapor. Heat transfer is by conduction across the vapor film and, at higher temperatures, by radiation from the surface to the liquid-vapor interface. Heat transfer to this interface produces the vapor bubbles seen in the photograph. The conduction heat transfer across the vapor film is relatively easy to analyze (29, 30).

For film boiling on tubes of diameter D, Bromley (29) recommends the following correlation equation for the heat transfer coefficient due to conduction only

$$\bar{h}_c = 0.62 \left\{ \frac{g(\rho_l - \rho_v)\rho_v k_v^3 [h_{fg} + 0.68c_{pv}\Delta T_x]}{D\mu_v \Delta T_x} \right\}^{1/4} \tag{10.6}$$

For very-large-diameter tubes and flat horizontal surfaces Westwater and Breen (30) recommend

$$\bar{h}_c = \left(0.59 + 0.69\frac{\lambda}{D} \right) \left\{ \frac{g(\rho_l - \rho_v)\rho_v k_v^3 [h_{fg} + 0.68c_{pv}\Delta T_x]}{\lambda\mu_v \Delta T_x} \right\}^{1/4} \tag{10.7}$$

where

$$\lambda = 2\pi \left[\frac{g_c \sigma}{g(\rho_l - \rho_v)} \right]^{1/2}$$

To account for radiation from the surface, Bromley (29) suggests combining the two heat transfer coefficients, that is,

$$\bar{h}_{total} = \bar{h}_c + 0.75\bar{h}_r \tag{10.8}$$

where $\bar{h}_c$ may be computed from Eq. (10.6) or (10.7). The radiation heat transfer coefficient $\bar{h}_r$ is calculated from Eq. (1.31) by assuming that the liquid-vapor interface and the solid surface are flat and parallel and that the interface has an emissivity of 1.0:

$$\bar{h}_r = \sigma\epsilon_s \left(\frac{T_s^4 - T_{sat}^4}{T_s - T_{sat}} \right) \tag{10.9}$$

Here ϵ_s is the surface emissivity and T_s is the absolute surface temperature.

EXAMPLE 10.2

Repeat Example 10.1 using a surface temperature of 400°C for the mechanically polished stainless steel surface.

Solution. From Eq. (10.2) we note that $q/A \sim \Delta T_x^3$, therefore

$$q'' = 29{,}289 \times \left(\frac{300}{6}\right)^3 = 3.6 \times 10^9 \text{ W/m}^2$$

This exceeds the critical heat flux (1.107×10^6 W/m²); therefore the system must be operating in the film boiling regime. From Appendix 2, Table 34, we find: $k_v = 0.0249$ W/m K, $c_{pv} = 2034$ J/kg K, $\mu_v = 1.21 \times 10^{-6}$ kg/m s. Using Eq. (10.7) for $D \to \infty$ we have

$$\lambda = 2\pi \left(\frac{58.8 \times 10^{-3}}{9.8 \times 962}\right)^{1/2} = 0.0157 \text{ m}$$

and

$$\bar{h}_c = (0.59)\left\{\frac{(9.8)(962)(0.60)(0.0249)^3[2250 + (0.68)(2034)(300)]}{(0.0157)(1.21 \times 10^{-6})(300)}\right\}^{1/4}$$

$$= 166.8 \text{ W/m}^2 \text{ K}$$

Since the surface is polished, $\epsilon_s \approx 0.05$, and from Eq. (10.9) we see that $\bar{h}_r$ is negligible. The heat flux is therefore

$$q'' = 166.8 \times 300 = 50{,}040 \text{ W/m}^2$$

10.3 BOILING IN FORCED CONVECTION

The heat transfer and pressure drop characteristics of forced-convection boiling play an important part in the design of boiling nuclear reactors, environmental control systems for spacecraft and space power plants, and other advanced power-production systems. Despite the large number of experimental and analytical investigations which have been conducted in the area of forced-convection boiling, it is not yet possible to predict all of the characteristics of this process quantitatively. This is due to the great number of variables upon which the process depends and the complexity of the various two-phase flow patterns which occur as the quality of the vapor-liquid mixture (defined as the percentage of the total mass which is in the form of vapor at a given station) increases during vaporization. However, the forced-convection vaporization process has been photographed (31, 32) and, based on these photographic observations, it is possible to give a qualitative description of the process.

In most practical situations, a fluid at a temperature below its boiling point at the system pressure enters a duct in which it is heated so that progressive vaporization occurs. Figure 10.11 shows schematically what happens in a vertical duct in which a liquid is vaporized with low heat flux. Part *e* of the figure is a qualitative graph on which the heat transfer coefficient at a specific location is plotted as a function of the local quality. In view of the fact that heat is continuously added to the fluid, the quality will increase with distance from the entrance.

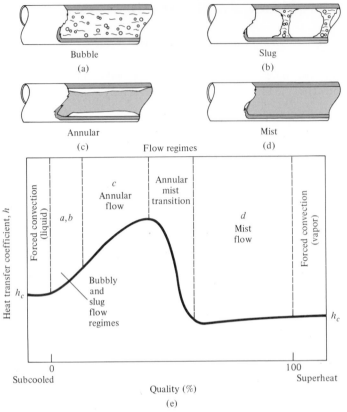

Figure 10.11 Characteristics of forced-convection vaporization: heat transfer coefficient versus quality and type of flow regime.

The heat transfer coefficient at the inlet can be predicted from Eq. (6.63) with satisfactory accuracy. However, as the fluid bulk temperature increases toward its saturation point, which usually occurs only a short distance from the inlet in a system designed to vaporize the fluid, bubbles will begin to form at nucleation sites and will be carried into the main stream as in nucleate pool boiling. This regime, known as the *bubbly-flow regime*, is shown schematically in Fig. 10.11a. Bubbly flow occurs only at very low quality and consists of individual bubbles of vapor entrained in the main flow. In the very narrow quality range over which bubbly flow exists, the heat transfer coefficient can be predicted by superimposing liquid-forced-convection and nucleate-pool-boiling equations as long as the wall temperature is not so large as to produce film boiling, see Section 10.3.1.

As the vapor volume fraction increases, the individual bubbles begin to agglomerate and form plugs or slugs of vapor as shown in Fig. 10.11b. Although in this regime, known as the *slug-flow* regime, the mass fraction of vapor is generally much less than 1 percent, as much as 50 percent of the volume fraction may be vapor and the fluid velocity in the slug-flow regime may increase appreciably. The plugs of vapor are compressible volumes which also produce flow oscillations within the duct even if the entering flow is steady. Bubbles may continue to nucleate at the wall, and it is probable that the heat transfer mechanism in plug flow is the same as in the bubbly regime: a superposition of forced convection to a liquid and nucleate pool boiling. Because of the increased liquid flow velocity the heat transfer coefficient rises, as can be seen in Fig. 10.11e.

While both the bubbly and slug-flow regimes are interesting, it should be noted that for density ratios of importance in forced-convection evaporators, the quality in these two regimes is too low to produce appreciable vaporization. These regimes become important in practice only if the temperature difference is so large as to cause film boiling, or if the flow oscillations produced in the slug-flow regime cause instability in a system.

As the fluid flows farther along in the tube and the quality increases, a third flow regime, commonly known as the *annular-flow regime*, appears. In this regime the wall of the tube is covered by a thin film of liquid and heat is transferred through this liquid film. In the center of the tube vapor is flowing at a higher velocity and, although there may be a number of active bubble nucleation sites at the wall, vapor is generated primarily by vaporization from the liquid-vapor interface inside the tube and not by the formation of bubbles inside the liquid annulus unless the heat flux is high. In addition to the liquid in the annulus at the wall, there may be a significant amount of liquid dispersed throughout the vapor core as droplets. The quality range for this type of flow is strongly affected by fluid properties and geometry, but it is generally believed that transition to the next flow regime, known as the *mist-flow regime*, occurs at qualities of about 25 percent or higher.

The transition from annular to mist flow is of great interest since this is presumably the point at which the heat transfer coefficient experiences a sharp decrease as shown in Fig. 10.11e. In systems with fixed heat flux a sharp increase in wall temperature results, while systems with fixed wall temperature will exhibit a sharp drop in heat flux. Generally, this point is referred to as the critical heat flux. Specifically, for low heat flux the condition is called dryout because the tube wall is no longer wetted by liquid. An important change takes place in the transition between annular and mist flow: In the former the wall is covered by a relatively high-conductivity liquid, whereas in the latter, due to complete evaporation of the liquid film, the wall is covered by a low-conductivity vapor. Berenson and Stone (31) observed that the wall-drying process occurs in the following manner: A small dry spot forms suddenly at the wall and grows in all directions as the liquid vaporizes because of conduction heat transfer through the liquid. The small strips of liquid remaining on the wall are almost stationary relative to the high-velocity vapor and the liquid droplets in the vapor core. The dominant heat transfer mechanism is conduction through the liquid film and, although nucleation may produce the initial dry spot on the wall, it has only a small effect on the heat transfer. It thus appears that the drying process in transition to mist flow is similar to that which occurs with a thin film of liquid in a hot pan whose temperature is not great enough to cause nucleate boiling.

Most of the heat transfer in mist flow is from the hot wall to the vapor, and after the heat has been transferred into the vapor core, it is transferred to the liquid droplets there. Vaporization in mist flow actually takes place in the interior of the duct, not at the wall. For this reason the temperature of the vapor in the mist-flow regime can be greater than the saturation temperature, and thermal equilibrium may not exist in the duct. While the volume fraction of the liquid droplets is small, they account for a substantial mass fraction because of the high liquid-to-vapor density ratio.

These observations are consistent with a theoretical stability analysis for a liquid film by Miles (34) which predicts that a liquid film is stable at sufficiently small Reynolds numbers irrespective of the vapor velocity. Since the Reynolds number of the liquid film in a forced-convection evaporator decreases as the quality increases, the liquid annulus will be stable at sufficiently high quality irrespective of the value of the vapor velocity.

The regimes of forced-convection boiling which depend upon the magnitude of the heat flux may be visualized with Fig. 10.12. At high heat fluxes the annular-flow regime is not exhibited. The critical heat flux under these conditions occurs because of a transition from saturated nucleate boiling in the bubble/slug-flow regime to saturated film boiling in the mist-flow regime and is known as departure from nucleate boiling (DNB). At still higher heat fluxes, the critical heat flux results from a transition from subcooled boiling in the bubble-flow regime to subcooled film boiling in the mist-flow regime. This transition is also known as DNB. At the higher heat fluxes which produce DNB to subcooled film boiling, there are very large temperature increases and actual

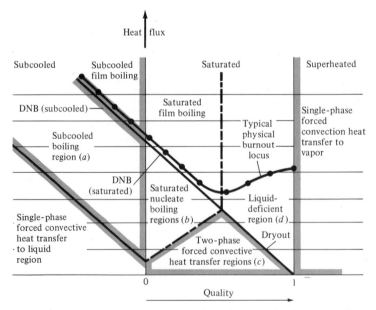

Figure 10.12   Regimes of two-phase forced-convective heat transfer as a function of quality with increasing heat flux as ordinate (46).

tube burnout may occur. At lower heat fluxes, where the transition is due to dryout, the temperature increase is much less and physical burnout is not likely.

For horizontal tubes, the situation is more complex due to stratification of the vapor and liquid phases by gravity, especially at low flow velocities. Much less data are available for the horizontal orientation than for the vertical orientation but it is clear that the critical heat flux is strongly influenced. In addition, stratification may lead to overheating of the upper portions of the tube where the vapor may become superheated before dryout occurs on the lower portion of the tube.

10.3.1 Nucleate Forced Convection Boiling

The method of correlating data for nucleate pool boiling described in Section 10.2.2 has also been applied successfully to boiling of fluids flowing inside tubes or ducts by forced (14) or natural (35) convection.

Figure 10.13 shows curves faired through boiling data, typical of sub-cooled forced convection in tubes or ducts (21, 36). The system in which these data were obtained consisted of a vertical annulus containing an electrically heated stainless steel tube placed centrally in tubes of various diameters. The heater was cooled by degassed distilled water flowing upward at velocities from 0.3 to 3.7 m/s and pressures from 207 to 620 kPa. The scale of Fig. 10.13 is logarithmic. The ordinate is the heat flux q/A, and the abscissa is ΔT, the temperature difference between the heating surface and the bulk of the fluid. The dotted lines represent forced-convection conditions at various velocities and various degrees of subcooling. The solid lines indicate the deviation from forced convection caused by surface boiling. We note that the onset of boiling caused by increasing the heat flux depends on the velocity of the liquid and the degree of subcooling below its saturation temperature at the prevailing pressure. At lower pressures the boiling point at a given velocity is reached at lower heat fluxes. An increase in velocity increases the effectiveness of forced convection, decreases the surface temperature at a given heat flux, and thereby delays the onset of boiling. In the boiling region the curves are steep and the wall temperature is practically independent of the fluid velocity. This shows that the agitation caused by the bubbles is much more effective than turbulence in forced convection without boiling. The heat flux data with surface boiling are plotted separately in Fig. 10.14 against the excess temperature. The resulting curve is similar to that for nucleate boiling in a saturated pool shown in Fig. 10.1 and emphasizes the similarity of the boiling processes and their dependence on the excess temperature; in particular, the heat flux increases approximately with ΔT^3. However, there are not yet sufficient data to suggest that the fully developed boiling curve for forced convection will always follow pool-boiling data correlations.

To apply the pool-boiling correlation to forced-convection boiling, the total heat flux must be separated into two parts, a *boiling flux* q_b/A and a *convective flux* q_c/A, and

$$q_{\text{total}} = q_b + q_c$$

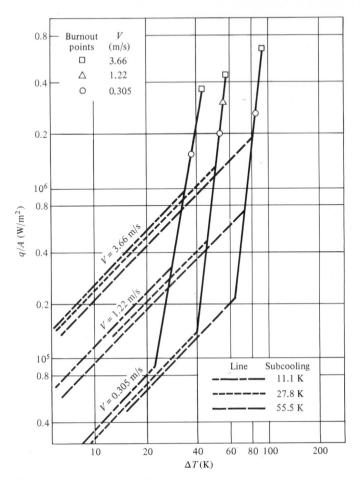

Figure 10.13 Typical boiling data for subcooled forced convection: heat flux versus temperature difference between surface and fluid bulk. [By permission from W. H. McAdams et al. (36).]

The boiling heat flux is determined by subtracting the heat flow rate, accountable for by forced convection alone, from the total flux, or

$$q_b = q_{total} - A\bar{h}_c(T_s - T_b) \tag{10.10}$$

where $\bar{h}_c$ is determined from Eq. (6.63)* using property values for the liquid phase. This value of q_b is to be determined by Eq. (10.2). The results of this method of correlating data for boiling superimposed on convection are shown in Fig. 10.15 for a number of fluid-surface combinations. Some of the data

---

* Rohsenow (14) recommends that the coefficient 0.023 in Eq. (6.66) be replaced by 0.019 in nucleate boiling.

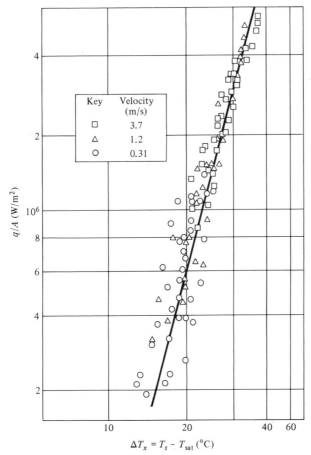

Figure 10.14 Approximate correlation of data for nucleate boiling with forced convection obtained by plotting heat flux versus excess temperature. [By permission from W. M. McAdams et al. (36).]

shown in Fig. 10.15 were obtained with subcooled liquids, others with saturated liquids containing various amounts of vapor.

10.3.2 Boiling with Net Vapor Production

Beyond the narrow range of quality in which bubble flow exists and Eq. (10.10) is valid, the bulk of the liquid will be at the saturation temperature. The heat transfer mechanism here is referred to as saturated nucleate boiling. Beyond this, in the annular regime, the heat is transferred across the film of liquid on the wall. In these flow regimes, Chen (37) has proposed a correlation which assumes that the convection and boiling heat transfer mechanisms play a role

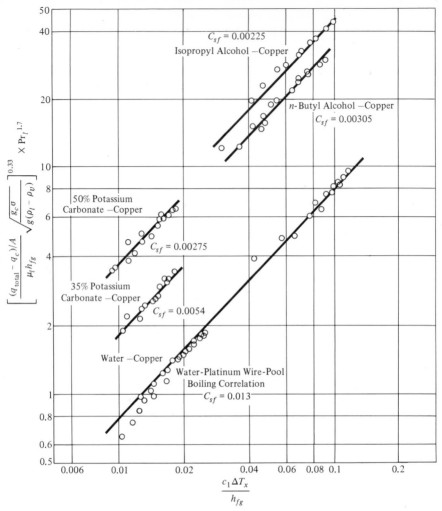

Figure 10.15 Correlation of data for subcooled convection boiling by the Rohsenow method. [From W. H. Jens and G. Leppert (1), with permission of the publishers, the Society of Naval Engineers.]

and that their effects are additive:

$$h = h_c + h_b \qquad (10.11)$$

where

$$h_c = 0.023\left[\frac{G(1-x)D}{\mu_l}\right]^{0.8} \mathrm{Pr}_l^{0.4}\frac{k_l}{D} F \qquad \text{contribution of annular region}$$

$$h_b = 0.00122\left(\frac{k_l^{0.79}c_l^{0.45}\rho_l^{0.49}g_c^{0.25}}{\sigma^{0.5}\mu_l^{0.29}h_{fg}^{0.24}\rho_v^{0.24}}\right)\Delta T_x^{0.24}\Delta p_{\mathrm{sat}}^{0.75} S \qquad (10.12)$$

contribution of nucleate boiling region

In Eq. (10.12), SI units are used with Δp_{sat} (the change in vapor pressure corresponding to a temperature change ΔT_x) expressed in N/m^2. The parameter F may be calculated (38) from

$$F = 1.0 \qquad \text{when } \frac{1}{X_{tt}} < 0.1$$

$$F = 2.35\left(\frac{1}{X_{tt}} + 0.213\right)^{0.736} \qquad \text{when } \frac{1}{X_{tt}} > 0.1$$

where

$$\frac{1}{X_{tt}} = \left(\frac{x}{1-x}\right)^{0.9}\left(\frac{\rho_l}{\rho_v}\right)^{0.5}\left(\frac{\mu_v}{\mu_l}\right)^{0.1}$$

and the parameter S is defined by

$$S = (1 + 0.12\ \mathrm{Re}_{TP}^{1.14})^{-1} \qquad \text{when } \mathrm{Re}_{TP} < 32.5$$
$$S = (1 + 0.42\ \mathrm{Re}_{TP}^{0.78})^{-1} \quad \text{in the range } 32.5 < \mathrm{Re}_{TP} < 70$$
$$S = 0.1 \qquad \text{when } \mathrm{Re}_{TP} > 70$$

with the Reynolds number defined as

$$\mathrm{Re}_{TP} = \frac{G(1-x)D}{\mu_l}F^{1.25} \times 10^{-4}$$

This correlation has been tested against data for several systems (water, methanol, cyclohexane, pentane, heptane, and benzene) for pressures ranging from 1/2 to 35 atm and quality X ranging from 1 to 71 percent, with an average deviation of 11 percent. Collier (19) describes how the Chen correlation can be extended to the subcooled boiling region.

EXAMPLE 10.3

Saturated liquid n-butyl alcohol, $C_4H_{10}O$, is flowing at 161 kg/h through a 1-cm-ID copper tube at atmospheric pressure. The tube wall temperature is held at 140°C by condensing steam at 361 kPa absolute pressure. Calculate the length of tube required to achieve a quality of 50%. The following property values may be used for the alcohol:

$\sigma = 0.0183$ N/m, surface tension

$h_{fg} = 591,500$ J/kg, heat of vaporization

$T_{sat} = 117.5°C$, atmospheric pressure boiling point

$P_{sat} = 2$ atm, saturation pressure corresponding to a saturation temperature of 140°C

$\rho_v = 2.3$ kg/m^3, density of the vapor

$\mu_v = 0.0143 \times 10^{-3}$ kg/m s, viscosity of the vapor

Solution.   Remaining property values taken from Appendix 2, Table 18 are:

$$\rho_l = 737 \text{ kg/m}^3$$
$$\mu_l = 0.39 \times 10^{-3} \text{ kg/m s}$$
$$c_l = 2876 \text{ J/kg K}$$
$$\text{Pr}_l = 6.9$$
$$k_l = 0.163 \text{ W/m K}$$
$$C_{sf} = 0.00305 \text{ from Table 10.2}$$

The mass velocity is

$$G = \frac{161}{3600} \frac{4}{\pi(0.01)^2} = 569 \text{ kg/m}^2 \text{ s}$$

and the Reynolds number for the liquid flow is

$$\text{Re}_D = \frac{GD}{\mu_l} = \frac{(569)(0.01)}{(0.39 \times 10^{-3})} = 14{,}590$$

The contribution to the heat transfer coefficient due to the two-phase annular flow is

$$h_c = (0.023)(14{,}590)^{0.8}(6.9)^{0.4}\left(\frac{0.163}{0.01}\right)(1-x)^{0.8}F$$
$$= 1741(1-x)^{0.8}F$$

Since the vapor pressure changes by 1 atm over the temperature range from T_{sat} to 140°C, we have $\Delta p_{\text{sat}} = 101{,}300 \text{ N/m}^2$. Therefore, the contribution to the heat transfer coefficient from nucleate boiling is

$$h_b = 0.00122\left[\frac{0.163^{0.79} \ 2876^{0.45} \ 737^{0.49} \ 1^{0.25}}{0.0183^{0.5}(0.39 \times 10^{-3})^{0.29} \ 591{,}300^{0.24} \ 2.3^{0.24}}\right]$$
$$\times (140 - 117.5)^{0.24}(101{,}300)^{0.75}S$$

or $h_b = 7754S$.

The calculation for $1/X_{tt}$ becomes

$$\frac{1}{X_{tt}} = \left(\frac{x}{1-x}\right)^{0.9}\left(\frac{737}{2.3}\right)^{0.5}\left(\frac{0.0143}{0.39}\right)^{0.1} = 12.86\left(\frac{x}{1-x}\right)^{0.9}$$

Since the liquid is at saturation temperature, the heat flux over a length Δl may be related to an increase in quality by

$$\dot{m}h_{fg}\Delta x = q\pi D \ \Delta l$$

Substituting in known quantities, we find

$$\Delta l = 842{,}031\frac{\Delta x}{q''}$$

where $q'' = h\Delta T_x$, and from Eq. (10.11) $h = h_c + h_b$.

We may now prepare a table showing stepwise calculations which track the increase in quality, from $x = 0$ to $x = 0.50$. We assume that the steps Δx are small enough that the heat flux and other parameters are reasonably constant in that step.

x	Δx	$\dfrac{1}{X_{tt}}$	F	h_c (W/m^2 K)	Re_{TP}	S	h_b (W/m^2 K)	h (W/m^2 K)	q'' (W/m^2)	Δl (m)	l (m)
0											0
0.01	0.01	0.206	1.24	2,138	1.94	0.797	6,179	8,317	187,129	0.048	0.045
0.05	0.04	0.909	2.56	4,273	4.49	0.601	4,658	8,931	200,952	0.168	0.213
0.10	0.05	1.78	3.90	6,247	7.19	0.468	3,626	9,873	222,142	0.190	0.403
0.20	0.10	3.69	6.41	9,330	11.90	0.331	2,568	11,898	267,706	0.315	0.718
0.30	0.10	6.00	9.01	11,796	15.94	0.262	2,030	13,826	311,093	0.271	0.989
0.40	0.10	8.93	11.98	13,857	19.51	0.220	1,705	15,562	350,135	0.240	1.229
0.50	0.10	12.86	15.59	15,585	22.60	0.192	1,492	17,077	384,239	0.219	1.448

The tube length required to reach 50% quality is 1.49 m.

Note the relative importance of the nucleate boiling contribution, h_b, and the two-phase flow contribution, h_c, along the tube.

10.3.3 Critical Heat Flux

Predictions of the critical heat flux for forced-convection systems are less accurate than those for pool boiling. This is primarily due to the number of variables involved and difficulties encountered in trying to perform controlled experiments to measure the critical heat flux or to determine its location.

A very large number of critical heat flux correlations have been proposed—primarily for boiling water in vertical round tubes with constant heat flux. An empirical critical heat flux correlation for forced covection has been developed by Griffith (39) which covers a wide range of conditions. Griffith correlated critical heat flux data for water, benzene, n-heptane, n-pentane, and ethanol at pressures varying from 0.5 to 96 percent of critical pressure, at velocities from 0 to 30.5 m/s, at subcooling from 0 to 138°C, and at qualities ranging from 0 up to 70 percent. The data used in this correlation were obtained in round tubes and rectangular channels. Figure 10.16 shows the correlated data, and an inspection of this figure suggests that the critical heat flux can apparently be predicted to within ± 33 percent for the conditions used in this study. In Fig. 10.16, h_g is the saturated vapor enthalpy and h_b is the bulk enthalpy of the fluid, which may be subcooled liquid, saturated liquid, or a two-phase flow mixture at some quality less than 70 percent.

The pressure drop in pipes and duct with two-phase flow has been investigated by numerous authors. The problem is quite complex and no entirely satisfactory method of calculation is available. A very useful summary of the state of the art has been prepared by Griffith (40), who concludes, as do several others, that the best available method for predicting the pressure loss is that

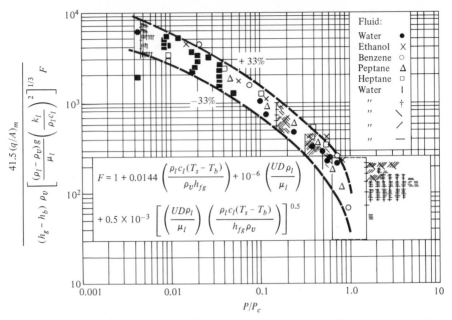

Figure 10.16 Peak heat flux correlation for forced-convection boiling and vaporization. (Courtesy of Dr. P. Griffith and the American Society of Mechanical Engineers.)

proposed by Lockhart and Martinelli (41). The reader interested in this problem is referred to the detailed treatments by Tong (42) and Collier (19).

A very effective method for increasing the peak heat flux attainable in low-quality forced-convection boiling is to insert twisted tapes into a tube to produce a helical flow pattern which generates a centrifugal force field corresponding to many g's (43). Gambill et al. (44) achieved a peak heat flux of 174 MW/m^2 in a swirl system with 61°C subcooled water at 5860 kPa flowing at a velocity of 30 m/s in a 0.5-cm-diameter tube; this is almost three times the energy flux emanating from the surface of the sun.

10.3.4 Heat Transfer Beyond the Critical Point

As suggested by Fig. 10.12, there are three critical transitions leading to a sudden increase in wall temperature for constant heat flux. Operation beyond the critical points involves (1) subcooled film boiling, (2) saturated film boiling, or (3) a liquid-deficient region (mist flow). For systems in which temperature is the independent variable a fourth critical transition, known as transition boiling, exists.

Film Boiling In the film-boiling regime a central liquid core is surrounded by an irregular vapor film. As in pool film boiling, the presence of the vapor film simplifies analysis of this boiling regime. These analyses generally follow that for filmwise condensation as described in Section 10.4. For film boiling in vertical tubes, a correlation which agrees reasonably well with the analyses is the one recommended for pool boiling on the outside of horizontal tubes by Bromley (29), that is, Eq. (10.6).

Liquid-Deficient Region   This regime results from a thinning of the annular liquid film on the heated surface, ultimately resulting in wall dryout. Note from Fig. 10.12 that at higher heat flux the liquid-deficient region results from a transition from saturated film boiling, that is, DNB.

In the saturated film-boiling regime the flow pattern is inverted from that in the annular regime, Fig. 10.11. That is, a central liquid core is surrounded by a vapor film. As the thermodynamic quality is increased, the liquid core breaks up into droplets and the resulting liquid-deficient flow is similar to that resulting from transition from annular flow.

Liquid droplets periodically hit the wall, thereby producing significantly higher heat transfer coefficients than in the saturated film-boiling regime—thus physical burnout is unlikely. Heat transfer from the wall to the vapor and then from the vapor to the droplets allows a nonequilibrium state to exist since the vapor can become superheated in the presence of the droplets. Correlations developed for heat transfer in this regime are of two types: (1) purely empirical and (2) empirical correlations which attempt to account for the nonequilibrium.

An empirical correlation developed by Groeneveld (46) is of the form of the Dittus-Boelter equation for single-phase, forced convection:

$$\frac{\bar{h}D}{k_v} = a\left\{\mathrm{Re}_v\left[x + \frac{\rho_v}{\rho_l}(1-x)\right]\right\}^b \mathrm{Pr}_v^c\left[1 - 0.1\left(\frac{\rho_l}{\rho_v} - 1\right)^{0.4}(1-x)^{0.4}\right]^d$$

(10.13)

Table 10.3 gives values of a, b, c, and d for various geometries and the range of operating conditions over which the correlation is valid.

Rohsenow (47) warns that all such purely empirical correlations should be used with caution. Collier (19) presents additional correlation equations which account for nonequilibrium in the liquid-deficient regime.

Transition Boiling   The transition boiling regime is difficult to characterize in a quantitative manner (2). Within this region the amount of vapor generated is not enough to support a stable vapor film, but is too large to allow sufficient liquid to reach the surface to support nucleate boiling. Berenson (11) suggests, therefore, that nucleate and film boiling occur alternately at a given location. The process is unstable and photographs show that liquid surges sometimes toward the heating surface and sometimes away from it. At times, this turbulent liquid becomes so highly superheated that it explodes into vapor (10). From an industrial viewpoint, the transition-boiling regime is of little interest; equipment designed to operate in the nucleate-boiling region may be sized with more assurance and operated with more reproducible results. Tong and Young (45) have proposed a correlation for heat flux in this region.

TABLE 10.3   CONSTANTS FOR EQ. (10.13)

Geometry	a	b	c	d	No. of points	RMS error, %
Tubes	1.09×10^{-3}	0.989	1.41	-1.15	438	11.5
Annuli	5.20×10^{-2}	0.688	1.26	-1.06	266	6.9
Tubes and annuli	3.27×10^{-3}	0.901	1.32	-1.50	704	12.4

Range of data on which correlations are based

Parameters and Units	Geometry	
	Tube	Annulus
Flow direction	Vertical and horizontal	Vertical
Inside diameter, cm	0.25 to 2.5	0.15 to 0.63
Pressure, atm	68 to 215	34 to 100
G, kg/m² s	700 to 5300	800 to 4100
x, fraction by weight	0.10 to 0.90	0.10 to 0.90
heat flux, kW/m²	120 to 2100	450 to 2250
$\hbar D/k_v$	95 to 1770	160 to 640
$\mathrm{Re}_v \left[x + \dfrac{\rho_v}{\rho_l}(1-x) \right]$	6.6×10^4 to 1.3×10^6	1.0×10^5 to 3.9×10^5
Pr	0.88 to 2.21	0.91 to 1.22
$1 - 0.1\left(\dfrac{\rho_l}{\rho_\mu} - 1\right)^{0.4}(1-x)^{0.4}$	0.706 to 0.976	0.610 to 0.963

10.4 CONDENSATION

When a saturated vapor comes in contact with a surface at a lower temperature, condensation occurs. Under normal conditions a continuous flow of liquid is formed over the surface and the condensate flows downward under the influence of gravity. Unless the velocity of the vapor is very high or the liquid film very thick, the motion of the condensate is laminar and heat is transferred from the vapor-liquid interface to the surface merely by conduction. The rate of heat flow depends, therefore, primarily on the thickness of the condensate film, which in turn depends on the rate at which vapor is condensed and the rate at which the condensate is removed. On a vertical surface the film thickness increases continuously from top to bottom, as shown in Fig. 10.17. As the plate is inclined from the vertical position, the drainage rate decreases and the liquid film becomes thicker. This, of course, causes a decrease in the rate of heat transfer.

10.4.1 Filmwise Condensation

Theoretical relations for calculating the heat transfer coefficients for filmwise condensation of pure vapors on tubes and plates were first obtained by Nusselt (48) in 1916. To illustrate the classical Nusselt approach we shall consider a plane vertical surface at a constant temperature T_s on which a pure vapor at

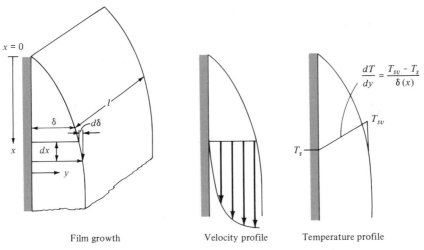

Film growth   Velocity profile   Temperature profile

Figure 10.17 Filmwise condensation on a vertical surface: film growth, velocity profile, and temperature distribution.

saturation temperature T_{sv} is condensing. As shown in Fig. 10.17, a continuous film of liquid flows downward under the action of gravity, and its thickness increases as more and more vapor condenses at the liquid-vapor interface. At a distance x from the top of the plate the thickness of the film is δ. If the flow of the liquid is laminar and is caused by gravity alone, we can estimate the velocity of the liquid by means of a force balance on the element $dx\,\delta l$. The downward force acting on the liquid at a distance greater than y from the surface is $(\delta - y)\,dx\,\rho_l g/g_c$. Assuming that the vapor outside the condensate layer is in hydrostatic balance $(dp/dx = \rho_v g/g_c)$, a partially balancing force equal to $(\delta - y)\,dx\,\rho_v g/g_c$ will be present as a result of the pressure difference between the upper and lower faces of the element. The other force retarding the downward motion is the drag at the inner boundary of the element. Unless the vapor flows at a very high velocity, the shear at the free surface is quite small and may be neglected. The remaining force will then simply be the viscous shear $(\mu_l\,du/dy)\,dx$ at the vertical plane y. Under steady-state conditions the upward and downward forces are equal, or

$$(\delta - y)(\rho_l - \rho_v)g = \mu_l\frac{du}{dy}$$

where the subscripts l and v denote liquid and vapor respectively. The velocity u at y is obtained by separating the variables and integrating. This yields the expression

$$u(y) = \frac{(\rho_l - \rho_v)g}{\mu_l}\left(\delta y - \frac{1}{2}y^2\right) + \text{const}$$

The constant of integration is zero because the velocity u is zero at the surface, that is, $u = 0$ at $y = 0$.

The mass rate of flow of condensate per unit breadth Γ_c is obtained by integrating the local mass flow rate at the elevation x, $\rho u(y)$, between the limits $y = 0$ and $y = \delta$, or

$$\Gamma_c = \int_0^\delta \frac{\rho_l(\rho_l - \rho_v)g}{\mu_l}\left(\delta y - \frac{1}{2}y^2\right)dy = \frac{\rho_l(\rho_l - \rho_v)\delta^3}{3\mu_l}g \qquad (10.14)$$

The change in condensate flow rate Γ_c with the thickness of the condensate layer δ is

$$\frac{d\Gamma_c}{d\delta} = \frac{g\rho_l(\rho_l - \rho_v)}{\mu_l}\delta^2 \qquad (10.15)$$

Heat is transferred through the condensate layer solely by conduction. Assuming that the temperature gradient is linear, the average enthalpy change of the vapor in condensing to liquid and subcooling to the average liquid temperature of the condensate film is

$$h_{fg} + \frac{1}{\Gamma_c}\int_0^\delta \rho_l u c_{pl}(T_{sv} - T)\,dy = h_{fg} + \frac{3}{8}c_{pl}(T_{sv} - T_s)$$

and the rate of heat transfer to the wall is $(k/\delta)(T_{sv} - T_s)$, where k is the thermal conductivity of the condensate. In the steady state the rate of enthalpy change of the condensing vapor must equal the rate of heat flow to the wall, or

$$\frac{q}{A} = k\frac{T_{sv} - T_s}{\delta} = \left[h_{fg} + \frac{3}{8}c_{pl}(T_{sv} - T_s)\right]\frac{d\Gamma_c}{dx} \qquad (10.16)$$

Equating the expressions for $d\Gamma_c$ from Eqs. (10.15) and (10.16) gives

$$\delta^3\,d\delta = \frac{k\mu_l(T_{sv} - T_s)}{g\rho_l(\rho_l - \rho_v)h'_{fg}}dx$$

where $h'_{fg} = h_{fg} + \frac{3}{8}c_{pl}(T_{sv} - T_s)$. Integrating between the limits $\delta = 0$ at $x = 0$ and $\delta = \delta$ at $x = x$ and solving for $\delta(x)$ yields

$$\delta = \left[\frac{4\mu_l kx(T_{sv} - T_s)}{g\rho_l(\rho_l - \rho_v)h'_{fg}}\right]^{1/4} \qquad (10.17)$$

Since heat transfer across the condensate layer is by conduction the local heat transfer coefficient h_x is k/δ. Substituting the expression for δ from Eq. (10.17) gives the unit-surface conductance as

$$h_x = \left[\frac{\rho_l(\rho_l - \rho_v)gh'_{fg}k^3}{4\mu_l x(T_{sv} - T_s)}\right]^{1/4} \qquad (10.18)$$

and the local Nusselt number at x is

$$\mathrm{Nu}_x = \frac{h_x x}{k} = \left[\frac{\rho_l(\rho_l - \rho_v)gh'_{fg}x^3}{4\mu_l k(T_{sv} - T_s)}\right]^{1/4} \qquad (10.19)$$

Inspection of Eq. (10.18) shows that the unit conductance for condensation decreases with increasing distance from the top as the film thickens. The thickening of the condensate film is similar to the growth of a boundary layer over a flat plate in convection. At the same time it is also interesting to observe that an increase in the temperature difference $(T_{sv} - T_s)$ causes a decrease in the surface conductance. This is caused by the increase in the film thickness as a result of the increased rate of condensation. No comparable phenomenon occurs in simple convection.

The average value of the conductance $\bar{h}$ for a vapor condensing on a plate of height L is obtained by integrating the local value h_x over the plate and dividing by the area. For a vertical plate of unit width and height L we obtain by this operation the average heat transfer coefficient

$$\bar{h}_c = \frac{1}{L} \int_0^L h_x \, dx = \frac{4}{3} h_{x=L} \tag{10.20}$$

or

$$\bar{h}_c = 0.943 \left[\frac{\rho_l(\rho_l - \rho_v)gh'_{fg}k^3}{\mu_l L(T_{sv} - T_s)} \right]^{1/4} \tag{10.21}$$

It can easily be shown that, for a surface inclined at an angle ψ with the horizontal, the average coefficient is

$$\bar{h}_c = 0.943 \left[\frac{\rho_l(\rho_l - \rho_v)gh'_{fg}k^3 \sin \psi}{\mu_l L(T_{sv} - T_s)} \right]^{1/4} \tag{10.22}$$

A modified integral analysis for this problem by Rohsenow (49), which is in better agreement with experimental data if $\mathrm{Pr} > 0.5$ and $c_{pl}(T_{sv} - T_s)/h'_{fg} < 1.0$, yields results identical to Eqs. (10.18)–(10.22) except that h'_{fg} is replaced by $[h_{fg} + 0.68c_{pl}(T_{sv} - T_s)]$.

Chen (50) considered the interfacial shear and momentum effects and computed a correction factor to Eq. (10.22)

$$\bar{h}'_c = \bar{h}_c \left(\frac{1 + 0.68A + 0.02AB}{1 + 0.85B - 0.15AB} \right)^{1/4} \tag{10.23}$$

where $\bar{h}'_c$ is the corrected heat transfer coefficient, $\bar{h}_c$ is the heat transfer coefficient from Eq. (10.22), and

$$A = \frac{c_l(T_{sv} - T_s)}{h_{fg}} < 2 \qquad \text{(upper limit of validity)}$$

$$B = \frac{k_l(T_{sv} - T_s)}{\mu_l h_{fg}} < 20 \qquad \text{(upper limit of validity)}$$

and

$$1 < \mathrm{Pr}_l < 0.05$$

Although the foregoing analysis was made specifically for a vertical flat plate, the development is also valid for the inside and outside surfaces of vertical tubes if the tubes are large in diameter compared with the film thickness. These results cannot be extended to inclined tubes, however. In such cases the film flow would not be parallel to the axis of the tube and the effective angle of inclination would vary with x.

The average heat transfer coefficient of a pure saturated vapor condensing on the outside of a horizontal tube can be evaluated by the method used to obtain Eq. (10.22). For a tube of outside diameter D it leads to the equation

$$\bar{h}_c = 0.725 \left[\frac{\rho_l(\rho_l - \rho_v)gh'_{fg}k^3}{D\mu_l(T_{sv} - T_s)} \right]^{1/4} \tag{10.23}$$

If condensation occurs on N horizontal tubes so arranged that the condensate sheet from one tube flows directly onto the tube below, the average unit-surface conductance for the system can be estimated by replacing the tube diameter D in Eq. (10.23) by DN. This method will in general yield conservative results because condensate does not fall in smooth sheets from one row to another.

Chen (50) suggested that, since the liquid film is subcooled, additional condensation occurs on the liquid layer between tubes. Assuming that all the subcooling is used for additional condensation, Chen's analysis yields

$$\bar{h}_c = 0.728 \left[1 + 0.2 \frac{c_p(T_{sv} - T_s)}{h_{fg}}(N - 1) \right] \left[\frac{g\rho_l(\rho_l - \rho_v)k^3 h'_{fg}}{ND\mu_l(T_{sv} - T_s)} \right]^{1/4} \tag{10.24}$$

which is in reasonably good agreement with experimental results, provided $[(N - 1)c_p(T_{sv} - T_s)/h_{fg}] < 2$.

In the preceding equations the unit-surface conductance will be in $W/m^2 \, °C$ if the other quantities are evaluated in the units listed below.

c_p = specific heat of vapor, J/kg °C

c_{pl} = specific heat of liquid, J/kg °C

k = thermal conductivity of liquid, W/m °C

ρ_l = density of liquid, kg/m³

ρ_v = density of vapor, kg/m³

g = acceleration of gravity, m/s²

h_{fg} = latent heat of condensation or vaporization, J/kg

$h'_{fg} = h_{fg} + \frac{3}{8}c_p(T_{sv} - T_s)$

μ_l = viscosity of liquid, N s/m²

D = tube diameter, m

L = length of plane surface, m

T_{sv} = temperature of saturated vapor, °C

T_s = wall surface temperature, °C

The physical properties of the liquid film in Eqs. (10.17)–(10.24) should be evaluated at an effective film temperature $T_{film} = T_s + 0.25(T_{sv} - T_s)$ (16). When used in this manner, Nusselt's equations are satisfactory for estimating surface conductances for condensing vapors. Experimental data are in general agreement with Nusselt's theory when the physical conditions comply with the assumptions inherent in the analysis. Deviations from Nusselt's film theory occur when the condensate flow becomes turbulent, when the vapor velocity is very high (51), or when a special effort is made to render the surface nonwettable. All of these factors tend to increase the surface conductance, and the Nusselt film theory will therefore always yield conservative results.

EXAMPLE 10.4 _____

A 0.013-m-OD, 1.5-m-long tube is to be used to condense steam at 40,000 N/m^2, $T_{sv} = 349$ K. Estimate the unit-surface conductances for this tube in (a) the horizontal position and (b) the vertical position. Assume that the average tube wall temperature is 325 K.

Solution.

(a) At the average temperature of the condensate film $[T_f = (349 + 325)/2 = 337$ K], the physical property values pertinent to the problem are

$$k_l = 0.661 \text{ W/m K} \qquad \mu_l = 4.48 \times 10^{-4} \text{ N s/m}^2$$
$$\rho_l = 980.9 \text{ kg/m}^3 \qquad c_{pl} = 4184 \text{ J/kg K}$$
$$h_{fg} = 2.349 \times 10^6 \text{ J/kg} \qquad \rho_v = 0.25 \text{ kg/m}^3$$

For the tube in the horizontal position Eq. (10.23) applies and the unit-surface conductance is

$$\bar{h}_c = 0.725 \left[\frac{(980.9)(980.6)(9.81)(2.417 \times 10^6)(0.661)^3}{(0.013)(4.48 \times 10^{-4})(349 - 325)} \right]^{1/4}$$
$$= 10{,}680 \text{ W/m}^2 \text{ K}$$

(b) In the vertical position the tube may be treated as a vertical plate of area πDL and, according to Eq. (10.21), the average unit-surface conductance is

$$\bar{h}_c = 0.943 \left[\frac{(980.9)(980.6)(9.81)(2.417 \times 10^6)(0.661)^3}{(4.48 \times 10^{-4})(1.5)(349 - 325)} \right]^{1/4}$$
$$= 4239 \text{ W/m}^2 \text{ K}$$

Effect of Film Turbulence   The preceding correlations show that, for a given temperature difference, the average unit conductance is considerably larger when the tube is placed in a horizontal position, where the path of the condensate is shorter and the film thinner, than in a vertical position, where the path is longer and the film thicker. This conclusion is generally valid when the length of the vertical tube is more than 2.87 times the outer diameter, as can be seen by a comparison of Eqs. (10.21) and (10.23). However, these equations are based on the assumption that the flow of the condensate film is laminar and consequently they do not apply when the flow of the condensate is turbulent. Turbulent flow is hardly ever reached on a horizontal tube, but may be established over the lower portion of a vertical surface. When this occurs, the average heat

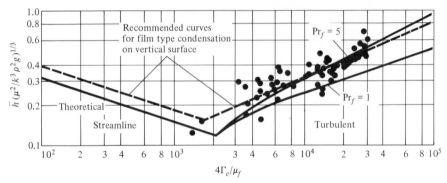

Figure 10.18 Effect of turbulence in film on heat transfer with condensation.

transfer coefficient becomes larger as the length of the condensing surface is increased because the condensate no longer offers as high a thermal resistance as it does in laminar flow. This phenomenon is somewhat analogous to the behavior of a boundary layer.

Just as a fluid flowing over a surface undergoes a transition from laminar to turbulent flow, so the motion of the condensate becomes turbulent when its Reynolds number exceeds a critical value of about 2000. The Reynolds number of the condensate film, Re_δ, when based on the hydraulic diameter [Eq. (6.2)], can be written as $Re_\delta = (4A/P)\Gamma_c/\delta\mu_f$, where P is the wetted perimeter, equal to πD for a vertical tube, and A is the flow cross-sectional area, equal to $P\delta$. According to an analysis by Colburn (52), the local heat transfer coefficient for turbulent flow of the condensate can be evaluated from

$$h_x = 0.056 \left(\frac{4\Gamma_c}{\mu_f}\right)^{0.2} \left(\frac{k^3 \rho^2 g}{\mu^2}\right)^{1/3} Pr_f^{1/2} \qquad (10.25)$$

To obtain average values of the conductance, integration of h_x over the surface by means of Eq. (10.18) for values of $4\Gamma_c/\mu_f$ less than 2000 and Eq. (10.25) for values larger than 2000 is necessary. The results of such calculations for two values of the Prandtl number are plotted as solid lines in Fig. 10.18, where some experimental data obtained with diphenyl in turbulent flow are also shown (53). The heavy dashed line shown on the same graph is an empirical curve recommended by McAdams (18) for evaluating the average unit-surface conductance of single vapors condensing on vertical surfaces.

EXAMPLE 10.5 —————————————————————————————

Determine whether the flow of the condensate in Example 10.4b is laminar or turbulent at the lower end of the tube.

Solution. The Reynolds number of the condensate at the lower end of the tube can be written with the aid of Eq. (10.14) as

$$Re_\delta = \frac{4\Gamma_c}{\mu_l} = \frac{4\rho_l^2 g \delta^3}{3\mu_l^2}$$

Substituting Eq. (10.17) for δ yields

$$\mathrm{Re}_\delta = \frac{4\rho_l^2 g}{3\mu_l^2}\left[\frac{4\mu_l k_l L(T_{sv} - T_s)}{gh_{fg}\rho_l^2}\right]^{3/4} = \frac{4}{3}\left[\frac{4k_l L(T_{sv} - T_s)\rho_l^{2/3}g^{1/3}}{\mu_l^{5/3}h'_{fg}}\right]^{3/4}$$

Inserting in the expression above the numerical values for the problem yields

$$\mathrm{Re}_\delta = \frac{4}{3}\left[\frac{4(0.661)(1.5)(349 - 325)(980.9)^{2/3}(9.81)^{1/3}}{(4.48 \times 10^{-4})^{5/3}(2.417 \times 10^6)}\right]^{3/4} = 564$$

Since the Reynolds number at the lower edge of the tube is below 2000, the flow of the condensate is laminar and the result obtained from Eq. (10.21) is valid.

Effect of High Vapor Velocity   One of the approximations made in Nusselt's film theory is that the frictional drag between the condensate and the vapor is negligible. This approximation ceases to be valid when the velocity of the vapor is substantial compared with the velocity of the liquid at the vapor-condensate interface. When the vapor flows upward, it adds a retarding force to the viscous shear and causes the film thickness to increase. With downward flow of vapor, the film thickness decreases, and surface conductances substantially larger than those predicted from Eq. (10.21) can be obtained. In addition, the transition from laminar to turbulent flow occurs at condensate Reynolds numbers of the order of 300 when the vapor velocity is high. Carpenter and Colburn (54) determined the heat transfer coefficients for condensation of pure vapors of steam and several hydrocarbons in a vertical tube, 2.44 m long and 1.27 cm ID, with inlet vapor velocities at the top up to 152 m/s. Their data are correlated reasonably well by the equation

$$\frac{\bar{h}_c}{c_{pl}G_m}\,\mathrm{Pr}_l^{0.50} = 0.046\sqrt{\frac{\rho_l}{\rho_v}}\,f \tag{10.26}$$

where Pr_l = Prandtl number of liquid

ρ_l = density of liquid, kg/m^3

ρ_v = density of vapor, kg/m^3

c_{pl} = specific heat of liquid, J/kg K

$\bar{h}_c$ = average unit conductance, W/m^2 K

f = pipe-friction coefficient evaluated at the average vapor velocity = $\tau_w/[G_m^2/2\rho_v]$

τ_w = wall shear stress, N/m^2

G_m = mean value of the mass velocity of the vapor, kg/s m^2

The value of G_m in Eq. (10.26) can be calculated from

$$G_m = \sqrt{\frac{G_1^2 + G_1 G_2 + G_2^2}{3}}$$

where G_1 = mass velocity at top of tube

   G_2 = mass velocity at bottom of tube

All physical properties of the liquid in Eq. (10.26) are to be evaluated at a reference temperature equal to $0.25T_{sv} + 0.75T_s$. These results may be used gennerally as an indication of the influence of vapor velocity on the heat tranfer coefficient of condensing vapors when the vapor and the condensate flow in the same direction.

   Soliman et al. (55) have modified the numerical coefficients in Eq. (10.26) on the basis of recent additional data

$$\frac{\bar{h}_c}{c_{pl}G_m} \Pr_l^{0.35} = 0.036 \sqrt{\frac{\rho_l}{\rho_v}} f \tag{10.27}$$

The effect of vapor velocity in a horizontal tube is complicated by the existence of several flow regimes created by the interaction of vapor and liquid within the tube. Collier (19) treats this problem in detail.

   For condensation on the outside of a horizontal tube when the effect of vapor velocity cannot be ignored, Shekriladze and Gomelauri (56) developed the following correlation equation:

$$\bar{h}_c' = \left[\frac{1}{2} \bar{h}_s^2 + \left(\frac{1}{4} \bar{h}_s^4 + \bar{h}_c^4 \right)^{1/2} \right]^{1/2} \tag{10.28}$$

where $\bar{h}_c'$ is the heat transfer coefficient corrected for the effect of vapor shear, $\bar{h}_c$ is the uncorrected heat transfer coefficient for condensation on horizontal tubes, Eq. (10.23), and $\bar{h}_s$, the contribution of the vapor shear to the heat transfer coefficient, is calculated from

$$\frac{\bar{h}_s D}{k_l} = 0.9 \left(\frac{\rho_l U_\infty D}{\mu_l} \right)^{0.5} \qquad \text{for} \qquad \frac{\rho_l U_\infty D}{\mu_l} < 10^6 \tag{10.29}$$

and $$\frac{\bar{h}_s D}{k_l} = 0.59 \left(\frac{\rho_l U_\infty D}{\mu_l} \right)^{0.5} \qquad \text{for} \qquad \frac{\rho_l U_\infty D}{\mu_l} > 10^6 \tag{10.30}$$

where U_∞ is the vapor velocity approaching the tube.

Condensation of Superheated Vapor   Although all of the preceding equations strictly apply only to saturated vapors, they can also be used with reasonable accuracy for condensation of superheated vapors. The rate of heat transfer from a superheated vapor to a wall at T_s will therefore be

$$q = A\bar{h}(T_{sv} - T_s) \tag{10.31}$$

where $\bar{h}$ = average value of the heat transfer coefficient determined from an equation appropriate to the geometric configuration with the same vapor at saturation conditions

   T_{sv} = *saturation temperature* corresponding to the prevailing system pressure

10.4.2 Dropwise Condensation

When a condensing surface material prevents the condensate from wetting the surface, such as is the case for a metallic (nonoxide) coating, the vapor will condense in drops rather than as a continuous film (57). This is known as dropwise condensation. A large part of the surface is not covered by an insulating film under these conditions, and the heat transfer coefficients are four to eight times as high as in filmwise condensation. The ratio of condensate mass flux for dropwise condensation, $\dot{m}_D$, from the outside of a horizontal tube of diameter D, to that for film condensation, $\dot{m}_f$, may be estimated (58) from

$$\frac{\dot{m}_D}{\dot{m}_f} = \left(\frac{\rho_l^2 D^2 g}{24.2\mu_l\dot{m}_f}\right)^{1/9} \tag{10.32}$$

For steam at atmospheric pressure and $\dot{m}_f = 0.014 \text{ kg/m}^2$ s, Eq. (10.32) predicts a ratio of 6.5.

For the purpose of calculating the unit conductance in practice, a conservative approach is to assume filmwise condensation because, even with steam, dropwise condensation can be expected only under carefully controlled conditions, which cannot always be maintained in practice (59, 60). Dropwise condensation of steam may, however, be a useful technique in experimental work when it is desirable to reduce the thermal resistance on one side of a surface to a negligible value.

10.5 CONDENSER DESIGN

The evaluation of the surface conductance of condensing vapors, as can be seen from Eqs. (10.21)–(10.23), presupposes a knowledge of the temperature of the condensing surface. In practical problems this temperature is generally not known because its value depends on the relative order of magnitudes of the thermal resistances in the entire system. The type of problem usually encountered in practice, whether it be a performance calculation for an existing piece of equipment or the design of equipment for a specific process, requires simultaneous evaluation of thermal resistances at the inner and outer surfaces of a tube or the wall of a duct. In most cases the geometric configuration is either specified, as in the case of an existing piece of equipment, or assumed, as in the design of new equipment. When the desired rate of condensation is specified, the usual procedure is to estimate the total surface area required and then to select a suitable arrangement for a combination of size and number of tubes that meets the preliminary area specification. The performance calculation can then be made as though one were dealing with an existing piece of equipment, and the results can later be compared with the specifications. The flow rate of the coolant is usually determined by the allowable pressure drop or the allowable temperature rise. Once the flow rate is known, the thermal resistances of the coolant and the tube wall can be computed without difficulty. The unit-surface conductance of the condensing fluid, however, depends on the

TABLE 10.4   APPROXIMATE VALUES OF UNIT-SURFACE
CONDUCTANCES FOR CONDENSATION OF PURE VAPORS

Vapor	System	Approximate range of $(T_{sv} - T_s)$ (K)	Approximate range of average unit conductance (W/m² K)
Steam	Horizontal tubes, 2.5–7.5 cm OD	3–22	11,400–22,800
Steam	Vertical surface 3.1 m high	3–22	5700–11,400
Ethanol	Vertical surface 0.15 m high	11–55	1100–1900
Benzene	Horizontal tube, 2.5 cm OD	17–44	1400–2000
Ethanol	Horizontal tube, 5 cm OD	6–22	1700–2600
Ammonia	Horizontal 5-to 7.5-cm annulus	1–4	1400–2600[a]

[a] Overall heat transfer coefficient U for water velocities between 1.2 and 24 m/s (71) inside the tube.

condensing-surface temperature, which can be computed only after the conductance is known. A trial-and-error solution is therefore necessary. One either assumes a surface temperature or, if more convenient, estimates the unit conductance on the condensing side and calculates the corresponding surface temperature. With this first approximation of the surface temperature, the unit-surface conductance is then recalculated and compared with the assumed value. A second approximation is usually sufficient for satisfactory accuracy.

The orders of magnitude of unit-thermal conductances for various vapors listed in Table 10.4 will aid in the initial estimates and reduce the amount of trial and error. We note that for steam the thermal resistance is very small, whereas for organic vapors it is of the same order of magnitude as the resistance offered to the flow of heat by water at a low turbulent Reynolds number. In the refrigeration industry and in some chemical processes, finned tubes have been used to reduce the thermal resistance on the condensing side. A method for dealing with condensation on finned tubes and tube banks is presented in (61). When repeated calculations of the conductance for condensation of pure vapors are to be made, alignment charts devised by Chilton, Colburn, Genereaux, and Vernon, and reproduced in (18), are convenient.

Mixtures of Vapors and Noncondensable Gases   The analysis of a condensing system containing a mixture of vapors, or a pure vapor mixed with noncondensable gas, is considerably more complicated than the analysis of a pure-vapor system. The presence of appreciable quantities of a noncondensable gas will in general reduce the rate of heat transfer. If high rates of heat transfer are desired, it is considered good practice to vent the noncondensable gas, which otherwise will blanket the cooling surface and add considerably to the thermal resistance. Noncondensable gases also inhibit the mass transfer by offering a diffusional resistance. A complete treatment of problems involving condensation of mixtures is beyond the scope of this text, and the reader is referred to (18) for a comprehensive summary of available information on this topic.

10.6* FREEZING AND MELTING

Problems involving the solidification or melting of materials are of considerable importance in many technical fields. Typical examples in the field of engineering are the making of ice, the freezing of foods, or the solidification and melting of metals in casting processes. In geology the solidification rate of the earth has been used to estimate the age of our planet. Whatever the field of application, the problem of central interest is the rate at which solidification or melting occurs.

We shall consider here only the problem of solidification, and it is left for the reader as an exercise to show that a solution to this problem is also a solution to the corresponding problem in melting. Figure 10.19 shows the temperature distribution in an ice layer on the surface of a liquid. The upper face is exposed to air at subfreezing temperature. Ice formation occurs progressively at the solid-liquid interface as a result of heat transfer through the ice to the cold air. Heat flows by convection from the water to the ice, by conduction through the ice, and by convection to the sink. The ice layer is subcooled except for the interface in contact with the liquid, which is at the freezing point. A portion of the heat transferred to the sink is used to cool the liquid at the interface SL to the freezing point and to remove its latent heat of solidification. The other portion serves to subcool the ice. Cylindrical or spherical systems may be described in a similar manner, but solidification may proceed either inward (as for freezing of water inside a can) or outward (as for water freezing on the outside of a pipe).

The freezing of a slab can be formulated as a boundary-value problem in which the governing equation is the general conduction equation for the solid phase

$$\frac{\partial^2 T}{\partial x^2} = \frac{1}{\alpha} \frac{\partial T}{\partial t}$$

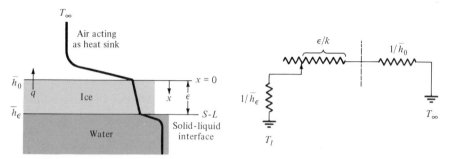

Figure 10.19 Temperature distribution for ice forming on water with air acting as heat sink, and simplified thermal circuit for the system with the heat capacity of the solid considered negligible.

subject to the boundary conditions that

$$-k\frac{\partial T}{\partial x} = \bar{h}_o(T_{x=0} - T_\infty) \qquad \text{at } x = 0$$

$$-k\frac{\partial T}{\partial x} = \rho L_f \frac{d\epsilon}{dt} + \bar{h}_\epsilon(T_l - T_{fr}) \qquad \text{at } x = \epsilon$$

where ϵ = distance to be solid-liquid interface, which is a function of time t

L_f = latent heat of fusion of the material

α = thermal diffusivity of the solid phase $(c\rho/k)$

ρ = density of the solid phase

T_l = temperature of the liquid

T_∞ = temperature of the heat sink

T_{fr} = freezing-point temperature

$\bar{h}_o$ = unit conductance at $x = 0$, the air-ice interface

$\bar{h}_\epsilon$ = unit conductance at $x = \epsilon$, the water-ice interface

The analytic solution of this problem is very difficult and has been obtained only for special cases. The reason for the difficulty is that the governing equation is a partial differential equation for which the particular solutions are unknown when physically realistic boundary conditions are imposed.

An approximate solution of practical value can, however, be obtained by considering the heat capacity of the subcooled solid phase as negligible relative to the latent heat of solidification. To simplify our analysis further we shall assume that the physical properties of the ice, ρ, k, and c, are uniform, that the liquid is at the solidification temperature (i.e., $T_l = T_{fr}$ and $1/\bar{h}_\epsilon = 0$), and that $\bar{h}_o$ and T_∞ are constant during the process.

The rate of heat flow per unit area through the resistances offered by the ice and the air, acting in series, as a result of the temperature potential $(T_{fr} - T_\infty)$ is

$$\frac{q}{A} = \frac{T_{fr} - T_\infty}{1/\bar{h}_o + \epsilon/k} \tag{10.33}$$

This is the heat flow rate which removes the latent heat of fusion necessary for freezing at the surface $x = \epsilon$, or

$$\frac{q}{A} = \rho L_f \frac{d\epsilon}{dt} \tag{10.34}$$

where $d\epsilon/dt$ is the volume rate of ice formation per unit area at the growing surface (m³/h m²) and ρL_f is the latent heat (J/m³). Combining Eqs. (10.33) and (10.34) to eliminate the rate of heat flow yields

$$\frac{T_{fr} - T_\infty}{1/\bar{h}_o + \epsilon/k} = \rho L_f \frac{d\epsilon}{dt} \tag{10.35}$$

which relates the depth of ice to the freezing time. The variables ϵ and t can now be separated, and we get

$$d\epsilon\left(\frac{1}{\bar{h}_o} + \frac{\epsilon}{k}\right) = \frac{T_{fr} - T_\infty}{\rho L_f}\, dt \tag{10.36}$$

To make this equation dimensionless, let

$$\epsilon^+ = \frac{\bar{h}_o \epsilon}{k}$$

and

$$t^+ = t\bar{h}_o^2 \frac{T_{fr} - T_\infty}{\rho L_f k}$$

Substituting these dimensionless parameters into Eq. (10.36) yields

$$d\epsilon^+(1 + \epsilon^+) = dt^+ \tag{10.37}$$

If the freezing process starts at $t = t^+ = 0$ and continues for a time t, the solution of Eq. (10.37), obtained by integration between the specified limits, is

$$\epsilon^+ + \frac{(\epsilon^+)^2}{2} = t^+ \tag{10.38}$$

or

$$\epsilon^+ = -1 + \sqrt{1 + 2t^+} \tag{10.39}$$

When the temperature of the liquids T_l is above the fusion temperature and the convective resistance at the liquid-solid interface is $\bar{h}_\epsilon$, the dimensionless equation corresponding to Eq. (10.37) in the foregoing simplified treatment becomes

$$\frac{(1 + \epsilon^+)\, d\epsilon^+}{1 + R^+ T^+(1 + \epsilon^+)} = dt^+ \tag{10.40}$$

where $\quad R^+ = \dfrac{\bar{h}_\epsilon}{\bar{h}_o}$

$$T^+ = \frac{T_l - T_{fr}}{T_{fr} - T_\infty}$$

and the other symbols represent the same dimensionless quantities used previously in Eq. (10.37).

For the boundary conditions that at $t^+ = 0$, $\epsilon^+ = 0$, and at $t^+ = t^+$, $\epsilon^+ = \epsilon^+$, the solution of Eq. (10.40) becomes

$$t^+ = -\frac{1}{(R^+ T^+)^2} \ln\left(1 + \frac{R^+ T^+ \epsilon^+}{1 + R^+ T^+}\right) + \frac{\epsilon^+}{R^+ T^+} \tag{10.41}$$

The results are shown graphically in Fig. 10.20, where the generalized thickness ϵ^+ is plotted against generalized time t^+ with the generalized potential-resistance ratio $R^+ T^+$ as parameter.

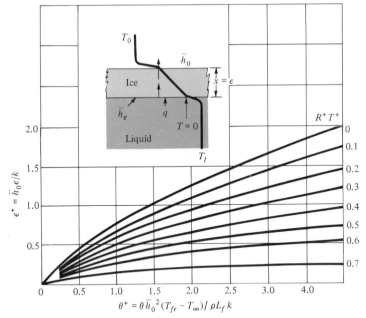

Figure 10.20 Solidification of slab: thickness versus time. [From A. L. London and R. A. Seban (70), with permission of the publishers, the American Society of Mechanical Engineers.]

EXAMPLE 10.6 _____

In the production of "Flakice," ice forms in thin layers on a horizontal rotating drum which is partly submerged in water. The cylinder is internally refrigerated with a brine spray at $-11°C$. Ice formed on the exterior surface is peeled off as the revolving-drum surface emerges from the water.

For the operating conditions listed below, estimate the time required to form an ice layer 0.25 cm thick.

Water liquid temperature, $4.4°C$

Liquid-surface conductance, $57 \ W/m^2 \ K$

Conductance between brine and ice
(including metal wall), $570 \ W/m^2 \ K$

Use the following properties for ice: latent heat of fusion, $333{,}700 \ J/kg$; thermal conductivity, $2.22 \ W/m \ K$

Solution. For the conditions stated above we have

$$R^+ = \frac{\bar{h}_\epsilon}{\bar{h}_o} = \frac{57}{570} = 0.1$$

$$T^+ = \frac{T_l - T_{fr}}{T_{fr} - T_\infty} = \frac{4.4 - 0}{0 - (-11)} = 0.4$$

$$\epsilon^+ = \frac{\bar{h}_o \epsilon}{k_{\text{ice}}} = \frac{(570 \ W/m^2 \ °C)(0.0025 \ m)}{2.32 \ W/m \ °C} = 0.614$$

We assume now that the ice is a sheet. This is justified because the thickness of the ice is very small compared to the radius of curvature of the drum. The boundary conditions of this problem are then the same as those assumed in the solution of Eq. (10.41). Hence, Eq. (10.41) is the solution to the problem at hand. Substituting numerical values for R^+, T^+, and ϵ^+ in Eq. (10.41) yields

$$t^+ = -\frac{1}{(0.04)^2}\ln\left(1 + \frac{0.0246}{1 + 0.04}\right) + \frac{0.614}{0.04} = 0.739$$

From the definition of t^+, the time t is

$$t = \frac{0.739\rho L_f k}{\bar{h}_o^2(T_{fr} - T_\infty)}$$
$$= \frac{(0.739)(918)(333,700)(2.22)}{(324,900)(11)}$$
$$= 141\ \text{s}$$

An estimate of the error caused by neglecting the heat capacity of the solidified portion has been obtained by means of an electrical network simulating the freezing of a slab originally at the fusion temperature (62). It was found that the error is not appreciable when $\epsilon\bar{h}_o/k$ is less than 0.1 (62) or when $L_f/(T_{fr} - T_\infty)c$ is larger than 1.5 (63). In the intermediate range, the freezing rates predicted by the simplified analysis are too large. The solutions presented here are valid for ice and other substances that have heats of fusion which are large compared to their specific heats. An approximate method for predicting the freezing rate of steel and other metals, where $L_f/(T_{fr} - T_\infty)c$ may be less than 1.5, is presented in (63). Numerical methods of solution for systems involving a change of phase are presented in (64) and (65). Melting and freezing in wedges and corners has been analyzed in (66). Melting in a spherical geometry has been treated in (67).

PROBLEMS

The problems for this chapter are related to the subject matter as shown below.

Topic	Section	Problem Number
Pool boiling	10.2	10.1 to 10.10
Convection boiling	10.3	10.11 to 10.13
Condensation	10.4	10.14 to 10.24
Freezing	10.6	10.25 to 10.29

10.1 Water at atmospheric pressure is boiling in a pot with a flat copper bottom on an electric range which maintains the surface temperature at 115°C. Calculate the boiling heat transfer coefficient.

10.2 Predict the nucleate-boiling heat transfer coefficient for water boiling at atmospheric pressure on the outside surface of a $\frac{5}{8}$-in.-OD vertical tube 5 ft long. Assume the tube-surface temperature is constant at 20°F above the saturation temperature.

10.3 Estimate the maximum heat flux obtainable with nucleate pool boiling on a clean surface for (a) water at 1 atm on brass, (b) water at 10 atm on brass, and (c) n-butyl alcohol at 3 atm on copper.

10.4 Determine the excess temperature at one-half of the maximum heat flux for the fluid-surface combinations in Problem 10.3.

10.5 For saturated pool boiling of water on a horizontal plate calculate the peak heat flux at pressures of 10, 20, 40, 60, and 80 percent of the critical pressure p_c and plot your results as q_{max} versus p/p_c. The surface tension of water may be taken as $\sigma = 5.3 \times 10^{-3} (1 - 0.0014T)$, where σ is in pounds per foot and T in degrees Fahrenheit.

10.6 A 0.6-cm-thick flat plate of stainless steel, 7.5 cm high and 0.3 m long, is immersed vertically at an initial temperature of 980°C in a large water bath at 100°C and at atmospheric pressure. Determine how long it will take this plate to cool to 540°C.

10.7 Calculate the maximum heat flux attainable in nucleate boiling with saturated water at 2 atm pressure in a gravitational field equivalent to one-tenth that of the earth.

10.8 Prepare a graph showing the effect of subcooling between 0 and 50°C on the maximum heat flux calculated in Problem 10.9.

10.9 Calculate the critical height at which film-boiling water at atmospheric pressure will undergo a transition on a vertical surface at 500°C. Calculate the average heat transfer coefficient for the laminar regime.

10.10 A thin-walled horizontal copper tube of 0.5 cm OD is placed in a pool of water at atmospheric pressure and 100°C. Inside the tube an organic vapor is condensing and the outside surface temperature of the tube is uniform at 232°C. Calculate the average unit-surface conductance at the outside of the tube.

10.11 Calculate the heat transfer coefficient for laminar film boiling of water in a $\frac{1}{2}$-in.-OD horizontal tube if the tube temperature is 1000°F and the system is placed under pressure of $\frac{1}{2}$ atm.

10.12 Calculate the maximum safe heat flux in the nucleate-boiling regime for water flowing at a velocity of 15 m/s through a 1.2-cm-ID tube 0.31 m long if the water enters at 1 atm pressure and 100°C and the heat flux in the tube is uniform at a rate of 15 MW/m² ft.

10.13 Calculate the peak heat flux for Problem 10.3c from Eq. (10.5) and from the correlation shown in Fig. 10.15, and compare the results.

10.14 Develop the Nusselt film-condensation relation for condensation inside small vertical tubes where the film builds up an annulus.

10.15 Consider a 1.3-cm-ID vertical tube at a surface temperature of 66°C with atmospheric saturated steam inside. Determine the tube length at which the condensate fills the tube and chokes the flow.

10.16 Calculate the average heat transfer coefficient for film-type condensation of water at pressures of 130 Pa and 101 kPa for (a) a vertical surface 1.5 m high; (b) the outside surface of a 1.6-cm-OD vertical tube 1.5 m long; (c) the outside surface of a 1.6-cm-OD horizontal tube 1.5 m long; and (d) a 10-tube vertical bank of 1.6-cm-OD horizontal tubes 1.5 m long. In all cases assume that the vapor velocity is negligible and that the surface temperatures are constant at 11°C below saturation temperature.

10.17 (a) Estimate the heat transfer surface area required and (b) suggest a suitable arrangement for the condenser of a 10-ton refrigeration machine. The working fluid is ammonia condensing on the outside of horizontal pipes at a pressure of 170 psia. The condenser is to be constructed with 1-in. steel pipes (1.00 in. OD, 0.834 in. ID), cooling water is available at 79°F, and the average water velocity in the pipes is not to exceed 6 ft/s.

10.18 The inside surface of a 3-ft-long vertical 2.0-in.-ID tube is maintained at 250°F. For saturated steam at 50 psia condensing inside, estimate the average unit-surface conductance and the condensation rate, assuming the steam velocity is small.

10.19 A horizontal 1.00-in.-OD tube is maintained at a temperature of 80°F on its outer surface. Calculate the average unit-surface conductance if saturated steam at 1.7 psia is condensing on this tube.

10.20 Repeat Problem 10.19 for a tier of six horizontal 1.0-in.-OD tubes under similar thermal conditions.

10.21 Saturated steam at 34 kPa condenses on a 1 m tall vertical plate whose surface temperature is uniform at 60°C. Compute the average unit-surface conductance and the value of the conductance $\frac{1}{3}$, $\frac{2}{3}$, and 1 m from the top. Also, find the maximum height for which the condensate film will remain laminar.

10.22 At a pressure of 490 kPa, the saturation temperature of sulfur dioxide (SO_2) is 32°C, the density is 1350 kg/m^3, and the latent heat of vaporization is 2419 kJ/m^3. If the SO_2 is to be condensed at 490 kPa on a 20-cm flat surface, inclined at an angle of 45°, whose temperature is maintained uniformly at 24°C, calculate: (a) the thickness of the condensate film 1.3 cm from the bottom, (b) the average heat transfer coefficient, and (c) the rate of condensation in kilograms per hour.

10.23 Repeat Problem 10.22 but assume that condensation occurs on a 5-cm-OD horizontal tube.

10.24 Saturated methyl chloride at 62 psia condenses on a horizontal bank of tubes, ten-by-ten, 2 in. OD, equally spaced, 4 in. apart center-to-center on rows and columns. At 62 psia the latent heat of vaporization of methyl chloride is 167 Btu/lb and the specific volume of the saturated liquid is 0.017 ft^3/lb$_m$. If the surface temperature of the tubes is maintained at 45°F by water pumped through them, calculate the rate of condensation of methyl chloride in lb$_m$/h.

10.25 Show that the dimensionless equation for ice formation at the outside of a tube of radius r_o is

$$t^* = \frac{r^{*2}}{2} \ln r^* \left(\frac{1}{2R^*} + \frac{1}{4} \right) (r^{*2} - 1)$$

where

$$r^* = \frac{\epsilon + r_0}{r_o} \qquad R^* = \frac{\bar{h}_o r_o}{k} \qquad t^* = \frac{(T_f - T_\infty)kt}{\rho L r_o^{\,2}}$$

Assume that the water is originally at the freezing temperature T_f, that the cooling medium inside the tube surface is below the freezing temperature at a uniform temperature T_∞, and that $\bar{h}_o$ is the total conductance between the cooling medium and the pipe-ice interface. Also show the thermal circuit.

10.26 In the manufacture of can ice, cans having inside dimensions of 11 by 22 by 50 in. with 1-in. inside taper are filled with water and immersed in brine at a temperature of 10°F. [For details of the process see (72).] For the purpose of a preliminary analysis, the actual ice can may be considered as an equivalent cylinder having the same cross-sectional area as the can, and end effects may be neglected. The overall conductance between the brine and the inner surface of the can is 40 Btu/h ft² °F. Determine the time required to freeze the water and compare with the time necessary if the brine circulation rate were increased to reduce the thermal resistance of the surface to one-tenth of the value specified above.

10.27 Estimate the time required to freeze vegetables in thin, tin cylindrical containers 15 cm in diameter. Air at -12°C is blowing at 4 m/s over the cans, which are stacked to form one long cylinder. The physical properties of the vegetables may be taken as those of water and ice, respectively.

10.28 Estimate the time required to freeze a 3 cm thickness of water due to nocturnal radiation with ambient air and initial water temperatures at 4°C. Neglect evaporation effects.

10.29 The temperature of a cooling pond is 7°C on a winter day. If the air temperature suddenly drops to -7°C, calculate the thickness of ice formed after three hours.

REFERENCES

1. W. H. Jens and G. Leppert, "Recent Developments in Boiling Research," parts I and II. *J. Am. Soc. Nav. Eng.*, vol. 66, pp. 437–456, 1955; vol. 67, pp. 137–155, 1955.
2. D. P. Jordan, "Film and Transition Boiling," in *Advances in Heat Transfer*, vol. 5, T. F. Irvine, Jr., and J. P. Hartnett, eds., pp. 55–125, Academic Press, New York, 1968.
3. G. Leppert and C. C. Pitts, "Boiling," in *Advances in Heat Transfer*, vol. 1, T. F. Irvine, Jr., and J. P. Hartnett, eds., pp. 185–265, Academic Press, New York, 1968.
4. W. M. Rohsenow, "Boiling Heat Transfer," in *Developments in Heat Transfer*, W. M. Rohsenow, ed., pp. 169–260, MIT Press, Cambridge, Mass., 1964.
5. E. A. Farber and R. L. Scorah, "Heat Transfer to Water Boiling under Pressure," *Trans. ASME*, vol. 70, pp. 369–384, 1948.
6. S. Nukiyama, "Maximum and Minimum Values of Heat Transmitted from a Metal to Boiling Water under Atmospheric Pressure," *J. Soc. Mech. Eng. Jpn.*, vol. 37, no. 206, pp. 367–394, 1934.

7. K. Engelberg-Forster and R. Greif, "Heat Transfer to a Boiling Liquid—Mechanism and Correlations," *Trans. ASME, Ser. C, J. Heat Transfer*, vol. 81, pp. 43–53, Feb. 1959.

8. R. F. Gaertner, "Photographic Study of Nucleate Pool Boiling on a Horizontal Surface," ASME Paper 63-WA-76.

9. R. Moissis and P. J. Berenson, "On the Hydrodynamic Transitions in Nucleate Boiling," *Trans. ASME, Ser. C, J. Heat Transfer*, vol. 85, pp. 221–229, Aug. 1963.

10. J. W. Westwater, "Boiling Heat Transfer," *Am. Sci.*, vol. 47, no. 3, pp. 427–446, Sept. 1959.

11. P. J. Berenson, "Experiments on Pool-Boiling Heat Transfer," *Int. J. Heat Mass Transfer*, vol. 5, pp. 985–999, 1962.

12. F. C. Gunther, "Photographic Study of Surface Boiling Heat Transfer with Forced Convection," *Trans. ASME*, vol. 73, pp. 115–123, 1951.

13. M. E. Ellion, "A Study of the Mechanism of Boiling Heat Transfer," Memorandum 20–88, Jet Propulsion Laboratory, Pasadena, Calif., March 1954.

14. W. M. Rohsenow, "A Method of Correlating Heat-Transfer Data for Surface Boiling Liquids," *Trans. ASME*, vol. 74, pp. 969–975, 1952.

15. R. I. Vachon, G. H. Nix, and G. E. Tanger, "Evaluation of Constants for the Rohsenow Pool-Boiling Correlation," *Trans. ASME, Ser. C, J. Heat Transfer*, vol. 90, pp. 239–247, 1968.

16. J. N. Addoms, "Heat Transfer at High Rates to Water Boiling outside Cylinders," D.Sc. thesis, Dept. of Chemical Engineering, Massachusetts Institute of Technology, 1948.

17. W. H. McAdams et al., "Heat Transfer from Single Horizontal Wires to Boiling Water," *Chem. Eng. Prog.*, vol. 44, pp. 639–646, 1948.

18. W. H. McAdams, *Heat Transmission*, 3d ed., McGraw-Hill, New York, 1954.

19. J. G. Collier, *Convective Boiling and Condensation*, 2d ed., McGraw-Hill, New York, 1981.

20. *Steam—Its Generation and Use*, Babcock & Wilcox Co., New York, 1955.

21. F. Kreith and M. J. Summerfield, "Heat Transfer to Water at High Flux Densities with and without Surface Boiling," *Trans. ASME*, vol. 71, pp. 805–815, 1949.

22. N. Zuber and M. Tribus, "Further Remarks on the Stability of Boiling Heat Transfer," Rept. 58–5, Dept. of Engineering, Univ. of Calif., Los Angeles, 1958.

23. S. S. Kutateladze, "A Hydrodynamic Theory of Changes in a Boiling Process under Free Convection," *Izv. Akad. Nauk SSSR Otd. Tekh. Nauk*, no. 4, p. 524, 1951.

24. N. Zuber, M. Tribus, and J. W. Westwater, "The Hydrodynamic Crisis in Pool Boiling of Saturated and Subcooled Liquids," in *Proceedings of the International Conference on Developments in Heat Transfer*, pp. 230–236, Am. Soc. of Mech. Eng. (ASME), New York, 1962.

25. J. H. Lienhard and V. K. Dhir, "Extended Hydrodynamic Theory of the Peak and Minimum Pool Boiling Heat Fluxes," NASA Contract. Rept. CR-2270, July 1973.

26. C. M. Usiskin and R. Siegel, "An Experimental Study of Boiling in Reduced and Zero Gravity Fields," *Trans. ASME, Ser. C*, vol. 83, pp. 243–253, 1961.

27. D. A. Huber and J. C. Hoehne, "Pool Boiling of Benzene, Diphenyl, and Benzene-Diphenyl Mixtures under Pressure," *Trans. ASME, Ser. C*, vol. 85, pp. 215–220, 1963.

28. L. Bernath, "A Theory of Local-Boiling Burnout and Its Application to Existing Data," *Chem. Eng. Prog. Symp. Ser.*, vol. 56, no. 30, pp. 95–116, 1960.

29. L. A. Bromley, "Heat Transfer in Stable Film Boiling," *Chem. Eng. Prog.*, vol. 46, pp. 221–227, 1950.

30. J. W. Westwater and B. P. Breen, "Effect of Diameter of Horizontal Tubes on Film Boiling Heat Transfer," *Chem. Eng. Prog.*, vol. 58, pp. 67–72, 1962.

31. P. J. Berenson and R. A. Stone, "A Photographic Study of the Mechanism of Forced-Convection Vaporization," AIChE Reprint No. 21, Symposium on Heat Transfer, San Juan, Puerto Rico, 1963.

32. K. Konmutsos, R. Moissis, and A. Spyridonos, "A Study of Bubble Departure in Forced Convection Boiling," *Trans. ASME, Ser. C., J. Heat Transfer*, vol. 90, pp. 223–230, 1968.

33. W. R. Gambill, "Generalized Prediction of Burnout Heat Flux for Flowing, Sub-cooled, Wetting Liquids," *Chem. Eng. Prog. Symp. Ser.*, vol. 59, pp. 71–87, 1963.

34. J. W. Miles, "The Hydrodynamic Stability of a Thin Film of Liquid in Uniform Shearing Motion," *J. Fluid Mech.*, vol. 8, pp. 592–610, 1961.

35. E. L. Piret and H. S. Isbin, "Natural Circulation Evaporation Two-Phase Heat Transfer," *Chem. Eng. Prog.*, vol. 50, p. 305, 1954.

36. W. H. McAdams, W. E. Kennel, C. S. Minden, R. Carl, P. M. Picarnell, and J. E. Drew, "Heat Transfer at High Rates to Water with Surface Boiling," *Ind. Eng. Chem.*, vol. 41, pp. 1945–1953, 1949.

37. J. C. Chen, "Correlation for Boiling Heat Transfer to Saturated Liquids in Convective Flow," *Ind. Eng. Chem. Proc. Des. Dev.*, vol. 5, p. 332, 1966.

38. J. G. Collier, "Forced Convective Boiling," in *Two Phase Flow and Heat Transfer*, chapter 8, pp. 247–248, Hemisphere, Washington, D.C., 1981.

39. P. Griffith, "Correlation of Nucleate-Boiling Burnout Data," ASME Paper 57-HT-21.

40. P. Griffith, "Two Phase Flow in Pipes," Course Notes, Massachusetts Institute of Technology, Cambridge, 1964.

41. R. W. Lockhart and R. C. Martinelli, "Proposed Correlation of Data for Isothermal Two-Phase Two-Component Flow in Pipes," *Chem. Eng. Prog.*, vol. 45, pp. 39–48, 1949.

42. L. S. Tong, *Boiling Heat Transfer and Two-Phase Flow*, Wiley, New York, 1965.

43. F. Kreith and M. Margolis, "Heat Transfer and Friction in Turbulent Vortex Flow," *Appl. Sci. Res.*, Sec. A, vol. 8, pp. 457–473, 1959.

44. W. R. Gambill, R. D. Bundy, and R. W. Wansbrough, "Heat Transfer, Burnout, and Pressure Drop for Water in Swirl Flow through Tubes with Internal Twisted Tapes," *Chem. Eng. Prog. Symp. Ser.*, no. 32, vol. 57, pp. 127–137, 1961.

45. L. S. Tong and J. D. Young, "A Phenomenological Transition and Film Boiling Heat Transfer Correlation," *Proc. 5th Int. Heat Transfer Conf.*, Tokyo, Sept. 1974.

46. D. C. Groeneveld, "Post-Dryout Heat Transfer at Reactor Operating Correlations," Paper AECL-4513, National Topical Meeting on Water Reactor Safety, ANS, Salt Lake City, Utah, 1973.

47. W. M. Rohsenow, J. P. Hartnett, and E. Ganic, eds. *Handbook of Heat Transfer*, 2d ed., McGraw-Hill, New York, 1985.

48. W. Nusselt, "Die Oberflächenkondensation des Wasserdampfes," *Z. Ver. Dtsch. Ing.*, vol. 60, pp. 541 and 569, 1916.

49. W. M. Rohsenow, "Heat Transfer and Temperature Distribution in Laminar-Film Condensation," *Trans. ASME*, vol. 78, pp. 1645–1648, 1956.

50. M. M. Chen, "An Analytical Study of Laminar Film Condensation," part 1, "Flat Plates," and part 2, "Single and Multiple Horizontal Tubes," *Trans. ASME, Ser. C*, vol. 83, pp. 48–60, 1961.

51. W. M. Rohsenow, J. M. Weber, and A. T. Ling, "Effect of Vapor Velocity on Laminar and Turbulent Film Condensation," *Trans. ASME*, vol. 78, pp. 1637–1644, 1956.

52. A. P. Colburn, "The Calculation of Condensation Where a Portion of the Condensate Layer is in Turbulent Flow," *Trans. AIChE*, vol. 30, p. 187, 1933.

53. C. G. Kirkbridge, "Heat Transfer by Condensing Vapors on Vertical Tubes," *Trans. AIChE*, vol. 30, p. 170, 1933.

54. E. F. Carpenter and A. P. Colburn, "The Effect of Vapor Velocity on Condensation Inside Tubes," in *Proceedings, General Discussion on Heat Transfer*, pp. 20–26, Inst. Mech. Eng. ASME, 1951.

55. M. Soliman, J. R. Schuster, and P. J. Berenson, "A General Heat Transfer Correlation for Annular Flow Condensation," *J. Heat Transfer*, vol. 90, pp. 267–276, 1968.

56. I. G. Shekriladze and V. I. Gomelauri, "Theoretical Study of Laminar Film Condensation of Flowing Vapor," *Int. J. Heat Mass Transfer*, vol. 9, pp. 581–591, 1966.

57. T. B. Drew, W. M. Nagle, and W. Q. Smith, "The Conditions for Dropwise Condensation of Steam," *Trans. AIChE*, vol. 31, pp. 605–621, 1935.

58. R. S. Silver, "An Approach to a General Theory of Surface Condensers," *Proc. Inst. Mech. Eng.*, vol. 178, part 1, no. 14, pp. 339–376, 1964.

59. J. W. Rose, "On the Mechanism of Dropwise Condensation," *Int. J. Heat Mass Transfer*, vol. 10, pp. 755–762, 1967.

60. P. Griffith and M. S. Lee, "The Effect of Surface Thermal Properties and Finish on Dropwise Condensation," *Int. J. Heat Mass Transfer*, vol. 10, pp. 697–707, 1967.

61. D. L. Katz, E. H. Young, and G. Balekjian, "Condensing Vapors on Finned Tubes," *Petroleum Refiner*, pp. 175–178, Nov. 1954.

62. F. Kreith and F. E. Romie, "A Study of the Thermal Diffusion Equation with Boundary Conditions Corresponding to Freezing or Melting of Materials at the Fusion Temperature," *Proc. Phys. Soc.*, vol. 68, pp. 277–291, 1955.

63. D. L. Cochran, "Solidification Application and Extension of Theory," Tech. Rept. 24, Navy Contract N6-onr-251, Stanford Univ., 1955.

64. W. D. Murray and F. Landis, "Numerical and Machine Solutions of Transient Heat Conduction Problems Involving Melting or Freezing," *Trans. ASME*, vol. 81, pp. 106–112, 1959.

65. A. Lazaridis, "A Numerical Solution of the Multi-Dimensional Solidification (or Melting) Problem," *Int. J. Heat Mass Transfer*, vol. 13, pp. 1459–1477, 1970.

66. H. Budhia and F. Kreith, "Melting or Freezing in a Wedge," *Int. J. Heat Mass Transfer*, vol. 15, 1972.

67. F. E. Moore and Y. Bayazitoglu, "Melting within a Spherical Enclosure," *J. Heat Transfer*, vol. 104, pp. 19–23, 1982.

68. M. T. Cichelli and C. F. Bonilla, "Heat Transfer to Liquids Boiling under Pressure," *Trans. AIChE*, vol. 41, pp. 755–787, 1945.

69. D. S. Cryder and A. C. Finalbargo, "Heat Transmission from Metal Surfaces to Boiling Liquids: Effect of Temperature of the Liquid on Film Coefficient," *Trans. AIChE*, vol. 33, pp. 346–362, 1937.

70. A. L. London and R. A. Seban, "Rate of Ice Formation," *Trans. ASME*, vol. 65, pp. 771–778, 1943.

71. A. P. Katz, H. J. Macintire, and R. E. Gould, "Heat Transfer in Ammonia Condensers," Bull. 209, Univ. Ill. Eng. Expt. Stn., 1930.

72. *The Refrigerating Data Book*, vol. II, pp. 9 and 56, American Society of Refrigeration Engineers (ASRE), 1940.

Chapter 11

Special Topics

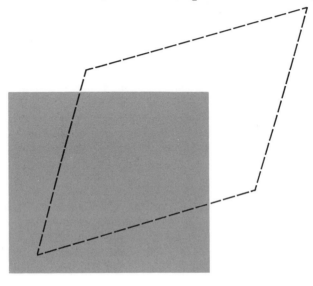

In this chapter we will consider a few special topics in heat transfer that cut across conventional lines and/or are still at a stage of research and development. The presentation here is not intended to be exhaustive, but rather to introduce some new concepts that are likely to grow in importance in the future. These topics are: 11.1 Second Law Analysis of Heat Transfer Processes, 11.2 Heat Transfer in High-Speed Flow, 11.3 Non-Newtonian Fluids, 11.4 Heat Pipes, 11.5 Solar Collectors, 11.6 Heat Transfer Enhancement.

11.1 SECOND LAW ANALYSIS OF HEAT TRANSFER PROCESSES

As a result of the growing awareness that fossil energy supplies are limited, increasing emphasis in thermal design is being placed on the efficient use of available energy. Consequently, the thermodynamic irreversibility of heat transfer processes has become a subject of current interest. In this section we will introduce the principles of a method of analysis that can identify the irreversibilities associated with various components in a system and thereby indicate how to improve a design by avoiding unnecessary irreversibilities that lead to a loss of the available energy or power. The method relies not only on the first but also on the second law of thermodynamics and is often called *second law analysis*.

In preceding chapters we have extensively used the first law of thermodynamics, the law of energy conservation. The second law of thermodynamics has so far been used only to show that the direction of heat flow must be from a higher to a lower temperature. Since thermal energy cannot spontaneously flow from a lower to a higher temperature, any heat transfer process is irreversible. Another way of expressing this phenomenon is to note that the energy that has been transferred to a lower temperature has become less available. The degree of irreversibility, or the loss of availability of the energy as a result of a process, is related to the rate of entropy generation, and in the rest of this section we will show how to minimize the generation of entropy in heat transfer and fluid flow systems.

11.1.1 Basic Equations for an Open System

Figure 11.1 illustrates schematically the control volume and the control surface of an open system. Observe that the control surface may have one or more openings through which mass may be transferred. The interior of the control volume contains certain amounts of mass M, energy E, and entropy S. For a complete analysis, conservation equations for each of the three system

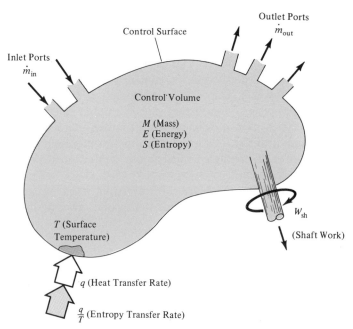

Figure 11.1 Nomenclature for open-system analysis.

properties are required. First, we write the *conservation of mass* in the form

$$\sum_{\text{in}} \dot{m} - \sum_{\text{out}} \dot{m} = \frac{\partial M}{\partial t} \tag{11.1}$$

This equation states that the difference between the net rate of mass flows into the control volume and the net rate of mass flows out equals the rate of mass accumulation within the control surface.

As shown in Chapter 1, the *first law of thermodynamics* can be written in the form

$$\sum_{\text{in}} \dot{m}(h + \tfrac{1}{2}U^2 + gz) - \sum_{\text{out}} \dot{m}(h + \tfrac{1}{2}U^2 + gz) + q - W_{\text{sh}} = \frac{\partial E}{\partial t} \tag{11.2}$$

The terms on the left-hand side represent energy transfer interaction of heat and shaft work, and energy transfers associated with the mass crossing the control surface. The specific enthalpy h, the fluid velocity U, and the height potential z should be evaluated at the boundary and, when necessary, averaged over a flow area cross section. The right-hand side is the time rate of change of energy within the control surface.

The final statement, the *second law of thermodynamics*, may be expressed in the form

$$\sum_{\text{in}} \dot{m}s - \sum_{\text{out}} \dot{m}s + \frac{q}{T} \leq \frac{\partial S}{\partial t} \tag{11.3}$$

$$\underset{\substack{\text{entropy} \\ \text{transfer}}}{} \qquad\qquad \underset{\substack{\text{entropy} \\ \text{change}}}{}$$

The first two terms in this equation represent the entropy transfer associated with the mass crossing the control surface. The term q/T is the entropy transfer associated with any heat transfer across the control surface; q is the heat transfer rate and T is the absolute temperature at the control surface across which the heat transfer takes place. Observe that *entropy transfer can only take place with the transfer of energy as heat*, whereas work transfer interactions are not accompanied by entropy transfer. The specific entropy s is representative of the thermodynamic state of each stream averaged across the flow area through which it enters or leaves the system.

For the purpose of evaluating the degree of irreversibility that occurs in engineering systems, it is convenient to rearrange the second law equality as stated in Eq. (11.3) in the form

$$\dot{S}_{\text{gen}} \equiv \frac{\partial S}{\partial t} - \frac{q}{T} + \sum_{\text{out}} \dot{m}s - \sum_{\text{in}} \dot{m}s \geq 0 \tag{11.4}$$

The degree of irreversibility is proportional to the entropy generation and the above equation defines the rate of entropy generation in the system in the units watts per kelvin (W/K). Observe that the rate of entropy generation $\dot{S}_{\text{gen}}$ must always be positive except in the case of a reversible interaction, where it approaches zero. The term $\dot{S}_{\text{gen}}$ can be used as a measure of a system's departure from reversibility. Hence, when two systems, A and B, operate under similar conditions, if $\dot{S}_{gen, A}$ is larger than $\dot{S}_{gen, B}$, system A operates more irreversibly than system B.

Equations (11.1)–(11.3) can be simplified when the system operates in steady state. Steady operation exists when the system state properties such as density, specific energy, and specific entropy do not change with time. In practice, local properties may vary spatially from one point to another within the control volume, but as long as the averaged system properties do not vary with time the operation is steady and the right-hand sides of Eqs. (11.1)–(11.3) equal 0.

A special kind of heat interaction occurs when the system exchanges heat with the environment, assumed to be at an absolute temperature T_o. In such a case the temperature of the boundary penetrated by heat transfer must be set equal to T_o. This means that the control surface must be drawn through regions of space that are not affected by the heat transfer interaction q and have no temperature gradients. Since any heat transfer interaction must be driven by a finite temperature difference, this temperature difference must be located within the control surface.

11.1.2 Exergy Loss and Entropy Generation in Heat Transfer

To quantify the dissipative phenomena it is often convenient to use the concept of *exergy*. A detailed exposition of the meaning of this concept is presented in (1), (2), and (3). In a simplified version suitable for analyzing the dissipation associated with a given energy source (4), the destruction of exergy can be explained by referring to Fig. 11.2. Suppose heat is released by a fuel at an absolute temperature T_h and transferred to a user (e.g., the working fluid in a power cycle), which receives it at a temperature T. If the heat Q were used to drive a Carnot engine operating between T_h and the ambient temperature T_o, the work output would be

$$W_{max, h} = \frac{Q(T_h - T_o)}{T_h} \tag{11.5}$$

According to the second law, since a Carnot engine is reversible and $T_h - T_o$ is the maximum temperature potential available, $W_{max,h}$ *is the maximum available*

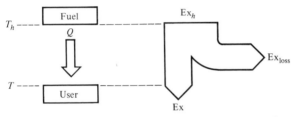

Figure 11.2 Destruction of exergy due to heat transfer.

work producible from Q at T_h. Similarly, the heat Q received by the user at temperature T has the maximum work-producing potential

$$W_{max} = \frac{Q(T - T_o)}{T} \tag{11.6}$$

In the simplified approach, the maximum work potential defined by Eq. (11.5) may be equated to *exergy*,* Ex. Since the user temperature T cannot exceed the fuel temperature T_h, Eqs. (11.5) and (11.6) show that the decrease in exergy, as this heat Q is transferred across the fuel-to-user temperature gap, is given by

$$Ex_{loss} = T_o\left(\frac{Q}{T} - \frac{Q}{T_h}\right) \geq 0 \tag{11.7}$$

The loss of energy, Ex_{loss}, as Q is transferred across the temperature gap is illustrated graphically in Fig. 11.2.

EXAMPLE 11.1 _____

Suppose that in a small community there is a need for an electric power plant taking in 100 MW of thermal energy at 850 K and rejecting 60 MW at 373 K, as well as for a district heating system that requires 60 MW of heat at 333 K. There are two options for supplying this energy: solar and coal. The sun may be taken as an energy source at 5800 K, while the flame temperature of a coal burner is 2300 K. One scheme envisions the use of a coal for the electric power plant and solar flat-plate collectors for the district heating system. The other scheme envisions the use of a solar power plant with the rejected heat from the plant used for district heating. Compare the total exergy loss for these two schemes. Assume that the environmental temperature T_o is 283 K.

Solution. Let Q_D be the heat for the district heating and Q_P be the heat for the power plant. Then for the first scheme, the exergy loss for the power plant is

$$Ex_{loss} = T_o\left(\frac{Q_P}{T} - \frac{Q_P}{T_h}\right) = T_o\left(\frac{100}{850} - \frac{100}{2300}\right)$$
$$= 283(0.118 - 0.043) = 21.2 \text{ MW}$$

and the exergy loss for the solar collector system is

$$Ex_{loss} = T_o\left(\frac{Q_D}{T} - \frac{Q_D}{T_h}\right) = 283\left(\frac{60}{333} - \frac{60}{5800}\right) = 48.1 \text{ MW}$$

The total exergy loss from both is 69.3 MW.

_____

* As shown in (1), (2), or (3), exergy is defined in general as $Ex = (h - h_o) - T_o(s - s_o)$, where h_o and s_o are the enthalpy and specific entropy of the ambient temperature and pressure, the dead state. When h_o and s_o are taken as zero, exergy and availability are numerically equal.

The exergy loss for the cascaded solar system is

$$Ex_{loss} = T_o\left[\left(\frac{Q_P}{850} - \frac{Q_P}{5800}\right) + \left(\frac{Q_D}{333} - \frac{Q_D}{850}\right)\right]$$
$$= 283(0.118 - 0.017 + 0.180 - 0.071) = 59.4 \text{ MW}$$

We see that using a cascaded solar system, often called cogeneration, reduces the exergy loss almost 20 percent compared to the first scheme, in which solar energy is used only for heating and coal is used to operate the electric power plant.

Any heat transfer process requires a temperature difference, and when heat is transferred from a higher to a lower temperature, entropy is generated. We will now determine the amount of entropy generated and the associated loss in available work when heat is transferred. The schematic diagram in Fig. 11.3 shows a hot body at temperature T_h transferring an amount of heat Q to a colder body at T_c. Since the two systems do not communicate directly, there exists a third system between. We will refer to this intermediate system as the temperature gap ΔT; the heat Q enters and leaves this intermediate system unchanged. If we apply the second law to the temperature gap, the entropy generation is

$$\dot{S}_{gen} = \frac{Q}{T_c} - \frac{Q}{T_h} = Q\left(\frac{T_h - T_c}{T_c T_h}\right) \tag{11.8}$$

Note that the entropy generation is always positive.

The available work lost as a result of this irreversibility is equal to the work that a reversible Carnot engine could deliver from the heat Q, if it operated between temperatures T_h and T_c:

$$W_{lost} = Q\left(\frac{T_h - T_c}{T_h}\right) = T_c\dot{S}_{gen} \tag{11.9}$$

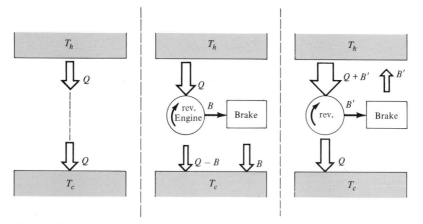

Figure 11.3 Entropy generation in heat transfer across a finite temperature difference.

As shown in Fig. 11.3, the transfer of heat from T_h to T_c is thermodynamically equivalent to a reversible engine operating between the two temperatures and dissipating its entire work output into a brake. The frictional heat thus generated can be rejected to either the hot or the cold temperature.

11.1.3 Entropy Generation for Convection Heat Transfer in Flow-Through Ducts

In this section we will treat the problem of entropy generation for flow in a duct with heat transfer through the wall. For this purpose consider a flow passage of cross section A and wetted perimeter P as shown in Fig. 11.4. The flow is steady at a mass flow rate $\dot{m}$ and bulk properties T, p, h, ρ, and s. Heat is transferred through the wall at a rate q' (W/m) per unit length and the frictional pressure gradient is $-dp/dx > 0$. If the wall-to-bulk temperature difference is ΔT, the rate of entropy generation in the differential element $A\,dx$ is

$$d\dot{S}_{gen} = \dot{m}\,ds - \frac{q'\,dx}{T + \Delta T} \tag{11.10}$$

If the first law is applied to the same system, we obtain

$$\dot{m}\,dh = q'\,dx \tag{11.11}$$

For a pure substance the enthalpy change can be cast into the form (3)

$$dh = T\,ds + v\,dp = T\,ds + \frac{1}{\rho}\,dp \tag{11.12}$$

Substituting for dh and ds from Eqs. (11.10) and (11.11) in the above yields

$$\frac{q'\,dx}{\dot{m}} = T\left[\frac{d\dot{S}_{gen}}{\dot{m}} + \frac{q'\,dx}{\dot{m}(T + \Delta T)}\right] + \frac{1}{\rho}\,dp \tag{11.13}$$

From Eq. (11.10) we can obtain the rate of entropy generation per unit length:

$$\frac{d\dot{S}_{gen}}{dx} = \frac{q'\,\Delta T}{T^2[(T + \Delta T)/T]} + \frac{\dot{m}}{\rho T}\left(-\frac{dp}{dx}\right) \tag{11.14}$$

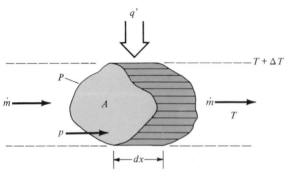

Figure 11.4 Nomenclature for entropy generation in forced-convection flow through a duct.

When $(T + \Delta T)/T$ is approximately unity, that is, $\Delta T \ll T$, Eq. (11.14) simplifies to

$$\frac{d\dot{S}_{gen}}{dx} = \frac{q' \Delta T}{T^2} + \frac{\dot{m}}{\rho T}\left(-\frac{dp}{dx}\right) \tag{11.15}$$

In Chapter 6 we derived expressions for convection heat transfer and pressure drop. The Stanton number St was defined as

$$St = \frac{h_c}{c_p \rho \bar{u}} = \frac{q'/P \Delta T}{c_p G} \tag{11.16}$$

where $h_c = q'/P \Delta T$, the heat transfer coefficient, and $G = \rho \bar{u} = \dot{m}/A$, the mass velocity. The friction factor f was defined in Chapter 6 as

$$f = \frac{\rho D_H}{2G^2}\left(-\frac{dp}{dx}\right) \tag{11.17}$$

where D_H is the hydraulic diameter $(=4A/P)$. The friction factor depends on the Reynolds number and on the geometry of the duct and its surface characteristics.

Using the above relations, we can obtain the dependence of entropy generations on St and f if the heat transfer rate per unit length of duct, q', and the mass flow rate $\dot{m}$ are specified:

$$\dot{S}'_{gen} = \frac{q'^2}{4T^2\dot{m}c_p}\frac{D_H}{St} + \frac{2\dot{m}^3}{\rho T}\frac{f}{D_H A^2} \tag{11.18}$$

Note that for a specified q' increasing the heat transfer coefficient (i.e., the Stanton number) reduces the entropy generation due to heat transfer, whereas increasing the friction coefficient increases the entropy generation due to viscous effects.

For flow through a round duct of internal diameter D, Eq. (11.18) becomes

$$\dot{S}'_{gen} = \frac{q'^2}{\pi k T^2 Nu_D} + \frac{32\dot{m}^2}{\pi^2 \rho^2 T}\frac{f}{D^5} \tag{11.19}$$

Using empirical relations for Nu_D and f, we can determine the tube diameter (or Re_D) that minimizes entropy generation. For turbulent flow the following empirical relations apply:

$$\overline{Nu}_D = 0.023 Re_D^{0.8} Pr^{0.4}$$

$$f = 0.046 Re_D^{-0.2}$$

Bejan (5) has shown that combining these expressions with Eq. (11.19) and setting $\partial \dot{S}_{gen}/\partial Re_D = 0$ gives the Reynolds number for minimum irreversibility, $Re_{D, opt}$, in the form:

$$Re_{D, opt} = \frac{4\dot{m}}{\pi \mu D_{opt}} = 2.023 Pr^{-0.071}\left[\frac{\rho \dot{m} q'}{\mu^{5/2}(kT)^{1/2}}\right]^{0.358} \tag{11.20}$$

The term in square brackets is sometimes called the *heat exchange duty parameter* B_0. It is determined once the rate of heat transfer, the mass flow rate, and the

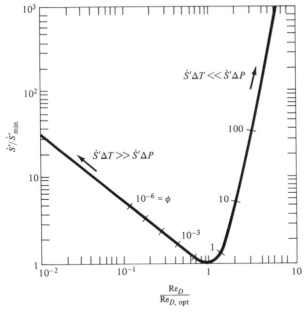

Figure 11.5 Relative entropy generation rate in a smooth tube. [From A. Bejan (26).]

working fluid are specified. Note that the optimum tube size decreases as B_0 decreases. The minimum entropy generation rate may be obtained by substituting Eq. (11.20) into Eq. (11.19). A convenient summary of the above analysis is shown in Fig. 11.5, where the ratio

$$\frac{\dot{S}'(\mathrm{Re}_D)}{\dot{S}'_{\min}(\mathrm{Re}_{D,\mathrm{opt}})}$$

is plotted against $\mathrm{Re}_D/\mathrm{Re}_{D,\mathrm{opt}}$. Note that the entropy generation increases sharply on either side of the optimum Reynolds number.

11.1.4 Entropy Generation for Forced-Convection Heat Transfer in Flow over Exterior Surfaces

As shown in Chapter 7, there are numerous important convection heat transfer problems in which a fluid flows over the surface of a solid body at a temperature different from that of the fluid. Typical examples are flow over a cylinder, an airplane wing, or a flat plate suspended in the stream. Also, fins protruding from a surface into the stream constitute such systems. We shall consider the entropy generation for such types of systems, but restrict our analysis to cases in which the body is at a uniform temperature T_s.

Consider a cylindrical body of arbitrary shape and surface area A suspended in a fluid stream of velocity U_∞ and temperature T_∞ as shown in Fig. 11.6. The total fluid drag force F_D consists of pressure and frictional forces. The

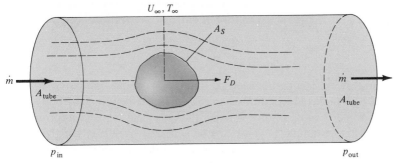

Figure 11.6 Nomenclature for entropy generation in flow over exterior surfaces.

stream tube shown by the solid line in Fig. 11.6 is sufficiently far from the solid body that disturbances are dampened out. If the heat flux between the surface of the body and the stream is q'' (W/m^2) and the flow is steady, the following three conservation relations apply to the stream tube:

$$\dot{m}_{\text{in}} = \dot{m}_{\text{out}} = \dot{m} \tag{11.21}$$

$$\dot{m}h_{\text{in}} + \iint_{A_s} q'' \, d\sigma - \dot{m}h_{\text{out}} = 0 \tag{11.22}$$

$$\dot{S}_{\text{gen}} = \dot{m}s_{\text{out}} - \dot{m}s_{\text{in}} - \iint_{A_s} \frac{q'' \, d\sigma}{T_s} \tag{11.23}$$

where $d\sigma$ stands for a differential surface area of the solid body whose surface area is A_s.

Assuming that the flow is incompressible and using the relation $dh = T \, ds + (1/\rho) \, dp$, we have

$$h_{\text{out}} - h_{\text{in}} = T_\infty(S_{\text{out}} - S_{\text{in}}) + \frac{1}{\rho_\infty}(p_{\text{out}} - p_{\text{in}}) \tag{11.24}$$

Combining Eqs. (11.21)–(11.24) yields the entropy generation rate

$$\dot{S}_{\text{gen}} = \iint_{A_s} q'' \left(\frac{1}{T_\infty} - \frac{1}{T_s}\right) d\sigma - \frac{\dot{m}}{\rho_\infty T_\infty}(p_{\text{out}} - p_{\text{in}}) \tag{11.25}$$

From Fig. 11.6 we see that

$$\dot{m} = A_{\text{tube}}\rho_\infty U_\infty$$

and

$$F_D = A_{\text{tube}}(p_{\text{in}} - p_{\text{out}})$$

Furthermore,

$$\iint_{A_s} q'' \, d\sigma = U A(T_s - T_\infty)$$

where U is the average overall heat transfer coefficient. Substituting the above three equations into Eq. (11.25), the rate of entropy generation becomes

$$\dot{S}_{gen} = \left(\frac{T_s - T_\infty}{T_\infty}\right)^2 UA + \frac{F_D U_\infty}{T_\infty} \qquad (11.26)$$

if the temperature difference between T_∞ and T_s is small compared to T_∞. Inspection of Eq. (11.26) shows that when the difference between the body temperature and ambient temperature is fixed, the only heat transfer approach to minimizing $\dot{S}_{gen}$ is to reduce $\bar{U}$. This can be achieved by insulating the body, decreasing the overall heat transfer coefficient, and thereby reducing the rate of heat loss.

Another important situation is that of constant heat flux over A_s. In that case (3)

$$\dot{S}_{gen} = \left(\frac{q''}{T}\right)^2 \iint\limits_{A_s} \frac{d\sigma}{h_{c,x}} + \frac{F_D U_\infty}{T_\infty} \qquad (11.27)$$

To reduce entropy generation when q'' is fixed, we must increase the heat transfer coefficient $h_{c,x}$. This can be achieved by enhancing or augmenting the heat transfer—for example, by increasing surface roughness or adding fins. These measures can, however, increase the pressure loss, and optimization must take the interplay of these two factors into account.

EXAMPLE 11.2 _____

Using the expressions for Nu_x and $C_{f,x}$ in Table 4.5, derive a dimensionless expression for the entropy generation in laminar flow of air over a flat plate fin of width W when the heat flux is uniform, and determine the fin length for minimum irreversibility.

Solution. The system is shown schematically in Fig. 11.7. From Table 4.5 we obtain for flow over a plate

$$Nu_x = \frac{h_{c,x} x}{k} = 0.332 \, Pr^{1/3} Re_x^{1/2}$$

$$C_{f,x} = \frac{\tau_x}{\frac{1}{2}\rho_\infty U_\infty^2} = 0.664 \, Re_x^{-1/2}$$

According to Eq. (11.27), the entropy generation is

$$\dot{S}_{gen} = \frac{q''^2 W}{T_\infty^2} \int_0^L \frac{dx}{h_{c,x}} + \frac{U_\infty W}{T_\infty} \frac{1}{2} \rho_\infty U_\infty^2 \int_0^L C_{f,x} dx$$

Substituting the relations for $h_{c,x}$ and $C_{f,x}$ multiplying both sides by $kT_\infty^2/WL^2 q''^2$ yields

$$\dot{S}_{gen} \frac{kT_\infty^2}{q''^2 L^2 W} = 2.0 \, Pr^{-1/3} Re_L^{-1/2} + 0.664 \frac{U_\infty^2 \mu k T_\infty}{(q''L)^2} Re_L^{1/2}$$

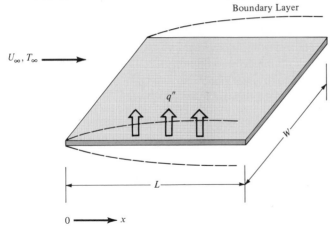

Figure 11.7 Schematic diagram for Example 11.2.

It can be shown (see Problem 11.9) that for minimum irreversibility the optimum Reynolds number is

$$\mathrm{Re}_{L,\,\mathrm{opt}} = 2.193 \; \mathrm{Pr}^{-1/3} \; \frac{(q''L)^2}{U_\infty \mu k T_\infty} \tag{11.28}$$

11.1.5 Application to Heat Transfer Augmentation and Fins

In previous chapters we have seen that the thermal performance of devices such as heat exchangers can be enhanced by various augmentation techniques, such as roughening the heat transfer surface, installing fins, or swirling the flow. Additional information on enhancement devices is presented in Section 11.6. The objective of these techniques is to improve the thermal performance without introducing a significant increase in the fluid pumping power necessary to exchange the heat. Unfortunately, these simultaneous objectives are in fundamental conflict because augmentation techniques that improve the heat transfer cause, in most cases, an increase in the pumping power, which is equivalent to mechanical power dissipation. In view of this conflict it is important to know which augmentation techniques will lead to a net improvement in thermodynamic performance. In making these calculations it is important to have a common conceptual basis because in the published literature there exists a great diversity in the techniques used to evaluate augmentation devices. We shall postulate that the objective of any augmentation technique should be the reduction of exergy destruction in order to improve the thermodynamic efficiency of the heat transfer equipment.

The entropy generation in a heat exchanger passage of length x, heat transfer per unit length q', and mass flow rate $\dot{m}$ is given by Eq. (11.15), which is reproduced for convenience below:

$$\frac{d\dot{S}_{\mathrm{gen}}}{dx} = \frac{q'\,\Delta T}{T^2} + \frac{\dot{m}}{\rho T}\left(-\frac{dP}{dx}\right) \tag{11.15}$$

The first term on the right-hand side in Eq. (11.15) represents the irreversibility as a result of heat transfer across the wall due to fluid temperature difference, and the second term is the irreversibility caused by fluid friction. To simplify the analysis, we follow Bejan (3) and define ϕ, the *irreversibility distribution ratio*:

$$\phi = \frac{S'_{\Delta P}}{S'_{\Delta T}} \tag{11.29}$$

where $\dot{S}'_{\Delta P} = -\dfrac{\dot{m}}{\rho T}\dfrac{dp}{dx}$

$\dot{S}'_{\Delta T} = \dfrac{q'\,\Delta T}{T^2}$

Equation (11.15) can then be expressed in terms of the irreversibility distribution ratio:

$$\dot{S}''_{gen} = S'_{\Delta T}(1 + \phi) \tag{11.30}$$

To evaluate the impact of an augmentation technique for a given heat exchanger passage we must calculate the entropy generation in the augmented passage, denoted by subscript a, and then compare it with the energy generation in the unaugmented passage, denoted by subscript u. This can be done conveniently by defining the *augmentation entropy generation number* $N_{s,a}$:

$$N_{s,a} = \frac{\dot{S}'_{gen,\,a}}{\dot{S}'_{gen,\,u}} \tag{11.31}$$

From a thermodynamic perspective, augmentation techniques that yield values of $N_{s,a}$ less than unity are advantageous because they not only enhance heat transfer but also reduce the irreversibility of the heat process. When the rate of heat flow per unit length and the mass flow are the same before and after augmentation and when the heat exchanger passage and its function are given, the entropy generation number can be recast in the form

$$N_{s,a} = \frac{N_T + \phi_u N_P}{1 + \phi_u} \tag{11.32}$$

where $N_T = \dfrac{St_u}{St_a}\dfrac{D_a}{D_u}$

$N_P = \dfrac{f_a}{f_u}\dfrac{D_u}{D_a}\left(\dfrac{A_u}{A_a}\right)^2$

$Re_a = \dfrac{D_a A_u}{D_u A_a} Re_u$

Inspection of Eq. (11.32) indicates that when the unaugmented passage is dominated by heat transfer reversibility, that is, when ϕ_u approaches 0, the entropy generation number $N_{s,a}$ approaches N_T, which is proportional to the ratio of the unaugmented to the augmented overall heat transfer coefficient.

In many designs of heat transfer passages the value of the unaugmented irreversibility distribution ratio is less than unity. This does not, however, imply necessarily that $N_{s,a}$ is proportional to N_T. It is therefore important to first calculate ϕ_u and, in conjunction with Eq. (11.29), to determine the impact of the augmentation technique on entropy generation. Bejan (3) has shown that the irreversibility distribution ratio for the unaugmented passage can be put in the simpler form

$$\phi_u = \left(\frac{T}{\Delta T}\right)_u^2 \left(\frac{U_\infty}{c_p T}\right)_u \frac{f_u/2}{\mathrm{St}_u} \tag{11.33}$$

As shown in Chapter 6, the ratio of the friction factor divided by 2 to the Stanton number is of the order of unity; thus, Eq. (11.33) permits a rapid estimate of the unaugmented irreversibility distribution ratio without the necessity of first calculating the Reynolds number.

To illustrate the above analysis, consider the influence of roughening the heat transfer surface on entropy generation in a duct. Since the wall roughness has a negligible influence on the flow cross section and the hydraulic diameter, the Reynolds number is the same in the rough and the smooth tube and the entropy generation number in Eq. (11.32) simplifies to

$$N_{s,a} = \left(\frac{\mathrm{St}_u}{\mathrm{St}_a}\right) + \frac{\phi_u}{1 + \phi_u}\left(\frac{f_a}{f_u} - \frac{\mathrm{St}_u}{\mathrm{St}_a}\right) \tag{11.34}$$

Bejan and Pfister (6) analyzed experimental data in the literature for air (Pr = 0.72) flowing through smooth and sand grain-roughened tubes, calculated $N_{s,a}$, and plotted the results against Re_u for various values of e/D (the ratio of roughness height to diameter). The results are shown in Fig. 11.8. Inspection of the

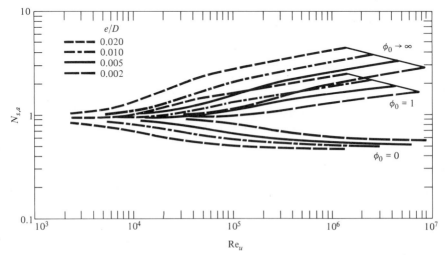

Figure 11.8 Effect of sand grain roughness on entropy generation. [From A. Bejan and P. A. Pfister (6).]

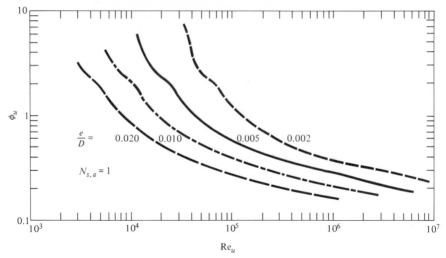

Figure 11.9   Critical irreversibility distribution ratio ϕ_u for sand grain roughness. [From A. Bejan and P. A. Pfister (6).]

resulting curves indicates that the ability of sand-roughening to reduce irreversibility depends on the values of ϕ_u and Re_u. By cross-plotting ϕ_u against Re_u for an entropy generation number of unity, as shown in Fig. 11.9, they obtained a contour that demonstrates the critical irreversibility distribution ratio ϕ_u for various roughnesses. Each of these marginal curves is for one roughness, and a reduction of entropy generation with increased roughness occurs whenever the operating conditions fall below the curve. For example, at a Reynolds number of 100,000 ($\mathrm{Re} = 10^5$) and $e/D = 0.002$, roughening will decrease $\dot{S}_{gen}$ only if ϕ_u is less than unity.

Similar analysis have been made for surfaces roughened with repeated ribs as well as for augmentation with various types of swirl promoters. The results are summarized in (3).

As shown in Chapter 2, a common method for increasing the rate of heat transfer per unit area is the use of fins. While addition of fins to a surface generally enhances heat transfer, it also increases drag and thereby generates irreversibility. Thus, there exists in general an optimum fin size or geometry for which the combined irreversibility due to thermal contact and fluid friction gives a minimum entropy generation rate.

According to Eqs. (11.22)–(11.24), the entropy generation rate for flow over a fin can be written as

$$(\dot{S}_{gen})_{ex} = \iint_A q'' \left(\frac{1}{T_\infty} - \frac{1}{T_s} \right) d\sigma + \frac{F_D U_\infty}{T_\infty} \tag{11.35}$$

Since the fin is not isothermal, it also generates entropy internally. Using Fig. 11.10 as a guide, the rate of internal entropy generation can be expressed as

$$(\dot{S}_{gen})_{int} = \iint_A \frac{q''}{T_s} d\sigma - \frac{q_B}{T_B} \tag{11.36}$$

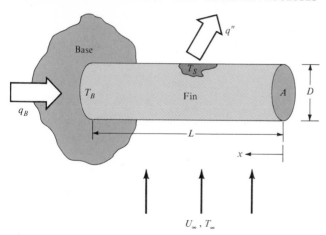

Figure 11.10   Pin fin nomenclature.

where q_B and T_B are the rate of heat transfer and the absolute temperature at the base of the fin.

Let $\theta_B = T_B - T_\infty$ and assume that $\theta_B \ll T_\infty$. Adding Eqs. (11.35) and (11.36) then yields

$$\dot{S}_{gen} = \frac{q_B \theta_B}{T_\infty^2} + \frac{F_D U_\infty}{T_\infty} \tag{11.37}$$

The optimum thermodynamic size of a fin can then be obtained by minimizing Eq. (11.37) subject to constraints such as flow velocity, fin shape and material, and base heat flux. For a pin fin, as shown in Fig. 11.10, we obtained in Chapter 2 the relation

$$\theta_B = \frac{4\,q_B}{\pi k D^2 m \, \tanh(mL)} \tag{11.38}$$

where $m = (4h/kD)^{1/2}$.

By combining Eqs. (11.37) and (11.38), it can be shown that the dimensionless entropy generation number N_s is given by

$$N_s = \frac{\dot{S}_{gen}}{\rho_\infty q_B^2 U_\infty / k_f \mu_\infty T_\infty^2}$$

$$= \frac{2(k_f/k_\infty)^{1/2}}{\pi \mathrm{Nu}^{1/2} \mathrm{Re}_D \tanh[2\mathrm{Nu}^{1/2}(k_f/k_\infty)^{1/2}\mathrm{Re}_L/\mathrm{Re}_D]} + \frac{BC_D \mathrm{Re}_L \mathrm{Re}_D}{2} \tag{11.39}$$

where $\mathrm{Re}_D = \rho_\infty U_\infty D/\mu_\infty$, $\mathrm{Re}_L = \rho_\infty U_\infty L/\mu_\infty$, $B = \mu_\infty^3 k_\infty T_\infty/(\rho_\infty q_B)^2$, and k_f/k_∞ is the ratio of fin material to fluid conductivity.

Empirical equations for the Nusselt number Nu_D and the drag coefficient for a cylinder in cross-flow (see Chapter 7) are:

$$\mathrm{Nu}_D = \frac{h_c D}{k_\infty} \simeq 0.683\,\mathrm{Re}_D^{0.466}\mathrm{Pr}^{0.33}$$

$$C_D = \frac{F_D}{\frac{1}{2}\rho_\infty U_\infty^2 D} \simeq \frac{5.484}{\mathrm{Re}_D^{0.246}} \qquad 40 < \mathrm{Re}_D < 10^3$$

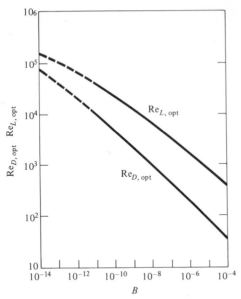

Figure 11.11 Optimum pin fin diameter and optimum length for minimum irreversibility ($M = 100$). [From D. Poulikakos and A. Bejan (7).]

The entropy generation number N_s is thus a function of fin geometry through Re_L and Re_D and design constraints through Pr, k_f/k_∞, and B. To obtain the optimum pin length we set $\partial N_s/\partial Re_L$ equal to zero and solve for $Re_{L,\,opt}$. This yields (see Problem 11.10):

$$Re_{L,opt} = \frac{Re_D}{2}\left(\frac{k_f}{Nu\,k_\infty}\right)^{1/2} \sinh^{-1}\left[\left(\frac{8}{\pi Re_D{}^3 C_D B}\right)^{1/2}\right] \qquad (11.40)$$

Poulikakos and Bejan (7) have calculated the optimum values of Re_L and Re_D for $M = (k_f/k_\infty)^{1/2}Pr^{-1/6} = 100$, with the results shown in Fig. 11.11.

An alternative condition of constraint would exist if the ratio L/D for the fin were specified. Then the entropy generation number would be a function of Re_D and L/D. The results of some numerical calculations to minimize N_s are shown in Fig. 11.12 for $B = 10^{-7}$ and $M = 100$. Approaches to thermodynamic optimization for other fin geometries are discussed by Poulikakos and Bejan (7).

11.1.6 Irreversibility in Heat Exchangers

This section presents a simplified analysis of the irreversibility in heat exchange devices. The irreversibility analysis of the heat transfer process in a heat exchanger presented here has been simplified by neglecting the pressure drop in the flow passages. Then the entropy is generated only as a result of the heat transfer between the hot and cold streams. Although neglecting irreversibility due to fluid friction introduces some error, the simplified approach provides

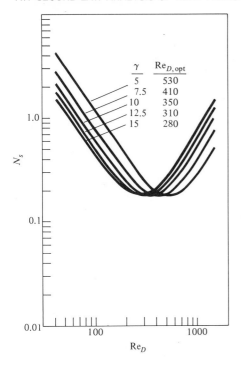

Figure 11.12 Optimization of a pin fin: entropy generation versus pin diameter ($M = 100$). [From D. Poulikakos and A. Bejan (7).]

insight into the relation between entropy generation and important heat exchanger design parameters without mathematical complexity. For a more complete second law analysis of heat exchangers that includes the effects of fluid friction, the reader is referred to (3) and (8).

Consider a counterflow heat exchanger such as the pipe-within-a-pipe design discussed in Chapter 8. Since heat is transferred from a warmer to a cooler fluid, the availability of energy in the warmer stream is decreased and entropy is generated. Conversely, the availability of energy in the cooler stream is increased. If the inlet and outlet temperatures are given, the rate of entropy generation in the control volume shown in Fig. 11.13 is

$$\dot{S} = \dot{m}_1(s_{\text{out}} - s_{\text{in}})_1 + \dot{m}_2(s_{\text{out}} - s_{\text{in}})_2 \tag{11.41}$$

$$= \int_{T_{1,\text{in}}}^{T_{1,\text{out}}} (mc_p)_1 \frac{dT_1}{T_1} + \int_{T_{2,\text{in}}}^{T_{2,\text{out}}} (mc_p)_2 \frac{dT_2}{T_2} \tag{11.42}$$

$$= \dot{C}_1 \ln \frac{T_{1,\text{out}}}{T_{1,\text{in}}} + \dot{C}_2 \ln \frac{T_{2,\text{out}}}{T_{2,\text{in}}} \tag{11.43}$$

where $\dot{C}_1 = (\dot{m}c_p)_1$ and $\dot{C}_2 = (\dot{m}c_p)_2$. Assuming that $\dot{C}_1 < \dot{C}_2$ and that the heat exchanger does not lose any heat to the environment, that is, the shell is perfectly insulated, a heat balance gives

$$\dot{C}_{\text{min}}(T_{1,\text{in}} - T_{1,\text{out}}) = \dot{C}_{\text{max}}(T_{2,\text{out}} - T_{2,\text{in}}) \tag{11.44}$$

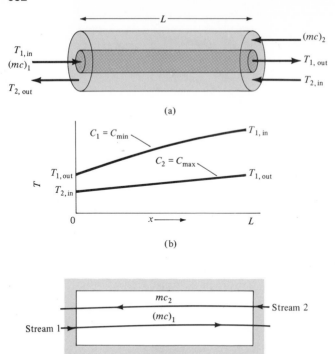

Figure 11.13 (a) Schematic diagram of counterflow heat exchanger. (b) Temperature distribution, $\dot{C}_1 < \dot{C}_2$. (c) Control volume for entropy generation analysis.

Using the above relation together with the definition of effectiveness from Chapter 8,

$$\epsilon = \frac{T_{1,\text{in}} - T_{1,\text{out}}}{T_{1,\text{in}} - T_{2,\text{in}}} = \left(\frac{\dot{C}_{\min}}{\dot{C}_{\max}}\right)\frac{T_{2,\text{out}} - T_{2,\text{in}}}{T_{1,\text{in}} - T_{2,\text{in}}}$$

we can eliminate the outlet temperatures $T_{1,\text{out}}$ and $T_{2,\text{out}}$ in Eq. (11.43) and obtain the rate of entropy generation in the form

$$\dot{S} = \dot{C}_{\min} \ln\left[\frac{\epsilon(T_{2,\text{in}} - T_{1,\text{in}})}{T_{1,\text{in}}} - 1\right]$$

$$+ \dot{C}_{\max} \ln\left[1 - \frac{\dot{C}_{\min}}{\dot{C}_{\max}} \epsilon\left(\frac{T_{2,\text{in}} - T_{1,\text{in}}}{T_{2,\text{in}}}\right)\right] \qquad (11.45)$$

Dividing both sides by $\dot{C}_{\max}$ yields, after some rearrangement,

$$\frac{\dot{S}}{\dot{C}_{\max}} = \frac{\dot{C}_{\min}}{\dot{C}_{\max}} \ln\left[1 - \epsilon\left(\frac{T_{2,\text{in}}}{T_{1,\text{in}}} - 1\right)\right] + \ln\left[1 - \epsilon\frac{\dot{C}_{\min}}{\dot{C}_{\max}}\left(1 - \frac{T_{1,\text{in}}}{T_{2,\text{in}}}\right)\right] \qquad (11.46)$$

The parameter $\dot{S}/\dot{C}_{\max}$ is referred to as the number of entropy production units, N_s.

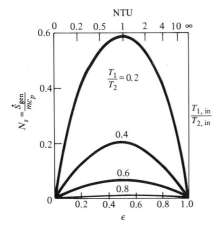

Figure 11.14 Entropy generation rate in a balanced counterflow heat exchanger with zero pressure drop.

If $\dot{C}_1 = \dot{C}_2$, a condition called balanced design, we get

$$N_{s,\,balanced} = \ln\left[1 + \epsilon\left(\frac{T_{2,\,in} - T_{1,\,in}}{T_{1,\,in}}\right)\right]\left[1 + \epsilon\left(\frac{T_{1,\,in} - T_{2,\,in}}{T_{2,\,in}}\right)\right] \quad (11.47)$$

Figure 11.14 shows the number of entropy production units in a balanced counterflow heat exchanger as a function of ϵ for various values of $T_{1,\,in}/T_{2,\,in}$. For this case, $\epsilon = NTU/(1 + NTU)$ and maximum entropy generation occurs when $\epsilon = 0.5$ or $NTU = 1$; $N_{s,\,max}$ can then be expressed in the form

$$\frac{\dot{S}}{\dot{m}c_p} = N_{s,\,balanced} = \ln\left[\frac{1}{2} + \frac{1}{4}\left(\frac{T_{1,\,in}}{T_{2,\,in}} + \frac{T_{2,\,in}}{T_{1,\,in}}\right)\right] \quad (11.48)$$

Note that for the balanced case the condition is symmetrical and T_1 can be interchanged with T_2 without changing the result. The dissipation in this case is entirely due to the finite size of the exchanger, because if $A \to \infty$ the outlet temperature of the colder stream equals the inlet temperature of the warmer stream.

When the capacity rates are not equal, the expression for N_s in Eq. (11.46) is the result of two dissipation effects:

1. The ΔT effect caused by the capacity rate imbalance, i.e., $C_{max} > C_{min}$.
2. The ΔT effect resulting from the finite heat transfer area, i.e., NTU is finite.

Bejan (3) has separated these two effects for the case of $(1 - \epsilon) \ll 1$, that is, for exchangers with high effectiveness. Using the relation between ϵ and NTU derived in Chapter 8, Bejan found that

$$N_s \approx C \ln\frac{T_{2,\,in}}{T_{1,\,in}} + \frac{1}{C}\ln\left[1 - C\left(1 - \frac{T_{2,\,in}}{T_{1,\,in}}\right)\right]$$
$$+ C^2(1 - C)\frac{(1 - T_{in}/T_{2,\,in})^2}{1 - C(1 - T_{in}/T_{2,\,in})} F(NTU, C) \quad (11.49)$$

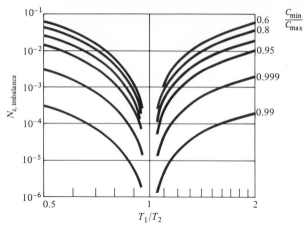

Figure 11.15   Entropy generation due to capacity rate imbalance in a counterflow heat exchanger.

where $\quad C = \dfrac{\dot{C}_{min}}{\dot{C}_{max}}$

$\qquad \text{NTU} = \dfrac{UA}{C_{min}}$

$F(\text{NTU}, C) = \dfrac{\exp[\text{NTU}(C - 1)]}{1 - C \exp[\text{NTU}(C - 1)]}$

The first two terms represent the contribution from capacity rate imbalance, while the third term is the contribution due to the finite NTU. The contribution to entropy generation due to the capacity rate imbalance is shown graphically in Fig. 11.15, where the sum of the first two terms in Eq. (11.49), $N_{s,\,imbalance}$, is plotted against the inlet temperature ratio $T_{1,\,in}/T_{2,\,in}$ for various values of C. As expected, the number of entropy generation units increases as the imbalance becomes more pronounced. Even if the area were increased to approach "ideal" thermal performance, Fig. 11.15 shows that this approach to increase performance reaches a point of diminishing returns as most of the dissipation is caused by imbalance rather than finite area, that is, $N_{s,\,imbalance} > N_{s,\,NTU}$. Additional optimization procedures—for example, calculating the minimum area or volume for a specified value of N_s—can be found in (3). The problem of minimizing irreversibility in solar collectors is treated in (9).

11.2 HEAT TRANSFER IN HIGH-SPEED FLOW

Convection heat transfer in high-speed flow is important for systems such as aircraft and missiles when the velocity approaches or exceeds the velocity of sound. For a perfect gas the accoustical velocity a can be obtained from

$$a = \sqrt{\frac{\gamma \mathscr{R} T}{\mathscr{M}}} \qquad\qquad (11.50)$$

where γ = specific heat ratio, c_p/c_v (about 1.4 for air)

         $\mathscr{R}$ = universal gas constant

         T = absolute temperature

         $\mathscr{M}$ = molecular weight of the gas

When the velocity of a gas flowing over a heated or cooled surface is of the order of the accoustical velocity or larger, the flow field can no longer be described solely in terms of the Reynolds number; instead, the ratio of the gas flow velocity to the accoustical velocity, i.e., the Mach number $M = U_\infty/a_\infty$, must also be considered. When the gas velocity in a flow system reaches a value of about one-half the speed of sound, the effects of viscous dissipation in the boundary layer become important. Under such conditions the temperature of a surface over which a gas is flowing can actually exceed the free-stream temperature. For flow over an adiabatic surface, such as a perfectly insulated wall, Fig. 11.16 shows the velocity and temperature distributions schematically. The high temperature at the surface is the combined result of the heating due to viscous dissipation and the temperature rise of the fluid as the kinetic energy of the flow is converted to internal energy while the flow decelerates through the boundary layer. The actual shape of the temperature profile depends on the relation between the rate at which viscous shear work increases the internal energy of the fluid and the rate at which heat is conducted toward the free stream.

Although the processes in a high-speed boundary layer are not adiabatic, it is general practice to relate them to adiabatic processes. The conversion of kinetic energy in a gas being slowed down adiabatically to zero velocity is described by

$$i_0 = i_\infty + \frac{U_\infty^2}{2\,Jg_c} \tag{11.51}$$

where i_0 is the stagnation enthalpy and i_∞ is the enthalpy of the gas in the free stream. For an ideal gas Eq. (11.51) becomes

$$T_0 = T_\infty + \frac{U_\infty^2}{2\,Jg_c c_p} \tag{11.52}$$

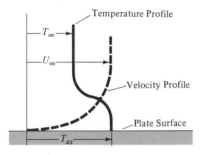

Figure 11.16 Velocity and temperature distribution in high-speed flow over an insulated plate.

or, in terms of the Mach number,

$$\frac{T_0}{T_\infty} = 1 + \frac{\gamma - 1}{2} M_\infty^2 \tag{11.53}$$

where T_0 is the stagnation temperature and T_∞ is the free-stream temperature.

In a real boundary layer the fluid is not brought to rest reversibly because the viscous shearing process is thermodynamically irreversible. To account for the irreversibility in a boundary-layer flow we define a recovery factor r as

$$r = \frac{T_{as} - T_\infty}{T_0 - T_\infty} \tag{11.54}$$

where T_{as} is the adiabatic surface temperature.

Experiments (10) have shown that in laminar flow,

$$r = \mathrm{Pr}^{1/2} \tag{11.55}$$

whereas in turbulent flow

$$r = \mathrm{Pr}^{1/3} \tag{11.56}$$

When a surface is not insulated, the rate of heat transfer by convection between a high-speed gas and that surface is governed by the relation

$$q_c'' = -k \left. \frac{\partial T}{\partial y} \right|_{y=0}$$

The influence of heat transfer to and from the surface on the temperature distribution is illustrated in Fig. 11.17. We observe that in high-speed flow heat can be transferred to the surface even when the surface temperature is above the free-stream temperature. This phenomenon is the result of viscous shear, often called aerodynamic heating. The heat transfer in high-speed flow over a flat surface can be predicted (11) from the boundary-layer energy equation

$$u \frac{\partial T}{\partial x} + v \frac{\partial T}{\partial y} = a \frac{\partial^2 T}{\partial y^2} + \frac{\mu}{\rho c_p} \left(\frac{\partial u}{\partial x} \right)^2$$

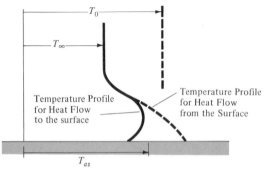

Figure 11.17 Temperature profiles in a high-speed boundary layer for heating and cooling.

where the last term accounts for the viscous dissipation. However, for most practical purposes the rate of heat transfer can be calculated with the same relations used for low-speed flow, if the average convection heat transfer coefficient is redefined by the relation

$$q_c'' = \bar{h}_c(T_s - T_{as}) \tag{11.57}$$

which will yield zero heat flow when the surface temperature T_s equals the adiabatic surface temperature.

Since in high-speed flow the temperature gradients in a boundary layer are large, variations in the physical properties of the fluid will also be substantial. Eckert (12) has shown, however, that the constant-property heat transfer equations can still be used if all the properties are evaluated at a reference temperature T^* given by

$$T^* = T_\infty + 0.5(T_s - T_\infty) + 0.22(T_{as} - T_\infty) \tag{11.58}$$

The local values of the heat transfer coefficient, defined by

$$h_{cx} = \frac{q''}{T_s - T_{as}} \tag{11.59}$$

can be obtained from the following equations:

Laminar Boundary Layer ($Re_x^* < 10^5$):

$$St_x^* = \left(\frac{h_{cx}}{c_p \rho u_\infty}\right)^* = 0.332(Re_x^*)^{-1/2}(Pr^*)^{-2/3} \tag{11.60}$$

Turbulent Boundary Layer ($10^5 < Re_x^* < 10^7$):

$$St_x^* = \left(\frac{h_{cx}}{c_p \rho u_\infty}\right)^* = 0.0288(Re_x^*)^{-1/5}(Pr^*)^{-2/3} \tag{11.61}$$

Turbulent Boundary Layer ($10^7 < Re_x^* < 10^9$):

$$St_x^* = \left(\frac{h_{cx}}{c_p \rho u_\infty}\right)^* = \frac{2.46}{(\ln Re_x^*)^{2.584}}(Pr^*)^{-2/3} \tag{11.62}$$

Eq. (11.62) is based on experimental data for local friction coefficients in high-speed gas flow (12) in the Reynolds number range 10^7 to 10^9 that are correlated by

$$C_{fx} = \frac{4.92}{(\ln Re_x^*)^{2.584}} \tag{11.63}$$

If the average value of the heat transfer coefficients is to be determined, the expressions above must be integrated between $x = 0$ and $x = L$. However, the integration may have to be done numerically in most practical cases because the reference temperature T^* is not the same for the laminar and turbulent portions of the boundary layer, as shown by Eqs. (11.55) and (11.56).

When the speed of a gas is exceedingly high, the boundary layer may become so hot that the gas begins to dissociate. In such situations Eckert (12) recommends that the heat transfer coefficient be based on the enthalpy difference between the surface and the adiabatic state, $(i_s - i_{as})$, and be defined by

$$q_c'' = h_{ci}(i_s - i_{as}) \tag{11.64}$$

If an enthalpy recovery factor is defined by

$$r_i = \frac{i_{as} - i_\infty}{i_0 - i_\infty} \tag{11.65}$$

the same relation used previously to calculate the reference temperature can be used to calculate a reference enthalpy, or

$$i^* = i_\infty + 0.5(i_s - i_\infty) + 0.22(i_{as} - i_\infty) \tag{11.66}$$

The local Stanton number is then redefined as

$$St_{x,i}^* = \frac{h_{c,i}}{\rho^* u_\infty} \tag{11.67}$$

and used in Eqs. (11.61), (11.62), and (11.63) to calculate the heat transfer coefficient. It should be noted that the enthalpies in the above relations are the total values, which include the chemical energy of dissociation as well as the internal energy. As shown in (12), this method of calculation is in excellent agreement with experimental data.

In some situations, for instance, at extremely high altitudes, the fluid density may be so small that the distance between gas molecules becomes of the same order of magnitude as the boundary layer. In such cases the fluid cannot be treated as a continuum and it is necessary to subdivide the flow processes into regimes. These flow regimes are characterized by the ratio of the molecular free path to a significant physical scale of the system, called the Knudsen number Kn. Continuum flow corresponds to small values of Kn, while at larger values of Kn molecular collisions occur primarily at the surface and in the

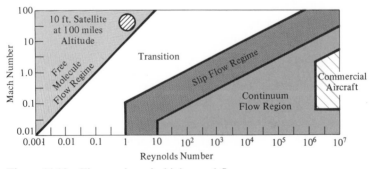

Figure 11.18 Flow regimes in high-speed flow.

main stream. Since energy transport is by free motion of molecules between the surface and the main stream, this regime is called the "free-molecule" regime. Between the free-molecule and the continuum regime is a transition range, called the "slip-flow" regime because it is treated by assuming temperature and velocity "slip" at fluid-solid interfaces. Figure 11.18 shows a map of these flow regimes. For a treatment of heat transfer and friction in these specialized flow systems the reader is referred to (13–15).

11.3 NON-NEWTONIAN FLUIDS

In previous chapters we have assumed that the hydrodynamic behavior of a fluid is Newtonian, that is, that the relation between the shear stress and the velocity gradient is linear. Thus, the shear stress τ is equal to zero when the velocity gradient is zero and is also independent of that gradient. Gases and liquids such as water, oil, and liquid metals generally behave as Newtonian fluids, but colloidal suspensions, polymeric solutions, blood, and other complex fluids do not obey linear relations between shear stress and shear rate and are referred to as *non-Newtonian fluids*. In this section an introduction to some fundamental characteristics of non-Newtonian fluids will be presented. For a more complete treatment, the specialized books by Metzner (16, 17) and surveys such as those in (18–20) are recommended.

Hydrodynamic Behavior   According to Metzner (16), all fluids can be classified into four categories, as shown in Table 11.1, with Newtonian fluids being a subclass of other viscous fluids. We will only treat fluids that fall in category 1, purely viscous fluids, but according to Metzner this may often also be used as a good approximation for fluids in categories 2 and 3.

The most common relation by which the shear stress behavior of purely viscous non-Newtonian flow can be described is the *power law*

$$\tau = K(u')^n \tag{11.68}$$

TABLE 11.1   TYPES OF FLUID BEHAVIOR

Category 1: Purely viscous fluids
- (a) μ = constant, independent of rate of strain (Newtonian fluid)
- (b) μ = decreasing function of the shear stress; includes "pseudoplastic," "Bingham plastic," and "shear-thinning" materials
- (c) μ = increasing function of the shear stress; includes the "dilatants" and "shear-thickening" materials

Category 2: Time-dependent shear stress fluids

Category 3: Viscoelastic fluids

Category 4: More complex materials

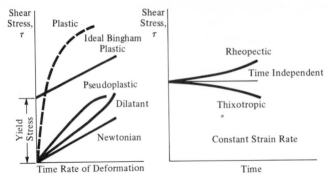

Figure 11.19 Viscous behavior of various materials listed in Table 11.1.

where u' is the shear rate, du/dy. For Newtonian fluids $n = 1$ and K is the viscosity, μ. Shear-thinning fluids whose behavior corresponds to an exponent n less than unity in the preceding relation have a local viscosity that decreases with increasing stress and are called "pseudoplastic," whereas those characterized by a value of n greater than one are shear-thickening and are called "dilatant." Values for the exponent n in the power law are presented in (20) for various fluids and some examples of non-Newtonian behavior are shown schematically in Fig. 11.19. Curves for true fluids that cannot resist shear must pass through the origin, but other substances, called yielding fluids, show a finite stress at zero strain rate.

According to Metzner (16) the friction factor for laminar flow in a pipe, corresponding to the friction factor for a Newtonian fluid, is given by

$$f = \frac{16}{\text{Re}'} \tag{11.69}$$

where

$$\text{Re}' = D^{n'}\bar{u}^{2-n'}\frac{\rho}{\gamma} \tag{11.70}$$

and

$$\gamma = K'8^{(n'-1)} \tag{11.71}$$

The coefficient n' is given by

$$n' = \frac{d(\log \tau_w)}{d[\log(8\bar{u}/D)]} \tag{11.72}$$

and the shear stress is

$$\tau_w = K'\left(\frac{8\bar{u}}{D}\right)^{n'} \tag{11.73}$$

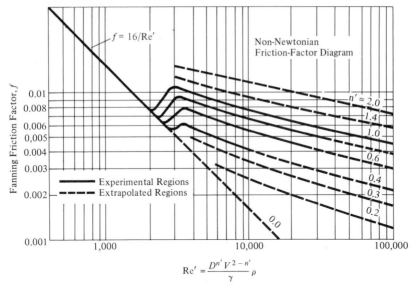

$$\mathrm{Re}' = \frac{D^{n'} V^{2-n'}}{\gamma} \rho$$

Figure 11.20 Friction factor-Reynolds number design chart for purely viscous non-Newtonian fluid. [From A. B. Metzner (16).]

Substituting the above expressions into the power law, one finds that

$$K' = K\left(\frac{3n'+1}{4n'}\right)^n\left(\frac{8\bar{u}}{D}\right)^{n-n'}\tag{11.74}$$

If the fluid obeys the power-law formulation, then $n = n'$ and Eqs. (11.74) and (11.70) simplify to

$$K' = K\left(\frac{3n+1}{4n}\right)^n\tag{11.75}$$

and

$$\mathrm{Re}' = \frac{D^n \bar{u}^{2-n}\rho}{(K/8)[(6n+2)/n]^n}\tag{11.76}$$

Note that $n = 1$ gives Newtonian fluid behavior.

For power-law fluids the laminar velocity distribution in a pipe of diameter $2r_w$ is (16)

$$\frac{u(r)}{\bar{u}} = \frac{1+3n}{1+n}\left[1-\left(\frac{r}{r_w}\right)^{(n+1)/n}\right]\tag{11.77}$$

where $\bar{u}$ is the average pipe velocity. For $n = 1$ this gives a parabolic velocity profile.

Results for the friction factor of power-law fluids in turbulent flow are shown in Fig. 11.20, where n' is given by Eq. (11.72). More detailed information is available in (16).

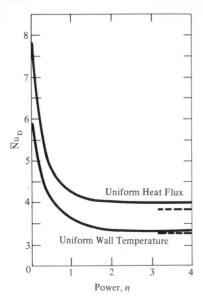

Figure 11.21 Downstream fully developed Nusselt number for laminar flow of a power-law, non-Newtonian fluid in round tubes (19, 21).

Convection heat transfer in laminar flow through a tube can be characterized by plotting the Nusselt number against the power n from Eq. (11.68). Figure 11.21 shows the results for fully developed flow with uniform heat flux as well as for uniform temperature along the tube. Observe that the Nusselt number decreases as the value of the exponent n' increases (37). A case of combined velocity and thermal entrance conditions has been treated in (22) for a power-law fluid with $n = 0.5$; the results are presented in Table 11.2 for various values of the modified Prandtl number Pr', defined as

$$\mathrm{Pr}' = \frac{Kc_p}{k}\left(\frac{D}{2\bar{u}}\right)^{1-n} \tag{11.78}$$

For heat transfer in turbulent flow of a non-Newtonian power-law fluid in a tube, the Stanton number can be estimated in the range of $3000 < \mathrm{Re}_a < 90{,}000$ from the relation

$$\mathrm{St} = \frac{0.0152}{\mathrm{Re}_a{}^{0.155}\mathrm{Pr}_a{}^{0.66}} \tag{11.79}$$

where $\mathrm{Re}_a = \rho\bar{u}D/y_a$
$\mathrm{Pr}_a = c_p y_a/k$
$y_a = \tau_w 4nD/(3n + 1)8\bar{u}$

According to (37), Eq. (11.79) correlates available data with a mean deviation of 2.3 percent for n between 0.2 and 0.9. Fig. 11.22 shows the experimental results.

TABLE 11.2 LOCAL AND MEAN NUSSELT NUMBERS FOR A POWER-LAW FLUID WITH $n = 0.5$ FOR COMBINED HYDRODYNAMIC AND THERMAL ENTRY LENGTH

| | $Pr' = 1.0$ | | | | $Pr' = 10$ | | | | $Pr' = 100$ | | | |
| | Uniform wall temp. | | Uniform heat flux | | Uniform wall temp. | | Uniform heat flux | | Uniform wall temp. | | Uniform heat flux | |
x_0 [a]	Nu_x	$\overline{Nu}_m$	Nu_x	$\overline{Nu}_m$	Nu_x	$\overline{Nu}_m$	Nu_x	$\overline{Nu}_m$	Nu_x	$\overline{Nu}_m$	Nu_x	$\overline{Nu}_m$
0.02	7.95	12.98	10.77	19.29	6.64	10.45	8.52	14.32	6.41	1.03	8.16	12.23
0.06	5.48	8.57	7.15	12.02	4.78	7.13	6.06	9.43	4.75	6.56	6.00	8.57
0.10	4.61	7.13	5.97	9.79	4.30	6.00	5.40	7.93	4.29	5.73	5.38	7.39
0.20	4.04	5.67	5.02	7.57	4.00	5.10	4.90	6.51	3.99	4.91	4.90	6.23
0.30	3.96	5.11	4.82	6.68	3.95	4.73	4.79	5.97	3.95	4.59	4.79	5.76

[a] $x_0 = 2x/D Re' Pr'$; $Pr' = (Kc_p/k)(D/2\bar{u})^{1-n}$, $Re' = \bar{u}^{2-n}(D/2)^n \rho/K$.

Source: W. L. Wilkinson (18).

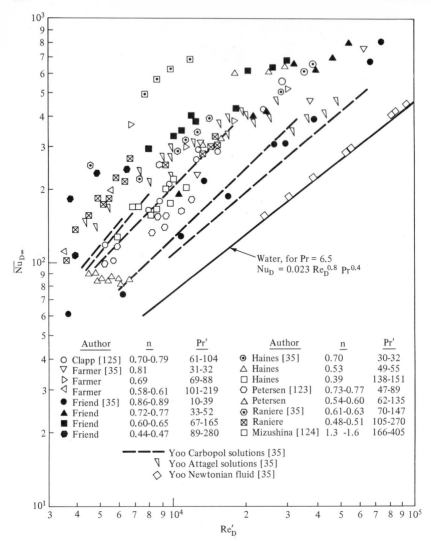

Figure 11.22 Experimental heat transfer measurements for purely viscous non-Newtonian fluids. [From S. S. Yoo, *Heat Transfer and Friction Factor for Non-Newtonian Fluids in Turbulent Pipe Flow*, Ph.D. Thesis, University of Illinois, Chicago, 1979.]

11.4 HEAT PIPES

One of the main objectives of energy conversion systems is to transfer energy from a receiver to some other location where it can be used to heat a working fluid. A novel device that can transfer large quantities of heat through small surface areas with small temperature differences is the heat pipe. The method of operation of a heat pipe is shown schematically in Fig. 11.23. The device consists of a circular pipe with an annular layer of wicking material covering

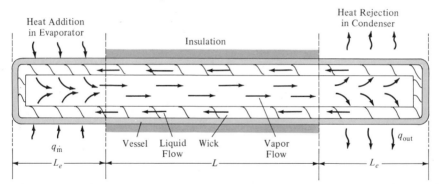

Figure 11.23 Schematic diagram of a heat pipe and the associated heat flow mechanisms.

the inside. The core of the system is hollow in the center to permit the working fluid to pass freely from the heat addition end on the left to the heat rejection end on the right. The heat addition end is equivalent to an evaporator, and the heat rejection end corresponds to a condenser. The condenser and the evaporator are connected by an insulated section of length L. The liquid permeates the wicking material by capillary action, and when heat is added to the evaporator end of the heat pipe, liquid is vaporized in the wick and moves through the central core to the condenser end, where heat is removed. Then the vapor condenses back into the wick and the cycle repeats.

A large variety of fluid and pipe material combinations have been used for heat pipes, and some typical working fluid and material combinations, as well as the temperature ranges over which they can operate, are presented in Table 11.3. In the fourth and fifth columns of the table measured axial heat fluxes and measured surface heat fluxes are listed, and it is apparent that very high heat fluxes can be obtained (23, 24).

In order for a heat pipe to operate, the maximum capillary pumping head, $(\Delta p_c)_{max}$, must be able to overcome the total pressure drop in the heat pipe. This pressure drop consists of three parts:

1. The pressure drop required to return the liquid from the condenser to the evaporator, Δp_e,
2. The pressure drop required to move the vapor from the evaporator to the condenser, Δp_v, and
3. The potential head due to the difference in elevation between the evaporator and the condenser, Δp_g.

The condition for pressure equilibrium can thus be expressed in the form

$$(\Delta p_c)_{max} \geq \Delta p_e + \Delta p_v + \Delta p_g \tag{11.80}$$

If this condition is not met, the wick will dry out in the evaporator region and the heat pipe will cease to operate.

TABLE 11.3 SOME TYPICAL OPERATING CHARACTERISTICS OF HEAT PIPES

Temperature range (K)	Working fluid	Vessel material	Measured axial heat flux[a] (W/cm²)	Measured surface heat flux[a] (W/cm²)	Comments
230–400	Methanol	Copper, nickel, stainless steel	0.45 at 373 K	75.5 at 373 K	Using threaded artery wick
280–500	Water	Copper, nickel	0.67 at 473 K	146 at 443 K	
360–850	Mercury	Stainless steel	25.1 at 533 K	181 at 533 K	Based on sonic limit in heat pipe
673–1073	Potassium	Nickel, stainless steel	5.6 at 1023 K	181 at 1023 K	
773–1173	Sodium	Nickel, stainless steel	9.3 at 1123 K	224 at 1033 K	

[a] Varies with temperature.

Source: Abstracted from C. H. Dutcher and M. R. Burke (27).

The liquid pressure drop in flow through a homogeneous wick can be calculated from the empirical relation

$$\Delta p_l = \frac{\mu_l L_{\text{eff}} \dot{m}}{\rho_l K_w A_w} \tag{11.81}$$

where μ_l = liquid viscosity

$\dot{m}$ = mass flow rate

ρ_l = liquid density

A_w = wick cross-sectional area

K_w = wick permeability or wick factor

L_{eff} = effective length between the evaporator and condenser, given by

$$L_{\text{eff}} = L + \frac{L_e + L_c}{2} \tag{11.82}$$

where L_e = evaporator length

L_c = condenser length

The pressure drop through longitudinal, grooved wicks or composite wicks can be obtained by minor modifications of Eq. (11.82), as shown in (23).

The vapor pressure drop is generally small compared to the liquid pressure loss. As long as the velocity of the vapor is small compared to the velocity of sound, say less than 30 percent, one can neglect compressibility effects and calculate the viscous pressure loss Δp_v from incompressible flow relations. For steady-state laminar flow (see Chapter 6)

$$\Delta p_v = f \frac{L_{\text{eff}}}{D} \rho \bar{u}^2 = \frac{64 \mu_v \dot{m} L_{\text{eff}}}{\rho_v \pi D_v^{\,4}} \tag{11.83}$$

where D_v is the inside wick diameter in contact with the vapor and the subscript v denotes vapor properties.

In addition to the viscous pressure drop, a pressure force is necessary to accelerate the vapor entering from the evaporator section to its axial velocity, but most of this loss is regained in the condenser, where the vapor stream is brought to rest. A more detailed treatment of the vapor pressure loss, including the pressure recovery in the evaporator, is presented in (23).

The pressure difference due to the hydrostatic or potential head of the liquid may be positive, negative, or zero, depending on the relative positions of the evaporator and condenser. The pressure difference Δp_g is given by

$$\Delta p_g = \rho_l g L \sin \phi \tag{11.84}$$

where ϕ is the angle between the axis of the heat pipe and the horizontal (positive when the evaporator is above the condenser).

The driving force in the wick is the result of surface tension. Surface tension is a force resulting from an imbalance of the natural attraction among a homogeneous assembly of molecules. For example, a molecule near the surface of a liquid will experience a force directed inward due to the attraction of neighboring molecules below. One of the consequences of surface tension is that the pressure on a concave surface is less than that on a convex surface. The resulting pressure difference Δp is related to the surface tension σ_l and the radius of curvature r_c. For a hemispherical surface the tension force action around the circumference is equal to $2\pi r_c \sigma_l$ and it is balanced by a pressure force over the surface equal to $\Delta p \pi r_c^2$. Hence

$$\Delta p = \frac{2\sigma_l}{r_c} \tag{11.84}$$

Another illustration of surface tension can be observed when a capillary tube is placed vertically in a wetting fluid; the fluid will rise in the tube due to capillary action, as shown in Fig. 11.24. A pressure balance then gives

$$\Delta p_c = \rho_l g h = \frac{2\sigma_l}{r_l} \cos \theta \tag{11.85}$$

where θ is the contact angle, which is between 0 and $\pi/2$ for wetting fluids. For a nonwetting fluid, θ is larger than $\pi/2$, and the liquid level in the capillary will be depressed below the surface. Hence, to obtain a capillary driving force only wetting fluids can be used in heat pipes.

Substituting Eqs. (11.81), (11.83), (11.84), and (11.85) for the pressure terms in the relation for the dynamic equilibrium, Eq. (11.80), yields one of the key design criteria for heat pipes:

$$\frac{2\sigma_l \cos \theta}{r_c} = \frac{\mu_l L_{\text{eff}} \dot{m}}{\rho_l K_w A_w} + \frac{64 \mu_v \dot{m} L_{\text{eff}}}{\rho_v \pi D_v^4} + \rho_l g L_{\text{eff}} \sin \phi \tag{11.86}$$

If $(64\mu_v/\rho_v\pi D_v^4) \ll (\mu_l/\rho_l K_w A_w)$ the pressure drop of the vapor is negligible and the second term in Eq. (11.80) can be deleted in a preliminary design.

The maximum heat transport capability of a heat pipe due to wicking limitations is given by the relation

$$q_{\max} = \dot{m}_{\max} h_{fg} \tag{11.87}$$

where $\dot{m}_{\max}$ can be obtained from Eq. (11.86). Assuming $\cos \theta = 1$ and a negligible vapor flow pressure drop, one can solve for $\dot{m}_{\max}$ and combine the result with Eq. (11.87) to obtain the following expression for the maximum heat trans-

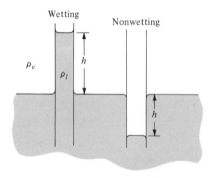

Figure 11.24   Capillary rise in a tube.

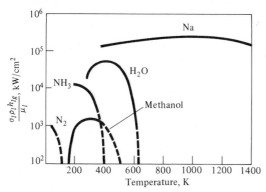

Figure 11.25 Figure of merit for several heat pipe working fluids as a function of temperature.

port capability:

$$q_{max} = \left(\frac{\rho_l \sigma_l h_{fg}}{\mu_l}\right)\left(\frac{A_w K_w}{L_{eff}}\right)\left(\frac{2}{r_c} - \frac{\rho_l g L_{eff} \sin \phi}{\sigma_l}\right) \tag{11.88}$$

In the above equation all the terms in the first parentheses $(\rho_l \sigma_l h_{fg}/\mu_l)$ are properties of the working fluid. This group is known as the figure of merit M

$$M = \frac{\rho_l \sigma_l h_{fg}}{\mu_l} \tag{11.89}$$

and is plotted in Fig. 11.25 as a function of temperature for a number of heat pipe fluids.

The wick geometric properties are functions of A_w, K, and r_c. Table 11.4 present data for pore size and permeability for a few wick materials and mesh sizes. They can be used for a preliminary design as shown in Example 11.3.

A widely used correlation between the maximum achievable power transfer by a heat pipe and its dominant dimensions and operating parameters is

$$q_{max} = \frac{A_w h_{fg} g \rho_l^2}{\mu_l}\left(\frac{l_w K_w}{L_{eff}}\right) \tag{11.90}$$

where A_w = wick area

g = gravitational acceleration

h_{fg} = heat of evaporization of liquid

ρ_l = liquid density

μ_l = liquid viscosity

l_w = wicking height of fluid in wick

The wicking height is given by

$$l_w = \frac{2\sigma_l}{r_c \rho_l g} \tag{11.91}$$

where σ_l = surface tension

r_c = effective pore radius

TABLE 11.4 WICK PORE SIZE AND PERMEABILITY DATA[a]

Material and mesh size	Capillary height[b] (cm)	Pore radius (cm)	Permeability (m²)	Porosity (%)
Glass fiber	25.4		0.061×10^{-11}	
Monel beads				
30–40	14.6	0.052[c]	4.15×10^{-10}	40
70–80	39.5	0.019[c]	0.78×10^{-10}	40
100–140	64.6	0.013[c]	0.33×10^{-10}	40
140–200	75.0	0.009	0.11×10^{-10}	40
Felt metal				
FM1006	10.0	0.004	1.55×10^{-10}	
FM1205		0.008	2.54×10^{-10}	
Nickel powder				
200 μm	24.6	0.038	0.027×10^{-10}	
500 μm	>40.0	0.004	0.081×10^{-11}	
Nickel fiber				
0.01 mm diameter	>40.0	0.001	0.015×10^{-11}	68.9
Nickel felt		0.017	6.0×10^{-10}	89
Nickel foam		0.023	3.8×10^{-9}	96
Copper foam		0.021	1.9×10^{-9}	91
Copper powder (sintered)	156.8	0.0009	1.74×10^{-12}	52
45–56 μm		0.0009		28.7
100–125 μm		0.0021		30.5
150–200 μm		0.0037		35
Nickel 50	4.8	0.0305	6.635×10^{-10}	62.5
Copper 60	3.0		8.4×10^{-10}	
Nickel				
100 (3.23)		0.0131	1.523×10^{-10}	
120 (3.20)	5.4		6.00×10^{-10}	
120[d] (3.20)	7.9	0.019	3.50×10^{-10}	
2[e] × 120 (3.25)			1.35×10^{-10}	
Nickel				
200	23.4	0.004	0.62×10^{-10}	
2 × 200			0.81×10^{-10}	
Nickel[d]				
2 × 250		0.002		
4[e] × 250		0.002		
325		0.0032		
Phosp./bronze		0.0021	0.296×10^{-10}	

[a] Abstracted from P. D. Dunn and D. A. Reay (23).
[b] Obtained with water, unless stated otherwise.
[c] Particle diameter.
[d] Oxidized.
[e] Denotes number of layers.

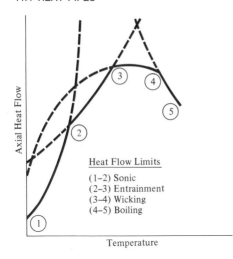

Figure 11.26 Limitations to heat transport in a heat pipe.

The maximum wicking height with sodium as the working fluid is about 38.5 cm, which is calculated by assuming an effective pore diameter of 8.6×10^{-3} cm. This is typical for a screen made with eight 4.1×10^{-3} cm diameter wires per square millimeter.

The most dominant parameters affecting the total power transfer capacity are the wick area, effective wicking height, and heat pipe length. For any effective wicking height a wicking area can be selected to achieve the desired total power transfer if the operating temperature as well as the temperature drops at the evaporator section and the condenser section can be freely selected. However, when a limit to the upper operating temperature as well as to the temperature of the heat pipe at the condenser section exists, the wicking thickness might be determined by these temperature considerations. In general, the temperature drops and the operating temperature increases with increasing wick thickness. If the wick thickness is based on temperature and temperature-drop considerations, the maximum heat pipe length for a given power transfer is determined.

Although a heat pipe behaves like a structure of very high thermal conductance, it has heat transfer limitations that are governed by certain principles of fluid mechanics. The possible effects of these limitations on the capability of a heat pipe with a liquid-metal working fluid are shown in Fig. 11.26. Individual limitations indicated in the figure are discussed below.

11.4.1 Sonic Limitation

When heat is transferred from the evaporator section of a heat pipe to the condenser section, the rate of heat transfer q between the two sections is given by

$$q = \dot{m}_v h_{fg} \tag{11.92}$$

where $\dot{m}_v$ is the rate of mass flow of vapor at evaporator exit and h_{fg} the latent heat of the fluid. Because the latent energy of the working fluid is used instead of its heat capacity, large heat transfer rates can be achieved with a relatively small mass flow. Furthermore, if the heat is transferred by high-density/low-velocity vapor, the transfer is nearly isothermal because only small pressure gradients are necessary to move the vapor.

To show the effect of vapor density and velocity on heat transfer, Eq. (11.92) can be modified by using the continuity equation

$$\dot{m}_v = \bar{\rho}_v \bar{u} A_v \tag{11.93}$$

where $\bar{\rho}_v$ is the radial average vapor density at the evaporator exit and A_v the cross-sectional area of vapor passage. By combining Eqs. (11.92) and (11.93) and rearranging, the result is

$$\frac{q}{A_v} = \bar{\rho}_v \bar{u} h_{fg} \tag{11.94}$$

where q/A_v is the axial heat flux based on the cross-sectional area of the vapor passage.

Equation (11.94) shows that the axial heat flux in a heat pipe can be held constant and the condenser environment adjusted to lower the pressure, temperature, and density of the vapor until the flow at the evaporator exit becomes sonic. Once this occurs, pressure changes in the condenser will not be transmitted to the evaporator. This sonic limiting condition is represented in Fig. 11.26 by the solid curve between points 1 and 2. Some values for sonic heat flux limits as a function of evaporator exit temperature are given in Table 11.5 for Cs, K, Na, and Li.

Although heat pipes are normally not operated at sonic flow, such conditions have been encountered during startup with the working fluids listed in Table 11.5. Temperatures during such startups are always higher at the beginning of the heat pipe evaporator than at the evaporator exit.

TABLE 11.5   SONIC LIMITATIONS OF HEAT PIPE WORKING FLUIDS

Evaporator exit temperature (°C)	Heat flux limits (kW/cm²)			
	Cs	K	Na	Li
400	1.0	0.5		
500	4.6	2.9	0.6	
600	14.9	12.1	3.5	
700	37.3	36.6	13.2	
800			38.9	1.0
900			94.2	3.9
1000				12.0
1100				31.1
1200				71.0
1300				143.8

11.4.2 Entrainment Limitation

Ordinarily, the sonic limitations just discussed do not cause dryout of the wick with attendant overheating of the evaporator. In fact, they often prevent the attainment of other limitations during startup. However, if the vapor density is allowed to increase without an accompanying decrease in velocity, some liquid from the wick return system may be entrained. The onset of entrainment can be expressed in terms of a Weber number,

$$\frac{\rho_v \bar{u}^2 L_c}{2\pi\sigma_l} = 1 \tag{11.95}$$

where L_c is a characteristic length describing the pore size. Equation (11.95) simply expresses the ratio of vapor inertial forces to liquid surface tension forces. When this ratio exceeds unity, a condition develops which is very similar to that of a body of water agitated by high-velocity winds into waves, which propagate until liquid is torn from their crests. Once entrainment begins in a heat pipe, fluid circulation increases until the liquid return path cannot accommodate the increased flow. This causes dryout and overheating of the evaporator.

Because the wavelength of the perturbations at the liquid-vapor interface in a heat pipe is determined by the wick structure, the entrainment limit can be estimated by combining Eqs. (11.94) and (11.95) to give

$$\frac{q}{A_v} = h_{fg}\left(\frac{\lambda\pi\sigma_l\rho_v}{L_c}\right)^{1/2} \tag{11.96}$$

Equation (11.96) can then be used to obtain the type of curve represented by the solid line between points 2 and 3 in Fig. 11.26.

11.4.3 Wicking Limitation

Fluid circulation in a heat pipe is maintained by capillary forces that develop in the wick structure at the liquid-vapor interface. These forces balance the pressure losses due to the flow in the liquid and vapor phases; they are manifest as many menisci which allow the pressure in the vapor to be higher than the pressure in the adjacent liquid in all parts of the system. When a typical meniscus is characterized by two principal radii of curvature (r_1 and r_2) the pressure drop ΔP_c across the liquid surface is given by

$$\Delta P_c = \sigma\left(\frac{1}{r_1} + \frac{1}{r_2}\right) \tag{11.97}$$

These radii, which are smallest at the evaporator end of the heat pipe, become even smaller as the heat transfer rate is increased. If the liquid wets the wick perfectly, the radii will be defined exactly by the pore size of the wick when a heat transfer limit is reached. Any further increase in heat transfer will cause the liquid to retreat into the wick, and drying and overheating will occur at the evaporator end of the system.

As indicated by Eq. (11.97), the capillary force in a heat pipe can be increased by decreasing the size of the wick pores that are exposed to vapor flow. However, if the pore size is also decreased in the remainder of the wick, the wicking limit might actually be reduced because of the increased pressure drop in the liquid phase. This is shown by *Poiseuille's equation* for the pressure drop through a capillary tube,

$$\Delta P_e = \frac{8\mu \dot{m}_l L}{\pi r^4 \rho} \tag{11.98}$$

where μ is the liquid viscosity, $\dot{m}_l$ the rate of mass flow of liquid, r the tube radius, ρ the liquid density, and L the tube length.

Equation (11.98) can be modified to obtain the liquid-pressure drop at a particular heat transfer rate q for various wick structures. The equations given below are for the examples shown in Fig. 11.27.

Artery:

$$\Delta P_L = \frac{8\mu q L_{\text{eff}}}{\pi r^4 \rho h_{fg}} \tag{11.98a}$$

Channels:

$$\Delta P_L = \frac{8\mu q L_{\text{eff}}}{\pi r_e{}^4 N \rho h_{fg}} \tag{11.98b}$$

Screen:

$$\Delta P_L = \frac{b\mu q L_{\text{eff}}}{\pi (R_W{}^2 - R^2)\epsilon r_c{}^2 \rho h_{fg}} \tag{11.98c}$$

Concentric annulus:

$$\Delta P_L = \frac{12\mu q L_{\text{eff}}}{\pi D w^3 \rho h_{fg}} \tag{11.98d}$$

Crescent annulus:

$$\Delta P_L = \frac{4.8\mu q L_{\text{eff}}}{\pi D w^3 \rho h_{fg}} \tag{11.98e}$$

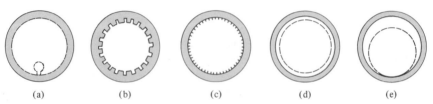

(a)   (b)   (c)   (d)   (e)

Figure 11.27 Cross sections of various wick structures: (*a*) artery, (*b*) channels, (*c*) screen, (*d*) concentric annulus, and (*e*) crescent annulus.

The quantities are defined as follows:

L_{eff} = effective length of heat pipe

r_e = effective channel radius

N = number of channels

b = screen tortuosity factor

R_W = outer radius of screen structure

R = radius of vapor passage

ϵ = screen void fraction

r_c = effective radius of screen openings

D = mean diameter of annulus

w = width of annulus

Although the artery wick system appears ideal, it requires an additional capillary network to distribute the liquid over surfaces which are used for heat addition and removal. Because of this complication, arteries are usually reserved for systems where boiling is likely to occur within the wick if the bulk of the liquid-return network is located in the path of the incoming heat. The consequences of such boiling will be discussed later.

Equation (11.98b) is essentially the same as Eq. (11.98c), except that it involves a number of channels, N, and an effective channel radius, r_e, which is obtained from the hydraulic diameter

$$\frac{D_H}{2} = r_e = 2\left(\frac{\text{flow area}}{\text{wetted perimeter}}\right)$$

Although open channels are subject to an interaction of vapor and liquid, which causes waves but no liquid entrainment, the interaction can be suppressed by covering the channels with a layer of fine-mesh screen. Because the screen is located at the interface of liquid and vapor, the fine pores of the screen provide large capillary forces for fluid circulation, while the channels provide a less restrictive flow path for liquid return. This general type of structure is called a *composite wick*.

All-screen composite wicks can be made by wrapping a layer of a fine screen around a mandrel followed by a second layer of coarse screen. The assembly can be placed in a container tube, the diameter of which is then drawn down until the inner wall makes contact with the coarse screen. The quantity $b/\epsilon r_c^2$ in Eq. (11.98c) can next be determined by liquid-flow measurements through the screen before the mandrel is removed.

An ideal wick system for liquid-metal working fluids consists of an inner porous tube separated from an outer container tube by a gap that provides an unobstructed annulus for liquid return. The pressure drop in a concentric annulus is obtained by deriving Poiseuille's equation for flow between two parallel plates. Although not as precise as the equation for flow between concentric cylinders, it is easier to handle and is fairly accurate provided the width of the annulus is small compared to its mean diameter. Equation (11.98e) for a crescent annulus is obtained by assuming the displacement obeys a cosine function—the width of the annulus doubles at the top of the tube, becomes zero at the bottom, and remains unchanged on the sides.

In Fig. 11.26, the wicking limitation is represented by the solid line between points 3 and 4. Although this limitation is shown to occur at temperatures where essentially all the pressure drop is in the liquid phase, the effect of a significant vapor-pressure drop is indicated by the dotted extension line at lower temperatures.

11.4.4 Boiling Limitations

In most two-phase flow systems the formation of vapor bubbles in the liquid phase (boiling) enhances convection, which is required for heat transfer. Such boiling is often difficult to produce in liquid-metal systems because the liquid tends to fill the nucleation sites necessary for bubble formation. In a heat pipe, convection in the liquid is not required because heat enters the pipe by conduction through a thin saturated wick. Furthermore, the formation of vapor bubbles is undesirable because they could cause hot spots and destroy the action of the wick. Therefore, heat pipes are usually heated isothermally before being used to allow the liquid to wet the inner heat pipe wall and to fill all but the smallest nucleation sites.

Boiling may occur at high input heat fluxes and high operating temperatures. The curve between points 4 and 5 in Fig. 11.26 is based on the equations

$$p_i - p_l = \frac{2\sigma}{r} \tag{11.99}$$

$$\frac{q}{A} = \frac{k(T_w - T_v)}{t} \tag{11.100}$$

where p_i is the vapor pressure inside the bubble, p_l the pressure in adjacent liquid, r the radius of the largest nucleation site, A the heat input area, k the effective thermal conductivity of saturated wick, T_w the temperature at the inside wall, T_v the temperature at the liquid-vapor interface, and t the thickness of the first layer in the wick (23).

Since the sizes of nucleation sites in a system are usually unknown, it is not possible to predict when boiling will occur. However, Eqs. (11.99) and (11.100) show how various factors influence boiling. For example, if nucleation sites are small, a large pressure difference will be required for bubbles to grow. For a given heat input flux, this pressure difference will depend on the thickness and thermal conductivity of the wick, on the saturation temperature of the vapor, and on the pressure drop in the vapor and liquid phases. This pressure drop is often overlooked because it is not a factor in the ordinary treatment of boiling.

Boiling is not a limitation with liquid metals, but when water is used as the working fluid, boiling may be a major heat transfer limitation because the thermal conductivity of the fluid is low and because it does not readily fill nucleation sites. Unfortunately, little experimental information is available concerning this limitation.

EXAMPLE 11.3 ───────────────────────────────────

Determine the maximum heat transport capability and the liquid flow rate of a water heat pipe operating at 100°C and atmospheric pressure. The heat pipe is 30 cm long and has an inner diameter of 1 cm. The heat pipe is inclined at 30° with the evaporator above the condenser. The wick consists of four layers of 250-mesh wire screen (wire diameter of 0.045 mm) on the inner surface of the pipe as shown in Fig. 11.27d.

Solution. The pressure balance relation to prevent dryout is

$$(\Delta p_c)_{max} \geq \Delta p_l + \Delta p_v + \Delta p_g$$

As a first approximation in the analysis we will neglect the vapor pressure drop Δp_v. Substituting Eq. (11.85) for the capillary pumping head Δp_c and Eqs. (11.81) and (11.84) for the liquid pressure drop Δp_l and the gravitational head Δp_c, respectively, gives

$$\frac{2\sigma_l \cos \theta}{r_c} = \frac{\mu_l q L_{eff}}{\rho_l h_{fg} A_w K_w} + \rho_l g L_{eff} \sin \phi$$

The area of the wick A_w is approximately

$$A_w = \pi D t = \pi(1 \text{ cm})(4)(0.0045 \text{ cm})$$
$$= 0.057 \text{ cm}^2$$

where t is the thickness of the four layers of wire mesh. The effective flow length L_{eff} is approximately 0.30 m. From Table 11.4, the pore radius r_c is 0.002 cm and the permeability K is 0.3×10^{-10} m². The water properties at 100°C are, from Table 13 in Appendix 2, and Table 10.2:

$$h_{fg} = 2.26 \times 10^6 \text{ J/kg}$$
$$\rho_l = 958 \text{ kg/m}^3$$
$$\mu_l = 279 \times 10^{-6} \text{ N s/m}^2$$
$$\sigma_l = 58.9 \times 10^{-3} \text{ N/m}$$

The maximum liquid flow rate through the wick can be obtained from the pressure balance equation. Assuming perfect wetting with $\theta = 0$, substituting $\dot{m}_{max} h_{fg}$ for q_{max}, and solving for $\dot{m}_{max}$ yields

$$\dot{m}_{max} = \left(\frac{2\sigma_l}{r_c} - \rho_l g L_{eff} \sin \phi\right) \frac{\rho_l h_{fg} A_w K}{\mu_l L_{eff} h_{fg}}$$

$$= \frac{2 \times 58.9 \times 10^{-3} \text{ N/m}}{0.002 \times 10^{-2} \text{ m}} - 958 \text{ kg/m}^3 \times 9.81 \text{ m/s}^2 \times 0.31 \text{ m} \times 0.5$$

$$= \frac{958 \text{ kg/m}^3 \times 0.057 \times 10^{-4} \text{ m}^2 \times 0.3 \times 10^{-10} \text{ m}^2}{279 \times 10^{-6} \text{ N s/m}^2 \times 0.30 \text{ m}}$$

$$= (5890 - 1362) \text{ N/m}^2 \times 0.20 \times 10^{-10} \text{ kg m}^2/\text{N s}$$

$$= 9.0 \times 10^{-6} \text{ kg/s}$$

The maximum heat transport capability is then, from Eq. (11.87),

$$q_{\max} = \dot{m}_{\max} h_{fg}$$
$$= 9.0 \times 10^{-6} \text{ kg/s} \times 2.26 \times 10^{6} \text{ J/kg}$$
$$= 20 \text{ W}$$

Note that the heat transport capability could be increased significantly by adding two or three layers of 100-mesh screen.

For a more complete treatment of the heat-pipe theory and practice, the reader is referred to (23, 24, 25, and 26).

11.5 SOLAR COLLECTORS

A solar collector is a heat exchanger capable of using solar radiation to increase the internal energy and temperature of a working fluid. In its simplest form it consists of a tube exposed to solar radiation. The solar insolation is partly absorbed by the tube, the temperature of the tube wall increases, and if a fluid at ambient temperature passes through the tube, heat is transferred from the tube to the fluid and the temperature of the fluid increases until the heat loss from the tube to the surroundings is equal to the solar energy absorbed. To improve the thermal performance of this simple system, fins can be attached to the tube to increase the area exposed to solar insolation and the heat losses can be reduced by placing one or two layers of glass between the incoming solar energy and the surface absorbing it (27). Figure 11.28 shows the cross section of a typical flat-plate solar collector. If a fluid such as water passes through the tubes, the useful energy delivered to the working fluid, q_u, is

$$q_u = \dot{m}c_p(T_{f,\text{out}} - T_{f,\text{in}}) \tag{11.101}$$

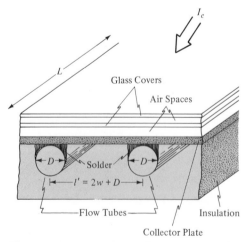

Figure 11.28   Flat-plate-collector cross section.

where $\quad \dot{m}$ = mass flow rate through collector

$\qquad c_p$ = constant pressure specific heat of working fluid

$(T_{f,\,\text{out}} - T_{f,\,\text{in}})$ = temperature rise of working fluid passing through collector

11.5.1 Energy Balance for a Flat-Plate Collector

The thermal performance of a solar collector can be evaluated by an energy balance that determines the portion of the incoming radiation delivered as useful energy to the working fluid. For a flat-plate collector of area A_c, this energy balance is

$$I_c A_c \bar{\tau}_s \alpha_{s,c} = q_u + q_{\text{loss}} + \frac{de_c}{dt} \qquad (11.102)$$

where $\quad I_c$ = solar irradiation on collector surface

$\qquad \bar{\tau}_s$ = effective solar transmittance of collector cover(s)

$\qquad \alpha_{s,c}$ = solar absorptance of collector absorber plate surface

$\qquad q_u$ = rate of heat transfer from collector absorber plate to working fluid

$\qquad q_{\text{loss}}$ = rate of heat transfer (or heat loss) from collector absorber plate to surroundings

$\qquad \dfrac{de_c}{dt}$ = rate of internal energy storage in collector

The instantaneous efficiency of a collector, η_c, is simply the ratio of the useful energy delivered to the total incoming solar energy:

$$\eta_c = \frac{q_u}{A_c I_c} \qquad (11.103)$$

In practice, the efficiency must be measured over a finite time period. In a standard performance test (28) the required period is of the order of 15 or 20 min, whereas for design the performance over some longer period t (e.g., a day or a month) is important. Then, the average efficiency is

$$\bar{\eta}_c = \frac{\displaystyle\int_0^t q_u\, dt}{\displaystyle\int_0^t A_c I_c\, dt} \qquad (11.104)$$

where t is the duration of the time period over which the performance is averaged.

A detailed and precise analysis of the efficiency of a solar collector is complicated by the nonlinear behavior of radiation heat transfer. However, a simple linearized analysis is usually sufficiently accurate at low temperature, and it illustrates the parameters of significance for a solar collector and how these parameters interact. Although for the design and economic evaluation of solar systems the results of standard performance tests are generally used, for a proper analysis and interpretation of these test results an understanding of the thermal analysis is imperative.

11.5.2 Collector-Heat-Loss Conductance

To gain an understanding of the parameters that determine the thermal efficiency of a solar collector, it is important to develop the concept of overall
collector-heat-loss conductance. Once the collector-heat-loss conductance U_c is
known, if the collector plate is at an average temperature T_c, the second right-
hand term in Eq. (11.102) can be written for a given ambient temperature T_a in
the simple form

$$q_{loss} = U_c A_c (T_c - T_a) \tag{11.105}$$

The simplicity of this relation is somewhat misleading because the collector-
heat-loss conductance cannot be specified without a detailed analysis of all the
heat losses. Figure 11.28 shows a schematic diagram of a double-glazed collector, and Fig. 11.29a shows the thermal circuit with all elements that must be
analyzed before they can be combined into a single conductance element, as
shown in Fig. 11.29b.

Figure 11.30 shows qualitatively the temperature distributions in a flat-
plate collector. Radiation impinges on the top of the plate connecting any two

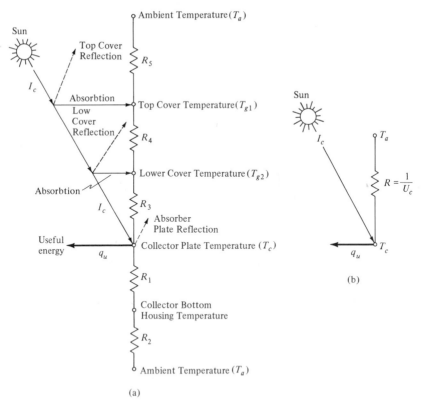

(a)

Figure 11.29   Thermal circuits for flat-plate collector shown in Fig. 11.28: (a) detailed
circuit; (b) approximate, equivalent circuit to (a). In both circuits, the absorber plate
absorbs incident energy equal to $\alpha_s I_s$, where $I_s = \bar{\tau}_s I_c$.

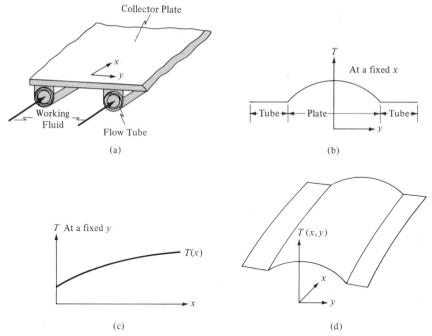

Figure 11.30 Temperature distribution in the absorber plate of a flat-plate collector: (a) Schematic diagram of absorber; (b) temperature profile in the direction of flow of the working fluid; (c) temperature profile at given y; (d) temperature distribution in the absorber plate.

adjacent flow tubes. It is absorbed uniformly by the plate and conducted later-ally toward the flow tubes, where it is then transferred by convection to the working fluid flowing through the ducts. It is apparent that at any cross section perpendicular to the flow direction the temperature is a maximum at the mid-point between two adjacent flow ducts and decreases along the plate toward the tube, as shown in Fig. 11.30b. Since heat is transferred to the working fluid, the temperature of the fluid as well as of the entire collector system will increase in the direction of flow. The increase in temperature at the midpoint between the two tubes is shown qualitatively in Fig. 11.30c. The temperature distribution in both the x and y directions is shown in three-dimensional view in Fig. 11.30d.

To construct a model suitable for a thermal analysis of a flat-plate collec-tor, the following simplifying assumptions will be made:

1. The collector is thermally in steady state.
2. The temperature drop between the top and bottom of the absorber plate is negligible.
3. Heat flow is one-dimensional through the covers as well as through the back insulation.
4. The headers connecting the tubes cover only a small area of the col-lector and provide uniform flow to the tubes.

5. The sky can be treated as though it were a blackbody source for infrared radiation at an equivalent sky temperature.

6. The irradiation on the collector plate is uniform.

For a quantitative analysis consider a location at x, y on a typical flat-plate collector, as shown in Fig. 11.30a. Let the plate temperature at this point be $T_c(x, y)$ and assume that solar energy is absorbed at the rate $I_s \alpha_s$. If the lower surface of the collector is well insulated, most of the heat loss occurs from the upper surface. The conductance for the upper surface of the collector can be evaluated by determining the thermal resistances R_3, R_4, and R_5 in Fig. 11.29a. Heat is transferred between the cover and the second glass plate and between the two glass plates by convection and radiation in parallel. Except for absorptance of solar energy by the second glass plate, the relations for the rate of heat transfer between T_c and T_{g2} and between T_{g2} and T_{g1} are the same. Thus, the rate of heat transfer per unit surface area of collector between the absorber plate and the second glass cover is

$$q_{\text{top loss}} = A_c \bar{h}_{c2}(T_c - T_{g2}) + \frac{\sigma(T_c^4 - T_{g2}^4)A_c}{1/\epsilon_{p,i} + 1/\epsilon_{g2,i} - 1} \tag{11.106}$$

where $\bar{h}_{c2}$ = heat transfer coefficient between plate and second glass cover

$\epsilon_{p,i}$ = infrared emittance of plate

$\epsilon_{g2,i}$ = infrared emittance of second cover

As shown in Chapter 1, if the radiation term is linearized, Eq. (11.106) becomes

$$q_{\text{top loss}} = (\bar{h}_{c2} + h_{r2})A_c(T_c - T_{g2}) = \frac{T_c - T_{g2}}{R_3} \tag{11.107}$$

where

$$h_{r2} = \frac{\sigma(T_c + T_{g2})(T_c^2 + T_{g2}^2)}{(1/\epsilon_{p,i}) + (1/\epsilon_{g2,i}) - 1} \tag{11.108}$$

A similar derivation for the rate of heat transfer between the two cover plates gives

$$q_{\text{top loss}} = (\bar{h}_{c1} + h_{r1})A_c(T_{g2} - T_{g1}) = \frac{T_{g2} - T_{g1}}{R_4} \tag{11.109}$$

where

$$h_{r1} = \frac{\sigma(T_{g1} + T_{g2})(T_{g1}^2 + T_{g2}^2)}{(1/\epsilon_{g1,i}) + (1/\epsilon_{g2,i}) - 1} \tag{11.110}$$

and h_{c1} = heat transfer coefficient between two transparent covers.

The emittances of the two covers will, of course, be the same if they are made of the same material. However, economic advantages can sometimes be achieved by using a plastic cover between an outer cover of glass and the plate, and in such a sandwich construction the radiative properties of the two covers may not be the same.

The equation for the thermal resistance between the upper surface of the outer collector cover and the ambient air has a form similar to the two preceding relations, but the heat transfer coefficient at the outer surface must be evaluated differently. If the air is still, natural-convection relations should be used, but when wind is blowing over the collector, forced-convection correlations apply as shown in Chapter 7. Radiation exchange occurs between the top cover and the sky at T_{sky}, whereas convection heat exchange occurs between T_{g1} and the ambient air at T_{air}. For convenience we shall refer both conductances to the air temperature. This gives

$$q''_{top\ loss} = (h_{c,\infty} + h_{r,\infty})(T_{g1} - T_{air}) = \frac{T_{g1} - T_{air}}{R_5} \tag{11.111}$$

where

$$h_{r,\infty} = \epsilon_{g1,i}\sigma(T_{g1} + T_{sky})(T_{g1}^2 + T_{sky}^2)\frac{T_{g1} - T_{sky}}{T_{g1} - T_{air}} \tag{11.112}$$

For a double-glazed flat-plate collector the total heat loss conductance $U_{c,total}$ can then be expressed in the form

$$U_{c,total} = \frac{1}{R_1} + \frac{1}{R_3 + R_4 + R_5} \tag{11.113}$$

where R_1 is the thermal resistance for the lower surface.

The evaluation of the collector-heat-loss conductance defined by Eq. (11.113) requires iterative solution of Eqs. (8.35) and (8.36) because the unit radiation conductances are functions of the cover and plate temperatures, which are not known a priori. An empirical procedure for calculating U_c for collectors with all covers of the same material, which is often sufficiently accurate and more convenient to use, has been suggested by Klein (29). For this approach the collector top loss, in watts, is written in the form

$$q_{top\ loss} = \frac{(T_c - T_a)A_c}{\dfrac{N}{(C/T_p)(T_c - T_a)/(N + f)^{0.33} + \dfrac{1}{h_{c,\infty}}}} + \frac{\sigma(T_c^4 - T_a^4)A_c}{\dfrac{1}{\epsilon_{p,i} + 0.05N(1 + \epsilon_{p,i})} + \dfrac{2N + f - 1}{\epsilon_{g,i}} - N} \tag{11.114}$$

where $f = (1 - 0.04h_{c,\infty} + 0.0005h_{c,\infty}^2)(1 + 0.091N)$

$C = 365.9(1 - 0.00883\beta + 0.00013\beta^2)$

N = number of covers

$h_{c,\infty} = 5.7 + 3.8\ V_\infty$ (V_∞ in m/s)

$\epsilon_{g,i}$ = infrared emittance of covers

The values of $q_{\text{top loss}}$ calculated from Eq. (11.114) agreed closely with the values obtained from Eq. (11.113).

To determine the efficiency of a solar collector, the rate of heat transfer to the working fluid must be calculated (30). If transient effects are neglected, the rate of heat transfer to the fluid flowing through a collector depends only on the temperature of the collector surface from which heat is transferred by convection to the fluid, the temperature of the fluid, and the heat transfer coefficient between the collector and the fluid. To calculate the rate of heat transfer, consider first the condition at a cross section of the collector shown in Fig. 11.28. Solar radiant energy impinging on the upper face of the collector plate is conducted in a transverse direction toward the flow channels. The temperature is a maximum at any midpoint between adjacent channels and the collector plate acts as a fin attached to the walls of the flow channel. The thermal performance of the fin plate can be expressed in terms of its fin efficiency η_f, defined as the ratio of the rate of heat flow through the real fin to the rate of heat flow through a fin of infinite thermal conductivity.

If U_c is the overall unit conductance from the collector plate surface to the ambient air, the rate of heat loss from a given segment of the collector plate at x, y in Fig. 11.30 is

$$q(x, y) = U_c[T_c(x, y) - T_a] \, dx \, dy \qquad (11.115)$$

where T_c = local collector plate temperature

T_a = ambient air temperature

U_c = overall unit conductance between plate and ambient air

If conduction in the x direction is negligible, a heat balance at a given distance x_0 for a cross section of the flat-plate collector per unit length in the x direction can be written in the form

$$\alpha_s I_s \, dy - U_c(T_c - T_a) \, dy + \left[\left(-kt \frac{dT_c}{dy} \bigg|_{y, x_0} \right) - \left(-kt \frac{dT_c}{dy} \bigg|_{y + dy, x_0} \right) \right] = 0$$

$$(11.116)$$

If the plate thickness t is uniform and the thermal conductivity of the plate is independent of temperature, Eq. (11.116) can be cast into the form of a second-order differential equation:

$$\frac{d^2 T_c}{dy^2} = \frac{U_c}{kt} \left[T_c - \left(T_a + \frac{\alpha_s I_s}{U_c} \right) \right] \qquad (11.117)$$

The boundary conditions for the system described above are:

1. At the center between any two ducts, the heat flow is zero, or at $y = 0$: $dT_c/dy = 0$.
2. At the duct the plate temperature is $T_b(x_0)$, or at $y = w = [(l' - D)/2]$: $T_c = T_b(x_0)$, where $T_b(x_0)$ is the temperature at the fin base.

If we let $m^2 = U_c/kt$ and $\Phi = T_c - (T_a + \alpha_s I_s/U_c)$, Eq. (8.42) becomes

$$\frac{d^2\Phi}{dy^2} = m\Phi \qquad (11.118)$$

subject to the boundary conditions

$$\frac{d\Phi}{dy} = 0 \quad \text{at} \quad y = 0$$

and

$$\Phi = T_b(x_0) - \left(T_a + \frac{\alpha_s I_s}{U_c}\right) \quad \text{at} \quad y = \frac{l' - D}{2}$$

The general solution of Eq. (11.118) is

$$\Phi = C_1 \sinh my + C_2 \cosh my \qquad (11.119)$$

The constants C_1 and C_2 can be determined by substituting the two boundary conditions and solving the two resulting equations for C_1 and C_2. This gives

$$\frac{T_c - (T_a + \alpha_s I_s/U_c)}{T_b(x_0) - (T_a + \alpha_s I_s/U_c)} = \frac{\cosh my}{\cosh mw} \qquad (11.120)$$

From the preceding equation the rate of heat transfer to the conduit from the portion of the plate between two conduits can be determined by evaluating the temperature gradient at the base of the fin per unit width, or

$$q_{\text{fin}} = -kt \left.\frac{dT_c}{dy}\right|_{y=w} = \frac{1}{m}\left[\alpha_s I_s - U_c(T_b(x_0) - T_a)\tanh mw\right] \qquad (11.121)$$

Since the conduit is connected to fins on both sides, the total rate of heat transfer is

$$q_{\text{total}}(x_0) = 2w[\alpha_s I_s - U_c(T_b(x_0) - T_a)]\frac{\tanh mw}{mw} \qquad (11.122)$$

If the entire fin were at the temperature $T_b(x)$, a situation corresponding physically to a plate of infinitely large thermal conductivity, the rate of heat transfer would be a maximum, $q_{\text{total, max}}$. As mentioned previously, the ratio of the rate of heat transfer with a real fin to the maximum rate obtainable is the fin efficiency η_f. Using this definition, Eq. (11.122) can be written in the form

$$q_{\text{total}}(x) = 2w\eta_f[\alpha_s I_s - U_c(T_b(x_0) - T_a)] \qquad (11.123)$$

where

$$\eta_f \equiv \frac{\tanh mw}{mw}$$

The fin efficiency η_f is plotted as a function of the dimensionless parameter $(U_c/kt)^{1/2}w$ in Fig. 11.31. When the fin efficiency approaches unity, the maximum

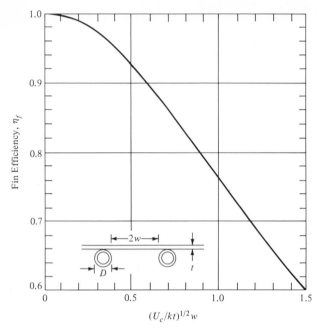

Figure 11.31 Fin efficiency for tube-and-sheet flat-plate solar collectors.

portion of the radiant energy impinging on the fin becomes available for heating the fluid.

In addition to the heat transferred through the fin, the energy impinging on the portion of the plate above the flow passage also provides useful energy. The rate of useful energy from this region available to heat the working fluid is

$$q_{duct}(x) = D[\alpha_s I_s - U_c(T_b(x_0) - T_a)]$$ (11.124)

Thus the useful energy per unit length in the flow direction becomes

$$q_u(x) = (D + 2w\eta_f)[\alpha_s I_s - U_c(T_b(x_0) - T_a)]$$ (11.125)

The energy $q_u(x)$ must be transferred as heat to the working fluid. If the thermal resistance of the metal wall of the flow duct is negligibly small and there is no contact resistance between the duct and the plate, the rate of heat transfer to the fluid is

$$q_u(x) = (\pi D_i)\bar{h}_{c,i}[T_b(x_0) - T_f(x_0)]$$ (11.126)

11.5.3 Collector Efficiency Factor

To obtain a relation for the useful energy delivered by a collector in terms of known physical parameters, the fluid temperature, and the ambient temperature, the collector temperature must be eliminated from Eqs. (11.125) and (11.126). Solving for $T_b(x_0)$ in Eq. (11.126) and substituting this relation in Eq.

TABLE 11.6 TYPICAL VALUES OF THE PARAMETERS
THAT DETERMINE THE COLLECTOR
EFFICIENCY FACTOR F' FOR A
FLAT-PLATE COLLECTOR IN
EQS. (11.127) AND (11.128)

U_c	two glass covers: 4 W/m² K one glass cover: 8 W/m² K
kt	copper plate, 1 mm thick: 0.4 W/K steel plate, 1 mm thick: 0.005 W/K
$h_{c,i}$	water in laminar flow forced convection: 300 W/m² K water in turbulent flow forced convection: 1500 W/m² K air in turbulent forced convection: 100 W/m² K

(11.125) gives

$$q_u(x) = (D + 2w)F'[\alpha_s I_s - U_c(T_f(x_0) - T_a)] \tag{11.127}$$

where F' is called the *collector efficiency factor*. It is given by

$$F' = \frac{1/U_c}{(D + 2w)\left[\dfrac{1}{U_c(D + 2w\eta_f)} + \dfrac{1}{\overline{h}_{c,i}(\pi D_i)}\right]} \tag{11.128}$$

Physically, the denominator in Eq. (11.128) is the thermal resistance between the fluid and the environment, whereas the numerator is the thermal resistance between the collector and the ambient air. The collector plate efficiency factor F' depends on U_c, $h_{c,i}$, and η_f. It is only slightly dependent on temperature and can for all practical purposes be treated as a design parameter. Typical values of the factors that determine the value of F' are given in Table 11.6.

The collector efficiency factor increases with increasing plate thickness and plate thermal conductivity but decreases with increasing distance between flow channels. Also, increasing the heat transfer coefficient between the walls of the flow channel and the working fluid increases F', but an increase in the overall conductance U_c will cause F' to decrease.

11.5.4 Collector-Heat-Removal Factor

Equation (11.127) yields the rate of heat transfer to the working fluid at a given point x along the plate for specified collector and fluid temperatures. However, in a real collector the fluid temperature increases in the direction of flow as heat is transferred to it. An energy balance for a section of flow duct dx can be written in the form

$$-\dot{m}c_p(T_f|_{x+dx} - T_f|_x) = q_u(x)\,dx \tag{11.129}$$

Substituting Eq. (11.127) for $q_u(x)$ and setting

$$\left[T_f(x) + \frac{dT_f(x)}{dx}\,dx\right] = T_f\Big|_{x+dx}$$

in Eq. (11.129) gives the differential equation

$$-\dot{m}c_p \frac{dT_f(x)}{dx} = (D + 2w)F'[\alpha_s I_s - U_c(T_f(x) - T_a)]$$

Separating the variables gives, after some rearranging,

$$\frac{dT_f(x)}{T_f(x) - T_a - (\alpha_s I_s/U_c)} = \frac{(D + 2w)F'U_c}{\dot{m}c_p} dx \qquad (11.130)$$

Equation (11.130) can be integrated and solved for the outlet temperature of the fluid, $T_{f,\text{out}}$, for a duct length L and fluid inlet temperature $T_{f,\text{in}}$, if we assume that F' and U_c are constant, or

$$\frac{T_{f,\text{out}} - T_a - \alpha_s I_s/U_c}{T_{f,\text{in}} - T_a - \alpha_s I_s/U_c} = \exp\left(-\frac{U_c(D + 2w)F'L}{\dot{m}c_p}\right) \qquad (11.131)$$

To compare the performance of a real collector with the thermodynamic optimum, it is convenient to define the heat removal factor F_R as the ratio of the actual rate of heat transfer to the working fluid, to the rate of heat transfer at the maximum temperature difference between the absorber and the environment. The thermodynamic limit corresponds to the condition of the working fluid remaining at the inlet temperature throughout the collector. This can be approached when the fluid velocity is very high. From its definition F_R can be expressed as

$$F_R = \frac{G_c c_p(T_{f,\text{out}} - T_{f,\text{in}})}{\alpha_s I_s - U_c(T_{f,\text{in}} - T_a)} \qquad (11.132)$$

where G_c is the flow rate per unit surface area of collector, $\dot{m}/A_c$. By regrouping the right-hand side of Eq. (11.132) and combining with Eq. (11.131), it can be easily verified that

$$F_R = \frac{G_c c_p}{U_c}\left[1 - \frac{(\alpha_s I_s/U_c) - (T_{f,\text{out}} - T_a)}{(\alpha_s I_s/U_c) - (T_{f,\text{in}} - T_a)}\right]$$

or

$$F_R = \frac{G_c c_p}{U_c}\left[1 - \exp\left(-\frac{U_c F'}{G_c c_p}\right)\right] \qquad (11.133)$$

Inspection of the relation above shows that F_R increases with increasing flow rate and approaches as an upper limit F', the collector efficiency factor. Since the numerator of the right-hand side of Eq. (11.132) is q_u, the rate of useful heat transfer can now be expressed in terms of the fluid inlet temperature:

$$q_u = A_c F_R[\alpha_s I_s - U_c(T_{f,\text{in}} - T_a)] \qquad (11.134)$$

This is a convenient form for design because the fluid inlet temperature to the collector is usually known or can be specified.

EXAMPLE 11.4

Calculate the averaged hourly and daily efficiency of a water solar collector on January 15, in Boulder, Colorado. The collector is tilted at an angle of 60° and has an overall conductance of 8.0 W/m²·K on the upper surface. It is made of copper tubes, 1 cm ID, 0.05 cm thick, connected by a 0.05-cm-thick plate at a center-to-center distance of 15 cm. The heat transfer coefficient for the water in the tubes is 1500 W/m²·K, the cover transmittance is 0.9, and the solar absorptance of the copper surface is 0.9. The collector is 1 m wide and 2 m long, the water inlet temperature is 330 K, and the water flow rate is 0.02 kg/s. The insolation, I_c, and the environmental temperature are tabulated below.

Time (h)	I_c (W/m²)	T_{amb} (K)
7–8	94	270
8–9	208	280
9–10	387	283
10–11	583	286
11–12	804	290
12–13	828	290
13–14	720	288
14–15	579	288
15–16	387	284
16–17	267	280

Solution. We shall assume that the collector operates at a "quasi" steady state during each hour as shown in the tabulation above. Using the analysis presented for the configuration as specified, first, the fin efficiency is obtained from Eq. (11.123):

$$\eta_f = \frac{\tanh mw}{mw}$$

With

$$m = \left(\frac{U_c}{kt}\right)^{1/2} = \left(\frac{8}{390 \times 5 \times 10^{-4}}\right)^{1/2} = 6.4$$

we get

$$\eta_f = \frac{\tanh[6.4(0.15 - 0.01)/2]}{6.4(0.15 - 0.01)/2} = 0.938$$

The collector efficiency factor F' is from Eq. (11.128):

$$F' = \frac{1/U_c}{(D + 2w)\left[\dfrac{1}{U_c(D + 2w\eta_f)} + \dfrac{1}{h_{c,i}\pi D_i}\right]}$$

$$= \frac{1/8.0}{0.15\left[\dfrac{1}{8.0(0.01 + 0.14 \times 0.938)} + \dfrac{1}{(1500\pi) \times (0.01)}\right]} = 0.920$$

Then we obtain the heat removal factor from Eq. (11.133):

$$F_R = \frac{G_c c_p}{U_c} [1 - e^{-(U_c F'/G_c c_p)}]$$

$$= \frac{0.01 \times 4184}{8} [1 - e^{-(8.0 \times 0.920/0.01 \times 4184)}] = 0.844$$

The useful heat delivery rate is, from Eq. (11.134),

$$q_u = A_c F_R [\alpha_s I_c - U_c (T_{f, in} - T_a)]$$

In the above relation I_c is the radiation incident on the collector cover glass. If the transmittance of the glass is 0.9, the radiation incident on the collector plate is

$$I_s = \tau I_c = 0.9 I_c \quad \text{and}$$
$$q_u = 2(0.844)[I_c(0.81) - 8.0(T_{f, in} - T_a)]$$

The efficiency of the collector is $\eta_c = q_u/A I_c$ and the hourly averages are tabulated below.

Hour	I_c (W/m^2)	q_u (W)	η_c
7–8	94	0	0
8–9	208	0	0
9–10	387	0	0
10–11	583	221	0.190
11–12	804	584	0.363
12–13	828	619	0.374
13–14	720	438	0.304
14–15	579	237	0.205
15–16	387	0	0
16–17	267	0	0
	$\Sigma I_{tot} = 4837$ W/m^2	$\Sigma q_u = 2099$ W	

Note that during the early morning and late afternoon the sun is too low on the horizon to deliver useful energy. The daily average is obtained by summing the useful energy for those hours during which the collector delivers heat and dividing by the total insolation between sunrise and sunset, as shown in the table above. This yields

$$\bar{\eta}_{b, day} = \frac{\Sigma q_u}{\Sigma A_c I_c} = \frac{2099}{4837} = 0.43\%$$

The presentation in this section is restricted to flat plate collectors that do not concentrate the solar radiation. Flat plate collectors are widely used for domestic hot water heating and for preheating water for commercial applications such as canning or hotel services. They constitute by far the largest percentage of solar installations worldwide. But they are limited to relatively low temperature service. To achieve higher temperatures it is necessary to concentrate the sunlight. This can be achieved by a variety of designs, but they are beyond the scope of an introductory text. For the design and analysis of concentrating solar collectors, the reader is referred to specialized books on solar thermal energy conversion. (36)

11.6 HEAT TRANSFER ENHANCEMENT

Heat transfer enhancement is the practice of modifying a heat transfer surface to increase the heat transfer coefficient between the surface and a fluid. In previous chapters we have treated some practical examples of heat transfer enhancement, e.g., fins, surface roughness. Heat transfer enhancement may also be achieved by surface or fluid vibration, electrostatic fields, or mechanical stirrers. These latter methods are often referred to as active techniques because they require the application of external power. The active techniques have received attention in the literature although practical applications have been limited. In this section we shall focus on the passive techniques, i.e., those based upon modification of the heat transfer surface.

Increases in heat transfer due to the surface treatment can be brought about by increased turbulence, increased surface area, improved mixing, or flow swirl. These effects generally result in an increase in pressure drop with the increase in heat transfer. The associated increase in pumping work is almost always greater than the increase in heat transfer relative to a smooth (untreated) heat transfer surface of the same nominal (base) heat transfer area. But heat transfer enhancement is gaining industrial importance because it gives one the opportunity to: (1) reduce the heat transfer surface area required for a given application and, therefore, reduce the heat exchanger size and cost, (2) increase the heat duty of the exchanger, or (3) permit closer approach temperatures. All of these can be visualized from the expression for heat duty for a heat exchanger, Eq. (8.16):

$$Q = U_0 A \text{ LMTD} \tag{8.16}$$

Any enhancement technique that increases the heat transfer coefficient also increases the overall conductance U_0. Therefore, one may either reduce the heat transfer area A, increase the heat duty Q, or decrease the temperature difference LMTD for fixed Q and LMTD, A and LMTD, or Q and A, respectively. Enhancement may also be used to prevent overheating of heat transfer surfaces in systems with a fixed heat generation rate.

In any practical application, a complete analysis is required to determine the economic benefit of enhancement. Such an analysis must include a possible increased first cost because of the enhancement, increased heat exchanger heat transfer performance, the effect on operating costs (especially a potential increase in pumping power because of roughness, turbulence promoters, and swirl devices), and maintenance costs. A major practical concern in industrial applications is the increased fouling of the heat exchange surface caused by the enhancement. Accelerated fouling can quickly eliminate any incease in the heat transfer coefficient achieved by enhancement of a clean surface.

11.6.1 Practical Embodiments

There is a very large, rapidly growing body of literature on the subject of heat transfer enhancement. Bergles et al. (31) cataloged 3045 papers and reports as of late 1983 and over 500 U.S. patents related to the technology. They logged the technical articles into a computerized information retrieval system that classifies each document according to the type of flow treated (single-phase natural convection, single-phase forced convection, pool boiling, flow boiling, condensation, etc.) and type of enhancement (rough surface, extended surface, displaced enhancement devices, swirl flow, fluid additives, vibration, etc.).

Table 11.7 shows how each enhancement technique applies to the different types of flow according to (31). Extended surfaces are probably the most common heat transfer enhancement technique. An example of an extended surface is shown in Fig. 11.32a. The fin was discussed in Chapter 2 as an extended surface with primary application in gas-side heat transfer. Effectiveness of the fin in this service is based on the poor thermal conductivity of the gas relative to that of the fin material. Thus, while the temperature drop along the fin reduces its effectiveness somewhat, overall an increase in surface area and thus heat transfer performance is realized. Recently, several manufacturers have made available tubing with integral internal fins. Extended surfaces may also take the form of interrupted fins where the objective is to force the redevelopment of boundary layers. Compact heat exchangers (32, 33) make use of extended surfaces to give a required heat transfer surface area in as small a volume as possible. This type of heat exchanger is important in applications such as automobile radiators, and gas turbine regenerators, where the overall size of the heat exchanger is of major concern.

Rough surfaces refer to small roughness elements approximately the height of the boundary-layer thickness. An example of a rough surface applied to the inner diameter of a tube is shown in Fig. 11.32b. The roughness elements provide only a small increase in surface area relative to the extended surfaces. Their

TABLE 11.7 APPLICATION OF ENHANCEMENT TECHNIQUES TO DIFFERENT TYPES OF FLOWS

	Single-phase natural convection	Single-phase forced convection	Pool boiling	Flow boiling	Condensation
Extended surfaces	c[a]	c	c	o	c
Rough surfaces	o[b]	c	o	c	c
Displaced enhancement devices		o		o	
Swirl flow devices		c		c	o
Treated surfaces		c	c	o	c

[a] c = commonly practiced.
[b] o = occasionally practiced.

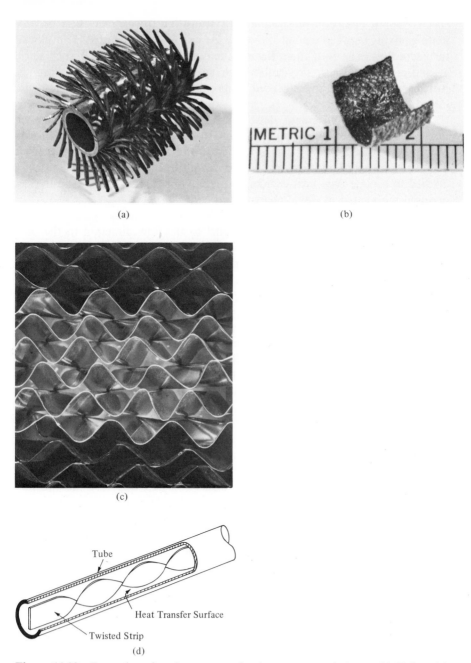

(a)

(b)

(c)

Tube

Heat Transfer Surface

Twisted Strip

(d)

Figure 11.32 Examples of various types of enhancement techniques. (*a*) Tube with exterior fins produced by machining process. (*b*) Section of a tube with sand-grained roughness. (*c*) Corrugated surface of a plate heat exchanger (view is down the flow channel). The corrugations strengthen the plates and promote mixing in the flow channels. The corrugations are considered a displacement device. (*d*) Swirl tape inserted into a tube.

effectiveness is based on promoting early transition to turbulent flow or promoting mixing between the bulk flow and the viscous sublayer in fully developed turbulent flow. The roughness elements may be randomly shaped, such as on a sand-grained surface, or regular such as machined grooves or pyramids. Rough surfaces are primarily used to promote heat transfer in single-phase forced convection.

Displaced enhancement devices are inserted into the flow channel to improve mixing between the bulk flow and the heat transfer surface. A common example is the static mixer that is in the form of a series of corrugated sheets meant to promote bulk flow mixing. An example of such displacement device applied to a plate-type heat exchanger is shown in Fig. 11.32c. These devices are used most often in single-phase forced convection.

Swirl flow devices inserted into the flow channel are designed to impart a rotational motion about an axis parallel to the flow direction to the bulk flow. An example of a twisted-tape swirl flow device inserted into a tube is shown in Fig. 11.32d. Enhancement arises due to increased flow velocity, secondary flows generated by the swirl, or increased flow path length in the flow channel. Swirl flow devices are used for single-phase forced flow and in flow boiling. The flow boiling application in Section 10.3.3 showed that a swirl flow device can increase the boiling critical heat flux to 174 MW/m^2.

Treated surfaces are used primarily in pool boiling and condensing applications. These consist of very small surface structures such as the surface inclusions discussed in Section 10.2.1, which promote nucleate boiling by providing bubble nucleation sites. Condensation may be enhanced by promoting the formation of droplets, rather than a film, on the condensing surface. This may be accomplished by coating the surface with a material that leaves the surface nonwetting.

Figure 11.33 compares the performance of four enhancement techniques for single-phase forced convection in a tube with that for a smooth tube (34). The basis of comparison is the heat transfer (Nusselt number) and pressure drop (friction factor) plotted as a function of the Reynold number. One can see that at a given Reynolds number, all four enhancement techniques provide an increased Nusselt number relative to the smooth tube but at the expense of an even greater increase in the friction factor.

11.6.2 Analysis of Enhancement Techniques

We have previously noted the need for a comprehensive analysis of any candidate enhancement technique to determine potential benefits. Since heat transfer enhancement may be used to accomplish several goals, no general procedure that would allow one to compare different enhancement techniques exists. A comparison such as that given in Fig. 11.33, which is limited to the thermal and hydraulic performance of the heat exchange surface, is often a useful starting point. Other factors that must be included in the analysis are the hydraulic diameter, length of the flow passages, and flow arrangement (crossflow or

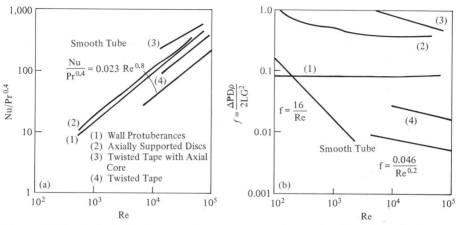

Figure 11.33 Typical data for turbulence promoters inserted inside tubes. (*a*) Heat transfer data; (*b*) friction data. (From W. M. Rohsenow and J. P. Hartnett, eds., *Handbook of Heat Transfer*, McGraw Hill, New York, 1985.)

counterflow, etc.). In addition to these geometric variables, the flow rate per passage or Reynolds number and the LMTD may be varied, or may be constrained for a given application. Those that can be varied must be adjusted in the analysis to produce the desired goal, e.g., increased heat duty, minimum surface area, or reduced pressure drop. Table 11.8 lists the variables that should be considered in a complete analysis.

Fortunately, many applications constrain one or more of these variables, thereby simplifying the analysis. As an example, consider an existing shell-and-tube heat exchanger being used to condense a hydrocarbon vapor on the shell

TABLE 11.8 VARIABLES IN THE ANALYSIS OF HEAT TRANSFER ENHANCEMENT

Symbol	Description	Comments
1. —	Type of enhancement technique	
2. $Nu(Re_D)$	Thermal performance of the enhancement technique	Determined by choice of technique
3. $f(Re_D)$	Hydraulic performance of the enhancement technique	Determined by choice of technique
4. Re_D	Flow Reynolds number	Probably an independent variable
5. D	Flow passage hydraulic diameter	May be determined by choice of technique
6. L	Flow passage length	Generally an independent variable with limits
7. —	Flow arrangement	May be determined by choice of technique
8. LMTD	Terminal flow temperatures	May be determined by the application
9. Q	Heat duty	Probably a dependent variable
10. A_s	Heat transfer surface area	Probably a dependent variable
11. Δp	Pressure drop	Probably a dependent variable

side with chilled water pumped through the tube side. It may be possible to increase the flow of vapor by increasing the water-side heat transfer since the vapor-side thermal resistance is probably negligible. Suppose the pressure drop on the water side is fixed due to pump constraints, and assume that it is necessary to keep the heat exchanger size and configuration the same to simplify installation costs. The water-side heat transfer could be increased by any of several devices placed in the tubes such as swirl tapes, static mixers, or roughness on the tube inner diameter. Assuming that thermal and hydraulic performance data are available for each enhancement technique to be considered, then items 1, 2, and 3 in Table 11.8, as well as 5, 6, 7, and 10, are known. We will adjust the Re_D, which will affect the water outlet temperature or LMTD, Q, and Δp. Since the LMTD is not important (within reason) we can determine which surface provides the largest Q (and hence vapor flow) at a fixed Δp.

Soland et al. (35) have developed a useful performance ranking methodology that incorporates the thermal/hydraulic behavior of the heat transfer surface with the flow parameters and the geometric parameters for the heat exchanger. For each heat exchange surface the method plots the fluid pumping power per unit volume of heat exchanger versus heat exchanger NTU per unit volume. These parameters are:

$$\frac{P}{V} = \frac{\text{pumping power}}{\text{volume}} \propto \frac{f \, Re_{D_H}{}^3}{D_H{}^3} \tag{11.135}$$

$$\frac{NTU}{V} = \frac{NTU}{\text{volume}} \propto \frac{j Re_{D_H}}{D_H{}^2} \tag{11.136}$$

Given the friction factor $f(Re)$ and the heat transfer performance $Nu(Re)$ or $j(Re)$ for the heat exchanger surface and the flow passage hydraulic diameter, D_H, one can easily construct a plot of the two parameters P/V and NTU/V.

In Eqs. (11.135) and (11.136) the Reynolds number is based upon the flow area A_f, which ignores any enhancement:

$$Re_{D_H} = \frac{G D_H}{\mu} \tag{11.137}$$

$$G = \frac{\dot{m}}{A_f} \tag{11.138}$$

where $\dot{m}$ is the mass flow rate in the flow passage of area A_f.

The friction factor is

$$f = \frac{\Delta p}{4(L/D_H)(G^2/2\rho)} \tag{11.138}$$

where Δp is the frictional pressure drop in the core.

The j factor is defined as

$$j = \frac{\bar{h}_c}{G c_p} Pr^{2/3} \tag{11.139}$$

where $\bar{h}_c$ is the heat transfer coefficient based on the base (without enhancement) surface area A_b. The hydraulic diameter is defined as in Chapter 6, but can be written more conveniently in the form

$$D_H = \frac{4V}{A_b} \qquad (11.140)$$

Using these definitions, a smooth tube of inside diameter D and a tube of inside diameter D with a twisted tape insert and with the same mass flow rate would have the same G, Re_D, A_b, and D but we would expect f and j to be larger for the latter tube.

Such a plot is useful for comparing two heat exchange surfaces because it allows a convenient comparison based on any of these constraints:

a. Fixed heat exchanger volume and pumping power
b. Fixed pumping power and heat duty
c. Fixed volume and heat duty

These constraints may be visualized with Fig. 11.34 in which the $f\mathrm{Re}_D{}^3/D^3$ and $j\,\mathrm{Re}_D/D^2$ data are plotted for the two surfaces to be compared. From the baseline point labeled "o" in Fig. 11.34, comparisons based on the three constraints are labeled.

A comparison based on constraint (a) may be made by constructing a vertical line through the baseline point. Comparing the two ordinate values where the vertical line intersects the curves allows one to compare the heat duty for each surface. The surface with the highest curve will transfer more heat. Constraint (b) may be visualized by constructing a line with slope $+1$.

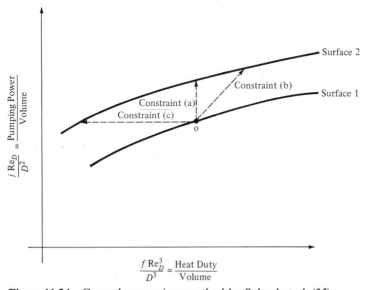

Figure 11.34   General comparison method by Soland et al. (35).

Comparing either the abscissa or ordinate where the line of slope $+1$ intersects the curves allows one to compare the heat exchanger volume required for each surface. The surface with the highest curve will require the least volume. Constraint (c) may be visualized by constructing a horizontal line. Comparing the abscissa where the line intersects the curves allows one to compare the pumping power for each surface. The surface with the highest curve will require the least pumping power.

EXAMPLE 11.5

Given the data in Fig. 11.33, compare the performance of wall protuberances and a twisted tape [surfaces (1) and (4) in Fig. 11.33] for a flow of air on the basis of fixed heat exchanger volume and pumping power. Assume that both surfaces are applied to the ID of a 1-cm-ID tube of circular cross section.

Solution. We must first construct the $f(\text{Re})$ and $j(\text{Re})$ curves for the two surfaces.

Curves (1) and (4) in Fig. 11.33a and b, can be represented by straight lines with good accuracy. From the data in Fig. 11.33a and b, these straight lines for the Nusselt numbers are

$$Nu_1/Pr^{0.4} = 0.054\ Re_D^{0.805}$$
$$Nu_4/Pr^{0.4} = 0.057\ Re_D^{0.772}$$

where the subscripts 1 and 4 denote surfaces 1 and 4.

Since $j = St\ Pr^{2/3} = Nu Re_D^{-1} Pr^{-1/3}$ we have

$$j_1 = 0.054\ Re_D^{-0.195} Pr^{1/15}$$

and

$$j_4 = 0.057\ Re_D^{-0.228} Pr^{1/15}$$

For the friction coefficient data we find

$$f_1 = 0.075\ Re_D^{0.017}$$
$$f_4 = 0.222\ Re_D^{-0.238}$$

In comparing the two surfaces we should restrict ourselves to the range

$$10^4 < Re_D < 10^5$$

where the data for both surfaces are valid.

Constructing the two comparison parameters, we have

$$\frac{f_1\ Re_D^3}{D_1^4} = \frac{0.075\ Re_D^{3.017}}{(0.01)^4} = 7{,}500{,}000\ Re_D^{3.017}\quad m^{-4}$$

$$\frac{f_4\ Re_D^3}{D_4^4} = \frac{0.222\ Re_D^{2.76}}{(0.01)^4} = 22{,}200{,}000\ Re_D^{2.76}\quad m^{-4}$$

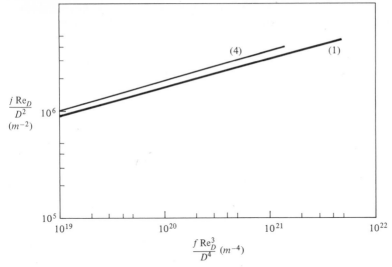

Figure 11.35 Comparison of wall protuberences and twisted tapes based on the method of Soland et al. (35).

$$\frac{j_1 \, \mathrm{Re}_D}{D_1^{\,2}} = \frac{0.054 \, \mathrm{Re}_D^{\,0.805} \mathrm{Pr}^{1/15}}{(0.01)^2} = 527.8 \, \mathrm{Re}_D^{\,0.805} \quad \mathrm{m}^{-2}$$

$$\frac{j_4 \, \mathrm{Re}_D}{D_4^{\,2}} = \frac{0.057 \, \mathrm{Re}_D^{\,0.772} \mathrm{Pr}^{1/15}}{(0.01)^2} = 557.1 \, \mathrm{Re}_D^{\,0.772} \quad \mathrm{m}^{-2}$$

These parameters are plotted in Fig. 11.35 for the Reynolds number range of interest. According to the specified constraint, a vertical line connecting the curves labeled (1) and (4) in Fig. 11.33 clearly demonstrates that surface 4, the twisted tape, is the better of the two surfaces. That is, for a fixed heat exchanger volume and at constant pumping power the twisted tape enhancement will transfer more heat.

PROBLEMS

The problems for this chapter are organized by subject matter as shown below

Topic	Section	Problem Number
Second law analysis	11.1	11.1 to 11.11
High-speed flow	11.2	11.12 to 11.16
Non-Newtonian flow	11.3	11.17 to 11.18
Heat pipes	11.4	11.19 to 11.21
Solar collectors	11.5	11.22 to 11.25
Heat transfer enhancement	11.6	11.26 to 11.29

11.1 Superheated steam at 1000 K and 2 MPa is considered for use in a steady-flow power conversion device. If the dead state is $T_o = 25°C$ and $P_o = 1$ atm, calculate the exergy in kilojoules per kilogram for the above conditions.

11.2 Derive an expression for the specific exergy of an ideal gas, Ex, under steady flow conditions in terms of the pressure and temperature ratios P/P_0 and T/T_0, where P_o and T_o are dead-state conditions, in the dimensionless form

$$\frac{Ex}{c_p T_o} = f\left(\frac{T}{T_o}, \frac{P}{P_o}\right)$$

11.3 In the Reynolds number range between 10^3 and 10^5 the drag coefficient for a sphere can be approximated by

$$\frac{F_D/\pi D^2}{\rho U_\infty^2/2} = 0.5$$

For a sphere of diameter D in a uniform flow field with velocity U_∞ and temperature T_∞ determine the rate of entropy generation if the rate of heat transfer from the sphere to the fluid is fixed at $\bar{q}$ (W/K).

11.4 Determine the sphere diameter for the conditions in Problem 11.3, D_{opt}, that would minimize the rate of entropy generation, and give a practical example to which this result could be applied.

11.5 Consider an isothermal solar collector operating with a solar flux q_s'' at temperature T_c and delivering heat at the rate q_u with the dead state at T_o (see diagram below). If the overall heat loss coefficient as defined in Chapter 8 is U_c and the area of the solar collector is A_c, show that the rate of entropy generation can be expressed in dimensionless form by

$$N_3 = \frac{S_{gen}'}{U_c A_c} = \frac{\theta_c + \theta_{max}}{\theta_c} - \frac{q_s''}{U_c T_s}$$

where T_s = sun absolute temperature

$$\theta = \frac{T_c}{T_o}$$

$$\theta_{max} = \frac{T_{c,max}}{T_o}$$

$T_{c,max}$ = stagnation temperature of collector when $q_u = 0$

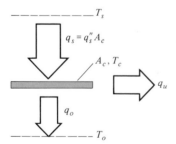

11.6 For the conditions of Problem 11.5 show that entropy generation is minimized when the collector operates at a temperature corresponding to $\theta_{max}^{1/2}$.

11.7 What is the thermal efficiency, $\eta_c = q_u/q_s$, at which entropy generation is a minimum for Problem 11.5.

11.8 Derive Eq. (11.20) in detail and then calculate the optimum Reynolds number $Re_{D, opt}$ for flow of water at 30°C and an average velocity of 0.1 m/s through an electrically heated tube at 100°C wall temperature. State your assumptions.

11.9 Derive the optimum Reynolds number for minimum irreversibility as given by Eq. (11.28) in detail from the basic equations.

11.10 Calculate the optimum length for a 200°C pin fin, made of aluminum, in an air stream at 25°C and velocity 2 m/s.

11.11 Show that when the Reynolds analogy (see Chapter 6) is valid Eq. (11.33) reduces to $f_u/2 = St_u$ and the irreversibility ratio ϕ_u can be estimated without first calculating the Reynolds number.

11.12 A flat plate is exposed to air with a temperature of 0°C, a pressure of 3500 N/m², and a velocity parallel to the plate of 800 m/s. How long is the laminar boundary layer, and what is the adiabatic wall temperature in the laminar region?

11.13 Air at a static temperature of 70°F and a static pressure of 0.1 psia flows at zero angle of attack over a thin electrically heated flat plate at a velocity of 800 fps. If the plate is 4-in. long in the direction of flow and 24 in. in the direction normal to the flow, determine the rate of electrical heat dissipation necessary to maintain the plate at an average temperature of 130°F.

11.14 Atmospheric air at 10°C flows at 250 ms^{-1} over a thermally nonconducting flat plate. What is the plate temperature 3 m downstream from the leading edge? How much does this temperature differ from that which exists 0.1 m from the leading edge?

11.15 Air at 15°C and 0.01 atmospheres pressure flows over a thin flat strip of metal, 0.1 m long in the direction of flow, at a velocity of 250 ms^{-1}. Determine (a) the surface temperature of the plate at equilibrium and (b) the rate of heat removal required per foot length if the surface temperature is to be maintained at 40°C.

11.16 An airplane model wing can be idealized as a flat plate, 2 ft long in the direction of flow and 3 ft wide. The wing is placed in an air flow at $M_\infty = 3.0$, $p_\infty = 0.05$ atm, and $T_\infty = 420$ R. Determine the temperature of this wing at a distance of 0.5 and 1.5 ft from the leading edge, if no cooling were provided, and estimate the rate at which heat must be removed from the surface of the wing to maintain its temperature at 100°F.   *Ans.* 1056 and 1093 R, 5×10^4 Btu/h.

11.17 Calculate the Nusselt number, the heat transfer coefficient, and the friction factor for a non-Newtonian fluid whose shear stress behavior follows a power law (see Eq. (11.68)) with $n = 0.5$. Assume the fluid flows through a 1 cm ID tube at a velocity of 1 m/s and has a thermal conductivity and density similar to water. The temperature of the tube is uniform at 200°C.

11.18 Compare the temperature rise and the pressure drop for the non-Newtonian fluid in the preceding problem with the temperature rise and the pressure loss for water. Assume the pipe is 2 m long and both fluids enter at 50°C.

11.19 Compare the axial heat flux achievable by a heat pipe using water as the working fluid with that of a silver rod. Assume that both are 20 cm long, that the temperature difference for the rod from end to end is 100°C and that the heat pipe operates at atmospheric pressure. State your other assumptions.

11.20 Estimate the cross-sectional area required for a 30 cm long methanol-nickel heat pipe to transport 10^5 Btu/hr at atmospheric pressure.

11.21 Design a heat pipe cooling system for a spherical satellite that generates 5×10^3 MW/m³, has a surface area of 5 m², and cannot exceed a temperature of 120°C. All the heat must be dissipated by radiation into space. State all your assumptions.

11.22 A flat plate solar collector, as shown schematically in the attached figure, has a selective surface with an emittance of 0.1 and a solar absorbance of 0.95. At 11:00 A.M. the absorber is at a temperature of 120°C, while the effective solar irradation received is 750 watts per square meter. The ambient air temperature at that time is measured to be 30°C, and the assumed effective sky temperature is $-10°C$. Assume that the air is calm and the convection heat transfer between the absorber surface and the environment may be estimated from the relationship $\bar{h}_c = 0.22 \, \Delta T^{1/3}$ where ΔT is the difference between the absorber and environmental temperature. Water is circulated through the tubes in the absorber plate. For the above conditions, calculate the rate at which the water is heated in the solar collector under steady state conditions. Also, defining the efficiency of the solar collector as the ratio of useful heat delivered to the water divided by the incident solar radiation, estimate the efficiency of the collector.

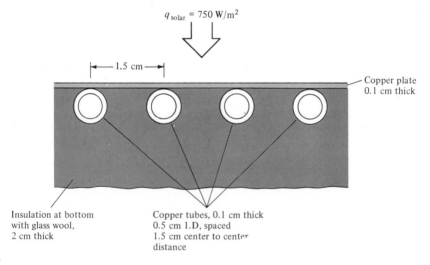

$q_{solar} = 750$ W/m²

1.5 cm

Copper plate
0.1 cm thick

Insulation at bottom
with glass wool,
2 cm thick

Copper tubes, 0.1 cm thick
0.5 cm I.D, spaced
1.5 cm center to center
distance

11.23 Repeat the calculation for a collector with a glass cover, spaced 1 cm above the absorber plate under the same conditions as in Problem 11.22.

11.24 Calculate the overall heat loss coefficient for a flat plate solar collector with a single glass cover having the following specifications:

plate to cover spacing	3 cm
plate emittance	0.2
plate absorptance	0.8
ambient air	283 K
wind speed	3 m/s
back insulation	2 cm
conductivity of insulation	0.04 W/mK
tilt angle	45 degrees
mean plate temperature	340 K

Note that this problem requires a trial and error solution.

11.25 Determine the efficiency of the collector described in Problem 11.24 if the insolation on the cover is 300 W/m². Assume that the working fluid is air entering at 50°C. The collector is 1 m wide and 5 m long, the air flow rate is 0.2 m³/min. The air flows in a duct 1.5 cm × 1 m, as shown below.

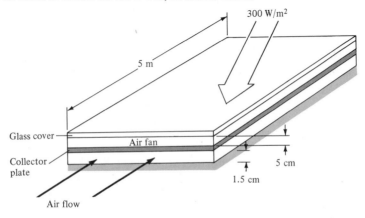

11.26 Extend Example 11.5 to compare the performance of all four turbulence promoters for which performance data are given in Fig. 11.33.

11.27 Supplement Problem 11.26 by adding results for a smooth tube.

11.28 An inventor claims to have developed an enhancement technique which can be applied to the inside diameter of a tube yielding the following performance.

$$Nu = 0.055 \, Re_D^{0.8} \, Pr^{0.4}$$
$$f = 0.80 \, Re_D^{-0.2}$$

Comment on the usefulness of this enhancement method.

11.29 A shell-and-tube heat exchanger is to be used to heat water from an inlet temperature of 25°C to 50°C with clean hot air available at 150°C and 5 atm pressure. The air is to be pumped through the tubes and the water will be pumped through the shell. The heat exchanger consists of 100 tubes of 2.54 cm ID and is arranged as a counterflow exchanger. Determine if it is economically feasible to add a twisted tape insert in the tubes. The air is to be supplied by an existing compressor and the cost of pumping the air through the heat exchanger is $0.05 per kWhr shaft input work at the compressor. Assume that the overall compressor efficiency is 30 percent (useful compression work per unit shaft input work). The cost of the heat exchanger is $330 per m^2 of tube surface area and adding the tape increases the cost by 15 percent. Assume a fixed charge rate of 15 percent which, when multiplied by the capital cost, yields the annual cost of capital. The plant will be operated 300 days per year. Neglect pressure drop due to inlet and outlet effects.

REFERENCES

1. J. Kestin, "Availability: The Concept and Associated Terminology," *Energy, Int. J.,* vol. 5, pp. 679–692, 1980.
2. J. E. Ahern, *The Exergy Method of Energy System Analysis,* Wiley, New York, 1980.
3. A. Bejan, *Entropy Generation through Heat and Fluid Flow,* Wiley, New York, 1982.
4. M. Bejan and A. Bejan, "Communications on Energy, a Supply Side Approach to Energy Policy," *Energy Policy,* pp. 153–157, June 1982.
5. A. Bejan, "A Study of Entropy Generation in Fundamental Convective Heat Transfer," *J. Heat Transfer,* vol. 101, p. 718, 1979.
6. A. Bejan and P. A. Pfister, *Lett. Heat Mass Transfer,* vol. 7, pp. 97–98, 1980.
7. D. Poulikakos and A. Bejan, "Fin Geometry for Minimum Entropy Generation in Forced Convection," *J. Heat Transfer,* vol. 105, 1983. See also D. Poulikakos, "Fin Geometry for Minimum Entropy Generation," M.S. thesis, Department of Mechanical Engineering, University of Colorado, Boulder, 1980.
8. A. Bejan, "The Concept of Irreversibility in Heat Exchanger Design: Counterflow Heat Exchangers for Gas to Gas Applications," *J. Heat Transfer,* vol. 99, pp. 374–379, 1977.
9. A. Bejan, D. W. Kearnly, and F. Kreith, "Second Law Analysis and Synthesis of Solar Collector Systems," *J. Solar Energy Eng.,* vol. 103, pp. 23–29, 1981.
10. J. Kaye, "Survey of Friction Coefficients, Recovery Factors, and Heat Transfer Coefficients for Supersonic Flow," *J. Aeronaut. Sci.,* vol. 21, no. 2, pp. 117–129, 1954.
11. H. Schlichting, *Boundary Layer Theory,* 6th ed., McGraw-Hill, New York, 1968.
12. E. R. A. Eckert, "Engineering Relations for Heat Transfer and Friction in High-Velocity Laminar and Turbulent Boundary Layer Flow over Surfaces with Constant Pressure and Temperature," *Trans. ASME,* vol. 78, pp. 1273–1284, 1956.
13. E. R. Van Driest, "Turbulent Boundary Layer in Compressible Fluids," *J. Aeronaut. Sci.,* vol. 18, no. 3, pp. 145–161, 1951.
14. A. K. Oppenheim, "Generalized Theory of Convective Heat Transfer in a Free-Molecule Flow," *J. Aeronaut. Sci.,* vol. 20, pp. 49–57, 1953.
15. W. D. Hayes and R. F. Probstein, *Hypersonic Flow Theory,* Academic Press, New York, 1959.
16. A. B. Metzner, *Handbook of Fluid Mechanics,* McGraw-Hill, New York, 1961.
17. A. B. Metzner, *Advances in Heat Transfer,* vol. 2, Academic Press, New York, 1965.
18. W. L. Wilkinson, *Non-Newtonian Fluids,* Pergamon, New York, 1960.
19. W. M. Rohsenow, *Developments in Heat Transfer,* MIT Press, Cambridge, Mass., 1964.

20. A. B. Metzner, *Advances in Chemical Engineering*, vol. 1, Academic Press, New York, 1956.

21. N. J. Beek and R. Eggink, *Ingenieur (Montreal)*, vol. 74, p. 81, 1962.

22. A. A. McKillop, *Int. J. Heat Mass Transfer*, vol. 7, p. 853, 1964.

23. P. D. Dunn and D. A. Reay, *Heat Pipes*, 3d ed., Pergamon, New York, 1982.

24. S. W. Chi, *Heat Pipe Theory and Practice*, Hemisphere, Washington, D.C., 1976.

25. R. Richter, "Solar Collector Thermal Power Systems," vol. 1, Rept. AFAPL-TR-74-89-I, Xerox Corp., Pasadena, Calif.; NTIS AD/A-000-940, National Technical Information Service, Springfield, Va., 1974.

26. C. H. Dutcher and M. R. Burke, "Heat Pipes: A Cool Way to Cool Circuits," *Electronics*, pp. 93–100, February 16, 1970.

27. J. F. Kreider and F. Kreith, *Solar Heating and Cooling*, 2nd ed., Hemisphere, New York, 1982.

28. J. E. Hill and T. Kusuda, "Methods of Testing for Rating Solar Collectors Based on Thermal Performance," National Bureau of Standards, U.S. Department of Commerce, Washington, D.C., Interim Rept. NBSIR 74-635, December 1974.

29. S. A. Klein, The Effects of Thermal Capacitance upon the Performance of Flat Plate Collectors, M. S. thesis, University of Wisconsin, Madison, 1973.

30. F. Kreith and J. Kreider, *Principles of Solar Engineering*, McGraw-Hill, New York, 1978.

31. A. E. Bergles, V. Nirmalan, G. H. Junkhan, and R. L. Webb, *Bibliography on Augmentation of Convective Heat and Mass Transfer-II*, Rept. HTL-31, ISU-ERI-Ames-84221, Iowa State University, Ames, Iowa.

32. R. K. Shah and R. L. Webb, "Compact and Enhanced Heat Exchangers," in *Heat Exchangers, Theory and Practice*, J. Taborek, G. F. Hewitt, and N. Afgan, eds., McGraw-Hill, New York, 1983.

33. W. M. Kays and A. L. London, *Compact Heat Exchangers*, 2d ed., McGraw-Hill, New York, 1964.

34. W. M. Rohsenow, J. P. Hartnett, and E. Ganic, eds., *Handbook of Heat Transfer Applications*, 2d ed., McGraw-Hill, New York, 1985. See Chapter 3, "Techniques to Augment Heat Transfer," by A. E. Bergles.

35. J. G. Soland, W. M. Mack, Jr., and W. M. Rohsenow, "Performance Ranking of Plate-Fin Heat Exchange Surfaces," *J. Heat Transfer*, vol. 100, pp. 514–519, 1978.

36. A. Rabl, *Active Solar Collectors and Their Applications*, Oxford Press, New York, 1985.

37. See Chapter 2, "Non-Newtonian Fluids" by Y. I. Cho and J. P. Hartnett, Ref. 3.

APPENDIXES

The International System of Units

The International System of Units (SI) has evolved from the MKS system, in which the meter is the unit of length, the kilogram is the unit of mass, and the second is the unit of time. The SI system is rapidly becoming the standard system of units throughout the industrialized world.

The SI system is based on seven units. Other derived units may be related to these seven base units through governing equations. The base units are listed in Table 1 along with the recommended symbols. Several defined units are listed in Table 2, while the derived units of interest in heat transfer and fluid flow are given in Table 3.

TABLE 1 SI BASE UNITS

Quantity	Name of unit	Symbol
Length	meter	m
Mass	kilogram	kg
Time	second	s
Electrical current	ampere	A
Thermodynamic temperature	kelvin	K
Luminous intensity	candela	cd
Amount of a substance	mole	mol

TABLE 2 SI DEFINED UNITS

Quantity	Unit	Defining equation
Capacitance	farad, F	$1\ F = 1\ A\ s/V$
Electrical resistance	ohm, Ω	$1\ \Omega = 1\ V/A$
Force	newton, N	$1\ N = 1\ kg\ m/s^2$
Potential difference	volt, V	$1\ V = 1\ W/A$
Power	watt, W	$1\ W = 1\ J/s$
Pressure	pascal, Pa	$1\ Pa = 1\ N/m^2$
Temperature	kelvin, K	$K = {}^{\circ}C + 273.15$
Work, heat, energy	joule, J	$1\ J = 1\ N\ m$

TABLE 3   SI DERIVED UNITS

Quantity	Name of unit	Symbol
Acceleration	meter per second squared	m/s^2
Area	square meter	m^2
Density	kilogram per cubic meter	kg/m^3
Dynamic viscosity	newton-second per square meter	$N\,s/m^2$
Force	newton	N
Frequency	hertz	Hz
Kinematic viscosity	square meter per second	m^2/s
Plane angle	radian	rad
Power	watt	W
Radiant intensity	watt per steradian	W/sr
Solid angle	steradian	sr
Specific heat	joule per kilogram-kelvin	J/kg K
Thermal conductivity	watt per meter-kelvin	W/m K
Velocity	meter per second	m/s
Volume	cubic meter	m^3

Standard prefixes can be used in the SI system to designate multiples of the basic units and thereby conserve space. The standard prefixes are listed in Table 4.

TABLE 4   SI PREFIXES

Multiplier	Symbol	Prefix	Multiplier	Symbol	Prefix
10^{12}	T	tera	10^{-2}	c	centi
10^9	G	giga	10^{-3}	m	milli
10^6	M	mega	10^{-6}	μ	micro
10^3	k	kilo	10^{-9}	n	nano
10^2	h	hecto	10^{-12}	p	pico
10^1	da	deka	10^{-15}	f	femto
10^{-1}	d	deci	10^{-18}	a	atto

Table 5 contains an alphabetical listing of physical constants that are frequently used in heat transfer and fluid flow problems, along with their values in the SI system of units.

TABLE 5   PHYSICAL CONSTANTS IN SI UNITS

Quantity	Symbol	Value
—	e	2.718281828
—	π	3.141592653
—	g_c	$1.00000\ kg\,m\,N^{-1}\,s^{-2}$
Avogadro constant	N_A	$6.022169 \times 10^{26}\ kmol^{-1}$
Boltzmann constant	k	$1.380622 \times 10^{-23}\,J\,K^{-1}$
First radiation constant	$C_1 = 2\pi hc^2$	$3.741844 \times 10^{-16}\ W\,m^2$
Gas constant	R_u	$8.31434 \times 10^3\ J\,kmol^{-1}\,K^{-1}$
Gravitational constant	G	$6.6732 \times 10^{-11}\ N\,m^2\,kg^{-2}$
Planck constant	h	$6.626196 \times 10^{-34}\ J\,s$
Second radiation constant	$C_2 = hc/k$	$1.438833 \times 10^{-2}\ m\,K$
Speed of light in a vacuum	c	$2.997925 \times 10^8\ m\,s^{-1}$
Stefan-Boltzmann constant	σ	$5.66961 \times 10^{-8}\ W\,m^{-2}\,K^{-4}$

TABLE 6A CONVERSION FACTORS

Physical quantity	Symbol	Conversion factor
Area	A	1 ft^2 = 0.0929 m^2
		1 in.^{2} = 6.452 × 10^{-4} m^2
Density	ρ	1 lb$_m$/ft^3 = 16.018 kg/m^3
		1 slug/ft^3 = 515.379 kg/m^3
Energy, heat	Q	1 Btu = 1055.1 J
		1 cal = 4.186 J
		1 (ft)(lb$_f$) = 1.3558 J
		1 (hp)(h) = 2.685 × 10^6 J
Force	F	1 lb$_f$ = 4.448 N
Heat flow rate	q	1 Btu/h = 0.2931 W
		1 Btu/s = 1055.1 W
Heat flux	q''	1 Btu/(h)(ft^2) = 3.1525 W/m^2
Heat generation per unit volume	$\dot{q}_G$	1 Btu/(h)(ft^3) = 10.343 W/m^3
Heat transfer coefficient	h	1 Btu/(h)(ft^2)(°F) = 5.678 W/m^2 K
Length	L	1 ft = 0.3048 m
		1 in. = 2.54 cm = 0.0254 m
		1 mile = 1.6093 km = 1609.3 m
Mass	m	1 lb$_m$ = 0.4536 kg
		1 slug = 14.594 kg
Mass flow rate	$\dot{m}$	1 lb$_m$/h = 0.000126 kg/s
		1 lb$_m$/s = 0.4536 kg/s
Power	$\dot{W}$	1 hp = 745.7 W
		1 (ft)(lb$_f$)/s = 1.3558 W
		1 Btu/s = 1055.1 W
		1 Btu/h = 0.293 W
Pressure	p	1 lb$_f$/in.^{2} = 6894.8 N/m^2 (Pa)
		1 lb$_f$/ft^2 = 47.88 N/m^2 (Pa)
		1 atm = 101,325 N/m^2 (Pa)
Specific energy	Q/m	1 Btu/lb$_m$ = 2326.1 J/kg
Specific heat capacity	c	1 Btu/(lb$_m$)(°F) = 4188 J/kg K
Temperature	T	$T(°R) = (9/5)T(K)$
		$T(°F) = [T(°C)](9/5) + 32$
		$T(°F) = [T(K) - 273.15](9/5) + 32$
Thermal conductivity	k	1 Btu/(h)(ft)(°F) = 1.731 W/m K
Thermal diffusivity	α	1 ft^2/s = 0.0929 m^2/s
		1 ft^2/h = 2.581 × 10^{-5} m^2/s
Thermal resistance	R_t	1 (h)(°F)/Btu = 1.8958 K/W
Velocity	U	1 ft/s = 0.3048 m/s
		1 mph = 0.44703 m/s
Viscosity, dynamic	μ	1 lb$_m$/(ft)(s) = 1.488 N s/m^2
		1 centipoise = 0.00100 N s/m^2
Viscosity, kinematic	v	1 ft^2/s = 0.0929 m^2/s
		1 ft^2/h = 2.581 × 10^{-5} m^2/s
Volume	V	1 ft^3 = 0.02832 m^3
		1 in.^{3} = 1.6387 × 10^{-5} m^3
		1 gal (U.S. liq.) = 0.003785 m^3

TABLE 6B TEMPERATURE CONVERSION TABLE

K	°C	°F	K	°C	°F	K	°C	°F
220	−53	−63	335	62	144	450	177	351
225	−48	−54	340	67	153	455	182	360
230	−43	−45	345	72	162	460	187	369
235	−38	−36	350	77	171	465	192	378
240	−33	−27	355	82	180	470	197	387
245	−28	−18	360	87	189	475	202	396
250	−23	−9	365	92	198	480	207	405
255	−18	0	370	97	207	485	212	414
260	−13	9	375	102	216	490	217	423
265	−8	18	380	107	225	495	222	432
270	−3	27	385	112	234	500	227	441
275	2	36	390	117	243	505	232	450
280	7	45	395	122	252	510	237	459
285	12	54	400	127	261	515	242	468
290	17	63	405	132	270	520	247	477
295	22	72	410	137	279	525	252	486
300	27	81	415	142	288	530	257	495
305	32	90	420	147	297	535	262	504
310	37	99	425	152	306	540	267	513
315	42	108	430	157	315	545	272	522
320	47	117	435	162	324	550	277	531
325	52	126	440	167	333	555	282	540
330	57	135	445	172	342	560	287	549

Tables

To facilitate conversion of property values from SI to English units, conversion factors have been incorporated into each table. For temperature-dependent data, temperature is listed in both system of units. Property values are given in SI units (with the exception of Table 37); to obtain a property in English units, the property in SI units must be multiplied by the conversion factor at the top. For example, suppose we wish to determine the absolute viscosity of water in English units, at 95°F. From Table 13 we have

$$\mu = (719.8 \times 10^{-6}) \quad \times \quad (0.6720) \quad = 4.84 \times 10^{-4} \ \text{lb}_m/\text{ft s}$$
$$\begin{array}{cc} \text{(SI value} & \text{(conversion factor} \\ \text{from table)} & \text{from column head)} \end{array}$$

Tables 41 and 42 are expressed in English units only, because pipe and tubing are not yet available in SI dimensional sizes.

PROPERTIES OF SOLIDS

TABLE 7 NORMAL EMITTANCE OF METALS

Substance	State of surface	Temperature (K)	Temperature (R)	Normal emittance, $\epsilon_n{}^a$
Aluminum	Polished plate	296	533	0.040
		498	896	0.039
	Rolled, polished	443	797	0.039
	Rough plate	298	536	0.070
Brass	Oxidized	611	1100	0.22
	Polished	292	526	0.05
		573	1031	0.032
	Tarnished	329	592	0.202
Chromium	Polished	423	761	0.058
Copper	Black oxidized	293	527	0.780
	Lightly tarnished	293	527	0.037
	Polished	293	527	0.030
Gold	Not polished	293	527	0.47
	Polished	293	527	0.025
Iron	Oxidized smooth	398	716	0.78
	Ground bright	293	527	0.24
	Polished	698	1256	0.144
Lead	Gray oxidized	293	527	0.28
	Polished	403	725	0.056
Molybdenum	Filament	998	1796	0.096
Nickel	Oxidized	373	671	0.41
	Polished	373	671	0.045
Platinum	Polished	498	896	0.054
		898	1616	0.104
Silver	Polished	293	527	0.025
Steel	Oxidized rough	313	563	0.94
	Ground sheet	1213	2183	0.520
Tin	Bright	293	527	0.070
Tungsten	Filament	3300	5940	0.39
Zinc	Tarnished	293	527	0.25
	Polished	503	905	0.045

[a] Hemispherical emissivity values ϵ may be approximated by: $\epsilon = 1.2\epsilon_n$ for bright metal surfaces; $\epsilon = 0.95\epsilon_n$ for other smooth surfaces; $\epsilon = 0.98\epsilon_n$ for rough surfaces.

Source: K. Raznjevič, *Handbook of Thermodynamic Tables and Charts,* McGraw-Hill, New York, 1976.

TABLE 8 NORMAL EMITTANCE OF NONMETALS

Substance	State of surface	Temperature (K)	Temperature (R)	Normal emittance, ϵ_n
Asbestos board		297	535	0.96
Brick	Red, rough	293	527	0.93
Carbon filament		1313		0.53
Glass	Smooth	293	527	0.93
Ice	Smooth	273	491	0.966
	Rough	273	491	0.985
Masonry	Plastered	273	491	0.93
Paper		293	527	0.80
Plaster, lime	White, rough	293	527	0.93
Porcelain	Glazed	293	527	0.93
Quartz	Fuzed, rough	293	527	0.93
Rubber				
Soft	Gray	297	535	0.86
Hard	Black, rough	297	535	0.95
Wood				
Beech	Planed	343	617	0.935
Oak	Planed	294	529	0.885

Source: K. Raznjevič, *Handbook of Thermodynamic Tables and Charts*, McGraw-Hill, New York, 1976.

TABLE 9 NORMAL EMITTANCE OF PAINTS AND SURFACE COATINGS

Substance	State of surface	Temperature (K)	Temperature (R)	Normal emittance, ϵ_n
Aluminum bronze		373	671	0.20–0.40
Aluminum enamel	Rough	293	527	0.39
Aluminum paint	Heated to 325°C	423–588	761–1058	0.35
Bakelite enamel		353	635	0.935
Enamel				
White	Rough	293	527	0.90
Black	Bright	298	536	0.876
Oil paint		273–473	491–851	0.885
Red lead primer		293–373	527–671	0.93
Shellac, black	Bright	294	529	0.82
	Dull	348–418	626–752	0.91

Source: K. Raznjevič, *Handbook of Thermodynamic Tables and Charts*, McGraw-Hill, New York, 1976.

TABLE 10 ALLOYS

Metal	Composition (%)	ρ (kg/m³) $\times 6.243 \times 10^{-2}$ = (lb$_m$/ft³)	c_p (J/kg K) $\times 2.388 \times 10^{-4}$ = (Btu/lb$_m$ °F)	k (W/m K) $\times 0.5777$ = (Btu/h ft °F)	$\alpha \times 10^5$ (m²/s) $\times 3.874 \times 10^4$ = (ft²/h)
		Properties at 293 K or 20°C or 68°F			
Aluminum					
Duralumin	94–96 Al, 3–5 Cu, trace Mg	2787	833	164	6.676
Silumin	87 Al, 13Si	2659	871	164	7.099
Copper					
Aluminum Bronze	95 Cu, 5 Al	8666	410	83	2.330
Bronze	75 Cu, 25 Sn	8666	343	26	0.859
Red brass	85 Cu, 9 Sn, 6 Zn	8714	385	61	1.804
Brass	70 Cu, 30 Zn	8522	385	111	3.412
German silver	62 Cu, 15 Ni, 22 Zn	8618	394	24.9	0.733
Constantan	60 Cu, 40 Ni	8922	410	22.7	0.612
Iron					
Cast iron	≈ 4 C	7272	420	52	1.702
Wrought iron	0.5 CH	7849	460	59	1.626
Steel	1 C	7801	473	43	1.172
Carbon steel	1.5 C	7753	486	36	0.970
	1 Cr	7865	460	61	1.665
Chrome steel	5 Cr	7833	460	40	1.110
	10 Cr	7785	460	31	0.867
	15 Cr, 10 Ni	7865	460	19	0.526
Chrome nickel steel	20 Cr, 15 Ni	7833	460	15.1	0.415
	10 Ni	7945	460	26	0.720
Nickel steel	20 Ni	7993	460	19	0.526
	40 Ni	8169	460	10	0.279
	60 Ni	8378	460	19	0.493
	80 Ni, 15 C	8522	460	17	0.444
Nickel chrome steel	40 Ni, 15 C	8073	460	11.6	0.305
	1 Mn	7865	460	50	1.388
Manganese steel	5 Mn	7849	460	22	0.637
	1 Si	7769	460	42	1.164
Silicon steel	5 Si	7417	460	19	0.555
	Type 304	7817	461	14.4	0.387
Stainless steel	Type 347	7817	461	14.3	0.387
	1 W	7913	448	66	1.858
Tungsten steel	5 W	8073	435	54	1.525

Source: E. R. G. Eckert and R. M. Drake, *Analysis of Heat and Mass Transfer*, McGraw-Hill, New York, 1972.

TABLE 11   INSULATIONS AND BUILDING MATERIALS

Material	Properties at 293 K or 20°C or 68°F			
	ρ (kg/m^3)	c_p (J/kg K)	k (W/m K)	$\alpha \times 10^5$ (m^2/s)
	$\times\ 6.243 \times 10^{-2}$ $= (lb_m/ft^3)$	$\times\ 2.388 \times 10^{-4}$ $= (Btu/lb_m\ °F)$	$\times\ 0.5777$ $= (Btu/h\ ft\ °F)$	$\times\ 3.874 \times 10^4$ $= (ft^2/h)$
Asbestos	383	816	0.113	0.036
Asphalt	2120		0.698	
Bakelite	1270		0.233	
Brick				
Common	1800	840	0.38–0.52	0.028–0.034
Carborundum (50% SiC)	2200		5.82	
Magnesite (50% MgO)	2000		2.68	
Masonry	1700	837	0.658	0.046
Silica (95% SiO$_2$)	1900		1.07	
Zircon (62% ZrO$_2$)	3600		2.44	
Cardboard			0.14–0.35	
Cement, hard			1.047	
Clay (48.7% moisture)	1545	880	1.26	0.101
Coal, anthracite	1370	1260	0.238	0.013–0.015
Concrete, dry	500	837	0.128	0.049
Cork, boards	150	1880	0.042	0.015–0.044
Cork, expanded	120		0.036	
Diatomaceous earth	466	879	0.126	0.031
Glass fiber	220		0.035	
Glass, window	2800	800	0.81	0.034
Glass, wool	50		0.037	
	100		0.036	
	200	670	0.040	0.028
Granite	2750		3.0	
Ice (0°C)	913	1830	2.22	0.124
Kapok	25		0.035	
Linoleum	535		0.081	
Mica	2900		0.523	
Pine bark	342		0.080	
Plaster	1800		0.814	
Plexiglas	1180		0.195	
Plywood	590		0.109	
Polystyrene	1050		0.157	
Rubber, Buna	1250		0.465	
Hard (ebonite)	1150	2009	0.163	0.0062
Spongy	224		0.055	

(continued)

TABLE 11 *(continued)*

Material	ρ (kg/m^3) $\times 6.243 \times 10^{-2}$ = (lb$_m$/ft^3)	c_p (J/kg K) $\times 2.388 \times 10^{-4}$ = (Btu/lb$_m$ °F)	k (W/m K) $\times 0.5777$ = (Btu/h ft °F)	$\alpha \times 10^5$ (m^2/s) $\times 3.874 \times 10^4$ = (ft^2/h)
Sand, dry			0.582	
Sand, moist	1640		1.13	
Sawdust	215		0.071	
Soil				
Dry	1500	1842	~ 0.35	~ 0.0138
Wet	1500		~ 2.60	0.0414
Wood				
Oak	609–801	2390	0.17–0.21	0.0111–0.0121
Pine, fir, spruce	416–421	2720	0.15	0.0124
Wood fiber sheets	200		0.047	
(celotex)	400		0.055	
Wool	200		0.038	

Source: E. R. G. Eckert and R. M. Drake, *Analysis of Heat and Mass Transfer*, McGraw-Hill, New York, 1972; K. Raznjevic, *Handbook of Thermodynamic Tables and Charts*, McGraw-Hill, New York, 1976.

TABLE 12 METALLIC ELEMENTS[a]

Element	Thermal conductivity k (W/m K)[b] ×0.5777 = (Btu/h ft °F)							Properties at 293 K or 20 °C or 68 °F				
	200 K −73°C	273 K 0°C 32°F	400 K 127°C 261°F	600 K 327°C 621°F	800 K 527°C 981°F	1000 K 727°C 1341°F	1200 K 927°C 1701°F	ρ (kg/m³) ×6.243×10⁻² = (lbₘ/ft³)	c_p (J/kg K) ×2.388×10⁻⁴ = (Btu/lbₘ °F)	k (W/m K) ×0.5777 = (Btu/h ft °F)	$\alpha \times 10^6$ (m²/s) ×3.874×10⁴ = (ft²/h)	Melting temperature (K)
Aluminum	237	236	240	232	220			2,702	896	236	97.5	933
Antimony	30.2	25.5	21.2	18.2	16.8			6,684	208	24.6	17.7	904
Beryllium	301	218	161	126	107	89	73	1,850	1750	205	63.3	1550
Bismuth[c]	9.7	8.2						9,780	124	7.9	6.51	545
Boron[c]	52.5	31.7	18.7	11.3	8.1	6.3	5.2	2,500	1047	28.6	10.9	2573
Cadmium[c]	99.3	97.5	94.7					8,650	231	97	48.5	594
Cesium	36.8	36.1						1,873	230	36	83.6	302
Chromium	111	94.8	87.3	80.5	71.3	65.3	62.4	7,160	440	91.4	29.0	2118
Cobalt[c]	122	104	84.8					8,862	389	100	29.0	1765
Copper	413	401	392	383	371	357	342	8,933	383	399	116.6	1356
Germanium	96.8	66.7	43.2	27.3	19.8	17.4	17.4	5,360		61.6		1211
Gold	327	318	312	304	292	278	262	19,300	129	316	126.9	1336
Hafnium	24.4	23.3	22.3	21.3	20.8	20.7	20.9	13,280		23.1		2495
Indium	89.7	83.7	74.5					7,300		82.2		430
Iridium	153	148	144	138	132	126	120	22,500	134	147	48.8	2716
Iron	94	83.5	69.4	54.7	43.3	32.6	28.2	7,870	452	81.1	22.8	1810
Lead	36.6	35.5	33.8	31.2				11,340	129	35.3	24.1	601
Lithium	88.1	79.2	72.1					534	3391	77.4	42.7	454
Magnesium	159	157	153	149	146			1,740	1017	156	88.2	923
Manganese	7.17	7.68						7,290	486	7.78	2.2	1517
Mercury[c]	28.9							13,546				234
Molybdenum	143	139	134	126	118	112	105	10,240	251	138	53.7	2883
Nickel	106	94	80.1	65.5	67.4	71.8	76.1	8,900	446	91	22.9	1726

(continued)

TABLE 12 (continued)

| Element | Thermal conductivity k (W/m K)[b] ×0.5777 = (Btu/h ft °F) | | | | | | | Properties at 293 K or 20 °C or 68 °F | | | | Melting temperature (K) |
	200 K −73°C	273 K 0°C 32°F	400 K 127°C 261°F	600 K 327°C 621°F	800 K 527°C 981°F	1000 K 727°C 1341°F	1200 K 927°C 1701°F	ρ (kg/m³) ×6.243×10⁻² = (lbm/ft³)	c_p (J/kg K) ×2.388×10⁻⁴ = (Btu/lbm °F)	k (W/m K) ×0.5777 = (Btu/h ft °F)	$\alpha \times 10^6$ (m²/s) ×3.874×10⁴ = (ft²/h)	
Niobium	52.6	53.3	55.2	58.2	61.3	64.4	67.5	8,570	270	53.6	23.2	2741
Palladium	75.5	75.5	75.5	75.5	75.5	75.5		12,020	247	75.5	25.4	1825
Platinum	72.4	71.5	71.6	73.0	75.5	78.6	82.6	21,450	133	71.4	25.0	2042
Potassium	104	104	52					860	741	103	161.6	337
Rhenium	51	48.6	46.1	44.2	44.1	44.6	45.7	21,100	137	48.1	16.6	3453
Rhodium	154	151	146	136	127	121	115	12,450	248	150	48.6	2233
Rubidium	58.9	58.3						1,530	348	58.2	109.3	312
Silicon	264	168	98.9	61.9	42.2	31.2	25.7	2,330	703	153	93.4	1685
Silver	403	428	420	405	389	374	358	10,500	234	427	173.8	1234
Sodium	138	135						971	1206	133	113.6	371
Tantalum	57.5	57.4	57.8	58.6	59.4	60.2	61	16,600	138	57.5	25.1	3269
Tin[c]	73.3	68.2	62.2					5,750	227	67.0	51.3	505
Titanium[c]	24.5	22.4	20.4	19.4	19.7	20.7	22	4,500	611	22.0	8.0	1953
Tungsten[c]	197	182	162	139	128	121	115	19,300	134	179	69.2	3653
Uranium[c]	25.1	27	29.6	34	38.8	43.9	49	19,070	113	27.4	12.7	1407
Vanadium	31.5	31.3	32.1	34.2	36.3	38.6	41.2	6,100	502	31.4	10.3	2192
Zinc	123	122	116	105				7,140	385	121	44.0	693
Zirconium[c]	25.2	23.2	21.6	20.7	21.6	23.7	25.7	6,570	272	22.8	12.8	2125

[a] Purity for all elements exceeds 99%.

[b] The expected percent errors in the thermal conductivity values are approximately within ±5% of the true values near room temperature and within about ±10% at other temperatures.

[c] For crystalline materials, the values are given for the polycrystalline materials.

Source: E. R. G. Eckert and R. M. Drake, *Analysis of Heat and Mass Transfer,* McGraw-Hill, New York, 1972; K. Raznjević, *Handbook of Thermodynamic Tables and Charts,* 3d ed., McGraw-Hill, New York, 1976; Y. S. Touloukian, ed., *Thermophysical Properties of Matter,* IFI/Plenum, New York, 1970.

THERMODYNAMIC PROPERTIES OF LIQUIDS

TABLE 13 WATER AT SATURATION PRESSURE

Temperature, T			Density, ρ (kg/m³) $\times 6.243 \times 10^{-2}$ = (lb$_m$/ft³)	Coefficient of thermal expansion, $\beta \times 10^4$ (1/K) $\times 0.5556$ = (1/R)	Specific heat, c_p (J/kg K) $\times 2.388 \times 10^{-4}$ = (Btu/lb$_m$ °F)	Thermal conductivity, k (W/m K) $\times 0.5777$ = (Btu/h ft °F)	Thermal diffusivity, $\alpha \times 10^6$ (m²/s) $\times 3.874 \times 10^4$ = (ft²/h)	Absolute viscosity, $\mu \times 10^6$ (N s/m²) $\times 0.6720$ = (lb$_m$/ft s)	Kinematic viscosity, $v \times 10^6$ (m²/s) $\times 3.874 \times 10^4$ = (ft²/h)	Prandtl number, Pr	$\frac{g\beta}{v^2} \times 10^{-9}$ (1/K m³) $\times 1.573 \times 10^{-2}$ = (1/R ft³)
°F	K	°C									
32	273	0	999.9	−0.7	4226	0.558	0.131	1794	1.789	13.7	—
41	278	5	1000	—	4206	0.568	0.135	1535	1.535	11.4	—
50	283	10	999.7	0.95	4195	0.577	0.137	1296	1.300	9.5	0.551
59	288	15	999.1	—	4187	0.585	0.141	1136	1.146	8.1	—
68	293	20	998.2	2.1	4182	0.597	0.143	993	1.006	7.0	2.035
77	298	25	997.1	—	4178	0.606	0.146	880.6	0.884	6.1	—
86	303	30	995.7	3.0	4176	0.615	0.149	792.4	0.805	5.4	4.540
95	308	35	994.1	—	4175	0.624	0.150	719.8	0.725	4.8	—
104	313	40	992.2	3.9	4175	0.633	0.151	658.0	0.658	4.3	8.833
113	318	45	990.2	—	4176	0.640	0.155	605.1	0.611	3.9	—
122	323	50	988.1	4.6	4178	0.647	0.157	555.1	0.556	3.55	14.59
167	348	75	974.9	—	4190	0.671	0.164	376.6	0.366	2.23	—
212	373	100	958.4	7.5	4211	0.682	0.169	277.5	0.294	1.75	85.09
248	393	120	943.5	8.5	4232	0.685	0.171	235.4	0.244	1.43	140.0
284	413	140	926.3	9.7	4257	0.684	0.172	201.0	0.212	1.23	211.7
320	433	160	907.6	10.8	4285	0.680	0.173	171.6	0.191	1.10	290.3
356	453	180	886.6	12.1	4396	0.673	0.172	152.0	0.173	1.01	396.5
392	473	200	862.8	13.5	4501	0.665	0.170	139.3	0.160	0.95	517.2
428	493	220	837.0	15.2	4605	0.652	0.167	124.5	0.149	0.90	671.4
464	513	240	809.0	17.2	4731	0.634	0.162	113.8	0.141	0.86	848.5
500	533	260	779.0	20.0	4982	0.613	0.156	104.9	0.135	0.86	1076.
536	553	280	750.0	23.8	5234	0.588	0.147	98.07	0.131	0.89	1360.
572	573	300	712.5	29.5	5694	0.564	0.132	92.18	0.128	0.98	1766.

(continued)

TABLE 13 (continued)

Saturation temperature T			Saturation pressure $p \times 10^{-5}$ (N/m²) $\times 1.450 \times 10^{-4}$ = (psi)	Specific volume of vapor v_g (m³/kg) $\times 16.02$ = (ft³/lb_m)	Enthalpy		
°F	K	°C			h_f (kJ/kg) $\times 0.430$ = (Btu/lb_m)	h_g (kJ/kg) $\times 0.430$ = (Btu/lb_m)	h_{fg} (kJ/kg) $\times 0.430$ = (Btu/lb_m)
32	273	0	0.0061	206.3	−0.04	2501	2501
50	283	10	0.0122	106.4	41.99	2519	2477
68	293	20	0.0233	57.833	83.86	2537	2453
86	303	30	0.0424	32.929	125.66	2555	2430
104	313	40	0.0737	19.548	167.45	2574	2406
122	323	50	0.1233	12.048	209.26	2591	2382
140	333	60	0.1991	7.680	251.09	2609	2358
158	343	70	0.3116	5.047	292.97	2626	2333
176	353	80	0.4735	3.410	334.92	2643	2308
194	363	90	0.7010	2.362	376.94	2660	2283
212	373	100	1.0132	1.673	419.06	2676	2257
248	393	120	1.9854	0.892	503.7	2706	2202
284	413	140	3.6136	0.508	589.1	2734	2144
320	433	160	6.1804	0.306	675.5	2757	2082
356	453	180	10.027	0.193	763.1	2777	2014
392	473	200	15.551	0.127	852.4	2791	1939
428	493	220	23.201	0.0860	943.7	2799	1856
464	513	240	33.480	0.0596	1037.6	2801	1764
500	533	260	46.940	0.0421	1135.0	2795	1660
536	553	280	64.191	0.0301	1237.0	2778	1541
572	573	300	85.917	0.0216	1345.4	2748	1403

Source: K. Raznjevič, *Handbook of Thermodynamic Tables and Charts*, McGraw-Hill, New York, 1976.

TABLE 14 FREON 12 (CCl_2F_2), SATURATED LIQUID

Temperature T °F	K	°C	Density, ρ (kg/m³) $\times 6.243 \times 10^{-2}$ = (lb$_m$/ft³)	Coefficient of thermal expansion, $\beta \times 10^3$ (1/K) $\times 0.5556$ = (1/R)	Specific heat, c_p (J/kg K) $\times 2.388 \times 10^{-4}$ = (Btu/lb$_m$ °F)	Thermal conductivity, k (W/m K) $\times 0.5777$ = (Btu/h ft °F)	Thermal diffusivity, $\alpha \times 10^8$ (m²/s) $\times 3.874 \times 10^4$ = (ft²/h)	Absolute viscosity, $\mu \times 10^4$ (N s/m²) $\times 0.6720$ = (lb$_m$/ft s)	Kinematic viscosity, $\nu \times 10^6$ (m²/s) $\times 3.874 \times 10^4$ = (ft²/h)	Prandtl number, Pr	$\frac{g\beta}{\nu^2} \times 10^{-10}$ (1/K m³) $\times 1.573 \times 10^{-2}$ = (1/R ft³)
−58	223	−50	1547	2.63	875.0	0.067	5.01	4.796	0.310	6.2	26.84
−40	233	−40	1519		884.7	0.069	5.14	4.238	0.279	5.4	
−22	243	−30	1490		895.6	0.069	5.26	3.770	0.253	4.8	
−4	253	−20	1461		907.3	0.071	5.39	3.433	0.235	4.4	
14	263	−10	1429		920.3	0.073	5.50	3.158	0.221	4.0	
32	273	0	1397	3.10	934.5	0.073	5.57	2.990	0.214	3.8	6.68
50	283	10	1364		949.6	0.073	5.60	2.769	0.203	3.6	
68	293	20	1330		965.9	0.073	5.60	2.633	0.198	3.5	
86	303	30	1295		983.5	0.071	5.60	2.512	0.194	3.5	
104	313	40	1257		1001.9	0.069	5.55	2.401	0.191	3.5	
122	323	50	1216		1021.6	0.067	5.45	2.310	0.190	3.5	

Source: E. R. G. Eckert and R. M. Drake, *Analysis of Heat and Mass Transfer*, McGraw-Hill, New York, 1972.

TABLE 15 AMMONIA (NH_3), SATURATED LIQUID

Temperature, T			Density, ρ (kg/m³) $\times 6.243 \times 10^{-2}$ = (lb$_m$/ft³)	Coefficient of thermal expansion, $\beta \times 10^3$ (1/K) $\times 0.5556$ = (1/R)	Specific heat, c_p (J/kg K) $\times 2.388 \times 10^{-4}$ = (Btu/lb$_m$ °F)	Thermal conductivity, k (W/m K) $\times 0.5777$ = (Btu/h ft °F)	Thermal diffusivity, $\alpha \times 10^8$ (m²/s) $\times 3.874 \times 10^4$ = (ft²/h)	Absolute viscosity, $\mu \times 10^4$ (N s/m²) $\times 0.6720$ = (lb$_m$/ft s)	Kinematic viscosity, $\nu \times 10^6$ (m²/s) $\times 3.874 \times 10^4$ = (ft²/h)	Prandtl number, Pr	$\dfrac{g\beta}{\nu^2} \times 10^{-10}$ (1/K m³) $\times 1.573 \times 10^{-2}$ = (1/R ft³)
°F	K	°C									
−58	223	−50	703.7		4463	0.547	17.42	3.061	0.435	2.60	
−40	233	−40	691.7		4467	0.547	17.75	2.808	0.406	2.28	
−22	243	−30	679.3		4476	0.549	18.01	2.629	0.387	2.15	
−4	253	−20	666.7		4509	0.547	18.19	2.540	0.381	2.09	
14	263	−10	653.6		4564	0.543	18.25	2.471	0.378	2.07	
32	273	0	640.1	2.16	4635	0.540	18.19	2.388	0.373	2.05	1.51
50	283	10	626.2		4714	0.531	18.01	2.304	0.368	2.04	
68	293	20	611.8	2.45	4798	0.521	17.75	2.196	0.359	2.02	18.64
86	303	30	596.4		4890	0.507	17.42	2.081	0.349	2.01	
104	313	40	581.0		4999	0.493	17.01	1.975	0.340	2.00	
122	323	50	564.3		5116	0.476	16.54	1.862	0.330	1.99	

Source: E. R. G. Eckert and R. M. Drake, *Analysis of Heat and Mass Transfer*, McGraw-Hill, New York, 1972.

TABLE 16 UNUSED ENGINE OIL, SATURATED LIQUID

Temperature T			Density, ρ (kg/m³) $\times 6.243 \times 10^{-2}$ = (lb$_m$/ft³)	Coefficient of thermal expansion, $\beta \times 10^3$ (1/K) $\times 0.5556$ = (1/R)	Specific heat, c_p (J/kg K) $\times 2.388 \times 10^{-4}$ = (Btu/lb$_m$ °F)	Thermal conductivity, k (W/m K) $\times 0.5777$ = (Btu/h ft °F)	Thermal diffusivity, $\alpha \times 10^{10}$ (m²/s) $\times 3.874 \times 10^4$ = (ft²/h)	Absolute viscosity, $\mu \times 10^3$ (N s/m²) $\times 0.6720$ = (lb$_m$/ft s)	Kinematic viscosity, $\nu \times 10^6$ (m²/s) $\times 3.874 \times 10^4$ = (ft²/h)	Prandtl number, Pr	$\dfrac{g\beta}{\nu^2}$ (1/K m³) $\times 1.573 \times 10^{-2}$ = (1/R ft³)
°F	K	°C									
32	273	0	899.1	0.70	1796	0.147	911	3848.	4280.	471.	8475
68	293	20	888.2		1880	0.145	872	799.	900.	104.	
104	313	40	876.1		1964	0.144	834	210.	240.	28.7	
140	333	60	864.0		2047	0.140	800	72.5	83.9	10.5	
176	353	80	852.0		2131	0.138	769	32.0	37.5	4.90	
212	373	100	840.0		2219	0.137	738	17.1	20.3	2.76	
248	393	120	829.0		2307	0.135	710	10.3	12.4	1.75	
284	413	140	816.9		2395	0.133	686	6.54	8.0	1.16	
320	433	160	805.9		2483	0.132	663	4.51	5.6	0.84	

Source: E. R. G. Eckert and R. M. Drake, *Analysis of Heat and Mass Transfer*, McGraw-Hill, New York, 1972.

TABLE 17 TRANSFORMER OIL (STANDARD 982-68)

Temperature, T °F	K	°C	Density, ρ (kg/m³) $\times 6.243 \times 10^{-2}$ = (lb$_m$/ft³)	Coefficient of thermal expansion, $\beta \times 10^3$ (1/K) $\times 0.5555$ = (1/R)	Specific heat, c_p (J/kg K) $\times 2.388 \times 10^{-4}$ = (Btu/lb$_m$ °F)	Thermal conductivity, k (W/m K) $\times 0.5777$ = (Btu/lb$_m$ °F)	Thermal diffusivity, $\alpha \times 10^{10}$ (m²/s) $\times 3.874 \times 10^4$ = (ft²/h)	Absolute viscosity, $\mu \times 10^3$ (N s/m²) $\times 0.6720$ = (lb$_m$/ft s)	Kinematic viscosity, $\nu \times 10^6$ (m²/s) $\times 3.874 \times 10^4$ = (ft²/h)	Prandtl number, Pr $\times 10^{-2}$
-58	223	-50	922		1700	0.116	742	29,320	31,800	4,286
-40	233	-40	916		1680	0.116	750	3,866	4,220	563
-22	243	-30	910		1650	0.115	764	1,183	1,300	170
-4	253	-20	904		1620	0.114	778	365.6	404	52
14	263	-10	898		1600	0.113	788	108.1	120	15.3
32	273	0	891		1620	0.112	778	55.24	67.5	8.67
50	283	10	885		1650	0.111	763	33.45	37.8	4.95
68	293	20	879		1710	0.111	736	21.10	24.0	3.26
86	303	30	873		1780	0.110	707	13.44	15.4	2.18
104	313	40	867		1830	0.109	688	9.364	10.8	1.57

Source: N. B. Vargaftik, *Tables on the Thermophysical Properties of Liquids and Gases*, 2nd ed., Hemisphere, Washington, D.C., 1975.

TABLE 18 *n*-BUTYL ALCOHOL ($C_4H_{10}O$)

Temperature T °F	K	°C	Density, ρ (kg/m³) $\times 6.243 \times 10^{-2}$ = (lb$_m$/ft³)	Coefficient of thermal expansion, $\beta \times 10^4$ (1/K) $\times 0.5556$ = (1/R)	Specific heat, c_p (J/kg K) $\times 2.388 \times 10^{-4}$ = (Btu/lb$_m$ °F)	Thermal conductivity, k (W/m K) $\times 0.5777$ = (Btu/h ft °F)	Thermal diffusivity, $\alpha \times 10^{10}$ (m²/s) $\times 3.874 \times 10^4$ = (ft²/h)	Absolute viscosity, $\mu \times 10^3$ (N s/m²) $\times 0.6720$ = (lb$_m$/ft s)	Kinematic viscosity, $\nu \times 10^6$ (m²/s) $\times 3.874 \times 10^4$ = (ft²/h)	Prandtl number, Pr	$\dfrac{g\beta}{\nu^2} \times 10^{-6}$ (1/K m³) $\times 1.573 \times 10^{-2}$ = (1/R ft³)
60	289	16	809		1305	0.168	901	3.36	4.16	26.1	
100	311	38	796	8.1	1392	0.166	816	1.92	2.41	16.1	1367
150	339	66	777	8.6	1502	0.164	743	1.00	1.29	9.16	5086
200	366	93	756		1609	0.163	666	0.57	0.76	5.6	
243.5	390.7	117.5	737		1706	0.163	769	0.39	0.53	4.1	
300	422	149			1830			0.28			

TABLE 19 COMMERCIAL ANILINE

Temperature T (°F)	(K)	(°C)	Density ρ (kg/m³)	Coefficient of thermal expansion β × 10⁴ (1/K)	Specific heat c_p (J/kg K)	Thermal conductivity k (W/m K)	Thermal diffusivity α × 10¹⁰ (m²/s)	Absolute viscosity μ × 10³ (N s/m²)	Kinematic viscosity ν × 10⁶ (m²/s)	Prandtl number, Pr	$\frac{g\beta}{\nu^2} \times 10^{-6}$ (1/K m³)
			× 6.243 × 10⁻² = (lbm/ft³)	× 0.5556 = (1/R)	× 2.388 × 10⁻⁴ = (Btu/lbm °F)	× 0.5777 = (Btu/h ft °F)	× 3.874 × 10⁴ = (ft²/h)	× 0.6720 = (lbm/ft s)	× 3.874 × 10⁴ = (ft²/h)		× 1.573 × 10⁻² = (1/R ft³)
60	289	16	1025		2011	0.173	839	4.84	4.72	56.0	
100	311	38	1009	8.82	2052	0.173	837	2.53	2.51	30.0	1373
150	339	66	985	8.86	2115	0.170	816	1.44	1.46	18.0	4100
200	366	93	961		2157	0.166	803	0.91	0.947	11.8	
300	422	149	921		2261	0.161	775	0.48	0.521	6.8	

TABLE 20 BENZENE (C_6H_6)

Temperature T (°F)	(K)	(°C)	Density ρ (kg/m³)	Coefficient of thermal expansion β × 10³ (1/K)	Specific heat c_p (J/kg K)	Thermal conductivity k (W/m K)	Thermal diffusivity α × 10¹⁰ (m²/s)	Absolute viscosity μ × 10³ (N s/m²)	Kinematic viscosity ν × 10⁶ (m²/s)	Prandtl number, Pr	$\frac{g\beta}{\nu^2} \times 10^{-6}$ (1/K m³)
			× 6.243 × 10⁻² = (lbm/ft³)	× 0.5556 = (1/R)	× 2.388 × 10⁻⁴ = (Btu/lbm °F)	× 0.5777 = (Btu/h ft °F)	× 3.874 × 10⁴ = (ft²/h)	× 0.6720 = (lbm/ft s)	× 3.874 × 10⁴ = (ft²/h)		× 1.573 × 10⁻² = (1/R ft³)
60	289	16	883	1.08	1675	0.161	1089	0.685	0.776	7.2	19,072
80	300	27	875		1759	0.159	1035	0.589	0.673	6.5	
100	311	38	865		1843	0.151	911	0.522	0.604	5.1	
150	339	66	857		1926			0.387	0.452	4.5	
200	366	93						0.302		4.0	

TABLE 21 ORGANIC COMPOUNDS AT 20 °C, 68 °F

Liquid	Chemical formula	Density, ρ (kg/m³) $\times 6.243\times10^{-2}$ = (lb$_m$/ft³)	Coefficient of thermal expansion $\beta\times10^4$ (1/K) $\times 0.5556$ = (1/R)	Specific heat, c_p (J/kg K) $\times 2.388\times10^{-4}$ = (Btu/lb$_m$ °F)	Thermal conductivity, k (W/m K) $\times 0.5777$ = (Btu/h ft °F)	Thermal diffusivity, $\alpha\times10^9$ (m²/s) $\times 3.874\times10^4$ = (ft²/h)	Absolute viscosity, $\mu\times10^4$ (N s/m²) $\times 0.6720$ = (lb$_m$/ft s)	Kinematic viscosity, $\nu\times10^6$ (m²/s) $\times 3.874\times10^4$ = (ft²/h)	Prandtl number, Pr	$\dfrac{g\beta}{\nu^2}\times10^{-8}$ (1/K m³) $\times 1.573\times10^{-2}$ = (1/R ft³)
Acetic acid	$C_2H_4O_2$	1049	10.7	2031	0.193	90.6				802.6
Acetone	C_3H_6O	791	14.3	2160	0.180	105.4	3.31	0.418	3.97	825.3
Chloroform	$CHCl_3$	1489	12.8	967	0.129	89.6	5.8	0.390	4.35	543.5
Ethyl acetate	$C_4H_8O_2$	900	13.8	2010	0.137	75.7	4.49	0.499	6.59	46.7
Ethyl alcohol	C_2H_6O	790	11.0	2470	0.182	93.3	12.0	1.52	16.29	
Ethylene glycol	$C_2H_6O_2$	1115		2382	0.258	97.1	199	17.8	183.7	
Glycerol	$C_3H_8O_3$	1260	5.0	2428	0.285	93.2	14,800	1175	12,609	0.0000355
n-Heptane	C_7H_{14}	684	12.4	2219	0.140	92.2	4.09	0.598	6.48	340.1
n-Hexane	C_6H_{14}	660	13.5	1884	0.137	11.02	3.20	0.485	4.40	562.8
Isobutyl alcohol	$C_4H_{10}O$	804	9.4	2303	0.134	72.4	39.5	4.91	67.89	3.82
Methyl alcohol	CH_4O	792	11.9	2470	0.212	108.4	5.84	0.737	6.80	214.9
n-Octane	C_8H_{18}	720	11.4	2177	0.147	93.8	5.4	0.750	8.00	198.8
n-Pentane	C_5H_{12}	626	16.0	2177	0.136	99.8	2.29	0.366	3.67	1171
Toluene	C_7H_8	866	10.8	1675	0.151	104.1	5.86	0.677	6.50	231.1
Turpentine	$C_{10}H_{16}$	855	9.7	1800	0.128	83.2	14.87	1.74	20.91	31.4

Source: K. Raznjević, *Handbook of Thermodynamic Tables and Charts*, McGraw-Hill, New York, 1976.

HEAT TRANSFER FLUIDS

TABLE 22 MOBILTHERM 600

Temperature, T °F	K	°C	Density, ρ (kg/m³) $\times 6.243 \times 10^{-2}$ = (lb$_m$/ft³)	Coefficient of thermal expansion, $\beta \times 10^3$ (1/K) $\times 0.5556$ = (1/R)	Specific heat, c_p (J/kg K) $\times 2.388 \times 10^{-4}$ = (Btu/lb$_m$ °F)	Thermal conductivity, k (W/m K) $\times 0.5777$ = (Btu/h ft °F)	Thermal diffusivity, $\alpha \times 10^{10}$ (m²/s) $\times 3.874 \times 10^4$ = (ft²/h)	Absolute viscosity, $\mu \times 10^3$ (N s/m²) $\times 0.6720$ = (lb$_m$/ft s)	Kinematic viscosity, $\nu \times 10^6$ (m²/s) $\times 3.874 \times 10^4$ = (ft²/h)	Prandtl number, Pr	$\dfrac{g\beta}{\nu^2} \times 10^{-6}$ (1/K m³) $\times 1.573 \times 10^{-2}$ = (1/R ft³)
50	283	10	953	0.621	1549	0.123	833				
122	323	50	929	0.637	1680	0.120	769	30.28	32.60	424	5.9
212	373	100	899	0.658	1859	0.116	694	5.48	6.10	87.9	173
302	423	150	870	0.680	2031	0.113	640	2.04	2.34	36.6	1218
392	473	200	839	0.705	2209	0.110	594	1.05	1.25	21.0	4425
482	523	250	810	0.730	2386	0.106	545	0.64	0.790	14.5	11,470

Source: P. L. Geiringer, *Handbook of Heat Transfer Media*, Krieger, New York, 1977.

TABLE 23 MOLTEN SALT (EQUIMOLAR KNO_3, $NaNO_3$)

Temperature T			Density, ρ (kg/m³) $\times 6.243 \times 10^{-2}$ = (lb$_m$/ft³)	Coefficient of thermal expansion, $\beta \times 10^4$ (1/K) $\times 0.5556$ = (1/R)	Specific heat, $c_p{}^a$ (J/kg K) $\times 2.388 \times 10^{-4}$ = (Btu/lb$_m$ °F)	Thermal conductivity, k (W/m K) $\times 0.5777$ = (Btu/h ft °F)	Thermal diffusivity, $\alpha \times 10^6$ (m²/s) $\times 3.874 \times 10^4$ = (ft²/h)	Absolute viscosity, $\mu \times 10^6$ (N s/m²) $\times 0.6720$ = (lb$_m$/ft s)	Kinematic viscosity, $\nu \times 10^6$ (m²/s) $\times 3.874 \times 10^4$ = (ft²/h)	Prandtl number, Pr	$\dfrac{g\beta}{\nu^2} \times 10^{-9}$ (1/K m³) $\times 1.573 \times 10^{-2}$ = (1/R ft³)
°F	K	°C									
675	630	357	1864	3.40	1645	0.512	0.167	2217	1.189	7.12	2.36
693	640	367	1858	3.47	1633	0.513	0.169	2100	1.130	6.68	2.67
711	650	377	1851	3.53	1621	0.515	0.172	1997	1.079	6.29	2.97
729	660	387	1845	3.60	1610	0.517	0.174	1906	1.033	5.94	3.31
747	670	397	1838	3.67	1598	0.519	0.177	1826	0.994	5.62	3.64
765	680	407	1831	3.74	1586	0.521	0.179	1755	0.959	5.34	3.99
783	690	417	1824	3.81	1574	0.523	0.182	1692	0.928	5.09	4.34
801	700	427	1817	3.88	1562	0.525	0.185	1636	0.900	4.87	4.70
819	710	437	1810	3.95	1551	0.527	0.188	1585	0.876	4.66	5.05
837	720	447	1803	4.02	1539	0.529	0.191	1540	0.854	4.48	5.41

a For the minimum melting point mixture, 54% KNO_3, 46% $NaNO_3$

Source: National Bureau of Standards Publication NSRDS-NBS 61, Part II, "Physical Properties Data Compilations Relevant to Energy Storage."

LIQUID METALS

TABLE 24 BISMUTH

Temperature T °F	Temperature T K	Temperature T °C	Density, ρ (kg/m³) $\times 6.243 \times 10^{-2}$ = (lb$_m$/ft³)	Coefficient of thermal expansion, $\beta \times 10^3$ (1/K) $\times 0.5556$ = (1/R)	Specific heat, c_p (J/kg K) $\times 2.388 \times 10^{-4}$ = (Btu/lb$_m$ °F)	Thermal conductivity, k (W/m K) $\times 0.5777$ = (Btu/h ft °F)	Thermal diffusivity, $\alpha \times 10^5$ (m²/s) $\times 3.874 \times 10^4$ = (ft²/h)	Absolute viscosity, $\mu \times 10^4$ (N s/m²) $\times 0.6720$ = (lb$_m$/ft s)	Kinematic viscosity, $\nu \times 10^7$ (m²/s) $\times 3.874 \times 10^4$ = (ft²/h)	Prandtl number, Pr	$\dfrac{g\beta}{\nu^2} \times 10^{-9}$ (1/K m³) $\times 1.573 \times 10^{-2}$ = (1/R ft³)
600	589	316	10,011	0.117	144.5	16.44	1.14	16.22	1.57	0.014	46.5
800	700	427	9,867	0.122	149.5	15.58	1.06	13.39	1.35	0.013	65.6
1000	811	538	9,739	0.126	154.5	15.58	1.03	11.01	1.08	0.011	106
1200	922	649	9,611		159.5	15.58	1.01	9.23	0.903	0.009	
1400	1033	760	9,467		164.5	15.58	1.01	7.89	0.813	0.008	

TABLE 25 MERCURY (SATURATED LIQUID)

Temperature, T °F	K	°C	Density, ρ (kg/m³) $\times 6.243 \times 10^{-2}$ = (lb$_m$/ft³)	Coefficient of thermal expansion, $\beta \times 10^4$ (1/K) $\times 0.5556$ = (1/R)	Specific heat, c_p (J/kg K) $\times 2.388 \times 10^{-4}$ = (Btu/lb$_m$ °F)	Thermal conductivity, k (W/m K) $\times 0.5777$ = (Btu/h ft °F)	Thermal diffusivity, $\alpha \times 10^{10}$ (m²/s) $\times 3.874 \times 10^4$ = (ft²/h)	Absolute viscosity, $\mu \times 10^4$ (N s/m²) $\times 0.6720$ = (lb$_m$/ft s)	Kinematic viscosity, $\nu \times 10^6$ (m²/s) $\times 3.874 \times 10^4$ = (ft²/h)	Prandtl number, Pr	$\dfrac{g\beta}{\nu^2} \times 10^{-10}$ (1/K m³) $\times 1.573 \times 10^{-2}$ = (1/R ft³)
32	273	0	13,628		140.3	8.20	42.99	16.90	0.124	0.0288	
68	293	20	13,579	1.82	139.4	8.69	46.06	15.48	0.114	0.0249	13.73
122	323	50	13,506		138.6	9.40	50.22	14.05	0.104	0.0207	
212	373	100	13,385		137.3	10.51	57.16	12.42	0.0928	0.0162	
302	423	150	13,264		136.5	11.49	63.54	11.31	0.0853	0.0134	
392	473	200	13,145		157.0	12.34	69.08	10.54	0.0802	0.0116	
482	523	250	13,026		135.7	13.07	74.06	9.96	0.0765	0.0103	
600	588.7	315.5	12,847		134.0	14.02	81.50	8.65	0.0673	0.0083	

Source: E. R. G. Eckert and R. M. Drake, *Analysis of Heat and Mass Transfer*, McGraw-Hill, New York 1972.

TABLE 26 SODIUM

Temperature, T			Density, ρ (kg/m³) $\times 6.243 \times 10^{-2}$ = (lb$_m$/ft³)	Coefficient of thermal expansion, $\beta \times 10^3$ (1/K) $\times 0.5556$ = (1/R)	Specific heat, c_p (J/kg K) $\times 2.388 \times 10^{-4}$ = (Btu/lb$_m$ °F)	Thermal conductivity k (W/m K) $\times 0.5777$ = (Btu/h ft °F)	Thermal diffusivity, $\alpha \times 10^5$ (m²/s) $\times 3.874 \times 10^4$ = (ft²/h)	Absolute viscosity, $\mu \times 10^4$ (N s/m²) $\times 0.6720$ = (lb$_m$/ft s)	Kinematic viscosity, $\nu \times 10^7$ (m²/s) $\times 3.874 \times 10^4$ = (ft²/h)	Prandtl number, Pr	$\dfrac{g\beta}{\nu^2} \times 10^{-9}$ (1/K m³) $\times 1.573 \times 10^{-2}$ = (1/R ft³)
°F	K	°C									
200	367	94	929	0.27	1382	86.2	6.71	6.99	7.31	0.0110	4.96
400	478	205	902	0.36	1340	80.3	6.71	4.32	4.60	0.0072	16.7
700	644	371	860		1298	72.4	6.45	2.83	3.16	0.0051	
1000	811	538	820		1256	65.4	6.19	2.08	2.44	0.0040	
1300	978	705	778		1256	59.7	6.19	1.79	2.26	0.0038	

THERMODYNAMIC PROPERTIES OF GASES

TABLE 27 DRY AIR AT ATMOSPHERIC PRESSURE

Temperature, T			Density ρ (kg/m³) $\times 6.243 \times 10^{-2}$ = (lb$_m$/ft³)	Coefficient of thermal expansion, $\beta \times 10^3$ (1/K) $\times 0.5556$ = (1/R)	Specific heat, c_p (J/kg K) $\times 2.388 \times 10^{-4}$ = (Btu/lb$_m$ °F)	Thermal conductivity k (W/m K) $\times 0.5777$ = (Btu/h ft °F)	Thermal diffusivity $\alpha \times 10^6$ (m²/s) $\times 3.874 \times 10^4$ = (ft²/h)	Absolute viscosity, $\mu \times 10^6$ (N s/m²) $\times 0.6720$ = (lb$_m$/ft s)	Kinematic viscosity, $\nu \times 10^6$ (m²/s) $\times 3.874 \times 10^4$ = (ft²/h)	Prandtl number, Pr	$\dfrac{g\beta}{\nu^2} \times 10^{-8}$ (1/K m³) $\times 1.573 \times 10^{-2}$ = (1/R ft³)
°F	K	°C									
32	273	0	1.252	3.66	1011	0.0237	19.2	17.456	13.9	0.71	1.85
68	293	20	1.164	3.41	1012	0.0251	22.0	18.240	15.7	0.71	1.36
104	313	40	1.092	3.19	1014	0.0265	24.8	19.123	17.6	0.71	1.01
140	333	60	1.025	3.00	1017	0.0279	27.6	19.907	19.4	0.71	0.782
176	353	80	0.968	2.83	1019	0.0293	30.6	20.790	21.5	0.71	0.600
212	373	100	0.916	2.68	1022	0.0307	33.6	21.673	23.6	0.71	0.472
392	473	200	0.723	2.11	1035	0.0370	49.7	25.693	35.5	0.71	0.164
572	573	300	0.596	1.75	1047	0.0429	68.9	39.322	49.2	0.71	0.0709
752	673	400	0.508	1.49	1059	0.0485	89.4	32.754	64.6	0.72	0.0350
932	773	500	0.442	1.29	1076	0.0540	113.2	35.794	81.0	0.72	0.0193
1832	1273	1000	0.268	0.79	1139	0.0762	240	48.445	181	0.74	0.00236

Source: K. Raznjević, *Handbook of Thermodynamic Tables and Charts*, McGraw-Hill, New York, 1976.

TABLE 28 CARBON DIOXIDE AT ATMOSPHERIC PRESSURE

Temperature, T °F	K	°C	Density, ρ (kg/m³) $\times 6.243 \times 10^{-2}$ = (lbm/ft³)	Coefficient of thermal expansion, $\beta \times 10^3$ (1/K) $\times 0.5556$ = (1/R)	Specific heat, c_p (J/kg K) $\times 2.388 \times 10^{-4}$ = (Btu/lbm °F)	Thermal conductivity, k (W/m K) $\times 0.5777$ = (Btu/h ft °F)	Thermal diffusivity, $\alpha \times 10^4$ (m²/s) $\times 3.874 \times 10^4$ = (ft²/h)	Absolute viscosity, $\mu \times 10^6$ (N s/m²) $\times 0.6720$ = (lbm/ft s)	Kinematic viscosity, $\nu \times 10^6$ (m²/s) $\times 3.874 \times 10^4$ = (ft²/h)	Prandtl number, Pr	$\dfrac{g\beta}{\nu^2} \times 10^{-6}$ (1/K m³) $\times 1.573 \times 10^{-2}$ = (1/R ft³)
−63	220	−53	2.4733		783	0.01080	0.0592	11.105	4.490	0.818	
−9	250	−23	2.1657		804	0.01288	0.0740	12.590	5.813	0.793	
81	300	27	1.7973	3.33	871	0.01657	0.1058	14.958	8.321	0.770	472
171	350	77	1.5362	2.86	900	0.02047	0.1480	17.205	11.19	0.755	224
261	400	127	1.3424	2.50	942	0.02461	0.1946	19.32	14.39	0.738	118
351	450	177	1.1918	2.22	980	0.02897	0.2481	21.34	17.90	0.721	67.9
441	500	227	1.0732	2.00	1013	0.03352	0.3084	23.26	21.67	0.702	41.8
531	550	277	0.9739	1.82	1047	0.03821	0.3750	25.08	25.74	0.685	26.9
621	600	327	0.8938	1.67	1076	0.04311	0.4483	26.83	30.02	0.668	18.2

Source: E. R. G. Eckert and R. M. Drake, Analysis of Heat and Mass Transfer, McGraw-Hill, New York, 1972.

TABLE 29 CARBON MONOXIDE AT ATMOSPHERIC PRESSURE

Temperature T			Density, ρ (kg/m³) $\times 6.243 \times 10^{-2}$ $=$ (lb$_m$/ft³)	Coefficient of thermal expansion, $\beta \times 10^3$ (1/K) $\times 0.5556$ $=$ (1/R)	Specific heat, c_p (J/kg K) $\times 2.388 \times 10^{-4}$ $=$ (Btu/lb$_m$ °F)	Thermal conductivity, k (W/m K) $\times 0.5777$ $=$ (Btu/h ft °F)	Thermal diffusivity, $\alpha \times 10^4$ (m²/s) $\times 3.874 \times 10^4$ $=$ (ft²/h)	Absolute viscosity, $\mu \times 10^6$ (N s/m²) $\times 0.6720$ $=$ (lb$_m$/ft s)	Kinematic viscosity, $\nu \times 10^6$ (m²/s) $\times 3.874 \times 10^4$ $=$ (ft²/h)	Prandtl number, Pr	$\dfrac{g\beta}{\nu^2} \times 10^{-6}$ (1/K m²) $\times 1.573 \times 10^{-2}$ $=$ (1/R ft³)
°F	K	°C									
−63		−53	1.554		1043	0.01906	0.1176	13.88	8.90	0.758	
−9		−23	0.841		1043	0.02144	0.1506	15.40	11.28	0.750	
81		27	1.139	3.33	1042	0.02525	0.2128	17.84	15.67	0.737	133
171		77	0.974	2.86	1043	0.02883	0.2836	20.09	20.62	0.728	65.9
261		127	0.854	2.50	1048	0.03226	0.3605	22.19	25.99	0.722	36.3
351		177	0.758	2.22	1055	0.04360	0.4439	24.18	31.88	0.718	21.4
441		227	0.682	2.00	1064	0.03863	0.5324	26.06	38.19	0.718	13.4
531		277	0.620	1.82	1076	0.04162	0.6240	27.89	44.97	0.721	8.83
621		327	0.569	1.67	1088	0.04446	0.7190	29.60	52.06	0.724	6.04

Source: E. R. G. Eckert and R. M. Drake, *Analysis of Heat Mass Transfer*, McGraw-Hill, New York, 1972.

TABLE 30 HELIUM AT ATMOSPHERIC PRESSURE

Temperature, T °F	K	°C	Density, ρ (kg/m³) $\times 6.243 \times 10^{-2}$ = (lb$_m$/ft³)	Coefficient of thermal expansion, $\beta \times 10^3$ (1/K) $\times 0.5556$ = (1/R)	Specific heat, c_p (J/kg K) $\times 2.388 \times 10^{-4}$ = (Btu/lb$_m$ °F)	Thermal conductivity, k (W/m K) $\times 0.5777$ = (Btu/h ft °F)	Thermal diffusivity, $\alpha \times 10^4$ (m²/s) $\times 3.874 \times 10^4$ = (ft²/h)	Absolute viscosity, $\mu \times 10^6$ (N s/m²) $\times 0.6720$ = (lb$_m$/ft s)	Kinematic viscosity, $\nu \times 10^6$ (m²/s) $\times 3.874 \times 10^4$ = (ft²/h)	Prandtl number, Pr	$\dfrac{g\beta}{\nu^2} \times 10^{-6}$ (1/K m³) $\times 1.573 \times 10^{-2}$ = (1/R ft³)
-454	3	-270			5200	0.0106		0.842			
-400	33	-240	1.466		5200	0.0353	0.04625	5.02	3.42	0.74	
-200	144	-129	0.3380	6.94	5200	0.0928	0.5275	12.55	37.11	0.70	49.4
-100	200	-73	0.2435	5.00	5200	0.1177	0.9288	15.66	64.38	0.694	11.8
0	255	-18	0.1906	3.92	5200	0.1357	1.3675	18.17	95.50	0.70	4.22
200	366	93	0.1328	2.73	5200	0.1691	2.449	23.05	173.6	0.71	0.888
400	477	204	0.1020	2.10	5200	0.197	3.716	27.50	269.3	0.72	0.284
600	589	316	0.08282	1.70	5200	0.225	5.215	31.13	375.8	0.72	0.118
800	700	427	0.07032	1.43	5200	0.251	6.661	34.75	494.2	0.72	0.0574
981	800	527	0.06023	1.25	5200	0.275	8.774	38.17	634.1	0.72	0.0305
1161	900	627	0.05286	1.11	5200	0.298	10.834	41.36	781.3	0.72	0.0178

Source: E. R. G. Eckert and R. M. Drake, *Analysis of Heat and Mass Transfer*, McGraw-Hill, New York, 1972.

TABLE 31 HYDROGEN AT ATMOSPHERIC PRESSURE

Temperature, T °F	K	°C	Density, ρ (kg/m³) ×6.243×10⁻² = (lbm/ft³)	Coefficient of thermal expansion, $\beta \times 10^3$ (1/K) ×0.5556 = (1/R)	Specific heat, c_p (J/kg K) ×2.388×10⁻⁴ = (Btu/lbm °F)	Thermal conductivity, k (W/m K) ×0.5777 = (Btu/h ft °F)	Thermal diffusivity, $\alpha \times 10^4$ (m²/s) ×3.874×10⁴ = (ft²/h)	Absolute viscosity, $\mu \times 10^6$ (N s/m²) ×0.6720 = (lbm/ft s)	Kinematic viscosity, $\nu \times 10^6$ (m²/s) ×3.874×10⁴ = (ft²/h)	Prandtl number, Pr	$\dfrac{g\beta}{\nu^2} \times 10^{-6}$ (1/K m³) ×1.573×10⁻² = (1/R ft³)
−369	50	−223	0.50955		10,501	0.0362	0.0676	2.516	4.880	0.721	
−279	100	−173	0.24572	10.0	11,229	0.0665	0.2408	4.212	17.14	0.712	333.8
−189	150	−123	0.16371	6.67	12,602	0.0981	0.475	5.595	34.18	0.718	55.99
−100	200	−73	0.12270	5.00	13,540	0.1282	0.772	6.813	55.53	0.719	15.90
−9	250	−23	0.09819	4.00	14,059	0.1561	1.130	7.919	80.64	0.713	6.03
81	300	27	0.08185	3.33	14,314	0.182	1.554	8.963	109.5	0.706	2.72
171	350	77	0.07016	2.86	14,436	0.206	2.031	9.954	141.9	0.697	1.39
261	400	127	0.06135	2.50	14,491	0.228	2.568	10.864	177.1	0.690	0.782
351	450	177	0.05462	2.22	14,499	0.251	3.164	11.779	215.6	0.682	0.468
441	500	227	0.04918	2.00	14,507	0.272	3.817	12.636	257.0	0.675	0.297
621	600	327	0.04085	1.67	14,537	0.315	5.306	14.285	349.7	0.664	0.134
800	700	427	0.03492	1.43	14,574	0.351	6.903	15.89	455.1	0.659	0.0677
981	800	527	0.03060	1.25	14,675	0.384	8.563	17.40	569	0.664	0.0379
1341	1000	727	0.02451	1.00	14,968	0.440	11.997	20.16	822	0.686	0.0145
2192	1200	927	0.02050	0.833	15,366	0.488	15.484	22.75	1107	0.715	0.00667

Source: E. G. Eckert and R. M. Drake, *Analysis of Heat and Mass Transfer*, McGraw-Hill, New York, 1972.

TABLE 32 NITROGEN AT ATMOSPHERIC PRESSURE

Temperature T			Density, ρ (kg/m³) $\times 6.243 \times 10^{-2}$ = (lb$_m$/ft³)	Coefficient of thermal expansion, $\beta \times 10^3$ (1/K) $\times 0.5556$ = (1/R)	Specific heat, c_p (J/kg K) $\times 2.388 \times 10^{-4}$ = (Btu/lb$_m$ °F)	Thermal conductivity, k (W/m K) $\times 0.5777$ = (Btu/h ft °F)	Thermal diffusivity, $\alpha \times 10^4$ (m²/s) $\times 3.874 \times 10^4$ = (ft²/h)	Absolute viscosity, $\mu \times 10^6$ (N s/m²) $\times 0.6720$ = (lb$_m$/ft s)	Kinematic viscosity, $\nu \times 10^6$ (m²/s) $\times 3.874 \times 10^4$ = (ft²/h)	Prandtl number, Pr	$\dfrac{g\beta}{\nu^2} \times 10^{-6}$ (1/K m³) $\times 1.573 \times 10^{-2}$ = (1/R ft³)
°F	K	°C									
−279	100	−173	3.4808		1072	0.00945	0.0253	6.86	1.97	0.786	
−100	200	−73	1.7108	5.00	1043	0.01824	0.1022	12.95	7.57	0.747	855.6
81	300	27	1.1421	3.33	1041	0.02620	0.2204	17.84	15.63	0.713	133.7
261	400	127	0.8538	2.50	1046	0.03335	0.3734	21.98	25.74	0.691	37.00
441	500	227	0.6824	2.00	1056	0.03984	0.5530	25.70	37.66	0.684	13.83
621	600	327	0.5687	1.67	1076	0.04580	0.7486	29.11	51.19	0.686	6.25
800	700	427	0.4934	1.43	1097	0.05123	0.9466	32.13	65.13	0.691	3.31
981	800	527	0.4277	1.25	1123	0.05609	1.1685	34.84	81.46	0.700	1.85
1161	900	627	0.3796	1.11	1146	0.06070	1.3946	37.49	91.06	0.711	1.31
1341	1000	727	0.3412	1.00	1168	0.06475	1.6250	40.00	117.2	0.724	0.714
1521	1100	827	0.3108	0.909	1186	0.06850	1.8591	42.28	136.0	0.736	0.482
	1200	927	0.2851	0.833	1204	0.07184	2.0932	44.50	156.1	0.748	0.335

Source: E. R. G. Eckert and R. M. Drake, *Analysis of Heat and Mass Transfer*, McGraw-Hill, New York, 1972.

TABLE 33 OXYGEN AT ATMOSPHERIC PRESSURE

Temperature, T °F	K	°C	Density, ρ (kg/m³) $\times 6.243 \times 10^{-2}$ = (lb$_m$/ft³)	Coefficient of thermal expansion, $\beta \times 10^{3}$ (1/K) $\times 0.5556$ = (1/R)	Specific heat, c_p (J/kg K) $\times 2.388 \times 10^{-4}$ = (Btu/lb$_m$ °F)	Thermal conductivity, k (W/m K) $\times 0.5777$ = (Btu/h ft °F)	Thermal diffusivity, $\alpha \times 10^{4}$ (m²/s) $\times 3.874 \times 10^{4}$ = (ft²/h)	Absolute viscosity, $\mu \times 10^{6}$ (N s/m²) $\times 0.6720$ = (lb$_m$/ft s)	Kinematic viscosity, $v \times 10^{6}$ (m²/s) $\times 3.874 \times 10^{4}$ = (ft²/h)	Prandtl number, Pr	$\dfrac{g\beta}{v^2} \times 10^{-6}$ (1/K m³) $\times 1.573 \times 10^{-2}$ = (1/R ft³)
−279	100	−173	3.992		948	0.00903	0.0239	7.768	1.946	0.815	
−189	150	−123	2.619	6.67	918	0.01367	0.0569	11.49	4.387	0.773	3398
−100	200	−73	1.956	5.00	913	0.01824	0.1021	14.85	7.593	0.745	850.5
−9	250	−23	1.562	4.00	916	0.02259	0.1579	17.87	11.45	0.725	299.2
80	300	27	1.301	3.33	920	0.02676	0.2235	20.63	15.86	0.709	129.8
171	350	77	1.113	2.86	929	0.03070	0.2968	23.16	20.80	0.702	64.8
261	400	127	0.9755	2.50	942	0.03461	0.3768	25.54	26.18	0.695	35.8
351	450	177	0.8682	2.22	957	0.03828	0.4609	27.77	31.99	0.694	21.3
441	500	227	0.7801	2.00	972	0.04173	0.5502	29.91	38.34	0.697	13.3
531	550	277	0.7096	1.82	988	0.04517	0.6441	31.97	45.05	0.700	8.79
621	600	327	0.6504	1.67	1004	0.04832	0.7399	33.92	52.15	0.704	6.02

Source: E. R. G. Eckert and R. M. Drake, *Analysis of Heat and Mass Transfer*, McGraw-Hill, New York, 1972.

TABLE 34 STEAM (H_2O) AT ATMOSPHERIC PRESSURE

Temperature, T			Density, ρ (kg/m³) $\times 6.243 \times 10^{-2}$ = (lb$_m$/ft³)	Coefficient of thermal expansion, $\beta \times 10^3$ (1/K) $\times 0.5556$ = (1/R)	Specific heat, c_p (J/kg K) $\times 2.388 \times 10^{-4}$ = (Btu/lb$_m$ °F)	Thermal conductivity, k (W/m K) $\times 0.5777$ = (Btu/h ft °F)	Thermal diffusivity, $\alpha \times 10^4$ (m²/s) $\times 3.874 \times 10^4$ = (ft²/h)	Absolute viscosity, $\mu \times 10^6$ (N s/m²) $\times 0.6720$ = (lb$_m$/ft s)	Kinematic viscosity, $\nu \times 10^6$ (m²/s) $\times 3.874 \times 10^4$ = (ft²/h)	Prandtl number, Pr	$\dfrac{g\beta}{\nu^2} \times 10^{-6}$ (1/K m³) $\times 1.573 \times 10^{-2}$ = (1/R ft³)
°F	K	°C									
212	373	100	0.5977		2034	0.0249	0.204	12.10	20.2	0.987	
225	380	107	0.5863		2060	0.0246	0.204	12.71	21.6	1.060	
261	400	127	0.5542	2.50	2014	0.0261	0.234	13.44	24.2	1.040	41.86
351	450	177	0.4902	2.22	1980	0.0299	0.307	15.25	31.1	1.010	22.51
441	500	227	0.4405	2.00	1985	0.0339	0.387	17.04	38.6	0.996	13.16
531	550	277	0.4005	1.82	1997	0.0379	0.475	18.84	47.0	0.991	8.08
621	600	327	0.3652	1.67	2026	0.0422	0.573	20.67	56.6	0.986	5.11
711	650	377	0.3380	1.54	2056	0.0464	0.666	22.47	66.4	0.995	3.43
800	700	427	0.3140	1.43	2085	0.0505	0.772	24.26	77.2	1.000	2.35
891	750	477	0.2931	1.33	2119	0.0549	0.883	26.04	88.8	1.005	1.65
981	800	527	0.2739	1.25	2152	0.0592	1.001	27.86	102.0	1.010	1.18
1071	850	577	0.2579	1.18	2186	0.0637	1.130	29.69	115.2	1.019	0.872

Source: E. R. G. Eckert and R. M. Drake, *Analysis of Heat and Mass Transfer*, McGraw-Hill, New York, 1972.

TABLE 35 METHANE AT ATMOSPHERIC PRESSURE

Temperature, T			Density, ρ (kg/m³) $\times 6.243 \times 10^{-2}$ = (lb$_m$/ft³)	Coefficient of thermal expansion, $\beta \times 10^3$ (1/K) $\times 0.5556$ = (1/R)	Specific heat, c_p (J/kg K) $\times 2.388 \times 10^{-4}$ = (Btu/lb$_m$ °F)	Thermal conductivity, k (W/m K) $\times 0.5777$ = (Btu/h ft °F)	Thermal diffusivity, $\alpha \times 10^4$ (m²/s) $\times 3.874 \times 10^4$ = (ft²/h)	Absolute viscosity, $\mu \times 10^6$ (N s/m²) $\times 0.6720$ = (lb$_m$/ft s)	Kinematic viscosity, $\nu \times 10^6$ (m²/s) $\times 3.874 \times 10^4$ = (ft²/h)	Prandtl number, Pr	$\dfrac{g\beta}{\nu^2} \times 10^{-6}$ (1/K m³) $\times 1.573 \times 10^{-2}$ = (1/R ft³)
°F	K	°C									
-112	193	-80	1.014	5.18		0.0207		7.4	7.30		954
-76	213	-60	0.9187	4.69		0.0230		8.1	8.82		592
-40	233	-40	0.8399	4.29		0.0260		8.8	10.48		383
-4	253	-20	0.7735	3.95		0.0278		9.5	12.28		257
32	273	0	0.7168	3.66	2165	0.0302	0.195	10.35	14.43		174
68	293	20	0.6679	3.41	2222	0.0332	0.224	10.87	16.27		126
122	323	50	0.6058	3.10	2307	0.0372	0.266	11.80	19.48		80.1
212	373	100	0.5246	2.68	2448			13.31	25.37	0.74	40.8
302	423	150	0.4626	2.36	2628			14.71	31.80	0.73	22.9
392	473	200	0.4137	2.11	2807			16.05	38.80	0.73	13.8
482	523	250	0.3742	1.91	2991			17.25	46.10		8.8
572	573	300	0.3415	1.75	3175			18.60	54.47		5.8

Source: N. B. Vargaftik, *Tables on the Thermophysical Properties of Liquid and Gases,* 2nd ed., Hemisphere, Washington, D.C., 1975.

TABLE 36 ETHANE AT ATMOSPHERIC PRESSURE

Temperature, T			Density, ρ (kg/m³) $\times 6.243 \times 10^{-2}$ = (lbm/ft³)	Coefficient of thermal expansion, $\beta \times 10^3$ (1/K) $\times 0.5556$ = (1/R)	Specific heat, c_p (J/kg K) $\times 2.388 \times 10^{-4}$ = (Btu/lbm °F)	Thermal conductivity, k (W/m K) $\times 0.5777$ = (Btu/h ft °F)	Thermal diffusivity, $\alpha \times 10^4$ (m²/s) $\times 3.874 \times 10^4$ = (ft²/h)	Absolute viscosity, $\mu \times 10^6$ (N s/m²) $\times 0.6720$ = (lbm/ft s)	Kinematic viscosity, $\nu \times 10^6$ (m²/s) $\times 3.874 \times 10^4$ = (ft²/h)	Prandtl number, Pr	$\dfrac{g\beta}{\nu^2} \times 10^{-6}$ (1/K m³) $\times 1.573 \times 10^{-2}$ = (1/R ft³)
°F	K	°C									
−103	198	−75	1.870	5.05	1647	0.0114		6.52	3.49		4066
32	273	0	1.356	3.66	1731	0.0183	0.0819	8.55	6.31	0.77	901
68	293	20	1.263	3.41	1815	0.0207	0.0947	9.29	7.36	0.78	617
104	313	40	1.183	3.19	1899	0.0235	0.109	9.86	8.33	0.76	451
140	333	60	1.112	3.00	1983	0.0265	0.126	10.50	9.44	0.75	330
176	353	80	1.049	2.83	2067	0.0296	0.142	11.11	10.66	0.75	244
212	373	100	0.992	2.68	2152	0.0328	0.160	11.67	11.76	0.74	190
248	393	120	0.942	2.54	2279			12.30	13.06		146
302	423	150	0.875	2.36				12.78	14.61		108
392	473	200	0.783	2.11	2490			14.09	17.99		63.9
482	523	250	0.708	1.91	2680			15.26	21.55		40.3

Source: N. B. Vargaftik, *Tables on the Thermophysical Properties of Liquid and Gases,* 2nd ed., Hemisphere, Washington, D.C., 1975.

TABLE 37 THE ATMOSPHERE[a]

Altitude, (ft)	Altitude, (m)	Absolute temperature (R) $\times \frac{5}{9}$ = (K)	Absolute pressure (lb_f/ft^2) $\times 47.88$ = (N/m^2)	Pressure ratio	Density (lb_f/ft^3) $\times 16.02$ = (kg/m^3)	Density ratio	Speed of sound (ft/s) $\times 0.3048$ = (m/s)
0	0	518	2,116	1.00	7.65×10^{-2}	1.00	1,120
5,000	1524	500	1,758	8.32×10^{-1}	6.60×10^{-2}	8.61×10^{-1}	1,100
10,000	3048	483	1,456	6.87×10^{-1}	5.66×10^{-2}	7.38×10^{-1}	1,080
20,000	6096	447	972	4.59×10^{-1}	4.08×10^{-2}	5.33×10^{-1}	1,040
30,000	9144	411	628	2.97×10^{-1}	2.88×10^{-2}	3.76×10^{-1}	997
40,000	12,192	392	392	1.85×10^{-1}	1.88×10^{-2}	2.45×10^{-1}	973
50,000	15,240	392	243	1.15×10^{-1}	1.16×10^{-2}	1.52×10^{-1}	973
60,000	18,288	392	151	7.13×10^{-2}	7.32×10^{-3}	9.45×10^{-2}	973
70,000	21,336	392	94.5	4.47×10^{-2}	4.51×10^{-3}	5.90×10^{-2}	974
80,000	24,384	392	58.8	2.78×10^{-2}	2.80×10^{-3}	3.67×10^{-2}	974
90,000	27,432	392	36.6	1.73×10^{-2}	1.67×10^{-3}	2.28×10^{-2}	974
100,000	30,480	392	22.8	1.08×10^{-3}	1.1×10^{-3}	1.4×10^{-2}	975
150,000	45,720	575	3.2	1.5×10^{-3}	9.7×10^{-4}	1.3×10^{-3}	1,190
200,000	60,960	623	0.73	3.6×10^{-4}	2.2×10^{-5}	2.9×10^{-4}	1,240
300,000	91,440	487	0.017	9.0×10^{-6}	6.9×10^{-7}	9.0×10^{-6}	1,110
400,000	121,920	695	0.0011	5.2×10^{-7}	2.7×10^{-8}	3.5×10^{-7}	1,430
500,000	152,400	910	1.2×10^{-4}	8.5×10^{-8}	3.1×10^{-9}	4.1×10^{-8}	
600,000	182,880	1,130	4.1×10^{-5}	1.9×10^{-8}	5.7×10^{-10}	7.5×10^{-9}	
700,000	213,360	1,350	1.3×10^{-5}	6.2×10^{-9}	1.5×10^{-10}	1.9×10^{-9}	
800,000	243,840	1,570	4.6×10^{-6}	2.2×10^{-9}	4.6×10^{-11}	6.0×10^{-10}	
900,000	274,320	1,800	1.9×10^{-6}	9.0×10^{-10}	1.7×10^{-11}	2.2×10^{-10}	

[a] *Sources of atmospheric property data*: C. N. Warfield, "Tentative Tables for the Properties of the Upper Atmosphere," *NACA TN* 1200, 1947; H. A. Johnson, M. W. Rubsein, F. M. Sauer, E. G. Slack, and L. Fossner, "The Thermal Characteristics of High Speed Aircraft," AAF, AMC, Wright Field, TR 5632, 1947; J. P. Sutton, *Rocket Propulsion Elements*, 2d ed., McGraw-Hill, New York, 1957.

MASS DIFFUSIVITIES

TABLE 38 BINARY GAS-PHASE MASS DIFFUSIVITIES AT
ATMOSPHERIC PRESSURE

Binary gas mixture	Temperature			$D_{AB} \times 10^5$ (m²/s)
	°F	K	°C	$\times 3.874 \times 10^4 = (ft^2/h)$
Air–ammonia	32	273	0	1.98
Air–aniline	77	298	25	0.726
Air–benzene	77	298	25	0.962
Air–carbon dioxide	32	273	0	1.36
Air–carbon disulfide	32	273	0	0.883
Air–chlorine	32	273	0	1.24
Air–ethyl alcohol	77	298	25	1.32
Air–hydrogen	32	273	0	5.472
Air–iodine	77	298	25	0.834
Air–mercury	597	614	314	4.73
Air–napthalene	77	298	25	0.611
Air–oxygen	32	273	0	1.75
Air–sulfur dioxide	32	273	0	1.22
Air–toluene	77	298	25	0.844
Air–water	77	298	25	2.60
CO_2–benzene	113	318	45	0.715
CO_2–carbon disulfide	113	318	45	0.715
CO_2–ethyl alcohol	32	273	0	0.693
CO_2–hydrogen	32	273	0	5.50
CO_2–nitrogen	77	298	25	1.58
CO_2–water	77	298	25	1.64
Hydrogen–nitrogen	55	286	13	7.376
Oxygen–ammonia	68	293	20	2.53
Oxygen–benzene	73	296	23	0.39
Oxygen–hydrogen	57	285	14	7.748
Oxygen–nitrogen	54	285	12	2.025

Source: R. D. Reid and T. K. Sherwood, *The Properties of Gases and Liquids*, McGraw-Hill, New York, 1966; *Handbook of Chemistry and Physics,* 39th ed., CRC Press, Cleveland, Ohio, 1957–58.

TABLE 39 BINARY LIQUID-PHASE MASS DIFFUSIVITIES (DILUTE SOLUTIONS; A IS SOLUTE, B IS SOLVENT)

Binary liquid mixture	Temperature			$D_{AB} \times 10^9$ (m^2/s)
	°F	K	°C	$\times 3.874 \times 10^4 = $ (ft^2/h)
Bromine in water	54	285	12	0.90
CO_2 in water	64	291	18	1.71
Chlorine in water	54	285	12	1.40
Glucose in water	59	288	15	0.52
Hydrogen in water	77	298	25	3.36
Iodine in water	77	298	25	1.25
Methanol in water	59	288	15	1.28
Nitrogen in water	72	295	22	2.02
Oxygen in water	77	298	25	2.60
Aniline in methanol	59	288	15	1.49
CCl_4 in methanol	59	288	15	1.70
Chloroform in methanol	59	288	15	2.07
Iodoform in methanol	59	288	15	1.33
Lactic acid in methanol	59	288	15	1.36
Acetic acid in benzene	59	288	15	1.92
Bromine in benzene	54	285	12	2.00
CCl_4 in benzene	77	298	25	2.00
Chloroform in benzene	59	288	15	2.11
Iodine in benzene	68	293	20	1.95

Source: R. D. Reid and T. K. Sherwood, *The Properties of Gases and Liquids*, McGraw-Hill, New York, 1966.

TABLE 40 BINARY SOLID-PHASE MASS DIFFUSIVITIES

Binary solid mixture	Temperature			$D_{AB} \times 10^{10}$ (m^2/s)
	°F	K	°C	$\times 3.874 \times 10^4 = $ (ft^2/h)
Aluminum in copper	68	293	20	1.30×10^{-24}
Antimony in silver	68	293	20	3.51×10^{-15}
Bismuth in lead	68	293	20	1.10×10^{-10}
Cadmium in copper	68	293	20	2.71×10^{-9}
Helium in silicon dioxide	68	293	20	$2.40 - 5.50 \times 10^{-4}$
Helium in Pyrex	932	773	500	2.00×10^{-2}
Helium in Pyrex	68	293	20	4.49×10^{-5}
Hydrogen in silicon dioxide	932	773	500	$0.573 - 2.10 \times 10^{-2}$
Hydrogen in nickel	185	358	85	1.16×10^{-2}
Hydrogen in nickel	329	438	165	0.105
Mercury in lead	68	293	20	2.50×10^{-9}

Source: R. M. Barrer, *Diffusion In and Through Solids*, Macmillan, New York, 1941.

TABLE 41 STEEL PIPE DIMENSIONS[a]

Nominal pipe size (in.)	Outside diameter (in.)	Schedule No.	Wall thickness (in.)	Inside diameter (in.)	Cross-sectional area metal (in.^{2})	Inside cross-sectional area (ft^2)
$\frac{1}{8}$	0.405	40[b]	0.068	0.269	0.072	0.00040
		80[c]	0.095	0.215	0.093	0.00025
$\frac{1}{4}$	0.540	40[b]	0.088	0.364	0.125	0.00072
		80[c]	0.119	0.302	0.157	0.00050
$\frac{3}{8}$	0.675	40[b]	0.091	0.493	0.167	0.00133
		80[c]	0.126	0.423	0.217	0.00098
$\frac{1}{2}$	0.840	40[b]	0.109	0.622	0.250	0.00211
		80[c]	0.147	0.546	0.320	0.00163
		160	0.187	0.466	0.384	0.00118
$\frac{3}{4}$	1.050	40[b]	0.113	0.824	0.333	0.00371
		80[c]	0.154	0.742	0.433	0.00300
		160	0.218	0.614	0.570	0.00206
1	1.315	40[b]	0.133	1.049	0.494	0.00600
		80[c]	0.179	0.957	0.639	0.00499
		160	0.250	0.815	0.837	0.00362
$1\frac{1}{4}$	1.660	40[b]	0.140	1.380	0.699	0.01040
		80[c]	0.191	1.278	0.881	0.00891
		160	0.250	1.160	1.107	0.00734
$1\frac{1}{2}$	1.900	40[b]	0.145	1.610	0.799	0.01414
		80[c]	0.200	1.500	1.068	0.01225
		160	0.281	1.338	1.429	0.00976
2	2.375	40[b]	0.154	2.067	1.075	0.02330
		80[c]	0.218	1.939	1.477	0.02050
		160	0.343	1.689	2.190	0.01556
$2\frac{1}{2}$	2.875	40[b]	0.203	2.469	1.704	0.03322
		80[c]	0.276	2.323	2.254	0.02942
		160	0.375	2.125	2.945	0.02463
3	3.500	40[b]	0.216	3.068	2.228	0.05130
		80[c]	0.300	2.900	3.016	0.04587
		160	0.437	2.626	4.205	0.03761
$3\frac{1}{2}$	4.000	40[b]	0.226	3.548	2.680	0.06870
		80[c]	0.318	3.364	3.678	0.06170
4	4.500	40[b]	0.237	4.026	3.173	0.08840
		80[c]	0.337	3.826	4.407	0.07986
		120	0.437	3.626	5.578	0.07170
		160	0.531	3.438	6.621	0.06447
5	5.563	40[b]	0.258	5.047	4.304	0.1390
		80[c]	0.375	4.813	6.112	0.1263
		120	0.500	4.563	7.953	0.1136
		160	0.625	4.313	9.696	0.1015

(*continued*)

TABLE 41   (*continued*)

Nominal pipe size (in.)	Outside diameter (in.)	Schedule No.	Wall thickness (in.)	Inside diameter (in.)	Cross-sectional area metal (in.²)	Inside cross-sectional area (ft²)
6	6.625	40ᵇ	0.280	6.065	5.584	0.2006
		80ᶜ	0.432	5.761	8.405	0.1810
		120	0.562	5.501	10.71	0.1650
		160	0.718	5.189	13.32	0.1469
8	8.625	20	0.250	8.125	6.570	0.3601
		30ᵇ	0.277	8.071	7.260	0.3553
		40ᵇ	0.322	7.981	8.396	0.3474
		60	0.406	7.813	10.48	0.3329
		80ᶜ	0.500	7.625	12.76	0.3171
		100	0.593	7.439	14.96	0.3018
		120	0.718	7.189	17.84	0.2819
		140	0.812	7.001	19.93	0.2673
		160	0.906	6.813	21.97	0.2532
10	10.75	20	0.250	10.250	8.24	0.5731
		30ᵇ	0.307	10.136	10.07	0.5603
		40ᵇ	0.365	10.020	11.90	0.5475
		60ᶜ	0.500	9.750	16.10	0.5185
		80	0.593	9.564	18.92	0.4989
		100	0.718	9.314	22.63	0.4732
		120	0.843	9.064	26.24	0.4481
		140	1.000	8.750	30.63	0.4176
		160	1.125	8.500	34.02	0.3941
12	12.75	20	0.250	12.250	9.82	0.8185
		30ᵇ	0.330	12.090	12.87	0.7972
		40	0.406	11.938	15.77	0.7773
		60	0.562	11.626	21.52	0.7372
		80	0.687	11.376	26.03	0.7058
		100	0.843	11.064	31.53	0.6677
		120	1.000	10.750	36.91	0.6303
		140	1.125	10.500	41.08	0.6013
		160	1.312	10.126	47.14	0.5592
14	14.0	10	0.250	13.500	10.80	0.9940
		20	0.312	13.376	13.42	0.9750
		30	0.375	13.250	16.05	0.9575
		40	0.437	13.126	18.61	0.9397
		60	0.593	12.814	24.98	0.8956
		80	0.750	12.500	31.22	0.8522
		100	0.937	12.126	38.45	0.8020
		120	1.062	11.876	43.17	0.7693
		140	1.250	11.500	50.07	0.7213
		160	1.406	11.188	55.63	0.6827

ᵃ Based on A.S.A. Standards B36.10.

ᵇ Designates former "standard" sizes.

ᶜ Former "extra strong."

TABLE 42 AVERAGE PROPERTIES OF TUBES

Diameter		Thickness		External			Internal				
External (in.)	Internal (in.)	BWG gage	Nom wall (in.)	Circumference (in.)	Surface per lineal foot (ft²)	Lineal feet of tube per square foot of surface	Transverse area (in.²)	Volume or capacity per lineal foot			Length of tube containing 1 ft³ (ft)
								(in.³)	(ft³)	U.S. Gal.	
$\frac{5}{8}$	0.527	18	.049	1.9635	0.1636	6.1115	0.218	2.616	0.0015	0.011	661
	0.495	16	.065	→	→	→	0.193	2.316	0.0013	0.010	746
	0.459	14	.083				0.166	1.992	0.0011	0.009	867
$\frac{3}{4}$	0.652	18	.049	2.3562	0.1963	5.0930	0.334	4.008	0.0023	0.017	431
	0.620	16	.065	→	→	→	0.302	3.624	0.0021	0.016	477
	0.584	14	.083				0.268	3.216	0.0019	0.014	537
	0.560	13	.095				0.246	2.952	0.0017	0.013	585
1	0.902	18	.049	3.1416	.2618	3.8197	0.639	7.668	0.0044	0.033	225
	0.870	16	.065	→	→	→	0.595	7.140	0.0041	0.031	242
	0.834	14	.083				0.546	6.552	0.0038	0.028	264
	0.810	13	.095				0.515	6.180	0.0036	0.027	280

Size											
1¼	1.152	18	.049	3.9270	.3272	3.0558	1.075	12.90	0.0075	0.056	134
	1.120	16	.065	→	→	→	0.985	11.82	0.0068	0.051	146
	1.084	14	.083				0.923	11.08	0.0064	0.048	156
	1.060	13	.095				0.882	10.58	0.0061	0.046	163
	1.032	12	.109				0.836	10.03	0.0058	0.043	172
1½	1.402	18	.049	4.7124	.3927	2.5465	1.544	18.53	0.0107	0.080	93
	1.370	16	.065	→	→	→	1.474	17.69	0.0102	0.076	98
	1.334	14	.083				1.398	16.78	0.0097	0.073	103
	1.310	13	.095				1.343	16.12	0.0093	0.070	107
	1.282	12	.109				1.292	15.50	0.0090	0.067	111
1¾	1.620	16	.065	5.4978	.4581	2.1827	2.061	24.73	0.0143	0.107	70
	1.584	14	.083	→	→	→	1.971	23.65	0.0137	0.102	73
	1.560	13	.095				1.911	22.94	0.0133	0.099	75
	1.532	12	.109				1.843	22.12	0.0128	0.096	78
	1.490	11	.120				1.744	20.92	0.0121	0.090	83
2	1.870	16	.065	6.2832	.5236	1.9099	2.746	32.96	0.0191	0.143	52
	1.834	14	.083	→	→	→	2.642	31.70	0.0183	0.137	55
	1.810	13	.095				2.573	30.88	0.0179	0.134	56
	1.782	12	.109				2.489	29.87	0.0173	0.129	58
	1.760	11	.120				2.433	29.20	0.0169	0.126	59

TABLE 43 THE ERROR FUNCTION

x	erf(x)	x	erf(x)	x	erf(x)
0.00	0.00000	0.76	0.71754	1.52	0.96841
0.02	0.02256	0.78	0.73001	1.54	0.97059
0.04	0.04511	0.80	0.74210	1.56	0.97263
0.06	0.06762	0.82	0.75381	1.58	0.97455
0.08	0.09008	0.84	0.76514	1.60	0.97635
0.10	0.11246	0.86	0.77610	1.62	0.97804
0.12	0.13476	0.88	0.78669	1.64	0.97962
0.14	0.15695	0.90	0.79691	1.66	0.98110
0.16	0.17901	0.92	0.80677	1.68	0.98249
0.18	0.20094	0.94	0.81627	1.70	0.98379
0.20	0.22270	0.96	0.82542	1.72	0.98500
0.22	0.24430	0.98	0.83423	1.74	0.98613
0.24	0.26570	1.00	0.84270	1.76	0.98719
0.26	0.28690	1.02	0.85084	1.78	0.98817
0.28	0.30788	1.04	0.85865	1.80	0.98909
0.30	0.32863	1.06	0.86614	1.82	0.98994
0.32	0.34913	1.08	0.87333	1.84	0.99074
0.34	0.36936	1.10	0.88020	1.86	0.99147
0.36	0.38933	1.12	0.88679	1.88	0.99216
0.38	0.40901	1.14	0.89308	1.90	0.99279
0.40	0.42839	1.16	0.89910	1.92	0.99338
0.42	0.44749	1.18	0.90484	1.94	0.99392
0.44	0.46622	1.20	0.91031	1.96	0.99443
0.46	0.48466	1.22	0.91553	1.98	0.99489
0.48	0.50275	1.24	0.92050	2.00	0.99532
0.50	0.52050	1.26	0.92524	2.10	0.997020
0.52	0.53790	1.28	0.92973	2.20	0.998137
0.54	0.55494	1.30	0.93401	2.30	0.998857
0.56	0.57162	1.32	0.93806	2.40	0.999311
0.58	0.58792	1.34	0.94191	2.50	0.999593
0.60	0.60386	1.36	0.94556	2.60	0.999764
0.62	0.61941	1.38	0.94902	2.70	0.999866
0.64	0.63459	1.40	0.95228	2.80	0.999925
0.66	0.64938	1.42	0.99538	2.90	0.999959
0.68	0.66378	1.44	0.95830	3.00	0.999978
0.70	0.67780	1.46	0.96105	3.20	0.999994
0.72	0.69143	1.48	0.96365	3.40	0.999998
0.74	0.70468	1.50	0.96610	3.60	1.000000

Tridiagonal Matrix Algorithm

Solution of a Tridiagonal System of Equations

The following BASIC program demonstrates an algorithm for solving tridiagonal matrices. Derivation of the algorithm is given by S. V. Patankar in *Numerical Heat Transfer and Fluid Flow*, Hemisphere Publishing Corporation, Washington, D.C., 1980. This program will accommodate any size matrix, but for demonstration purposes a 10×10 system has been used.

```
10*
20* READ IN DATA
30*
40  N=10//* SIZE OF SYSTEM = N × N
50  DIM A(N),B(N),C(N),D(N),P(N),Q(N),T(N)
60  A(1)=1//A(2)=.9//A(3)=.8//A(4)=1.1//A(5)=.95//* DIAGONAL ELEMENTS
70  A(6)=.85//A(7)=1.15//A(8)=.7//A(9)=.75//A(10)=1.2
80  B(1)=-.6//B(2)=-.5//B(3)=-.4//B(4)=-.7//B(5)=
     -.6//*NEGATIVE OF UPPER DIAGONAL ELEMENTS
90  B(6)=-.4//B(7)=-.6//B(8)=-.4//B(9)=-.8//B(10)=0
100 C(1)=0//C(2)=-.3//C(3)=-.2//C(4)=-.7//C(5)=
     -.5//* NEGATIVE OF LOWER DIAGONAL ELEMENTS
110 C(6)=-.1//C(7)=-.3//C(8)=-.2//C(9)=-.1//C(10)=-.5
120 D(1)=.1666//D(2)=.2022//D(3)=.2177//D(4)=.5155//D(5)=.5906
130 D(6)=.5489//D(7)=1.075//D(8)=.8755//D(9)=1.4728//D(10)=1.6056
140*
150* CALCULATE RECURSION VARIABLES
160*
170 P(1)=B(1)/A(1)
180 Q(1)=D(1)/A(1)
190 FOR I=2 TO N
200 P(I)=B(I)/(A(I)-C(I)*P(I-1))
210 Q(I)=(D(I)+C(I)*Q(I-1))/(A(I)-C(I)*P(I-1))
220 NEXT I
230*
240* BACK SUBSTITUTE FOR T(I)
250*
260 T(N)=Q(N)
270 FOR I=N-1 TO 1 STEP-1
280 T(I)=P(I)*T(I+1)+Q(I)
290 NEXT I
300*
310* PRINT RESULTS
320*
330 FOR I=1 TO N
```

```
340  PRINT # 1,TAB(5);'I = ';TAB(8);I;TAB(15);'A = ';INT(1000*A(I))/1000;
350;  # 1,TAB(25);'B = ';INT(10000*B(I))/10000;
360;  # 1,TAB(35);'C = ';INT(10000*C(I))/10000;
370;  # 1,TAB(45);'D = ';INT(10000*D(I))/10000;
380;  # 1,TAB(55);'T = ';INT(10000*T(I))/10000
390     NEXT I
```

PROGRAM OUTPUT

I = 1	A = 1.0	B = −.6	C = .0	D = .1666	T = .0999
I = 2	A = .9	B = −.5	C = −.3	D = .2022	T = .1111
I = 3	A = .8	B = −.4	C = −.2	D = .2177	T = .1443
I = 4	A = 1.1	B = −.7	C = −.7	D = .5155	T = .1999
I = 5	A = .95	B = −.6	C = −.5	D = .5906	T = .2778
I = 6	A = .85	B = −.4	C = −.1	D = .5489	T = .3777
I = 7	A = 1.15	B = −.6	C = −.3	D = 1.075	T = .5
I = 8	A = .7	B = −.4	C = −.2	D = .8755	T = .6443
I = 9	A = .75	B = −.8	C = −.1	D = 1.4728	T = .8111
I = 10	A = 1.2	B = .0	C = −.5	D = 1.6056	T = 1.0

The Heat Transfer Literature

A limitation of any textbook is the depth to which the material can be covered. Textbooks can only provide the background needed to understand principles and prepare the student to handle more complex "real-world" problems.

We have made an effort to present up-to-date information in this book. But before starting to solve a "real life" heat transfer problem, one should become familiar with work in the area performed by other experts. A few hours spent in a library can save many hours "reinventing the wheel." In addition to specialized textbooks and handbooks, several periodicals are devoted to heat transfer and provide the most current literature available. Conference proceedings are also a valuable source of information. Although the articles in these sources have been reviewed by specialists in the field before publication, it is important to critically evaluate each paper and not assume that the work is infallible.

The following list includes the most important English language heat transfer journals, with information on the publisher, frequency of publication, and method of indexing.

Journal of Heat Transfer. Published quarterly by the American Society of Mechanical Engineers. No annual indices after 1977.

International Journal of Heat and Mass Transfer. Published monthly by Pergamon Press. Annual indices.

Numerical Heat Transfer. Published quarterly by Hemisphere Publishing Corporation. Annual indices.

AIChE Journal. Published bimonthly by the American Institute of Chemical Engineers. No annual indices.

Journal of Fluid Mechanics. Published monthly by Cambridge University Press. No annual indices.

Heat Transfer and Fluid Flow Digest. Titles and keywords only. Published monthly by Hemisphere Publishing Corporation.

Heat Transfer Engineering. Published quarterly by Hemisphere Publishing Corporation. No annual indices.

Advances in Heat Transfer. Published annually until 1978 and now every 3 years by Academic Press.

Advances in Chemical Engineering. Published by Academic Press. Frequency varies.

Proceedings of the International Heat Transfer Conferences. Published every four years by Hemisphere Publishing Corporation.

INDEX

Absolute temperature
  monochromatic emissive power and, 431–433
  Stefan-Boltzmann law and, 18, 432
  thermal conductivity and, 9
Absorbing and transmitting media, enclosures with, 485–487
Absorptance
  defined, 442
  gas, 489, 494
  monochromatic, versus wavelength, 443–445
  source temperature and, 453–454
Accoustical velocity, 584–589
Achenbach, E., 361
Addoms, J. N., 519
Adiabatic surface
  boundary conditions for, 162–164
  gas flow over, 585–586
  radiation to or from, 469
Aerodynamic heating, 586
Aiba, S., 375
Al-Arabi, M., 261
Al-Fakhri, A. A. M., 357
Algebraic shape-factor, 464–466
Alternating direction implicit (ADI) method, 159
Aluminum
  absorptance of, 454

hemispherical emittance of, 450–455
  thermal contact resistance for, 13
American engineering system, 42–44
Analogic method, in steady-state conduction, 86
Analogy between heat, mass, and momentum transfer, 202
Analytic solution
  finite-difference method compared with, 148
  for laminar boundary-layer flow over flat plate, 209–214
  for steady-state conduction, 85–89
Anderson, J. T., 266
Annular-flow regime, 531
Apparent turbulent shear, 226
Approximate analysis of boundary layer, 202
Approximate integral boundary-layer analysis, 219–225
Arpaci, V., 123
Artery wick system, 604–605
Augmentation entropy generation number, 576, 580
Augmentation techniques, 575–580
August, S. E., 315
Average bulk temperature, 288, 300, 304
Average film temperature, in forced convection, 295, 366